T0344794

Chemometrics for Pattern Recognition

Chemometrics for Pattern Recognition

RICHARD G. BRERETON

Centre for Chemometrics, School of Chemistry, Bristol University, Bristol, UK

A John Wiley and Sons, Ltd., Publication

This edition first published 2009

Registered office
John Wiley & Sons Ltd, The Atrium, Southern Gate, Chichester, West Sussex, PO19 8SQ, United Kingdom

For details of our global editorial offices, for customer services and for information about how to apply for permission to reuse the copyright material in this book please see our website at www.wiley.com.

2 2010

Library of Congress Cataloging-in-Publication Data

Brereton, Richard G.
Chemometrics for pattern recognition / Richard G. Brereton.
p. cm.
Includes bibliographical references and index.
ISBN 978-0-470-98725-4
1. Chemometrics. 2. Pattern perception. I. Title.
QD75.4.C45B74 2009
543.01′5195—dc22
2009009689

A catalogue record for this book is available from the British Library.

ISBN 978-0470-987254

Set in 10/12pt Times by Integra Software Services Pvt. Ltd, Pondicherry, India
Printed and bound in Singapore by Fabulous Printers Pte Ltd

Contents

Acknowledgements xi
Preface xv

1 Introduction **1**
1.1 Past, Present and Future 1
1.2 About this Book 9
Bibliography 12

2 Case Studies **15**
2.1 Introduction 15
2.2 Datasets, Matrices and Vectors 17
2.3 Case Study 1: Forensic Analysis of Banknotes 20
2.4 Case Study 2: Near Infrared Spectroscopic Analysis of Food 23
2.5 Case Study 3: Thermal Analysis of Polymers 25
2.6 Case Study 4: Environmental Pollution using Headspace Mass Spectrometry 27
2.7 Case Study 5: Human Sweat Analysed by Gas Chromatography Mass Spectrometry 30
2.8 Case Study 6: Liquid Chromatography Mass Spectrometry of Pharmaceutical Tablets 32
2.9 Case Study 7: Atomic Spectroscopy for the Study of Hypertension 34
2.10 Case Study 8: Metabolic Profiling of Mouse Urine by Gas Chromatography of Urine Extracts 36
2.11 Case Study 9: Nuclear Magnetic Resonance Spectroscopy for Salival Analysis of the Effect of Mouthwash 37
2.12 Case Study 10: Simulations 38
2.13 Case Study 11: Null Dataset 40
2.14 Case Study 12: GCMS and Microbiology of Mouse Scent Marks 42
Bibliography 45

3 Exploratory Data Analysis **47**
3.1 Introduction 47
3.2 Principal Components Analysis 49
3.2.1 Background 49
3.2.2 Scores and Loadings 50

3.2.3 Eigenvalues 53
3.2.4 PCA Algorithm 57
3.2.5 Graphical Representation 57
3.3 Dissimilarity Indices, Principal Co-ordinates Analysis and Ranking 75
3.3.1 Dissimilarity 75
3.3.2 Principal Co-ordinates Analysis 80
3.3.3 Ranking 84
3.4 Self Organizing Maps 87
3.4.1 Background 87
3.4.2 SOM Algorithm 88
3.4.3 Initialization 89
3.4.4 Training 90
3.4.5 Map Quality 93
3.4.6 Visualization 95
Bibliography 105

4 Preprocessing 107
4.1 Introduction 107
4.2 Data Scaling 108
4.2.1 Transforming Individual Elements 108
4.2.2 Row Scaling 117
4.2.3 Column Scaling 124
4.3 Multivariate Methods of Data Reduction 129
4.3.1 Largest Principal Components 129
4.3.2 Discriminatory Principal Components 137
4.3.3 Partial Least Squares Discriminatory Analysis Scores 145
4.4 Strategies for Data Preprocessing 150
4.4.1 Flow Charts 150
4.4.2 Level 1 153
4.4.3 Level 2 161
4.4.4 Level 3 162
4.4.5 Level 4 175
Bibliography 176

5 Two Class Classifiers 177
5.1 Introduction 177
5.1.1 Two Class Classifiers 178
5.1.2 Preprocessing 180
5.1.3 Notation 180
5.1.4 Autoprediction and Class Boundaries 181
5.2 Euclidean Distance to Centroids 184
5.3 Linear Discriminant Analysis 185
5.4 Quadratic Discriminant Analysis 192
5.5 Partial Least Squares Discriminant Analysis 196

5.5.1 PLS Method 196
5.5.2 PLS Algorithm 198
5.5.3 PLS-DA 199
5.6 Learning Vector Quantization 201
5.6.1 Voronoi Tesselation and Codebooks 206
5.6.2 LVQ1 207
5.6.3 LVQ3 209
5.6.4 LVQ Illustration and Summary of Parameters 211
5.7 Support Vector Machines 213
5.7.1 Linear Learning Machines 214
5.7.2 Kernels 218
5.7.3 Controlling Complexity and Soft Margin SVMs 223
5.7.4 SVM Parameters 228
Bibliography 231

6 One Class Classifiers 233
6.1 Introduction 233
6.2 Distance Based Classifiers 235
6.3 PC Based Models and SIMCA 236
6.4 Indicators of Significance 239
6.4.1 Gaussian Density Estimators and Chi-Squared 239
6.4.2 Hotelling's T^2 241
6.4.3 *D*-Statistic 243
6.4.4 *Q*-Statistic or Squared Prediction Error 248
6.4.5 Visualization of *D*- and *Q*-Statistics for Disjoint PC Models 249
6.4.6 Multivariate Normality and What to do if it Fails 263
6.5 Support Vector Data Description 266
6.6 Summarizing One Class Classifiers 275
6.6.1 Class Membership Plots 275
6.6.2 ROC Curves 279
Bibliography 286

7 Multiclass Classifiers 289
7.1 Introduction 289
7.2 EDC, LDA and QDA 291
7.3 LVQ 295
7.4 PLS 298
7.4.1 PLS2 298
7.4.2 PLS1 300
7.5 SVM 304
7.6 One against One Decisions 304
Bibliography 309

8 Validation and Optimization 311
8.1 Introduction 311
8.1.1 Validation 311
8.1.2 Optimization 315

8.2 Classification Abilities, Contingency Tables and Related Concepts 315
8.2.1 Two Class Classifiers 315
8.2.2 Multiclass Classifiers 318
8.2.3 One Class Classifiers 318
8.3 Validation 320
8.3.1 Testing Models 320
8.3.2 Test and Training Sets 321
8.3.3 Predictions 324
8.3.4 Increasing the Number of Variables for the Classifier 331
8.4 Iterative Approaches for Validation 335
8.4.1 Predictive Ability, Model Stability, Classification by Majority Vote and Cross Classification Rate 335
8.4.2 Number of Iterations 348
8.4.3 Test and Training Set Boundaries 352
8.5 Optimizing PLS Models 361
8.5.1 Number of Components: Cross-Validation and Bootstrap 361
8.5.2 Thresholds and ROC Curves 374
8.6 Optimizing Learning Vector Quantization Models 377
8.7 Optimizing Support Vector Machine Models 380
Bibliography 390

9 Determining Potential Discriminatory Variables 393
9.1 Introduction 393
9.1.1 Two Class Distributions 394
9.1.2 Multiclass Distributions 395
9.1.3 Multilevel and Multiway Distributions 396
9.1.4 Sample Sizes 399
9.1.5 Modelling after Variable Reduction 401
9.1.6 Preliminary Variable Reduction 405
9.2 Which Variables are most Significant? 405
9.2.1 Basic Concepts: Statistical Indicators and Rank 405
9.2.2 T-Statistic and Fisher Weights 407
9.2.3 Multiple Linear Regression, ANOVA and the F-Ratio 417
9.2.4 Partial Least Squares 431
9.2.5 Relationship between the Indicator Functions 434
9.3 How Many Variables are Significant? 440
9.3.1 Probabilistic Approaches 440
9.3.2 Empirical Methods: Monte Carlo 442
9.3.3 Cost/Benefit of Increasing the Number of Variables 447
Bibliography 450

10 Bayesian Methods and Unequal Class Sizes 453
10.1 Introduction 453
10.2 Contingency Tables and Bayes' Theorem 453
10.3 Bayesian Extensions to Classifiers 458
Bibliography 467

11 Class Separation Indices 469
11.1 Introduction 469
11.2 Davies Bouldin Index 470
11.3 Silhouette Width and Modified Silhouette Width 475
11.3.1 Silhouette Width 475
11.3.2 Modified Silhouette Width 475
11.4 Overlap Coefficient 477
Bibliography 478

12 Comparing Different Patterns 479
12.1 Introduction 479
12.2 Correlation Based Methods 481
12.2.1 Mantel Test 481
12.2.2 R_V Coefficient 483
12.3 Consensus PCA 484
12.4 Procrustes Analysis 487
Bibliography 492

Index 493

Acknowledgements

Case Study 1 (Forensic)

Rich Sleeman and Jim Carter of Mass Spec Analytical and their team are thanked for providing the mass spec dataset. This is described in the following paper:

- S.J. Dixon, R.G. Brereton, J.F. Carter, R. Sleeman, Determination of cocaine contamination on banknotes using tandem mass spectrometry and pattern recognition, *Anal. Chim. Acta*, **559**, 54–63 (2006).

Case Study 2 (NIR Spectroscopy of Food)

Camo ASA including Sabyasachi Goswami and Kenneth Olafsson are thanked for providing the NIR spectra of four essential oils, which is described in the Camo multivariate analysis training manual.

Case Study 3 (Polymers)

John Duncan and colleagues of Triton Technology Ltd are thanked for providing the polymer DMA analyses. These are described in the following papers:

- B.M. Lukasiak, S. Zomer, R.G. Brereton, R. Faria, J.C. Duncan, Pattern recognition and feature selection for the discrimination between grades of commercial plastics, *Chemometrics Intell. Lab. Systems*, **87**, 18–25 (2007).
- B.M. Lukasiak, S. Zomer, R.G. Brereton, R. Faria, J.C. Duncan, Pattern Recognition for the Analysis of Polymeric Materials, *Analyst*, **131**, 73–80 (2006).
- R. Faria, J.C. Duncan, R.G. Brereton, Dynamic Mechanical Analysis and Chemometrics for Polymer Identification, *Polym. Testing*, **26**, 402–412 (2007).
- G. Lloyd, R. Faria, R.G. Brereton, J.C. Duncan, Learning vector quantization for multiclass classification: application to characterization of plastics, *J. Chem. Information Modeling*, **47**, 1553–1563 (2007).
- G.R. Lloyd, R.G. Brereton, J.C. Duncan, Self Organising Maps for distinguishing polymer groups using thermal response curves obtained by dynamic mechanical analysis, *Analyst*, **133**, 1046–1059 (2008).

Case Study 4 (Pollution)

Miguel Sanchez and colleagues of the University of Salamanca are thanked for providing the MS of soils. This is described in the following paper:

- S. Zomer, M. Sánchez, R.G. Brereton, J.L. Pérez Pavón, Active Learning Support Vector Machines for Optimal Sample Selection in Classification, *J. Chemometrics*, **18**, 294–305 (2004).

Case Study 5 (Sweat)

Milos Novotny and Helena Soini (University of Indiana), and Dustin Penn and Alexandra Katzer (Konrad Lorentz Institute Vienna), are thanked for providing the GCMS of sweat; Karl Grammer and Liza Oberzaucher (University of Vienna) for field data collection. The experimental data collection in this work was sponsored by ARO Contract DAAD19-03-1-0215. Opinions, interpretations, conclusions, and recommendations are those of the authors and are not necessarily endorsed by the United States Government. This is approved for public release, distribution unlimited. This is described in the following papers:

- S.J. Dixon, R.G. Brereton, H.A. Soini, M.V. Novotny, D.J. Penn, An Automated Method for Peak Detection and Matching in Large Gas Chromatography-Mass Spectrometry Data Sets, *J. Chemometrics*, **20**, 325–340 (2006).
- D.J. Penn, E. Oberzaucher, K. Grammer, G. Fischer, H.A. Soini, D. Wiesler, M.V. Novotny, S.J. Dixon, Y. Xu, R.G. Brereton, Individual and gender fingerprints in human body odour, *J. R. Soc. Interface*, **4**, 331–340 (2007).
- H.A. Soini, K.E. Bruce, I. Klouckova, R.G. Brereton, D.J. Penn, M.V. Novotny, *In-situ* surface sampling of biological objects and preconcentration of their volatiles for chromatographic analysis, *Anal. Chem.*, **78**, 7161 –7168 (2006).
- S.J. Dixon, Y. Xu, R.G. Brereton, A. Soini , M.V. Novotny, E. Oberzaucher, K. Grammer, D.J. Penn, Pattern Recognition of Gas Chromatography Mass Spectrometry of Human Volatiles in Sweat to distinguish the Sex of Subjects and determine potential Discriminatory Marker Peaks, *Chemometrics Intell. Lab. Systems*, **87**, 161–172 (2007).
- Y. Xu, F. Gong, S.J. Dixon, R.G. Brereton, H.A. Soini, M.V. Novotny, E. Oberzaucher, K. Grammer, D.J. Penn, Application of Dissimilarity Indices, Principal Co-ordinates Analysis and Rank Tests to Peak Tables in Metabolomics of the Gas Chromatography Mass Spectrometry of Human Sweat, *Anal. Chem.*, **79**, 5633 –5641 (2007).

Case Study 6 (LCMS of Tablets)

Richard Escott, Christine Eckers and Jean-Claude Wolff of Glaxo Smith Kline and their team, are thanked for providing the data. This is described in the following paper:

- S. Zomer, R.G. Brereton, J.-C. Wolff, C.Y. Airiau, C. Smallwood, Component Detection Weighted Index of Analogy: Similarity Recognition on Liquid Chromatographic Mass

Spectral Data for the Characterisation of Route/Process Specific Impurities in Pharmaceutical Tablets, *Anal. Chem.*, **77**, 1607–1621 (2005).

Case Study 7 (Clinical)

Shahida Waheed and Sohaila Rahman of the Pakistan Institute of Nuclear Science and Technology are thanked for the obtaining the samples and atomic spectroscopy measurements. The data are described in Dr Rahman's PhD thesis:

- S. Rahman, Significance of blood trace metals in hypertension and diseases related to hypertension in comparison with normotensive, *PhD Thesis*, Pakistan Institute of Science and Technology, Islamabad, Pakistan.

Case Study 8 (Mouse Urine)

Nina Heinrich, Maria Holmboe and Jose Trevejo of the Charles Stark Draper Laboratory (Cambridge, MA) and their team are thanked for obtaining GCMS measurements; Michele Schaeffer and Randall Reed (Johns Hopkins University, Baltimore, MD) for mouse experiments. The experimental data collection in this work was sponsored by ARO Contract DAAD19-03-1-0215. Opinions, interpretations, conclusions and recommendations are those of the authors and are not necessarily endorsed by the United States Government. This is approved for public release, distribution unlimited. This is described in the following paper:

- S.J. Dixon, N. Heinrich, M. Holmboe, M.L. Schaefer, R.R. Reed, J. Trevejo, R.G. Brereton, Use of cluster separation indices and the influence of outliers: application of two new separation indices, the modified silhouette index and the overlap coefficient to simulated data and mouse urine metabolomic profiles, *J. Chemometrics* **23**, 19–31 (2009).
- K. Wongravee, G.R. Lloyd, J. Hall, M.E. Holmboe, M.L. Schaefer, R.R. Reed, J. Trevejo, R.G. Brereton, Monte-Carlo methods for determining optimal number of significant variables. Application to mouse urinary profiles, *Metabolomics* (2009) doi 10.1007/s11306-009-0164-4.
- K. Wongravee, N. Heinrich, M. Holmboe, M.L. Schaefer, R.R. Reed, J. Trevejo, R.G. Brereton, Variable Selection using Iterative Reformulation of Training Set Models for Discrimination of Samples: Application to Gas Chromatography Mass Spectrometry of Mouse Urinary Metabolites, *Anal. Chem.* (2009) doi 10.1021/ac900251c.

Case Study 9 (NMR Spectroscopy)

Martin Grootveld of the University of Bolton, and Chris Silwood of South Bank University (London) are thanked for providing the NMR measurements.

Case Study 12 (Mouse Scent Marks)

Dustin Penn (Konrad Lorentz Institute, Vienna) and Morris Gosling (University of Newcastle) and their teams led also by Tony Clare (GCMS) and Tony O'Donnell (microbiology) are thanked for providing the data. The experimental data collection in this work was sponsored by ARO Contract DAAD19-03-1-0215. Opinions, interpretations, conclusions, and recommendations are those of the authors and are not necessarily endorsed by the United States Government. This is approved for public release, distribution unlimited. This is described in the following paper:

- S. Zomer, S.J. Dixon, Y. Xu, S.P. Jensen, H. Wang, C.V. Lanyon, A.G. O'Donnell, A.S. Clare, L.M. Gosling, D.J. Penn, R.G. Brereton, Consensus Multivariate methods in Gas Chromatographic Mass Spectrometry and Denaturing Gradient Gel Electrophoresis: MHC-congenic and other strains of mice can be classified according to the profiles of volatiles and microflora in their scent-marks, *Analyst*, **134**, 114–123 (2009).

Preface

In chemometrics there is much less of a tradition of writing books as a primary mechanism for the advancement of ideas than in many other parallel areas. In subjects such as statistics and machine learning this is a well known way of expounding ways of thinking that subsequently become widespread. Whereas there are many books on chemometrics (although perhaps not so much in the area of pattern recognition within a chemical context), the majority involve cataloguing existing well established methods or as tutorials often associated with a software package. Yet in chemometrics, ideas probably fit well into book format, rather than become fragmented in numerous papers, and it is hoped that this text will inspire others to produce books that are not only a list of already published methods (which of course is certainly also a valuable service) but also that introduce new ways of thinking about their data. In the area of chemometrics there have been enormous advances in, for example, multivariate calibration and signal processing, but pattern recognition has perhaps lagged behind intellectually over the last few years, especially in terms of fundamental understanding. There are several early books in the chemometrics literature addressing this topic, mainly from the 1980s or before, but these tend to describe methods algorithmically rather than discuss their statistical and intellectual basis, and so many users of pattern recognition methods often get confused by what the underlying motivation is. With the advent of fast desktop computer power and good user friendly graphics, together with a huge explosion of data primarily from the biological interface, especially in the area of metabolomic profiling, there is an urgent need to revisit this area of chemometrics and place it in a new, and modern, intellectual framework.

This book is the product of many influences on myself over several years. Although I am the guy hunched up in front of the PC at 2 in the morning, nevertheless, many people over the last few years have acted as significant influences, and really this text is the product of a large team, possibly a hundred or more.

First and foremost I am very fortunate to have had the opportunity to work with a large and diverse team of collaborators within my research group in the University of Bristol. It is primarily the research students that spur work on, whose enthusiasm and skills are what powers most academic research. Since 2002 there have been 5 senior visitors/postdocs, 28 internal research students mainly studying for PhDs, and 31 visiting students and project students within the Chemometrics Group in Bristol. This book is a product of a tremendously diverse range of activity over the past few years. It is impossible to highlight every student that has contributed to the advance in understanding and application of pattern recognition, but I wish to specifically acknowledge 7 key researchers. During the 6 months working on this book I have been fortunate to have a small team of excellent co-workers helping me in developing code and testing ideas and obtaining the results presented in this text. Gavin Lloyd has been an outstanding programmer and many of the diagrams in this text owe to his ability to rapidly develop good graphical software, as well as contributing many ideas especially in the area of machine learning. Kanet Wongravee has worked within this and a large biological pattern recognition project over 3 years and helped to obtain many of the results. Sim Fong worked unbelievably hard, often through the night, to document and catalogue a vast amount of results. Sila Kittiwachana has helped especially with the multiblock methods. Prior to this, several pioneering research students in the group helped to established many of the approaches that have metamorphosed since into the body of this text. These include Yun Xu, who worked in particular on one class classifiers and biological pattern recognition. Simeone Zomer pioneered pharmaceutical applications of pattern recognition and our first applications of Support Vector Machines. Sarah Dixon helped in leading a team in two large international pattern recognition projects in quick succession. These and many other team members have had a large influence on the rapid development of ideas in Bristol, which have rapidly led to ideas that have influenced this text.

Numerous collaborators from outside Bristol have also contributed to joint projects which have resulted in us working on many diverse applications: these range from biology, to medicine, to pharmaceuticals, to forensic, environmental and materials science, and are thanked for their faith in the work we are doing. These collaborations have given myself and my colleagues in Bristol an enormously diverse flavour of how pattern recognition methods can be applied to a vast array of scientific problems. Collaboration is essential for ideas to grow in applied areas: chemometrics is about developing tools to solve real world problems. Sometimes it can be frustrating if data are not perfect but by working over many years with a wide variety of organizations, some very small, some enormous, working on projects both large and small, we get a flavour of why pattern recognition methods are so important, but also why they are somewhat misunderstood and what to do to ensure they are better appreciated. Most of my main collaborators have been generous with access to data and are thanked in the previous section (Acknowledgements) with respect to the various case studies, and have been very understanding of the constraints on my time over the period of preparation of this book.

Being able to write a book also requires the ability to organize one's time in a suitably relaxing environment. In the area of applied science, there are two needs. The first is for the 'buzz' of collaboration and of daily contact with co-workers – one cannot work as an island – and the second is the need for quiet concentration. I was fortunate to be able to spend much of the summer of 2008 in Naples, and have many excellent memories of the wonderful archaeology and fantastic museums of the region, as well as the world class islands. Visiting Pompeii, or Herculaneum, or Capri, or Vesuvius, or the National Archaeological Museum, is a good preparation for an evening reading about pattern recognition. I have many pleasant memories of walking down past the unique historic centre through the Via Toledo via the Opera House and Castle on a baking hot day to watch the many boats came to land in the Port. Plus, of course, budget airlines allow one to return home every ten days or so, at no more than the cost of a train fare from Bristol to London. Finally, good wireless Internet means that one can access electronic libraries and desktop computers and networks from hundreds of miles away, just as if one were back home. So there are ways of working now that were impossible – perhaps inconceivable – ten years ago.

Finally thanks must go to the publisher John Wiley & Sons, Ltd, Chichester, UK who have worked successfully with me on three books in six years. Jenny Cossham is thanked for her continuing support and Richard Davies for progressing this book towards production.

Supplementary material related to this book is available on the www.spectroscopynow.com website.

Richard Brereton

1
Introduction

1.1 Past, Present and Future

Chemometrics has been with us, as a recognizable name, for around 35 years, its origins being traced back to the early 1970s. In those heydays many of the pioneers were programmers, good at writing software in Fortran, or Basic, that would be used to perform, what in those days were considered intensive calculations, often on large mainframe computers. The concept of the laboratory based chemist sitting at his or her desk manipulating data from an NMR or HPLC instrument in real time on a cheap micro was probably not even contemplated at the time. Indeed in the early 1970s most people had little idea that computers would shrink and become user friendly and on everyone's desk, the future of computers was 'big', to be developed and used by large institutes such as for defence and for space exploration. People who used computers were in the most part programmers and so a computationally based discipline that can look back to the early 1970s is a discipline that was founded primarily by people that could program computers.

It is my view that the development and aspiration of chemometricians and most computer based disciplines changed during the late 1970s, when the economic catalyst for the development of computing started to move in new directions. With the oil crisis and global inflation in Western countries of the 1970s, there was no longer an inexorable march to larger and more expensive computing. In the early 1970s many people might have envisaged that we would have established colonies on the moon by now, powered by huge, but somewhat impersonal and clunky, mainframes. View any science fiction film from this era, and computer screens are full of numbers and letters, not ergonomic mice and windows, but although the vision of this future computer interface was primitive, in our modern view, the computer power envisaged was enormous, way beyond what we might normally come across nowadays. In science fiction films and television programmes of that era, most space ships and cities had a large central computer control room filled with big cables, consoles, rather old fashioned keyboards, and lineprinters churning out printouts on large perforated sheets of paper: that is what people thought the future would be. So with early chemometrics,

Chemometrics for Pattern Recognition Richard G. Brereton

did the pioneers realize that chemometric methods would end up on a desktop, and that the click of a mouse and the ability to rotate a plot in real time would be considered more important than the development of computer intensive pattern recognition algorithms? The answer is probably not.

The first methods to be marketed specifically for chemometrics were reported in those early years, methods such as NIPALS (for performing Principal Components Analysis), k Near Neighbours (for pattern recognition), and SIMCA (also for pattern recognition). Early texts such as Sharaf *et al.* from the mid 1980s still illustrated diagrams by using lineprinter output. That was the norm.

One of the planks of early chemometrics was undoubtedly pattern recognition. This area in itself also is very broad, and includes numerous journals and conferences associated with it. Pattern recognition can mean a lot of different things to different people, and in the context of chemometrics there are two broadly based sets of approaches, one influenced by the machine learning community, that emphasizes Neural Networks, Kernel Methods, Self Organising Maps, Support Vector Machines etc. and another influenced by the statistical community emphasizing Discriminant Analysis, Method Validation, Bayesian approaches and so on. Whereas both traditions are important ones they tend not to meet much, almost existing as some parallel universes that occasionally collide: to the chemist this seems strange almost looking as an outsider, but both types of community really feels they are separate. Each 'tribe' has different beliefs probably based on prior education and experience. It is well known that people with different underlying psychologies choose to work (or succeed) in different disciplines. Hence someone successful as a statistician will probably have a different psychological makeup to someone that is successful in machine learning, and both will approach similar problems often from very different perspectives. They will often be reluctant to cross into each others' territories, and prefer to discuss scientific problems with their own 'tribe' who will organize their own conferences and journals. Chemometrics, too, has developed its own 'tribe', the foundations being set in the 1970s. I remember when starting my academic career, being asked to attend a course on teaching, and being told that people choosing chemistry are in themselves very much outliers against all other sciences. If one performs certain psychological tests and then performs some form of multivariate analysis, I was told, biologists, physicists and mathematicians would cluster together, with chemists as a separate outlying group. This at first may seem strange but walk into any department of biology or mathematics. The 'big professor' often does hands-on work himself or herself: there are less research students and smaller groups but more long term junior researchers; research students often pursue quite independent lines with encouragement; research groups are often less well defined, or at least are not defined around the individual academic but more as a 'sub-department' with common themes and several leaders. In a chemistry department there tend to be more research students and larger groups; the job of the research professor is primarily that of a manager with very few working hands-on in the laboratory; a group is primarily connected to the group leader and research students feel their strongest affiliation to their professor and his or her fellow students. So, some of the pioneering chemometricians, by psychology chemists, were good organizers as well as having strong mathematical ability. In contrast in different disciplines, such as computer science or statistics, senior scientists often publish sole author papers throughout their career, and in many cases their research students also publish independently; in chemistry most senior scientists publish mainly collaborative papers often with large teams. Hence many of the early chemometrics pioneers had the aptitude of mathematicians and/or computer scientists but the psychology of chemists. This probably flavoured the early development.

So what has this to do with pattern recognition? Over the past decades there have been enormous developments in statistical and computational pattern recognition, but despite many of the original seeds of chemometrics being within the application of pattern recognition to chemistry these theoretical developments are rarely understood in modern chemometrics practice. What happened is that there were indeed pioneering ideas in the 1970s and early 1980s but at the time chemometrics was developing as a recognizable subject. Good organizers and managers look for developing their teams and their funding. Only a few of those prominent in the 1970s regularly published books or papers as sole authors after the initial seeds in the 1970s: their job was to build up students, departments or companies to market their ideas because that was what they were very good at doing. Although in fact there were (and continue to be) people that have not moved in this direction within chemometrics, several of those 'feeling the water' in the 1970s and 1980s have moved back to other disciplines such as statistics, where one can be of the highest distinction without having set up a company or a large research group, leaving behind a somewhat large gap between a growing and excellent theoretical base, but hardly understood or even acknowledged by analytical chemistry users of chemometrics, and a slower intellectually developing base primarily focused on software that is much better known by users of chemometrics techniques.

Chemometrics in the 1980s onwards focused on dollars. Newcomers to the field were successful newcomers if their ideas could be quickly converted to funds. This meant that some areas moved faster than others. Areas in particular where there were specific challenges to the chemist, particularly in industry, and where chemometrics could help, developed fast. Multivariate calibration, especially in the context of NIR (Near Infrared) spectroscopy, is a definite success story from this era, catalysing the development of methods especially of benefit to the food and the pharmaceutical industry, metamorphosing over the last few years into initiatives such as the PAT (Process Analytical Technology) initiative. The more theoretical chemometrics still moved forward, but mainly in areas that are specific to chemistry such as multivariate curve resolution (MCR), e.g. in spectroscopy or coupled chromatography, encompassing approaches such as Factor Analysis or Alternating Least Squares, mainly applied in the interface of physical and analytical chemistry. Although there were many papers published applying pattern recognition to analytical chemistry problems over this period, look at any major chemometrics conference, and the topics emphasized tend to be areas such as calibration, multiway methods, signal resolution, experimental design, etc. – and NIR spectroscopy of course.

So pattern recognition has got left behind within the chemometrics context. Pattern recognition though provided one of the main original motivations of chemometrics. What happened? Some of those publishing in the early days of chemometrics pattern recognition have moved back into core disciplines such as statistics or machine learning. The reason for this 'divorce' is two fold. The first is that the 'audience', primarily of practicing chemists, found (and still find) many of the more intense statistical papers very tough to understand. Theoreticians mainly have a very intense way of writing – which they are trained in and which referees will often insist on – but that makes many of their papers inaccessible to all but a small handful of readers outside the field, so although their ideas may be fantastic they are not communicated in a way that the mainstream user of chemometrics methods might like. The second is that many of the chemists that took over chemometrics in the 1970s and 1980s were often very good at selling themselves – to get funds and develop their own groups and companies – and because they were more in tune with the psychology of their market, the way they communicated ideas won out, and being able to communicate ideas is a very important skill especially in a subject such as chemometrics which involves talking to a wide variety of collaborators. Hence much of the accepted 'philosophy' of chemometric pattern recognition is based on principles that have hardly changed for nearly thirty years, whilst other areas have progressed much faster. One reason why areas such as calibration and multivariate curve resolution could develop at such a pace is that the methods were very specific to chemical problems and so could thrive within an 'island' and a fairly isolated community. Plus, as we will see below, for many of the problems encountered by the early chemometrics pioneers, such as in NIR spectroscopy, the existing methods were good enough at the time. Finally, there were many more dollars at the time associated with development of multivariate calibration and signal analysis compared to pattern recognition within an analytical chemical context, although with the growing interface to metabolic profiling this is now starting to change.

Since the 1980s, though, there have been huge changes in scientific data analysis. The first and foremost is the continual explosion of computer power. In several places in this text we will cite Moore's law. This can be formulated in a variety of ways, but is based in an exponential increase in computing power per unit cost, variously put at doubling every one to two years (dependent on the yardstick). If we say it doubles every 18 months, then it increases around 100 times every 10 years, or has increased 10 million times in the 35 years or so that chemometrics has been around. Naturally 'at the beginning' people could not foresee the power of desktop PCs and used mainframes, so if one estimates a thousand fold difference between a mainframe and a desktop, this still results in an impressive 100 000 fold increase in capabilities since the 'beginning'. This means that computationally intense methods that were impossible several years ago can now be applied routinely. Many of the earliest chemometrics packages were based around being a very economical use of computer power: that is what a good programmer and algorithm developer will learn; still these skills are needed for good systems programmers but now it is possible to contemplate more computationally intense approaches – if these are useful. Why use cross-validation if one can use the bootstrap even if the latter has to be repeated 200 times and so is 200 times slower? People tell me they remember when they would go out for a long cup of coffee to wait for their 'cross-validation' to creek through their micro – they now sit in front of a PC for a few seconds and the bootstrap is completed. Why divide into a single test set when one can repeatedly divide into 100 different test sets and average the models? In the past each test set split was computationally expensive and so took time – one could not wait a week for the answer. However there is now no need to work within the straightjacket of methods developed thirty years ago. The old methods were good for their time, and indeed there is nothing wrong with them in their days. We may use energy efficient light bulbs rather than old wasteful ones now, but for one hundred years the old light bulbs served their purpose. We use DVDs now rather than VHS tapes – the latter technology seems very primitive in this day and age but in its time VHS was a revolution and served people well for twenty years. We buy CDs or download music rather than purchase vinyl gramophone records but gramophone records served an important purpose in civilisation for a century. Hence there is always a need to evaluate what methods are most appropriate according to the needs of the time: we no longer live in caves – give a caveman a flatscreen TV and they would not know what to do with it, but chemometrics should adapt with the times even though the pioneering work of the 1970s and 1980s made the subject what it is.

There is however a problem in chemometrics in that lots of people, often without a good mathematical or computational background, want to 'use' it. Often I am surprised that people without any prior knowledge of this subject feel that they can pick it up in a workshop that 'should not last too long'. They want to walk in, then walk out and understand how to do pattern recognition in a couple of afternoons. This desire, unfortunately, is an important economic driving force in this subject. I say to my students that it may take a year or so just working through examples to learn the basis and gain sufficient feel for the subject. They accept this, but that is why they are giving up so much of their time to learn chemometrics. If they did not accept this, they would not be my students. The dilemma though is that for chemometrics to become widespread there should be a big user base. This is where the subject, and especially the application of pattern recognition, has a problem – almost like a split personality. Keep the subject theoretical and to an elite who are really good at maths and computing, and it is not widespread. Tell an analytical chemist in the lab that he or she cannot do any pattern recognition and many will turn round, download a package, and put some data through and go away, even if he or she cannot understand the results. A few will get interested and learn but then they need to be in an environment where they have a lot of time – and many employers or even research supervisors will not allow this. So most will either give up or try to cut corners. They will pay money for chemometrics, but not to spend a year or two learning the ropes, but rather to buy a package, that they believe does what they want, and go on a course that will teach them how to enter data and print out results in a couple of afternoons. The people that market these packages will make it easy for someone to take a series of spectra, import them into a package, view a graph on a screen, change the appearance with a mouse or a menu and incorporate into a nice report in Word that will be on their boss's (or their sponsor's) desk within a few hours. They won't gain much insight, but they will spread the word that chemometrics is a useful discipline. The course they go on will not really give them an insight into chemomertics (how can one in an afternoon?) but will teach them how to put data through a package and learn to use software and will catalyse the wider name recognition of the subject.

Many of the early successes in chemometrics were in quite narrowly defined areas, NIR spectroscopy being one of them. In fact in NIR spectroscopy many of the challenges are not so much with the multivariate statistics, but with the preparation of data. Baseline correction, multiplicative scatter correction, derivatives, smoothing etc. are all tools of the trade and have been developed to a high degree and incorporated into most commercially available NIR chemometrics software. The user spends more time on this than on the front end multivariate analysis – and indeed NIR spectroscopy is a technique for which, if the data are correctly processed, excellent results can be obtained, but this is more in the area of visualization of multivariate trends that could usually be determined using other methods. Still this provided an impetus, and most NIR software contains some chemometrics capability, particularly in calibration. NIR spectroscopists often feel they understand chemometrics but maybe this is not so, particularly in pattern recognition. However since many NIR problems are quite straightforward from the chemometrics point of view (see in this text Case Study 2 – NIR of Food), in many cases no harm has come, and this at least demonstrates the tremendous power of multivariate methods for simplifying and visualizing data, even if the difficult part is the spectroscopic data handling, and so NIR spectroscopy can be considered correctly as an early success story and an important historic driving force of the subject.

The problem is that over the past decade new sources of data have come on-stream, and this is particularly confusing many analytical chemists. The development of metabolic profiling, e.g. using coupled chromatography, mass spectrometry and nuclear magnetic resonance spectroscopy, has had a very fast development, with improved, more sensitive, and automated instruments. It looks easy, but it is not. The problem is that datasets have now become vastly more difficult to handle, and are no longer the easy NIR spectra that have comforted chemometricians for a generation. The potential application of chemometrics to analytical data arising from problems in biology and medicine is enormous, but often the experimentalists have little understanding of how to acquire and handle these data. They want to learn but have only the odd afternoon or downloaded package with which to learn. They are funded to obtain data not to spend a year learning about Matlab. They usually want quick fixes. A few turn to collaborators, but sometimes the collaborator will say their data are not good enough, or the experiments need to be designed differently, or that it will take a long time and lots of resources and be very expensive, and so in many cases these collaborations don't develop or stop at a small pilot study or a paper. The biologists are anxious to be first to publish their 'marker compounds' and to claim that their work is a success and see data analysis as the afterthought that can be done on a Friday afternoon once all experiments are complete. So they will turn to the user-friendly packages and afternoon workshops and learn how to use the mouse and the menu and get a graph for incorporation into their report and then move on to the next project.

Many do not realize that the methods they are using probably were developed for different purposes. Most chemometrics methods have their origins in traditional analytical chemistry, where there are often underlying certainties, for example in calibration we know what answer we are aiming for and as such just want to get our multivariate method as close as possible to the known answer. In some of the original applications of chemical pattern recognition such as spectroscopy we know what the underlying groups of compounds are and want our methods to classify spectra as effectively as possible into these groupings. We aim for 100 % accuracy and the original algorithms were considered to be better the more accurate the answer. With nice reproducible spectra, a known solution, and no hidden factors, this was possible. But there often is no certain answer in biology, for example, we cannot be sure that by measuring some compounds in a patient's serum that we can predict whether they will develop kidney disease within the next five years: we are uncertain whether there will be an answer or not. We are testing hypotheses as well as trying to obtain accurate predictions, and now do not just want to predict properties with a high degree of accuracy, but also to determine whether there really is sufficient information in the analytical data to detect the desired trend. Overfitting involves overinterpreting data and seeing trends that are not really there. Many biologists do not have a feel for whether data are overfitted or not. One can start with purely random data and by a judicious choice of variables end up with graphs that look as if two arbitrarily selected groups are separate. Most people when submitting a paper for publication will actively seek out the graph that 'looks better' even if it is misleading. Variable selection prior to pattern recognition is common in chemometrics even though it can be dangerous. A reason is that if we know there is a certain answer we want the graphs to look their best. If we have a series of NMR spectra of compounds differing in only a small region, for clarity we often show only that part of the spectrum that differs, to emphasize why we are confident in our assignment – this is not dishonest, it is normal scientific practice as the rest of the spectrum may be irrelevant. However in pattern recognition it is like wiping evidence that we disagree with so that the final picture confirms our prejudice (which, if correct, may result in publishing a paper, getting another grant or obtaining a PhD). Many people do this unwittingly because they are not aware of the problem and many packages try their best to present the data in a way they think the user likes – and people pay money for this. With an inadequate background it is often hard to assess what comes out of the 'black box'. A final problem is that there often is enormous variability in biological samples. In traditional analytical chemistry the variability is often much more limited, and so groups of samples are much easier to define. In biomedical studies there are often a large number of factors that can influence the chemical signal, and so groups can be much harder to distinguish. This often means one may need tens or hundreds or even thousands more samples to obtain an adequate result. Many traditional papers on chemical pattern recognition are published on quite small sample sets – in some cases this is justified – for example it is probably quite reasonable to try to distinguish ten ketones from ten esters spectroscopically – but once translated into metabolomic applications (or indeed in other related areas such as environmental or forensic investigations), there are so many other factors that can influence variability that some papers using limited sample sizes are quite misleading. If one has many variables, e.g. hundreds of GCMS peaks, on small sample sizes, it is always possible to find some variables that appear good discriminators – just by chance – like tossing a coin ten times and repeating this experiment over and again hundreds of times – there will always be a few occasions where we obtain eight or nine 'heads' and if these are selected in the absence of the other tosses, it will look like the coin is biased.

How to solve this is not easy as the problem is not technical but about persuasion and education. In areas such as bioinformatics and chemoinformatics it is now generally accepted that there are specially trained informaticians who do nothing all day except process data and mine databases. The need for these specialists is accepted by the market. One reason why chemometrics has to function differently, is that in the informatics areas, most data analysis is done post event, over five, ten or even twenty years after the data have been obtained. There are many databases available, and the aim of the informatician is to interpret trends that might have been lurking in public (or company) databases accumulated over many years. Chemical structures, pKas, quantum mechanical descriptors, do not change much over time, and so can be gradually built up: and many sponsors insist that these data are made available to the scientific community, and so there are not too many barriers. But most chemometrics is a 'here and now' subject. The experiments and analysis are performed in real time, and often answers are required during or immediately after the project, examples being process analysis or clinical experiments. One cannot afford to wait years for the results. Plus although there are databases available, often there are incompatibilities between instrumental techniques, for example chromatography varies according to instrument and conditions, and also over the years methods improve, old machines are discarded and so on; therefore really large international and compatible databases are often hard to set up, and most databases relate to specific applications performed by particular groups. However, with the 'here and now' need for chemometrics expertise and often the commercially sensitive nature of large databases, only limited information is available in the public domain and there is much less emphasis on long term data mining and interpretation, and more on integrating chemometrics into the daily life of a laboratory. Hence most chemometrics groups or companies or institutes are primarily geared up to solving other people's problems in real time as this is where the funding comes from, and there is comparably little long term investment in development, in contrast to areas such as bioinformatics where sponsors are patient and it is accepted that the product may be many years down the line, so long as the product is a good one, of course. So although there are several groups that do train chemometricians, and give them time to pick up fundamental skills, the market emphasis is on producing something that can be used on the spot in the laboratory.

The problem is that software development costs money and takes time, especially user friendly software. It can take a hundred times longer to produce a graphical user interface than to develop the underlying algorithm. Many companies are reluctant to invest such resources. The market likes colour graphics and menu driven screens, and prefers this to some advanced statistical output that probably cannot easily be understood. Many of the main chemometrics software companies were founded in the 1980s, where, as we have seen, there were quite different needs, and are locked into methods that were excellent then but maybe not so appropriate now. The cost of software development can be enormous, and the market for chemometrics software is not enormous. If we estimate that one billion copies of Windows have been sold, we might estimate the total global market in chemometrics software to be between 10 000 and 100 000 or between 1/(100 000) and 1/(10 000) of that for Windows. This limited market means that to sell one has either to have a cheap product that is not very sophisticated or a front end product that is extremely expensive but with a narrow market. The problem here is that this limits capacity for radical redevelopment of the underlying philosophy of a software package. To produce a new package that trashes the old one is often not very good for the established market, but also a huge amount of work and a big risk. Twenty years ago the emphasis on a good user interface was much less – the market would be for people that in themselves would tolerate teleprinter output and awkward ways of typing in data and the occasional crash, and so the cost of getting something accepted on the market would be much lower. Nowadays, even with the best algorithms in the world, if one wants to sell, especially to the upper end of the market, the interface must be excellent. So there have been only modest changes in the underlying algorithms in the best established packaged software over many years, with just a slow and conservative evolution in the nature of the underlying algorithms, despite a revolution in the way chemometrics pattern recognition can be applied, in the quantity of data and breadth of problems, and the capability of computers.

So who should be doing chemometrics? One additional feature of chemometrics is that the number of variables often far exceeds the number of samples – something traditional statisticians rarely encountered. Plus often there is a lot of interest in interpreting the variables and finding out which are important – after all, variables relate to chemicals and so provide insight into the chemical process. Many traditional statistical and machine learning procedures were not designed for this – mainly we are interested in distinguishing samples (or objects or subjects) and not why they can be distinguished. Hence we cannot just slot into the existing data analysis community, as these people are not experienced in dealing with chemometrics projects that have become increasingly multivariate over time.

Of course ideally practitioners of chemometrics should have good programming ability and be able to understand the basis of the algorithms and statistical methods and this would ideally solve the dilemma of the occasional user of packaged software trying to navigate through the minefield of pattern recognition. But this is never likely to be and the community will always be very diverse. A major aim of this text is to bridge the gap between application scientists, and the more theoretical literature that is excellent but often inaccessible to the practicing user of chemometrics methods, and illustrate the principles primarily graphically. This text is therefore aimed at gaining insight into the use and ideas behind pattern recognition within a chemometrics context, without delving too far into the theory. It is not however a software package, although we have developed a Matlab code for the methods in this text, but not in the form of a user friendly GUI (Graphical User Interface). The need to bridge the gap between theory and practice in the area of chemometric pattern recognition is particularly pressing with the rapid growth of applications in the area of biology and medicine, and perhaps was not so evident ten years ago when the focus was more on process monitoring and on spectroscopy which pose different challenges.

It is hoped that readers of this text will gain much more understanding of the 'whats and whys' of pattern recognition within a modern chemometric context. Some readers will decide to continue this self learning and turn into programmers themselves (this book is not a programming manual though), developing and applying methods to their own and other people's data. For a graduate student or consultant or independent researcher, this text will provide an important springboard. Other readers may feel they have not the time, or cannot cope with, the necessary technical programming, but still will gain from the insight and use pattern recognition methods with care in the future. They may decide to team up with a colleague who is good at developing numerical methods, or form a collaboration, or go to a consultant, or even hire someone to do this for them if there is enough work. There will be a group of people from disciplines such as statistics, machine learning and chemical engineering that will gain insight into which methods are most applicable in chemometrics and how these can be applied in real world practice, and it is hoped that it will attract some such people back into the subject, even at a periphery as many have unfortunately deserted over the years. Finally there will be some who are applications scientists and not maybe at the moment able to spend the time learning about chemometrics, who will have heard of pattern recognition in a conference or read a paper or be thinking whether chemometrics can help their work, who will benefit – as a highly visual case study based text, skimming the chapters and looking at the diagrams and diving in and out of different sections will give a flavour of what is possible and allow planning as to whether to invest time or resources into applying methods to their research in the future.

1.2 About this Book

There have been a few books about pattern recognition in chemometrics, as listed in the Bibliography, but most of these are quite old, and are primarily algorithmic descriptions. Much of the development of chemometrics in the 1980s and 1990s involved cataloguing methods – formally describing algorithms, listing them and often applying them to a few benchmark datasets. Many of those using chemometrics would describe themselves first and foremost as analytical chemists, and indeed, this is where the majority of papers in chemometrics have been published: indeed the two subjects have developed hand in hand with some saying that chemometrics is the theoretical basis of analytical chemistry. But analytical chemists tend to like to list things. Previously we discussed the unusual psychology of chemists, and many chemists, whilst having many other skills, are scientific 'stamp collectors', and so a lot of the early pattern recognition books read like a list of methods, often though without the statistical motivation.

Surprisingly, despite a huge increase in the number of chemometrics texts over the past few years, with several well established ones in niche areas such as calibration, signal resolution and experimental design, and the large growth of pattern recognition texts in areas such as machine learning, there has been very limited development of such texts in the subject of chemometrics. Some of the historic reasons why this may be so are discussed above.

This current book does not aim to be an encyclopaedia. There are probably a hundred or more named 'methods' in the chemometrics literature for classification in the chemometrics literature. In order to publish a paper, win a grant or obtain a PhD it is often useful to propose a 'named' method. Some such methods become widespread, others are reported in one or two papers, cited half a dozen times and forgotten. Most named methods do indeed involve one or more steps that are different to existing published methods, but if every defined method for pattern recognition contained ten steps, and each step had five alternatives, there would be ten million combinations of alternatives, and so plenty of possibilities of announcing a new set of alternatives, for example, one could study all existing methods and find the combination of alternatives that is most different and publish a paper on this, and so on. This plethora of 'named' methods can be quite confusing, and it is not the intention of this author to describe all in detail. However there are only a few underlying principles and so it has been decided in this book to focus on a smaller number of generic methods and approaches and to discuss the motivations behind the methods rather than focus on a comprehensive list of methods. All the widespread ones including Principal Components Analysis, Linear Discriminant Analysis, Partial Least Squares Discriminant Analysis, Support Vector Machines, SIMCA and Self Organizing Maps are discussed, plus some others, but the choice is very much based on illustrating generic principles rather than being comprehensive.

The methods are illustrated by their use in analysing a variety of case studies chosen for their applicability in many branches of science, including biology, medicine, materials characterization, food, environment, pharmaceuticals and forensics, and many analytical techniques including MS, LCMS, GCMS, NIR Spectroscopy, NMR Spectroscopy, Thermal Analysis and Atomic Spectroscopy. For readers unfamiliar with these techniques (such as those from the machine learning or statistics community) it is not necessary to have an in depth appreciation of the analytical techniques to understand this text. The techniques and applications are biased a little towards this author's experience but attempt to be broad enough to be of interest to a wide variety of readers. For reasons of brevity only the most common techniques (such as PCA) are illustrated on all case studies, and in most cases we choose specific examples where it is most useful to illustrate the technique.

There also needs to be a decision as to the depth with which each method is discussed. Some, such as Partial Least Squares, can quite easily be described in adequate detail in a few pages, but others such as Support Vector Machines, could occupy a full text in its own right to adequately describe the maths. So the policy in this book is to allocate approximately equal space to techniques according to a judgement as to their relative importance, whilst in all cases trying to aid understanding using diagrams. In most cases statistically based methods are described in sufficient detail that a good programmer can reproduce the results if need be. All results in this text were obtained using in-house software written in Matlab which aided our understanding of the methods.

Decisions have also been made as to which topics to discuss in detail. The main emphasis of this book is on classification methods which are the focus of Chapters 5 to 11, although exploratory data analysis and data preprocessing are also discussed in some detail. Areas that could be expanded include Multiblock approaches (Chapter 12) and ANOVA based methods for looking at the significance of variables (Chapter 9). Both these could form books in their own rights, and for specific groups of chemometricians are considered very important as they pose interesting potential development areas. But practicing laboratory based chemists or biologists only rarely come across the need for such methods and so they are introduced only in a limited form although in sufficient detail to provide readers with a springboard and complete description of methods where needed. Multiway methods which are hotly topical in certain circles are not described in this book as they are already the subject of another excellent and dedicated text, and this book could not do justice to them except in a very peripheral way.

The references associated with each chapter are not aimed to be comprehensive but to allow readers a springboard should they wish to explore certain topics further. They are primarily chosen as references that my coworkers and I have found useful when trying to understand topics in more depth. They also include precursor papers published from the Bristol group which often expand on explanations and examples in this text around certain key topics.

Readers of a book such as this are unlikely to work through in a linear fashion (from beginning to end), and will probably come from different backgrounds, dipping in and out of the book. One good way of understanding methods (if you are a programmer) is to try to reproduce results or apply methods to your favourite datasets and see what happens. Some methods are iterative and not completely reproducible and so it is not always possible to obtain identical results each time, but the general principles should be comprehensible and it is hoped that this book offers insights into what is possible and what the fundamental assumptions are behind this.

Finally whereas we recommend certain protocols (for example repeatedly splitting datasets into test and training sets many times), this is not prescriptive, and a good scientist may come up with his or her own combination of techniques: this book gives important information about the building blocks and what we recommend, but other people may piece together their house differently. Furthermore some of the illustrations using PC scores plots are for purpose of visualization to understand the basis of methods and in practice one may use more PCs for a model but then not be able to visualize the results; however by experience a simple visual representation of a method makes it much easier to understand than by describing it in several pages of equations, although of course, we do not neglect the necessary algebra where appropriate.

Bibliography

General Chemometrics Texts

R.G. Brereton, *Applied Chemometrics for Scientists*, John Wiley & Sons, Ltd, Chichester, UK, 2007.
R.G. Brereton, *Chemometrics: Data Analysis for the Laboratory and Chemical Plant*, John Wiley & Sons, Ltd, Chichester, UK, 2003.
D.L. Massart, B.G.M. Vandeginste, L.M.C. Buydens, S. De Jong, P.J. Lewi, J. Smeyers-Verbeke, *Handbook of Chemometrics and Qualimetrics*, Part A, Elsevier, Amsterdam, The Netherlands, 1997.
B.G.M. Vandeginste, D.L. Massart, L.M.C. Buydens, S. De Jong, P.J. Lewi, J. Smeyers-Verbeke, *Handbook of Chemometrics and Qualimetrics*, Part B, Elsevier, Amsterdam, The Netherlands, 1998.
M. Otto, *Chemometrics: Statistics and Computer Applications in Analytical Chemistry*, Wiley-VCH, Weinheim, Germany, 2007.
H. Mark, J. Workman, *Chemometrics in Spectroscopy*, Elsevier, London, UK, 2007.
K. Varmuza, P. Filmoser, *Introduction to Multivariate Statistical Analysis in Chemometrics*, CRC Press, Boca Raton, FL, USA, 2009.
K.R. Beebe, R.J. Pell, M.B. Seasholtz, *Chemometrics: A Practical Guide*, John Wiley & Sons, Inc., New York, NY, USA, 1998.
R. Kramer, *Chemometrics Techniques for Quantitative Analysis*, Marcel Dekker, New York, NY, USA, 1998.
P.J. Gemperline (editor), *Practical Guide to Chemometrics*, Second Edition, CRC Press, Boca Raton, FL, USA, 2006.
D.L. Massart, B.G.M. Vandeginste, S.N. Deming, Y. Michotte, L. Kaufman, *Chemometrics: A Textbook*, Elsevier, Amsterdam, The Netherlands, 1988.
M.A. Sharaf, D.L. Illman, B.R. Kowalski, *Chemometrics*, John Wiley & Sons, Inc., New York, NY, USA, 1986.
R.G. Brereton, *Chemometrics: Applications of Mathematics and Statistics to Laboratory Systems*, Ellis Horwood, Chichester, UK, 1990.
M. Meloun, J. Militky, M. Forina, *Chemometrics for Analytical Chemistry*, Vols 1 and 2, Ellis Horwood, Chichester, UK, 1992 and 1994.
B.R. Kowalski (editor), *Chemometrics: Mathematics and Statistics in Chemistry*, Reidel, Dordrecht, The Netherlands, 1984.
D.L. Massart, R.G. Brereton, R.E. Dessy, P.K. Hopke, C.H. Spiegelman, W. Wegscheider, (editors), *Chemometrics Tutorials*, Elsevier, Amsterdam, The Netherlands, 1990.
R.G. Brereton, D.R. Scott, D.L. Massart, R.E. Dessy, P.K. Hopke, C.H. Spiegelman, W. Wegscheider, (editors), *Chemometrics Tutorials II*, Elsevier, Amsterdam, The Netherlands, 1992.
J.N. Miller., J.C. Miller, *Statistics and Chemometrics for Analytical Chemistry*, Fifth Edition, Pearson, Harlow, UK, 2005.
M.J. Adams, *Chemometrics in Analytical Spectroscopy*, Second Edition, The Royal Society of Chemistry, Cambridge, UK, 2004.
K.H. Esbensen, *Multivariate Data Analysis in Practice*, CAMO, Oslo, Norway, 2002.
L. Eriksson, E. Johansson, N. Kettaneh-Wold, S. Wold, *Multi- and Megavariate Data Analysis: Principles and Applications*, Umetrics, Umeå, Sweden, 2001.
S.D. Brown, R. Tauler, B. Walczak, (editors), *Comprehensive Chemometrics*, Elsevier, Amsterdam, 2009.
NIST Engineerig Statistics Handbook [http://www.itl.nist.gov/div898/handbook/].

Chemometrics Texts Focused on Pattern Recognition

R.G. Brereton (editor), *Multivariate Pattern Recognition in Chemometrics, Illustrated by Case Studies*, Elsevier, Amsterdam, The Netherlands, 1992.
D.L. Massart, L. Kaufmann, *The Interpretation of Analytical Chemical Data by the Use of Cluster Analysis*, John Wiley & Sons, Inc., New York, NY, USA, 1983.

O. Strouf, *Chemical Pattern Recognition*, Research Studies Press, Letchworth, UK, 1986.
K. Varmuza, *Pattern Recognition in Chemistry*, Springer-Verlag, Berlin, Germany, 1980.
P.C. Jurs, T.L. Isenhour, *Chemical Applications of Pattern Recognition*, John Wiley & Sons, Inc., New York, NY, USA, 1975.
D. Coomans, I. Broeckaert, *Potential Pattern Recognition in Chemical and Medical Decision Making*, Research Studies Press, Letchworth, UK, 1986.

2

Case Studies

2.1 Introduction

Most practical pattern recognition is based around analysing datasets, to use them to answer questions about the underlying process or experiment. Before using computational methods, it is important to examine the structure of a dataset, and determine what information is required.

In some cases the information may be exploratory, for example, we may be interested in whether there is enough information in a sample of urine from a patient to determine whether he or she has advanced diabetes, or whether by using chromatography a series of samples from similar geographical areas group together. We do not know what the answer is, and so the aim of data analysis is first to explore the data (does it look as if there are the trends we hope for?) and then to test hypotheses (how sure are we that there is sufficient information in the data to be able to determine these trends?). Even if we know that these trends must be there (for example, we know that we can distinguish humans by gender) we do not always know whether the method of analysis (e.g. extraction followed by chromatography, mass spectrometry or NMR spectroscopy) is adequate to detect these trends.

In other cases we may want to produce a predictive model. Under such circumstances, we are sure (or at least believe) that the analytical data are sufficient to model the trends, and so our aim is not so much to determine whether there is a trend but to obtain as accurate as possible a prediction of the properties of a sample. This latter type of modelling is called predictive modelling and is common in traditional analytical chemistry: the lower the error, the better the method. It is important not to confuse predictive modelling where the aim is to decrease errors and increase accuracy, with hypothesis based testing where there is no underlying certainty that we can obtain a

Chemometrics for Pattern Recognition Richard G. Brereton

specific result. Because most current chemometrics practice derives from traditional analytical chemistry where there are often underlying certainties, there is often an unfortunate confusion: if we are sure that a result can be obtained then we are entitled to try to optimize our method but until then, otherwise this could be dangerous.

In this text we will use several case studies for illustration of the methods, where each dataset has been obtained for a different purpose and as such contains characteristic features and poses specific problems. This text, to be limited in length, is primarily about pattern recognition and not signal analysis, peak deconvolution, calibration or experimental design, all of which though also play a key role in chemometrics.

When dealing with analytical chemistry data we are rarely directly interested in raw signals coming from an instrument (which may, for example, be in the form of voltages recorded as a function of time), but in chemically interpretable information such as the area of a chromatogram corresponding to a particular compound or the intensity of a spectrum at a known wavelength. Usually there is a significant number of steps required to convert this raw data into information that is suitable as input to a pattern recognition algorithm, such as data smoothing, Fourier transformation, resolution enhancement, deconvolution, peak detection, calibrating peak areas to concentrations and so on. We will describe briefly, for each case study, how the raw numbers are converted to a meaningful dataset that can be used for pattern recognition, but not provide detailed guidance on choosing these early steps, which would require an additional volume in its own right.

In addition, data have to be designed so that there is enough information to study the desired trends or factors. There is often a great deal of argument between statisticians and chemometricians as to the importance of experimental design. Many datasets available to chemometrics are not perfectly designed in the formal statistical sense. This may mean that they cannot be used to make definitive predictions, but nevertheless often something can be obtained from the data, and it is important for the chemometrics expert to realize the limitations of a dataset. For example we will often use the analogy of the toss of a coin. How times do I need to toss a coin to determine that it is biased? We could formulate this in another way, i.e. what sort of evidence is required to be 99.999 % sure that a coin is biased? This of course depends on the number of times the coin turns up H ('heads') and T ('tails') as well as the number of times the coin is tossed. Let us say the coin turns up H 90 % of the time. By calculating probabilities we find that we can be 99.999 % certain that a coin is not unbiased (i.e. it is biased) if we obtain this result after tossing the coin 21 times. Less than that, then either there must be more than 90 % H or we are less confident the coin is biased. However our experiment may be limited in size, and we may only have managed 10 tosses, and the coin comes up H 8 times. What can we say about bias? We are somewhat less sure it is biased, but nevertheless we can state something. Our design is not perfect and if we want to avoid making a mistake we do need more conclusive results or bigger datasets. But in advance of experimentation this is unknown, and our laboratory may only have limited resources. A more conclusive experiment may cost ten times the amount. Even if such experiments can be visualized, new problems may emerge, for example, samples may have to be analysed on an instrument such as a chromatograph that is not completely stable with time: columns need changing, the source of the instrument may need cleaning, there may be retention time drifts and downtime; the number of samples may be such that high throughput analysis is difficult or impossible because a chromatogram under optimal conditions may take an hour, so there may emerge a problem about storage as the number of samples may be too many to run on one instrument at a rate equal to the sampling; using more than one instrument may introduce errors relating to different characteristics of each instrument. So there has to be a compromise and in some cases experiments cannot be perfectly designed using classical statistical methods, and a chemometrician must be pragmatic. Nevertheless he or she should be aware of the limitations of an experimental design, what can safely be deduced, and what conclusions are dangerous.

2.2 Datasets, Matrices and Vectors

In order to understand how data are represented we need a basic understanding of matrices. In order to represent numerical information, we use the IUPAC recommended notation as follows:

- A scalar (or a single number for example a single concentration), is denoted in italics, e.g. x.
- A vector (or row/column of numbers, for example, a single spectrum) is denoted by lower case bold italics, e.g. $\boldsymbol{x}$.
- A matrix (or two dimensional array of numbers, for example, a set of spectra) is denoted by upper case bold italics, e.g. $\boldsymbol{X}$. Scalars and vectors can be regarded as matrices, one of both of whose dimensions are equal to 1.

A matrix can be represented by a region of a spreadsheet and its dimensions are characterized by the number of rows (for example, the number of samples or spectra) and by the number of columns (for example, the number of chromatographic peaks or biological measurements). Figure 2.1 is of a 3×5 matrix. The inverse of a matrix is denoted by $^{-1}$ so that $\boldsymbol{X}^{-1}$ is the inverse of $\boldsymbol{X}$. Only square matrices (where the number of rows and columns are equal) can have inverses. A few square matrices do not have inverses; this happens when the columns are correlated. The product of a matrix and its inverse is the identity matrix, a square matrix whose diagonals are equal to 1, and its off-diagonal elements equal to 0.

7	3	5	4	9
2	1	6	0	4
3	5	5	11	1

Figure 2.1 A 3×5 matrix represented by a region of a spreadsheet

The transpose of a matrix is often denoted by $'$ and involves swapping the rows and columns: so, for example, the transpose of a 5×8 matrix is an 8×5 matrix.

Most people in chemometrics use the dot ('.') product of matrices (and vectors). Usually although this is called the dot product the '.', or multiplication sign is left out (this varies according to the author), which we will do in this text. For the dot product $\boldsymbol{X}\,\boldsymbol{Y}$ to be viable, the number of columns in $\boldsymbol{X}$ must equal the number of rows in $\boldsymbol{Y}$ and the dimensions of the product equal the number of rows in the first matrix and number of columns in the second one. Note that in general $\boldsymbol{Y}\,\boldsymbol{X}$ (if allowed) is usually different to $\boldsymbol{X}\,\boldsymbol{Y}$.

The concept of a pseudoinverse of a matrix is often employed in chemometrics. To understand this, remember that only square matrices can have inverses, and so we need a different method when matrices are rectangular. This can be derived as follows.

If:

$$\boldsymbol{Z} = \boldsymbol{X}\,\boldsymbol{Y}$$

then:

$$\boldsymbol{X}'\,\boldsymbol{Z} = \boldsymbol{X}'\,\boldsymbol{X}\,\boldsymbol{Y}$$

and so:

$$(\boldsymbol{X}'\,\boldsymbol{X})^{-1}\,\boldsymbol{X}'\,\boldsymbol{Z} = (\boldsymbol{X}'\,\boldsymbol{X})^{-1}\,\boldsymbol{X}'\,\boldsymbol{X}\,\boldsymbol{Y} = \boldsymbol{Y}$$

where we call $(\boldsymbol{X}'\,\boldsymbol{X})^{-1}\,\boldsymbol{X}'$ the left 'pseudoinverse' of $\boldsymbol{X}$ which has dimensions $J \times I$ where $\boldsymbol{X}$ has dimensions $I \times J$ and will denote this by $\boldsymbol{X}^{+}$ for simplicity, in this text. There is also a right 'pseudoinverse', which in the case above is of $\boldsymbol{Y}$ denoted by $\boldsymbol{Y}'\,(\boldsymbol{Y}\,\boldsymbol{Y}')^{-1}$ which will also be denoted $\boldsymbol{Y}^{+}$ as appropriate.

In this text we will describe datasets in the form of matrices or vectors. Most experimental data can be arranged in a matrix format, where the columns represent variables (e.g. chromatographic peaks or spectroscopic wavelengths) and the rows represent samples (e.g. an extract from a soil or a sample of urine from a patient). The numbers in a matrix are called the elements of a matrix, and consist of individual measurements (e.g. the chromatographic peak height of a specific compound in a specific urine sample). In this text we use a number of definitions as follows:

- We denote an experimental datamatrix by $\boldsymbol{X}$.
- The matrix consists of I samples (e.g. extracts of urine) and J variables (e.g. chromatographic peak heights) and so is of dimensions $I \times J$.
- Elements of the matrix are labelled x_{ij} where i is the sample number and j the variable number, so that x_{35} is the measurement of the 3rd sample and 5th variable.
- Analytical measurements on individual samples are represented by row vectors, so that $\boldsymbol{x}_i$ corresponds to the vector of measurements for sample i. Note that in this text we use row vectors without a transpose to represent the measurement on a sample (e.g. a spectrum or a chromatogram), and some authors transpose these vectors.

This is illustrated in Figure 2.2.

Often there is some complementary information about the samples, for example a classifier, that is a label attached to each sample relating to its origin: we may record samples of human sweat from males and females and so an extra 'block' of information is available according to gender. We will call this block the c block, and it is usually represented by a vector which we will denote $\boldsymbol{c}$ (for classifier or calibrant). A major aim of pattern recognition is to see whether the two types of information ($\boldsymbol{X}$ and $\boldsymbol{c}$) can be related and if so, how, by forming a mathematical model between each type of data as

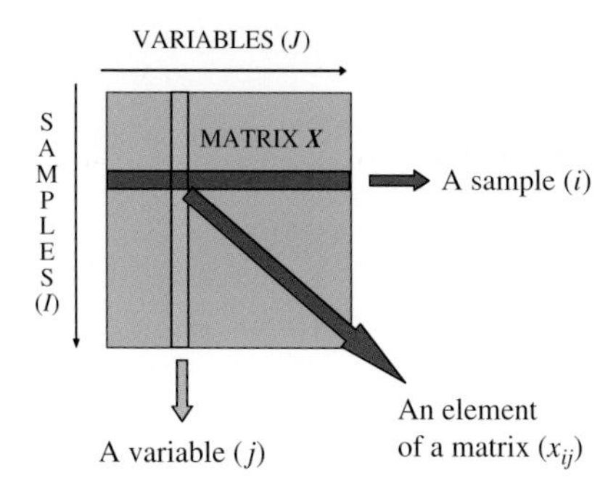

Figure 2.2 Representing analytical chemical data in matrix form

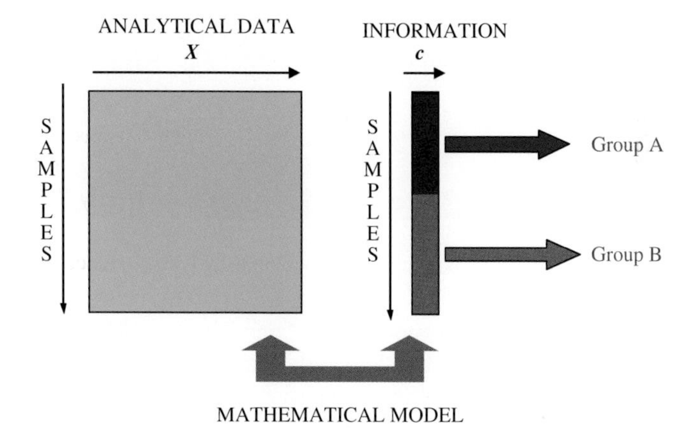

Figure 2.3 Including information about class membership

illustrated in Figure 2.3. Although in simpler cases this is a single vector (e.g. a value that can take on +1 if a sample is a member of one class or −1 if another class), sometimes it is more complex. If there are more than two classes, often $\boldsymbol{C}$ is a matrix, each of whose columns represent a single class, with +1 for samples that are a member of that class and −1 for samples that are not. There are of course a variety of other approaches for multiclass classification which we will discuss in Chapter 7.

Below we will look at the structure of each dataset in turn. Most will be quite simple as, in fact, the majority of datasets encountered in chemometrics are; however, a few will illustrate particular challenges requiring specific methodology. Where necessary preprocessing is described – for more detail, see Chapter 4 where the influence of the preprocessing method on the data is discussed; for the purpose of this chapter we list the default methods that are applied to each dataset later in the text, unless otherwise stated; it is important to recognize that other approaches are possible but for brevity we will illustrate the application of pattern recognition algorithms using a single type of preprocessing appropriate to each dataset. One transformation that is always applied is to remove any columns that contain just zero (this only relates to Case Studies 1 and 9).

2.3 Case Study 1: Forensic Analysis of Banknotes

In most countries in the world, illicit drug use is common. The proceeds of this business are purchased using cash in hand due to its illegal nature. It is known that banknotes that are in close proximity to certain substances such as drugs, contain small traces of contamination, as the compounds rub off and are trapped in the holes in the fibres (see Figure 2.4 for cocaine). A small percentage of all banknotes in the UK is likely to have been at some stage used for drug related transactions. However some compounds such as cocaine also are known to rub off from one banknote to another and so almost all banknotes in the UK (and indeed most Western countries) contain traces of cocaine, as banknotes are usually stored in close proximity to each other. Therefore nearly any banknote taken from a person's pocket will exhibit some traces of contamination.

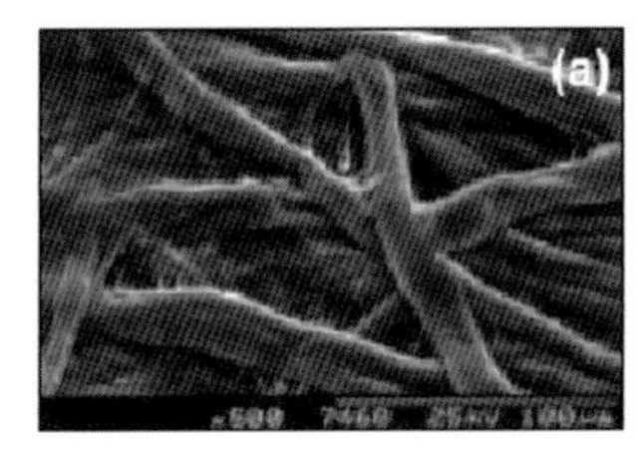

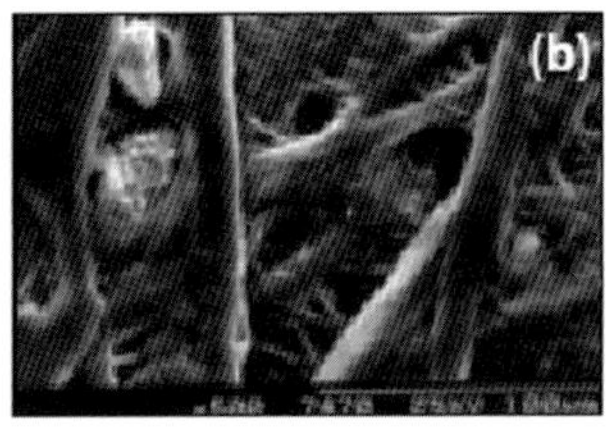

Figure 2.4 Electron micrographs of banknote fibres: (a) new; (b) with cocaine trapped in the holes

In the modern day, drug syndicates consist of a large number of street vendors, who are only minor players in what is a large scale illicit crime, and a smaller number of dealers who are rarely seen on the streets but who collect money and distribute drugs to the street vendors. Evidence against dealers may not involve direct CCTV pictures of them selling drugs, but may involve large cash hauls. Whereas cash in hand business is not illegal (e.g. it is possible to pay for building works in used banknotes), clearly drug dealing is illegal, and so in addition to large cash finds, additional evidence involves determining whether the cash is contaminated with drugs. If heavily contaminated it is likely to be a primary proceed of drug dealing. The difficulty is that all banknotes are to some extent contaminated (at least with cocaine) and so finding large cash deposits and finding that most banknotes contain detectable traces of cocaine is not sufficient. However if the levels of contamination are high then this is much stronger evidence – for banknotes in general circulation we may expect a few highly contaminated notes as these derive directly from the original dealers' banknotes 'diluted' into general circulation, but the majority of banknotes will be contaminated only to a very low extent, due to secondary rubbing off from the source. Hence if we analyse a set of banknotes we expect a 'distribution' of contamination, and we can use pattern recognition techniques to see whether this distribution arises from banknotes that come from general circulation or from potential drug dealers.

In order to study this all banknotes were analysed using tandem mass spectrometry and thermal desorption, and the peak area of the transition, which is very characteristic of cocaine, m/z 304 $\rightarrow$ 182 recorded. For readers unfamiliar with this technology, the more intense the peak, the more cocaine is found on the banknote. Using a series of banknotes it is possible to measure the intensity of the cocaine peak as illustrated in

Figure 2.5. For this case study, a sample or run consists of a set of banknotes, typically around 100. These are analysed for cocaine contamination, the cocaine peaks identified in the mass spectrometric trace, and integrated (several steps such as smoothing and baseline correction are necessary for this procedure to provide numerically valid results which are not discussed here), and a distribution can be obtained of contamination. This is most easily presented on a logarithmic scale since the contamination patterns follow approximately a log-normal distribution as illustrated in Figure 2.6(a). Each sample or batch of banknotes follows a different distribution, and the question is whether we can use pattern recognition to distinguish a set of banknotes obtained from general circulation and so innocent sources, and defendants that are likely to be guilty of drug dealing and money laundering – see Figure 2.6(b). Whereas visual inspection of the graphs may be possible if there are very few samples, it is important to realize that there will be a series of graphs, one per sample, and so superimposing many such distributions would become complicated and may not necessarily lead to an easily interpretable answer.

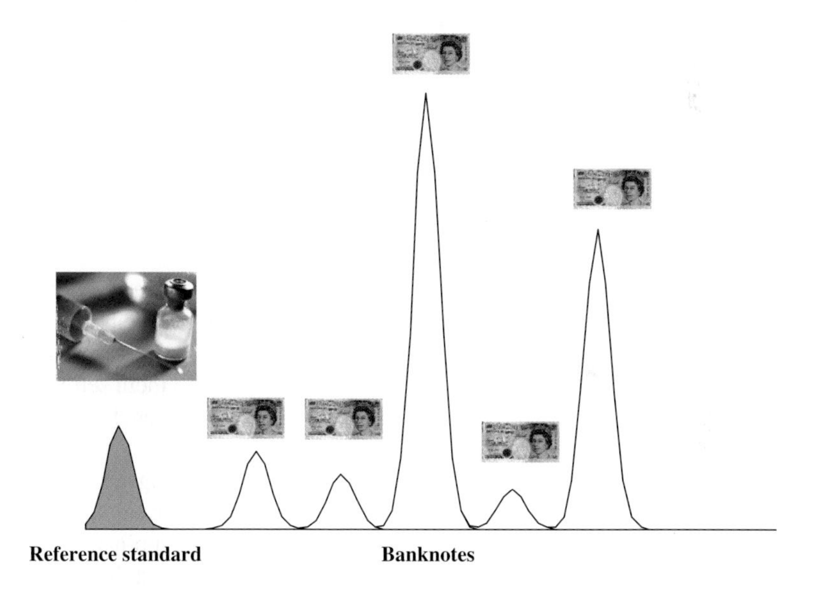

Figure 2.5 Five banknotes being analysed for cocaine using tandem mass spectrometry; first peak represents a reference standard

In this dataset, there are two classes of samples:

1. The first, class A, consists of samples obtained from defendants that are suspected drug dealers. Each run relates to a specific bundle of notes that has been seized by police. 46 such samples are analysed, consisting of 4826 banknotes in total, representing a mean of 105 notes per sample, also with a median of 105 notes.
2. The second, class B, consists of batches of banknotes withdrawn from banks in the UK at different times, consisting of 49 runs, each of which contained a minimum of 50 banknotes, consisting of a total of 7157 banknotes in total, representing a mean of 146 notes per sample with a median of 152 notes. These represent notes that come from innocent sources (banks).

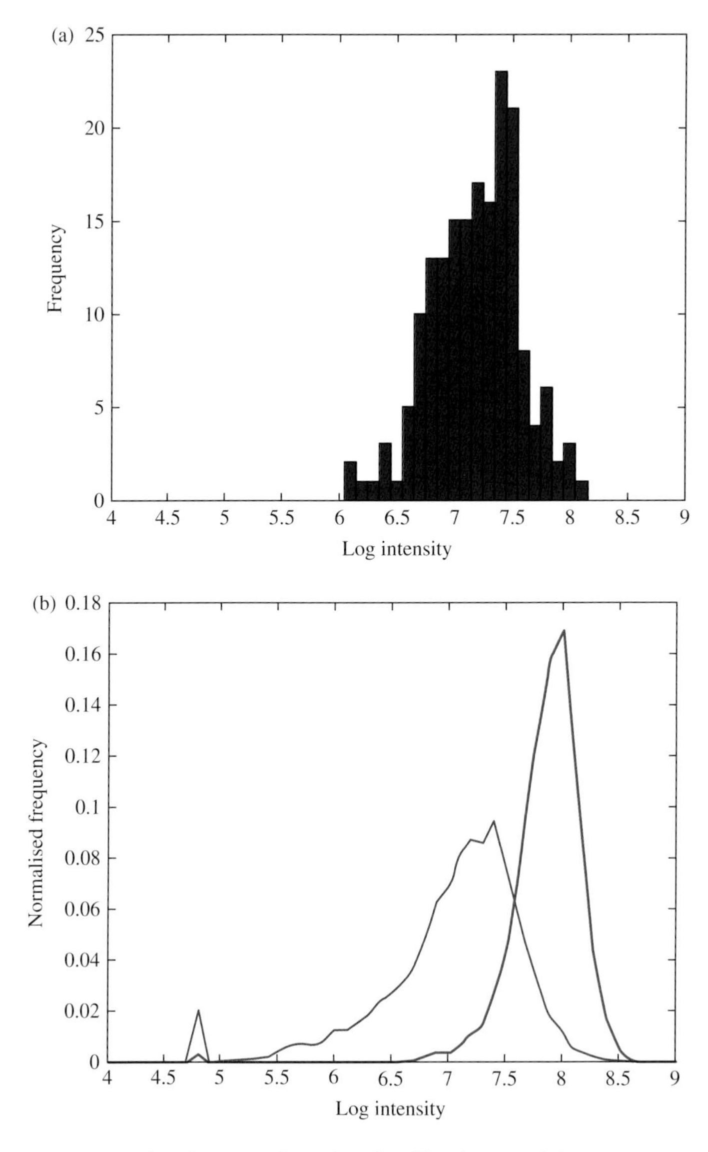

Figure 2.6 (a) Intensity distribution of one batch of banknotes. (b) Comparing two intensity distributions, blue = background general circulation, red = defendant. All logarithms are to the base 10

Pattern recognition is used to distinguish banknotes from both sources, and so potentially ask whether a batch of banknotes comes from general circulation or potential drug dealers. In order to formulate this problem in matrix terms the log-intensity distribution of peak areas is divided into 51 equally spaced bins from 4.0 to 9.0 (on a log10 scale). For a very few notes no contamination was detected and these were given a value that is half that of the minimum detected amount which results in a log-peak area of 4.8. As an example, if a run consisted of 200 banknotes, and the number of notes with log-contamination levels between greater than 7.25 and less than 7.35 is 8, then the intensity reading for the bin number 34 (bin number 1 corresponding a log-intensity value of 4.0, bin 2 to 4.1 etc.) is 8. Because the number of notes in each sample differs, in order to compare the frequency distributions, these numbers were divided by the total number of notes in the run. This would make an 'intensity' reading of 8/200 or 0.04 for bin 34 in this run.

Therefore each run can be represented as a vector containing 51 elements, which is row scaled to equal one. The overall dataset consists of 46 (class A) + 49 (class B) = 95 samples. Hence the data can be represented by a 95×51 matrix, the aim being to determine first whether there is a significant difference between the two sets of distributions and if so how well the membership of each group can be predicted. Below we describe the methods for preparing or preprocessing each dataset prior to pattern recognition. More about preprocessing is discussed in Chapter 4.

Prior to pattern recognition the following default transformations are performed:

1. All columns that contain only '0s' are removed, as otherwise there can be problems with matrices that have determinants of '0'. This leaves 39 columns, reducing the size of the data matrix to 95×39.
2. Columns are centred, but not standardized.

As usual (we will not repeat this below), there are other possible approaches to preprocessing which will be discussed as appropriate, but the aim of this chapter is to describe the defaults.

2.4 Case Study 2: Near Infrared Spectroscopic Analysis of Food

This study involves trying to assign samples of vegetable oils into one of four classes, using NIR (Near Infrared) spectroscopy, a traditional technique for the application of chemometrics. Because NIR spectroscopy is one of the earliest techniques to benefit from chemometrics, there is in fact an enormous literature on the use of multivariate methods in this area. Many approaches though are very specific to NIR spectroscopy and are not generally applicable in all areas of pattern recognition, and this text is not primarily one on infrared signal analysis, which is well covered elsewhere.

The data consist of 72 spectra from the following:

1. 18 samples of Corn Oils (class A).
2. 30 samples of Olive Oils (class B).
3. 16 samples of Safflower Oils (class C).
4. 8 samples of Corn Margarines (class D).

Note that the number of Corn Margarine samples is quite low and there can be problems in modelling groups with few samples.

In this dataset the following steps are used to prepare the data:

1. The NIR data are baseline corrected using an approach called MSC (Multiplicative Scatter Correction).
2. A region of the spectrum between 600 and 1500 nm wavelength is used for pattern recognition (the remainder is uninformative and can lead to degradation of results).

3. The data are mean centred because some regions are more intense than others, but the variability at each wavelength is very similar. No standardizing is required providing that mean centring has been performed.

The MSC corrected spectra are illustrated in Figure 2.7. It can be seen that there are some small differences between spectra of the groups, for example, at around 700 nm the safflower oils appear to exhibit more intense absorbance followed by the corn oil; however these differences are quite small and there is a little bit of spread within each group (as expected), and so it would be quite hard to identify an unknown oil, by eye, using a single NIR spectrum, and pattern recognition techniques can be employed to determine whether the groups can be distinguished, which spectral features are best for discrimination and how well an unknown can be assigned to a specific group.

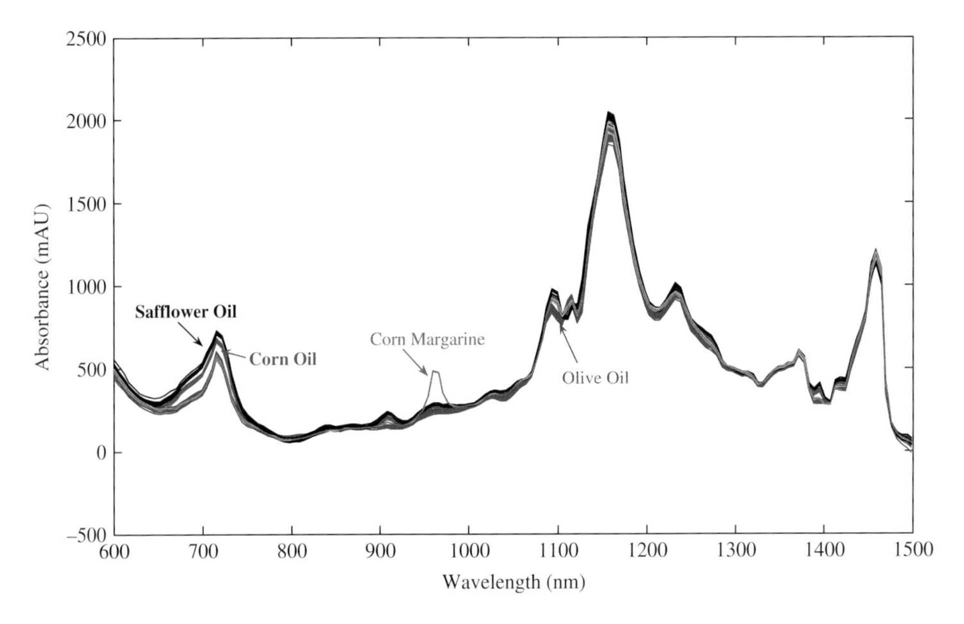

Figure 2.7 MSC corrected NIR spectra of the four groups of oils

This particular application is very much one of classical chemometrics and one for which the pattern recognition techniques perform successfully. It is a classification problem, but unlike that of Case Study 1, there are 4 rather than 2 classes, and there are particular issues about dealing with data when there are more than two groups in the data; usually there is no sequential meaning to the class labels in the ***C*** block, and rather than using a vector (with a number for each class), the classifier is often best presented by a matrix as shown in Figure 2.8, with each column representing one class. In addition to having a multiclass structure, there also is a problem in that the number of corn margarines is very small. However on NIR datasets chemometrics techniques usually tend to work very well, as will be shown in later chapters.

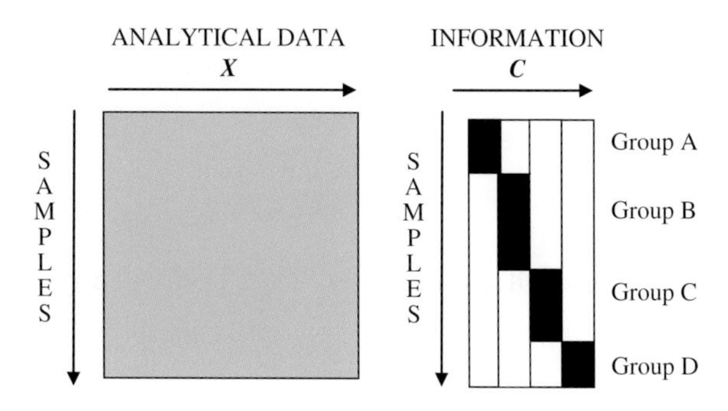

Figure 2.8 Structure of Case Study 2

2.5 Case Study 3: Thermal Analysis of Polymers

Most commercial plastics are polymers. The aim of this study is to be able to determine the group a plastic belongs to using its thermal properties. Commercial plastics have different properties, as their structure changes when heated, and each type of plastic has a different use and so will have different characteristics. The changes involve going from a solid to a glass to a liquid state. By applying an oscillating force and measuring the resulting displacement, the stiffness of the sample can be determined, which will change as the polymer is heated, using the technique of Dynamic Mechanical Analysis (DMA). Several parameters can be measured but in this study we use the Loss Modulus (E''), which in this text is measured in units of 10^7 Pascals, which is related to the proportion of the energy dissipated or non-recoverable per cycle, as a force is applied. For those unfamiliar with the concept an analogy involves throwing a ball onto the ground and seeing how far it will bounce back – the less it bounces the more the Loss Modulus, and the less elastic and more stiff it is. Hence high values of E'' correspond to temperatures at which the plastic is stiff. A typical trace is illustrated in Figure 2.9.

The temperature range studied is from −51 °C until the minimum stiffness of the polymer is reached, after which no further meaningful data can be collected. Measurements are made approximately every 1.5 °C. Each raw trace curve consists of between 99 and 215 data points dependent on the highest recordable datapoint of the polymer. After the highest recordable temperature, in order to ensure a similar temperature range for each sample, values (which were not recorded) are replaced by the value of E'' obtained at the highest measurable temperature: this is a relatively small number and makes very little difference to the analysis. Because the measurements for different samples are not performed at the same equally spaced temperatures the data are linearly interpolated to 215 equally spaced data points corresponding to an interpolated E'' value for each of the temperatures between −51 °C and 270 °C in increments of 1.5 °C.

293 samples are used for this case study to give a data matrix $\mathbf{X}$ of dimensions 293 × 215, which is first centred. The dataset is not preprocessed further as there is approximately similar interest in variability at all temperatures. An interest in this data is that there are two ways in which the polymers can be classified, either into polymer type by their main physical properties (amorphous or semi-crystalline), in two main classes, or into nine groups

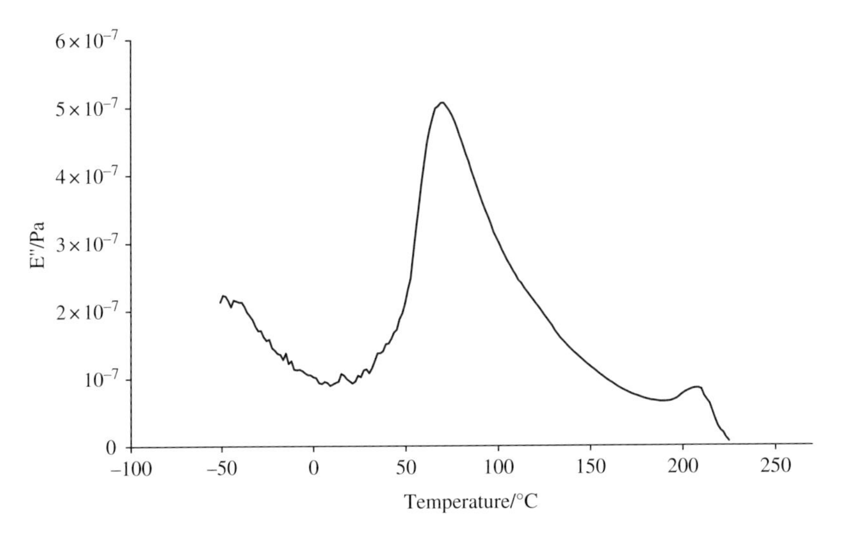

Figure 2.9 A typical DMA trace as obtained in Case Study 3

according to the polymeric material as listed in Table 2.1. Note the class lettering system we will use in this text, that class A (polymer type) is different to class A (polymer group). The classification is hierarchical in nature, as a particular polymeric group is all of one type. This dataset can be viewed both as a two class problem (as in Case Study 1) or a multiclass problem (as in Case Study 2).

Table 2.1 *Samples for Case Study 3*

Type			Group		
Amorphous	A	92	Polystyrene (PS)	A	35
			Acrylonitrile–Butadiene–Styrene (ABS)	B	47
			Polycarbonate (PCarb)	C	10
Semi-crystalline	B	201	Low-density polyethylene (LDPE)	D	56
			Polypropylene (PP)	E	45
			High-density polyethylene (HDPE)	F	30
			Polyamide 6 (PA6)	G	20
			Poly(butylene terephthalate) (PBT)	H	10
			Poly(ethylene terephthalate) (PET)	I	40

The aim of this study is to determine the origins of an unknown polymer using DMA traces from a dataset of polymers of known origins. There are many potential reasons for this, where one possible area is waste recycling of industrial plastics, where plastics have to be treated in different ways for environmentally safe disposal. In addition, the methods can also be extended to look at grades (which we will not discuss in this text) or for Quality Control of manufacturing processes.

2.6 Case Study 4: Environmental Pollution using Headspace Mass Spectrometry

The aim here is to study pollution by detecting hydrocarbons in different natural habitats, especially in zones where crude oil is extracted and spillages are frequent. Although gas chromatography methods are traditionally used here, they are time consuming and for automated analysis require alignment of peaks or manual interpretation. MS (Mass Spectrometric) techniques are a faster alternative, and although, unless coupled to chromatography or used in tandem techniques, they may not provide information about individual components in a mixture they can be used as a 'fingerprint' with enough information, suitably processed with the appropriate chemometric techniques, to make decisions about the proposed problem. For Case Study 4, a headspace sampler coupled to a mass spectrometer (HS–MS) was used for the analysis of soils polluted by crude oil and derivatives.

The dataset consisted of 213 samples of soil and sand. Of these 179 were spiked with oil in the laboratory at different levels, representing polluted samples, and 34 were 'clean', representing unpolluted samples, as listed in Table 2.2. For the purpose of this text we are primarily concerned with determining whether a sample comes from the polluted group (class A) or the unpolluted group (class B), rather than the extent of pollution, which is a calibration problem; however we would expect some of the more lightly polluted samples to be close to the boundary between the classes. We are trying to ask whether it is possible to distinguish polluted from unpolluted samples using MS data and pattern recognition and then to determine how well we can classify samples into one of these two groups.

MS data are recorded from m/z 49 to 160, and a typical spectrum is illustrated in Figure 2.10. In a small number of cases (Figure 2.11) the spectrum is dominated by a small number of large peaks. Data preprocessing has to take this into account and three steps are performed as follows:

1. The MS intensities are first square rooted. This is not always necessary for handling of MS data, but if it precedes row scaling it protects against a few intense peaks over-influencing the result; it is an alternative to log-scaling, suitable when there are zero or very low intensity values.
2. Each square rooted mass spectrum is then row-scaled to a total of 1. This is because the amount of sample introduced to the MS instrument cannot easily be controlled, and it is hard to find internal standards for HS–MS.
3. Finally the columns (each variable) are standardized to allow each m/z value to have equal influence on the resultant pattern recognition. Sometimes low intensity ions can be diagnostic of interesting compounds that are present in low quantities.

More details of these steps are discussed in Chapter 4. For this particular dataset quite a variety of preprocessing options could be employed with most giving comparable answers (in other cases a correct choice of preprocessing is essential), but for this text we stick to one protocol for the majority of results.

One feature of the environmental dataset is the very different group sizes – we will discuss approaches for modifying classifiers for unequal class sizes in Chapter 10. Some approaches do not require significant modification for group sizes but others are quite sensitive.

Table 2.2 *Samples for Case Study 4*

Sample	Spiked levels (mg/kg)	Number
Polluted samples (Class A)		
Commercial sand spiked with gas oil	1.2	3
	2.4	3
	5.9	3
	12	3
	24	3
	59	3
	74	3
	148	3
	297	3
	371	3
Commercial sand spiked with Iran light crude oil	1.4	3
	2.8	3
	6.9	3
	14	3
	28	3
	69	3
	86	3
	172	3
	345	3
	431	3
Commercial sand spiked with Brass River light crude oil	1.3	3
	2.6	3
	6.5	3
	13	3
	26	3
	65	3
	82	3
	164	3
	327	3
	409	3
Santander beach sand 1 spiked with gas oil	1.2	3
	2.4	3
	5.9	3
	12	3
	24	3
	59	3
	74	3
	148	3
	297	3
	371	3
Santander beach sand 1 spiked with Iran light crude oil	1.4	2
	2.8	3
	6.9	3
	14	3
	28	3
	69	3
	86	3
	172	3
	345	3
	431	3

Table 2.2 (*continued*)

Sample	Spiked levels (mg/kg)	Number
Santander beach sand 1 spiked with Brass River light crude oil	1.3	3
	2.6	3
	6.5	3
	13	3
	26	3
	65	3
	82	3
	164	3
	327	3
	409	3
Unpolluted samples (Class B)		
Santander beach sand 2	none	3
Coruña beach sand 4	none	3
Soil 1	none	3
Soil 2	none	3
Soil 3	none	3
Soil 4	none	3
Commercial sand	none	10
Santander beach sand 1	none	3
Coruña beach sand 3	none	3

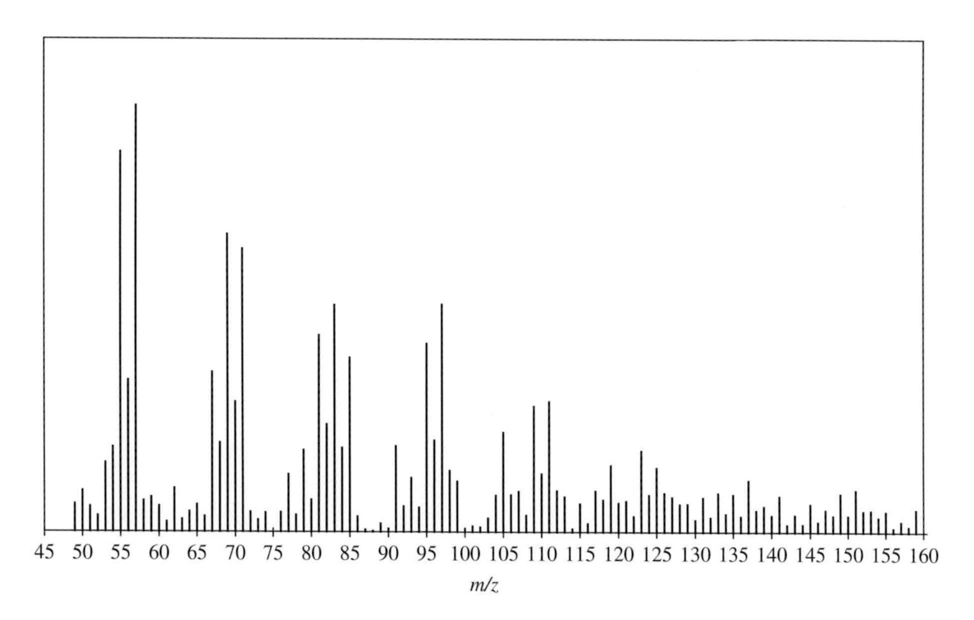

Figure 2.10 Typical mass spectrum (Santander beach sand spike with 2.4 mg/kg gas oil) for Case Study 4

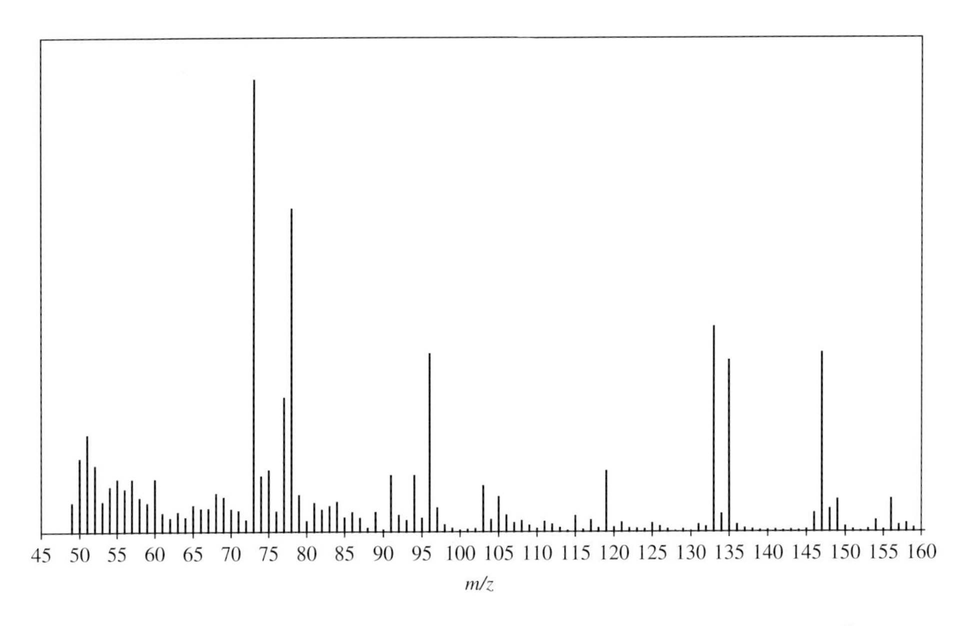

Figure 2.11 Mass spectrum of soil 3, dominated by a few very intense peaks

2.7 Case Study 5: Human Sweat Analysed by Gas Chromatography Mass Spectrometry

The aim of this study is to determine whether one can classify humans by gender from their sweat, and also whether there is a signal due to individuality within their sweat. There could be various reasons behind the signal, for example, genetics, personal habits, behaviour that is characteristic of sexes, but this is not the primary aim of this study, to determine why, but simply whether, and with what confidence.

The dataset studied consisted of human sweat samples of five repeats from subjects, sampled once a fortnight for five fortnights between June and August 2005 obtained from an isolated population in Carinthia, Southern Austria. 182 subjects were sampled five times over the period of the study including 83 males (class A – total 415 samples) and 99 females (class B – total 495 samples).

In metabolomic profiling it is important to recognize that samples taken from an individual over time are not strictly replicates as they will be influenced by different habits, different times of the year and so on, which can be quite confusing for analytical chemists used to dealing with replicate samples, and are likely to show substantial variability from one fortnight to the next. Although we may primarily be interested in asking some questions about the data, for example, "can we distinguish subjects by gender from their sweat?", these questions are unlikely to relate to the main or only factors that influence the metabolic signal: there may be twenty or more factors, many of which will be unknown to the investigator, and so although we might be able to form such a model to relate sweat to gender, it is likely that only a small portion of the compounds detected in sweat help us to answer such questions. This results in a major dilemma, especially in the area of metabolomics, in variable selection (see Chapter 9). If, for example, we identify ten thousand compounds and suspect only ten or twenty are relevant to our study, should we preselect the variables we think are most important and then form models using this subset of compounds (often called markers)? We can show that this often leads to flawed and overoptimistic predictions, as the models are being prejudiced in advance. Techniques to overcome this problem and produce safe predictions without the risk of overfitting are discussed in several chapters of the text, particularly Chapter 8, in the context of model validation.

The samples were extracted and then analysed by GCMS (Gas Chromatography Mass Spectrometry). A typical TIC (Total Ion Current) is presented in Figure 2.12. Several steps were then necessary:

1. Peaks were detected in each GCMS trace.
2. Peaks from each GCMS trace that had a common chemical origin as based on chromatographic Retention Time and Mass Spectral similarity were identified.
3. Peaks known to come from siloxanes (due to the analytical process) were removed.
4. Peaks were retained that were found in at least one subject at least four out of five fortnights: this is to look at peaks that are constant in individuals and more likely to be due to factors such as individuality and gender. This retained 337 peaks of unique origins. The integrated intensity over all masses was calculated for each peak.
5. Finally the data were presented as peak table of dimensions 910×337, where the columns refer to unique compounds and the rows to individuals. This peak table can also be presented as sub-tables for each fortnight.

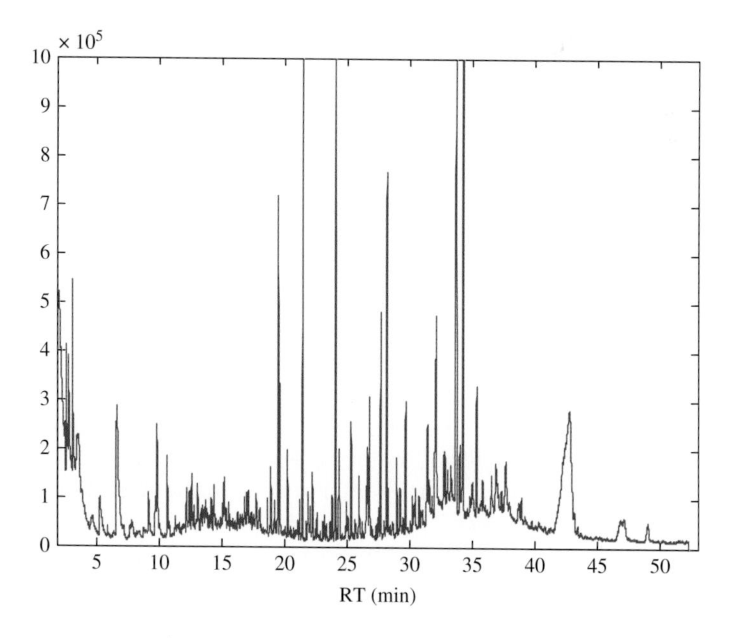

Figure 2.12 Typical GCMS trace of sweat

In order to analyse the peak tables, the data were further processed as follows:

6. The areas of peaks were square rooted. This is because there were a few very large peaks, and these would otherwise dominate the analysis. Logarithmic scaling, although an alternative, suffers from the fact that there are many zero intensity measurements in the data that have to be handled in a special way.
7. The square rooted peak table was row scaled to 1 for each chromatogram over all 337 possible identified peaks (no chromatogram contains all these peaks).
8. The data are standardized for the overall dataset.

One feature of this data, like many metabolomic datasets, is that there are often peaks that are not detected in each sample, that is the raw datamatrix contains a lot of zero values. Another feature is that there are likely to be many factors influencing the observed GCMS and as such it is unlikely that simple visualization, for example, of the first Principal Components (see Chapter 3) will reveal much, and so that statistical pattern recognition methods are necessary to get the best of the data.

2.8 Case Study 6: Liquid Chromatography Mass Spectrometry of Pharmaceutical Tablets

Chemometrics has a major role to play in the pharmaceutical industry, especially with the PAT (Process Analytical Technology) initiative, particularly to look at quality and origins of products. For most pharmaceutical products of the same drug, the main ingredient (API – Active Pharmaceutical Ingredient) and for tablets the excipient (which binds the tablet together), are the same, but small differences in their composition are indicative of differences in manufacturing processes. These small impurities are often diagnostic of specific problems in manufacturing, and can be used as fingerprints of the origin of a batch of product, for example, which factory did it come from, who was the supplier, is one plant behaving in a similar way to another plant, etc.

In LCMS (Liquid Chromatography Mass Spectrometry) these impurities that provide signatures of how a tablet (for example) has been manufactured can be detected but as small peaks. However in addition to the manufacturing route, also the instrument on which the sample was analysed and when it was analysed will play a part in influencing the appearance of the LCMS trace. This is because instruments (and chromatographic columns) age, that is their performance may change with time, as they get dirty, they are cleaned, columns are changed, etc.: and no two instruments perform identically; this is a real problem because instruments will be replaced over time so an analytical method developed one year may have to be transferred to a new instrument the next year as a laboratory is upgraded, and especially in a regulated industry such as pharmaceuticals there will always be constant change as old equipment becomes obsolete or fails or simply wears out. Hence we expect the LCMS signal to be influenced both by the origin of the tablet and by the instrumental conditions. In Case Study 6, we are primarily interested in determining the origins of tablets, that come from three manufacturing routes, and seeing if there is a process specific signature that can be detected in the impurities. However imposed upon this is an instrumental signal and so two primary factors will influence the appearance of the LCMS signal, the route and the instrumental conditions. We study 79 samples as listed in Table 2.3, divided into origins (different manufacturing routes) and batches (different instrumental conditions). It is of interest both to determine whether it is possible to classify samples due to their origin (which is the primary aim of this investigation) but also how important the instrumental conditions or batch are, and what features are primarily due to origin and what are due to the instrument.

Table 2.3 *Samples for Case Study 6*

	Batch 1	Batch 2	Batch 3	Batch 4	Batch 5		
Origin 1	8	7	8	8	8	Class A	39
Origin 2	3	3	3	3	3	Class B	15
Origin 3	5	5	5	5	5	Class C	25
	Class A	Class B	Class C	Class D	Class E		
	16	15	16	16	16		

The raw LCMS trace (Figure 2.13(a)) does not look very promising because the main peaks are due to the excipient and API. However in order to obtain a more informative LCMS trace the following were performed:

1. The regions below 3 min, where the excipient elutes, and above 25 min, were removed.
2. Component stripping was used to remove the large peak due to the API eluting at around 10 min; this procedure allows the retention of co-eluents and is an alternative to simply removing this region, and is a way of removing the large peaks but retaining peaks underneath if they exhibit different spectra.
3. The data were smoothed using a procedure called CODA that results in chromatograms that are less noisy. A resultant LCMS trace is seen in Figure 2.13(b).

The next stage is to form a peak table, as follows:

4. Peaks were detected in each LCMS trace.
5. Peaks from each LCMS trace that had a common chemical origin as based on chromatographic Retention Time and Mass Spectral similarity; these identified 90 possible peaks. The integrated intensity over all masses was calculated for each peak.
6. Finally the data was presented as peak table of dimensions 79×90, where the columns refer to unique compounds and the rows to samples.

The data are subjected to square root scaling (to reduce the influence of large peaks when row scaling), row scaling to a constant total of 1, and then standardizing down each column so that small peaks have equal influence on the model to large peaks.

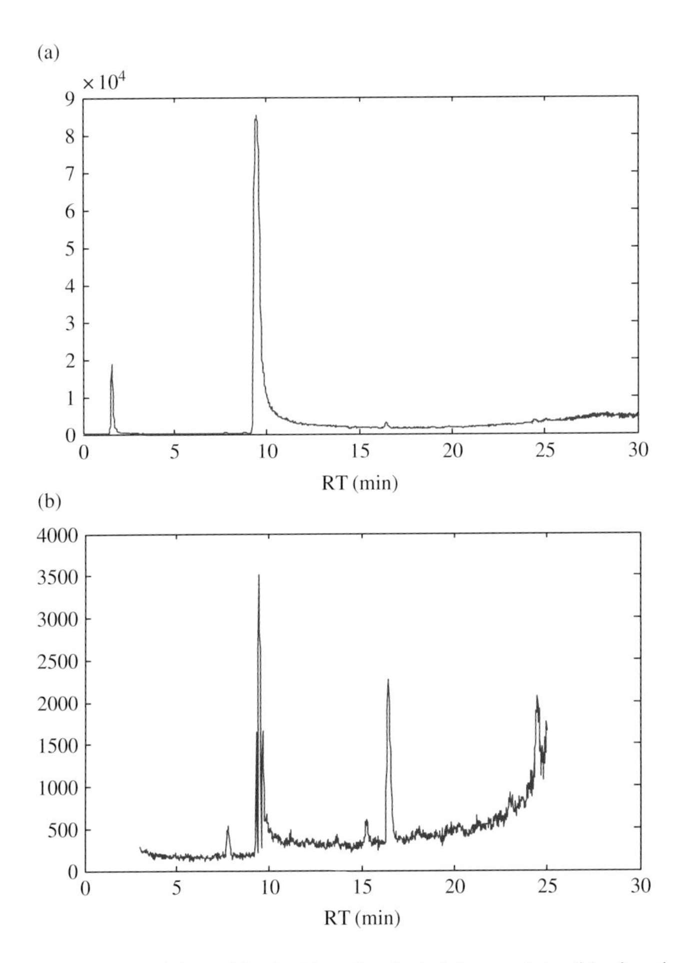

Figure 2.13 LCMS traces of the tablet in Case Study 6: (a) raw data; (b) after the main peaks have been removed and 3–25 min are retained

2.9 Case Study 7: Atomic Spectroscopy for the Study of Hypertension

This dataset contains 540 samples that can be separated into 94 samples of patients (class A) that have Hypertension disease (High blood pressure) and 446 samples of controls (class B) that are known not to have hypertension. Class A can be further split into 4 different subgroups, group C – cardiovascular hypertension (CV, 31 samples), group D – cardiovascular accident (CA, 19 samples), group E – renal hypertension (RH, 21 samples) and group F – malegnial hypertension (MH, 23 samples) studied. The structure of this dataset is thus partially hierarchical, as illustrated in Figure 2.14.

In the dataset, there are only 5 variables, which are the amounts of Cu, Zn, Mg, Pb and Cd in blood in units of μg/mL. These are all detected over quite different ranges as indicated in Table 2.4, and so the data are standardized prior to pattern recognition.

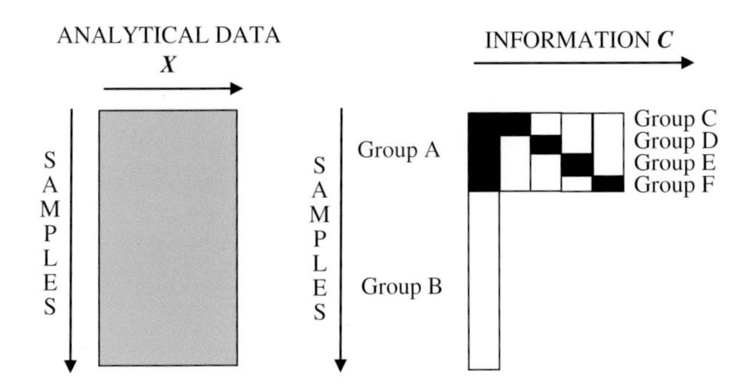

Figure 2.14 Structure of Case Study 7

Table 2.4 *Ranges in which five elements are detected in blood for Case Study 7, in μg/mL*

Element	Cu	Zn	Mg	Pb	Cd
Maximum	1.8	13.2	79.5	321	171.1
Minimum	0.42	4.9	14.5	0.078	0.08

There are a number of features of this dataset, as follows:

1. There are very few variables.
2. The class sizes (A and B) are very different, and one class has subgroups.
3. Some of the trends (between control and diseased patients) are very easy to see by eye, and could be determined using a univariate measure – see Table 2.5 for Cd and Pb – so pattern recognition is mainly used for quick data visualization (although this is still very useful); however for the different types of the disease there are overlapping ranges for each variable.

Whereas this dataset is quite simple compared to most in this text, it is very typical of the sort of data for which easy visualization of trends is helpful in reports and papers.

Table 2.5 *Ranges for the five elements of Case Study 7, in μg/mL*

Element	Cu	Zn	Mg	Pb	Cd		Cu	Zn	Mg	Pb	Cd
A	1.6	9.3	52.1	0.8	9.9						
	0.6	4.9	14.5	0.078	0.08						
B	1.8	13.2	79.5	321	171.1	C	1.8	9.9	79.5	321	132
	0.42	6.1	32.4	131	60		0.62	6.3	32.4	195	81.3
						D	0.89	10.2	56.9	307.1	98.3
							0.42	6.5	36.2	165.7	60
						E	0.91	13.2	51.2	314	171.1
							0.57	6.9	36.5	198	102.9
						F	1.7	11.7	57.5	297	106
							0.59	6.1	37.5	131	79

2.10 Case Study 8: Metabolic Profiling of Mouse Urine by Gas Chromatography of Urine Extracts

This case study consists of 2 datasets: 8(a) involving a study of the effect of diet on mice and 8(b) genetics.

Dataset 8(a) consists of the following:

1. 59 samples from mice fed on a special high fat diet (class A).
2. 60 samples from mice fed on a normal diet (class B).

All mice were of the same strain (B6H2b) and age (11–12 weeks) and all were male. The aim is to determine whether by analysing the chemical profile of urine one can determine whether a mouse is on a high fat diet on not.

Dataset 8(b) consists of the following:

1. 71 samples from mice of strain and haplotype AKRH2k (class A).
2. 59 samples from mice of strain and haplotype B6H2k (class B).
3. 75 samples from mice of strain and haplotype B6H2b (class C).

All mice were of the same age (12 weeks). Class A is a different strain to classes B and C, whereas class B and C differ in one gene (denoted k and b) – called a haplotype – which class A shares in common with class B. All mice were kept in the same environment and all were male. We might anticipate somewhat better separation between strains than haplotypes. It is expected that there are some differences in the urine according to genetic origin as this is one way mice communicate. The aim is to determine whether a mouse can be classified into a genetic group according to its urinary chemosignal. This particular dataset consists of three classes; however class A is likely to differ more from the other two. There are likely to be a few characteristics in common with classes A and B (as these share a specific gene that is thought to be important in the urinary signal) and a lot of characteristics in common between classes B and C. However classes A and C will be the most different.

Similar to Case Study 5, the samples were extracted and then subjected to GCMS analysis. The following steps were followed for each dataset:

1. Peaks were detected in each GCMS trace.
2. Peaks from each GCMS trace that had a common chemical origin as based on chromatographic Retention Time and Mass Spectral similarity were identified.
3. Peaks known to come from siloxanes (due to the analytical process) were removed.
4. Rare peaks were eliminated: for dataset 8(a) a threshold of being detected in at least 20 samples retaining 307 peaks, and for 8(b) at least 40 samples were employed retaining 316 peaks. This threshold is to reduce the size of the peak table, but other criteria could be employed; however in practice rare peaks are unlikely to be good markers or discriminators. The thresholds for occurrence is always below the minimum group size, and so a perfect marker that is found just in one of the groups would always be retained.
5. The integrated intensity over all masses was calculated for each peak.

There are two resultant peak tables of dimensions 119×307 for Case Study 8(a) and 205×316 for Case Study 8(b). The final steps are similar to steps 6 to 8 of Case Study 5.

2.11 Case Study 9: Nuclear Magnetic Resonance Spectroscopy for Salival Analysis of the Effect of Mouthwash

NMR spectroscopy is commonly employed in metabolomics. This study aims to use NMR spectroscopy to determine the effect of an oral mouth rinse on salival samples. A series of volunteers (16 donors) without any form of active dental disease were selected. Each donor provided a saliva specimen at daily 'waking' time in the morning. Each sample was split in two and was treated with either 0.5 mL of an oral rinse formulation (class A) or water (class B) and each donor was studied over three days with one sample being obtained per day. The structure of the dataset is illustrated in Figure 2.15, resulting in 16 (donors) $\times$ 2 (treatments) $\times$ 3 (days) $=$ 96 samples.

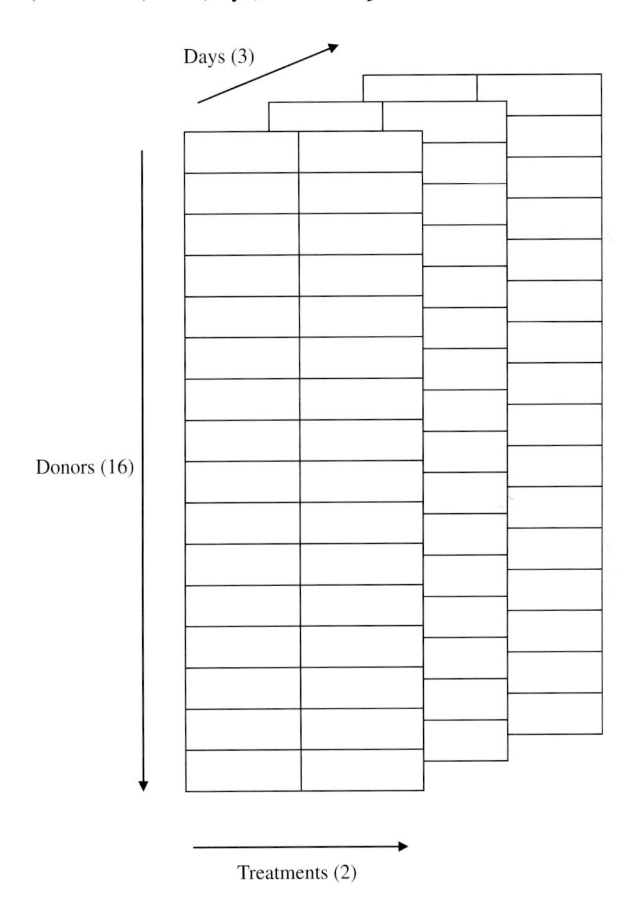

Figure 2.15 Structure of Case Study 9

^{1}H NMR spectroscopy was used to analyse each sample, with the data divided into 146 'intelligent buckets' which is a way of dividing spectra up over specific ppm regions. The intensity in each bucket is summed up in each spectrum over the corresponding regions. The reason for doing this rather than using the raw Fourier transformed NMR data is that there sometimes are small shifts between spectra (of a few datapoints), but by taking an integrated area over several datapoints, these small shifts are averaged out and reduce the errors that could otherwise be introduced; the buckets are unequal in size. Some uninformative regions of the spectra are removed by this procedure; all buckets where the summed intensity is less than 1 % of the bucket with the maximum summed intensity are removed, as these in most case just contain noise. Apart from a couple of small buckets in the aromatic region all chemical shifts were less than 5.44 ppm as the main interest is in hydrocarbons. This leaves 49 buckets. A typical NMR spectrum is illustrated in Figure 2.16. A datamatrix of 96×49 is created for subsequent pattern recognition.

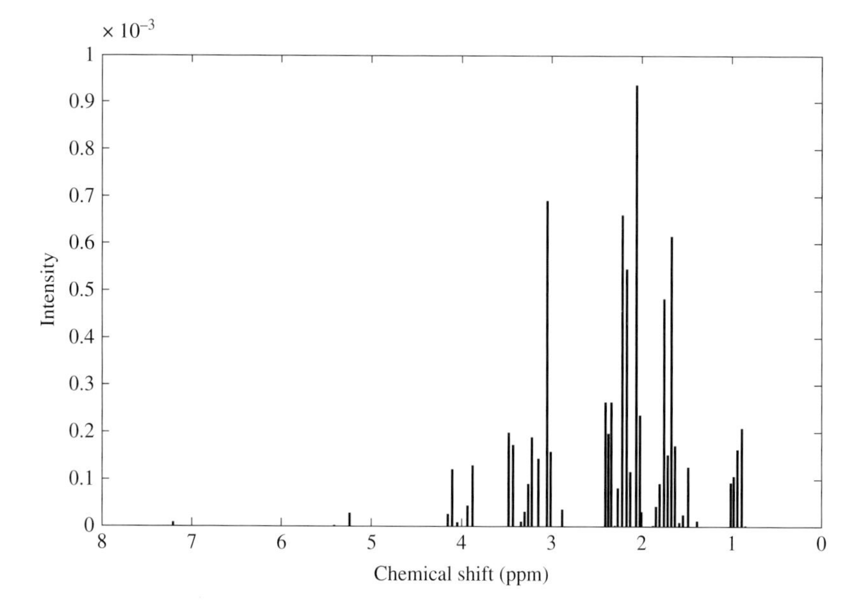

Figure 2.16 Typical bucketed NMR spectrum used in this study

Two further preprocessing steps are performed as follows:

1. Square root scaling. This is because there are some large peaks in certain samples. In this case the reason is not for the purpose of normalization but because these occur in occasional places in each column for certain variables. An alternative of log scaling suffers because there are some zero intensity readings.
2. Centring. Since the variability an each of the selected regions is approximately the same there is no need to standardize.

2.12 Case Study 10: Simulations

The aim of this case study is to simulate datasets consisting of two classes with different degrees of overlap. In addition only certain variables are discriminatory, that is potential markers for each group. Simulations play a role in validating methods as the results are known in advance but often have to be used with caution because sometimes there are different features in real datasets (e.g. outliers, asymmetric distributions, noise, etc.) that are not taken into account in the simulations, and so results should be interpreted with care. For example, the performance of classification methods depends on the underlying data distribution, and so one simulation may suggest that method A is better than method B, but for a different simulation the reverse. Unless the simulations relate quite closely to the real life problem under consideration they can be misleading.

Case Study 10 is primarily used to determine how easy it is to determine which variables are good potential discriminators (Chapter 9) between two groups of samples, according to degree of overlap, and whether these are correct, and also to demonstrate the use of class separation indices (Chapter 11).

Three simulations 10(a) to 10(c) were generated as follows. Each consisted of 200 samples, 100 from class A and 100 from class B, characterized by 100 variables and are each of dimensions 200×100. Twenty of the variables were induced to be potential discriminators, with the remaining variables being non-discriminatory.

The distributions of the variables and data matrices were obtained as follows:

1. In many situations, it is found that the mean intensities of variables (over an entire dataset) often follow an approximate log-normal distribution, and so this is generated as follows for each variable j.
   (a) λ_j is generated using a random normal distribution generator with mean 11.5, and standard deviation 1.5 or N (11.5,1.5) where N stands for a set of measurements obtained from an underlying normal distribution.
   (b) The underlying population mean (μ_j) of this variable is given by $\mu_j = e^{\lambda_j}$. This value is used both for discriminatory and non-discriminatory variables, with modification for the former as described below.
2. The underlying population standard deviation of each variable is generated as follows.
   (a) For non-discriminators, $\sigma_j = v_j\ \mu_j$ where v_j is generated using a random uniform distribution between limits of 0.1 and 0.3.
   (b) For discriminators there are two values σ_{jA} and σ_{jB} for each class A and B which are independently generated by a uniform distribution as described in step (a), and so reflect somewhat different spreads in values for each variable as often happens in practice.
3. The underlying population means of the discrimatory variables are generated by setting

$$\mu_{j\mathrm{A}} = \mu_j + s\ p\ \sigma_{j\,\mathrm{pooled}} \text{ and } \mu_{j\mathrm{B}} = \mu_{\mathrm{j}} - s\ p\ \sigma_{j\,\mathrm{pooled}}$$

   where:

   (a) $\sigma_{j\ \mathrm{pooled}}$ is the pooled standard deviation of the variable j over both classes.
   (b) s takes the value of $+1$ or -1 with 50 % probability, where if positive, the mean of the variable is greater in class A to class B and vice versa.
   (c) p takes one of three values 1.5 (Case Study 10(a)), 1.0 (Case Study 10(b)), and 0.5 (Case Study 10(c)), with a lower value of p resulting in more overlap between the classes.

For the non discriminators the mean generated in step 1 is employed.

4. Once the population means and standard deviations for each variable have been determined, the intensity values are obtained using a random normal distribution with underlying population mean and standard deviations as described in steps 2 and 3 above to give three matrices of dimensions 200×100 (according to 3(c)). For the discriminators the elements of the matrix were generated separately for each class. In a small number of cases this leads to negative intensities (3.22 % of samples over all samples) and these were set to 0.

Because each variable is on a very different scale due to the log-normal distrbution of intensities, the three datasets from Case Study 10 are standardized prior to pattern recognition.

The 20 variables chosen to be discriminatory in each of the three simulations are presented in Table 2.6 by variable number. This can be used to determine the effectiveness of methods for variable selection as discussed in Chapter 9. Note that there are several approaches for simulating datasets, some of which involve artificially adding correlations between variables, and this is a potentially large subject. In fact the method above does introduce some correlation between discriminatory variables although different results can be obtained according to how a simulation is set up.

Table 2.6 *Discriminatory variables for Case Studies 10a, 10b and 10c*

10a	10b	10c
1	9	7
3	10	10
12	13	11
16	16	12
17	17	18
31	24	20
35	26	22
43	27	42
44	30	43
48	33	52
49	45	54
51	58	59
55	59	61
56	67	63
70	77	64
80	78	65
86	79	68
91	89	79
96	92	81
98	100	89

2.13 Case Study 11: Null Dataset

This case study is also a simulation and one that will be employed throughout this book. In hypothesis driven science we often do not know in advance of analysis whether there is a trend or not. It is in fact easy to make mistakes, and many methods of data analysis can falsely suggest that classes are separable or trends are present when, in fact, they are not. In order to check for overfitting a useful technique is to simulate a null dataset that contains random numbers and follow all data analysis techniques through using this case study. If, for example, the case study consists of two classes, then we expect approximately 50 % classification ability, that is about equal chances of a sample assigned to each class using techniques that protect against overfitting. If a technique results in, for example, 80 % classification ability we know that this method provides an overoptimistic view of the data and should not normally be employed for assessments of real datasets.

For the null dataset, we simulate a 400 × 100 data matrix, with each element consisting of either 1 or 0 generated by a random number generator with equal probability for each state. Figure 2.17 illustrates the number of '1s' for each variable: the average is 200.91 with a minimum of 177 and a maximum of 235. A few variables may, by chance, appear discriminatory, because the number of '1s' in each class differs by a large amount; an example is variable 19 which has 108 '1s' in class A and 84 in class B. Figure 2.18 illustrates the distribution of variables between each class. We will see that this difference in distribution if inappropriately handled can result in an apparent false prediction that class A and B can be discriminated, for example it is possible to take the variables that show maximum difference between the classes and form a classification model just using these variables: in many cases especially in biomarker discovery and metabolomic profiling it is common to select the variables that appear best discriminators and then form models using these variables, which result in apparently falsely good separations, even if there is no underlying trend.

Because the variables are all on the same scale, the only preprocessing we will perform involves centring the columns.

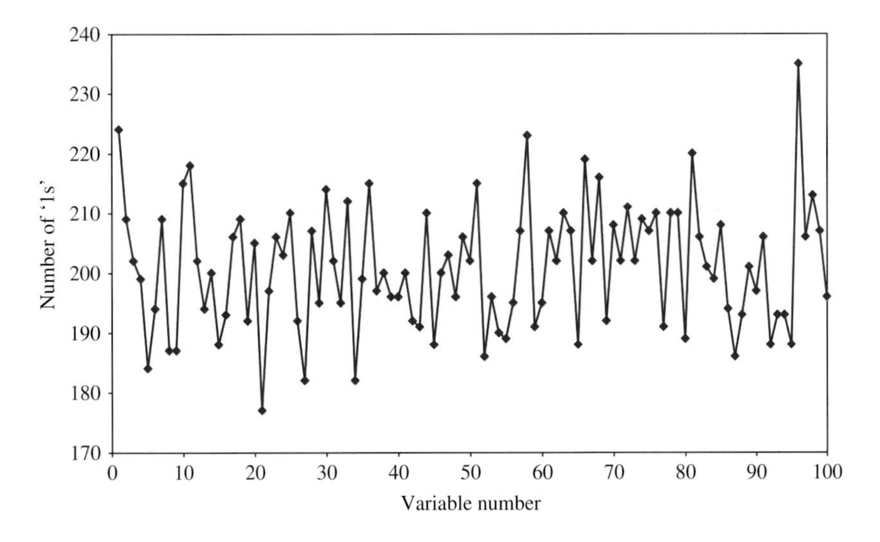

Figure 2.17 Number of '1s' for each variable of Case Study 11

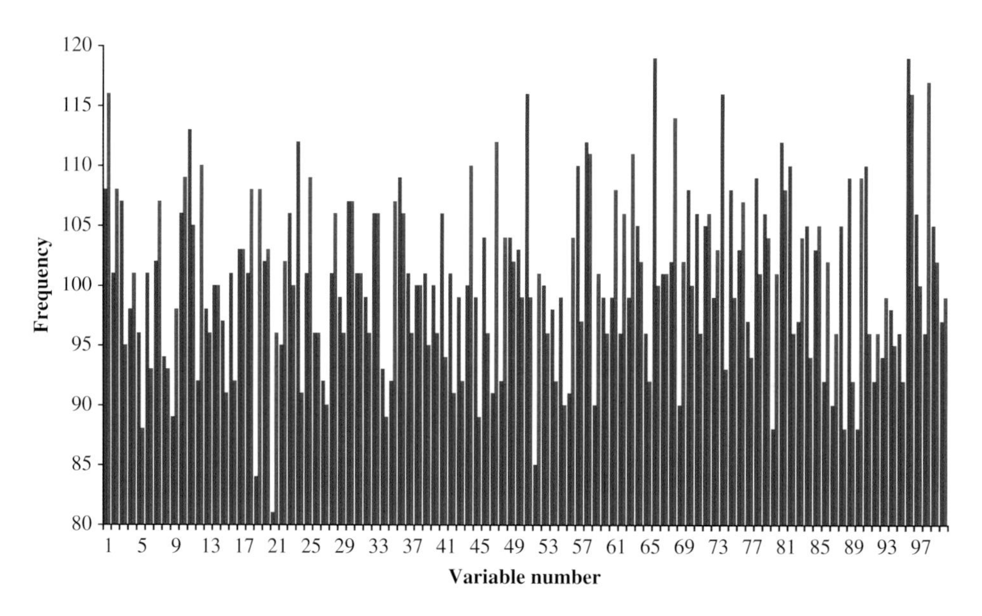

Figure 2.18 Frequency of '1s' in class A (blue) and class B (red) for each of the 100 variables in Case Study 11

2.14 Case Study 12: GCMS and Microbiology of Mouse Scent Marks

Case Studies 1 to 11 all have one block of experimental data, for example GCMS or MS or NIR spectroscopy or DMA measurements, the aim usually being to see if the samples can be classified into two or more groups according to the analytical measurements. However in some situations more than one type of measurement can be performed on a set of samples. This provides what is often called multiblock data. In more theoretical chemometrics circles there is great excitement about such types of data, and the potential for many methodological advances. In practice this type of data, however, is still quite rare in modern laboratory practice, and only found in quite specialist situations, with particularly important application areas being industrial process monitoring, where several types of spectroscopies and process variables can be acquired. This book would be incomplete without some discussion of how to handle such datasets (Chapter 12), although not as the main focus of this text: a full volume could be written about multiblock methods which of course are interesting technically in their own rights.

In order to demonstrate the application of these methods we use data obtained from the analysis of mouse scent marks. It is postulated that mice produce very specific compounds in their scent that allows individuals to identify not only each other but also main genetic groups, so that the scent marks of different groups of cloned mice should have distinct characteristics that can be studied by chemical analysis such as GCMS. However it is also likely that these chemicals are obtained from the action of microbes which are specific to each group of mice. Hence the aim is to distinguish groups of mice by both GCMS and microbiology and to determine if there are common trends between the two types of data, thus suggesting a link between microbiology and chemistry.

The dataset consists of 34 samples from extracts of mouse scent marks. The mice are in four groups, consisting of two strains (BALB and C57) and two haplotypes (d and b) as listed in Table 2.7, and denoted as in the table (so S1H1 represents strain 1 – BALB and haplotype 1 – d). A haplotype involves a single background gene, whereas the strains are congenic apart from this single gene. The aim is to see if we can distinguish these four groups.

Table 2.7 *Sample numbers and types for Case Study 12*

	Haplotype	
Strain	d	b
BALB	S1H1 (9) Class A (samples 1 to 9)	S1H2 (9) Class B (samples 19 to 26)
C57	S2H1 (8) Class C (samples 10 to 18)	S2H2 (8) Class D (samples 27 to 34)

There are two blocks of data:

1. GCMS peak detection and alignment was performed in a similar manner to Case Studies 5 and 8, and only peaks found in at least 10 samples were retained to reduce the peak table to 334 peaks.
2. The microbiological data were obtained using DGGE (Denaturing Gradient Gel Electrophoresis) in which each sample forms a lane on a plate, and each microbe a band at a characteristic position. 30 different bands (corresponding to different microbes) were identified and where present (according to a spot on a gel) its intensity is recorded, although this value is semiquantitive and not linearly related to the number of microbes. The microbiological data are represented symbolically in Figure 2.19.

Hence two matrices of dimensions 34×30 (microbiology) and 34×334 (GCMS) are obtained, each of which can be considered a block, as illustrated in Figure 2.20. These can then be related to the $\boldsymbol{c}$ block which contains information about their origins.

Both blocks are preprocessed as follows:

1. Each variable is square root scaled to remove the influence of very large values.
2. The rows are scaled to constant total of 1 (for each block separately).
3. All variables are standardized.

As shown in Chapter 12 some multiblock methods such as consensus PCA can be used to adjust the relative importance of each block.

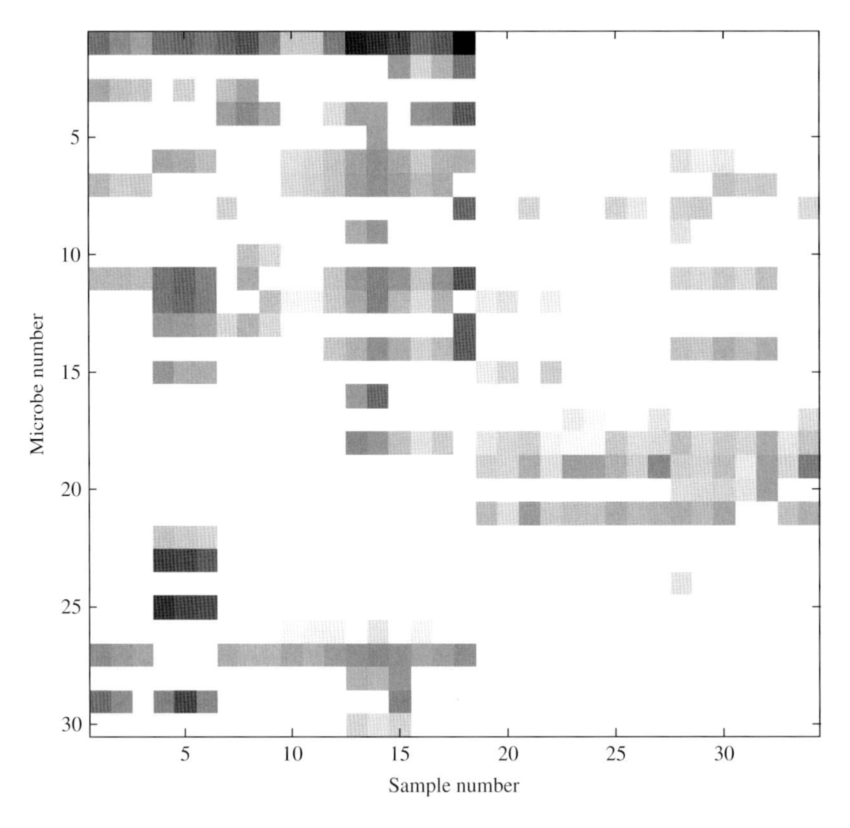

Figure 2.19 Diagrammatic representation of the DGGE data for Case Study 12. The intensity of the pixels represent the intensity of the band on the gel. Bands not detected are represented in white

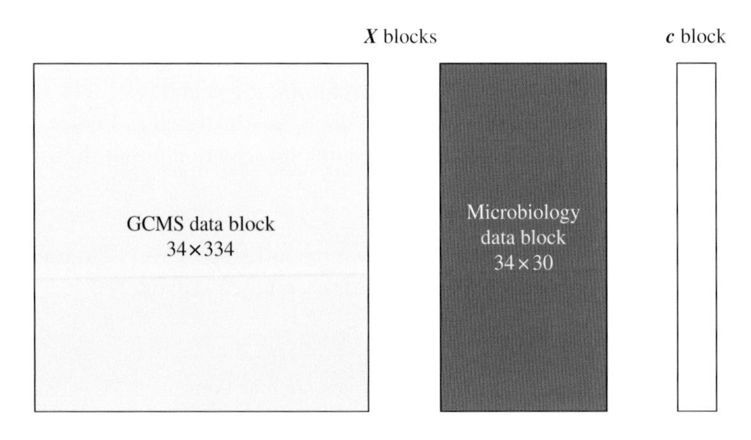

Figure 2.20 Structure of Case Study 12

Bibliography

Case Study 1

S.J. Dixon, R.G. Brereton, J.F. Carter, R. Sleeman, Determination of cocaine contamination on banknotes using tandem mass spectrometry and pattern recognition, *Anal. Chim. Acta*, **559**, 54–63 (2006).

Case Study 2

CAMO training exercise MVA II [www.camo.com].

Case Study 3

B.M. Lukasiak, S. Zomer, R.G. Brereton, R. Faria, J.C. Duncan, Pattern recognition and feature selection for the discrimination between grades of commercial plastics, *Chemometrics Intell. Lab. Systems*, **87**, 18–25 (2007).

B.M. Lukasiak, S. Zomer, R.G. Brereton, R. Faria, J.C. Duncan, Pattern Recognition for the Analysis of Polymeric Materials, *Analyst*, **131**, 73–80 (2006).

R. Faria, J.C. Duncan, R.G. Brereton, Dynamic Mechanical Analysis and Chemometrics for Polymer Identification, *Polym. Testing*, **26**, 402–412 (2007).

G.R. Lloyd, R. Faria, R.G. Brereton, J.C. Duncan, Learning vector quantization for multi-class classification: application to characterization of plastics, *J. Chem. Inform. Modeling*, **47**, 1553–1563 (2007).

G.R. Lloyd, R.G. Brereton, J.C. Duncan, Self organising maps for distinguishing polymer groups using thermal response curves obtained by dynamic mechanical analysis, *Analyst*, **133**, 1046–1059 (2008).

Case Study 4

S. Zomer, M. Sánchez, R.G. Brereton, J.L. Pérez Pavón, Active Learning Support Vector Machines for Optimal Sample Selection in Classification, *J. Chemometrics*, **18**, 294–305 (2004).

Case Study 5

S.J. Dixon, R.G. Brereton, H.A. Soini, M.V. Novotny, D.J. Penn, An Automated Method for Peak Detection and Matching in Large Gas Chromatography–Mass Spectrometry Data Sets, *J. Chemometrics*, **20**, 325–340 (2006).

D.J. Penn, E. Oberzaucher, K. Grammer, G. Fischer, H.A. Soini, D. Wiesler, M.V. Novotny, S.J. Dixon, Y. Xu, R.G. Brereton, Individual and gender fingerprints in human body odour, *J. R. Soc. Interface*, **4**, 331–340 (2007).

H.A. Soini, K.E. Bruce, I. Klouckova, R.G. Brereton, D.J. Penn, M.V. Novotny, *In-situ* surface sampling of biological objects and preconcentration of their volatiles for chromatographic analysis, *Anal. Chem.*, **78**, 7161–7168 (2006).

S.J. Dixon, Y. Xu, R.G. Brereton, A. Soini , M.V. Novotny, E. Oberzaucher, K. Grammer, D.J. Penn, Pattern Recognition of Gas Chromatography Mass Spectrometry of Human Volatiles in Sweat to Distinguish the Sex of Subjects and Determine Potential Discriminatory Marker Peaks, *Chemometrics Intell. Lab. Systems*, **87**, 161–172 (2007).

Y. Xu, F. Gong, S.J. Dixon, R.G. Brereton, H.A. Soini, M.V. Novotny, E. Oberzaucher, K. Grammer, D.J. Penn Application of Dissimilarity Indices, Principal Co-ordinates Analysis and Rank Tests to Peak Tables in Metabolomics of the Gas Chromatography Mass Spectrometry of Human Sweat, *Anal. Chem.*, **79**, 5633–5641 (2007).

Case Study 6

S. Zomer, R.G. Brereton, J.-C. Wolff, C.Y. Airiau, C. Smallwood, Component Detection Weighted Index of Analogy: Similarity Recognition on Liquid Chromatographic Mass Spectral Data for the Characterisation of Route/Process Specific Impurities in Pharmaceutical Tablets, *Anal. Chem.*, **77**, 1607–1621 (2005).

Case Study 7

S. Rahman, Significance of blood trace metals in hypertension and diseases related to hypertension in comparison with normotensive, PhD Thesis, Pakistan Institute of Nuclear Science and Technology, Islamabad, Pakistan.

Case Study 8

S. J. Dixon, N. Heinrich, M. Holmboe, M.L. Schaefer, R.R. Reed, J. Trevejo, R.G. Brereton, Use of cluster separation indices and the influence of outliers: application of two new separation indices, the modified silhouette index and the overlap coefficient to simulated data and mouse urine metabolomic profiles, *J. Chemometrics*, **23**, 19–31 (2009).

K. Wongravee, G.R. Lloyd, J. Hall, M.E. Holmboe, M.L. Schaefer, R.R. Reed, J. Trevejo, R.G. Brereton, Monte-Carlo methods for determining optimal number of significant variables. Application to mouse urinary profiles, *Metabolomics* (2009) doi 10.1007/s11306-009-0164-4.

K. Wongravee, N. Heinrich, M. Holmboe, M.L. Schaefer, R.R. Reed, J. Trevejo, R.G. Brereton, Variable Selection using Iterative Reformulation of Training Set Models for Discrimination of Samples: Application to Gas Chromatography Mass Spectrometry of Mouse Urinary Metabolites, *Anal. Chem.* (2009) doi 10.1021/ac900251c.

Case Study 12

S. Zomer, S.J. Dixon, Y. Xu, S.P. Jensen, H. Wang, C.V. Lanyon, A.G. O'Donnell, A.S. Clare, L.M. Gosling, D.J. Penn, R.G. Brereton, Consensus Multivariate methods in Gas Chromatographic Mass Spectrometry and Denaturing Gradient Gel Electrophoresis: MHC-congenic and other strains of mice can be classified according to the profiles of volatiles and microflora in their scent-marks, *Analyst*, **134**, 114–123 (2009).

3
Exploratory Data Analysis

3.1 Introduction

Most pattern recognition is either supervised (usually involving seeing whether samples fall into groups, how well, and what causes this separation), or Exploratory Data Analysis (EDA). The latter methods primarily involve visualizing the relationship between samples and between variables and do not necessarily require them to be assigned to groups. Usually EDA is a preliminary stage prior to supervised modelling such as classification (or calibration in other contexts) and can answer whether there are any groupings in the data, whether there are outliers, whether samples from a similar source are related, or even whether there are trends such as temporal trends in the data. In this text we will mainly be concerned with the ability to detect groups in data.

Most users of chemometrics methods have heard of PCA (Principal Components Analysis) which is regarded by many as the most widespread method for visualization of samples, but not of other methods for EDA. Many, however, also are often not well aware of the influence of variable selection (Chapter 9) or data preprocessing (Chapter 4) on the appearance of these graphs.

In this chapter we will describe a variety of approaches for data visualization. PCA is the most flexible general purpose approach, but suffers when there are many different groups and also often can use the visualization space quite inefficiently. Furthermore there are sometimes problems choosing the most appropriate PCs for visualization, especially if the trends of interest are minor. PCA also has problems if the variables are categorical, especially if the categories are not numerically related (e.g. three species or types of material that have no specific relationship to each other), and if there are a lot of undetected variables (e.g. in metabolomic studies when metabolites may be found only in some of the samples). Nevertheless PCA is still an excellent method for preliminary visualization in many cases, and will be employed widely in this text, although a purpose of this chapter is also to describe alternatives to PCA which could be employed in certain circumstances.

Chemometrics for Pattern Recognition Richard G. Brereton

Before describing methods for visualization in detail, it is important to understand that if inappropriately used they can mislead and result in erroneous conclusions about the data. A simple example involves performing variable selection prior to PCA, involving choosing those variables that appear to be able to discriminate two classes best and forming a model just on these. In Figure 3.1, the plots of PC2 versus PC1 of the Case Study 11 (null) are presented both using all the 100 raw variables and using just the top 20 variables selected using the *t*-statistic (Chapter 9) by choosing the 10 best discriminators for each of the classes. It can be seen that after variable selection, class separation appears to improve even though the raw data are completely random: the reason is that we have chosen the 20

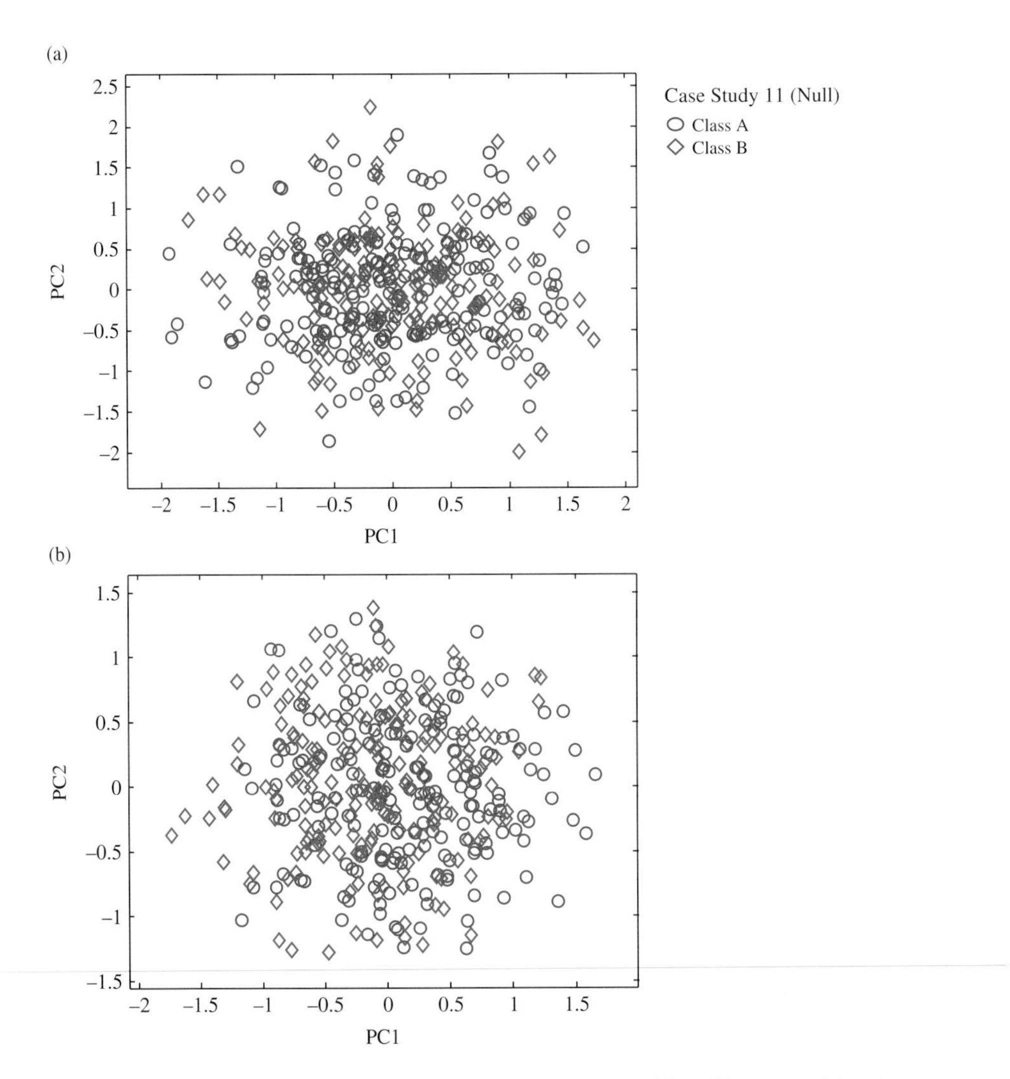

Figure 3.1 Plots of PC2 versus 1 for dataset 11 (null): (a) all variables; (b) 20 variables that appear best discriminators (10 for each class)

out of 200 variables that appear the top discriminators. If we toss an unbiassed coin 10 times, occasionally we might obtain 8 'Heads' or more just by chance, this should happen over 5 % of the time. If we repeated this experiment 200 times, by chance we expect a result of 8 'Heads' or more over 10 times out of the 200 coin tosses – if we then selected the 10 sets of coin tosses that gave 10 or more 'Heads', and rejected the others, it would look like the coin is biased, as we are rejecting the rest of the evidence. Hence if there are a large number of variables, retaining just the variables that appear best discriminators and reforming the model on these can result in anomalously good separations, even when starting with random data, and we are prejudicing the model in advance. Therefore caution is required when using exploratory methods and a nice looking graph may not necessarily accurately represent the underlying data. It is a common mistake to perform variable selection in a biased manner prior to EDA and then by visual inspection of the resultant graphs, falsely assume that there are underlying trends for which there is no particular evidence, and so caution should be employed when using approaches described in this chapter to ensure that the underlying variables have been recorded or selected in an unbiased manner if one is to use visual methods for determining whether there are underlying relationships between samples based on analytic measurements. This problem is sometimes called overfitting and we will regularly discuss this in the text primarily by reference to Case Study 11 (null).

3.2 Principal Components Analysis

3.2.1 Background

PCA is probably the most widespread multivariate chemometric technique, because of the importance of multivariate measurements in chemistry, and certainly the most widespread technique in multivariate EDA. In other disciplines it often goes by different names, for example, factor analysis (although this has a very specific meaning in some circles, often synonymous with Principal Components Regression) and eigenanalysis. There are numerous claims to the first description of PCA in the literature. Probably the most famous early paper was by Pearson in 1901. However, the fundamental ideas are based on approaches well known to physicists and mathematicians for much longer. An early description of the method in physics was by Cauchy in 1829. It has been claimed that the earliest non-specific reference to PCA in the chemical literature was in 1878, although the author of the paper almost certainly did not realise the potential, and was dealing mainly with a simple problem of linear calibration. It is generally accepted that the revolution in the use of multivariate methods in experimental science took place in psychometrics in the 1930s and 1940s of which Hotelling's work is regarded as a classic. Psychometrics is well understood to most students of psychology and one important area involves relating answers in tests to underlying factors, for example, verbal and numerical ability as illustrated in Figure 3.2.

Natural scientists of all disciplines, from biologists, geologists and chemists have caught on to these approaches over the past few decades. Within the chemistry community the first widespread applications of PCA as a method for data visualization were reported in the 1970s.

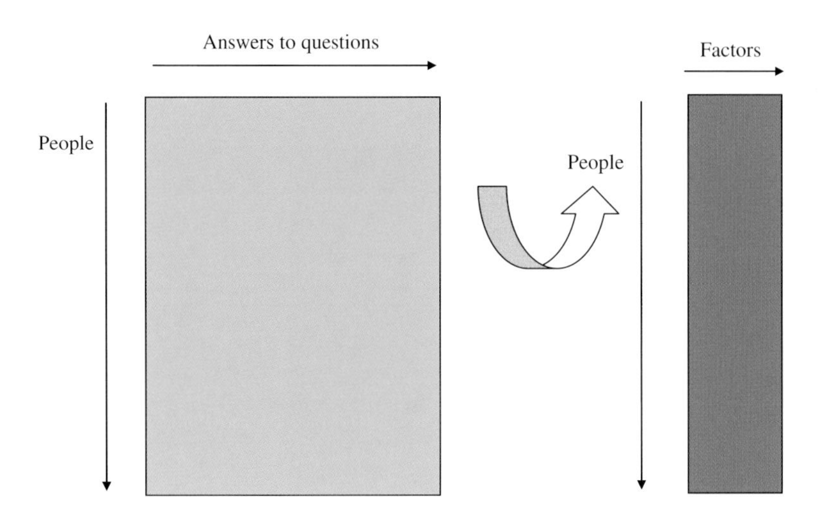

Figure 3.2 Factor analysis in psychology

3.2.2 Scores and Loadings

PCA involves an abstract mathematical transformation of the original data matrix, which, can be represented by the equation:

$$\boldsymbol{X} = \boldsymbol{T}\boldsymbol{P} + \boldsymbol{E}$$

where:

- $\boldsymbol{X}$ is the original data matrix of dimensions $I \times J$ whose rows represent samples, whose columns represent variables and whose elements represent the values of the measurements on these samples.
- The number of columns in the matrix $\boldsymbol{T}$ equals the number of rows in the matrix $\boldsymbol{P}$ and equals the number of significant components, which we will define by A.
- $\boldsymbol{T}$ are called the scores, and have as many rows as the original data matrix, and are represented by a matrix of dimensions $I \times A$.
- $\boldsymbol{P}$ are called the loadings and have as many columns as the original data matrix and are represented by a matrix of dimensions $A \times J$.
- The ath column of $\boldsymbol{T}$ and ath row of $\boldsymbol{P}$ can be represented by vectors $\boldsymbol{t}_a$ and $\boldsymbol{p}_a$ and are the vector representations of the ath PC.
- The product $\boldsymbol{TP}$ can be regarded as a model of the data that is an approximation to the original dataset, the error being represented by matrix $\boldsymbol{E}$.
- The first scores vector and the first loadings vector are often called the eigenvectors of the first principal component. Each successive component is characterized by a pair of eigenvectors for both the scores and loadings. This is illustrated in Figure 3.3.

It is possible to calculate scores and loadings matrices as large as desired, providing the 'common' dimension is no larger than the smallest dimension of the original data matrix,

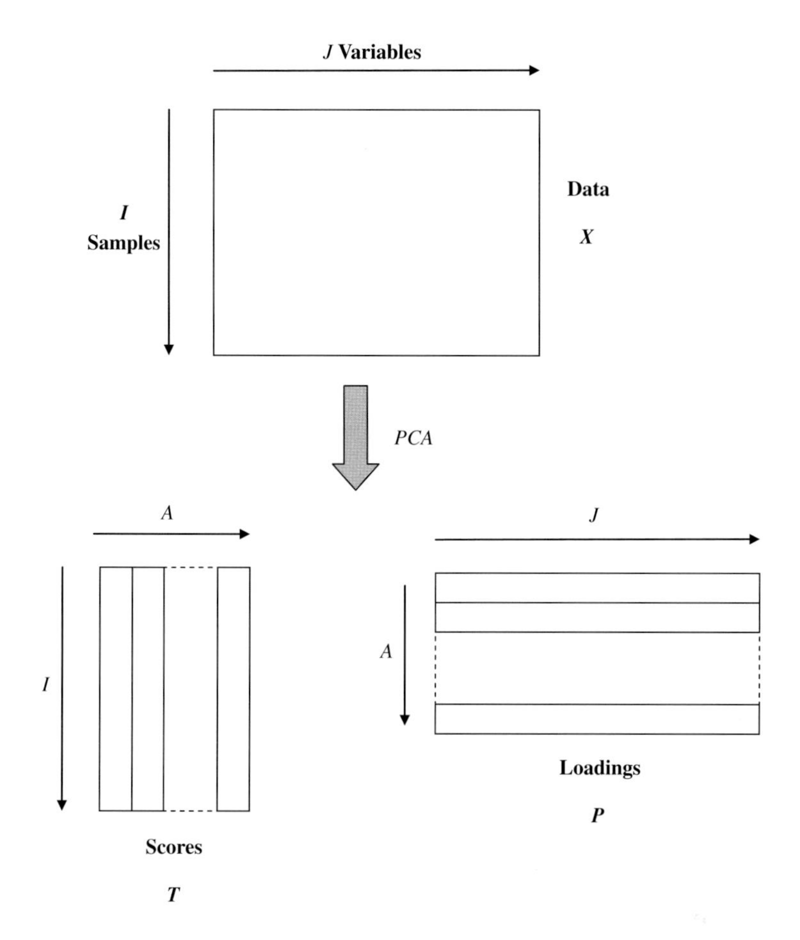

Figure 3.3 PCA method

and this corresponds to the number of PCs that are used in the model. Using A PCs, it is possible to establish a model for each element of $\boldsymbol{X}$ of the form:

$$x_{ij} = \sum_{a=1}^{A} t_{ia} \, p_{aj} + e_{ij}$$

which is the non-matrix version of the fundamental PCA equation above.

Hence if the original data matrix is of dimensions 50×100 (or $I \times J$), involving obtaining 50 samples and measuring 100 variables on each of these samples, no more than 50 (non-zero) PCs can be calculated. If 5 PCs are calculated then:

- the dimensions of $\boldsymbol{T}$ will be 50×5;
- the dimensions of $\boldsymbol{P}$ will be 5×100.

In traditional analytical chemistry there are many methods available for determining the value of A which primarily involves modelling the data to determine as low a difference as possible between the observed (***X***) and predicted (***TP***) data. We will discuss methods for determining the optimum number of PCs in Section 4.3. In many situations in analytical chemistry there is an exact physical interpretation of the matrix ***X***, for example it may consist of a series of spectra obtained during a reaction of a known number of chemical species whose individual spectra overlap and as such may be directly related to the underlying chemistry and even the kinetics and concentrations of each reactant: the residual error ***E*** is ideally related to instrumental noise or measurement error and the number of components A to the number of reactants. However in pattern recognition and for data visualization the precise value of A is often not so crucial, and ***E*** often does not have a physically interpretable meaning, and so the determination of the optimum value of A is not so important in the context of the problems discussed in this text. A simple example may involve studying the factors that influence the composition of a mammalian urine sample: these may be gender, age, disease state, for females hormonal cycle, genetics, etc. The number and relative importance of factors do not necessarily correlate with the number of chemicals observed in a series of urine samples and it is not the main aim of PCA to provide a physical prediction of the chemical composition of a urine sample by knowing all these factors, in fact this would probably be impossible as there would be a large number of hidden factors and enormous variability among populations.

There are a number of important features of scores and loadings. It is important to understand that PCs are simply abstract mathematical entities, and in themselves have no particular physical meaning, although they can sometimes be interpreted graphically and can be used for example as input to classifiers as a first step in modelling.

All scores and loadings vectors have the property that the sums $\Sigma_{i=1}^{I} t_{ia} t_{ib} = 0$ and $\Sigma_{j=1}^{J} p_{aj} p_{bj} = 0$ where $a \neq b$, and ***t*** and ***p*** correspond to the corresponding eigenvectors. Some authors state that the scores and loadings vectors are mutually orthogonal, since some of the terminology of chemometrics originated from multivariate statistics where people like to think of PCs as vectors in multidimensional space, with each variable representing an axis, and so some of the geometric analogies have been incorporated into the mainstream literature. If the columns are mean-centred, then also the correlation coefficient between any two scores vectors is equal to 0.

In addition, each loadings vector is also normalized. There are various different definitions of a normalized vector, but we use $\Sigma_{j=1}^{J} p_{aj}^2 = 1$. The loadings are often called orthonormal because they obey this property.

Note that there are several algorithms for PCA: using the SVD (Singular Value Decomposition) method the scores are also normalized, while for the NIPALS method this is not so; in this book for simplicity we will use NIPALS notation and methodology in most places unless inconvenient, although users of SVD and other methods should be aware of the relationship between the different approaches. The relationship between the matrices obtained by NIPALS and SVD is given by the following:

$$\boldsymbol{TP} = \boldsymbol{USV}$$

where ***U*** corresponds to the scores matrix ***T*** but is orthonormal, ***V*** to the loadings matrix, and ***S*** is a diagonal matrix of dimensions $A \times A$ that contains the square root of the eigenvalues ($\sqrt{\lambda_a}$) (or the square root of the sum of squares of the NIPALS scores of each successive component), as defined in Section 3.2.3.

It is possible to calculate the square matrix $\boldsymbol{T}'\boldsymbol{T}$ which has the properties that all elements are zero except along the diagonals, the value of the diagonal elements relating to the size (or importance) of each successive PC or the eigenvalue. The square matrix $\boldsymbol{P}\boldsymbol{P}'$ has the special property that it is an identity matrix, the dimensions equal to the number of PCs.

After PCA, the original variables (e.g. intensities of 100 peaks measured by GCMS) are reduced to A significant PCs (e.g. 3). PCA can be used as a form of variable reduction, reducing the large original dataset (involving the measurements of 100 chemicals) to a much smaller more manageable dataset (e.g. consisting of 3 PCs) which can be visualized more easily. The loadings represent the means to this end but also help us interpret which variables are likely to be most responsible for clustering or separation between the samples, as discussed below. Note that PCA can be used both for EDA (primarily data vizualization as in this chapter) and modelling – in later chapters we will explore the use of PCs as numerical inputs to classifiers.

A final point to note is that it is not possible to control the sign of a PC, and different algorithms, or even the same algorithm with a different starting point, can result in reflection in the PCs; this reflection is both in the scores and loadings simultaneously.

3.2.3 Eigenvalues

Many chemometricians use the concept of the rank of a matrix, which is a mathematical concept that relates to the number of significant components in a dataset, or the number of PCs required to obtain an adequate reconstruction. In traditional analytical chemistry this is often of great importance, for example it may correspond to the number of absorbing compounds in a series of spectra, which in turn may equal the number of unique compounds in a reaction mixture. During the first two decades when chemometrics existed there were many hundreds or even thousands of papers discussing this issue, and perhaps one hundred or more methods available, as much of the focus was on spectroscopic and chromatographic problems where the correct answer is often known in advance. It is important to distinguish the chemical rank of a matrix, which is the number of PCs needed to approximate a good reconstruction of a data matrix (small $\boldsymbol{E}$ according to a variety of criteria), from the mathematical rank. The former relates to the number of PCs that approximately model the data – and is usually considerably lower than the mathematical rank, which exactly reconstructs a matrix with zero error (all elements of $\boldsymbol{E}$ are 0). In traditional chemometrics the later PCs that contribute to the mathematical but not chemical rank of a matrix are usually considered to originate from errors such as sampling and instrumental noise.

In chemical pattern recognition, although it is still useful to determine the number of significant components in a series of samples, this no longer has a great physical significance and as the later PCs are smaller in size, often not very crucial. However it is still important to determine the size of each PC which is often called an eigenvalue: the earlier (and more significant) the components, the larger their size. There are a number of definitions in the literature, but a simple one defines the eigenvalue of a PC as the sum of squares of the scores, so that:

$$\lambda_a = \sum_{i=1}^{I} t_{ia}^2$$

where λ_a is the ath eigenvalue.

The sum of all non-zero eigenvalues for a datamatrix, equals the sum of squares of the entire data-matrix so that:

$$\sum_{a=1}^{K} g_a = \sum_{i=1}^{I} \sum_{j=1}^{J} x_{ij}^2$$

where K is the smaller of I or J. Note that if the data are preprocessed prior to PCA (see Chapter 4), x must likewise be preprocessed for this property to hold; if mean-centring has been performed K cannot be larger than $I-1$ where I equals the number of samples.

Frequently eigenvalues are presented as percentages, for example of the sum of squares of the entire (preprocessed) dataset, which is often called percent variance, although strictly speaking this is only when columns have been centred first, defined by:

$$\boldsymbol{V}_a = 100 \frac{\lambda_a}{\sum_{i=1}^{I} \sum_{j=1}^{J} x_{ij}^2}$$

Successive eigenvalues correspond to smaller percentages. It is sometimes useful to determine the percentage variance of each successive PC. If the first few PCs correspond to a small percentage variance they do not model much of the data, but also this means that there are probably large number of factors influencing the observations. In Figure 3.4 we see plots for %variance modelled by the first 20 PCs for Case Studies 1, 8a and 11. There are substantial differences between the appearance of these graphs. For Case Study 11 (null) successive eigenvalues are very similar in size to each other (initially representing about 2 % of the variance), descending in size very slowly, with no sharp drop off; when represented using a logarithmic scale the descent is almost linear. Case Study 1 (forensic), however shows a sharp drop after the first few PCs; PCs 1 and 2 model over 70 % of the data, in contrast to Case Study 11 where it is around 4 %, and so probably we can visualize the main trends easily with 2 or 3 PCs. We would expect Case Study 1 to be a relatively straightforward dataset as there are likely to be two main types of distribution, from the background and defendants that are in most cases clearly distinguishable. Case Study 8a (the mouse urine diet study) appears deceptively similar but if we look at the vertical scale, the first two PCs model far less of the variance (16 %) as we would expect many more factors to influence the urinary signal of the mice, even though there are likely to be some recognizable patterns in the first few components relating to the main experimental factor.

The cumulative percentage eigenvalue is often used to determine (approximately) what proportion of the data has been modelled using PCA and is given by $\Sigma_{a=1}^{A} \lambda_a$. The closer to

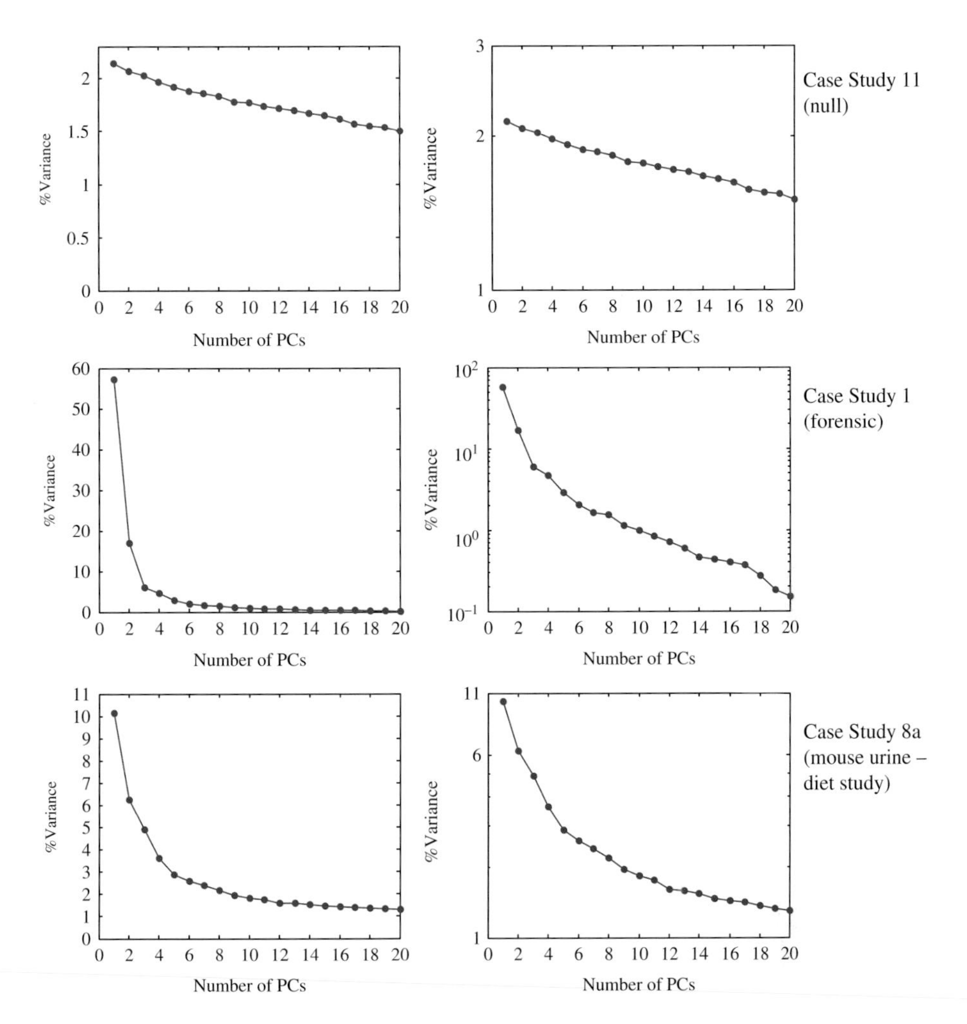

Figure 3.4 Plot of %variance against PC number for Case Studies 11, 1 and 8a: left, using a linear scale; right using a logarithmic scale

100 % of the total sum of squares of the data, the more faithful the model. The percentage can be plotted against the number of eigenvalues in the PC model. The graph of cumulative percentage eigenvalue for Case Studies 1, 8a and 11 is plotted as a function of PC number in Figure 3.5, which makes very clear the difference between the structure of the data in the three case studies.

The residual sum of squares after A PCs have been calculated is defined by:

$$RSS_A = \sum_{i=1}^{I} \sum_{j=1}^{J} x_{ij}^2 - \sum_{a=1}^{A} \lambda_a$$

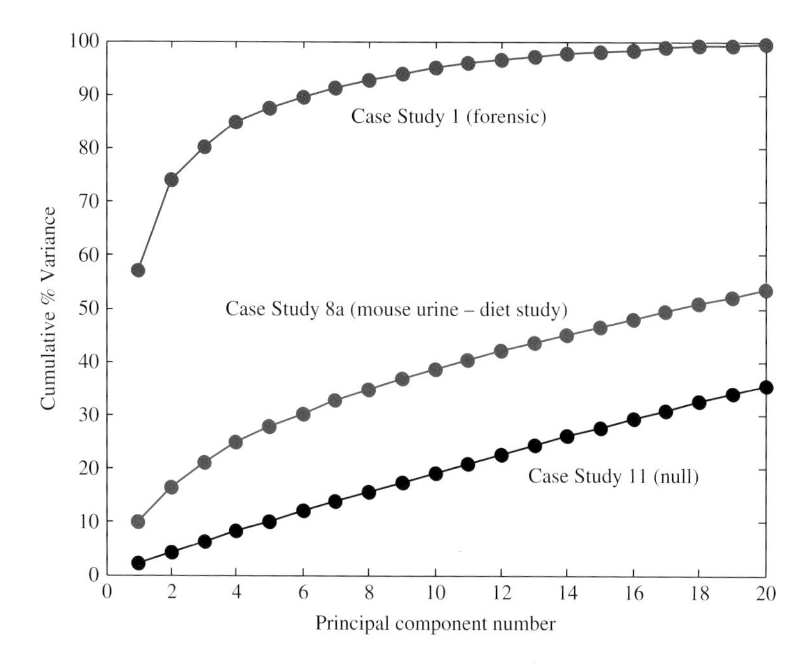

Figure 3.5 Cumulative percentage variance data for Case Studies 1, 8a and 11

after A eigenvalues, and also equals the sum of squares for the error matrix, between the PC model and the raw data, whose elements are defined by:

$$e_{ij} = x_{ij} - \hat{x}_{ij} = x_{ij} - \sum_{a=1}^{A} t_{ia} p_{aj}$$

where $\hat{x}_{ij}$ is the predicted value of x_{ij} using a model based on A PCs, or:

$$\sum_{i=1}^{I} \sum_{j=1}^{J} x_{ij}^2 - \sum_{a=1}^{A} \lambda_a = \sum_{i=1}^{I} \sum_{j=1}^{J} e_{ij}^2$$

This is because the product of the elements of any two different eigenvectors is 0 as discussed in Section 3.2.2.

Sometimes the size of successive eigenvalues is used to determine how many PCs are best used to model a datamatrix and there are a particularly large number of so called indicator functions, most developed by Malinowski in the 1970s and 1980s whose context was primarily spectroscopy in physical and analytical chemistry. However whereas quite valuable in spectroscopy this is perhaps less important in pattern recognition and approaches such as cross validation or the bootstrap are more usual as described in Section 4.3. In this chapter we are principally using PCs to visualize the data, although taking a note of the size of PCs (as determined by their eigenvalues) is often useful, especially visualizing the relative importance of successive PCs.

3.2.4 PCA Algorithm

We describe here the NIPALS algorithm. Although one of many algorithms for PCA, it is a common one in chemometrics and should be compared to the PLS1 algorithm (Section 5.5.2) and CPCA algorithm (Section 12.3). Data must have been preprocessed first (Chapter 4). The notation we use in this text is based on NIPALS unless otherwise stated.

1. Take a column of matrix $\boldsymbol{X}$ (often the column with greatest sum of squares) as the first guess of the scores first principal component – call it ${}^{initial}\hat{\boldsymbol{t}}$.
2. Calculate the following:

$$\frac{{}^{unnorm}\hat{\boldsymbol{p}} = {}^{initial}\hat{\boldsymbol{t}}'\boldsymbol{X}}{\sum \hat{t}^2}$$

3. Normalise the guess of the loadings, so that:

$$\hat{\boldsymbol{p}} = \frac{{}^{unnorm}\hat{\boldsymbol{p}}}{\sqrt{\sum {}^{unnorm}\hat{p}^2}}$$

4. Now calculate a new guess of the scores:

$${}^{new}\hat{\boldsymbol{t}} = \boldsymbol{X}\,\hat{\boldsymbol{p}}'$$

5. Check if this new guess differs from the first guess; a simple approach is to look at the size of the sum of square difference in the old and new scores, i.e. $\Sigma({}^{initial}\hat{t} - {}^{new}\hat{t})^2$. If this is small the PC has been extracted; set the PC scores ($\boldsymbol{t}$) and loadings ($\boldsymbol{p}$) for the current PC to $\hat{\boldsymbol{t}}$ and $\hat{\boldsymbol{p}}$. Otherwise return to step 2, substituting the initial scores by the new scores.
6. Subtract the effect of the new PC from the datamatrix to get a residual datamatrix

$${}^{resid}\boldsymbol{X} = \boldsymbol{X} - \boldsymbol{tp}$$

7. If it is desired to compute further PCs, substitute the residual data matrix obtained in step 6 for $\boldsymbol{X}$ and go to step 1.

3.2.5 Graphical Representation

Once scores (relating to the samples) and loadings (relating to the variables) have been calculated there are several ways of representing these graphically.

We will first discuss scores.

1. *1D scores plot*. This is generally represented as a bar chart. Each score can be plotted against sample number, although the samples can be reordered in any way that is desirable. If a particular PC is influenced by a specific grouping then this should be clear in the bar chart, providing samples are ordered suitably. If the samples are divided into groups, this can be indicated in colour or by vertical lines between groups.
2. *2D scores plot*. This involves plotting the scores of one PC against those of another. The most common involves the scores of PC 2 versus 1. The samples are then plotted according to the values of the scores. It is possible to plot the scores

of any PC against any other PC, for example choose the PCs that appear best discriminators. Each sample grouping can be represented by a different symbol and/or colour. In Chapters 5 to 7 we will show how the scores plots can be divided up into different regions corresponding to different groups, if the data are represented as a two dimensional graph.

3. *3D scores plot*, analogous to a 2D scores plot, but with each axis representing one PC. Most graphics software allows for rotation of these plots.

In Figure 3.6 to Figure 3.17 we present the scores plots of the largest PCs for all the case studies in this text, as a visual summary of each dataset. In some cases (e.g. Case Study 3 – the polymers) there are different ways of dividing the samples (in this case by type and by group), which we illustrate. Below we interpret the scores plots of each dataset in turn. In each case we use default preprocessing as discussed in Chapter 2.

1. Case Study 1 (forensic – Figure 3.6) is fairly straightforward. There is good separation in PCs 1 and 2, especially PC2 as demonstrated by the 1D loadings bar chart. PCs 3 and 4 are particularly influenced by one outlying sample in class B (background) which has a high negative score in PC3 and a high positive score in PC4. However from the first two PCs we see that it is easy to separate the two groups of samples, and the origin of the samples is the main trend that differentiates the classes. There are a few

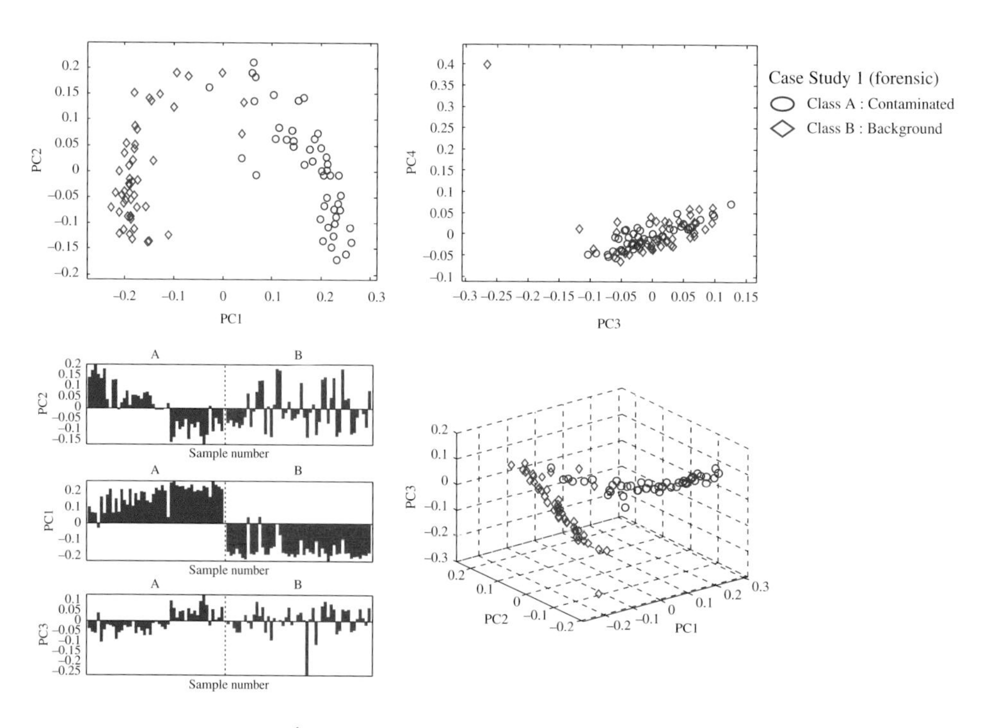

Figure 3.6 Scores plots for Case Study 1

samples between the groups, either background samples with slight contamination or case samples that are not heavily contaminated.

2. Case Study 2 (NIR spectroscopy of food – Figure 3.7) shows excellent separation between all four groups. NIR is a classical application of chemometrics and multivariate methods often work extremely well. PCA is useful as a method for visualization although the classification is unambiguous in this case. Two PCs are required to separate out the groups, classes B and D having high scores on PC1 and so being indistinguishable in PC1. Class D is well distinguished in PC2 with very negative scores. Classes A and C are distinguishable along PC1 according to the size of their scores, with Class C having the most negative scores. Note that the PC plots would change appearance considerably if the number of samples in each class were different. However this case study shows how multivariate methods can be used dramatically to distinguish spectra that are quite hard to pull apart by 'eyeballing', but is not a very challenging pattern

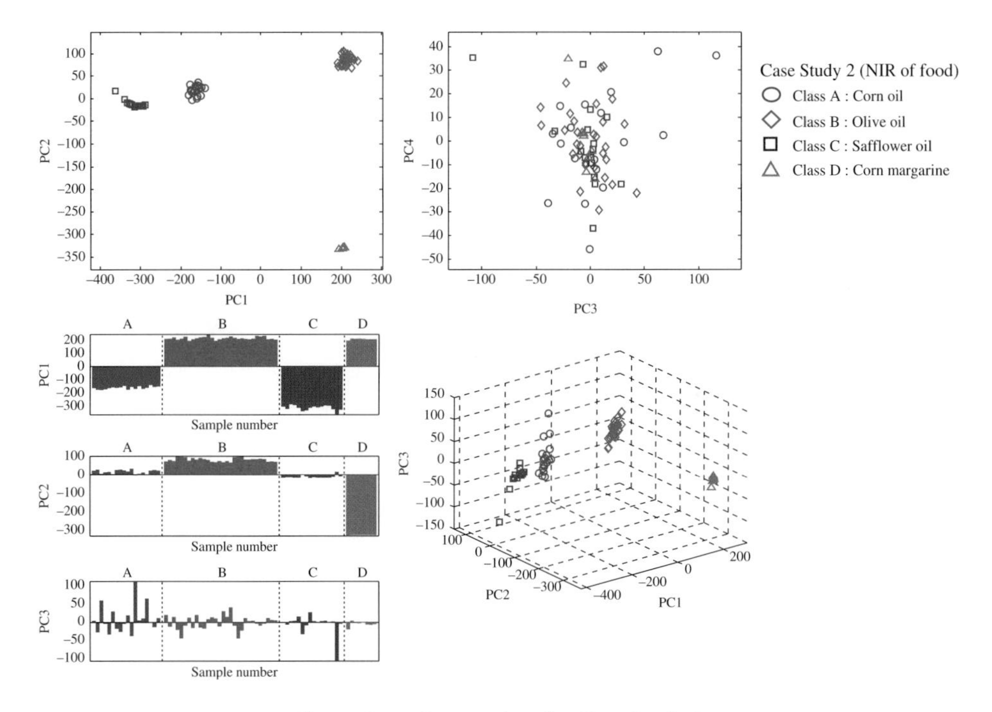

Figure 3.7 Scores plots for Case Study 2

recognition problem. The earlier steps of preparing the NIR spectroscopic data, such as MSC (Multiplicative Scatter Correction) have a bigger role.

3. For Case Study 3 (polymers – Figure 3.8) we first illustrate the two main types of polymers for which there is a good separation between the types in PCs 1 and 2. Class A has mainly positive scores for PC1 and class B mainly negative. There are a few samples that are exceptions to this rule and in fact these samples (small negative scores

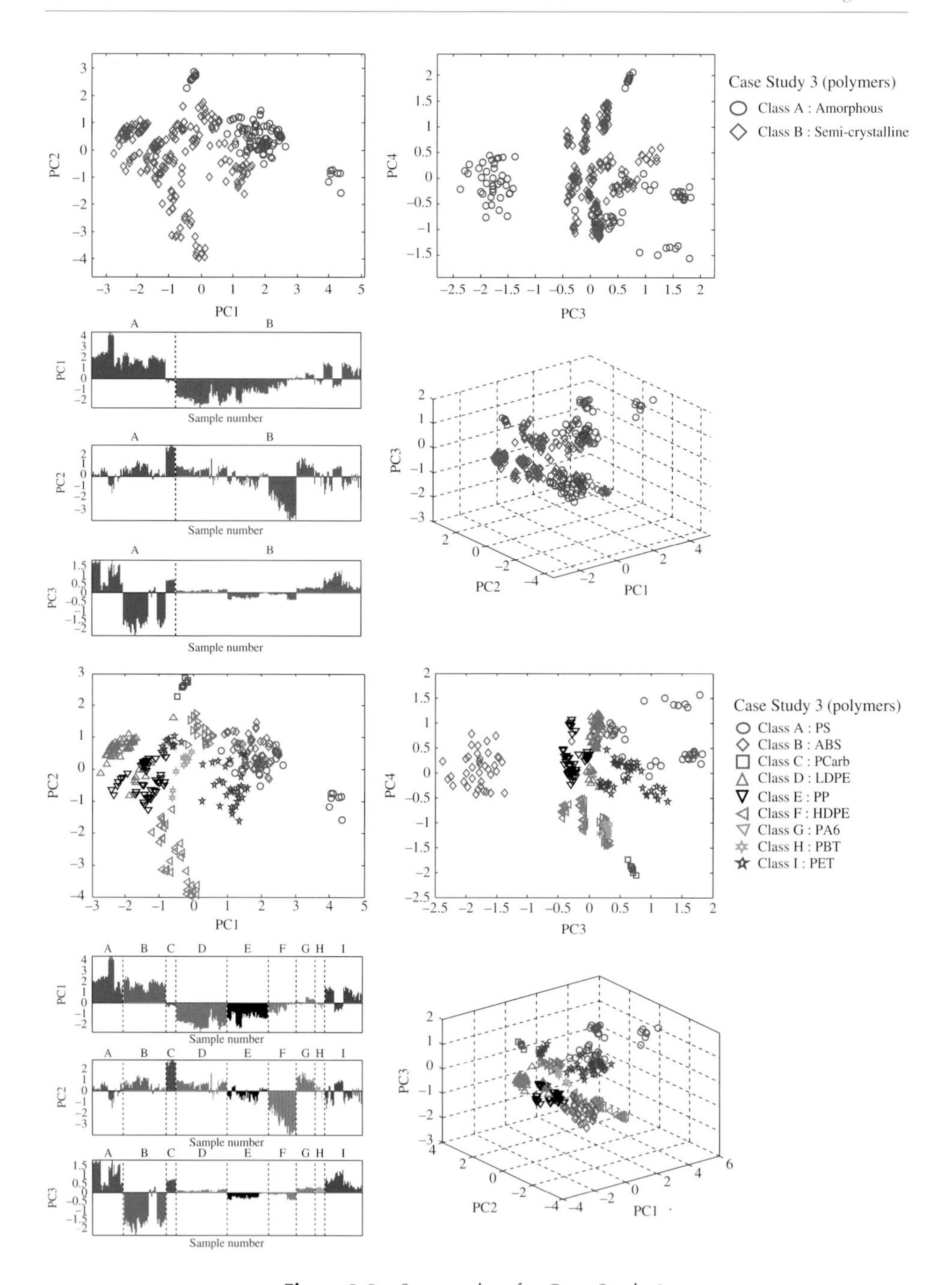

Figure 3.8 Scores plots for Case Study 3

for class A and positive scores for class B in PC2 belong mainly to specific groups of polymer) come from two types of polymer (group G – PA6 and group I – PET). In fact using a cut-off of 0 for the bar charts is rather arbitrary and will depend in part on how many samples we have from each group, with PA6 having less positive scores than the amorphous polymers (PS and ABS). The only ambiguous samples appear to be from PET which are hard to distinguish from PS in either of the first PCs. Note that the samples are ordered according to grade as well as type: a grade involves taking a polymer from a different manufacturer, and in fact only certain grades of PET are difficult to distinguish from PS. We would, however, expect that we can quite easily classify most polymers into two types using the first two or three PCs, with only a small number of misclassifications, primarily due to confusing groups A and I.

4. At first glance the PC scores plots of Case Study 4 (pollution – Figure 3.9) seem strange, especially the 1D scores plots. In the 2D and 3D projections we can see that there are three samples from the non-polluted class that appear mixed with the polluted samples. These are in fact all from soil 4 and so possibly this soil contained a small amount of pollutants that were not added artificially: there can be problems with labelling samples or collecting samples that originally appear to be clean but are in fact polluted. Although the 1D scores seem strange this is because the data are centred, and the number of samples in class A (polluted) is very much larger than class B and so the overall mean of the data is within class A, and the negative scores from class A (PC 1) are primarily caused by this shift in the origin; however apart from the three samples of soil 4, all samples from class B exhibit more positive scores in PC1 than class A. The pattern of the 1D scores plot could be changed by calculating a

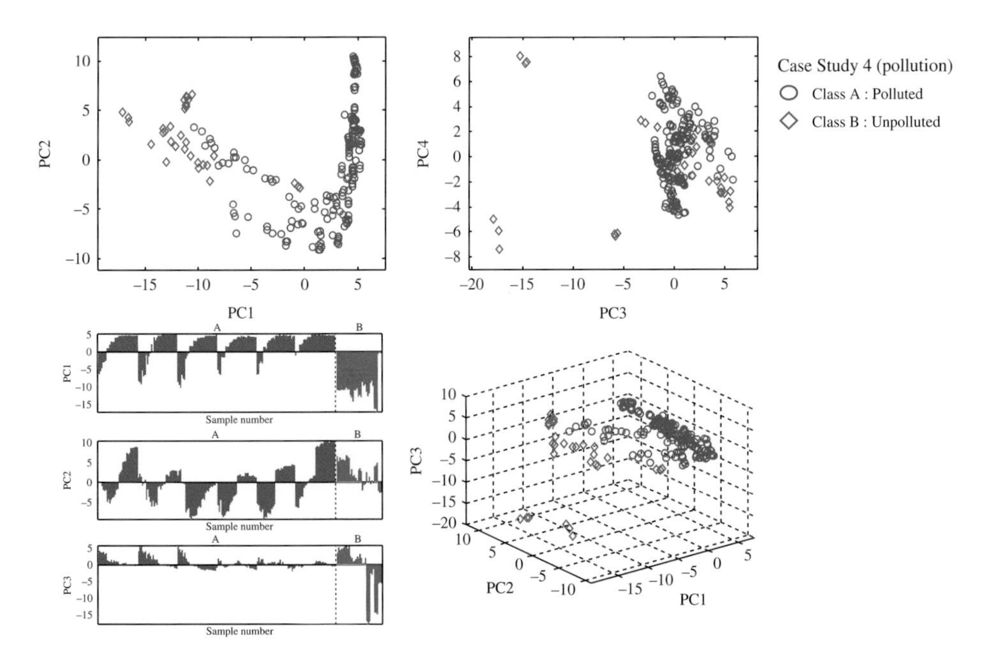

Figure 3.9 Scores plots for Case Study 4

weighted centring of each column where the average of the centroids of classes A and B is subtracted from each column. The repeating pattern in the 1D scores bar chart is primarily because the level of pollution is different in each sample. The samples are listed in Table 2.2 (see earlier), in the order they are displayed in Figure 3.9. It can be seen that the first 30 samples consist of commercial sand spiked with varying quantities of gas oil, and these correspond to the first 30 bars in the PC scores plot; the first 6 of these samples have had a lower level of hydrocarbon pollutant added and as such exhibit negative scores, as the amount of pollutant increases the scores become more positive. The scores plots suggest it is quite easy to separate the two classes using the first two or three PCs but also that the amount of pollutant added influences the spectrum and is reflected in the scores plots.

5. In Case Study 5 (sweat – Figure 3.10), it looks very difficult to distinguish the groups. Note that the class B (female) symbols are drawn over the class A (male) symbols, and so these hide a large cluster of symbols. It looks as if the first few PCs are dominated by a few outlying samples, although some visual separation between the male and female groups is discernable in the 3D PC plot. This may not be surprising: a person's sweat will be influenced by very many factors and we cannot expect gender to be the prime reason why people's sweat differs. There are over 300 compounds used in the model and only a few might be markers for gender. This does not however mean we cannot obtain good class models, and in fact it is precisely this sort of problem where pattern recognition becomes a valuable tool. Case Studies 1 to 4 are typical of what a laboratory

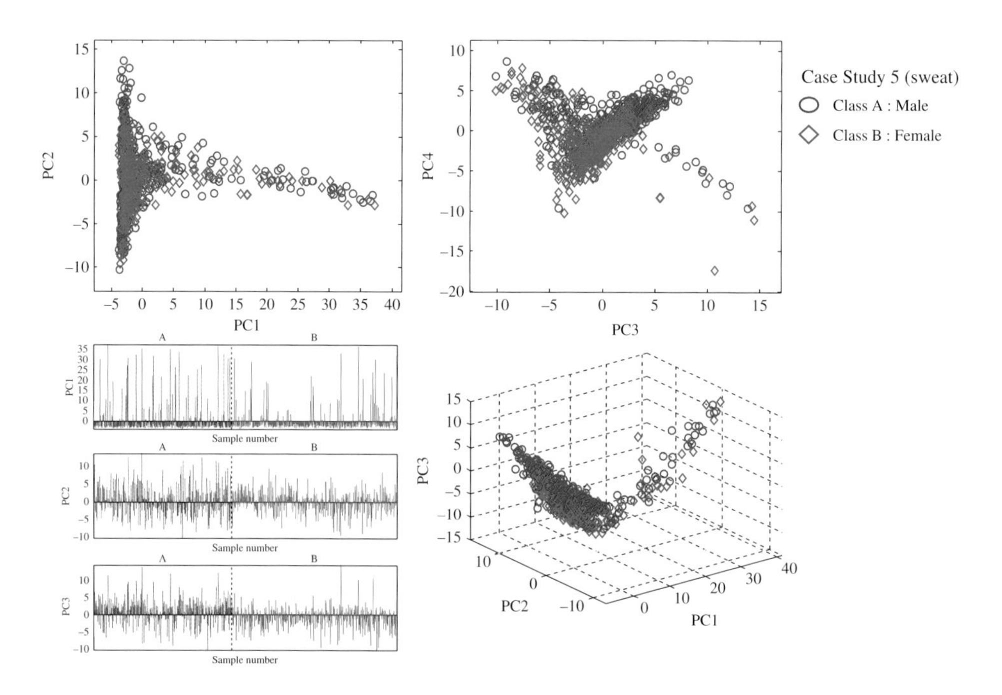

Figure 3.10 Scores plots for Case Study 5

based analytical chemist might encounter, often with some quite strong and dominant factors such as the type of material influencing the first PCs. In the case of human metabolomic profiling there may be 20 or more factors that influence the signal, and we cannot necessarily expect the earlier components to reflect the factor we are primarily interested in; this does not mean we cannot use these data for our intended purpose but it does suggest that it is hard to visualize – at least without overfitting and prior variable selection which can bias the picture. In this situation numerical methods are essential for the determination whether there is sufficient evidence to classify samples into groups. We will discuss ways of validating models in Chapter 8 and show that although graphically the scores of the first PCs look disappointing, we can, in fact, obtain reasonable classification models.

6. Case Study 6 (LCMS of tablets – Figure 3.11) appears quite hard to tease out. In this situation we are primarily interested in the origins of the tablets, but the instrument on which the analysis was performed and when it was performed also have an influence (the batch). In fact in many areas involving chromatography, instrumental performance is often very important for the effect on the appearance of small peaks; the pattern of trace contaminants change when chromatographic columns are changed, or chopped (in GC), or instruments cleaned, or instruments get dirty, or columns age. However there does seem some separation between the three classes if the scores of the first 3 PCs are plotted and so, although we do not expect this problem to be as easy as the first few case studies, it looks as if a model based on several PCs may succeed in separating the groups. The batches however have quite a strong influence on the first few PCs with batches 4 and 5 appearing quite distinct from the others. In fact these were run in August with the remaining batches run in April. Batch 5, furthermore, was run on a different instrument. The graphs suggest that both factors (batch and origin) might be distinguishable but the observed chromatogram is influence by both factors. It also demonstrates that instrumental effects can in practice be quite significant in a real dataset.
7. In Case Study 7 (clinical – Figure 3.12), we can see that the controls are easily separable from the diseased subjects, although this is obvious by inspection of the original numerical data and the PC scores plot do not add much additional insight, but are an easy way of visualizing the separation. The four different groups of diseased patients are harder to separate but inspection of the PC2 scores suggests that classes C and F (cardial hypertension and renal hypertension) are easy to distinguish from classes D and H (cardiovascular accident and malegnial hypertension) in PC2, whereas class F is quite well separated in PC3. Because there are only 5 variables and because the data are standardized, a maximum of 4 PCs can be computed. We might ascertain from these graphs that whereas we can separate some of the subgroups of diseased patients quite well, we cannot obtain perfect models from these data.
8. Case Study 8a (mouse urine – diet study – Figure 3.13) suggests that the two classes of mice on a diet and controls are reasonable well separable using 2 PCs. We do not expect perfect separation, but diet is clearly a major factor in distinguishing samples, as might be expected as all mice are clones, live under identical conditions, are of a similar age and we analyse urine which is a bodily

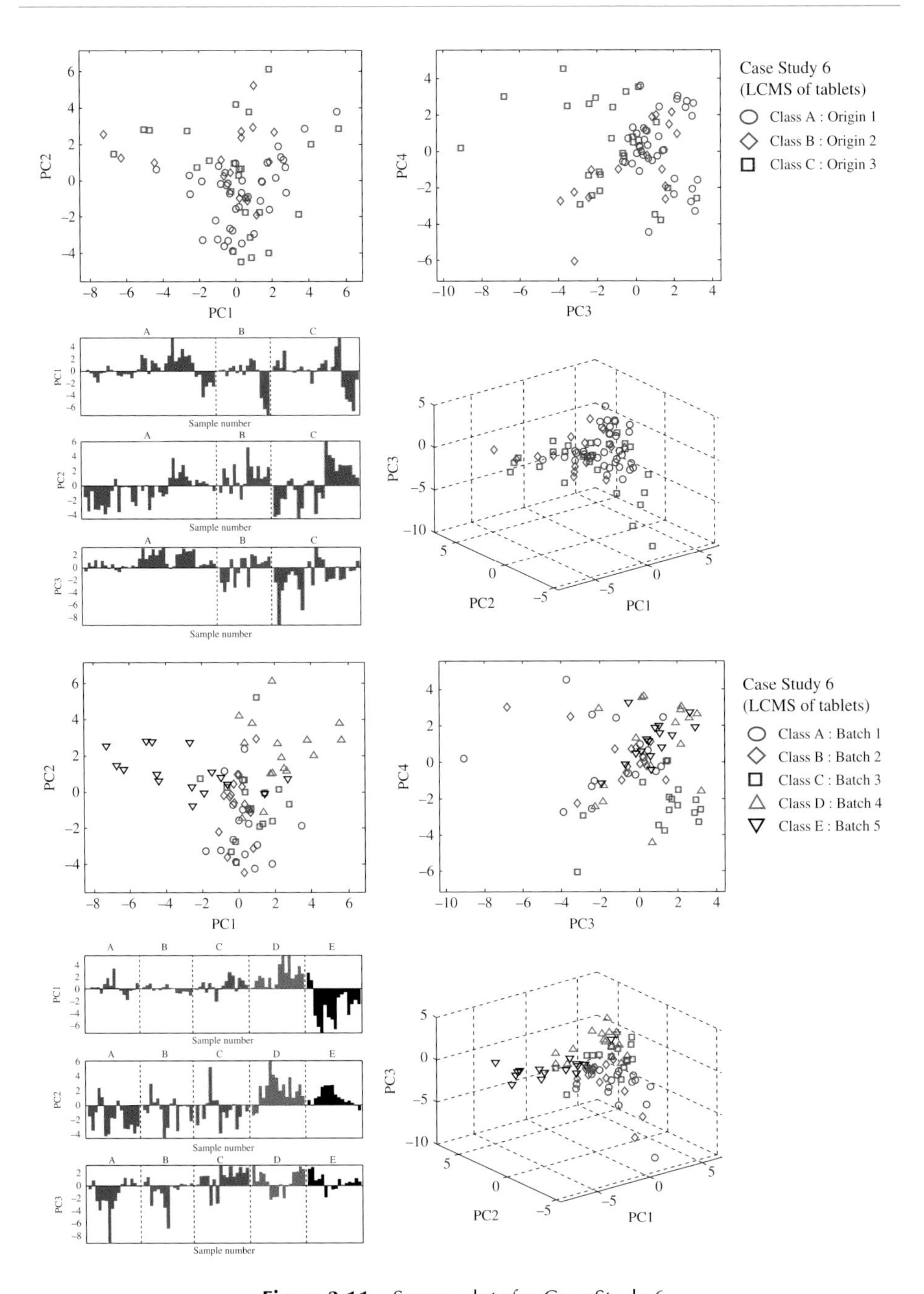

Figure 3.11 Scores plots for Case Study 6

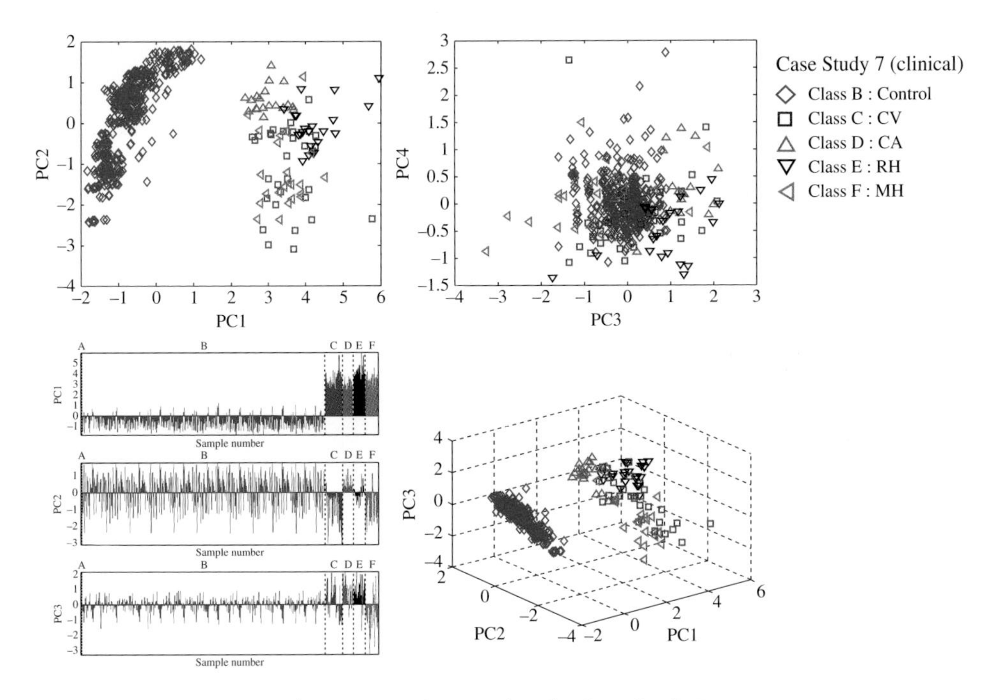

Figure 3.12 Scores plots for Case Study 7

fluid that contains waste products. Case Study 8b seems somewhat harder to separate, especially along the first components. In PC1 there appears some grouping within each of the three classes, and in fact the samples are ordered first according to class, and second according to analysis time on the instrument – so the aging of an instrument also has a major effect on the data. In fact the data of Case Studies 8a and 8b were analysed in a random order (that is, the samples from each group were interspersed) and the PC scores plots show that the change in the instrumental conditions has the largest influence on the analysis of the samples from mice with different genetics. For Case Study 8a there is also an effect due to sample order as can be seen from the scores of PC1 (the samples are ordered first by class and then by sample time), but it is less extreme because the effect of diet is less pronounced. For Case Study 8b (mouse urine – genetics – Figure 3.13), the genetic signal becomes more evident at later PCs. The PC scores plot of PC4 versus PC3 shows some level of separation. Interestingly, class B, which has characteristics both of class A and C, is in the middle. These graphs suggest that there is a genetic signal but it is weaker than the diet signal and the instrumental effect, and may require several PCs to model adequately, but will not be as difficult as distinguishing humans by gender from their sweat (Case Study 5).

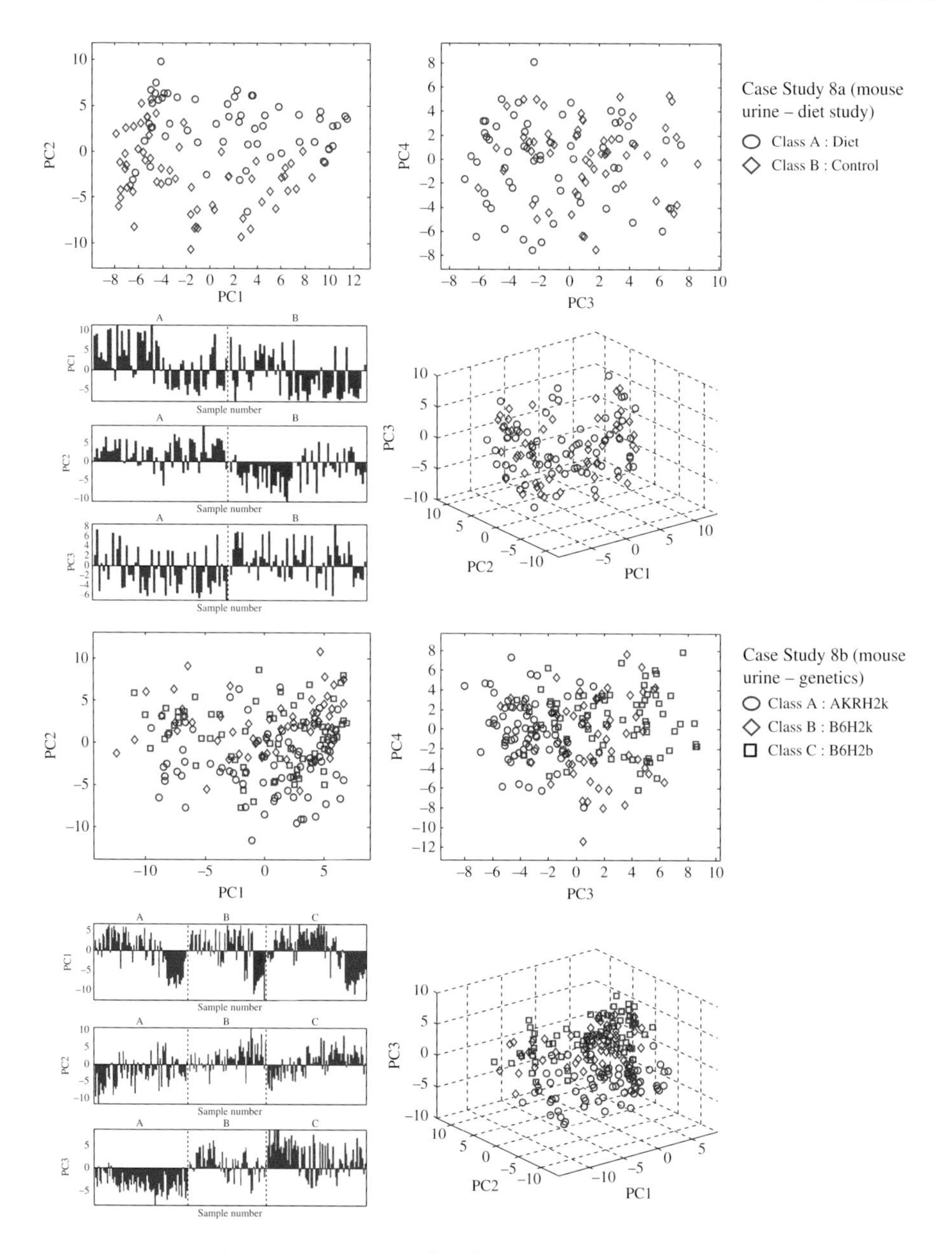

Figure 3.13 Scores plots for Case Studies 8a and 8b

9. Case Study 9 (NMR – Figure 3.14) suggests it will be fairly easy to distinguish salival samples according to treatment by mouthwash, with PC2 showing quite good separation. This problem appears of approximate difficulty to Case Study 8a, although there is, in fact, not such a clear instrumental effect

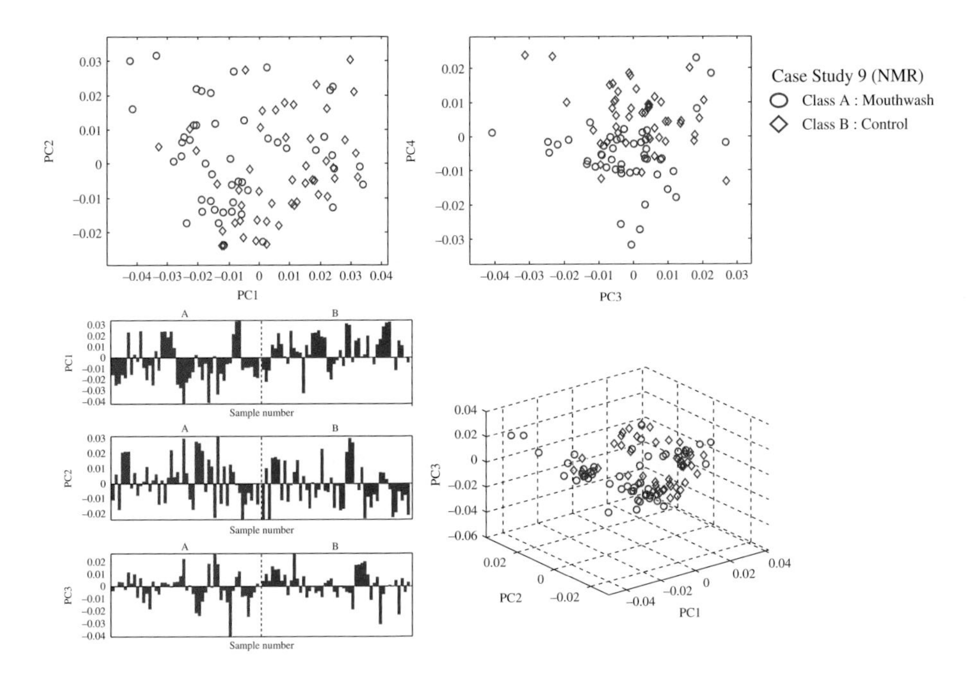

Figure 3.14 Scores plots for Case Study 9

10. The scores plots of Case Studies 10a to 10c (simulations – Figure 3.15) illustrate increasing overlap between the two groups.
11. Case Study 11 (null – Figure 3.16) encouragingly shows no easily visualized separation for either class. However randomness is not the same as uniformity, for example a sequence of numbers 10101010 is not random, but very ordered, and we actually expect some clumping in a sequence of random numbers. Hence there should be some clumping of samples from the same groups in the PC scores plots. For example in the graph of PC2 versus PC1 there is a small group of samples originating from class A in the bottom left, in contrast to a small group of samples originating from class B in the bottom right. These patterns are not unexpected but may confuse some pattern recognition methods, which will try to focus on these small regions to try to improve classification abilities.
12. There are not many samples in Case Study 12 (mouse scent marks – Figure 3.17), insufficient for classification studies but sufficient for data visualization. It can be seen that there is a quite strong separation between class A and B (Strain S1) and classes C and D (Strain S2), especially in the microbial scores plot – this is due in part to quite strong patterns of presence and absence of microbes. There is a less pronounced trend due to haplotypes in the samples studied. It is important to recognize that because not many samples have been studied, if the trend is not very pronounced, it is often hard to get good separation as the influence of noise will be relatively large.

Loadings plots can be used to see how the variables relate. They can be represented in 1D, 2D and 3D just as scores plots, although the 3D representation can be a bit crowded or

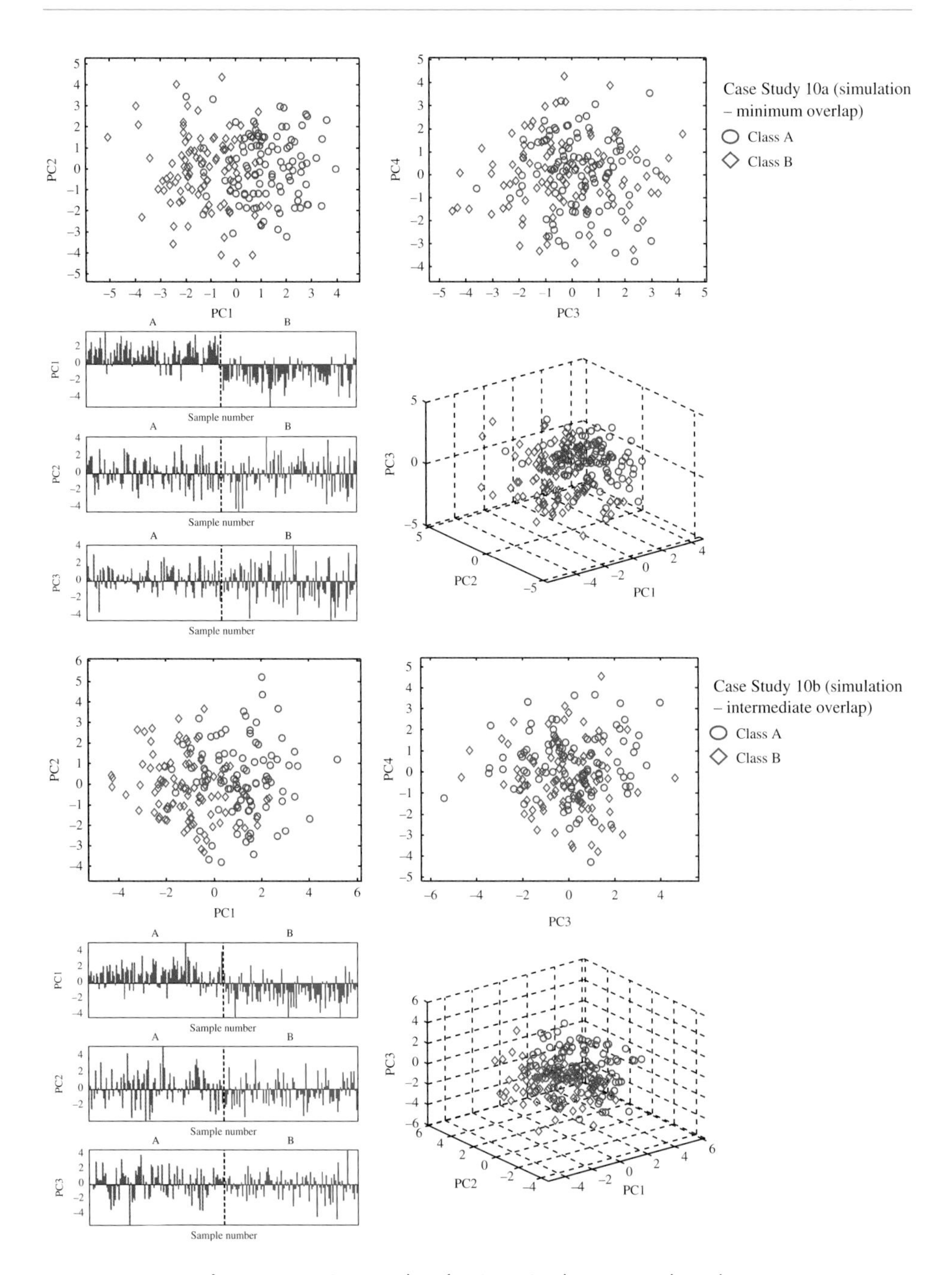

Figure 3.15 Scores plots for Case Studies 10a, 10b and 10c

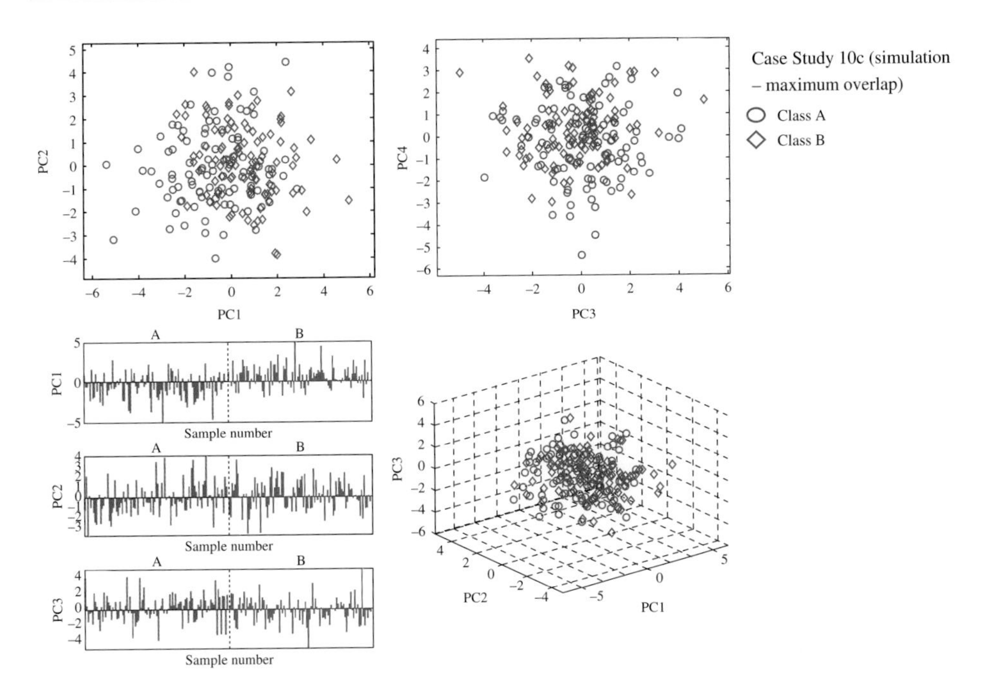

Figure 3.15 (continued)

complex as in most situations there are typically many hundreds or thousands of variables. However in many areas of chemometrics, such as NIR or NMR spectroscopies there often is a sequential meaning to the loadings and they can be related to a spectrum. Under such circumstances it is usual to join successive points. They should be compared to the scores plots and give clues as to which variables are most diagnostic of trends in the data. In Figure 3.18 we present the loadings plots for Case Studies 8a (mouse urine – diet study) and 2 (NIR spectroscopy of food).

Although there seems no easily interpretable pattern to the 1D loadings plot for Case Study 8a (mouse urine – diet study), this is because there is no specific sequence in the variables; however, these should be compared to the corresponding scores plots for the same case study (Figure 3.13). For example, a positive score for PC2 general indicates membership of class A (mice on a diet) and a negative score class B (controls) – of course, the fact that the dividing line is approximately at 0 is a function of several factors one of which is because each class contains nearly the same number of samples. However variables in the PC loadings plots with very positive values of PC2 are most likely to be markers for class A unlike those with very negative values of class B, and so one could look at the loadings plots and identify the extreme variables in the 1D PC2 loadings plot, and propose these as likely to be markers for each group. Of course it is a bit more complicated than this, and the relationship between PCs and class separation not quite as simple, as the PCs do not in themselves have a directly interpretable physical meaning but in many situations where the factor(s) of interest is/are the main ones that influence the simple visual interpretation provides good exploratory clues. In the 2D

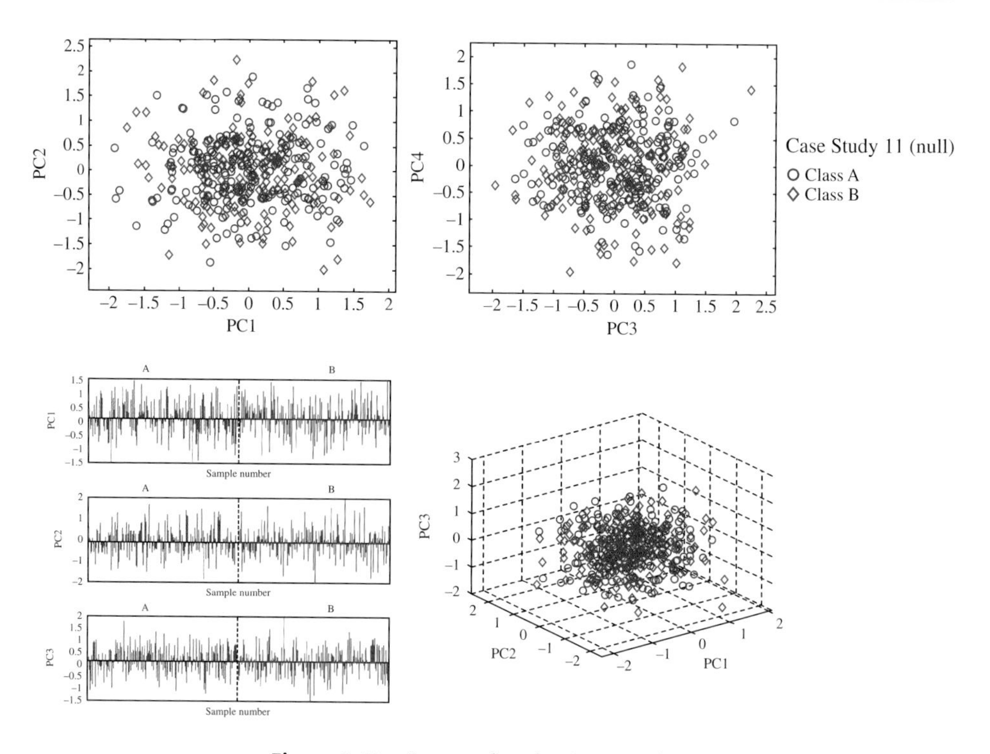

Figure 3.16 Scores plots for Case Study 11

loadings plots, the 20 variables that have the highest discriminatory power between the two groups, as assessed by the t-statistic (Section 9.2.2), are marked, and it can be seen that these cluster mainly in the top right or bottom left, which regions are most closely associated with one or other of the groups in the scores plots. An additional feature of the loadings plots is that variables that cluster most closely together are usually well correlated, so that for example, variables 55, 149 and 238 probably exhibit similar trends in the samples; if one is at a high level the others are too, and if one is at a low level the others are also. This is a somewhat simplified interpretation but on the whole such graphs are often used to provide a general idea of the basic relationship between variables and also between samples and variables.

The loadings from Case Study 2 (NIR spectroscopy of food) can likewise be interpreted in terms of the underlying classes of samples, but in this case there is sequential meaning to the horizontal scale which relates to wavelength in nm. As an example, we see that classes A and C (corn oil and safflower oil) exhibit high negative scores in PC1. Examining the loadings plot we can see that there is a prominent negative peak around 700 nm. This suggests that the region of the spectrum around 700 nm is correlated with these two oils – that is, it will be more intense for these oils and less for the others. It is important not to get confused by the negative sign of the loadings: if the scores and loadings are similar in sign,

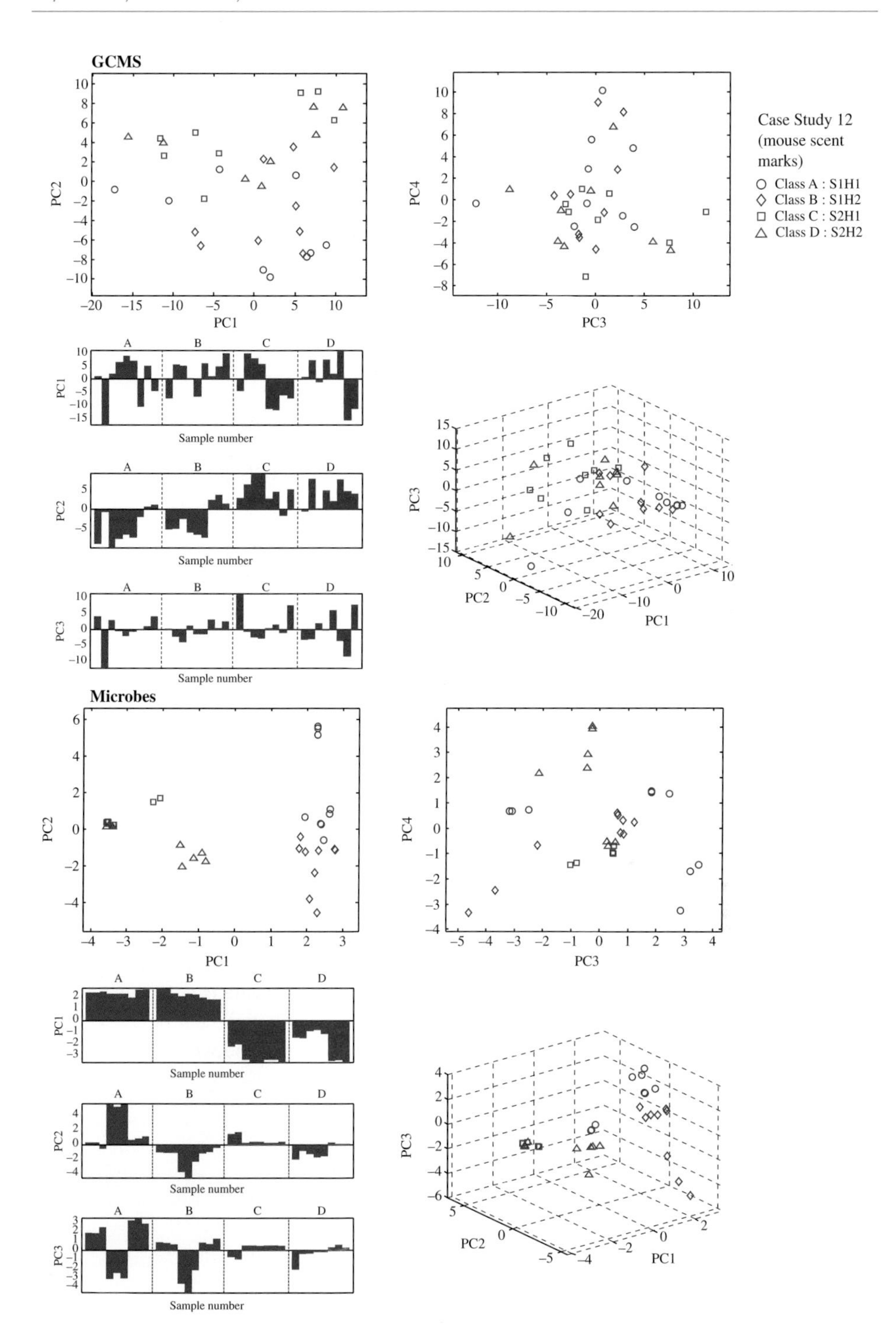

Figure 3.17 Scores plots for Case Study 12

then they group in samples, and spectral features are correlated. Returning to Figure 2.7 we see that this is indeed so. Corn margarine is characterized by negative scores in PC2 and so spectral regions with negative loadings in PC2 are likely to be diagnostic of this group: it appears that the region 954 to 971 nm is also

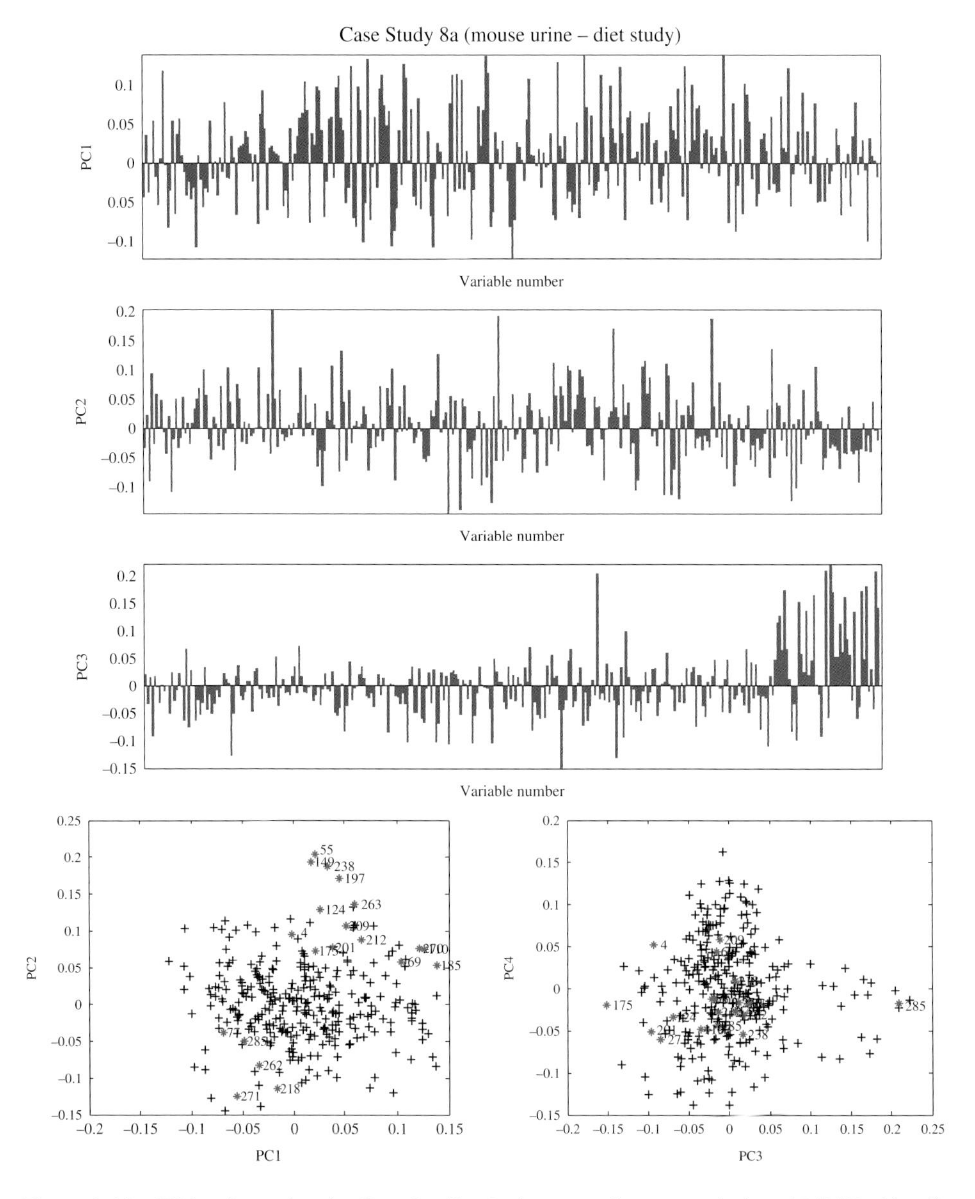

Figure 3.18 PC loadings plots for Case Studies 8a (mouse urine – genetics) and 2 (NIR of food). Variables found to be most discriminatory using the *t*-statistic (Case Study 8a – variable number; see Table 9.3) or Fisher weight (Case Study 2 – wavelength in nm) are indicated in the 2D loadings plots. NIR loadings plots are labelled by wavelength in nm

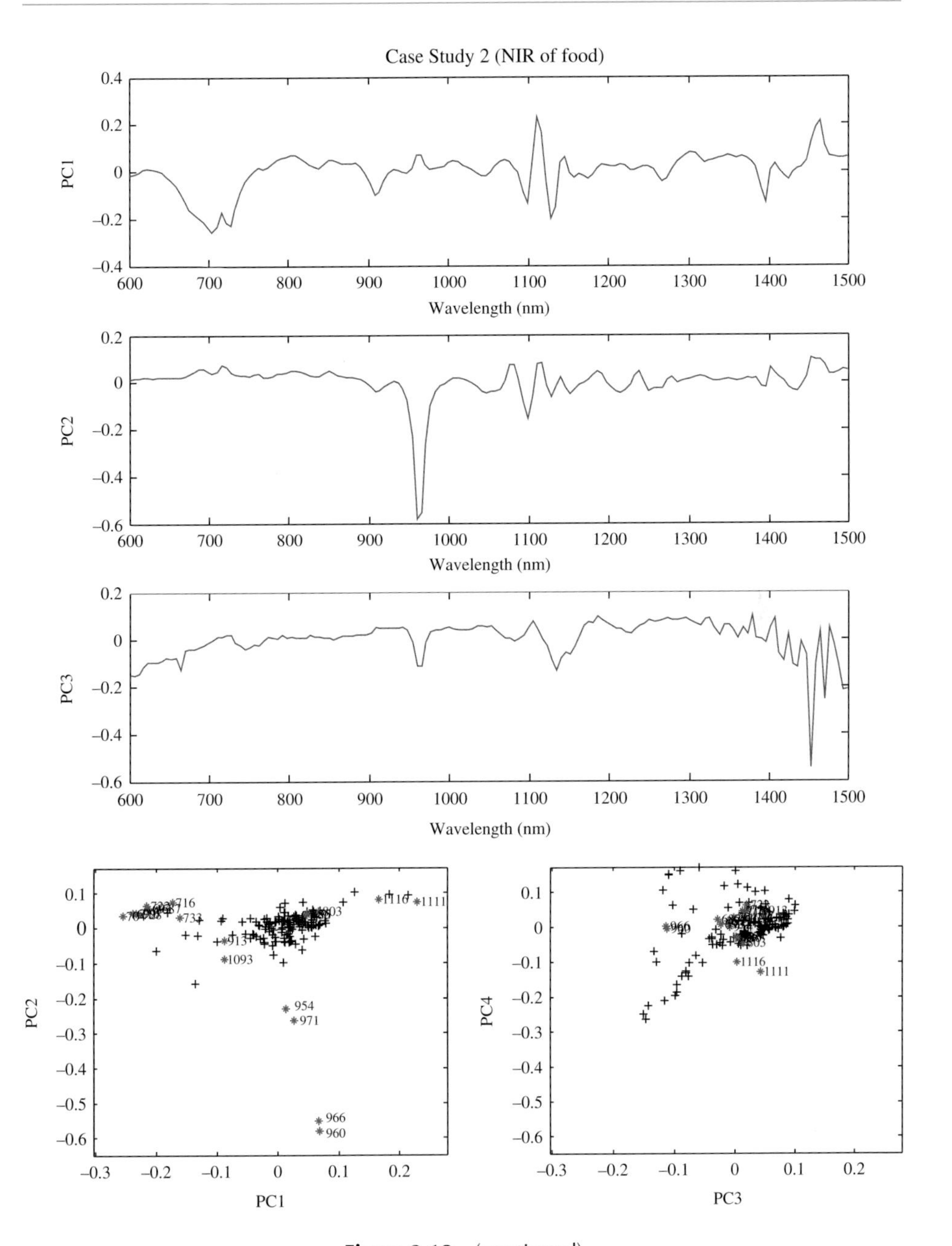

Figure 3.18 (continued)

characterized by negative loadings, and this is indeed a diagnostic spectral region for this group of oils.

Loadings plots are complementary to the more formal methods described in Chapter 9 for selecting the most discriminatory variables. In cases such as spectroscopy they do provide valuable information about, for example, which regions of the spectra are most useful, e.g. to discriminate between two products, but in other situations (e.g. Case Study 8a) they can sometimes be confusing: this is because there are often a large number of variables compared to samples and so the loadings can be rather crowded, consisting of a mass of points; for situations where there are not too many variables, however they are always quite useful.

Biplots are another method of data display that involve superimposing scores and loadings plots. Normally it is sufficient to compare these visually by looking at each type of graph separately, and as sometimes the loadings plots, as discussed above, can be very crowded. However in some cases it is useful, as in Case Study 7, i.e. the clinical study using atomic spectroscopy which is represented in Figure 3.19. We can see that Pb and Cd are most associated with the diseased patients as we expect from the data of Table 2.5 (see earlier). Cu seems to yield very limited diagnostic information for either groups. We can see visually which of the other elements are most diagnostic of a specific group. Although such information can be gained by inspecting tables of numbers, often it is easier to present this graphically. Note that there are several ways of scaling the loadings plot to fit the scores plot, but this is generally done automatically in most biplot routines. Further details of how to superimpose and scale both types of information is available in the technical literature.

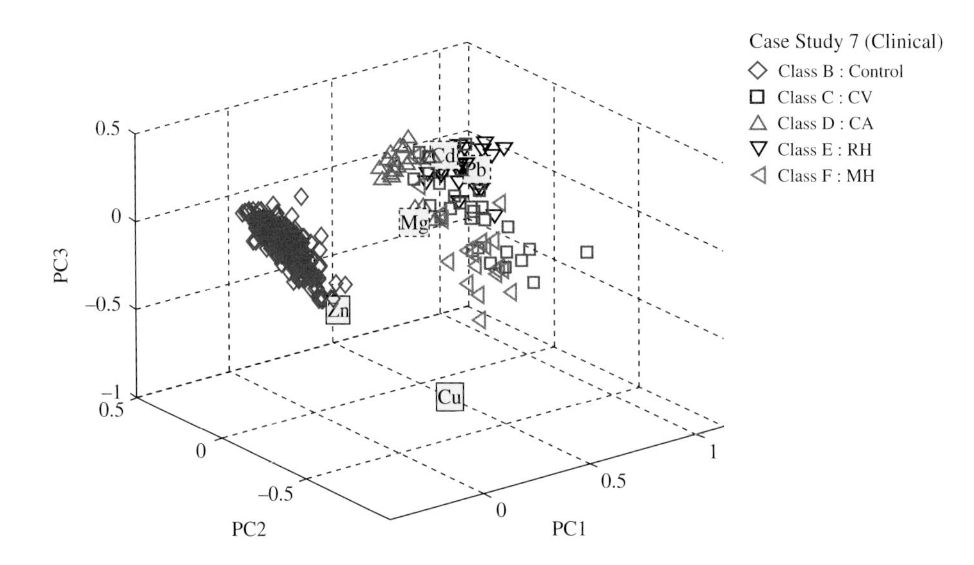

Figure 3.19 Biplot for Case Study 7

3.3 Dissimilarity Indices, Principal Co-ordinates Analysis and Ranking

3.3.1 Dissimilarity

One aim of EDA is to determine how (dis)similar each sample is from each other. The less similar, the less well two samples are related. The dissimilarity between each pair of samples in a data matrix can be measured, the larger its value, the less similar they are. A measure for similarity has the opposite property – the larger it is, the more similar the samples are. Any similarity measure can be transformed to a dissimilarity measure and vice versa if required.

In some cases these (dis)similarity measures can then be used as the first step in cluster analysis involving linking samples which are most similar together ultimately to form a dendrogram. We will not discuss the use of dendrograms in this text, which, although common in applications areas such as taxonomy, are not so frequently used in the chemometrics literature. Whereas there are many strengths in dendrograms for visual representation of data there are two principal weaknesses. The first is that the samples are ordered in a one dimensional sequence, yet often similarity between samples can be much more complex. The second is that it is possible to reflect branches of dendrograms, often resulting in quite different presentation of the same results according to the method of graphical display, with no general agreement on how the branches should be ordered. Nevertheless, cluster analysis does have a role and there are numerous texts and packages available for construction of dendrograms available, and the input is usually a (dis)similarity matrix as described below.

There are numerous dissimilarity measures that can be calculated, most also obeying the metric property and sometimes also called distances: many chemometricians visualize samples as projected on points in space. If we look at a cluster of samples in a diagram, the basis is that the farther the samples are apart, the less they are related and so the more dissimilarity they possess. To obey the metric property, the dissimilarity measure must obey the following rules:

1. The distance must always be ≥ 0.
2. $d_{ik} = d_{ki}$, that is the distance or dissimarity between samples i and k must be the same however they are measured.
3. $d_{ii} = 0$, that is the distance between a sample and itself is 0.
4. $d_{kl} \leq d_{ik} + d_{il}$, that is when considering three samples the distance between any two cannot exceed the sum of the distances between the other two pairs. This is often called the 'triangle' rule.

A dissimilarity measure that does not obey rule 4 is not strictly a distance, but nevertheless can be employed in pattern recognition, and is often called a 'semi-metric' to distinguish it from a pure metric or distance measure. Note that not all metric distances obey Euclidean properties, that they can be projected onto a Euclidean space; this is a rather esoteric distinction but all that it is necessary to remember is that distances that obey these rule 4 properties are not necessarily Euclidean in nature.

If a coefficient of dissimilarity can be measured, then it is possible to compute a dissimilarity matrix which consists of pairwise dissimilarities between all samples i and k where d_{ik} is the measure of dissimilarity between these samples. This is a symmetric

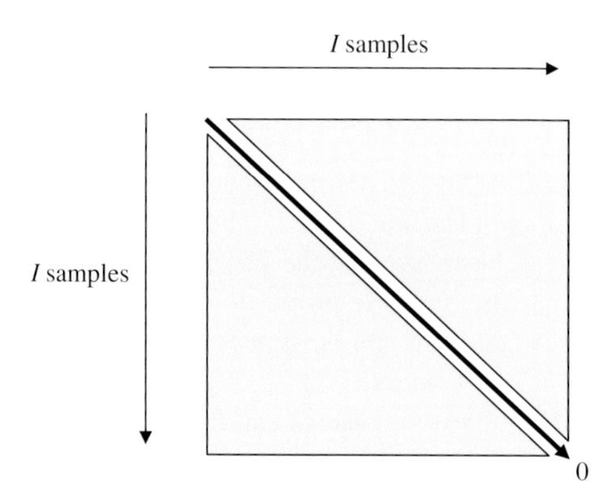

Figure 3.20 Dissimilarity matrix

square matrix of dimensions $I \times I$, since $d_{ik} = d_{ki}$ for all dissimilarity measures, and the diagonal values are equal to 0, since the dissimilarity between any sample and itself is 0. This is illustrated in Figure 3.20.

Below we will list the main types of (dis)similarity measure. Note that in some cases the basic measurement is one of similarity but this can be easily formulated into a dissimilarity measure if required. For cluster analysis (dendrograms) the main input is a similarity measure, whereas for Principal Co-ordinates Analysis (Section 3.3.2) it is a dissimilarity measure.

Binary Dissimilarity Measures

In traditional analytical chemistry, normally there is a measurement for each variable in all samples. For example, if we obtain a series of NIR spectra we measure intensities at all wavelengths and as such we obtain a spectrum over the full recorded wavelength range for each sample. There are no missing values, although at some spectral wavelengths the intensity may be lower.

Biologists in contrast have been faced with the problem of situations where a feature is present or absent for many years, for example we might be trying to determine whether several dogs are similar and might determine whether a dog has black spots or not, or a dog has a bushy tail or not; by building up a list of features (or variables), we can group dogs according to how many features they have in common, and those that have more in common with each other are likely to be more closely related. We can therefore set up a similarity matrix between each dog according to which features they can have in common. In some areas of modern chemical pattern recognition, especially metabolic profiling, matrices based on the presence or absence of features are becoming increasingly common. If we perform GCMS to detect the compounds in an emanation such as urine or sweat of several subjects, often only certain compounds will be detected in each subject. The presence of a compound may relate to disease state, to diet, to environment, etc. and there is no reason to expect that we will find the same set of compounds in each subject, and significant differences between individuals may be determined by studying how many compounds are detected in common to each pair of samples.

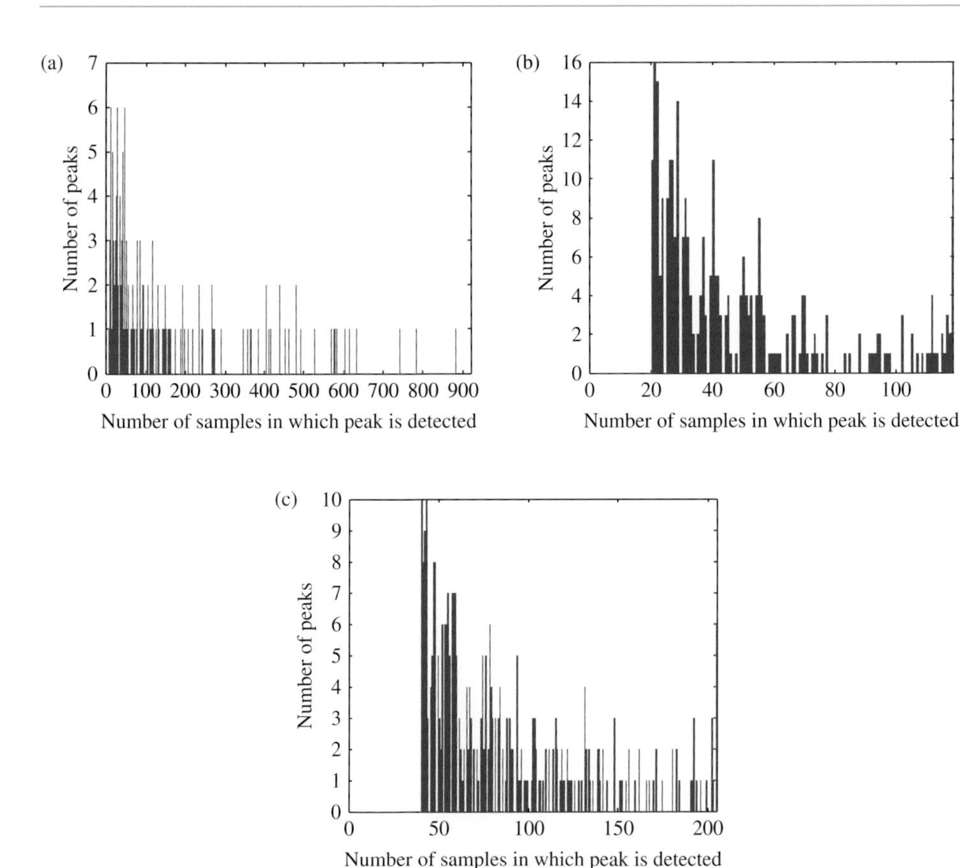

Figure 3.21 Number of times each compound is detected in different samples from (a) Case Study 5, (b) Case Study 8a and (c) Case Study 8b

Figure 3.21 illustrates how many times each compound is detected in samples for Case Study 5 (human sweat) and Case Studies 8a and 8b (mouse urine for the study of the effect of diet and genetics, respectively), after rare peaks are removed, as discussed in Sections 2.7 and 2.10. For Case Study 5, we see that of the 337 peaks, most are quite rare, none being detected in all samples (note that compounds known to arise from the analytical process such as siloxanes have already been removed). For Case Studies 8a and 8b, there are a few peaks detected in all samples (8 out of 307 in 8a and 6 out of 316 in 8b), but most are found in just a subset of samples. Good marker compounds may be detected only in a subset of samples and so determining which samples have most markers detected in common provides us with one way of deciding which are most similar, based on the detection of chemicals in their emanations. In analytical chemistry we cannot always distinguish whether the failure to detect a compound is due to it being absent or below the detection limit, but so long as the analytical procedure (e.g. extraction, chromatography) is similar for a series of samples, we will regard the failure to detect a compound as equivalent to its absence.

Dissimilarity measures determined by the presence or absence of features are usually known as binary dissimilarity indices. There are a very large number of such measures in the literature, and we will list a few of the most common, together with their properties. Table 3.1 illustrates a situation in which a number of features are detected in two samples i and k. In this case a features are detected in common to both samples, b are found only in sample i and so on. If J features or variables are measured over the entire set of samples, then $a + b + c + d = J$, and in sample i, $a + b$ features are detected. In the context of chemometrics, 1000 peaks (= J) may be detected over a series of chromatograms, but only 300 (= $a + b$) detected in chromatogram i.

Table 3.1 *Four binary cases for comparison of presence and absence of the features in samples* i *and* k

Sample i	Sample k	
+	+	a
+	−	b
−	+	c
−	−	d

Table 3.2 lists a number of pairwise (dis)similarity indices. S is a measure of similarity; the higher its value, the more similar the samples, with in all cases the maximum value equals 1. Each index either has a commonly accepted name (e.g. the Jaccard index which is widespread) or is given the name of the authors of the first paper where it has been introduced in the table. The simplest dissimilarity index that can be derived is $(1-S)$. Some of these indices have metric properties, that is, they obey the 4 rules of a distance, while others do not; it is not necessary to have an index that is metric, for example, in quantitative similarity measures, the correlation coefficient is not metric, but is often used. In no case do any of these indices have Euclidean properties Another dissimilarity index can be computed by $\sqrt{(1-S)}$. In all cases listed in Table 3.2 this index has metric properties even where $(1-S)$ does not and in all cases it has Euclidean properties: this

Table 3.2 *Binary (dis)similarity indices. All values of $\sqrt{(1-S)}$ are metric and Euclidean. No values of (1 – S) are Euclidean*

Number	Similarity index S	Definition	$(1-S)$ Metric
1	Jaccard	$a/(a+b+c)$	Y
2	Simple matching	$(a+d)/(a+b+c+d)$	Y
3	Sockal and Sneath (a)	$a/[a+2(b+c)]$	Y
4	Rogers and Tanimoto	$(a+d)/[a+2(b+c)+d]$	Y
5	Sorensen	$2a/(2a+b+c)$	N
6	Gower and Legendre	$[a-(b+c)+d]/(a+b+c+d)$	Y
7	Ochiai	$a/\sqrt{[(a+b)(a+c)]}$	N
8	Snockal and Sneath (b)	$ad/\sqrt{(a+b)(a+c)(d+b)(d+c)}$	N
9	Pearson's ϕ	$(ad-bc)/\sqrt{(a+b)(a+c)(d+b)(d+c)}$	N
10	Russell and Rao	$a/(a+b+c+d)$	Y

should not be confused with the Euclidean distance as discussed in Section 3.3.1 above, but implies that the distances can be projected onto a Euclidean surface.

In many cases where many unique compounds are detected over a dataset, but most are not detected in any specific sample, indices that include the double zero (*d*) are not best suited. By this logic, probably the most useful dissimilarity measures are given by the value of $\sqrt{(1-S)}$ for indices 1, 3, 5 and 7. However it is not necessary to be restricted to indices that are Euclidean or even metric.

Quantitative Dissimilarity Measures

Quantitative dissimilarity measures are more familiar in chemometric pattern recognition and well established. The main methods are summarised in Table 3.3, for computing

Table 3.3 *Quantitative (dis)similarity indices between samples* i *and* k

Number			
	Dissimilarity measure		
1	Euclidean	$\sqrt{\sum_{j=1}^{J}(x_{ij}-x_{kj})^2}$	Used in all cases
2	City block	$\sum_{j=1}^{J}\lvert x_{ij}-x_{kj}\rvert$	Used in all cases
3	Chebyshev	$max\lvert x_{ij}-x_{kj}\rvert$	Used in all cases
4	Minkowski (order *n*)	$\sqrt[n]{\sum_{j=1}^{J}(x_{ij}-x_{kj})^n}$	General for order *n*, equal to Euclidean distance if $n=2$
5	Canberra	$\sqrt{\sum_{j=1}^{J}\frac{\lvert x_{ij}-x_{kj}\rvert}{x_{ij}+x_{kj}}}$	Only useful for non-negative values
	Similarity measure		
6	Pearson correlation coefficient	$\frac{\sum_{j=1}^{J}(x_{ij}-\bar{x}_i)(x_{kj}-\bar{x}_k)}{\sqrt{\sum_{j=1}^{J}(x_{ij}-\bar{x}_i)^2\sum_{j=1}^{J}(x_{kj}-\bar{x}_k)^2}}$	Varies from +1 to −1
7	Squared Pearson correlation coefficient	$\left[\frac{\sum_{j=1}^{J}(x_{ij}-\bar{x}_i)(x_{kj}-\bar{x}_k)}{\sqrt{\sum_{j=1}^{J}(x_{ij}-\bar{x}_i)^2\sum_{j=1}^{J}(x_{kj}-\bar{x}_k)^2}}\right]^2$	Varies from +1 to 0
8	Cosine	$\frac{\sum_{j=1}^{J}x_{ij}x_{kj}}{\sqrt{\sum_{j=1}^{J}x_{ij}^2\sum_{j=1}^{J}x_{kj}^2}}$	Varies from +1 to −1

the (dissimilarity) between samples i and k where there are J variables in total measured for each sample. The indices can be divided into those that measure dissimilarity and those that measure similarity.

Of the dissimilarity measures, the Euclidean distance measure is by far the most common, but of course the use of this depends in part on how the data are preprocessed (Chapter 4), and so the results should be interpreted with this in mind. The City block and Chebyshev distances are computationally much less intense especially if there are several variables. The Minkowski distance generalizes to the City block distance when $n = 1$ and the Euclidean distance when $n = 2$. Interesting when n becomes large it tends towards the Chebyshev distance. The Canberra distance can only be used for positive values, and involves using a ratio between the difference between two points and their average, and so weights distances close to the origin as more significant; there are some advantages here particularly as this will be less susceptible to outliers that may result in readings that have an undue influence on the overall distance matrix.

Similarity based measures can also be obtained, but need to be converted to dissimilarities to have a comparable meaning to the distances described above, typically by subtracting from the highest achievable value ($= +1$ in the three measures described below), so that a value of 0 is obtained if two samples are identical. The cosine distance is often described as the cosine of the angle between the two vectors x_i and x_k especially by the more geometrically minded members of the chemometric community. It is identical to the Pearson correlation coefficient if two vectors are row centred; note that although it is usual to mean centre vectors, it is not usual to row centre. If the means of each row (or sample) vector change greatly these two measures may give quite different results.

There are a large number of other quantitative indices proposed in the literature, far exceeding the number of binary indices, but for brevity in this text we list only the most widespread and useful for the purpose of pattern recognition.

3.3.2 Principal Co-ordinates Analysis

After a dissimilarity matrix has been obtained, the next step involves visualizing this matrix. Principal Co-ordinates Analysis (PCO) is a generalized approach for visualizing such matrices. It is sometimes also called Multidimensional Scaling (MDS), although there are additional types of non-linear MDS which we will not discuss in this text. A major advantage of PCO over PCA is that it can cope with any (dis)-similarity measure and so is more flexible. PCA only uses the Euclidean distances between samples, and although it is possible to preprocess the data to influence the appearance of the PC plots, there are limitations, and so a more general approach is sometimes called for. The disadvantage of PCO is that it is designed only to examine similarity between samples and it is not easy to show the influence or significance of the variables. The input to PCO is an $I \times I$ dissimilarity matrix $\boldsymbol{D}$ that whose elements contain dissimilarity indices calculated using one of the methods described in Section 3.3.1.

The following steps are performed:

1. The $I \times I$ matrix $\boldsymbol{D}^{(2)}$ is calculated, by squaring the elements of the original dissimilarity matrix.
2. A matrix $\boldsymbol{A}$ is computed, whose elements are given by:

$$a_{ij} = -1/2d_{ij}^{(2)}$$

3. Finally a column and row centred matrix $\boldsymbol{H}$ is computed so that:

$$h_{ij} = a_{ij} - \overline{a}_i - \overline{a}_j + \overline{a}$$

   where $\overline{a}_i$ and $\overline{a}_j$ represent the row and column means for sample i and variable j of matrix A and $\overline{a}$ the overall mean.
4. PCA is performed so that:

$$\boldsymbol{H} = \boldsymbol{TP} + \boldsymbol{E}$$

   The expression $\boldsymbol{TP}$ can be reformulated as $\boldsymbol{TP} = \boldsymbol{P}'\Lambda^{1/2}\boldsymbol{P}$ where $\Lambda^{1/2}$ is a matrix whose diagonals contain the square root of the eigenvalues of the PCs. Hence the scores and loadings matrices are the same except in magnitude (if using Singular Value Decomposition they are identical).

Visualization can be done by plotting one row of $\boldsymbol{P}$ against another. There are some limitations as to which similarity measures can be used; however all the $\sqrt{(1-S)}$ dissimilarity measures of Table 3.2 and all those of Table 3.3 should provide meaningful answers.

An interesting result is that PCO using Euclidean distance measures is the same as the scores obtained from PCA apart from some reflectional ambiguity (note that the signs of PCs cannot be controlled), and hence PCO can be considered a generalized version of PCA. Although PCO is a powerful technique we stick primarily to PCA in this text partly because it is more widespread and so accepted and partly because information on the variables can also be obtained (in PCO the scores and loadings are the same apart from scaling issues if NIPALS is employed). However PCO allows additional flexibility particularly in the case of binary variables, and permits a wider range of approaches for visualizing data, and can be employed as an exploratory method and also the components obtained from PCO can be used as input to classifiers just as for the scores from PCA.

We will illustrate the performance of PCO using several different dissimilarity metrics using Case Study 8b (mouse urine – genetics) where there a number of the variables are characterized by the presence and absence, presented in Figure 3.22. For the quantitative measures, an undetected compound is set at 0, prior to standardization. Using the Euclidean distance the graph, in fact, is identical to that obtained by PCA (Figure 3.13) except that PC1 is reflected. We can see that in this case the Chebyshev distance is not very helpful – it is dominated by samples that have large differences in intensities for a few variables. The Jaccard and Sorensen distances appear less encouraging to the Euclidean distance; however when extended to more PCs, there are in fact differences (not illustrate for brevity). There is no one dissimilarity measure that is best, although some such as the Chebyshev distance appear inappropriate. Which similarity measure is most appropriate will depend on the dataset being studied, and a useful approach is to try several such dissimilarity measures and choose the one that is most appropriate for the study in hand.

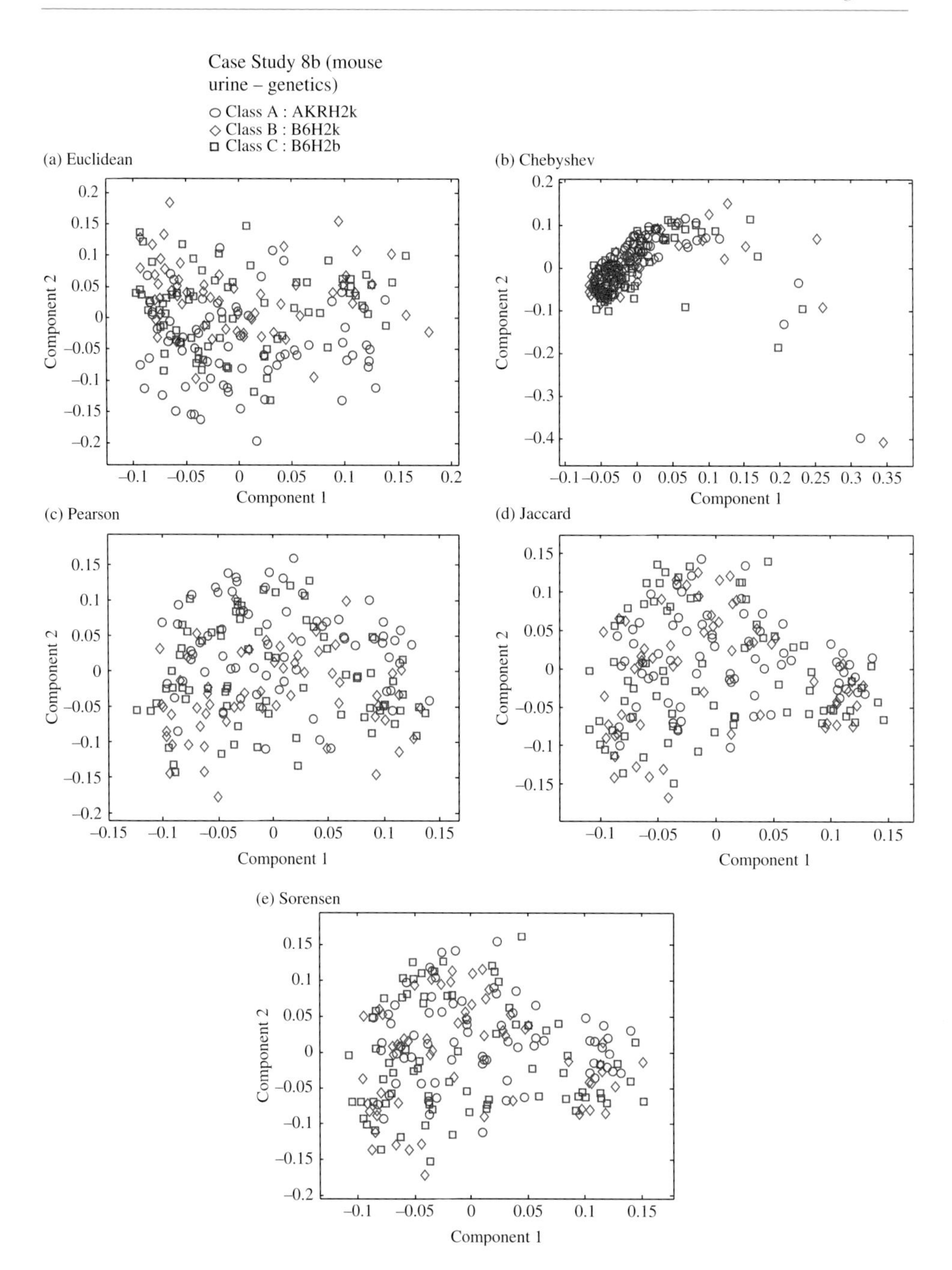

Figure 3.22 PCO on Case Study 8b (mouse urine – diet study) using (a) Euclidean, (b) Chebyshev, (c) Pearson correlation coefficient, (d) Jaccard $\sqrt{(1-S)}$ and (e) Sorensen $\sqrt{(1-S)}$ dissimilarity indices, using the first three components. The data for the quantitative indices (a) to (c) have first been preprocessed by square root scaling, row scaling to constant total and standardizing. The first two components are illustrated

Although for Case Study 8b only the Chebyshev distance seems to show significant differences from the other measures, for Case Study 7 (clinical) there are dramatic differences, as illustrated in Figure 3.23, in particular the cosine and the Pearson correlation coefficient measures give plots that are very different. The reason for this is that the average of each row (or sample) differs significantly between samples, as the amount of Pb and Cd detected in the diseased patients are in orders of magnitude larger than in the control group. This difference will be retained even in standardized samples. The Pearson correlation coefficient involves centring each row, whereas the cosine measure does not; hence both will have a different influence on the results of PCO. Note that for these

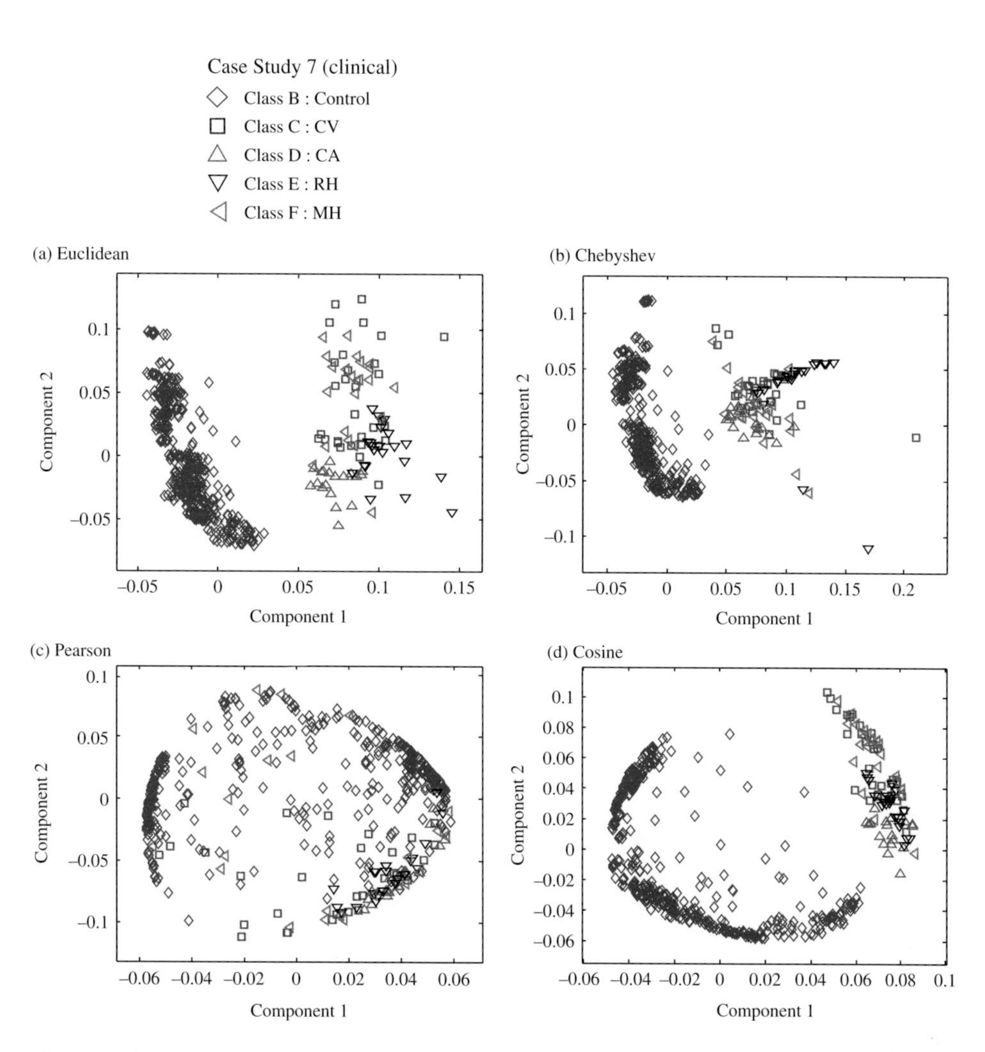

Figure 3.23 PCO on Case Study 7 (clinical study using atomic spectroscopy) using (a) Euclidean, (b) Chebyshev, (c) Pearson correlation coefficient and (d) cosine distance using the first two components, on standardized data

samples, since there is a positive value for each of the elements measured, qualitative dissimilarity measures cannot be used.

3.3.3 Ranking

Dissimilarity measures as described in Section 3.3.1 can be used for PCO but can also be used for rank based approaches for determining whether there is any grouping between samples. These look at which samples are most similar, involving ranking each pairwise similarity between samples according to how much they resemble each other. If we have two groups of samples, we would expect the 'within-group' similarity of samples to be greater than the 'between-group' similarity. The pairwise similarities can be ranked in order, from most to least similar. If there are I samples there are $I(I-1)/2$ pairwise similarity measures excluding the similarity of a sample with itself, e.g. for 100 samples, there will be 4950 similarities. The two samples whose similarity measure is ranked at $I(I-1)/2$ are the least similar (lowest rank).

If there are two classes, we may want to explore whether the similarity of samples within the classes is greater than samples between classes. If, for example, there is a dataset consisting of 50 samples of class A and 50 of class B, of the 4950 similarity measures there will be a population of 2450 within class (dis)similarity measures (corresponding to 2 sets of $50 \times 49/2$ measures for each class) and a population of 2500 between class (dis)similarity measures ($= 50 \times 50$). The question posed is whether the ranks of the 2450 'within class' (dis)similarity measures are significantly different (more similar) than the 2500 'between class' similarity measures? The way to do this is to create a separate list of the ranks of 'between' and 'within' class dissimilarities, and see if these lists are significantly different. The simplest approach is to represent these using a rank graph: the principles are illustrated in Figure 3.24. In the top graph all possible pairwise similarities are grouped together, for example if there are 4950 pairwise similarities we look at what proportion of the samples have a similarity exceeding rank 1, rank 2, and so on up to rank 4950; this must be a linear graph. If however the pairwise similarities are separated into two populations, and one population includes similarities of that are, on the whole, of higher rank than the other, we can plot the proportion of 'within class' and 'between class' (dis)similarity measures against their rank, called a cdf (cumulative distribution function). If there is a difference, we expect these to form significantly different populations and the two curves will diverge, as shown in Figure 3.24(b). Using the Euclidean distance for ranking, the rank graphs are presented for Case Studies 1, 3, 8a and 11 in Figure 3.25. We can see for the Case Study 11 (null) that the two lines are in practice indistinguishable, suggesting that the two classes are to all intense and purposes indistinguishable. Case Study 3 (polymers) can be visualized in one of two ways. Either way we can see whether the two main types are distinguishable from each other or the nine groups. We can see that the latter are much easier to distinguish. This is easy to rationalize. Within each type there are several groups, some of which are very different to others, and so these will result in large and distinct dissimilarities. However if each type in turn consists of several groups of samples, which have different characteristics, then there will be some high dissimilarities within each type, meaning that the average within type dissimilarity

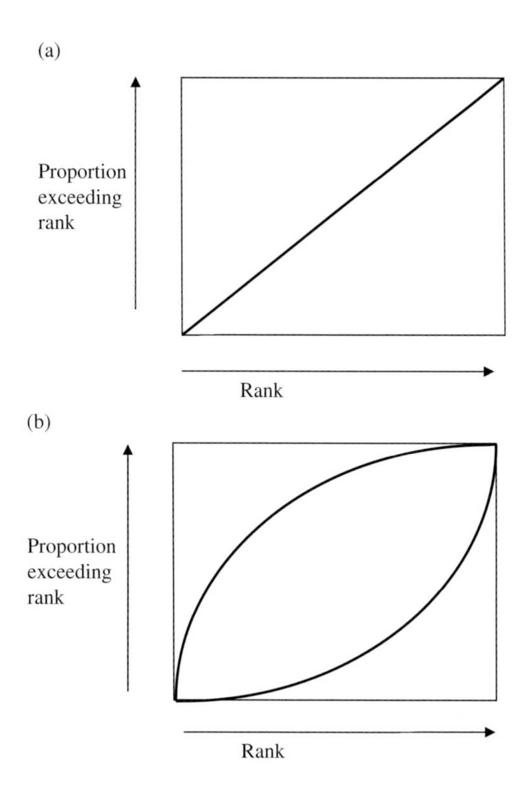

Figure 3.24 Illustration of rank charts: (a) a graph of proportion of dissimilarity measures exceeding a given rank against rank for an entire population; (b) separating the population of dissimilarity measures into two groups of pairwise comparisons, one which is more similar to each other (top line), and the other which is less similar (bottom line)

will be more than the average within group dissimilarity for this reason: this suggests that the two main types are not homogeneous groups and the polymers are more appropriately treated as nine different groups than just two main types. Case Study 1 (forensic) is very clear as already suggested by the PC scores plots, showing extremely strong separation between groups. Case Study 8a (distinguishing between mice on a diet and controls) shows some separation but is not as clear as Case Studies 1 and 3 as anticipated from the PC scores plots. Where the two lists clearly diverge there is likely to be a significant separation between classes, and so supervised methods described in Chapters 5 to 7 can be safely employed. The use of ranking can be extended to visualizing almost any type of similarity, for example we may have a series of individuals and ask whether chemicals obtained from repeat samples of a single individual are more similar to those from different individuals.

Some people also like to attach a probability to whether the two sets of ranks are different; much classical statistics is about computing probability or p values, and so we

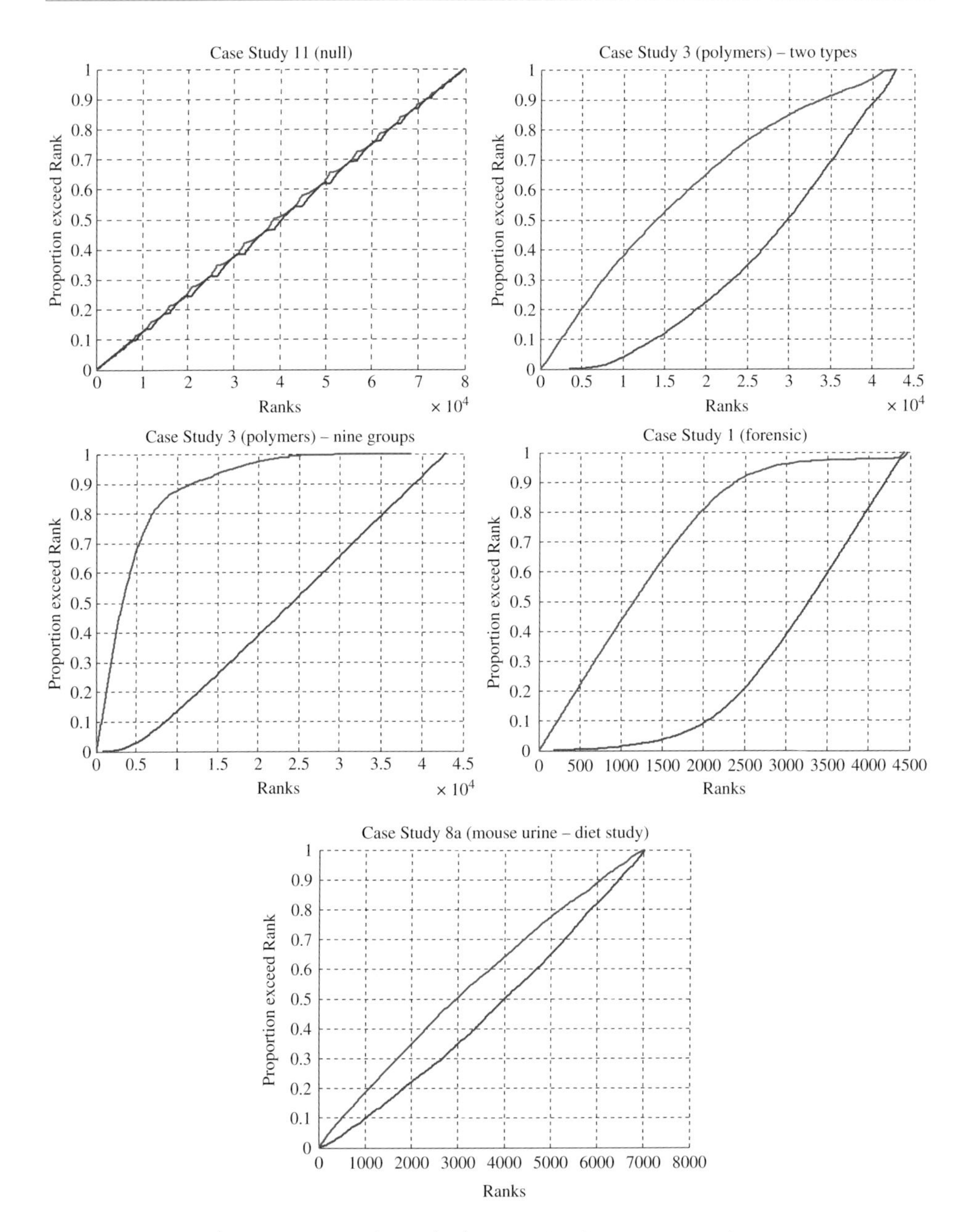

Figure 3.25 Rank graphs for Case Studies 11, 3, 1 and 8a

include a description for completeness. If we call the 'within class' rank list the AA list, and the 'between class' the AB list, then we can define $G_{AA}(\theta)$ and $G_{AB}(\theta)$ as the two-sample number based cumulative distribution functions from the rank lists AA and AB respectively (the proportion of the rank values less than or equal to θ, where θ varies from 0 to the maximum possible rank over all pairwise (dis)similarity measures). The two-sample Kolomorogov–Smirnov (K–S) statistic ($S_{\text{K–S}}$) can be calculated as follows:

$$S_{\text{K–S}} = \max\left(|G_{AA}(\theta) - G_{AB}(\theta)|\right)$$

We want to test the hypothesis that samples from the same group are more similar to each other than samples from different groups or that the AA rank list contains more high ranking comparisons than the AB list; the null hypothesis is that both two lists arise from the same underlying distribution. If the null hypothesis cannot be rejected (that is, there is no real evidence that the two rank lists are different), $F(\theta)$ and $G(\theta)$ would be fairly close for all values of θ and $S_{\text{K–S}}$ is almost zero. Thus, the two-sample K–S statistic can be used to test the hypothesis that the ranks in the AA list are significantly lower than those in the AB list. The p value for the one sided test can be approximated by using the following equations:

$$\lambda = \left(\sqrt{K} + 0.12 + \frac{0.11}{\sqrt{K}}\right) \times S_{\text{K–S}} \quad \text{and} \quad p = e^{-2\cdot\lambda^2}$$

where $K = K_1 \times K_2/(K_1 + K_2)$ and K_1 and K_2 are the number of measurements in the two distribution functions. This approximation is reasonably accurate when $K \geq 4$.

3.4 Self Organizing Maps

3.4.1 Background

Self Organizing (or Kohonen) maps (SOMs) are an alternative approach for data visualization to PCA and PCO. SOMs were first described by Kohonen in the 1980s as a method of visualizing the relationship between different speech patterns. There have, however, been relatively few descriptions of the applications of SOMs in the context of chemometric pattern recognition compared to PCA, probably due to the lack of well established packages. Yet SOMs can be a powerful means of visualization. In addition to maps that provide information similar to scores plots about the relationship between samples, component planes can be used to visualize characteristic variables, specific samples or groups of samples. There are many tools associated with SOMs, which can be employed to provide detailed insight into a dataset including numerous approaches for displaying maps, the most common approach used being the U-matrix. SOMs which are an important alternative to PCA and are particularly effective where there are several classes: most multivariate visualization methods perform most effectively when there are two or a small number of classes characterizing a dataset.

The SOM is a type of Artificial Neural Network which uses unsupervised learning to produce a low dimensional representation of training samples whilst preserving the topological properties of the original input space. This makes SOMs suitable for visualizing data that have a large number of variables. SOMs can be particularly effective if the data structure is quite complex and there are a large number of datapoints which would overlap unacceptably in a scores plot. An overview of the SOM visualization methods is described in Section 3.4.6.

3.4.2 SOM Algorithm

An SOM involves a map, or two dimensional grid, made up of a number of units (equivalent to pixels in a low resolution map) and the aim is to position the samples on the map such that relative distances between them in the original data-space is preserved as far as possible within a 2D approximation. To do this, each 2D map unit is described by a vector of weights for each variable in the original dataset, resulting in a weight vector for each unit. A map can therefore be thought of as a two dimensional collection of map units, with a layer of weights for each variable making up the third dimension; the number of weights for each unit (or dimension of the weight vector associated with each unit) corresponds to the number of variables measured in the original dataset. Figure 3.26 shows the structure of a typical map for a dataset described by $J = 3$ variables. In this text, maps will be represented by two dimensional hexagonal lattices with P columns and Q rows consisting of a total of K $(= P \times Q)$ map units; there is, of course, no requirement for the units to be hexagonal and sometimes square or rectangular geometries are employed as alternatives but the hexagonal representation is quite common. Each map unit is given a label (p, q) corresponding the row and column in which the unit is located: this label has no direct relationship to a unit's geometric distance to other units (e.g. there is a greater distance between units (2,2) and (3,1) than there is between units (2,1) and (3,2), though this may not be immediately evident by the labels) and is primarily used to locate the unit on the grid in a uniform way. The corresponding weights for each map unit can be represented by a $K \times J$ matrix $\boldsymbol{W}$ where there are J variables in the original dataset (e.g. spectral wavelengths or compounds analysed by chromatography). All calculations relating to the distance between map units are carried out on the 2 dimensional Cartesian coordinates of the map unit centres, which are described in terms of the $K \times 2$ matrix $\boldsymbol{M}$ shown in Figure 3.26(b). If the centre of unit (1,1) has Cartesian co-ordinates (1,1) and unit (1,2) has co-ordinates (2,1), where the first number refers to the rows, then unit (2,1) has co-ordinates (1.5,1.866) ($0.866 = \sqrt{3}/2$ and ensures the distance between unit (1,1) and unit (2,1) equals 1), the distance between the map units labelled (1,1) and (1,2) using Cartesian co-ordinates is $1\left(= \sqrt{(1/2)^2 + (\sqrt{3}/2)^2}\right)$.

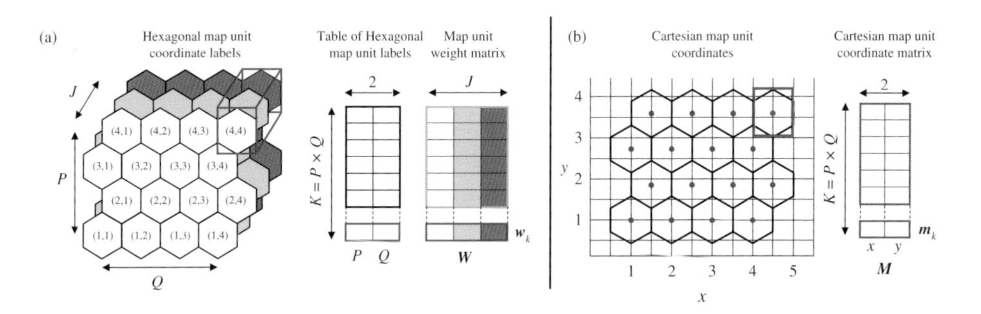

Figure 3.26 (a) $P \times Q$ map with J weights containing a total of K map units (pixels) and the corresponding weight matrix $\boldsymbol{W}$. (b) The same $P \times Q$ map showing the map unit centres and the corresponding map coordinate matrix $\boldsymbol{M}$. In the example shown, $P = 4$, $Q = 4$ and $J = 3$

Once initialized (see Section 3.4.3 for details), training of the map unit weights can begin. For SOMs the term 'training' refers to the process of iteratively updating the map unit weights to become more similar to vectors representing the original samples rather than supervised modelling. Once training is complete samples can then be mapped onto the units according to how similar they are to the corresponding units' weight vectors as follows. Training of the weights is carried out using a competitive learning technique. In each iteration, t a sample $\boldsymbol{x}_r$ is randomly selected from the data matrix $\boldsymbol{X}$, where $\boldsymbol{r}$ is a randomly selected integer usually obtained from a uniform distribution in the interval 1 to I (where I is the number of samples) and is newly generated for each iteration t. The sample vector $\boldsymbol{x}_r$ is then compared to the weight vectors of each map unit $\boldsymbol{w}_k$. The map unit whose weight vector is most similar to the response vector of the currently selected sample is designated the 'winner' or Best Matching Unit (BMU) of the selected sample and becomes the centre of learning for that iteration. Once the weight vectors have been updated the entire process is repeated for $t = 1, 2, \ldots, T$ iterations, using a randomly selected sample vector $\boldsymbol{x}_r$ for each value of t.

The SOM algorithm has several stages and a number of parameters that need to be set and these are described below.

3.4.3 Initialization

Prior to map training, each map unit k must first have a $1 \times J$ weight vector $\boldsymbol{w}_k$ initialized, usually by a random number generator which takes values within the range of the observed data. Each variable j is also therefore associated with a $K \times 1$ weight vector $\boldsymbol{w}_j$, where there are K units in a map (it is important to realize that the map cannot be considered as a matrix and the values of the co-ordinates are labels for a unit or cell rather than elements of a matrix as described in Section 3.4.2). Since both the initial weights of the map units are random, and the order in which the samples are selected is also random, the final representation of the sample space by the map also depends on this initial choice and is therefore not reproducible. However, the training process used for each map is the same, and so similar conclusions should be drawn from any of the maps generated. It is possible to 'force' a particular orientation of the data onto the map by initializing the map unit weights with values corresponding to the Principal Components of the data for example, and if being able to produce identical maps is important then the samples can be selected in an ordered fashion, e.g. cyclically, but this is not normally necessary. Note that although it is not necessary to force a particular orientation it is generally advisable for maps to be rectangular in shape (i.e. $P \neq Q$) so that there is a preferred direction for the data to organise themselves onto the map. For any given number of map units K there will be a number of combinations of P and Q that may or may not give a rectangular map. A dynamic adaptation of the SOM algorithm, often referred to as a Growing SOM, starts with a small map and iteratively adjusts the values of P and Q to try and find an optimal solution; however in most applications the optimal values of P and Q are chosen empirically, primarily according to the number of samples and complexity of data. For illustration purposes many of the example figures in this text used to illustrate principles of the method-used square maps, but when training maps are used for real data only rectangular maps are used.

3.4.4 Training

Important for the training process are the neighbourhood width σ and the learning rate α, described below, both of which are dependent on the dimensions of the map and the total number of iterations T.

Neighbourhood Width

The neighbourhood width σ determines which map units are close enough to the current sample's BMU in the map to be updated. For a map unit to be updated its distance from the current BMU must be less than the neighbourhood width, or otherwise it is not updated. The neighbourhood width is normally a non integer number, whereas the distance is the largest integer number that is less than the neighbourhood width. The neighbourhood width is generally initialized with a value σ_0 such that the first learning steps will update a large proportion of the map. As learning continues, the neighbourhood width should then monotonically decrease until, at the end of training, only the BMU and its immediate neighbours are updated. The exact form of the relationship between iteration t and the neighbourhood width σ is not critical provided it monotonically decreases, but for maps with a large number of units it is preferable to use a function inversely proportional to t to ensure that a large portion of the map is updated during the initial learning steps. This means that all units of the map can be trained to represent the input samples, and the map becomes globally ordered. A recommendation involves using an exponential function, as follows:

$$\sigma_t = \sigma_0\, exp\left[-\frac{t\ln(\sigma_0)}{T}\right]$$

where T is the total number of iterations and the distance between a unit and its immediate neighbours is equal to 1. A good value for the initial value σ_0 is the length of the smallest dimension of the map, $\min\{P, Q\}$ and the final value of σ after T iterations is 1. Figure 3.27(a) shows how the value of σ decreases with t for an 11×11 map. The effect of σ on the number of units being updated is shown in Figure 3.27(b).

Learning Rate

The learning rate α, which is a number between 1 and 0, determines the maximum amount a map unit can learn to represent a sample, or how similar it can make its weight vector to the sample vector $\boldsymbol{x}_r$ in each iteration. The learning rate for each iteration decreases monotonically with t. The exact starting value is not important, but an initial value close to 1 results in an increased number of iterations required for the map to stabilize. Conversely, an initial value that is too small (close to zero) can result in the weight vectors in the neighbourhoods of the BMU not being updated sufficiently, and so regions of the map never learn to represent one type of sample. The exact relationship between t and the learning rate does not have a large effect on the learning process provided it monotonically decreases with t.

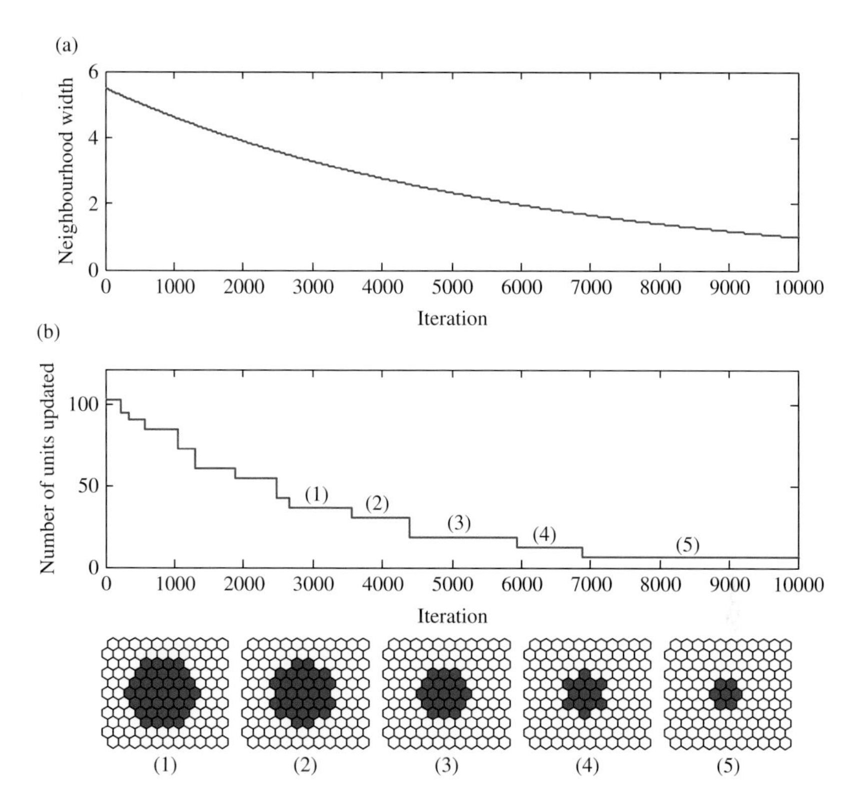

Figure 3.27 (a) Graph showing how the neighbourhood width decreases non-linearly with t for an 11×11 map. (b) Graph showing the number of units being updated in each iteration as defined by the neighbourhood width. A visual representation of the units within the neighbourhood width is shown at five different points during training. (1) to (5) are five different stages in training

A recommendation is to calculate α for each training step using the following exponential:

$$\alpha_t = \alpha_0 \exp\left[-\frac{t\ln(\sigma_0)}{T}\right]$$

where T is the total number of iterations and α_0 the initial learning rate.

Note that the learning rate relates to the initial map width in the implementation discussed in this text. The amount a map unit can learn is controlled not only by the learning rate but by its distance from the BMU. Units in close proximity to the BMU learn to represent the input sample more than units a greater distance away. The amount a map unit can learn, based on its distance from the BMU, is described in the form of a neighbourhood weight, ω. The overall amount a map unit is able to learn is then proportional to both the learning rate and the neighbourhood weight. A number of functions have been described for the neighbourhood weight, but one of the more common choices is a Gaussian function, as follows:

$$\omega_{kt} = \exp\left[-\frac{d^2_{(\boldsymbol{m}_b,\boldsymbol{m}_k)}}{2\sigma_t^2}\right]$$

where ω_{kt} is the neighbourhood weight of map unit k for iteration t, $\boldsymbol{m}_k$ is the Cartesian coordinate vector of map unit k, $\boldsymbol{m}_b$ is the Cartesian co-ordinate vector of the BMU and d is the geometric distance between the two co-ordinate vectors.

Iterations

Once the map and starting parameters have been initialized, training of the map weights can begin, as follows:

1. A sample vector $\boldsymbol{x}_r$ is selected from the $I \times J$ data matrix $\boldsymbol{X}$, where r is a randomly selected integer, e.g. obtained from a uniform distribution in the interval 1 to I and newly generated for each iteration t. Providing the number of iterations far exceeds the number of samples, it is highly likely that all samples will be tested many times.
2. The dissimilarity between the selected sample $\boldsymbol{x}_r$ and the weight vector $\boldsymbol{w}_k$ for each of the map units k is calculated as follows:

$$s_{(\boldsymbol{x}_r,\boldsymbol{w}_k)} = \sqrt{\sum_{j=1}^{J}(x_{rj} - w_{kj})^2}$$

   where x_{rj} is the value of variable j for the randomly selected sample x_r and w_{kj} is the weight of the jth variable for map unit k.
3. The map unit whose weight vector is most similar to the sample vector is declared as the Best Matching Unit (BMU) of the selected sample for the current iteration:

$$s_{(\boldsymbol{x}_r,\boldsymbol{w}_b)} = \sqrt{\sum_{j=1}^{J}(x_{rj} - w_{bj})^2} = \min_k\left\{s_{(\boldsymbol{x}_r,\boldsymbol{w}_k)}\right\}$$

   where b is defined as the value of k with the most similar weight vector w_k to the randomly selected sample vector $\boldsymbol{x}_r$ so that:

$$b = \arg\min_k\left\{s_{(\boldsymbol{x}_r,\boldsymbol{w}_k)}\right\}$$

   (or the co-ordinates of BMU)
4. The neighbourhood width σ_t, learning rate α_t and neighbourhood learning weight ω_t are calculated for this iteration t as described above.
5. The Euclidean distance between the Cartesian coordinates of each map unit $\boldsymbol{m}_k$ and the Cartesian coordinates of the BMU $\boldsymbol{m}_b$ is calculated as follows:

$$d_{(\mathbf{m}_b,\mathbf{m}_k)} = \sqrt{\sum_{j=1}^{J}(m_{bj} - m_{kj})^2}$$

6. Map units close enough to the BMU are declared neighbours of the BMU:

$$N_b = \{k : d_{(\mathbf{m}_b, \mathbf{m}_k)} < \sigma_t\}$$

   where N_b contains the values of k for which the kth map units distance from the best matching unit b is less than the neighbourhood width in iteration t.
7. Map units that have been declared neighbours of the BMU have their weights updated proportionally to the learning rate α_t and the neighbourhood learning weight ω_t for this iteration. All other map units remain unchanged.

$$\mathbf{w}_k = \begin{cases} \mathbf{w}_k + \omega_t \alpha_t (\mathbf{x}_r - \mathbf{w}_k) & k \in \mathbf{N}_b \\ \mathbf{w}_k & k \notin \mathbf{N}_b \end{cases}$$

The entire process is repeated until $t = T$. The exact value for the number of iterations is recommended to be at least 500 × the total number of the map units K as a first estimate. Since a sample is randomly selected in each iteration, it is also important to ensure that T is much larger than the number of samples in order to ensure that the map has a sufficient number of chances to learn about each sample. This is generally dependent on the data however, and however around 10 000 iterations are generally adequate unless the sample sizes are very large.

Recommended ranges for each of the training parameters to ensure that the learning process is sufficient have been published by Kohonen. In addition, it is sometimes recommended that training is done in two stages. In stage 1 the initial learning rate is higher than for stage 2. The second stage of training is a 'fine tuning' stage and can be initialized with a much smaller initial learning value as the map has already been partially trained; however in this text we use only the first stage, as there is little difference in the fine tuning for our datasets. Table 3.4 contains a list of recommended parameter values and the parameters used in this text and a schematic of the training process is shown in Figure 3.28.

Table 3.4 *Recommended values of the training parameters for SOMs, together with the parameters used in this text*

Parameter	Recommended value	Used in this text
Total number of iterations, T	~ 500 × Number of map units	10 000
Initial neighbourhood width, σ_0	~ Half width of map	Half width of map
Initial learning rate (Stage 1), α_0	~ 0.1	0.1
Initial learning rate (Stage 2), α_0	~ 0.01	Not applied

3.4.5 Map Quality

As with any unsupervised method for data visualization, outliers can have an adverse effect on the results obtained by using SOMs. There are a number of quality measures that can be employed to detect outliers and also to ensure that the map obtained is a good quality representation of the actual data. Two common measures are used, namely the Mean Quantization Error (MQE) and the Topographic Error (TE).

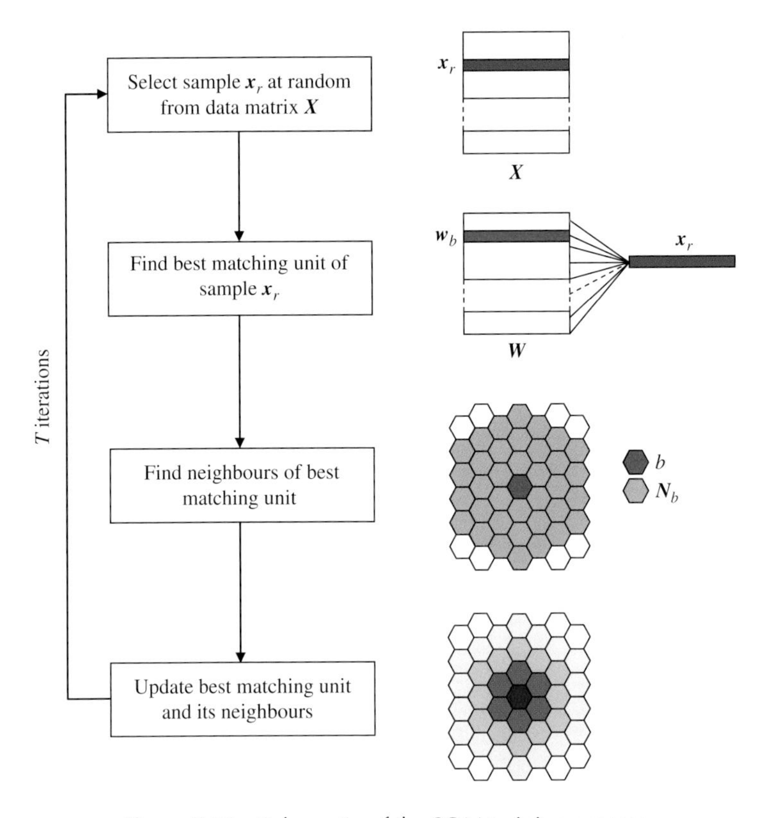

Figure 3.28 Schematic of the SOM training process

The MQE represents the average similarity between the map and the samples it was trained on, and hence can be used as a measure of the resolution, or quality, of the map. It is calculated as the mean similarity between each sample and its BMU when $t = T$:

$$MQE = \frac{1}{I}\sum_{i=1}^{I} s_{(\mathbf{x}_i,\mathbf{w}_b)} \qquad b = \arg\min_{k}\{s_{(\mathbf{x}_i,\mathbf{w}_k)}\}$$

where b is the value of k with the most similar weight vector w_k to the randomly selected sample vector $\mathbf{x}_r$.

The TE measures the topology preservation of the map, and indicates the proportion of the map that is mapped correctly. A TE is defined as a local discontinuity in the map. If $\mathbf{w}_b$ is the weight vector of the BMU of a sample, and $\mathbf{w}_{b+1}$ is the second most similar weight vector to the sample, then the map is locally continuous if $\mathbf{w}_b$ and $\mathbf{w}_{b+1}$ are from adjacent units. If not, then there is a local discontinuity and this contributes to the total TE:

$$TE = \frac{1}{I}\sum_{i=1}^{I} g_i$$

$$g_i = \begin{cases} 1 & \text{if } \mathbf{w}_b \text{ and } \mathbf{w}_{b+1} \text{ are adjacent} \\ 0 & \text{otherwise} \end{cases}$$

A high TE indicates that there is not a smooth transition between a map unit and its neighbours, which in turn suggests that samples that are very similar to each other in input space are not necessarily neighbours on the map. Since the aim of training an SOM is to represent the data and maintain the topography of the input space, if the map has lots of discontinuities then it has not achieved this aim.

Both the MQE and TE typically give the best results when the data has been over-fitted, and so caution is advised when considering these values if a map has more units than there are samples.

3.4.6 Visualization

Once an SOM has been trained there are numerous methods for visualization, dependent on application. We will describe four methods, namely the Unified Distance Matrix (U-Matrix), Hit Histograms and Component Planes, and a fourth method to allow class information to be visualized.

Unified Distance Matrix

The aim of a U-Matrix is to show the similarity of a unit to its neighbours and hence reveal potential clusters present in the map. If there are classes present in the data, then the border between neighbouring clusters could be interpreted as a class border. The unified distance of each unit is calculated as the sum of the similarities between the weight vector of a map unit and the weight vectors of its immediate neighbours N_k:

$$u_k = \sum_{h=1}^{H_k} s_{(w_k, w_n)} \qquad n \in N_k$$

where H_k is the number of units neighbouring map unit k (for a centralized unit on a hexagonal map $H_k = 6$, but for units at the edges of a map this is reduced). This gives map units whose neighbours are represented by similar weight vectors (e.g. in the middle of a cluster) a low value, and map units with very dissimilar neighbours (e.g. at the border of different clusters), a high value. Generally, the same similarity measure used to train the SOM is used, which in most cases is the Euclidean distance between two vectors. The U-Matrix values can be converted to greyscale (or any other equivalent colour scale) and the SOM grid can be plotted, shading each map unit with the scaled colour value. An illustration of how the U-Matrix is employed for visualization can be seen in Figure 3.29. As examples, Figure 3.30 shows U-matrices for Case Studies 2 (NIR of food) and 3 (polymers). The dark blue background is where most of the samples lie, the borders between each group being light in colour. The colouring is scaled between the lowest and highest values of u encountered in the map, the actual numerical value of the maximum and minimum u may differ according to map. For Case Study 2 we can see that the SOM is clearly divided into four regions, the two regions on the right being most distinct, whereas the two regions on the left are somewhat less distinct. Note an important feature of the map in that most cells are dark blue, representing samples, and the portion of the map that is empty, i.e. a barrier between the classes, is quite small, and so the map makes most use of the space. For the polymers we clearly see that the map is divided into several distinct regions, more than the nine groups, and so there may be some unsuspected

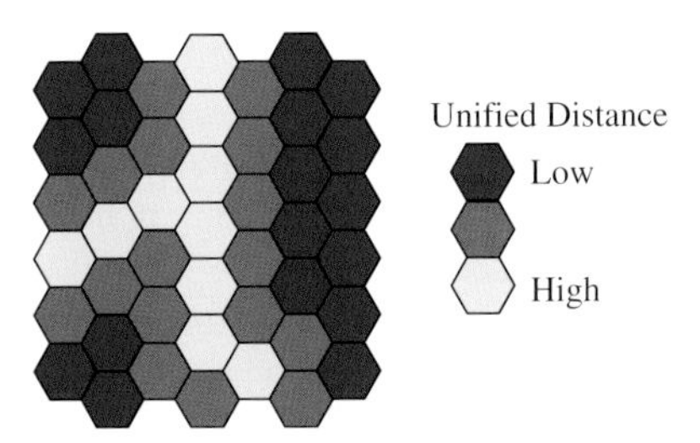

Figure 3.29 Example of the U-matrix method where each unit is shaded according to its value of u. Three blue clusters can be seen, separated by the main boundaries in yellow. For clarity this figure only has three levels of u, but for actual data the scale will be continuous

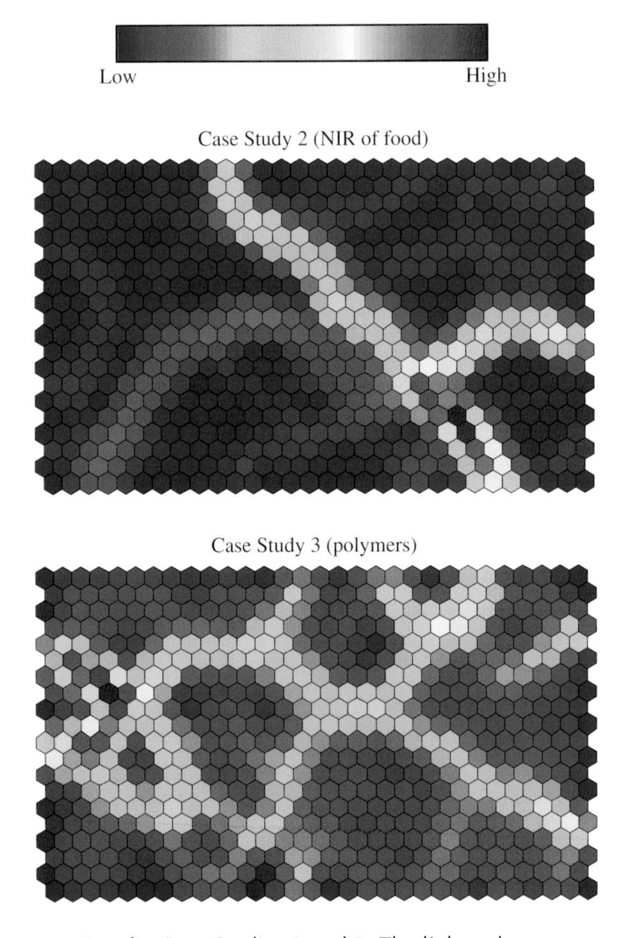

Figure 3.30 U-matrices for Case Studies 2 and 3. The light colours represent boundaries

Case Study 2 (NIR of food)

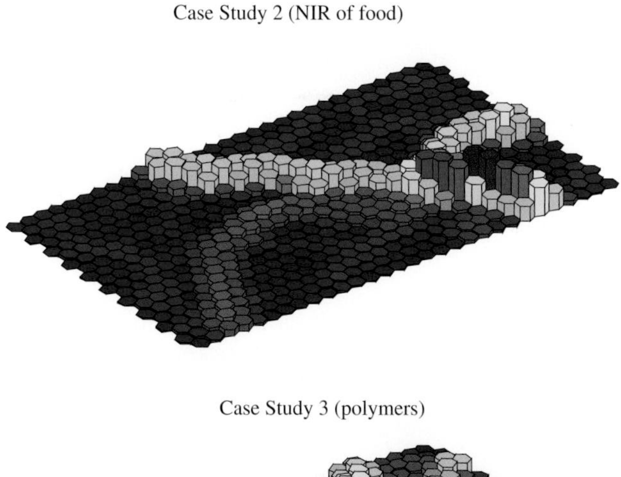

Case Study 3 (polymers)

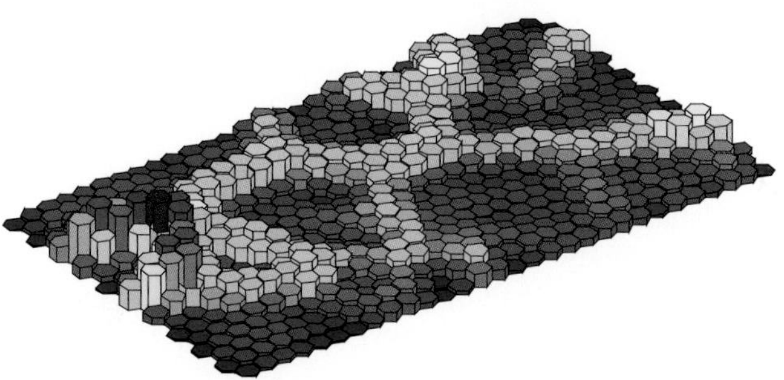

Figure 3.31 3D representations of the U matrices in Figure 3.31

structure. These U matrices can also be represented in 3 dimensions (Figure 3.31), clearly showing the boundaries between different regions of the map.

Hit Histograms

Hit histograms are a method of visualizing the BMU for each sample at the end of training. The value of each map unit is the number of times that the map unit was the BMU of any sample at the end of training. If there are several classes present in a dataset, then an ideal situation might be where one or a small number of map units correspond to the BMU of all samples from the same class and so correspond to a high number of 'hits'. This can be represented as a 3-dimensional plot, as shown in Figure 3.32(a), where the height of the map unit represents the frequency or, more commonly, the size of the map unit on the grid is scaled to represent frequency. The frequency with which a map unit is the BMU of a sample in different classes can also be represented by shading each hexagon by a different amount (Figure 3.32 (b)), according to how frequently a map unit is chosen, and also separate hit histograms could be generated for each class if required.

The 3D hit histograms from Case Studies 1, 2 and 3 are presented in Figure 3.33. In these representations we choose to scale the height of the bars for each class independently to the maximum number of hits found in each class. The hit histogram for Case Study 2 (NIR

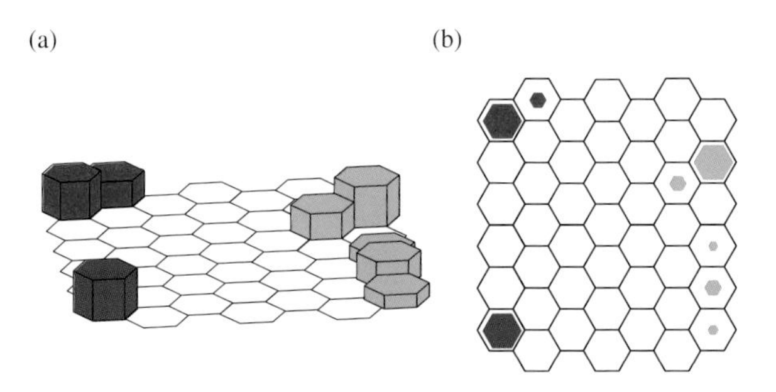

Figure 3.32 Example visualizations for the Hit Histograms method using the same data as for Figure 3.29. Regions with lots of hits correspond to regions of similarity shown in the U-Matrix method. (a) 3D representation, where the height of each unit is proportional to the number of hits. (b) 2D representation of the same map where the size of the shaded hexagons relates to the number of hits for each class

spectroscopy of food) is the clearest. There are 72 samples in this dataset and each sample has one corresponding BMU, where 72 out of the 600 ($= 20 \times 30$) cells are shaded and in fact are similar to Figure 3.37(a) (except we choose to represent the latter in 2D), the SOM class structure visualization. This is because there are relatively few samples and they are well spread out. For Case Study 1 (forensic) there is a cell in class B (red) – top centre that corresponds to two hits – whereas all of the others correspond to one hit, and so are less intense for that class. Case Study 3 (polymers) is less straightforward as the number of hits in each cell is quite different, unlike for Case Studies 1 and 2 where there is one hit per cell

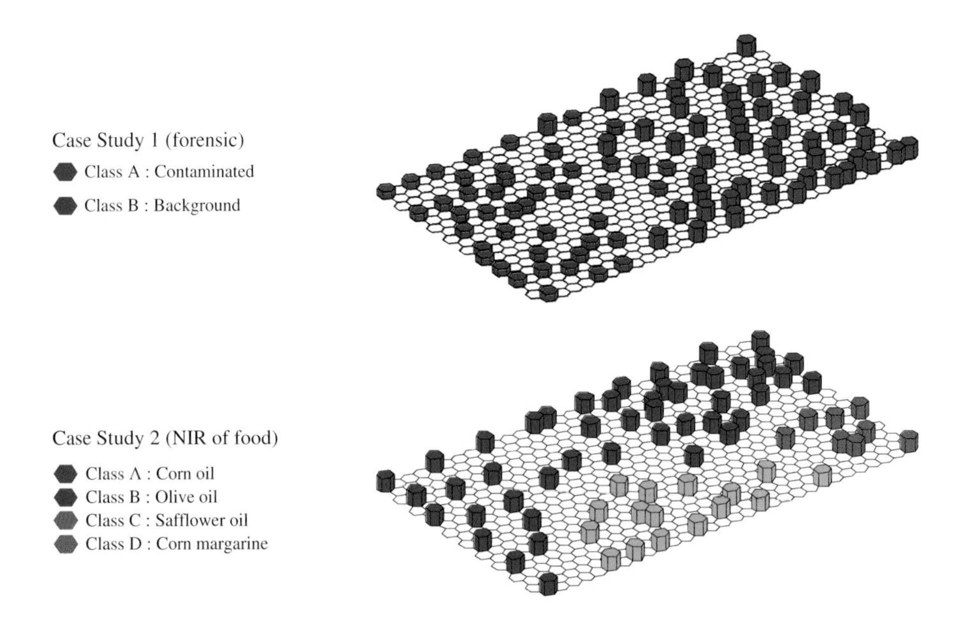

Figure 3.33 Hit histograms for Case Studies 1, 2 and 3

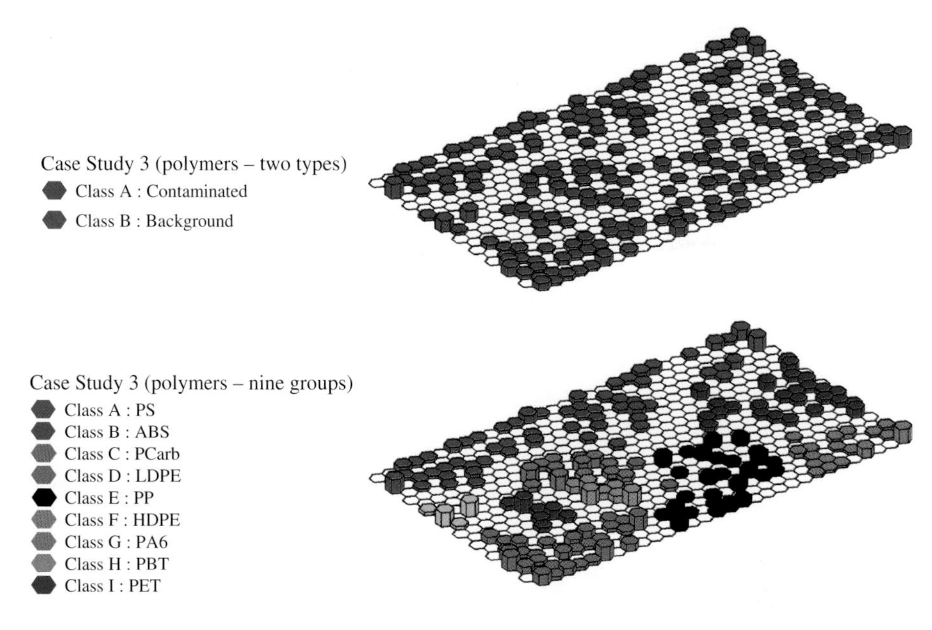

Figure 3.33 (continued)

with a single exception. This might be expected as there are 293 samples to place in the 600 cell grid, rather than just 95 (Case Study 1) or 72 (Case Study 2) and the class structure is more complex and so we do not expect such even distribution of samples. For the polymer groups we can see that there are usually one or two cells that have a high number of hits for each group thus suggesting a concentration of samples around that cell, and these cells are mainly on the periphery of the map, as different groups will try to get as far away from each other as possible to make maximum use of the map space.

Component Planes

In analogy to loadings, it is possible to see how each variable influences the map and which of the samples a variable is most associated with. Component Planes can be employed for this purpose and have some analogy to loadings in PCA, except normally a single component plane is calculated per variable and involves converting the weights for a specific plane (or variable) into greyscale or colour coded according to the importance for describing a given region of the map, analogous to the U matrix, for example, and means that the relationship between samples and variables can be visualized by shading each map unit k proportionally to the weight w_{kj} of map unit k for the chosen variable j. An example of component planes is shown in Figure 3.34. Note that for component planes we are primarily interested in whether a variable can describe a class rather than whether it can discriminate between two classes. If there are only two classes the variables that best describe a class are also good discriminators between the classes, but if there are more than two classes the results may be different.

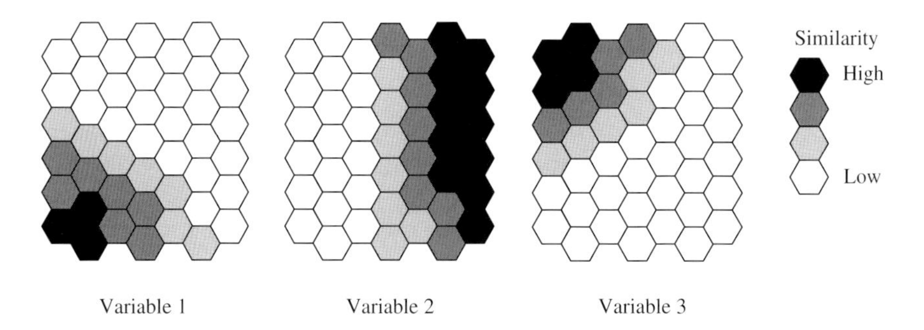

Figure 3.34 Example of possible component plane visualization for the data of Figure 3.29, showing the magnitude of each map unit's weight for the selected variable. Each variable has large values for regions of the map corresponding to regions of high similarity when using the U-matrix visualization. For clarity, the variables in this figure only have four levels. Note that a high similarity corresponds to a low unified distance and vice versa

As an example of application to case studies in the book we provide component planes of 3 of the variables for Case Study 2 (NIR of food) in Figure 3.35. In the top diagram, the darker the shading, the less important the variable is for describing the group. We can clearly see that there are four groups in the map as expected, which is especially obvious in case of variable 1. These should be compared to Figure 3.37. The graphs at the bottom involve superimposing a colour code for each class, then the lighter the colour, the more important the variable. We see that variable 1 is important for describing classes B and D (olive oil and corn margarine), and is quite useful for modelling class C (safflower oil) but does not model class A well. Variable 2, in contrast, models class D very well, and may correspond to a wavelength highly diagnostic of these groups.

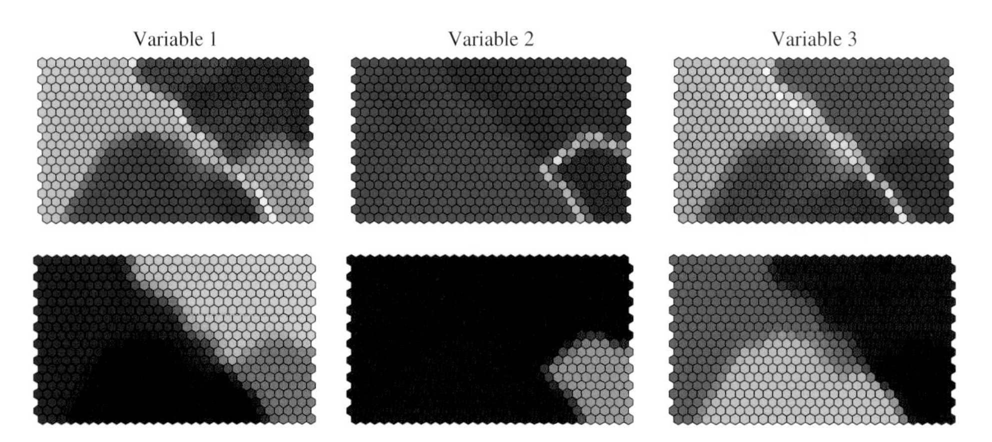

Figure 3.35 Component planes for three variables in Case Study 2: bottom, shaded according to class with black meaning not found; top, shaded using the scheme in Figure in 3.30

Including Class Information

Although the maps are usually trained using an unsupervised method that does not make prior use of any classifier; it is still useful to visualize the class membership of samples represented in the map. For SOMs as an exploratory technique, the most appropriate way to include class information is to create an additional set of variables, one for each class, that contain a value of 1 if the sample is in that class, and 0 if not. For example, if a sample is in the 3rd class (C) and there are 7 (A to G) classes, then the additional variables are [0,0,1,0,0,0,0] for that sample. It is possible to use SOMs in a supervised manner, but this is outside the scope of this text, in which case the additional variables are scaled appropriately and the map is then trained in the normal way, with the additional weight vectors for each map unit containing class information. However, when using SOMs as an exploratory technique, class membership can be included in the same way as for supervised SOMs, but not used when locating the BMU of the sample and therefore not allowed to influence the learning process. The process of updating the vectors remained the same for all variables however, but each map unit contains an additional set of weights containing class information. By plotting the Component Planes that correspond to the class variables it is possible to visualize how class membership is represented in the map.

If the number of groups is small enough, the method described above can be modified further so that rather than representing the class membership of samples as a vector of '1s' and '0s', a colour vector can be included with each sample that corresponds to the colour being used to represent its particular class. Each map unit will therefore have additional weights corresponding to a trained colour that can be used to shade the map unit for visualization purposes. This allows all class regions to be displayed on the same map, in contrast to using Component Planes which require a separate visualization for each class. A limitation of this method is that it is not appropriate for a very large number of classes where the distinction between colours used to represent classes is not clear. By visualizing the shading of the maps units in each iteration of the algorithm it is also possible to watch the development of a map in real time as the training progresses.

Figure 3.36 represents various ways of visualizing the class information for the two classes in Case Study 1 (forensic). Figure 3.36(a) represents just the BMUs for each class. In Figure 3.36(b) every cell is coloured according to their nearest BMUs; however cells that have the nearest BMUs from more than one class (equidistant) are shaded in a combination of colours. Figure 3.36(c) superimposes both together. When there are several classes and types of grouping, SOMs become very powerful. Figure 3.37 illustrates the visualization of SOMs for Case Study 2 (NIR spectroscopy of food). It can be seen that the map is clearly divided into four regions. The area of each region is roughly proportional to group size, because there is an equal probability of each sample being trained: if one group contains just a few samples it will be chosen less frequently, and hence class D (corn margarine) occupies a smaller area to the others, being represented by just 8 samples. Figure 3.38 illustrates the use of SOMs for Case Study 3 (the polymers). In this case we have two main types whose BMUs are illustrated in Figure 3.38(a) and nine groups (Figure 3.38(b)). We can see, for example, that PS and ABS are quite similar as they are neighbours, but very different to HDPE, and so such a map gives a good idea of the similarity between the types of polymer, which may relate to their thermal properties and

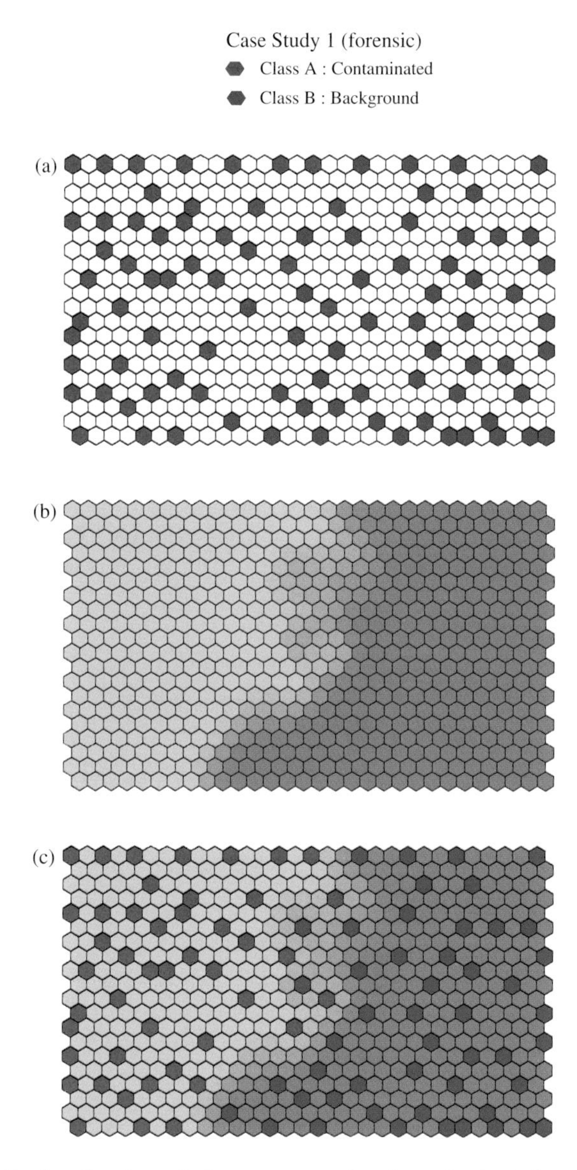

Figure 3.36 SOM visualization of class structure for Case Study 1 (forensic): (a) with the BMUs indicated; (b) with all units shaded according to the nearest class (c) superimposing (a) and (b)

physical structure. If samples are misclassified they are likely to be mistaken for samples in neighbouring groups. We can also see that there is a small isolated group of PET that appears similar to PA6 among other polymer groups which is probably a specific grade of the polymer. It is possible to further superimpose the grades on the types (Figure 3.38(c)) and see which are close to the boundaries. We will not discuss grades in detail in this text.

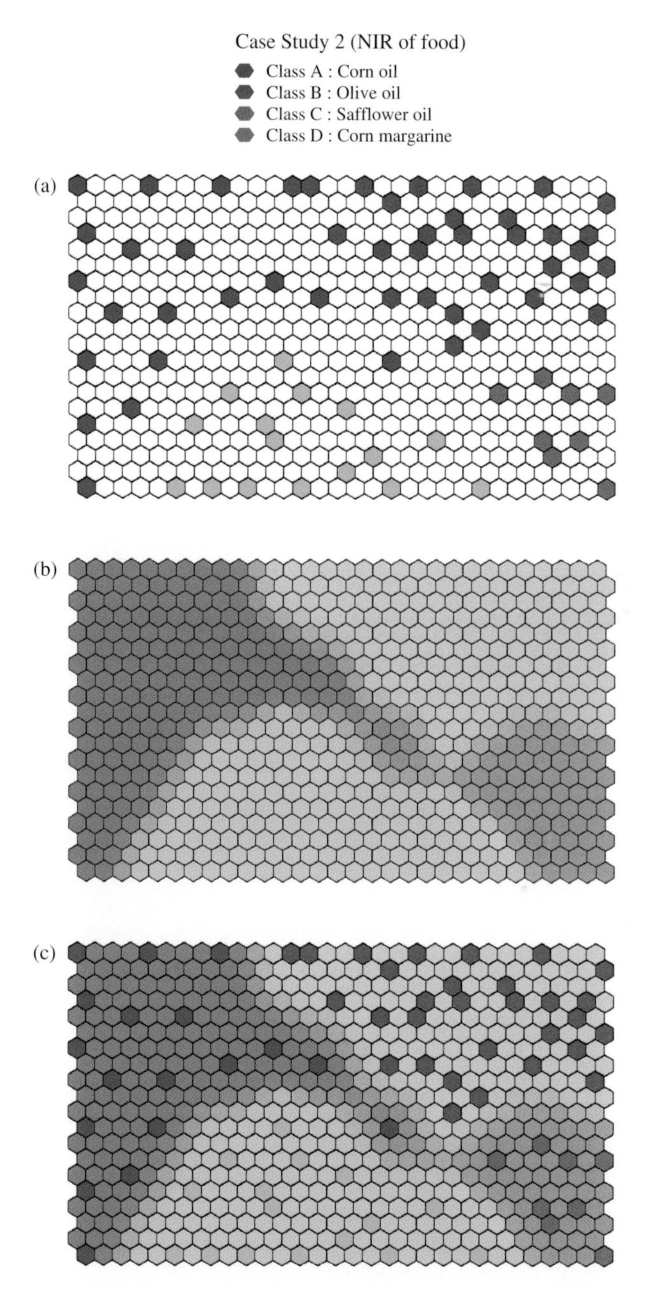

Figure 3.37 SOM visualization of class structure for Case Study 2 (NIR of food): (a) with the BMUs indicated; (b) with all units shaded according to the nearest class (c) superimposing (a) and (b)

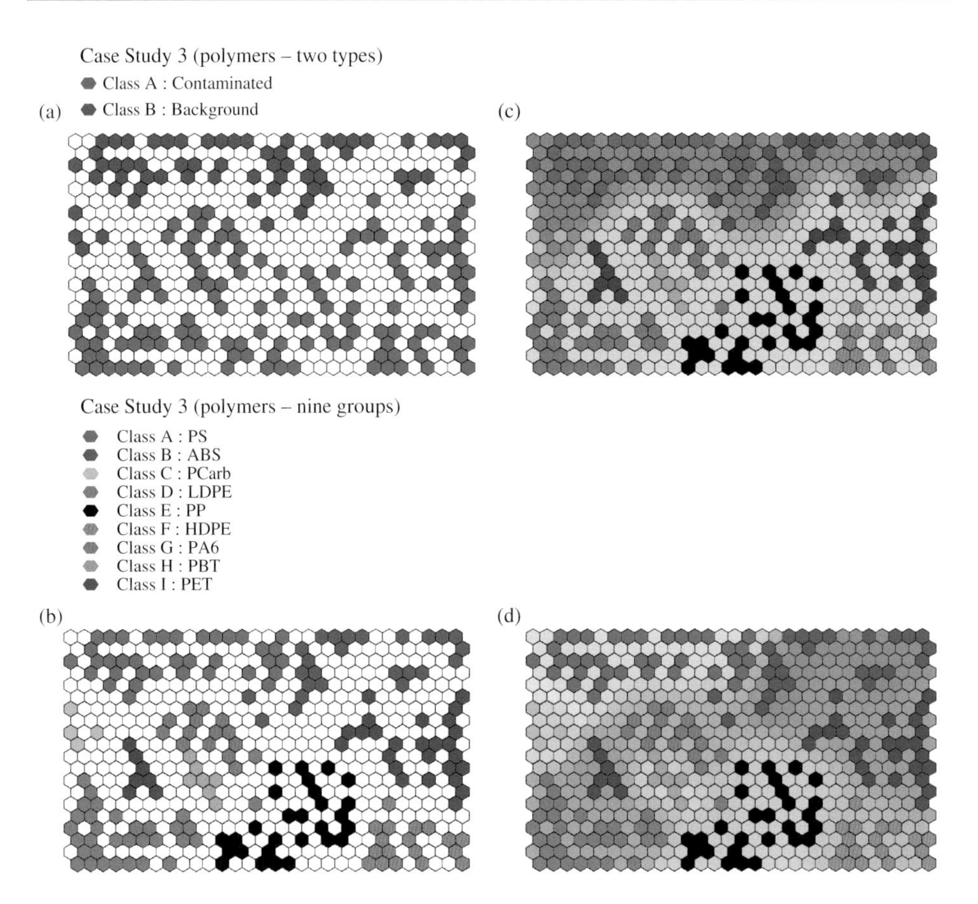

Figure 3.38 SOM visualization of class structure for Case Study 3 (polymers): (a) with the BMUs indicated for the two polymer types; (b) BMUs indicated for the nine polymer types; (c) superimposing BMUs for polymer groups over background shading for the nearest type; (d) superimposing BMUs for polymer groups over background shading for the nearest group

Note that that samples from more than one group may fall onto the same cell as for Case Study 7 (clinical) – see Figure 3.39. This is hardly surprising as there are 540 samples, and 600 cells. Better separation could be obtained using a larger grid; however one problem with this dataset is that the majority of samples are from class B (446 samples) and so fill most of the map, crowding the remaining four classes into a small corner. An alternative way round this is to change the probability of selection of each sample, and so there is equal chance that a sample from each class is chosen in training. This changes the map so that each class occupies roughly equal areas, as illustrated in Figure 3.40. Class B (controls) occupy a very compact region on the bottom right. There is some mixing between the classes as they are not completely distinguishable, but nevertheless the four patient classes do occupy roughly equal regions. Probably better distinction could be obtained if the number of samples in the diseased groups was increased.

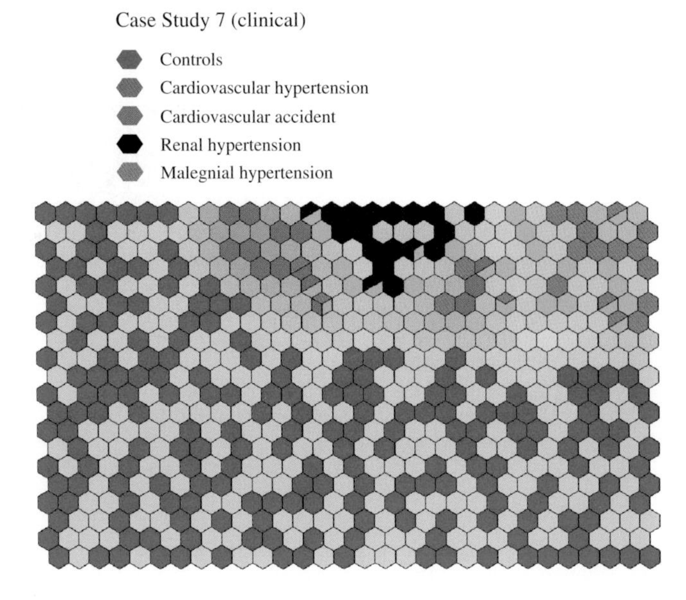

Figure 3.39 SOM visualization of the class structure of Case Study 7 (clinical)

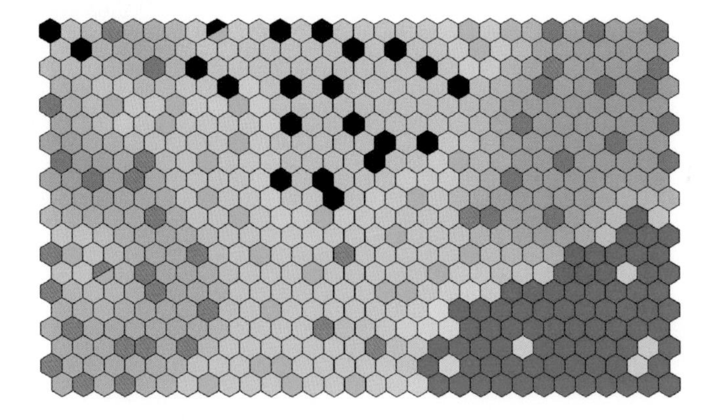

Figure 3.40 SOM visualisation of the class structure of Case Study 7 (clinical) but with equal probability of selecting samples from each class whilst training

Bibliography

PCA

S. Wold, K. Esbensen, P. Geladi, Principal Component Analysis, *Chemometrics Intell. Lab. Systems*, **2**, 37 (1987).

I.T. Jolliffe, *Principal Components Analysis*, Second Edition, Springer-Verlag, Berlin, Germany, 2002.

K.V. Mardia, J.T. Kent, J. Bibby, *Multivariate Analysis*, Academic Press, London, UK, 1979.

B.F.J. Manly, *Multivariate Statistical Method: A Primer*, Third Edition, Chapman & Hall/CRC Press, New York, NY, USA, 2004.
J.E. Jackson, *A User's Guide to Principal Components*, John Wiley & Sons, Inc., New York, NY, USA, 1991.

PCO, Rank Tests, Similarity

Y. Xu, F. Gong, S.J. Dixon, R.G. Brereton, H.A. Soini, M.V. Novotny, E. Oberzaucher, K. Grammer, D.J. Penn Application of Dissimilarity Indices, Principal Co-ordinates Analysis and Rank Tests to Peak Tables in Metabolomics of the Gas Chromatography Mass Spectrometry of Human Sweat, *Anal. Chem.*, **79**, 5633–5641 (2007).
J.C. Gower, P. Legendre, Metric and Euclidean Properties of Dissimilarity Coefficients, *J. Classification*, **3**, 5–48 (1986).
J.C. Gower, Some Distance Properties of Latent Root and Vector Methods used in Multivariate Analysis, *Biometrika*, **53**, 325–338 (1966).
B. Zhang and S.N. Srihari, Properties of Binary Vector Dissimilarity Measures, in *JCIS CVPRIP 2003*, Cary, NC, USA, September 26–30, 2003.

Self Organizing Maps

G.R. Lloyd, R.G. Brereton, J.C. Duncan, Self organising maps for distinguishing polymer groups using thermal response curves obtained by dynamic mechanical analysis, *Analyst*, **133**, 1046–1059 (2008).
T. Kohonen, *Self-Organizing Maps*, Third Edition, Springer-Verlag, Berlin, Germany, 2001.
T. Kohonen, Self-Organized Formation of Topologically Correct Feature Maps, *Biol. Cybernetics* **43**, 59–69 (1982).

4 Preprocessing

4.1 Introduction

Prior to exploratory data analysis or to classification it is necessary to prepare the data. There is a large area of signal analysis and experimental design which we will not be discussing in detail in this text, although in Chapter 2 the steps employed to handle the datasets obtained from the case studies prior to pattern recognition have been described in outline. We will, however, be concerned about how the raw data matrix, which may consist of, for example, chromatographic peak heights, or baseline corrected spectroscopic intensities, can be scaled or reduced in size prior to pattern recognition. In this chapter, in addition to data scaling, we discuss how to create new variables such as PCs prior to exploratory or supervised data analysis. Variable selection, as opposed to creating new variables (which former method involves choosing some of the original measurements) will be discussed in Chapter 9 in the context of determining which are the most significant variables for discrimination, e.g. in the area of biomarker discovery, and can also be employed as a form of preprocessing providing care is taken.

Data preprocessing should be considered part of the overall strategy for pattern recognition and it is especially important to understand the consequences on the results of model validation as discussed in Chapter 8, when data are divided into training and test (or validation) sets; the test set is used to determine how well a classifier performs: if there are two groups (A and B), if the test set %CC (percentage correctly classified) is much greater than 50 % it suggests that the classifier is able to discriminate between two groups. However if preprocessing and classification is incorrectly performed it can lead to false conclusions about the data, and it is important to understand that if the test set is used together with the training set for preprocessing, when this can lead to erroneous conclusions about the data, when this is safe (but optional) and when this is necessary. A way of checking this is to perform all steps of data analysis on a null dataset, e.g. case study 11 in this text: if a %CC much above 50 % is obtained on the test set we know that we are overfitting the data and making mistakes.

Chemometrics for Pattern Recognition Richard G. Brereton

We will illustrate the influence of data preprocessing primarily using scores plots of the first two PCs, on data that are characterized by two classes. The principles however are quite general and preprocessing influences all steps of data analysis; in the case of PLS-DA there can be quite considerable differences especially as a consequence of column scaling, and this will be illustrated in Section 5.5.3, but we describe all the main approaches to data preprocessing in this chapter. In Chapter 2 we discuss default methods for preprocessing and in this chapter we will usually change one of these steps to illustrate the effect of the methods described below unless specified.

4.2 Data Scaling

There are three principle ways in which a matrix can be scaled as illustrated in Figure 4.1. Scaling is normally performed in the following order:

1. transforming individual elements of a matrix
2. row scaling
3. column scaling

although if it is desired to optimize the method for scaling it is sometimes necessary to perform steps (in the correct order) to see what the result is and then return to the earlier steps and so on, and there are a few exceptions to the rule (e.g. ratioing to landmark peaks as discussed in Section 4.2.2).

4.2.1 Transforming Individual Elements

It is often useful to scale individual measurements, usually represented as elements of a matrix. If done on the training set, this must also be done on the test set.

In traditional analytical chemistry, often we anticipate a linear model between one set of data and another, for example in calibration or resolution, since spectra, if within a certain absorbance range, obey the Beer–Lambert law and so are likely to be linearly additive: there is then often an exact physical relationship between the observed spectrum or chromatogram and the underlying factors of interest, such as the concentrations of

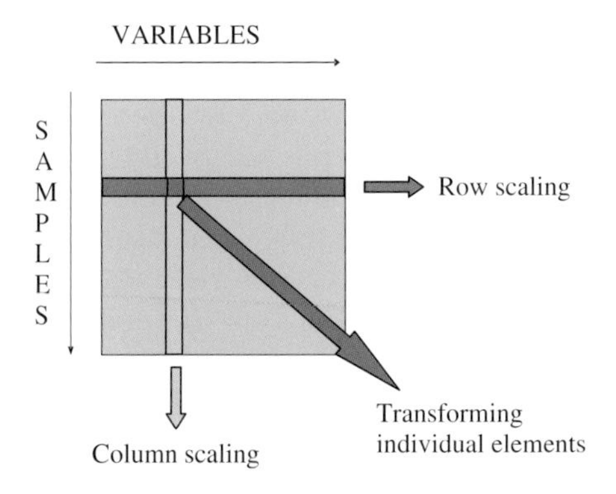

Figure 4.1 Types of scaling

analytes, and so great efforts are often made to develop linear models. However in pattern recognition this may not necessarily be so, for example, let us say we want to classify two groups of subjects into diseased and controls, using the concentration of a compound found in their urine. The controls may have a relatively low level that varies over a small range, whereas the diseased may have a much wider range of concentrations of the compound, and a few (perhaps very ill ones) might have very high levels. We do not necessarily expect a linear relationship between the amount of compound in the urine and whether a person has a disease or not, and so the concentration measurements are used primarily as guides to whether a person is diseased rather than to develop a linear model between the disease state and concentration of a compound. So in pattern recognition, there should be no significant barrier to transforming data in a non-linear manner where appropriate.

The most common problems encountered are:

1. Occasional large values, e.g. peaks in GCMS or NMR spectroscopy that are very high relative to others in a few samples and would otherwise dominate the analysis if not suitably scaled.
2. Values that are below the limit of detection (LoD) and are entered as 0.

Transforming individual measurements is usually only necessary if distributions of values are skew; note, however, that this serves a different purpose to methods for column scaling such as standardization (Section 4.2.3) which ensure that the variation in each variable has an influence that is approximately equal on the result of pattern recognition. Note also that it may, in exceptional cases where variables are of different types, be possible to transform each type of variable in different ways, e.g. log transforming one set of variables, and keeping the raw values of others: however such datasets are quite rare and we do not illustrate this situation below.

An additional and separate problem arises when there are missing data, i.e. no information was recorded for a specific variable in a specific sample – this may have been because the instrument was not working properly on a particular day which sometimes happens, for example, in environmental research, and this should be distinguished from case 2 above which involves measurements that are below the detection limit. The datasets in this text do not contain missing values, but several contain measurements of value 0 (either below the detection limit or absent). If many measurements are missing altogether it is often necessary to question whether the data are suitable for chemometric analysis or whether samples with missing values should be removed from the dataset. If a sample is retained with a small number of missing values, then one possible approach is to replace the missing value with the average (usually after transformation of the variables) of the variable over the entire dataset.

There are several common approaches to transformation of the elements of a matrix.

Logarithmic Transformation

This involves replacing x_{ij} by log (x_{ij}).

The main advantages are two fold:

- The influence of large values such as outliers or occasional large peaks is reduced, which could otherwise have a disproportionate influence on least squares solutions. These large or outlying peaks may be scattered around different samples, and so may be found in several samples.

- In some situations distributions of intensities are log-normal, that is they form a distorted distribution which is not symmetrical: this often happens in metabolomics and in environmental monitoring, and log scaling converts these to a more symmetric distribution. Most multivariate approaches work best when distributions are symmetric rather than skew. In addition, statistical significance tests often assume approximate normality.

The main drawbacks encountered with the use of log scaling in the context of chemometrics are as follows:

- If there are values very close to the LoD or the baseline they may have a disproportionate influence on the resultant scaled data matrix. Especially when close to the detection limit, the uncertainty, or errors in measurement, can be relatively large compared to the measurement even if small in absolute terms, and will also have a disproportionately large influence on the value of the logarithm.
- In cases where there are variables entered as 0 (absent or below the LoD) their logarithms are undefined.
- All values must be positive. Usually if there are many negative values this either is a consequence of noise in the dataset or baseline problems, which we will not discuss in detail in this text, although there can be a few situations such as the use of derivatives in spectroscopy where negative intensities may be encountered as a matter of course; however on the whole in the majority of pattern recognition studies, raw intensities should ideally be positive.

There are several solutions to these problems involving refining the method:

- In order to solve the problems of measurements that are below the LoD, such measurements can be replaced by a positive value that is small relative to other values in the dataset. There are several guidelines for choosing the value of this number but usually it is a fraction of the smallest value detected in the entire dataset (providing all other values are positive in the dataset of interest), and is usually the same value for all measurements if they have been recorded on roughly the same scale; typical choices are the minimum positive value of $\boldsymbol{X}$ or, quite commonly, half the minimum positive value of $\boldsymbol{X}$. If measurements are on very different scales (see Case Study 7) then values of 0 can be replaced by a fraction of the minimum observed value over all samples for each variable separately. The smaller the value used for replacing numbers below the LoD, the more important we weight the evidence that a variable is not detected, and if there are quite a few zero values (as in Case Studies 8 (mouse urine) and 9 (NMR spectroscopy)), the appearance of the scores plots can vary considerably according to what we use to choose to replace the undetected measurements. For example, we may be interested in metabolic profiling and attach some significance to the fact that a compound is not detected. Replacing '0s' by an extremely small value prior to taking logarithms has nearly the same effect as replacing all detected values by 1 and all those not detected by 0: this is easy to demonstrate, if the range of detected values varies from 1 to 100 (a logarithmic range of 2 to the base 10), then choosing a value of 10^{-6} for values below the LoD means that the full range in the logarithmically scaled dataset will be from –6 to 2, and the influence of these values on the resultant patterns is quite considerable; so as this replacement value gets very small in magnitude the effect after performing logarithmic transformation is similar to replacing undetected measurements by 0 and detected measurements

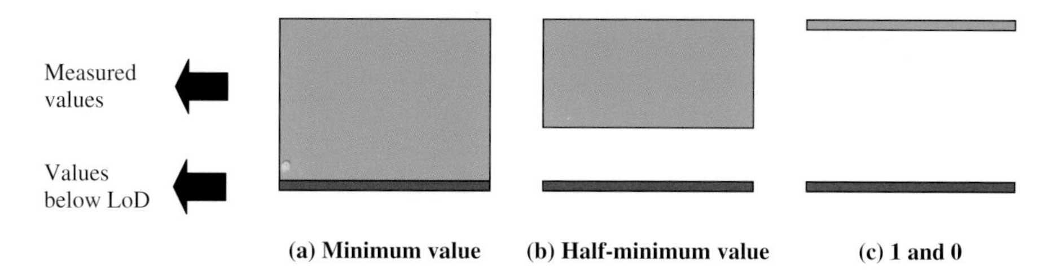

Figure 4.2 Weighting the importance of values below the LoD using replacing (a) by the minimum value (b) by half the minimum value, both prior to taking logarithms and (c) measured values by 1 and others by 0 without taking logarithms

by 1. Hence the decision as to what to replace values below the LoD by relates in part to how important we consider the evidence that a variable is not detected in a sample; this is illustrated in Figure 4.2. If there are few such values and no particular significance placed on undetected variables (just that they are small) using half the minimum measurement may be a good default choice. Quite commonly this approach might be used where a matrix consists of, for example, chromatographic peak areas, all of which are positive, and which have been obtained often after peak detection and alignment, as in Case Study 8 (mouse metabolic profiling).

- An alternative is to replace log (x_{ij}) by log (x_{ij} + *offset*) where *offset* is chosen to ensure that all values are positive. The value of the offset can of course also have an influence on the result in a similar manner to the choice of the value used to replace measurements below the LoD. If there are baseline or noise problems that might otherwise result in negative values, e.g. in Case Study 9 (NMR spectroscopy) and we are dealing with raw spectroscopic data, this procedure may be a good choice.

In Figure 4.3 we illustrate the influence of log scaling on the PC scores of Case Studies 8a (mouse urine – diet study) and 9 (NMR spectroscopy). For Case Study 8a, logarithmic scaling only has a small influence on the appearance of the PC scores plot: partly this is because the data are standardized, meaning that a variable with a very large peak will not be so influential on the overall model. In addition, there is only a small change in appearance of the scores plots when the value with which the '0s' are replaced is changed which is encouraging: the reason for this is that there is a large dynamic range of values in Case Study 8a, from 6.92×10^3 to 4.42×10^8 or nearly 5 orders of magnitude; hence replacing the 0 value by one that is one tenth the smallest value adds only one order of magnitude or about one sixth of the range after log scaling. Replacing by '1s' and '0s', however, is much more dramatic and loses the separation between groups almost completely. For Case Study 9, where the data are not standardized, logarithmic scaling has a much greater influence on the appearance of the scores plots. Surprisingly, replacement of the data matrix by '1s' and '0s' is quite effective for Case Study 9, with good separation for the two groups and clear identification of outliers; this may be because there are quite strong trends associated with presence and absence of metabolites; note that some form of data scaling is essential for Case Study 9 to get acceptable separation in the first two PCs. The way in which the '0s' are handled, prior to log scaling, is particularly important in metabolic profiling as there are often many peaks that are undetected in samples.

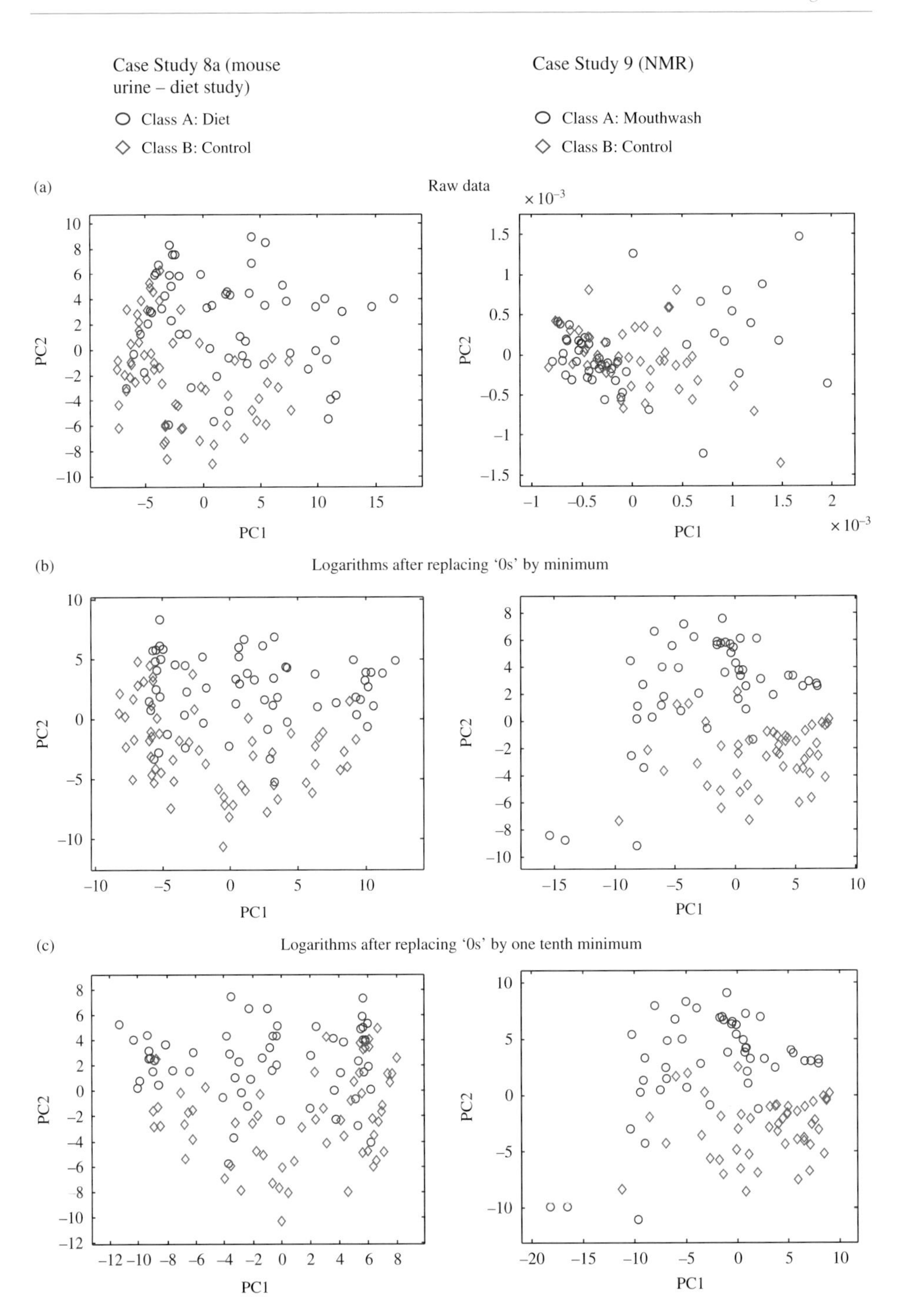

Figure 4.3 Influence of log scaling on the scores plots of Case Studies 8a and 9: (a) raw data; (b) replacing '0s' by the minimum value; (c) replacing zeroes by 1/10 of the minimum value; (d) using '0s' for absence and '1s' for presence. After log scaling the data are row scaled and standardized for Case Study 8a and centred for Case Study 9

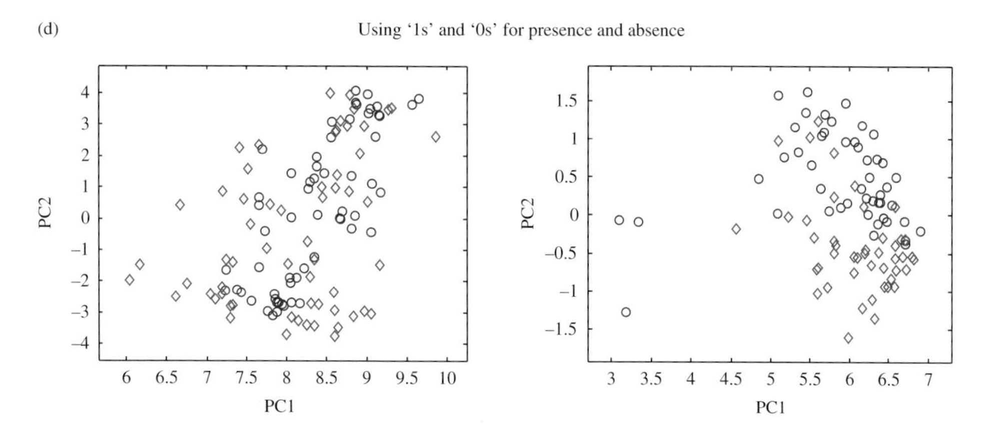

Figure 4.3 (continued)

Power Transformation

An alternative to using logarithms is to replace x_{ij} by $(x_{ij})^{1/n}$; commonly $n = 2$ and so a square root.

The main advantages of this method are as follows:

- As per logarithmic scaling, this reduces the influence of large values.
- It can cope with values that are zero, unlike logarithms, so there is no need to replace values below the LoD.
- Unlike logarithms, uncertainties in small values do not have a major influence on the resultant data analysis, whereas a small change in value close to the LoD could have a substantial influence on the value of a logarithm;, the smaller a value relative to others, the smaller its influence on the *n*th root transformed data.

In Figure 4.4 we illustrate a comparison between log and square root scaling, using the data of Table 4.1, where there are three quite large measurements in each case. The main result of data scaling is to change the distribution of points, so in order to understand this clearly the transformed datapoints are visualized on the same scale, with 1 for the highest and 0 for the lowest value. Tables 4.1(a) and (b) represent the same underlying data but with a small amount of uncertainty added to each measurement, which could be due to sampling or instrumental irreproducibility, and so that the two columns are slightly different. In both cases the distributions of the raw data look fairly similar. When computing the logarithms we can see that the uncertainly has a significant influence on the spread of values at low intensities, whereas the influence of the square root is less. The logarithmic transformation, however, does make the data more symmetric, whereas the influence of the high values is still evident (although to a lesser degree) when using square root scaling.

The main drawbacks are as follows:

- All values must be positive, at least for even powers.
- Sometimes data are approximately log-normally distributed and power scaling will not transform these to a symmetric distribution.

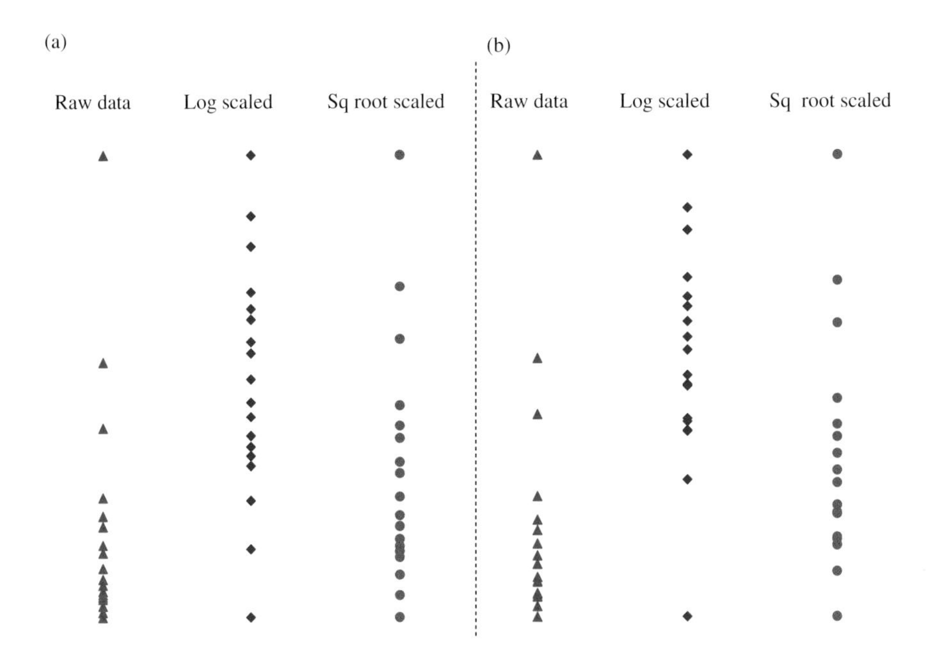

Figure 4.4 Raw data, log scaled and square root scaled data, represented on a scale of 0 to 1, for Table 1(a) and (b)

Table 4.1 *Data for Figure 4.4*

(a)	(b)
23.69	23.78
13.21	13.41
9.87	10.54
6.35	6.33
5.42	5.14
4.89	4.60
3.95	3.92
3.54	3.31
2.76	2.88
2.21	2.19
1.92	1.95
1.6	1.99
1.44	1.21
1.32	1.37
1.2	1.33
0.86	1.20
0.54	0.71
0.28	0.16

- The value of n must be chosen. A default is to 'square root' which is probably adequate in most cases. There are no agreed rules for choosing n but one can examine visually the spread of values as the power is changed, and if desired see what values creates a symmetric distribution of transformed values. A difficulty here can occur in multivariate data in that each variable may have a different distribution, which may be distorted further by other factors such as row scaling (Section 4.2.2).

The influence on the PC scores plots of using different powers of n for Case Studies 8a (mouse urine – diet study) and 3 (polymers) is illustrated in Figure 4.5. For data of Case Study 8a, once the square root is used, the two groups become more distinct and of approximately similar shape. This is because the influence of some large peaks in class A has been reduced. Using higher powers make a little difference, but visually from the first 2 PCs it looks as if square root scaling is optimal. For Case Study 3 (polymers) smaller distinct sub-clusters separate out when using higher powers. However the two main groups (amorphous and semicrystalline) are probably best separated when there is no power scaling, and we choose not to power scale this dataset as a default.

In this text when we need to reduce the influence of large peaks we will use square root scaling as a default; however using logarithmic scaling so long as the 0 values are replaced by some small number that is appropriately chosen and the use of higher powers are alternatives and in specific cases should be investigated.

Box Cox Transformation

Another transformation that can be employed is the Box Cox transformation for which:

- x_{ij} is transformed to $(x_{ij}^{\lambda} - 1/\lambda \text{ if } \lambda \neq 0)$
- x_{ij} is transformed to $\log(x_{ij})$ if $\lambda = 0$.

The value of λ is usually non-integer (typically it might take a value of 0.2, for example). If x_{ij} contains negative values, an offset is added to x (see Section 4.2.1 on logarithmic transforms).

The main rationale is to transform measurements into a normal distribution. The majority of traditional statistical tests assume normality and so great efforts are often made to first transform the data into a distribution that is approximately normal. The value of λ has to be optimized which can be done by many statistical packages.

However the use of this transform in chemometrics is rare, probably reflecting the different philosophy of the discipline compared to traditional statistics. One of the problems is that in many multivariate datasets, each variable might have a different distribution – this is especially so in areas such as metalobomics or forensics or environmental monitoring, and so the parameters that might transform the distribution of one variable into a normal distribution might be quite different from those for another variable. The consequences of using such a transformation then can become quite complicated to understand if further preprocessing such as row scaling is performed subsequently. So long as the results of pattern recognition are not interpreted in the form of probabilities based on an underlying normal distribution, such transformations are probably unnecessarily complicated and if row scaling and variable weighting are also introduced can have unpredictable consequences. However in the context of univariate data or multivariate matrices where each variable has a similar distribution and there are little or no further

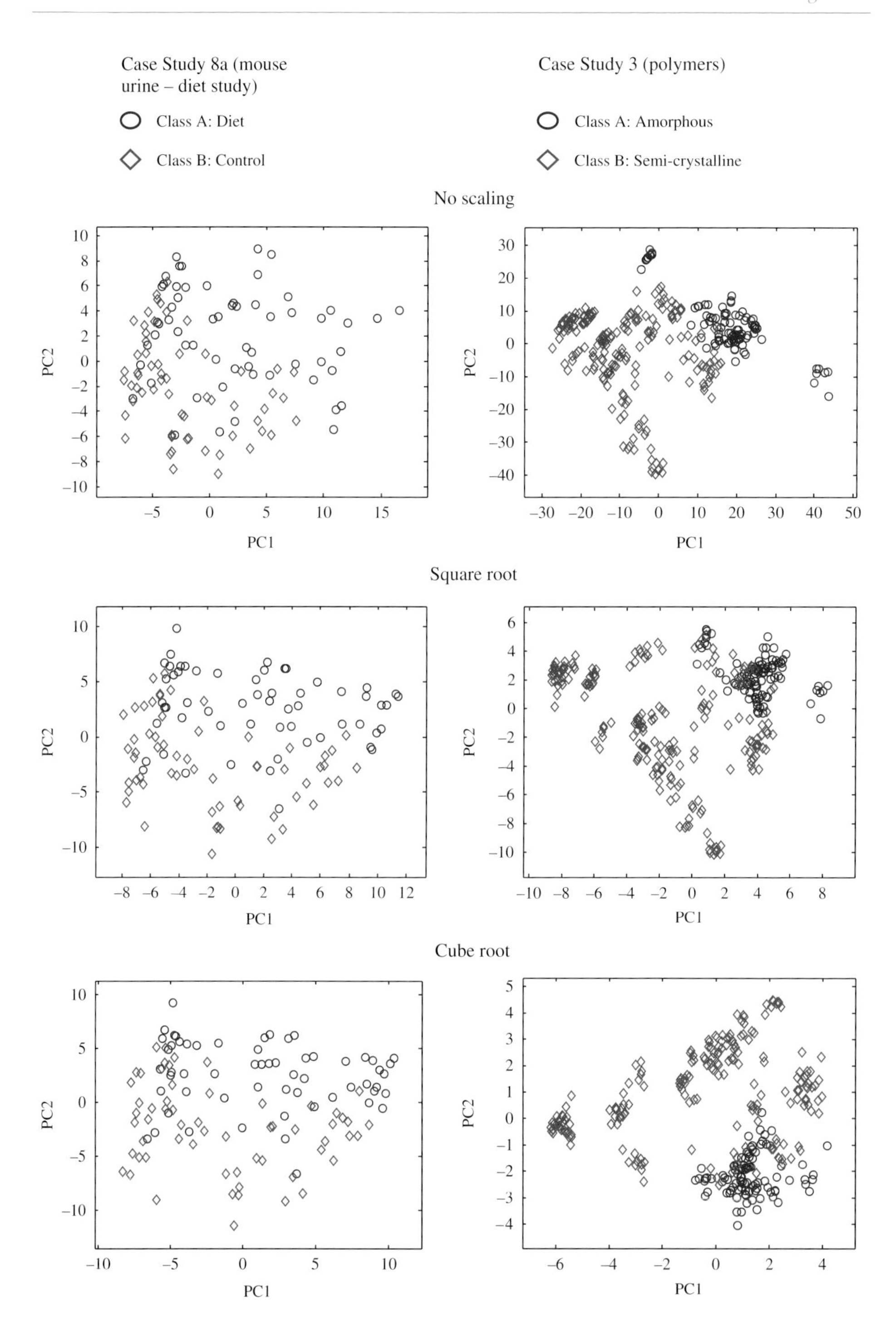

Figure 4.5 Influence of the use of different power scaling on the PC scores plots of Case Studies 8a and 3 (in all cases the later steps of mean centring (Case Study 1) and row scaling plus standardizing (Case Study 8a) are performed subsequent to transforming individual elements of the data matrices)

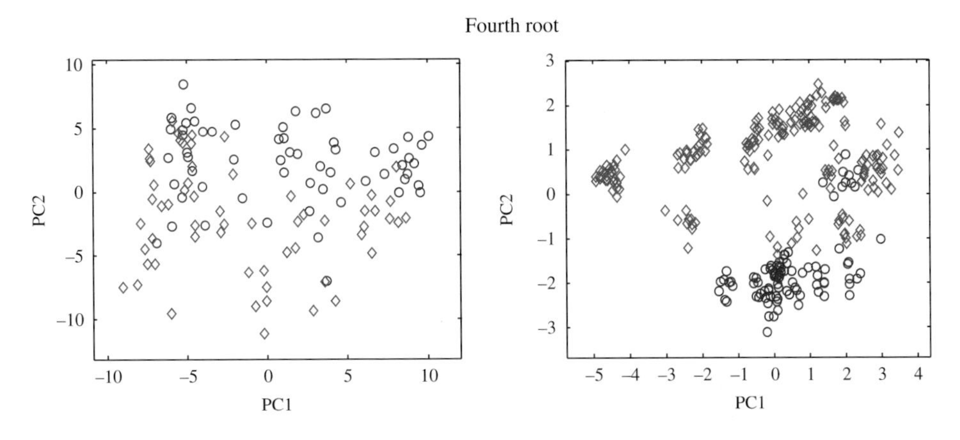

Figure 4.5 (continued)

steps required for preprocessing the data, the Box Cox transformation has potential value. It is important to realize that every transformation has its use, but should always be viewed in context.

It is also useful to understand that if the dataset is split into training and test sets (Chapter 8), if Box Cox transformation parameters are determined from the training set, they will be dependent on which samples are chosen for the training set, and may be unstable if a few samples contain large values of a variable (e.g. high concentrations of a compound) which may not always be included in the training set.

4.2.2 Row Scaling

Row scaling is an important procedure particularly when the amount of sample to be analysed is hard to control, involving scaling each sample independently. A typical example may involve human sweat which is hard to obtain in a controlled and quantitative manner, and so each sample may contain an unknown amount of sweat. Row scaling can, however, distort data and lead to a property called closure which means that measurements become proportions: in practice we are now analysing ratios between variables rather than absolute quantities, and it is important to watch out for unwanted side effects. Row scaling should usually be done after transforming individual variables, although sometimes if there is a set method for row scaling and it is desirable to optimize the transformation of individual elements (Section 4.2.1), this stage (and column scaling of Section 4.2.3 where appropriate) needs to be performed to see how the distribution of variables changes, each iteration in the optimization.

Ratioing to Landmarks

One of the simplest procedures in quantitative analytical chemistry involves adding reference standards to a mixture; in chromatography these are usually called internal standards (ISs), for example, one may add a certain known amount of a compound to a urine sample and then record a chromatogram. We will call the IS a landmark variable. The intensities of each analyte are then ratioed to the intensity of the landmark, to give an absolute measure that is proportional to its concentration, x_{ij}/y_i where y_i is the intensity (e.g. chromatographic peak area) of the landmark peak in sample i. This approach, whilst well established, and in some cases entirely appropriate, depends on the how accurately the IS can be measured, and so it would be normal to perform some independent tests on the quantitative accuracy of measuring the intensity of the IS, which may depend on, for example, weighing, dilution, making up standards, etc. In some cases it is quite easy to determine the accuracy of an IS, but in other cases, a good example being sweat, this can be very difficult and ratioing to the IS could introduce additional unwanted errors.

A somewhat more sophisticated alternative is to add several internal standards, preferably independently (not as a single reference mixture), to create a series of landmark variables or peaks. A first test is whether the ratio of intensities of these landmark variables is constant; often a limit can be established in advance. If not, is one variable an outlier in a particular sample, or is the sample itself an outlier and should it be discarded? If the ratios of intensities of these landmarks are within given limits (often this has to be established for a specific type of instrument and analytical technique and we will not provide generalized recommendations in this text), then the intensities of other variables can be corrected in various ways, a simple approach being to take the geometric average of the ratio to the internal standards transforming to $\left(\prod_{s=1}^{S} x_{ij}/y_{is}\right)^{1/S}$ where there are S internal standards of which y_{is} is the intensity of the internal standard s in sample i.

The IS method mainly used in applications such as chromatography, NMR spectroscopy and MS. Usually ratioing to landmarks is performed as the first step on raw intensity data prior to any of the transformations of Section 4.2.1 and is a rare case where transformation of individual elements (e.g. square root or logarithm) can be performed after row scaling.

Scaling Rows to Constant Total

If the amount of sample is hard to control then scaling rows to a constant total is often useful. This is sometimes called normalizing, but the word has different connotations so we will use the terminology 'row scaling to constant total' in this text.

In its simplest form this involves dividing each variable by the sum of all variables in the sample, replacing x_{ij} by $x_{ij}/\sum_{j=1}^{J} x_{ij}$, so that $\sum_{j=1}^{J} x_{ij} = 1$ after transformation; this is generally done after scaling of individual elements of the matrix.

There are three principle problems with this approach:

- The most serious is that high intensity variables can have an undue influence on the resultant row scaled data. If there are a few intense variables (e.g. they may be metabolites that are present in high concentrations in all samples but not be very relevant to pattern recognition), the resultant sample vector after row scaling may mainly be influenced by changes in intensity of these variables. However in some cases smaller variables are more diagnostic of differences between groups and their variability will be lost if care is not taken; especially if data are standardized subsequently, as discussed in Section 4.2.3, there could be nonsensical results in the resultant patterns.

- This technique is not applicable if there are negative values that are large in magnitude and expected to have a significant influence on the pattern recognition.
- Peaks may come from various sources, including the data that are of direct interest (e.g. a biological factors that influence the composition of a sample) and the experimental and analytical procedures. Background peaks from the analytical procedure may be of more or less constant intensity but once row scaled may appear to vary considerably and could be mistaken for genuine markers that have an influence over separation between groups: this is hard to avoid if there is one factor, e.g. the analytical procedure that is relatively constant and another, e.g. the amount of sample extracted to be analysed which varies substantially, unless the variables due to the analytical procedure are known and removed from the analysis in advance.

There are various ways of overcoming these limitations:

- Providing data have been transformed in an appropriate way it is possible to remove the influence of large variables on row scaling. This is why power or logarithmic transformation prior to row scaling is sometimes important, and so each step in data preparation must be considered together.
- If variables are known to arise from the analytical procedure or factors that are not directly interesting for pattern recognition, it is recommended they are removed prior to row scaling (and in all subsequent data analysis). A common example involves identifying siloxanes that are from plastics that might be used to cap vials; these are easy to identify in mass spectrometry based analysis, and their characteristic peaks can be removed.
- There is very little that can be done about negative values, unless a specific transformation is performed prior to row scaling; however if negative values are mainly due to instrumental noise these can be set to 0 or retained providing they are not too large; baseline correction is often important under such circumstances.

An alternative to summing the row to a constant total involves what is called by some vector normalization. We may represent a series of measurements on a sample by a vector, so that $\boldsymbol{x}_i$ is a row vector for sample i of dimensions $1 \times J$ where there are J variables: $\boldsymbol{x}_i$ is transformed to $(x_i/\|x_i\|)$ where $\|$ stands for the norm of a vector and is the square root of the sum of its elements. In other words x_{ij} is transformed to $x_{ij}/\sqrt{\sum_{j=1}^{J} x_{ij}^2}$, and so the sum of squares after transformation of the elements of $\boldsymbol{x}_i$ equals one. Note that this transformation is not the same as summing the elements of a vector to a constant total, and the results of the two methods are illustrated for a simple four element vector in Table 4.2. Whereas for a

Table 4.2 *(a) Vector consisting of four measurements, (b) transformed so that the sum of the elements comes to a constant total of 1 and (c) transformed so that the sum of the squares of the elements comes to a constant total of 1*

(a)	(b)	(c)
1	0.0714	0.1195
2	0.1429	0.2390
4	0.2857	0.4781
7	0.5000	0.8367

single sample it makes no difference which approach is employed, when several samples are placed together in a data matrix it will influence the relative values of measurements in each sample as to whether the data are row scaled to a constant total or each sample as represented by a vector is normalized to a constant length.

There is no general guidance as to whether it is more appropriate to sum the elements of each sample to a constant total, or to use the sum of squares criterion. In this book we will, where it is necessary, use the straight summation, but this does not imply specific advocacy of either approach.

Some datasets (Case Studies 2 – NIR spectrometry of food and 4 – environmental pollution) are already row scaled to constant total, although the data can be transformed, e.g. by square root scaling and row scaling performed again.

The influence of row scaling is best seen in combination with other transformations and is illustrated in Figure 4.6 for Case Studies 3, 8b and 9. For Case Study 3 (polymers) there are unusual effects, probably because the intensity of peaks as well as their position and shape is diagnostic of the group they belong to, and row scaling loses some key information, for example, we can see a small group of semi-crystalline polymers is mixed in with the amorphous polymers. The force in Pa has a direct physical meaning which will be lost if the data are row scaled and we would expect the strength of a force to have some bearing on the properties of a material: in the case of spectroscopy the vertical scale relates primarily to the concentration of a compound and so has limited diagnostic value for composition of samples unlike in DMA. Square root scaling and row scaling has the effect of splitting the semi-crystalline polymers into one main and two small groups. For Case Study 8b (mouse urine – genetics), in contrast, the amount of sample cannot easily be controlled and so if we do not row scale we are also looking at effects due to extraction and analysis in addition to those we are primarily interested in, and from the first two PCs we see most samples are clustered together with a few outliers due primarily to inconsistencies in the analytical procedure. Once this effect is removed by row scaling we obtain some separation on PCs1 and 2, especially for class A. For Case Study 9 (NMR spectroscopy), it should not be necessary to row scale as each saliva sample is extracted only once and split into aliquots of the same amount. When row scaling, one sample (number 70) exhibits two very intense peaks that dominate the spectrum and as such appears an outlier; this problem is somewhat reduced when performing square root transformation, but best separations are found when using either raw data or vector normalized data.

Usually the most appropriate form of row scaling, if any, can be determined by thinking about the underlying problem in hand, how the data are obtained, e.g. the analytical procedure; however sometimes the relationship between row scaling and other transformations should be investigated before deciding on the best approach.

Block Scaling

Occasionally it is useful to divide the data into blocks, for example, we may be interested in different regions of a chromatogram, different wavelength regions or different sets of marker compounds.

Block scaling, which can also be called selective normalization, involves taking all measurements in each block and summing these to a constant total, and so each block of measurements in each sample adds up to a constant total: these totals may be different for

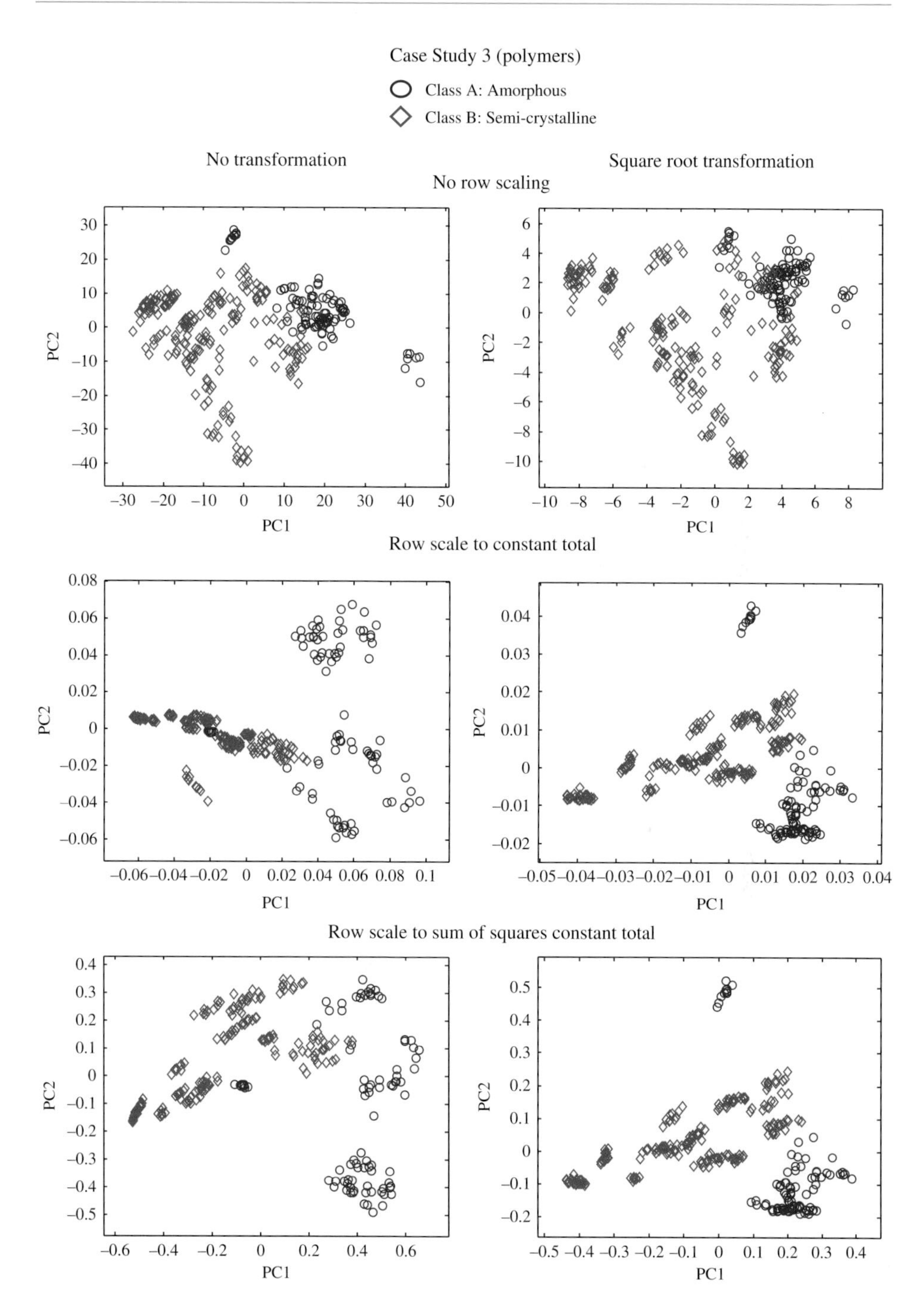

Figure 4.6 Influence of different types of row scaling on the PC scores plot of Case Studies 3, 8b and 9. For Case Studies 3 and 9 data are centred after row scaling and for Case Study 8b they are standardized

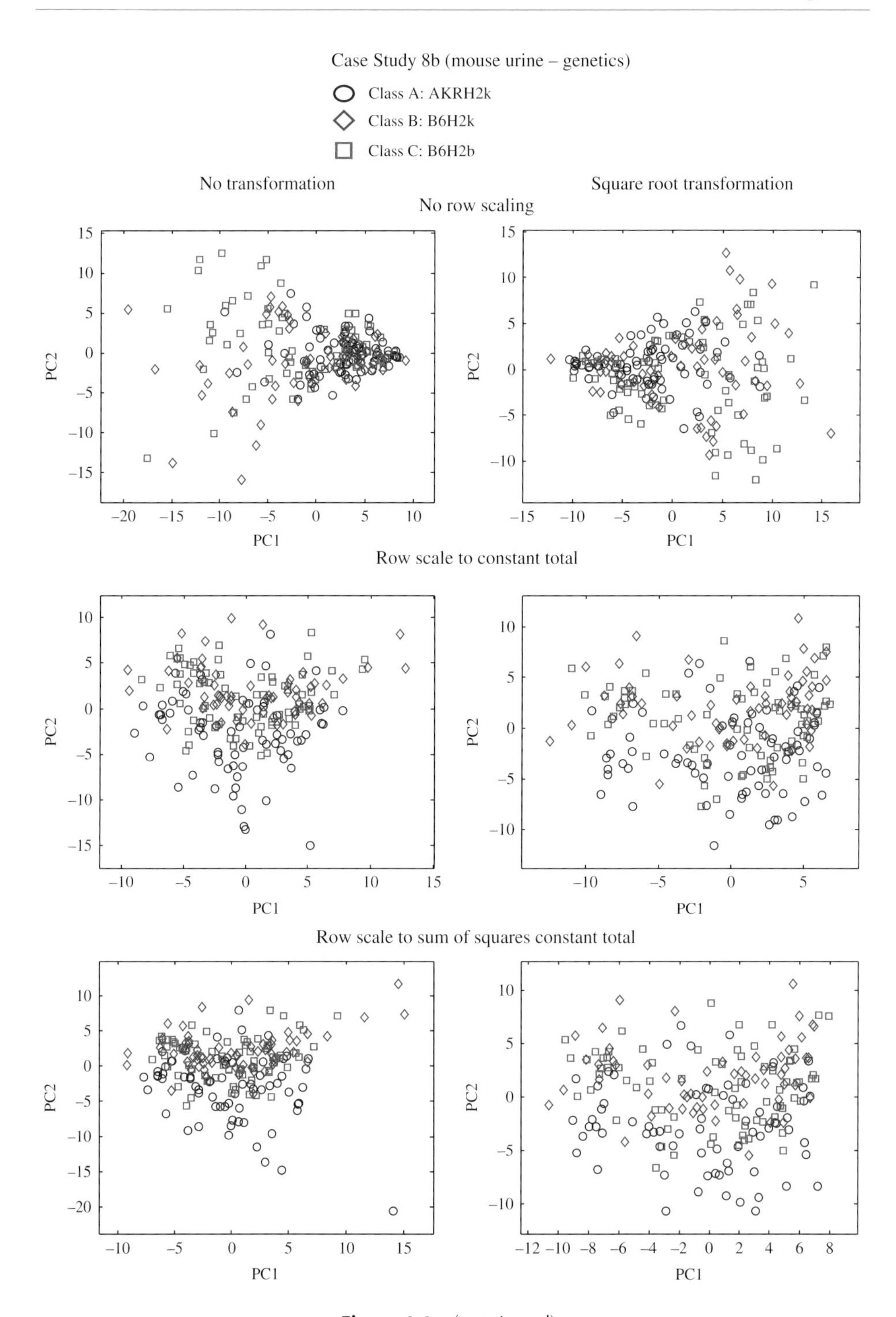

Figure 4.6 (continued)

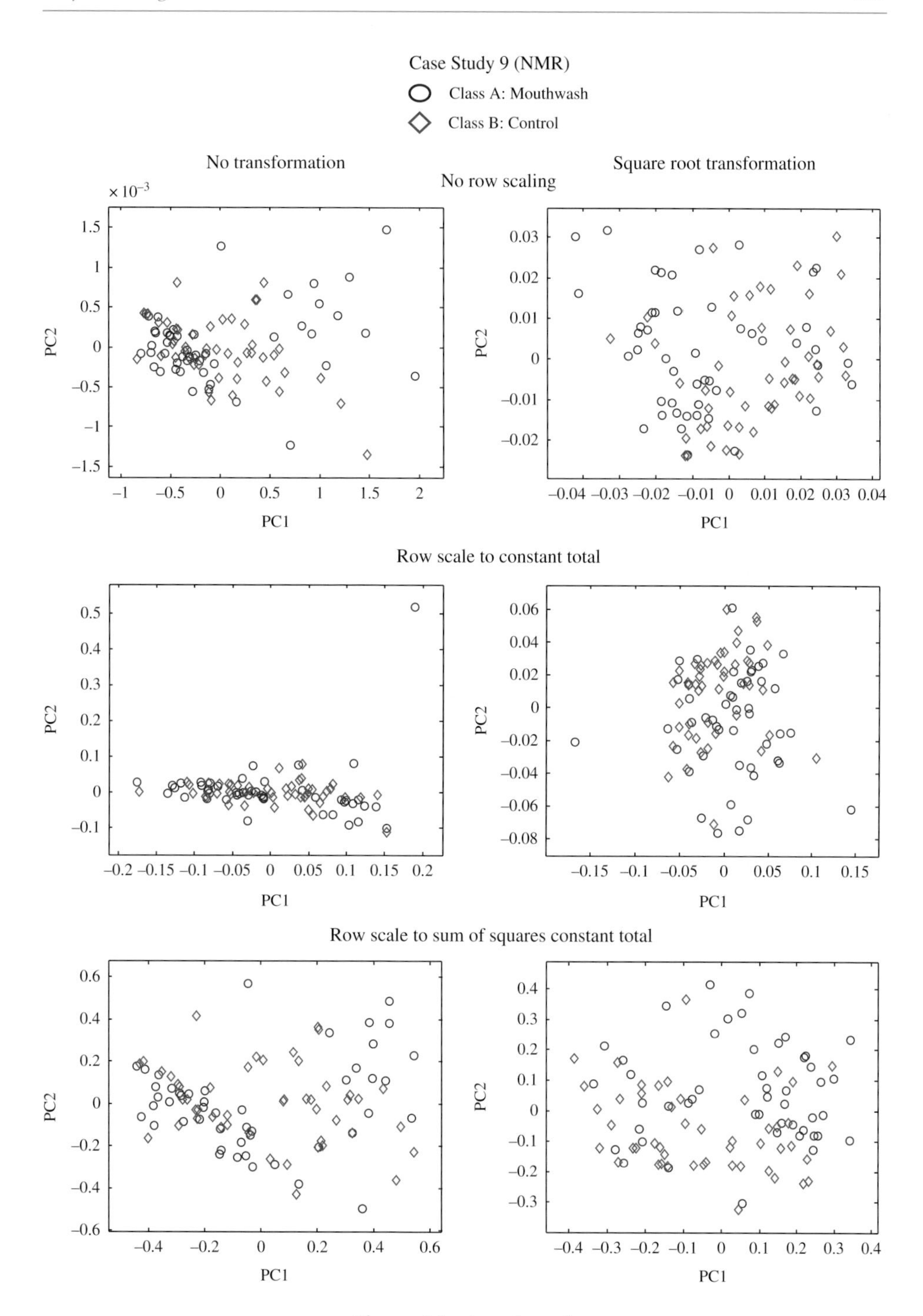

Figure 4.6 (continued)

each block if we feel that one part of the data is more important than another. The advantage of this is that different regions or blocks, for example, of a spectrum, can be weighted as of equal importance to each other, even if some regions contain many more measurements than others; it also means that for the purpose of row scaling if one region contains variables that are probably not very relevant to the study in hand but vary considerably over all the samples, they do not influence the row scaling of the rest of the data, for example, in spectroscopy there may be a wavelength range that contains a lot of variability which we want to retain but not influence the rest of a spectrum, or in chromatography we may be interested in looking at different groups of compounds separately.

However, block weighting can also be achieved by column scaling as discussed in Section 4.2.3 and, if used in the context of row scaling, is primarily useful when we want to normalize measurements from each sample.

4.2.3 Column Scaling

The final decision involves whether to scale the columns. If variables are on very different scales, it is usually necessary to do this to ensure that all variables have a similar influence on the outcome. For example in metabolomic profiling, we may measure several hundred metabolites but a few may be very intense in magnitude; if we do not column scale, these intense variables will dominate the analysis and the variation in the smaller variables will have very little influence on the results. In other areas such as NIR spectroscopy, all wavelengths may be of approximately equal importance and so standardizing is not so critical, and in some cases where there are regions of a spectrum where there is not much information, column scaling can actually make the result worse.

The application of column scaling is a little more complex than row scaling in pattern recognition, especially if data are separated into training and test sets and or the bootstrap (Chapter 8). In all cases column scaling should be performed as the last step prior to pattern recognition, and we will assume that x_{ij} has already been row scaled using the methods of Section 4.2.2 and the matrix elements transformed using the methods of Section 4.2.1 as appropriate to the problem in hand. In Section 4.4 we will look a little more closely at how column scaling is best incorporated into the data analysis strategy, especially when separating into training and test sets (Chapter 8) or using one class models (Chapter 6). For the most appropriate results the part of the data that is used to form the model (e.g. training set and/or one class) is column scaled, and the rest of the data are scaled by the column parameters (means or standard deviations) of the (one class) training set. However when we visualize data in 2D scores plots this is not feasible as the positions of the points will change according to which are used for the model, and so if we, for example, repeatedly generate test and training set models (Section 8.4), boundaries that may appear linear with one training set, will appear distorted in the 'space' created by another training set, and so we adopt the approximation that the overall dataset column parameters apply to all points.

In this section we primarily list the main approaches and discuss their application when there are 'splits' of the data in Section 4.4.

Mean Centring

This relatively straightforward approach involves subtracting column means, so that x_{ij} is transformed into $x_{ij} - \bar{x}_j$ where $\bar{x}_j$ is the column mean of variable j. It can be useful if different variables have different means, but usually does not have a major influence on pattern recognition and classification results. However in this text we will always mean centre the $\boldsymbol{X}$ block, except where stated otherwise.

In the case of the bootstrap (Sections 4.3.1 and 8.5.1) some samples are represented more than once in the bootstrap training set: these are counted as multiple incidences and so should be included more than once in the calculation of the mean for the bootstrap training set.

Weighted Centring

When performing PLS-DA classification, there may be different numbers of samples in each class. We recommend (Chapter 8) for the training set that equal numbers are selected from each class, but for autoprediction (that is performing model building on the entire dataset) it cannot be avoided that there are different sample sizes in each class. If there are two groups rather than mean centre the entire $\boldsymbol{X}$ block using the overall mean, which may be biased in favour of one of the classes, it is preferable to create a global mean $\bar{\boldsymbol{x}} = (\bar{\boldsymbol{x}}_A + \bar{\boldsymbol{x}}_B)/2$ and subtract this from the data. The effect of weighted centring as in this section and mean centring on PLS-DA will be discussed in Section 5.5.3.

This concept can be extended when there are several classes, all of which where there is a separate vector $\boldsymbol{c}_g$ vector for group g. The $\boldsymbol{X}$ matrix can be transformed by subtracting the weighted mean $\bar{x} = (\bar{x}_g + (\sum_{h \neq g} \bar{x}_h)/(G-1))/2$, from each column where there is a total of G groups, although we do not use this in this text; however the effect of this type of centring could be studied.

When samples are divided into test and training sets (Chapter 8) we recommend equal numbers of samples from each class, and so there is no requirement for weighted centring. However when performing the bootstrap (Section 8.5.1) there may be unequal representatives of each class, because the samples are chosen randomly, and this can have a small effect, but on average the number of representatives of each class should be equal so long as the data come from a training set of equal representations of each class and so we do not use weighted centring for the bootstrap, but normal centring, as the adjustments will be very minor. However if the bootstrap is performed on training set data of unequal class sizes (which is not done in this text), weighted adjustment of column means as described above may be important.

Standardization

Standardization is sometimes also called autoscaling or normalization (where the latter word is ambiguous and so best avoided), and involves first mean centring and then dividing each column (or variable) by its standard deviation, so that x_{ij} is transformed into $(x_{ij} - \bar{x}_j)/s_j$ where s_j is the standard deviation of variable j. Usually the population standard deviation given by $s_j = \sqrt{\sum_{i=1}^{I} (x_{ij} - \bar{x}_j)^2 / I}$ is computed rather than the sample standard deviation, as the purpose of the divisor is to scale columns and not to estimate statistical parameters, and we adopt this approach in this text. Note that many computational packages use sample rather than population standard deviations as defaults and so it is often necessary to specify the population equation carefully.

Standardization is important if different variables are measured over very different ranges and ensures that each variable has a similar influence on the resultant EDA or classifier. Whether standardization is useful or not depends on the specific problem in hand. In the metabolomic profiling Case Studies 5 (sweat) and 8 (mouse urine) it clearly has an advantage as otherwise the data will be swamped by intense variables as there are several orders of magnitude differences between the variables. However in other situations there may be variables or regions of the spectra that consist primarily of noise and standardizing makes the separation worse, an example being Case Study 1 (forensic) where there are regions of the distribution graph that contain very limited information.

Figure 4.7 illustrates the effect of standardization on the PC scores plot on Case Studies 1 and 8a. In Case Study 1 (forensic), standardizing worsens the separation, leading

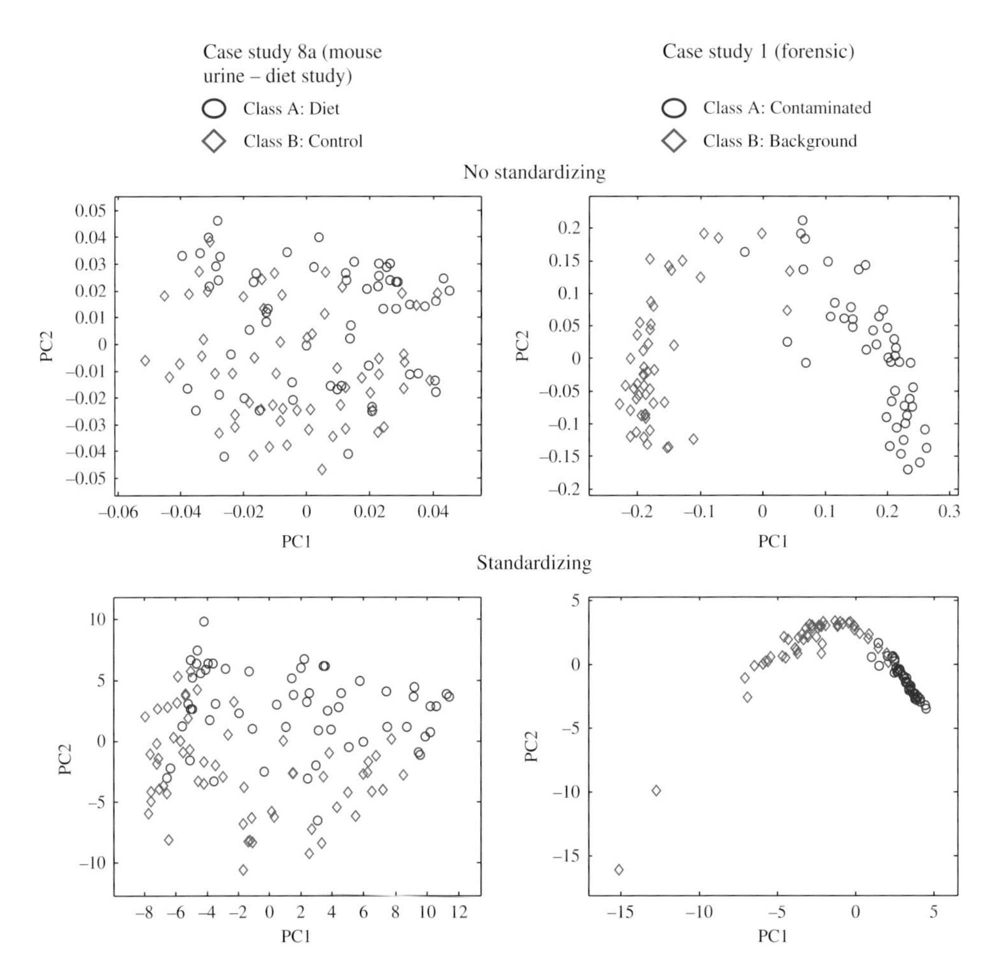

Figure 4.7 Influence of standardizing on the PC scores plots of Case Studies 8a and 1. For Case Study 8a, the data are row scaled to constant total and square rooted; for Case Study 1 they are row scaled to constant total by default. In all cases data are mean centred. Standardization is always the last step prior to PCA

to two outliers, with the two groups being very close together. This is primarily because there are many variables that are not very useful, representing primarily noise, whose influence is amplified during standardization. The two outliers happen to have non-zero values for these noisy variables (the other samples having a value of 0). However for Case Study 8a (mouse urine – diet study), standardization definitely improves separation. This is because there is a large dynamic range of five orders of magnitude, and if the data are not standardized the observed patterns are dominated by a few intense peaks, that may not be ones of direct interest to the study in hand.

An additional important consideration about standardization is its application when data are split into training and test sets, as discussed in Chapter 8 and Section 4.4. There are two approaches, as follows:

1. The overall data (including training and test set) is standardized using the overall mean and standard deviation of the entire dataset studied, and there is no re-standardization of the training set. As commented above, when visualizing boundaries between groups this is often the most appropriate approach.

The alternative is as follows:

2. The training set is standardized and the test set samples are adjusted for the training set mean and standard deviation, that is x_{ij} is transformed to $(x_{ij} - \bar{x}_{jtrain})/s_{jtrain}$ where $\bar{x}_{jtrain} = \sum_{i \in train} x_{ij}/I_{train}$ and $s_{jtrain} = \sqrt{\sum_{i \in train} (x_{ij} - \bar{x}_{jtrain})^2/I_{train}}$ and there are I_{train} samples in the training set. Strictly speaking this is statistically most correct and when making or testing quantitative predictions, e.g. in analytical chemistry calibration, probably essential. However if there are several test and training set splits which are generated this can become quite complex for book keeping, especially if there are two training sets (if the bootstrap is employed for model optimisation – see Section 8.5.1) and it is often hard to define what is meant by an overall model, and in many cases it makes very little difference at the cost of computational complexity. A major reason for standardization is to scale data so that all variables have a similar influence on the classifier, and usually the small differences that occur when a different subset of samples are included within a training set are not very important, unless there are a small number of clear outliers.

Note that all standardized data is automatically mean centred. In addition, in the case of calibrating of two blocks of data as in PLS (an 'X' block and a 'c' block), it is never necessary to standardise $\boldsymbol{c}$ if it contains one classifier (PLS1), although it can make a difference if PLS2 is employed (Section 7.4.1).

Range Scaling

A simple alternative to standardization is range scaling. This involves scaling each variable to the same range, e.g. between 1 and 0. The transformation is as follows. The value of x_{ij} is transformed to $(x_{ij} - min_j(x_{ij}))/(max_j(x_{ij}) - min_j(x_{ij}))$, where $max_j(x_{ij})$ is the maximum value of variable j over all samples and $min_j(x_{ij})$ the corresponding minimum. If there are potential outliers, an alternative is to take the quartiles rather than the maximum and minimum.

Standardization is normally used preferentially but if range scaling is employed many of the similar caveats and comments apply as for standardization. In this text we will employ standardization, which has the additional advantage that the columns are centred.

Block Scaling and Weighting

In Section 4.2.2 we discussed block scaling of rows, which mainly is useful to control how influential each set of variables is when row scaling. However block scaling by columns has a much more important role to play.

Consider the situation of Figure 4.8 which may be, for example, of process data which are monitored by IR spectroscopy and UV spectroscopy and for which we also have some process variables (e.g. temperature, pH, etc.). A typical example may involve recording data at 200 IR wavelengths, 50 UV wavelengths and measuring 5 process variables associated with each sample. If all variables are standardized, the variation in the IR spectra will be 4 times more influential than the UV spectra and 40 times more influential than the process variables. It may be that we feel that all three types of information are of approximately similar importance, but we just happen to have a detector with 200 IR wavelengths and another with 50 UV wavelengths, and only 5 process variables. In order to compensate for this and to make each block equally important we can divide the value of each type of variable by a further factor equal to the number of variables in each block b, J_b which equals 200 for the IR spectra and so on, and so that x_{ij} is transformed to $(x_{ij} - \bar{x}_j)/(s_j \times J_b)$.

This reasoning can be extended and variables or blocks of variables can be given any weight according to the relative importance attached to them, although it is necessary to be relatively unprejudiced as to how to choose the weighting. If the aim is to classify samples, and it is chosen with the classifier in mind it could bias the

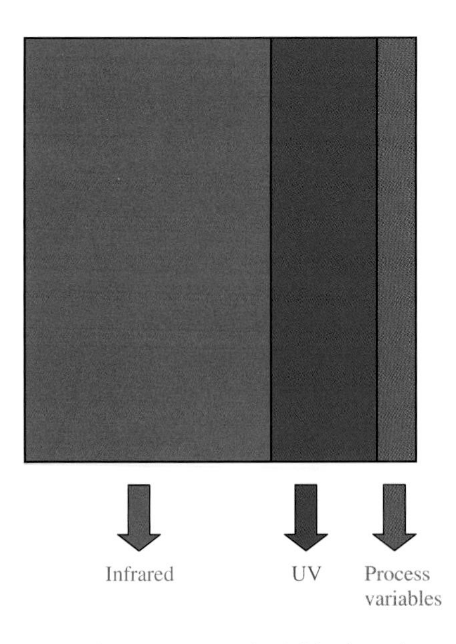

Figure 4.8 Typical situation in which block scaling is important

classification and result in artefactually good predictions, and so care should be taken into how far to go down this route, and it is recommended that block scaling is primarily used to adjust the relative importance of several types of measurement, particularly if each block consists of different numbers of variables.

4.3 Multivariate Methods of Data Reduction

In addition to scaling individual measurements it is sometimes useful to reduce the number of variables prior to exploratory or supervized data analysis. One approach is to choose a subset of the original variables as discussed in Chapter 9. An alternative is to create new variables that contain information obtained from the original variables, by multivariate methods. The advantage is that often the classification and exploratory procedures are much simpler computationally, for example, there may be several hundred peaks recorded in a chromatogram, but many will be from noise, the analytical process etc. and will offer little insight for the problem in hand, and so reducing to a few new variables makes it easier to visualize what is happening. In some cases such as classifiers based on the Mahalanobis distance, such as LDA (Linear Discriminant Analysis) or QDA (Quadratic Discriminant Analysis), it is essential to reduce the number of variables if their number exceeds the number of samples (Sections 5.3 and 5.4). However the disadvantage is that it is harder to interpret the results of classifiers in terms of the original variables, for example, to determine which compounds or variables are potential markers (see Chapter 9). PLS-DA based methods in fact give identical classification rates to LDA under certain circumstances of data scaling, but contain additional diagnostics of which variables are important. In classical statistics there often were very few variables to samples, and so problems with coping with situations where the sample to variable ratio was less than 1 were not encountered, but in chemometrics special techniques are needed if it is desired to interpret discrimination in terms of original variables (often related to known compounds in the samples).

In this section we will focus primarily on how to choose new variables as input to a classifier, and illustrate several of the methods by how they influence classification abilities, in contrast to Section 4.2 where we primarily illustrate the influence on the appearance of PC plots. In order to understand some of the results presented in this section it is useful to refer to Chapter 8 where the division of data into test and training sets is described. An ideal model, whereas it may result in high predictions for a training set even on a randomly generated dataset (e.g. Case Study 11), should provide realistic predictions on the test set and so for Case Study 11, the overall proportion of samples correctly classified in the test set should be close to 50 %. If it is substantially higher than this there may be problems with the methods employed.

4.3.1 Largest Principal Components

The PCA algorithm has been described previously in the context of EDA (Section 3.2). Principal Components can be used for many different purposes and in addition to visualizing data can also be used for data reduction. Instead of the original J raw variables in matrix $\boldsymbol{X}$ we only need to use A orthogonal variables or PCs as represented by a scores matrix $\boldsymbol{T}$ as input to the classifier. An important question is how many PCs should be chosen to model the data, i.e. what value of A should be used, to simplify the data, without losing essential features.

Using most algorithms PCs are ordered according to their size, or their eigenvalues, with PC1 being the largest (whose scores have the largest sum of squares or largest eigenvalue) with PC A the smallest. One important feature of PCA in the context of supervised pattern recognition is that the calculation of PCs does not take group membership of samples into account: therefore it is an unsupervised method for reducing the data. This has many advantages when used for data reduction prior to testing models because it means that there is no risk of over-fitting if PCA is performed on the overall dataset including the test and training sets together prior to classification (we will see in Section 4.3.3, that the same is not true for PLS components). Hence one can safely replace the original sample vector $\mathbf{x}_i$ by a $1 \times A$ vector of its scores without prejudicing the classifier. Incorrect applications of variable reduction are discussed in Sections 4.3.3 and 9.1.5 where it is shown that if supervised methods are used to reduce the data, for example, by choosing the best markers or PLS components, and this is performed on the entire dataset including a test set, a high %CC (Percentage Correctly Classified) can be obtained even when using the Case Study 11 (null dataset). However, reducing the overall dataset (test and training set together) by PCA and retaining the 2 largest PCs results in a %CC of 47.00 % on the average test set of 100 iterations using EDC (Euclidean Distance to Centroids) and 46.87 % using LDA (Linear Discriminant Analysis) (see Chapter 8 for details); retaining the 10 largest PCs results in 49.37 % and 49.37 %, respectively, sufficiently close to 50 % to give us confidence that PC reduction is not overfitting the data even if test set samples are included. There is a large literature about how many PCs should be retained to adequately represent a dataset. In this text we will just explore a few of these.

The simplest, of course, is just to select a specified number of PCs. To visualize class boundaries and separations we cannot use more than 3 PCs, and it is easier just to use 2PCs; however, this is mainly for illustrative purposes which we will adopt in this text to demonstrate the use of different classifiers. Having a set cut off number of PCs, e.g. 10, is another alternative. It is important to remember that we are not trying to produce exact numerical predictions as in classical chemometrics such as multivariate calibration, but primarily to simplify the data, on the basis that the classifier we are most interested in is likely to be the largest factor; other factors that influence the data (for example individuality) that may be important to model the data fully are not so important for the task in hand. Consider the analogy of trying to distinguish one species of fish from another. There may be many hundreds of measurements that could be made on each fish, and to distinguish each individual fish adequately from the others we probably need these measurements, but to distinguish one species from another it may not be necessary and just a few of the more important or distinguishing measurements are adequate to get on with the job, as it is not our primary purpose to build up a full description of each fish (which may have unique individual distinguishing features). So having a somewhat arbitrary cut off is often adequate, although it may not describe the data fully, it describes the group trends that are more important than the individual trends; these should be reflected in more significant PCs. Sometimes it is useful to determine how much of the variability is described when an increasing number of PCs is employed, which is presented in Table 4.3 for the case studies in this text. Graphs for Case Studies 1, 8a and 11 were

Table 4.3 *Percent variance for the first 20 PCs for the case studies described in this book*

Case study	Principal component																			
	1	2	3	4	5	6	7	8	9	10	11	12	13	14	15	16	17	18	19	20
1 (forensic)	57.30	16.90	5.98	4.65	2.88	2.03	1.63	1.54	1.15	1.00	0.85	0.71	0.60	0.46	0.43	0.40	0.37	0.27	0.18	0.15
2 (NIR spectroscopy of food)	75.44	22.08	1.10	0.43	0.33	0.17	0.12	0.07	0.06	0.04	0.03	0.02	0.02	0.01	0.01	0.01	0.01	0.01	0.01	0.01
3 (polymers)	39.82	23.30	10.76	9.11	5.73	5.14	2.50	1.21	0.78	0.40	0.37	0.27	0.16	0.09	0.07	0.06	0.05	0.04	0.03	0.02
4 (pollution)	34.09	23.96	10.29	7.45	6.05	4.09	1.89	1.32	0.87	0.81	0.70	0.53	0.47	0.46	0.42	0.37	0.33	0.32	0.30	0.26
5 (sweat)	14.20	3.93	3.36	2.68	1.96	1.55	1.38	1.35	1.29	1.25	1.22	1.16	1.07	0.94	0.90	0.88	0.87	0.84	0.83	0.80
6 (LCMS of tablets)	5.63	5.32	4.86	4.53	4.17	3.54	3.27	3.23	3.07	2.96	2.82	2.73	2.61	2.57	2.44	2.37	2.31	2.16	2.08	1.95
7 (clinical)	62.17	23.37	8.97	4.70	0.79															
8a (mouse – diet)	10.15	6.27	4.89	3.63	2.88	2.58	2.39	2.18	1.96	1.83	1.76	1.60	1.58	1.54	1.47	1.43	1.41	1.37	1.33	1.30
8b (mouse – genetics)	7.56	5.34	4.66	3.67	2.91	2.33	2.18	2.14	1.86	1.81	1.59	1.46	1.40	1.28	1.26	1.25	1.16	1.13	1.12	1.06
9 (NMR spectroscopy)	26.35	16.45	10.88	8.74	6.39	4.79	3.60	2.79	2.18	1.78	1.64	1.55	1.27	1.09	0.99	0.89	0.80	0.77	0.69	0.67
10a (simulation)	3.05	2.66	2.65	2.50	2.34	2.31	2.24	2.21	2.13	2.10	2.04	1.97	1.96	1.89	1.84	1.83	1.78	1.72	1.70	1.68
10b (simulation)	3.03	2.71	2.60	2.40	2.35	2.26	2.25	2.21	2.18	2.08	2.03	2.00	1.94	1.89	1.83	1.81	1.78	1.75	1.70	1.69
10c (simulation)	2.82	2.66	2.55	2.48	2.39	2.31	2.25	2.17	2.15	2.13	2.05	1.98	1.97	1.94	1.88	1.84	1.83	1.78	1.72	1.69
11 (null)	2.14	2.07	2.03	1.97	1.92	1.88	1.86	1.83	1.78	1.77	1.74	1.72	1.70	1.67	1.65	1.62	1.57	1.55	1.54	1.50
12 (GCMS)	19.06	9.11	5.81	5.08	4.85	4.19	3.63	3.16	3.03	2.79	2.71	2.55	2.44	2.36	2.24	2.11	2.09	1.94	1.92	1.87
12 (microflora)	21.72	15.37	11.97	9.96	8.05	6.87	5.18	4.48	3.80	3.53	2.31	1.62	1.36	0.92	0.78	0.67	0.50	0.33	0.24	0.17

presented in Figure 3.4 in Chapter 3. The relationship between the number of PCs and explained variance differs substantially between datasets – for Case Study 11 each PC accounts for approximately similar variance with only a very small decrease, as the data are essentially random. This contrasts with Case Study 1 (forensic) for which the first two PCs correspond to 74.20 % of the variance. Even more extreme is Case Study 2 (NIR spectroscopy of food) for which the first 2 PCs correspond to 97.52 % of the variance and as can be seen in Chapter 3 excellent separation is obtained by using just the first 2 PCs for that dataset.

However there is a large literature on determining the optimal number of PCs needed to correctly model a dataset. Although the simplest approach might be to include only the number of PCs or the number of PCs that represent a set percentage variance, in classical chemometrics determining the number of PCs that adequately describe a dataset is equivalent to determining the number of compounds in a mixture – for example, in coupled chromatography this could correspond to the number of compounds that characterize a peak cluster in a chromatogram. The basis is that significant PCs model 'data', whilst later (and redundant) PCs model 'noise', e.g. instrumental or sampling error, and the idea is to determine how many PCs are necessary to model the data and to throw away the remainder. Autopredictive models, which will be discussed in more detail in Chapters 5 to 7, involve fitting PCs to the entire dataset, and always provide a closer fit to the data the more the components are calculated. Hence, the residual error will be smaller if 10 rather than 9 PCs are calculated. However, this does not necessarily imply that it is correct to retain all 10 PCs.

The significance of the each PC can be test out by seeing how well an 'unknown' sample is predicted. Cross-validation is an important classical technique that has been employed for the determination of the number of significant components. In the most widely used LOO (Leave One Out) form of cross-validation, each sample is removed once from the dataset and then the properties of the remaining samples are predicted. For example, if there are 25 samples, perform PC on 24 samples, and see how well the remaining sample is predicted. The algorithm is as follows:

1. Initially leave out sample 1 ($= i$).
2. Perform PCA on the remaining $I-1$ samples, e.g. samples 2 to 25. For efficiency it is possible to calculate several PCs ($= A$) simultaneously. Obtain the scores $\boldsymbol{T}$ and loadings $\boldsymbol{P}$. Notice that there will be different scores and loadings matrices according to which sample i is removed from the dataset.
3. Next determine the what the scores would be for sample i by:

$$\hat{\boldsymbol{t}}_i = \boldsymbol{x}_i \boldsymbol{P}'$$

   Notice that this equation is quite simple, and is obtained from standard multiple linear regression $\hat{\boldsymbol{t}}_i = \boldsymbol{x}_i \boldsymbol{P}'(\boldsymbol{P}\boldsymbol{P}')^{-1}$, but since the loadings are orthonormal, $(\boldsymbol{P}\boldsymbol{P}')^{-1}$ is a unit matrix, where the "^" or "hat" sign indicates an estimate.
4. Then calculate the estimate for sample i using an a PC model by

$${}^{a,cv}\hat{\boldsymbol{x}}_i = {}^{a}\hat{\boldsymbol{t}}_i {}^{a}\boldsymbol{P}$$

   where the superscript a refers to the model formed from the first a PCs, so that ${}^{a}\hat{\boldsymbol{x}}_i$ has the dimension $1 \times J$, and ${}^{a}\hat{\boldsymbol{t}}_i$ has dimensions $1 \times a$ (i.e. is a scalar if only 1 PC is

retained) and consists of the first a scores obtained in step 3, and ${}^{a}\boldsymbol{P}$ has dimensions $a \times J$ and consists of the first a rows of the loadings matrix.
5. Next, repeat this by leaving another sample out and going to step 2 until all samples have been removed once.
6. The error, often called the Predicted Residual Error Sum of Squares or PRESS, is then calculated as follows:

$$\text{PRESS}_a = \sum_{i=1}^{I} \sum_{j=1}^{J} ({}^{a,cv}\hat{x}_{ij} - x_{ij})^2$$

This is the sum of the square difference between the observed and true values for each object using an a PC model. PRESS can be calculated for different numbers of PCs.

The PRESS errors can then be compared to the RSS (residual sum of square) errors for each object for straight PCA on the entire dataset (sometimes called the autoprediction error), given by the following:

$$\text{RSS}_a = \sum_{i=1}^{I} \sum_{j}^{J} x_{ij}^2 - \sum_{k=1}^{a} g_k$$

or:

$$\text{RSS}_a = \sum_{i=1}^{I} \sum_{j=1}^{J} ({}^{a,auto}\hat{x}_{ij} - x_{ij})^2$$

All equations presented above assume no column scaling, if there is mean centring or standardization (Sections 4.2.3) or other forms of column scaling, further steps are required involving usually requiring subtracting the mean of $I-1$ samples each time a sample is left out. If the data are preprocessed prior to cross-validation, it is essential that both PRESS and RSS are calculated using the same preprocessing. A problem is that if one takes a subset of the original data, the mean and standard deviation will differ for each group, and so it is safest to convert all the data to the original units for calculation of errors. The computational method can be quite complex and there are no generally accepted conventions but we recommend the following.

1. Preprocess the entire dataset.
2. Perform PCA on the entire dataset, to give predicted $\hat{\boldsymbol{X}}$ in preprocessed units (e.g. mean-centred or standardized).
3. Convert this matrix back to the original units.
4. Determine the RSS in the original units.
5. Next, take one sample out and determine statistics such as means or standard deviations for the remaining $I-1$ samples.
6. Then preprocess these remaining samples and perform PCA on these data.
7. Then scale the remaining sample using the mean and standard deviation (as appropriate) obtained in step 5.

8. Then obtain the predicted scores $\hat{t}_i$ for this sample, using the loadings in step 6 and the scaled vector x_i obtained in step 7.
9. Then predict the vector $\hat{x}_i$ by multiplying $\hat{t}_i$ by p where the loadings have been determined from the $I-1$ preprocessed samples in step 6.
10. Now rescale the predicted vector to the original units.
11. Next remove another sample, and repeat steps 6 to 11 until each sample is removed once.
12. Finally calculate PRESS values in the original units.

There are various ways of interpreting the PRESS and RSS errors, but a common approach is to compare the PRESS error using $a + 1$ PCs to the RSS using a PCs, so computing $PRESS_{a+1}/RSS_a$. If the latter error is significantly larger, then the extra PC is modelling only noise, and so is not significant, so once this exceeds 1, take a as the number of PCs. A modification is to use a value $\left(Q = \sqrt{(J-a)(I-a-1)/(J-a-1)(I-a)}\right)$ rather than 1 for determining the threshold, although this makes little difference if there are many samples and variables. An alternative is to compute the ratio $PRESS_{a+1}/PRESS_a$ and if this exceeds 1, use a PCs in the model. The principle is that if the errors are quite close in size, it is safest to continue checking further components and PRESS should start increasing after the optimum number of components have been calculated; however, this in practice does not always happen for PCA (different results are obtained when using PLS for calibration which we do not discuss in this text). We illustrate this for Case Studies 1, 2, 3, 4, 8a and 11 in Figure 4.9. It can be seen that different conclusions can be obtained by using each of the two criteria. On the whole $PRESS_{a+1}/RSS_a$ gives a clearer picture with a clear point above which the ratio exceeds 1. We model Case Study 1 (forensic) by 2 PCs and Case Study 2 (NIR spectroscopy of food) by 3 PCs: both of these involve readily separable groups of samples. For Case Study 3 (polymers) it looks as that all of the 14 PCs are needed: this is not surprising as there are several different groups of polymers, each in themselves formed of grades or subgroups, and so we expect a lot of variability. Case Study 4 (pollution) exhibits unusual behaviour allowing us to choose either 3 or 6 PCs, where probably the former is the most appropriate. Case Study 8a (mouse urine – diet study) seems to be modelled well by using just 2 PCs, which is rather surprising given the amount of biological variability; however in order to use a comparable scale for errors, we transformed back into the original units which may have the effect that a few influential variables (of high intensity) dominate the error analysis; there are other ways of scaling the data which are a bit more complex and which could result in different conclusions – Case Study 8a is the only one where the data are standardized prior to analysis amongst those illustrated. In Case Study 11 (null) it looks as if 0 PCs are necessary to model the data, according to the $PRESS_{a+1}/RSS_a$ which suggests that the data can be modelled by its mean and there is no significance in further variability – which as it is a randomly generated dataset might be a possible answer. The $PRESS_{a+1}/PRESS_a$ graphs are not as clear as anticipated, and some do not even reach 1 – slightly different conclusions can be obtained with varying adjustments for degrees of freedom, but we recommend $PRESS_{a+1}/RSS_a$.

The bootstrap can also be employed for a similar purpose. This procedure is discussed in detail in Section 8.5.2 in the context of PLS-DA, and only briefly outlined below. This is an iterative procedure that involves taking a group of samples out for

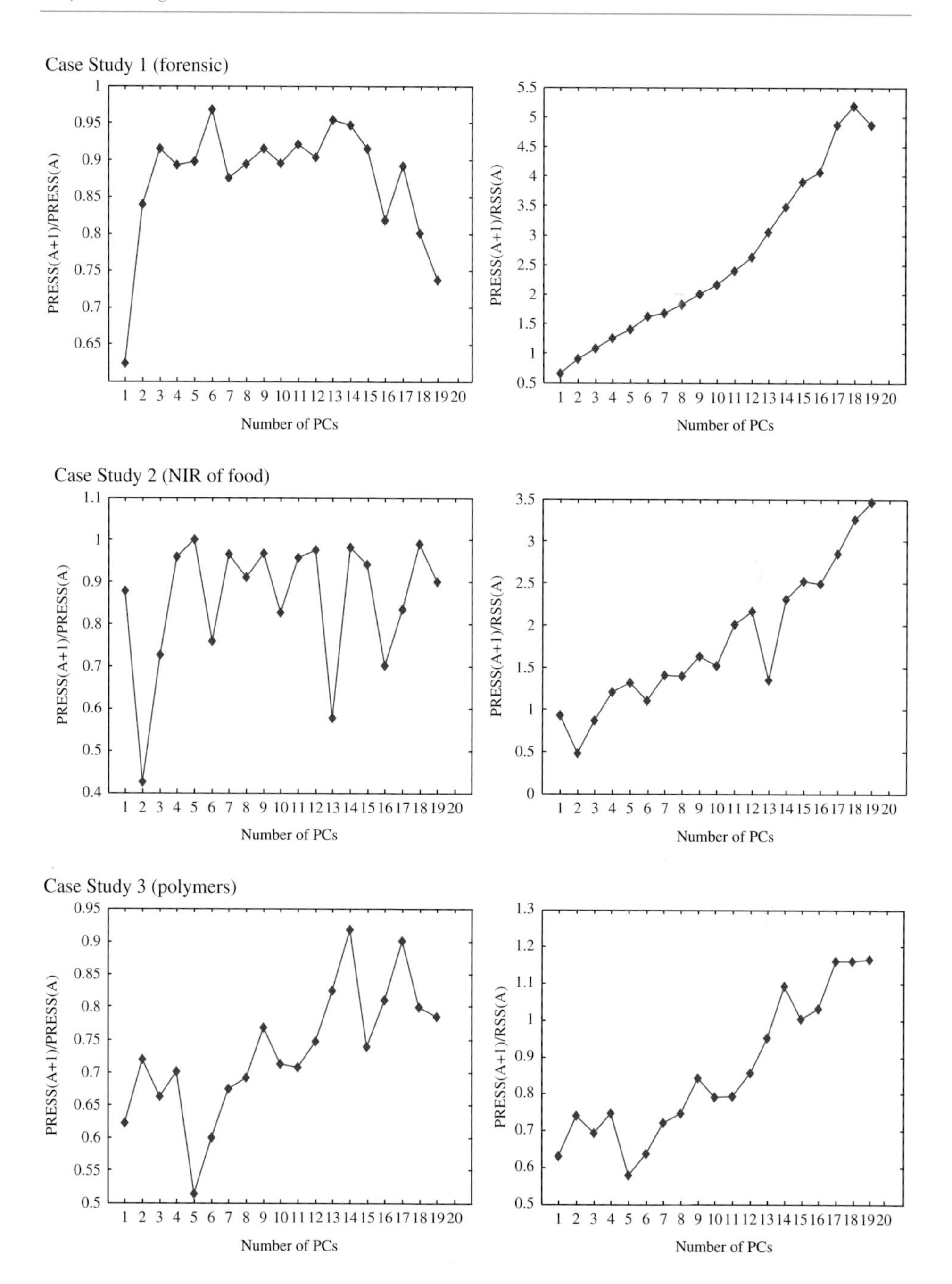

Figure 4.9 Cross-validation to determine the number of significant PCs for Case Studies 1, 2, 3, 4, 8a and 11

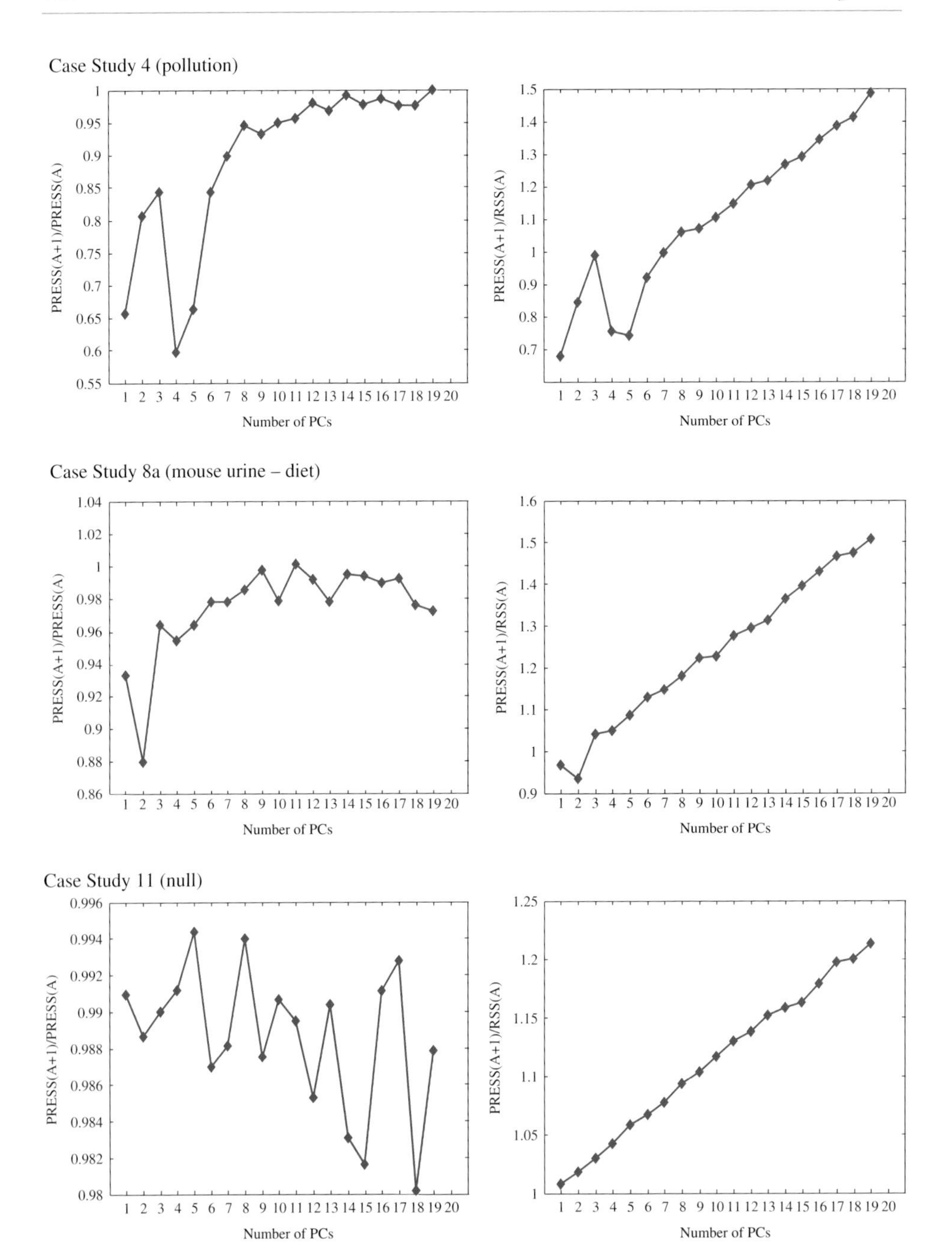

Figure 4.9 (continued)

testing each time; we use 200 bootstrap iterations. An abbreviated description is as follows:

1. Split the data into bootstrap training and test. The bootstrap training set has repeat samples, while the bootstrap test set does not, using repeat re-sampling, as discussed in Section 8.5.1
2. Preprocess the bootstrap training set.
3. Preprocess the bootstrap test set using the bootstrap training set column parameters.
4. Do PCA on the preprocessed boostrap training set, and predict the bootstrap training set data using 1 to A PCs where A is the maximum number of PCs to be tested for the model.
5. Predict the bootstrap test set data using 1 to A PCs using the test set model.
6. Un-scale the predicted training set and predicted test set data.
7. Calculate the training and test set residuals, square, and sum for each sample.
8. Repeat R times (200 is recommended in this text)
9. For each sample calculate the average squared residual over all 200 iterations for both the bootstrap training set and test set, taking into account repeats and when the sample(s) was/were not selected for the bootstrap training or test set.
10. Calculate the RSS value for each PC as the sum of the average squared auoprediction residuals for each sample and the PRESS value for each PC as the sum of the average squared bootstrap test set residuals.

The results of the bootstrap are presented in Figure 4.10, and it can be seen that they are clearer, than for cross-validation, probably because of the averaging effect. Especially for Case Study 11 (null) we see a much clearer trend.

It is usually acceptable to perform PCA on the combined training and test sets and then use the same PC scores for input to the classifier even if new training and test sets are formed providing that PCs are chosen according to size rather than classification ability, as discussed above. This is because PCs are unsupervised and have no direct relation to the classifier. An important test of whether any proposed method is valid involves applying it to the null dataset (Case Study 11); if the classification of the Case Study 11 is around 50 % this implies that PCA has not prejudiced the classifier and is safe.

In Section 4.3.2 we will look the alternative approach, where the PCs are selected not according to size but according to classification ability. Under such circumstances it is dangerous to select PCs on both the test and training set together.

4.3.2 Discriminatory Principal Components

In Section 4.3.1 we reduced the data by selecting PCs according to their size, choosing to retain the largest PCs. The question we asked was not which PCs we would retain, but how many, the only decision being about what value of A to use. Of course if we are interested, for example, in classifying samples into two groups, there may be many other factors that influence variability. Take, for example, a situation where we are interested primarily in determining whether a subject is diseased or a control by analysing their urine. Whereas disease may be one factor that influences the urinary chemosignal, so will age, gender, recent diet, environment and so on. Even if the sampling is performed under controlled conditions, these factors will be present, and there is no reason to assume that the disease state of a patient is the major factor influencing the compounds found in their urine. PCs primarily look at overall variability in the data and not why there is such variability. If we are lucky and the main factor is the one we are interested in, or if the experiments are very

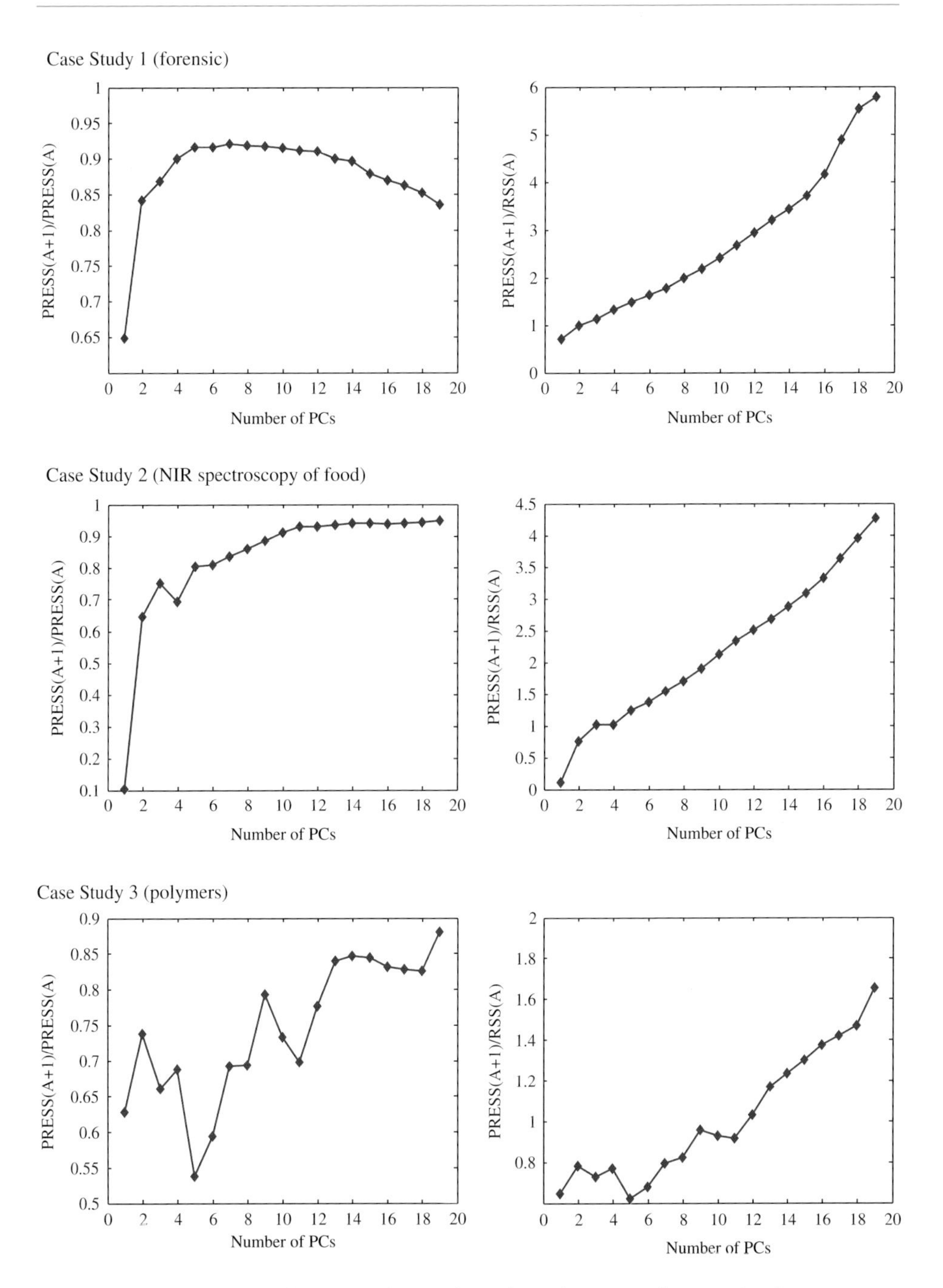

Figure 4.10 Bootstrap to determine the number of significant PCs for Case Studies 1, 2, 3, 4, 8a and 11

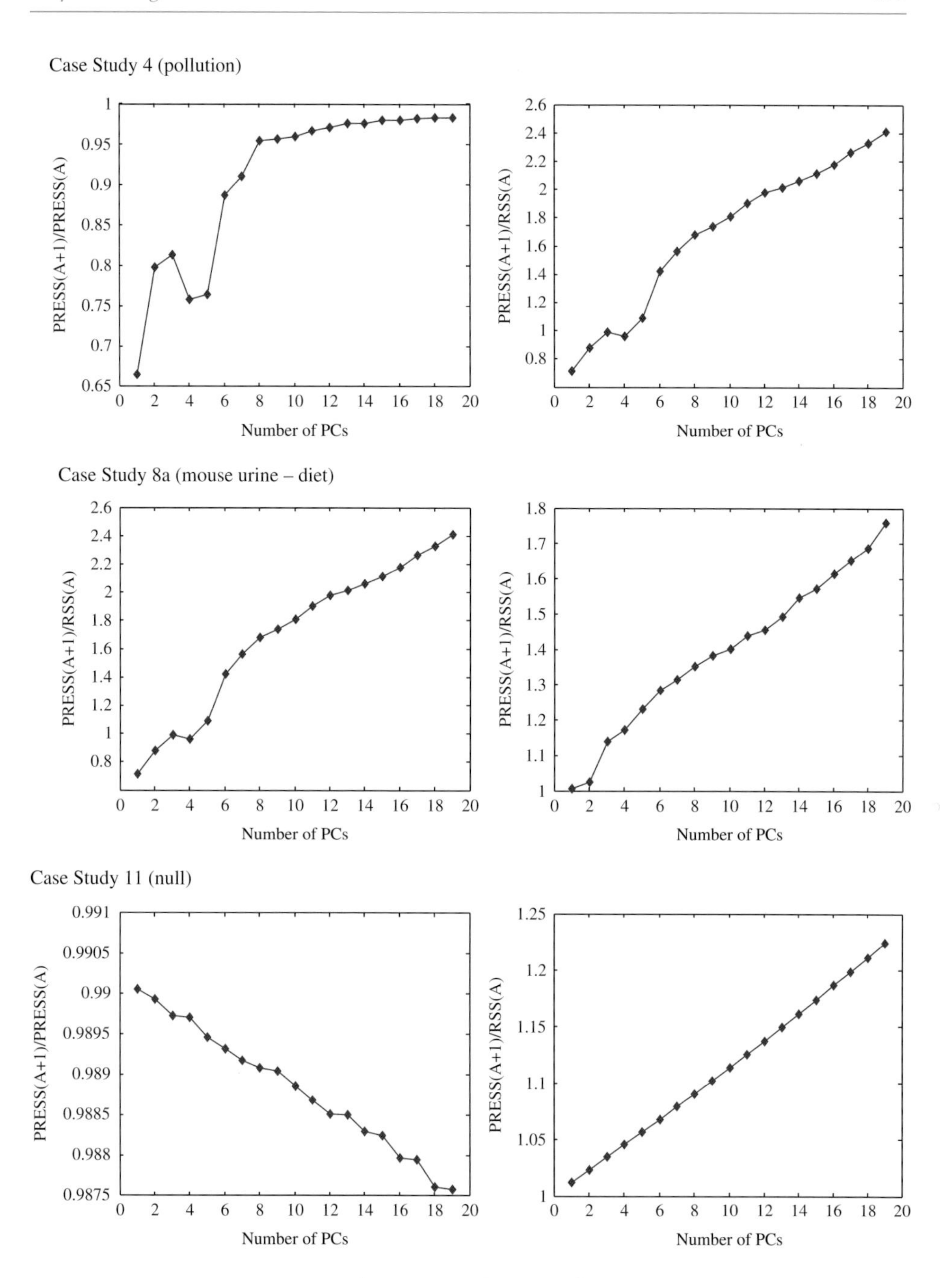

Figure 4.10 (continued)

controlled, we would anticipate the variability we are interested in to be reflected in the earlier (larger) PCs, but we cannot guarantee this, and there is no guarantee that the larger the PC, the more relevant it is to the factor we are most interested in.

Hence PCs can also be selected in a supervised way, that is, we can find out which PCs are most related to the factor (usually as in this book a discriminator or classifier) of prime interest, and instead of selecting the largest PCs we can select the most discriminatory one.

The simplest approach is to determine which PCs are most discriminatory and rank these in order. In Chapter 9, we discuss a number of approaches for determining which variables from the raw dataset are best discriminators, and many of these approaches can also be applied to PCs. For brevity we discuss only one approach in this chapter, i.e. the t-statistic which can be used to look at the significance of how well each variable can discriminate between two groups: similar principles can be applied by using a variety of other statistics for ranking the discriminatory power of the variables. The t-statistic can be calculated for each PC and then those with t-statistics that have the highest magnitude are taken as the most significant, with the sign relating to which class they are markers for (so a positive sign indicates that the scores of that PC are most positive for class A which is labelled +1, as discussed in Section 5.1.3: note that the sign of the PCs can vary according to algorithm). For Case Studies 1, 4, 8a and 11 these are calculated in Table 4.4 for the 30 largest PCs: the higher the rank, then the more relevant the PC is to the classification study, and the greater the size the larger the PC as assessed by its eigenvalue. It can be seen for Case Study 11 (null) that there is no relationship between rank according to discriminatory power and size, the largest

Table 4.4 *t-statistics of the first 30 PCs by size of Case Studies 11, 8a, 1 and 4 together with their ranks. The rank (size) relates to the sum of squares or eigenvalue of the corresponding PC*

Case Study 11 (null)				Case Study 8a (mouse urine – diet study)			
Rank (discriminatory)	Rank (size)	t	abs(t)	Rank (discriminatory)	Rank (size)	t	abs(t)
1	5	−2.6184	2.6184	1	2	7.7993	7.7993
2	11	−1.7141	1.7141	2	3	−3.6428	3.6428
3	9	−1.4947	1.4947	3	1	3.6053	3.6053
4	21	1.2873	1.2873	4	13	2.2179	2.2179
5	14	1.2129	1.2129	5	12	−2.1084	2.1084
6	23	1.1286	1.1286	6	6	1.6807	1.6807
7	16	−1.1160	1.1160	7	9	1.6399	1.6399
8	8	1.0319	1.0319	8	21	1.4227	1.4227
9	3	1.0072	1.0072	9	16	−1.3943	1.3943
10	24	0.9390	0.9390	10	11	1.3261	1.3261
11	7	0.7806	0.7806	11	25	1.2356	1.2356
12	27	0.7713	0.7713	12	17	1.2227	1.2227
13	19	0.7098	0.7098	13	29	−1.1474	1.1474
14	28	−0.7027	0.7027	14	27	−1.1231	1.1231
15	29	0.7023	0.7023	15	20	−0.9373	0.9373
16	6	0.6497	0.6497	16	24	0.9182	0.9182
17	17	0.5729	0.5729	17	4	0.8140	0.8140
18	15	0.5370	0.5370	18	19	−0.6881	0.6881
19	1	−0.5263	0.5263	19	30	0.6136	0.6136
20	20	−0.4466	0.4466	20	22	0.5456	0.5456
21	25	−0.4378	0.4378	21	18	−0.4611	0.4611

Table 4.4 *(continued)*

Case Study 11 (null)				Case Study 8a (mouse urine – diet study)			
Rank (discriminatory)	Rank (size)	*t*	abs(*t*)	Rank (discriminatory)	Rank (size)	*t*	abs(*t*)
22	12	0.4286	0.4286	22	26	−0.4353	0.4353
23	10	0.4154	0.4154	23	28	−0.3724	0.3724
24	13	0.3494	0.3494	24	15	0.3586	0.3586
25	18	0.3320	0.3320	25	7	−0.1077	0.1077
26	26	−0.2497	0.2497	26	8	0.0639	0.0639
27	22	0.2011	0.2011	27	14	−0.0461	0.0461
28	4	−0.1729	0.1729	28	5	−0.0219	0.0219
29	2	0.0641	0.0641	29	23	−0.0178	0.0178
30	30	−0.0291	0.0291	30	10	−0.0077	0.0077

Case Study 1 (forensic)				Case Study 4 (pollution)			
Rank (discriminatory)	Rank (size)	*t*	abs(*t*)	Rank (discriminatory)	Rank (size)	*t*	abs(*t*)
1	1	26.3159	26.3159	1	1	17.5930	17.5930
2	6	0.8076	0.8076	2	7	−5.7903	5.7903
3	3	0.7215	0.7215	3	3	−3.2780	3.2780
4	4	−0.6385	0.6385	4	2	−2.9373	2.9373
5	2	0.5497	0.5497	5	12	2.4139	2.4139
6	22	0.4702	0.4702	6	4	−2.1752	2.1752
7	9	0.4613	0.4613	7	19	1.7261	1.7261
8	13	−0.4163	0.4163	8	9	1.6865	1.6865
9	8	−0.4028	0.4028	9	6	1.4513	1.4513
10	28	0.3728	0.3728	10	30	−1.3678	1.3678
11	7	−0.3131	0.3131	11	18	1.2692	1.2692
12	5	−0.2623	0.2623	12	10	1.0899	1.0899
13	18	−0.2452	0.2452	13	24	−1.0148	1.0148
14	27	0.2208	0.2208	14	25	0.9509	0.9509
15	25	−0.1993	0.1993	15	20	0.8137	0.8137
16	11	−0.1832	0.1832	16	8	0.7933	0.7933
17	29	0.1830	0.1830	17	28	0.7469	0.7469
18	23	0.1697	0.1697	18	14	−0.6224	0.6224
19	10	−0.1531	0.1531	19	27	−0.6164	0.6164
20	26	−0.1356	0.1356	20	15	0.5260	0.5260
21	12	−0.1285	0.1285	21	22	0.4855	0.4855
22	19	0.1242	0.1242	22	16	0.3026	0.3026
23	24	−0.0896	0.0896	23	29	0.2615	0.2615
24	16	−0.0644	0.0644	24	13	0.2442	0.2442
25	20	0.0587	0.0587	25	26	−0.2250	0.2250
26	14	0.0510	0.0510	26	23	0.1639	0.1639
27	30	0.0496	0.0496	27	5	−0.1503	0.1503
28	17	0.0304	0.0304	28	11	−0.0873	0.0873
29	21	0.0137	0.0137	29	17	0.0870	0.0870
30	15	−0.0067	0.0067	30	21	0.0223	0.0223

PC being ranked 19th: most of the PCs for Case Study 11 are very similar in size and so their ability to discriminate is almost random relative to size (for pictorial purposes we will illustrate the effects of classifiers on the first two PCs but in reality there is no good answer as to how many PCs to use for this case study). For Case Studies 1 (forensic) and 4 (pollution) the largest PC is also the most discriminatory PC, thus suggesting that this is the main factor that influences the data; examining the value of the *t*-statistic we see for the forensic dataset that the first PC really does dominate the discrimination, whereas for Case Study 4 there is not such a sharp cut-off. Case Study 8a (mouse urine – diet study) is interesting as the most discriminatory PC is number 2. Although we certainly expect that whether mice are on a diet or not will have a major influence, in fact there are other factors such as when the samples

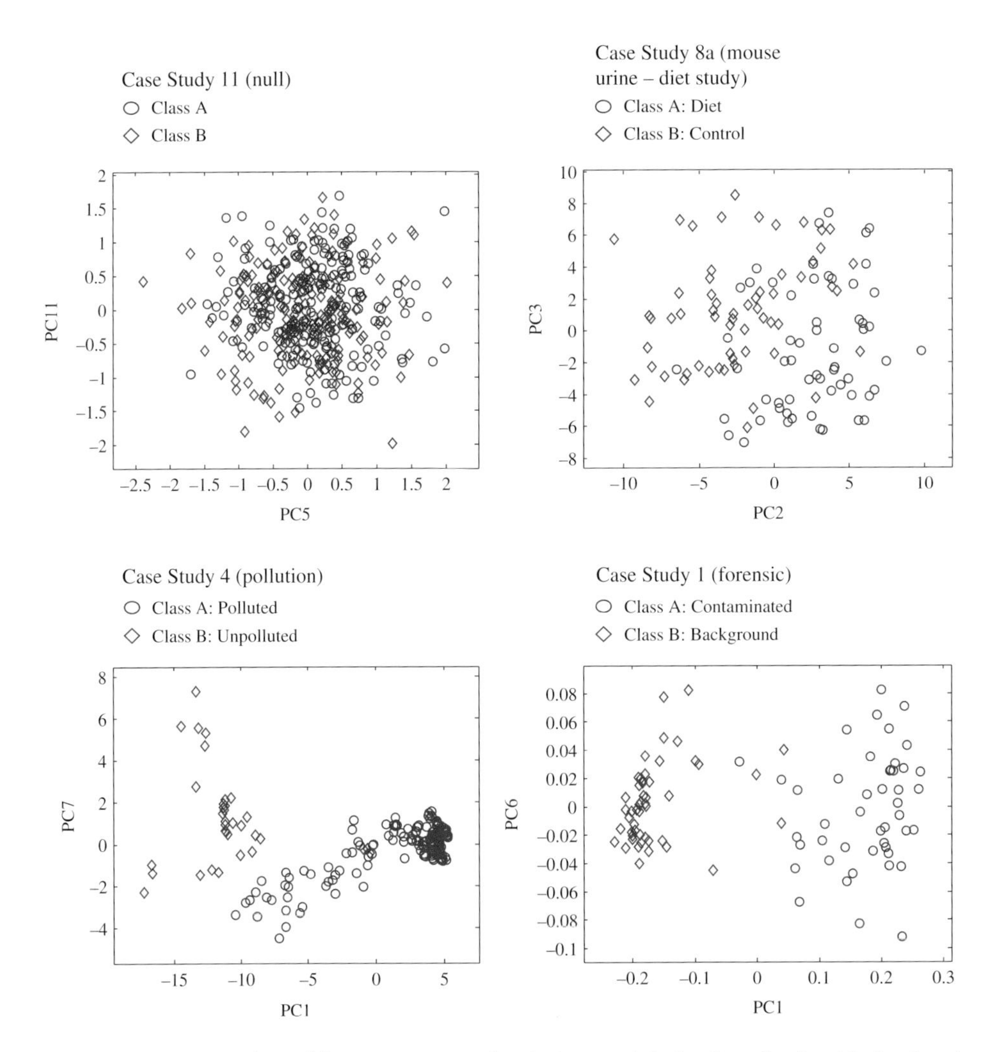

Figure 4.11 Scores plots of the top two most discriminatory PCs for Case Studies 11, 8a, 4 and 1 using the default data preprocessing methods appropriate to each dataset

were run on the instrument that influence the largest PCs as well; PC2 has a t-statistic around twice that of PCs 3 and 1, the next most discriminatory. However it is interesting that the three most discriminatory PCs are also the three largest in size. Note that cross-validation and the bootstrap suggested less than three PCs are required to describe the dataset fully but that was probably because of the 'unscaling' to compute RSS and PRESS in the original units (a comparison of how scaling influences the calculation of the number of significant components is outside the scope of this text). The score plots of the two most discriminatory PCs are presented in Figure 4.11. For Case Study 11 (null) it appears that there is some discrimination; this is by chance that we have picked the two PCs that show a little difference between the classes, and this would lead to a false deduction that there are some differences between the classes. It is important to understand that randomness (which is quite a difficult concept) usually involves some 'clumping' and so we may expect to find PCs which appear discriminatory (by accident). The discrimination of Case Study 8a (mouse urine – diet) is improved a little. However, there is only slight improvements for Case Studies 1 (forensic) and 4 (pollution): this is because PC 1 is already very discriminatory for these two datasets.

An alternative approach for determining which combination of PCs is optimal is stepwise discriminant analysis. There are a number of variations on the theme (it can also be used for choosing the most discriminatory variables), but a simple approach is as follows:

1. Perform PCA on the overall dataset.
2. Select A PCs, so long as A is not too small (typically, for example, 10 PCs). It is not critical that this is the optimal number for fully modelling the data obtained by cross-validation.
3. Test a 1 PC model on each of the A PCs in turn, using the bootstrap or cross-validation to determine the classification error (in this text, we recommend the bootstrap but most classical applications of stepwise discriminant analysis use cross-validation; note that we recommend a variation that involves selecting an equal number of samples from each class in the model in the bootstrap training set as this is a supervised approach – rather than for unsupervised PCA, as in Section 4.3.1).
4. Choose the PC with the minimum misclassification error. If there is a tie use a secondary criterion such as the autopredictive error. This PC will not necessarily be the PC that has the highest magnitude. Denote this PC as r_1 which will be a number between 1 and A.
5. Then using the PC chosen in step 4, repeat steps 3 and 4 but with a 2 PC models, containing PC r_1 and each of the remaining A–1 PCs in turn. Choose the 2 PC model that minimizes misclassification according to the criteria used above.
6. Continue adding PCs in turn until a desired stopping criterion is met, a common one being when the test set error no longer decreases, although this may depend a little on dataset and convergence.

The results of stepwise discriminant analysis using QDA for Case Studies 1, 4, 8a and 11 are listed in Table 4.5 by using the 30 largest PCs and the bootstrap for determining classification errors. These should be compared to Table 4.4. For Case Study 1 (forensic) there is quite good agreement. Four PCs are selected using stepwise discriminant analysis and of these (PCs 1, 4, 11 and 8 in order of importance), PCs 1 and 4 also appear in the top four using the t-statistic. PC1 is by far the largest PC in variance and has by far the largest t-statistic. For Case Study 4 (pollution) there is also good agreement. 13 PCs are selected. Looking at

Table 4.5 *Choosing the most discriminatory PCs using stepwise discriminant analysis using QDA . Only the ranks of the PCs selected are listed*

Case Study 11 (null)	6 PCs selected	Case Study 8a (mouse urine – diet study)	9 PCs selected	Case Study 1 (forensic)	4 PCs selected	Case Study 4 (pollution)	13 PCs selected
Rank (size)	Rank (discriminatory)	Rank (size)	Rank (discriminatory)	Rank (size)	Rank (discriminatory)	Rank (size)	Rank (discriminatory)
1	—	1	5	1	1	1	1
2	—	2	1	2	—	2	2
3	—	3	3	3	—	3	11
4	—	4	—	4	2	4	3
5	—	5	6	5	—	5	4
6	—	6	—	6	—	6	—
7	3	7	—	7	—	7	10
8	—	8	—	8	4	8	6
9	—	9	7	9	—	9	5
10	—	10	—	10	—	10	8
11	—	11	—	11	3	11	13
12	—	12	8	12	—	12	9
13	5	13	4	13	—	13	7
14	—	14	—	14	—	14	—
15	—	15	—	15	—	15	—
16	—	16	—	16	—	16	—
17	—	17	9	17	—	17	—
18	4	18	—	18	—	18	—
19	—	19	—	19	—	19	—
20	6	20	—	20	—	20	—
21	2	21	—	21	—	21	—
22	—	22	—	22	—	22	—
23	—	23	—	23	—	23	—
24	—	24	—	24	—	24	—
25	—	25	2	25	—	25	—
26	—	26	—	26	—	26	—
27	1	27	—	27	—	27	—
28	—	28	—	28	—	28	12
29	—	29	—	29	—	29	—
30	—	30	—	30	—	30	—

the *t*-statistic, although the first PC has a much larger *t*-statistic than the next 12 PCs, the change in size is gradual and PC1 is not so overwhelmingly large. Of the PCs chosen to be discriminatory by stepwise discriminant analysis (1, 2, 4, 5, 8, 9, 13, 10, 12, 7, 3, 28 and 11), eight are within the top thirteen for the *t*-statistic, and seven out of the eight variables with the largest *t*-statistic are also within the group of discriminators as assessed by stepwise discriminant analysis. For Case Study 8a (mouse urine – diet study) there is a suggestion that nine PCs are discriminatory (2, 25, 3, 13, 1, 5, 9, 12, 17). Six of these also are within the nine PCs that have the highest *t*-statistic, including six out of the top seven: several PCs with relatively large variances, such as 4, 7 and 8, make neither list. For Case Study 11 (null) there is almost no correspondence between the two lists, with only PC 21 making it into the top six of the *t*-statistic as well as the PCs selected by stepwise discriminant analysis, thus suggesting that the choice of PCs is almost completely random and differs according to the criterion chosen.

Both the approaches discussed in this section will select the PCs that appear most discriminatory, rather than the largest PCs. A problem arises here that these PCs could then bias the classification model and so this should only be done on the test set if it is an aim of the study to validate the model correctly as discussed in Chapter 8. If selected using both the training and test set together it can lead to artificially high %CCs which have little meaning even if the correct validation splits of test and training sets are performed subsequently, although it is still useful for visualization and if it is know that there is a relationship between the classifier and the analytical data it is likely to improve classification ability. In Table 4.6 we tabulate the classification abilities using Linear Discriminant Analysis (LDA) (Section 5.3) on test and training sets for Case Study 11 (null) when selecting PCs according to their apparent discriminatory power over the entire dataset. We see that the %CC for the test set is consistently above 50 %, in some cases almost 60 %. In contrast, selecting PCs according to size gives 51.13 % for the test set when using a 2 PC model and 49.63 % for 10 PCs (these numbers vary slightly according to the random split of data – see Section 8.4.2 – as the test set results are based on the average of 100 iterations), which is much closer to the expected 50 %. The PCs that are chosen are only weakly discriminating and so do not offer a very major advantage (unlike PLS as discussed in Section 4.3.3) but in this case do lift the overall apparent classification rates by over 5 % – in other cases this could be more – and the PC scores appear to show some small separation when it is not there. Hence selecting discriminatory PCs and then performing classification on these can be dangerous if done inappropriately.

4.3.3 Partial Least Squares Discriminatory Analysis Scores

An alternative is to perform PLS-DA on the data and choose as input to the classifier, the most significant PLS components. We will discuss this method in more detail in Section 5.5, but unlike PCA, PLS is a supervised method and the PLS components are new variables that include information about the classifier. Instead of PC scores we use PLS scores.

This has led to a number of methods in the literature where PLS is first used as a data reduction method and then a second step involves classification, e.g. PLS-LDA which involves first using PLS, picking the top PLS components and then performing LDA on

Table 4.6 *Classification ability for LDA (Linear discriminant analysis) when (a) the most discriminatory PCs (using the t-statistic) have been chosen over the entire dataset for Case Study 11 (null) and (b) the largest PCs by size are chosen*

(a)

Number of components selected	Training (%CC)			Test (%CC)		
	Overall	Class A	Class B	Overall	Class A	Class B
1	55	55.5	54.5	54.63	54.75	54.5
2	56.13	53.5	58.75	55	53	57
3	54.63	54.75	54.5	52.88	52.75	53
4	55.38	54.75	56	52.75	52	53.5
5	57	56	58	54.88	54.75	55
6	57.75	54.5	61	54.38	52	56.75
7	60.75	61.25	60.25	54.38	53.25	55.5
8	59.88	60.25	59.5	56.63	58.25	55
9	61	60	62	57	57.75	56.25
10	61.88	64.25	59.5	55.63	56.25	55

(b)

Number of components selected	Training (%CC)			Test (%CC)		
	Overall	Class A	Class B	Overall	Class A	Class B
1	51.25	53.5	49	51.13	53.25	49
2	52.75	54.75	50.75	49	51.5	46.5
3	55	56.25	53.75	49.63	50.5	48.75
4	55.5	56	55	44.63	45	44.25
5	56.38	55.25	57.5	51.33	53.5	49.25
6	57	58.25	55.75	51.13	52	50.25
7	57.5	58	57	49.63	52	47.25
8	57.88	60.5	55.25	51.25	53.75	48.75
9	58.63	59.5	57.75	50.13	52	48.25
10	57.5	59	56	49.63	51	48.25

these. The problem here is that the input to the classifier has already incorporated information about the origins of samples and is not therefore already prejudiced, if PLS is performed on the entire dataset (including the test and training set). Whereas this can lead to apparent improvement in %CCs, if the training set has been included in the computation of PLS scores, the resultant classification has little meaning. As an example we will consider Case Study 11 (null), where ideally the test set %CC should be close to 50 %. In Table 4.7 we list the %CCs for an increasing number of PLS components using these as inputs to three common classifiers, EDC, LDA and QDA (see Chapter 5 for the formal description of these methods). These components are selected by size, as the most significant components should be most diagnostic of the classifier, since the larger the PLS component, then the more important it is for distinguishing groups (which is not necessarily true for PCA). An iterative strategy of forming 100 test and training set splits, as

Table 4.7 *%CC for training and test sets when PLS components are selected on the entire dataset as input to the classifier with test and training set %CCs for three common classifiers (EDC, LDA and QDA), for Case Study 11 (null dataset)*

PLS-EDC

	Training set			Test set		
PLS comp	Overall	Class A (%CC)	Class B (%CC)	Overall	Class A (%CC)	Class B (%CC)
2	71.00	69.00	73.00	70.50	68.50	72.50
3	70.50	69.00	72.00	70.00	68.50	71.50
4	71.50	70.50	72.50	69.00	68.00	70.00
5	71.00	71.00	71.00	69.75	70.00	69.50
6	71.50	71.50	71.50	69.75	69.50	70.00
7	72.00	71.50	72.50	69.50	69.00	70.00
8	72.00	71.50	72.50	69.00	69.00	69.00
9	72.00	71.50	72.50	69.25	68.00	70.50
10	72.25	71.00	73.50	67.75	66.50	69.00

PLS-LDA

	Training set			Test set		
PLS comp	Overall	Class A (%CC)	Class B (%CC)	Overall	Class A (%CC)	Class B (%CC)
2	70.25	69.00	71.50	69.75	68.00	71.50
3	70.75	70.00	71.50	70.00	69.00	71.00
4	71.25	72.00	70.50	69.50	69.50	69.50
5	71.00	71.50	70.50	70.50	71.00	70.00
6	71.75	72.00	71.50	70.00	70.50	69.50
7	72.25	72.00	72.50	69.75	70.00	69.50
8	71.75	72.50	71.00	69.25	70.50	68.00
9	72.00	72.50	71.50	70.00	70.50	69.50
10	72.00	72.50	71.50	69.00	69.00	69.00

PLS-QDA

	Training set			Test set		
PLS comp	Overall	Class A (%CC)	Class B (%CC)	Overall	Class A (%CC)	Class B (%CC)
2	70.75	69.00	72.50	69.00	66.50	71.50
3	71.25	69.00	73.50	69.25	67.50	71.00
4	72.00	66.00	78.00	69.50	64.50	74.50
5	72.75	67.00	78.50	68.75	64.00	73.50
6	73.25	67.50	79.00	67.75	62.50	73.00
7	74.50	68.00	81.00	68.50	62.00	75.00
8	74.00	71.00	77.00	66.75	64.00	69.50
9	76.50	73.00	80.00	67.25	63.50	71.00
10	76.50	71.50	81.50	64.00	58.50	69.50

discussed in Section 8.4, is employed to give an average %CC. The effect of selecting the most significant PLS components is dramatic and lifts the classification ability for Case Study 11 from an expected 50 % to nearly 70 %, even on the test set. Only when many components are selected does this start to decline a little, as later PLS components may not be very relevant to the classifier. In order to illustrate what is happening, we illustrate the scores of the first two PLS components for this case study, together with the autopredictive boundary in Figure 4.12. It can be seen that the PLS apparently separates out the two groups, even though there is no underlying difference, because it creates a combined variable that includes the classifier. The three classification methods create a boundary between the groups which apparently suggests some discrimination. This demonstrates that methods for data reduction can often play a much greater role on the apparent classification abilities than the classification algorithm itself, unless great care is taken.

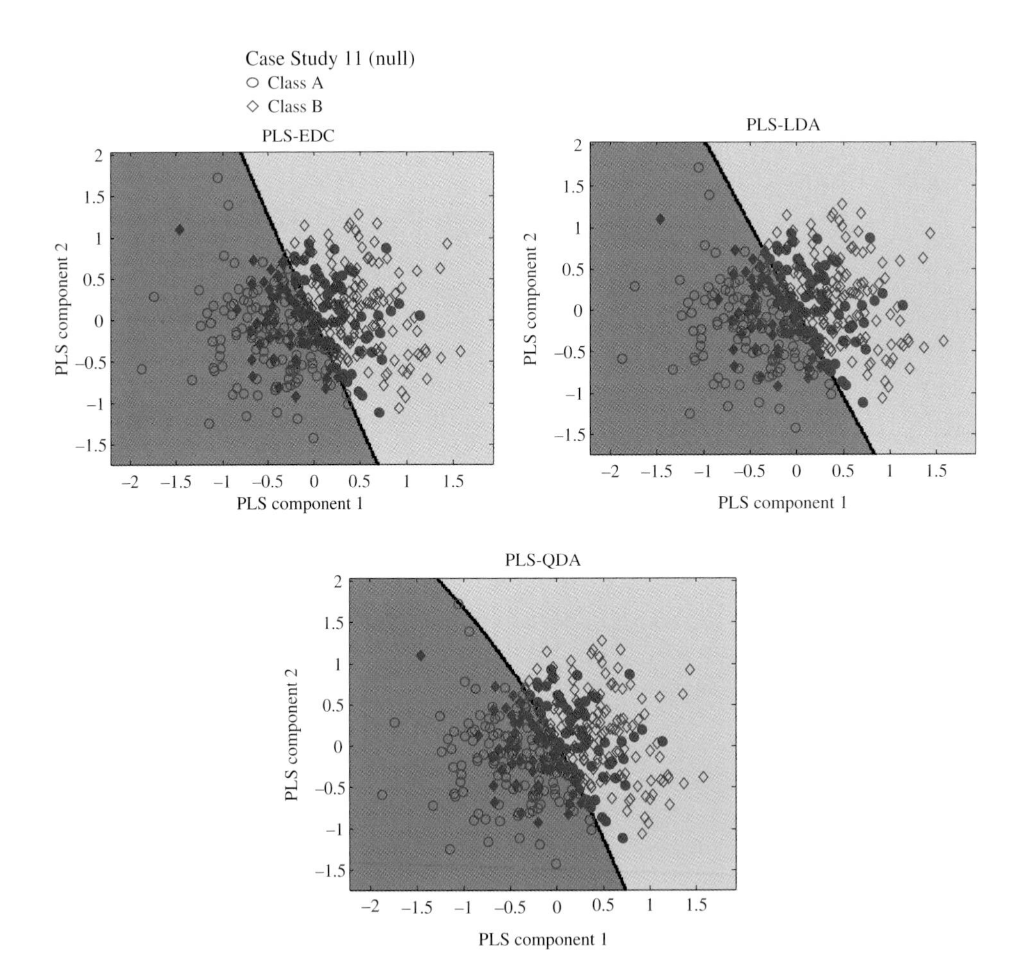

Figure 4.12 Autopredictive results of PLS followed by three common classification methods, represented graphically, for Case Study 11 (null dataset). Open circles represent correctly classified samples and filled circles misclassifications

It is, however, still possible to select PLS components as inputs to a classifier, and safe results can be obtained, but only if this is done on the training set (the samples used to form a model) and not on the test set (the samples used to determine how well the model performs). The results of this alternative strategy are presented in Table 4.8 as applied to

Table 4.8 *%CC for training and test sets when PLS components are selected on just on the training test set as input to the classifier with test and training set %CCs for three common classifiers (EDC, LDA and QDA), for Case Study 11 (null dataset)*

PLS-EDC

	Training			Test		
PLS comp	Overall	Class A (%CC)	Class B (%CC)	Overall	Class A (%CC)	Class B (%CC)
2	81.25	82.50	80.00	48.75	45.50	52.00
3	81.00	80.50	81.50	49.75	38.00	61.50
4	83.00	84.00	82.00	51.75	48.50	55.00
5	82.00	80.00	84.00	48.25	43.00	53.50
6	83.75	83.00	84.50	46.75	50.00	43.50
7	83.50	84.50	82.50	47.25	51.00	43.50
8	82.50	83.00	82.00	48.00	48.50	47.50
9	83.75	83.00	84.50	49.50	36.50	62.50
10	83.50	82.50	84.50	48.00	48.00	48.00

PLS-LDA

	Training			Test		
PLS comp	Overall	Class A (%CC)	Class B (%CC)	Overall	Class A (%CC)	Class B (%CC)
2	83.00	84.50	81.50	49.25	43.50	55.00
3	82.75	83.00	82.50	49.00	55.50	42.50
4	84.50	84.00	85.00	52.00	51.50	52.50
5	83.75	82.50	85.00	46.50	45.00	48.00
6	84.50	84.50	84.50	48.00	40.00	56.00
7	84.00	84.00	84.00	43.00	32.50	53.50
8	84.00	84.00	84.00	48.50	38.50	58.50
9	85.25	85.50	85.00	46.50	44.00	49.00
10	83.75	83.50	84.00	44.75	44.50	45.00

PLS-QDA

	Training			Test		
PLS comp	Overall	Class A (%CC)	Class B (%CC)	Overall	Class A (%CC)	Class B (%CC)
2	81.50	83.50	79.50	49.25	23.50	75.00
3	82.50	82.00	83.00	49.25	29.00	69.50
4	84.25	84.50	84.00	46.25	30.00	62.50
5	84.75	84.00	85.50	45.50	28.00	63.00
6	88.00	88.00	88.00	48.00	24.50	71.50
7	87.25	87.00	87.50	45.00	33.00	57.00
8	89.00	89.00	89.00	47.00	9.50	84.50
9	90.25	90.50	90.00	48.25	23.00	73.50
10	90.00	89.00	91.00	47.50	20.00	75.00

Case Study 11 (null) where we see that the overall %CC is much closer to 50 % in all cases (for QDA there is a big difference in how each class is modelled, which will be discussed in Chapter 5, but the average is nevertheless around 50 %). Hence although PLS scores can legitimately be employed as inputs to the classifier, in order not to obtain misleading results they must not be computed including the test set. In this way the use of PLS as a data reduction tool contrasts to PCA where calculating and using the PCs with highest variance (Section 4.3.1) as input to the classifier is acceptable without any undesirable side effects, and in some cases such as LDA and QDA, may be necessary, if the number of variables exceeds the number of samples.

4.4 Strategies for Data Preprocessing

In this text we will adopt a variety of methods for data preprocessing according to the method we wish to illustrate. There are very few basic methods in chemometrics and the differences between the results obtained often depends on how the data are preprocessed, especially when performing classification. The decisions can be quite complex when data are split into different groups such as test and training sets, and the training set is further split, e.g. via the bootstrap. In this section we will define the basic methods for data scaling and reduction, primarily in the context of preparing the data as inputs to a classifier, in the form of flow charts. In subsequent chapters, for clarity we will reference these named flow charts where appropriate and we feel there may be ambiguities so that it is clear to the reader how the data were prepared. Readers may wish to refer to this section when reading subsequent chapters of the text. It is important to recognize that there are a large variety of possible ways of preprocessing data and we describe, for brevity, only the main approaches we employ in this text, although other approaches could legitimately be conceived and in some cases are illustrated.

4.4.1 Flow Charts

The main flow charts are summarized in Figure 4.13.

- There are four levels of flow chart, each defining a process.
- Every method of calculation is initiated by a process defined by a flow chart at the highest level (1).
- Flow charts consist of a series of tasks (or computer instructions) that may call other processes represented by flow charts (often in the form of procedure calls). In our scheme, a flow chart can only call a process defined by a lower level flow chart, never the same or higher level.
- Each flow chart is denoted by a number and a letter. The number refers to the level and the letter to the type of flow chart; there is no relationship between letters at different levels, and so, for example, flow chart 1D has no specific relationship to flow charts 2D or 3D.
- There are several possible routes from one level to another; in no case is there a unique route from one level to another; for example, a flow chart will either be called from more than one flow chart of a higher level or call more than one flow chart at a lower level.

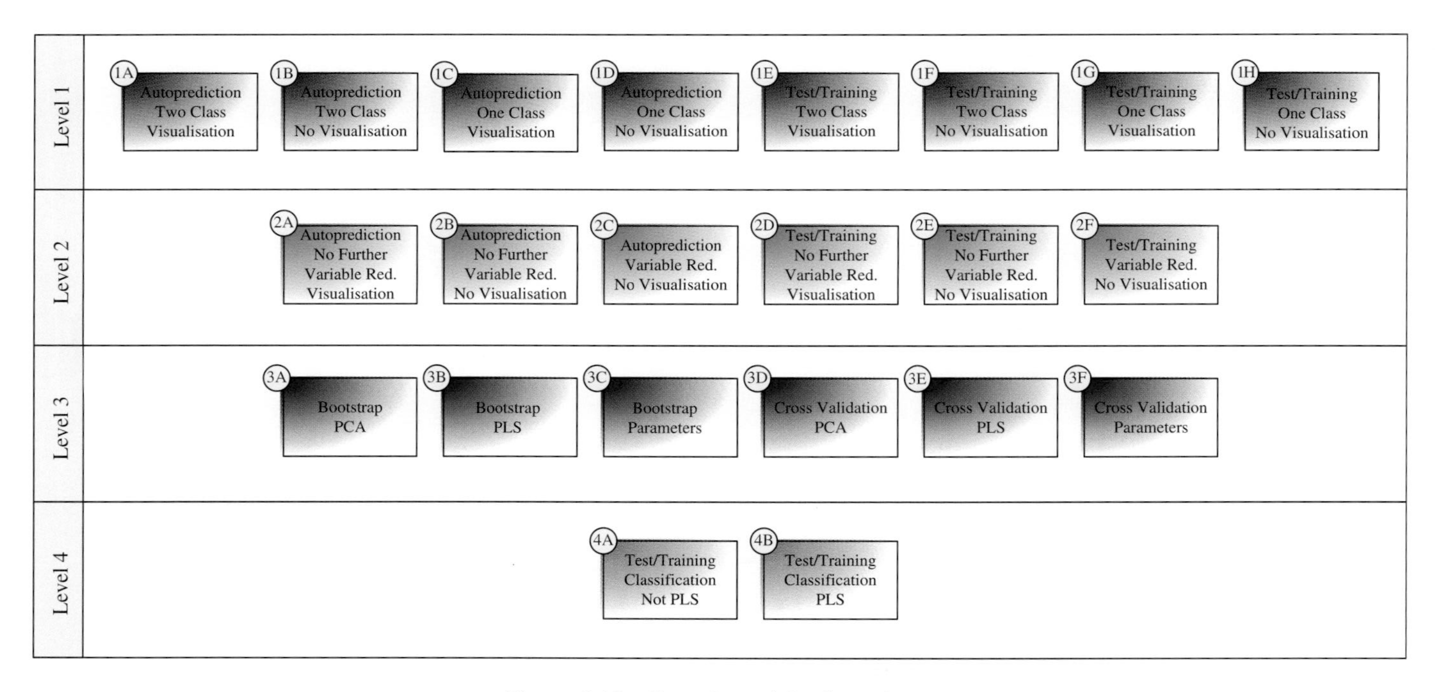

Figure 4.13 Overview of the flow charts

Symbols in flow charts have formal meanings; the main symbols that we use are defined in Figure 4.14. A higher level flow chart 'calls' a lower level, i.e. one using the 'Predefined Process' symbol.

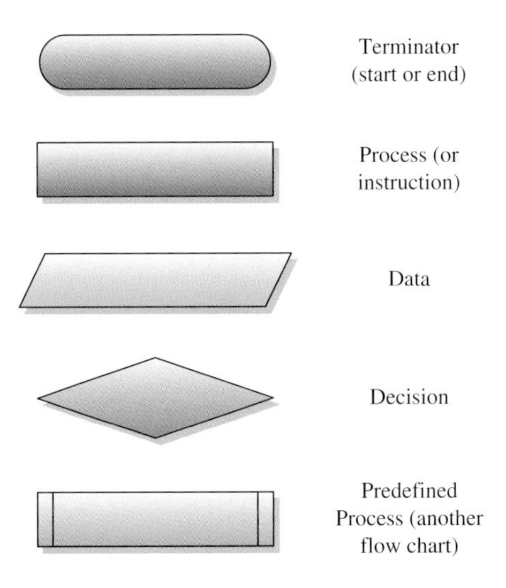

Figure 4.14 Main flow chart symbols

It is good practice not to make flow charts too complicated; around 20 boxes is approximately the maximum that can be taken in by visual inspection, and so it is preferable to split these into a number of smaller flow charts than have few very large ones that are hard to comprehend.

A typical summary calculation is illustrated in Figure 4.15. This involves starting with the process defined by flow chart 1F (test/training set prediction using a two class classifier without visualization) which calls 2F (variable reduction, e.g. by PCA) which then calls in sequence 3C (bootstrap to optimize parameters in a classifier) and 4A (the classifier which is not PLS-DA). These summaries will be placed near tables and figures in subsequent chapters where it is important to indicate how a result was obtained, for clarity. The flow charts primarily relate to the method of data scaling employed at different steps of the procedure. Not all tables and figures are appropriately illustrated by these symbols which we will use

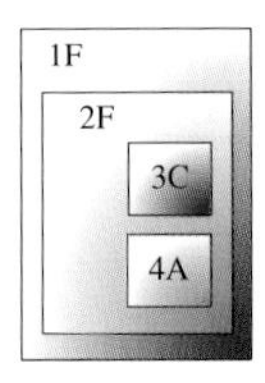

Figure 4.15 A typical calculation

primarily where it is felt useful to clarify the preprocessing steps where there may be ambiguity or it saves describing minor adjustments to algorithms in the text. These are primarily used for supervised classification. Some of the simpler, exploratory, methods, such as those of Chapter 3, can be understand without this detail. The motivation behind these flow charts can best be understood by reference to later chapters, especially Chapter 8 on validation and optimization, and represent detailed recommendations for data scaling, especially along the columns, under various different circumstances. For readers interested in general principles, rather than reproducing results computationally, some of the discussion below may be unnecessary.

4.4.2 Level 1

Level 1 consists of 8 flow charts, 4 for autoprediction and 4 when the aim is to test a model by splitting into test and training sets. These are further divided into two class classifiers (Chapter 5) and one class classifiers (Chapter 6). Finally we will illustrate the application of these classifiers using 2 dimensional PC scores plots, for visualization as to how the classes are separated and to illustrate boundaries and confidence limits. The detailed flow charts are illustrated in Figure 4.16.

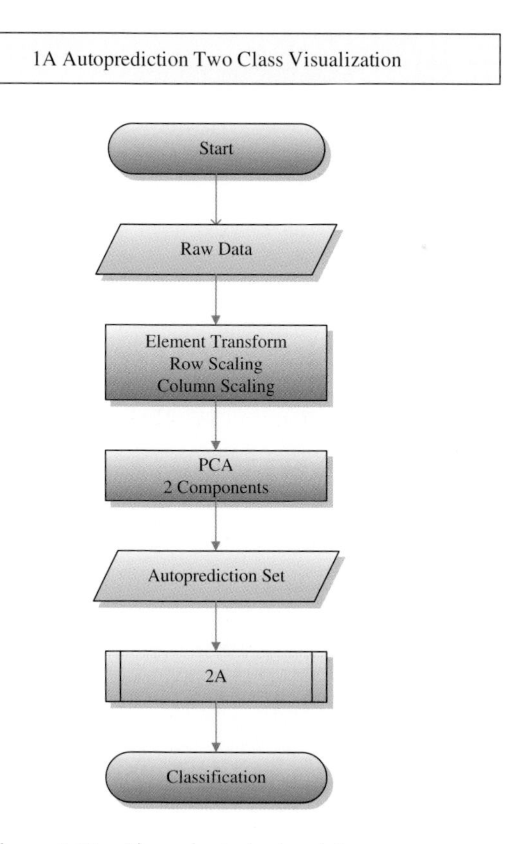

Figure 4.16 Flow charts for level 1 processes

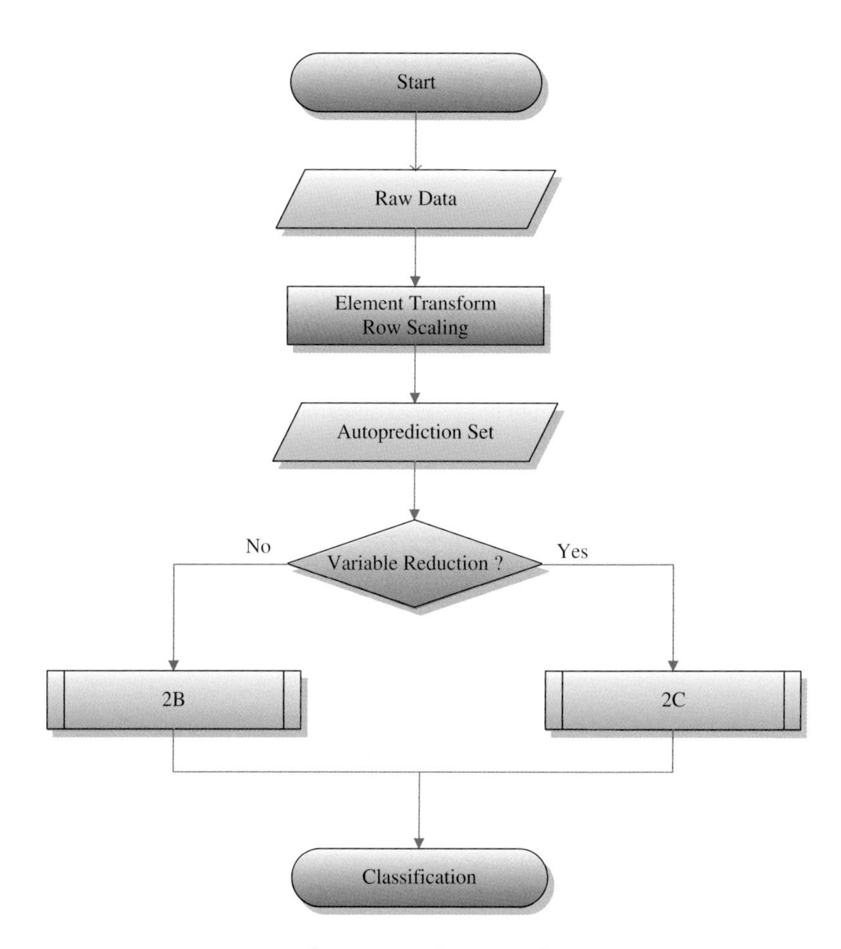

Figure 4.16 (continued)

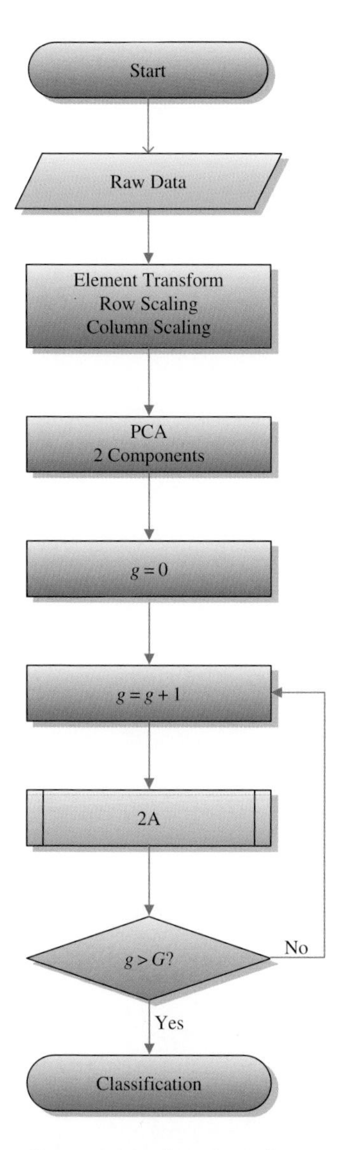

Figure 4.16 (continued)

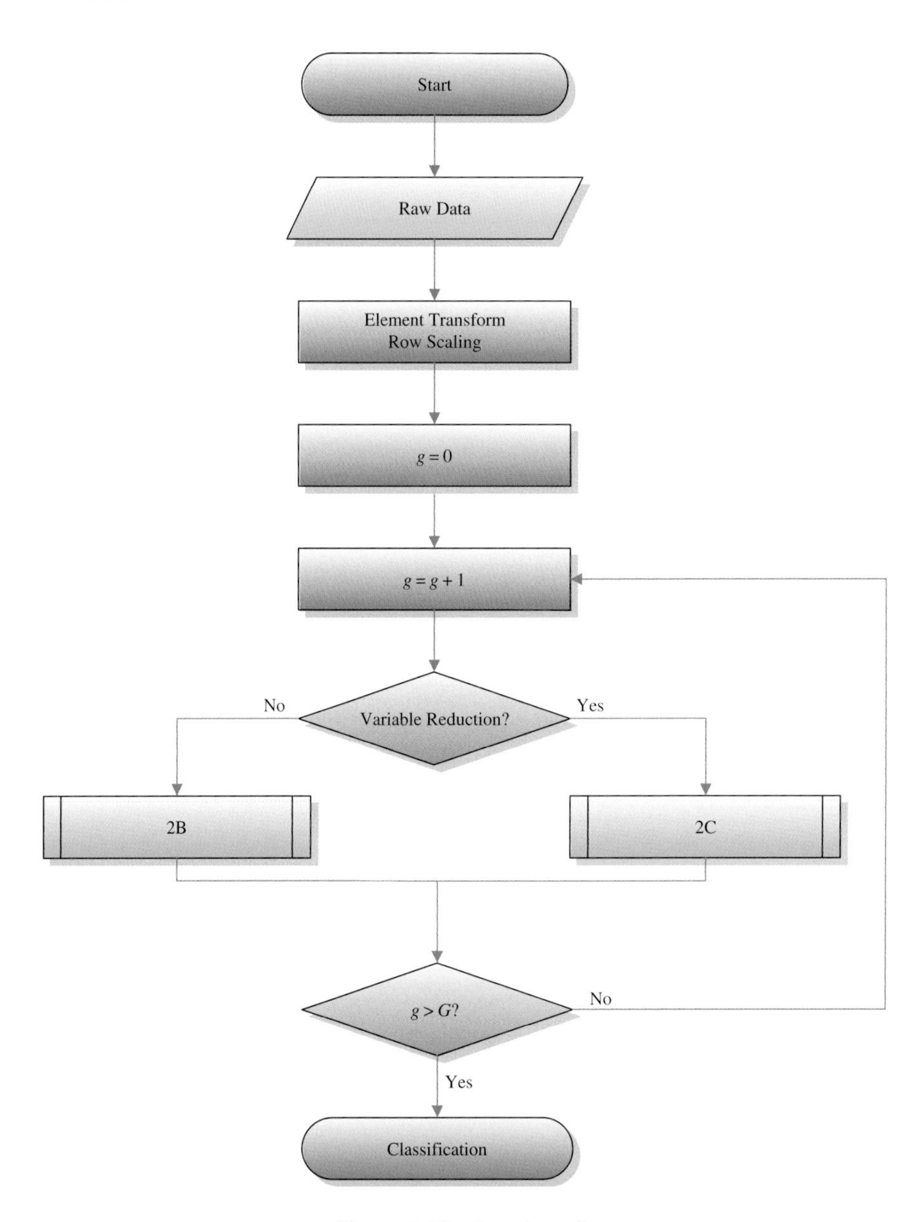

Figure 4.16 (continued)

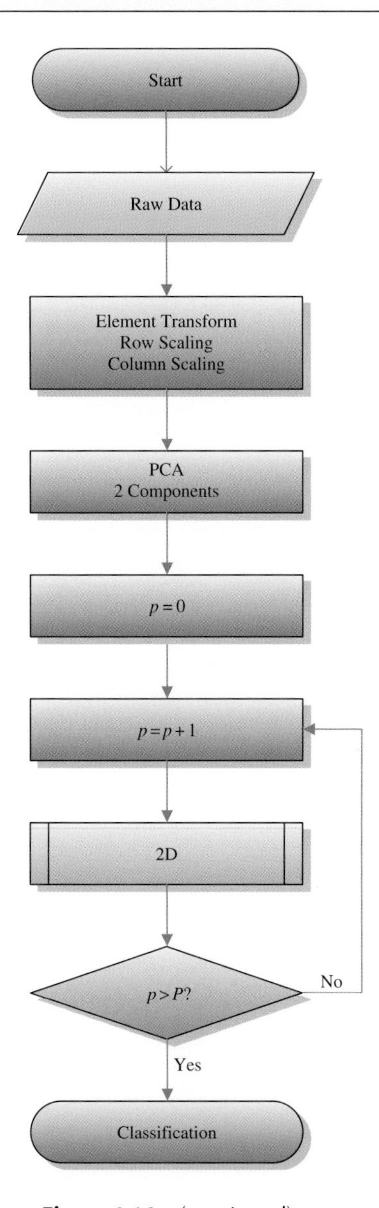

Figure 4.16 (continued)

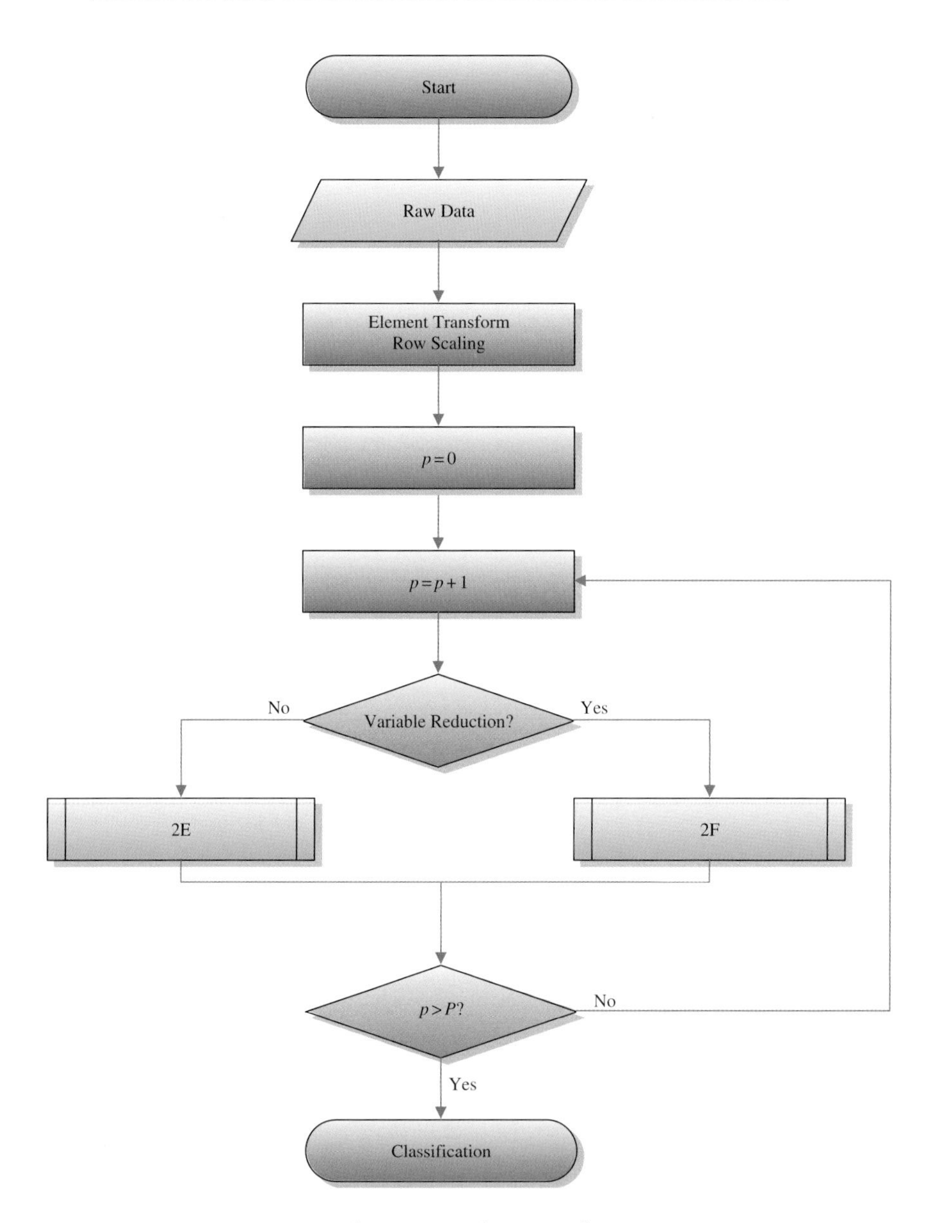

Figure 4.16 (continued)

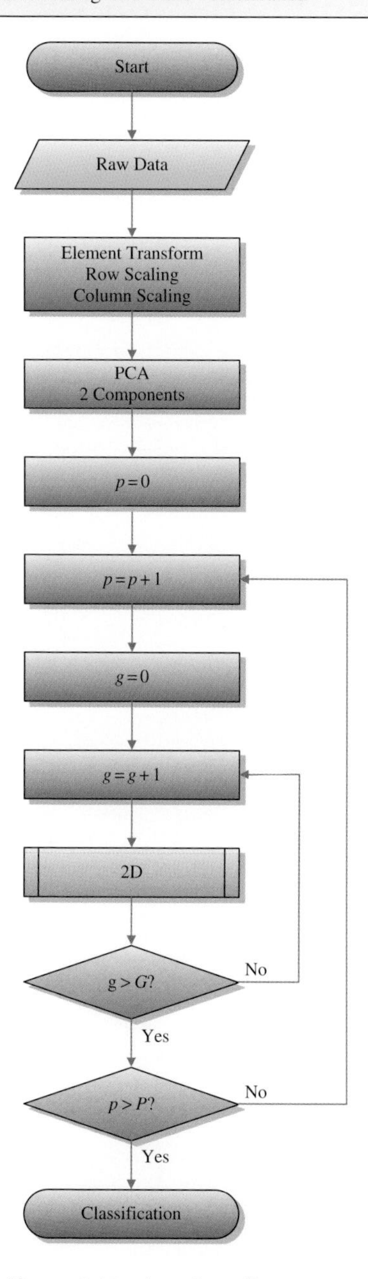

Figure 4.16 (continued)

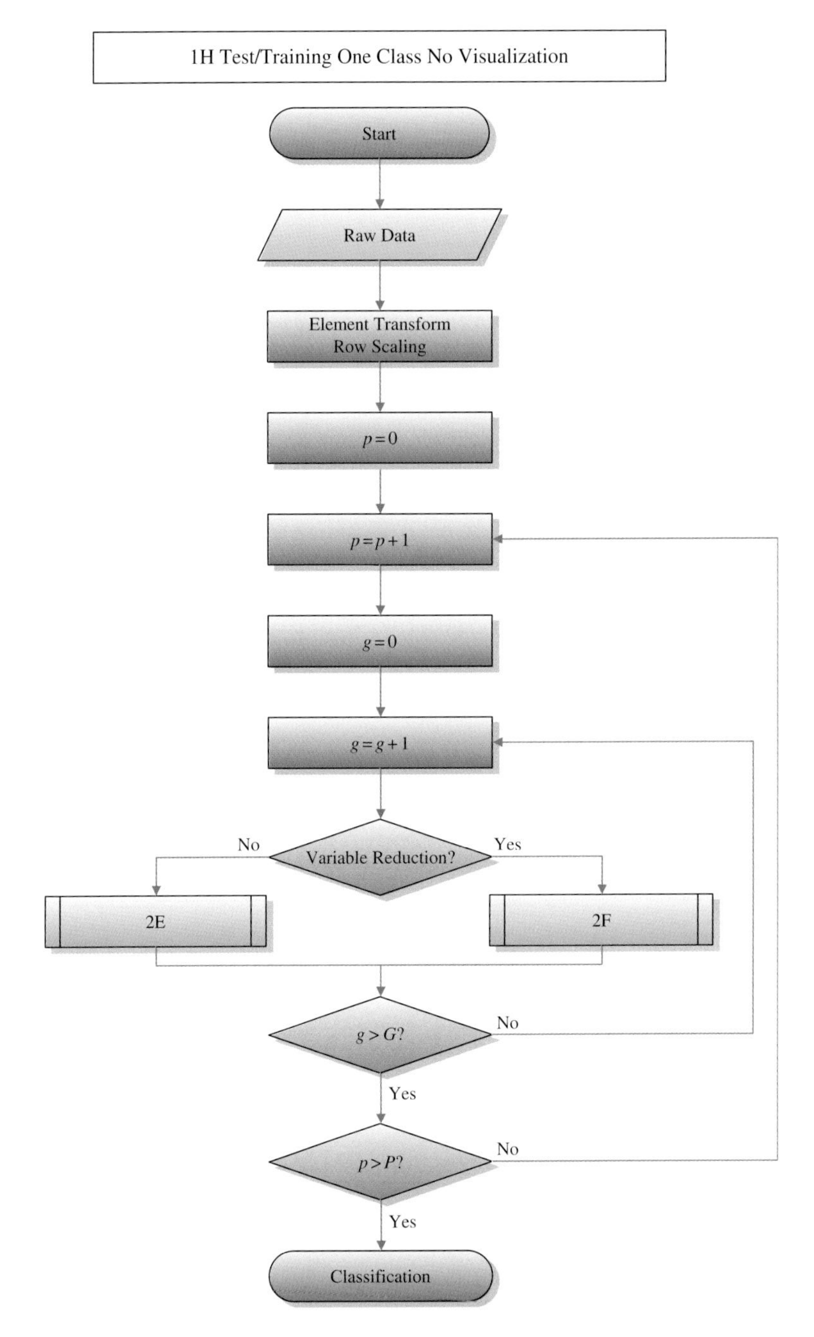

Figure 4.16 (continued)

For flow charts 1A, 1C, 1E and 1G, data are initially reduced to the scores of the first two PCs, so that they can be visualized graphically. Note that other approaches of visualization are possible, for example, choosing two of the original variables or two PLS components, but in most cases in this text we use the scores of the first two PCs; however this is not mandatory and primarily for illustrating how different methods work graphically. Flow charts 1B, 1D, 1F and 1H are more usual when trying to predict the origin of a sample except in rare cases where the data are best represented by 2 PCs.

In the case of one class classifiers (1C, 1D, 1G and 1H) a model can be formed separately for each of the G classes (Chapter 6) if desired. For two class classifiers, a single model is formed, dividing the data into two groups (Chapter 5). When validating models using a test set (1E, 1F, 1G, 1H), we recommend that the data are repeatedly split into training and test sets P times, in each case involving a different random split of the data. Conventionally only one test set is formed but in this book we advocate that this procedure is repeated many times; by default we use $P = 100$ but we discuss the effect of changing this value in Section 8.4.2. When there is no visualization, models can either be performed on the raw data without further variable reduction (although for LDA and QDA the number of variables must be less than the number of samples) or else variables can be reduced by PCA, as discussed in Section 4.3.1.

Note that when there are several classes, one either uses a series of one class classifiers in which case the data are scaled separately according to each class, or an extension of two class classifiers, in which case the entire data are scaled using information from all samples included in the model building, which will come from more than one group. Chapters 6 and 7 discuss these situations.

4.4.3 Level 2

The level 2 flow charts are presented in Figure 4.17. They are called from level one, the 'start' being the call from this higher level.

There are three for autoprediction (2A, 2B and 2C) and three for validation by splitting into test and training sets (2D, 2E and 2F). When visualizing data no further data reduction is possible. The reason is that all calculations are performed, usually using the scores of PCs 1 and 2 as inputs to the classifier, and in order to visualize, for example, different class boundaries it is not possible for example to rescale the data each time a new training set is obtained by splitting the data.

When PLS-DA is performed we recommend using class weighted centring (Section 4.2.3); the importance of this will be discussed in Section 5.5.3, although this is not a formal requirement of the algorithm. We will show that if there are different numbers of samples in each class without this, the PLS-DA prediction boundary will be biased towards the smaller class using a default decision threshold of $c = 0$. However when the data are split into test and training sets, we recommend equal numbers of samples in each class for the training set and so there is no difference (in the case of a two class classifier) between normal centring and weighted centring.

Flow charts 2E and 2F are complicated by the fact that the test set should be column scaled according to the training set column parameters (the mean if just mean centring or the mean and standard deviation if standardizing; other approaches such as range scaling are also feasible). For autoprediction this is not a necessary consideration, and when visualizing the test/training set boundaries using flow chart 2D we want all the data to be on the same scale each time a boundary is formed.

In the case of the processes represented by flow charts 2C and 2F the raw data are reduced using PCA prior to input to the classifier. Either an optimum number of PCs is determined or the number *N* is fixed in advance. For these flow charts we will distinguish the number of principal components (*N*) from the number of PLS components (*M*).

4.4.4 Level 3

The level 3 flow charts are represented in Figure 4.18. The processes they represent are all related to optimization which is discussed in more detail in Chapter 8.

Flow charts 3A, 3B and 3C refer to the bootstrap which is an iterative way of splitting the training set further into a bootstrap training and bootstrap test set. The method is discussed in the context of optimizing the number of PLS components for the PLS-DA classifier in

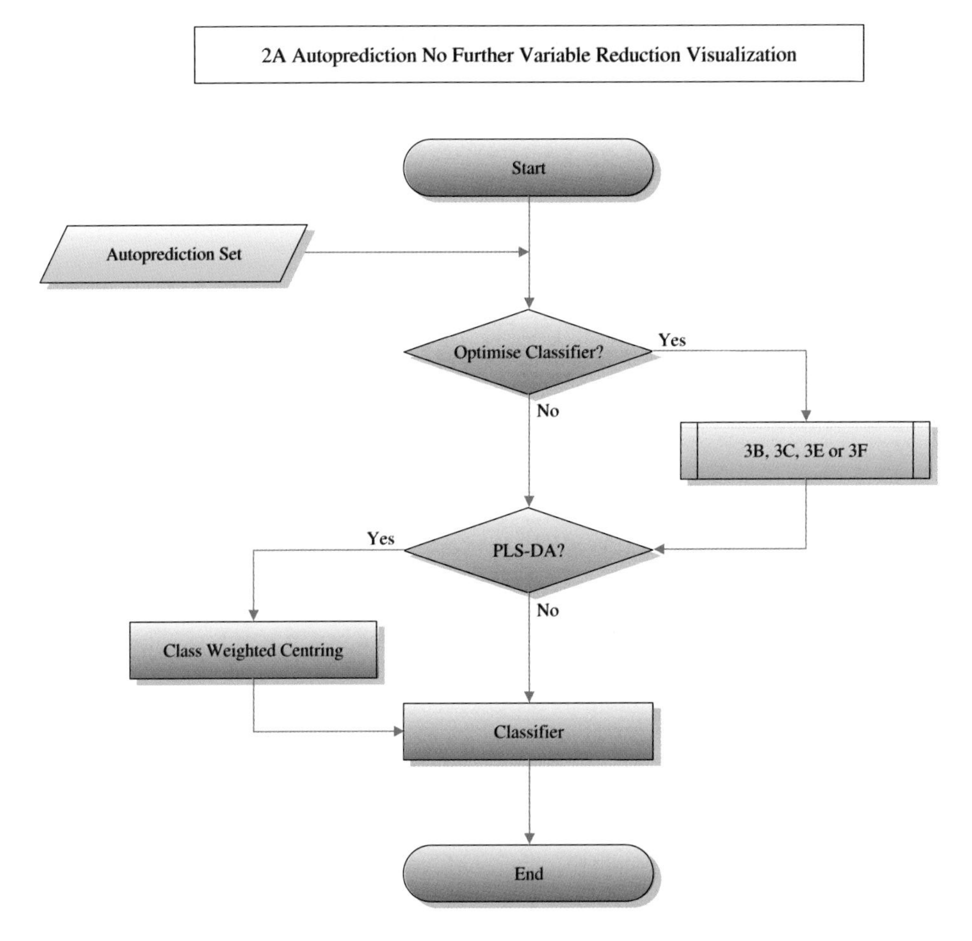

Figure 4.17 Flow charts for level 2 processes

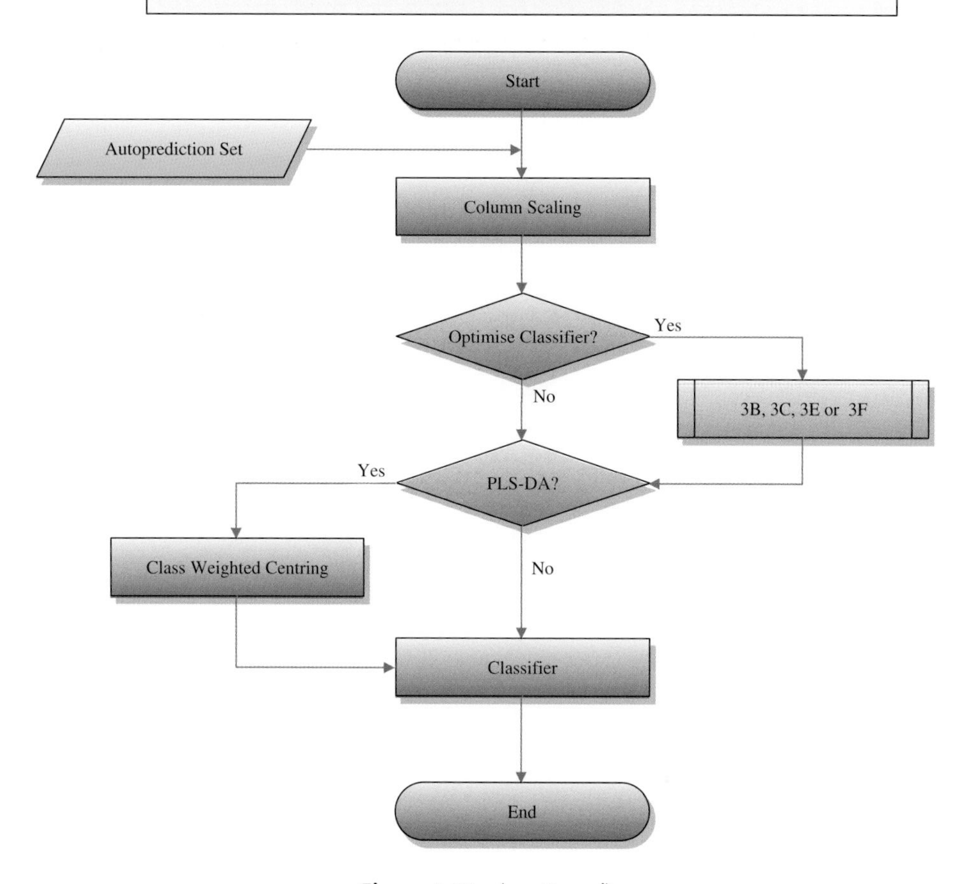

Figure 4.17 (continued)

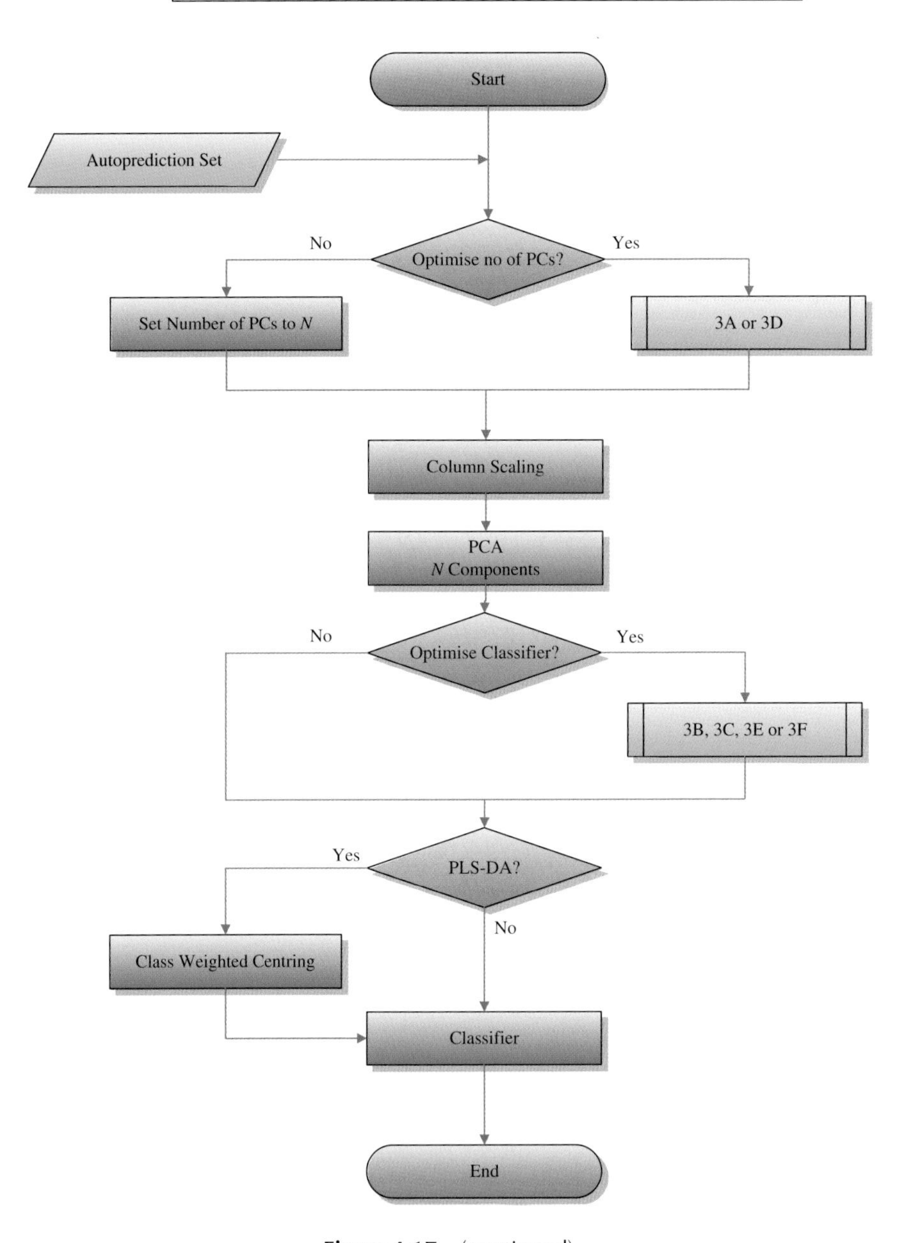

Figure 4.17 (continued)

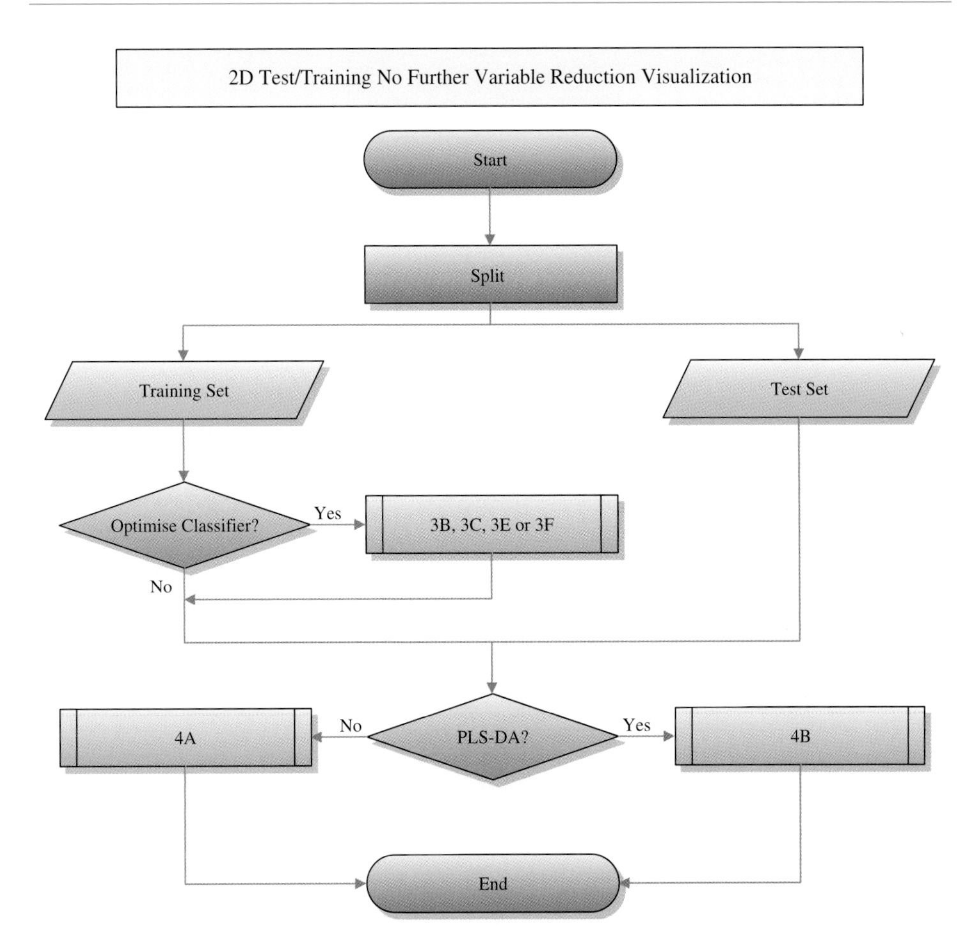

Figure 4.17 (continued)

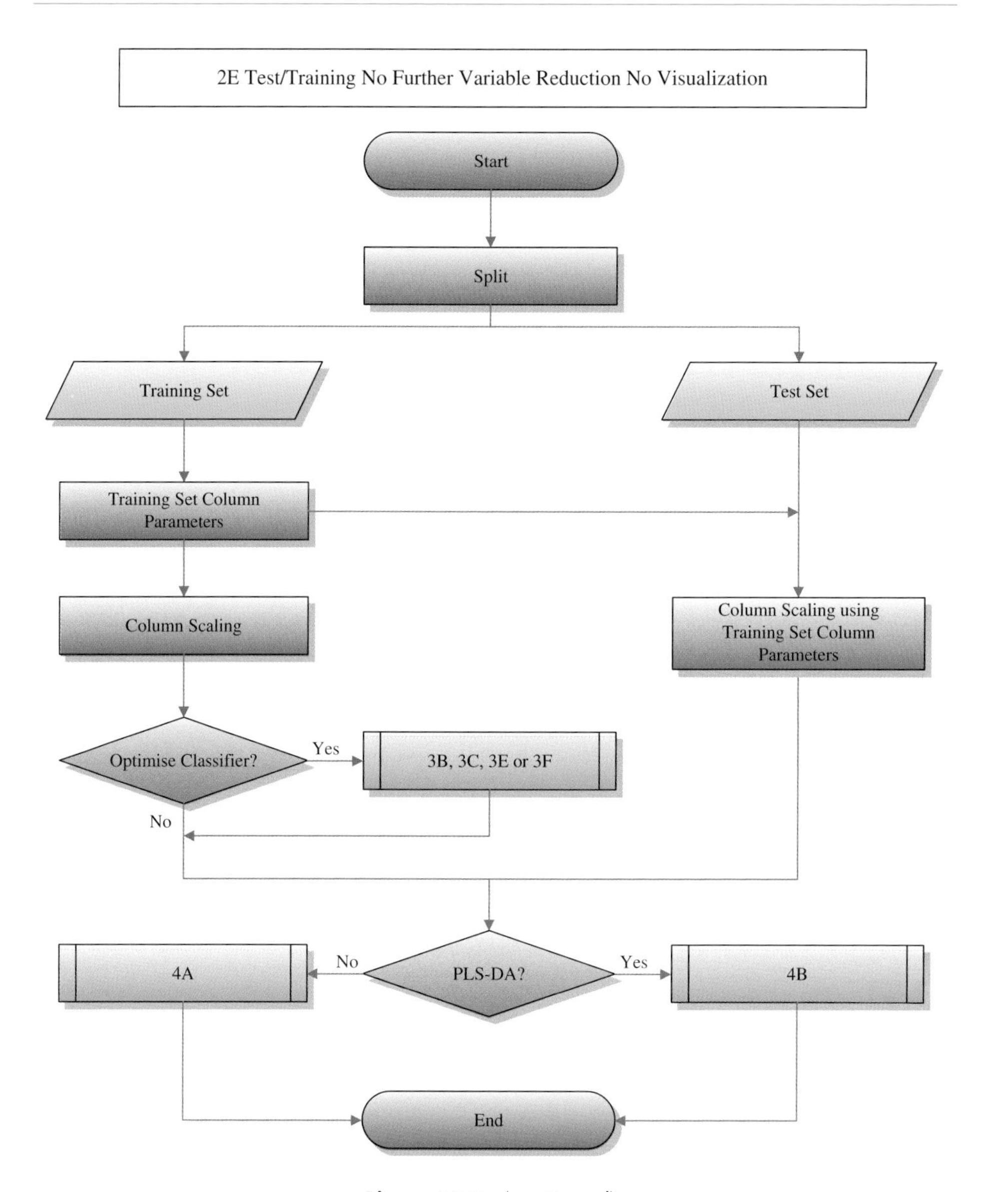

Figure 4.17 (continued)

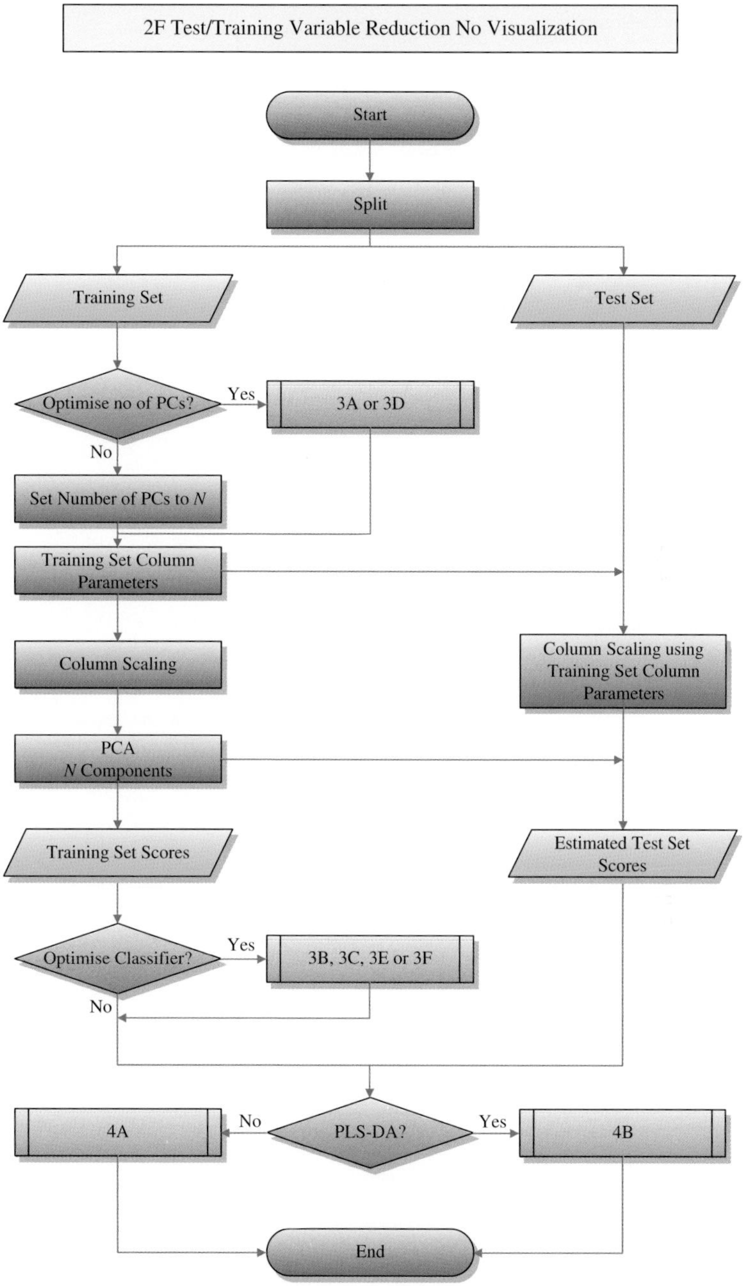

2F Test/Training Variable Reduction No Visualization

Start

Split

Training Set

Test Set

Optimise no of PCs?

Yes

3A or 3D

No

Set Number of PCs to *N*

Training Set Column Parameters

Column Scaling

Column Scaling using Training Set Column Parameters

PCA
N Components

Training Set Scores

Estimated Test Set Scores

Optimise Classifier?

Yes

3B, 3C, 3E or 3F

No

4A

No

PLS-DA?

Yes

4B

End

Figure 4.17 (continued)

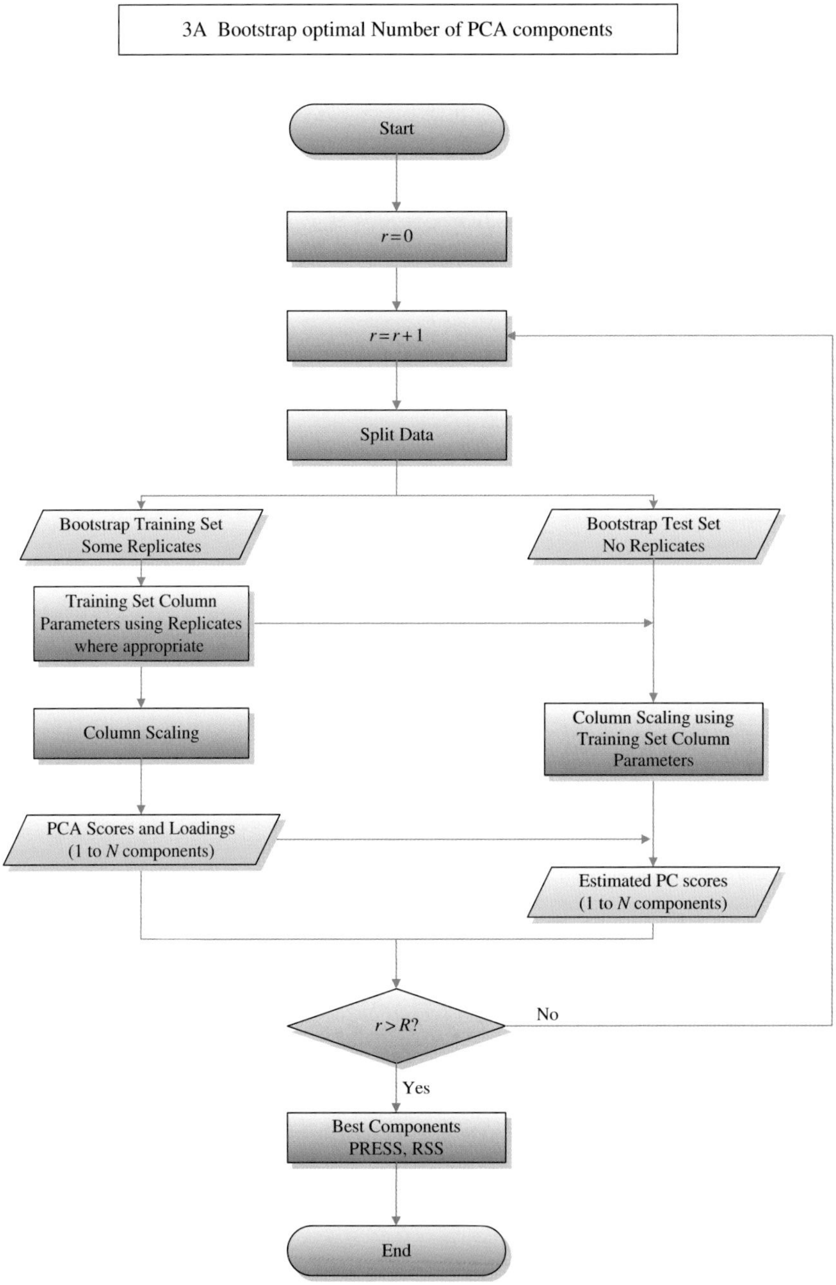

Figure 4.18 Flow charts for level 3 processes

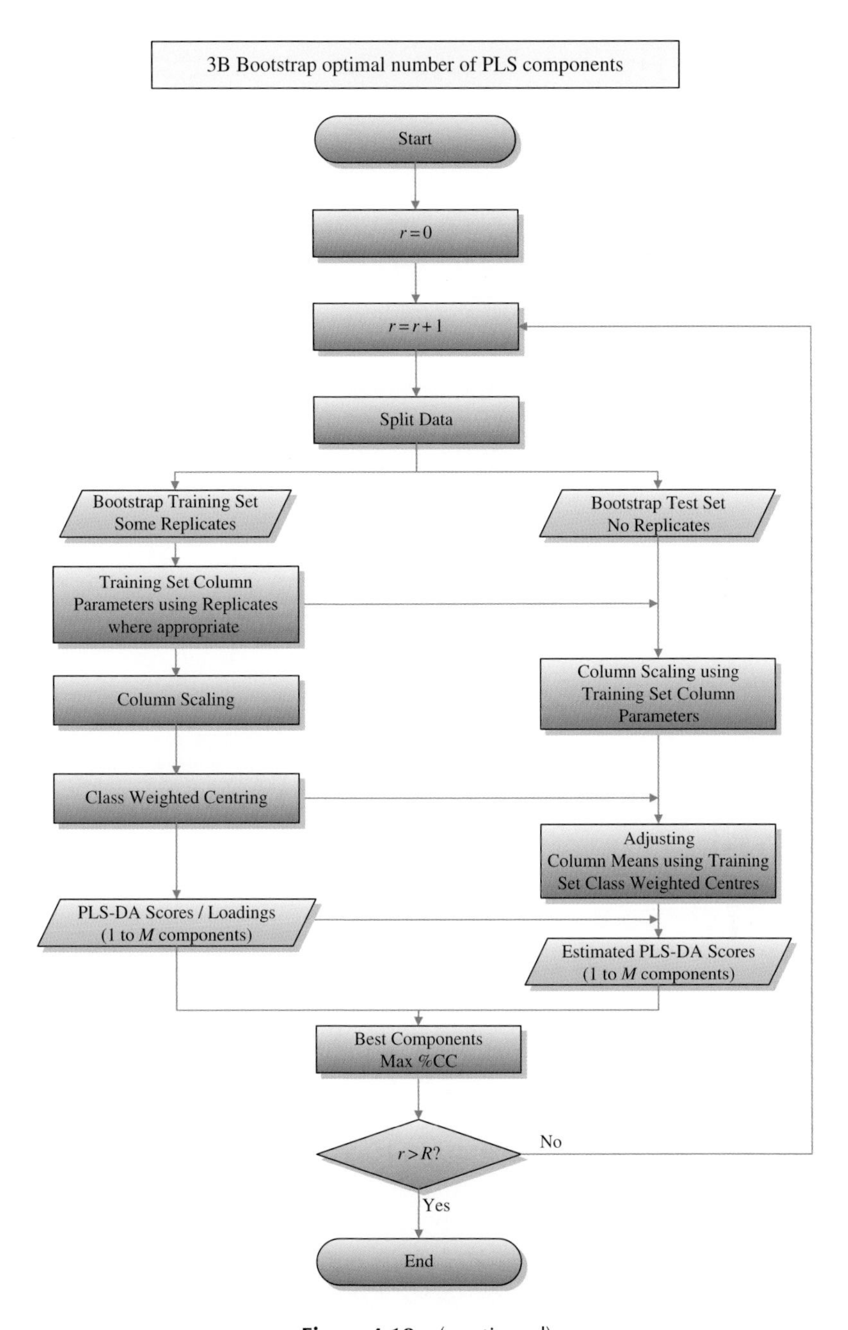

Figure 4.18 (continued)

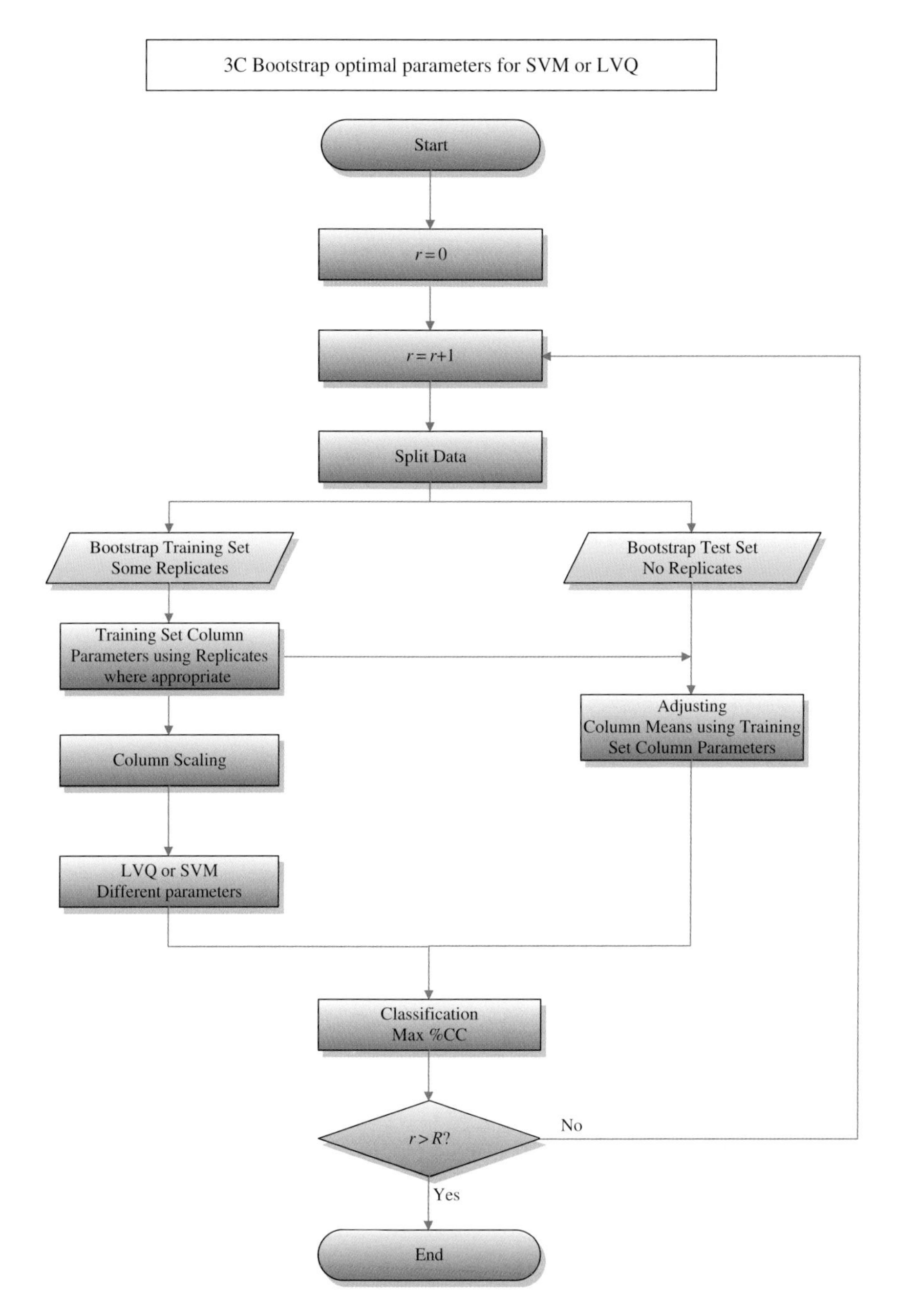

Figure 4.18 (continued)

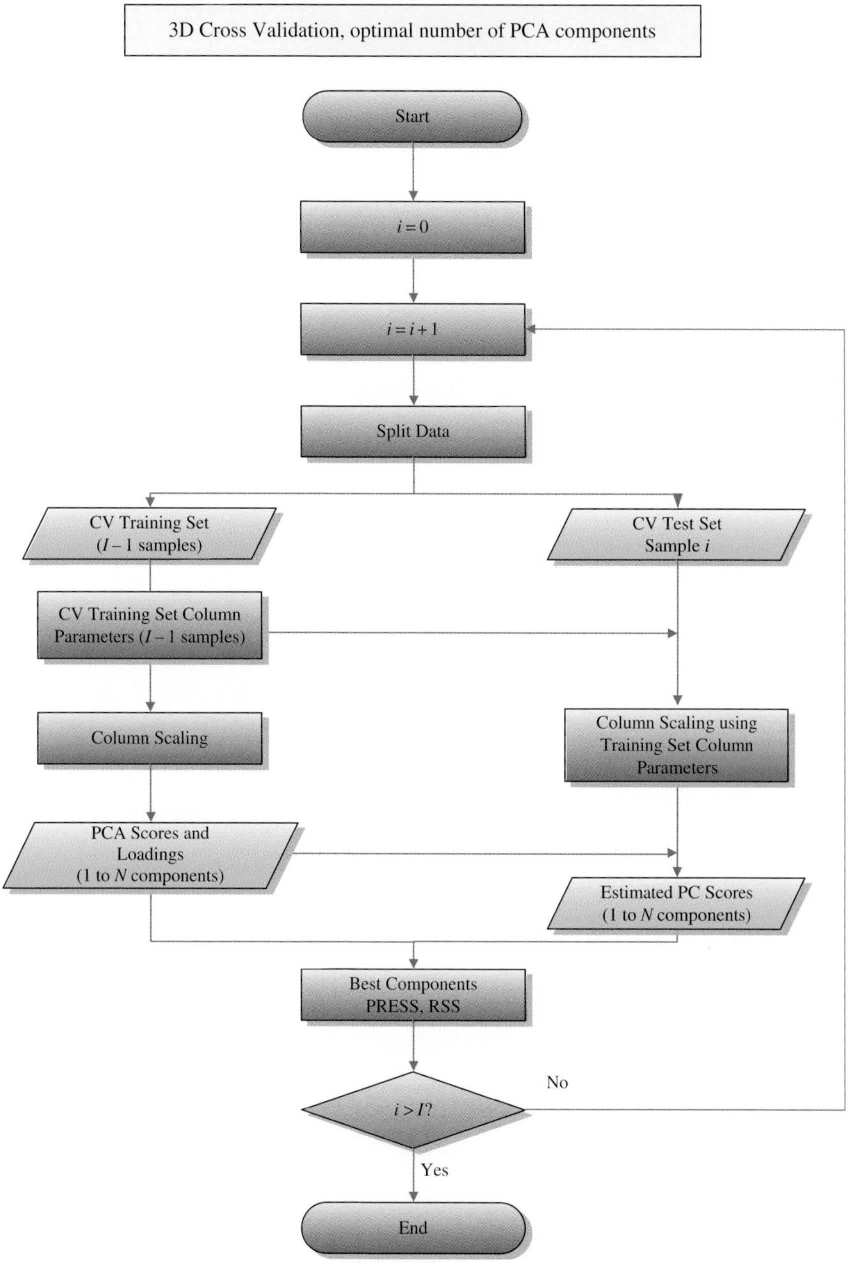

3D Cross Validation, optimal number of PCA components

Start

$i = 0$

$i = i + 1$

Split Data

CV Training Set
($I - 1$ samples)

CV Test Set
Sample i

CV Training Set Column
Parameters ($I - 1$ samples)

Column Scaling

Column Scaling using
Training Set Column
Parameters

PCA Scores and
Loadings
(1 to N components)

Estimated PC Scores
(1 to N components)

Best Components
PRESS, RSS

$i > I$?

No

Yes

End

Figure 4.18 (continued)

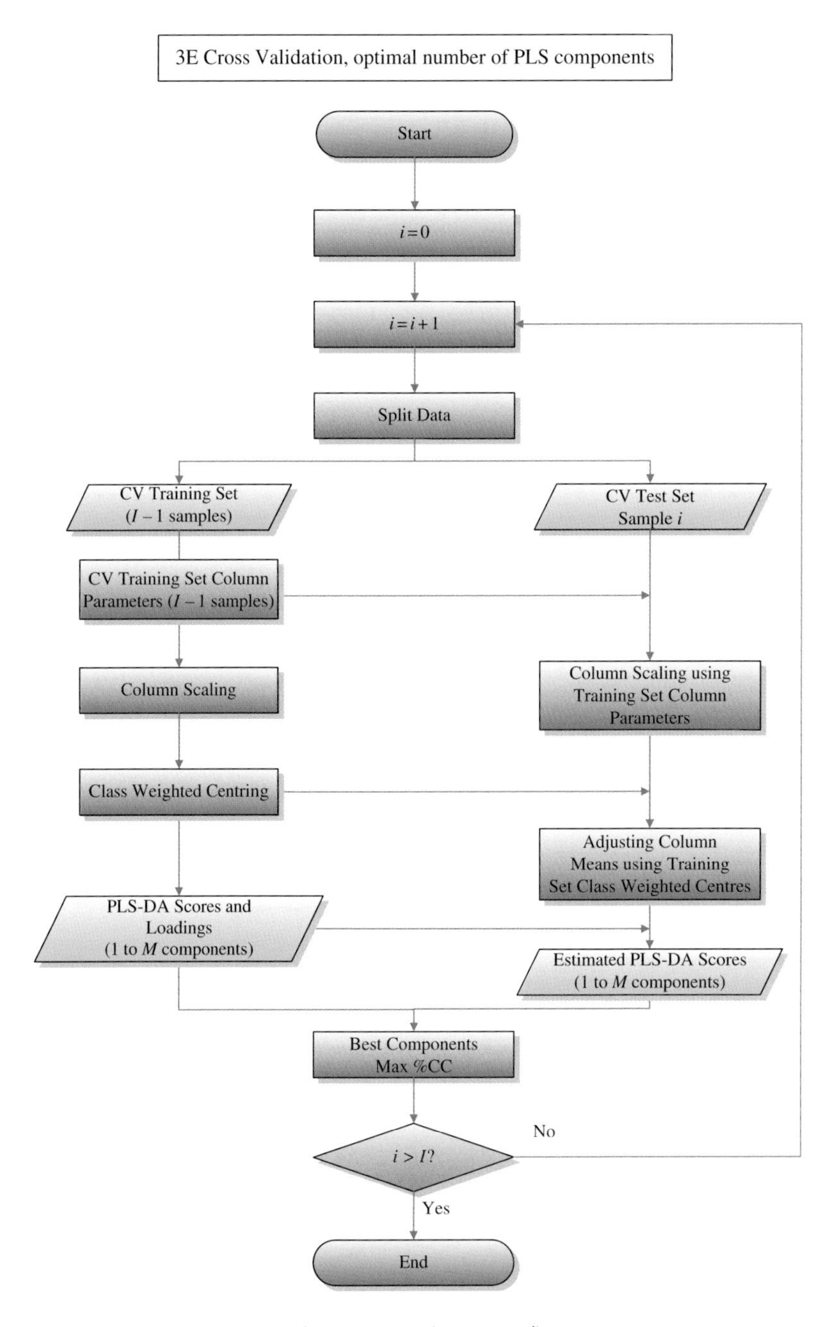

Figure 4.18 (continued)

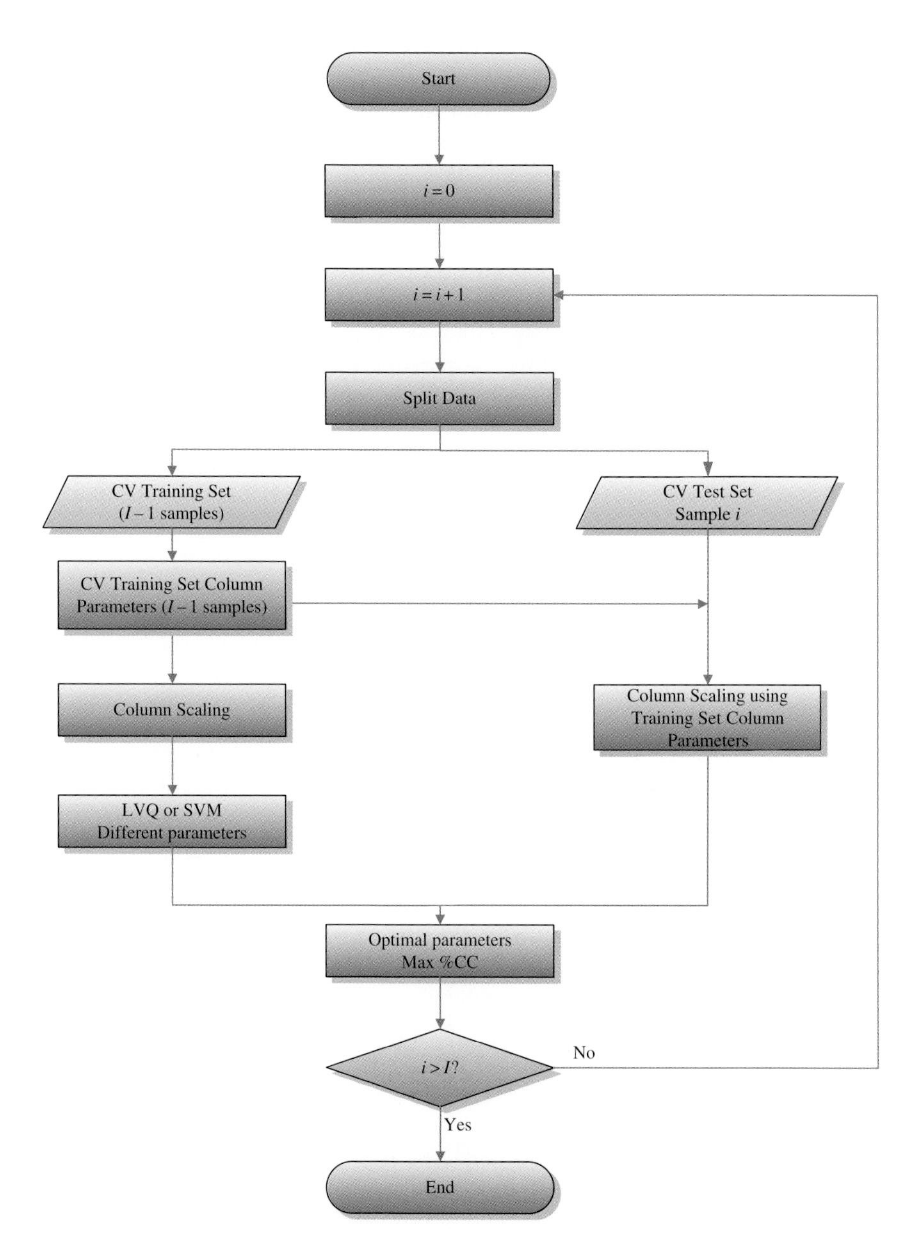

Figure 4.18 (continued)

Section 8.5.1 (see flow chart 3B) but can also be employed for PCA (flow chart 3A) as discussed in Section 4.3.1 or for optimizing the parameters in more complex classifiers (flow chart 3C). We define the number of times the bootstrap training set is generated by R which in this book we usually set to 200. An unusual feature of the bootstrap training set (Section 8.5.1) is that some samples may be replicated in the training set, that is, occur more than once. In this text we recommend that if this is so, these samples are included more than once in the computation of bootstrap training set column parameters such as the mean and standard deviation, so for example if sample 1 is included 3 times and sample 2 is included 2 times, then the calculation of the standard deviation and mean of each column in the bootstrap training set contains multiple instances of each sample. However when optimizing the number of components in PLS-DA we can then readjust the mean of each column by the class weighted mean of each class in the bootstrap training set (Section 4.2.3) prior to model building, although all other scaling (e.g. dividing by the standard deviation) is performed using the parameters from the bootstrap training set without taking class information into account. The reason for this is that unlike the overall training set where we recommend (Section 8.3.2) there are equal representatives of each class to be distinguished, for the bootstrap training set this may not necessarily be so. Of course there could be alternative strategies for column scaling but in this text we use only one method which we recommend is quite robust.

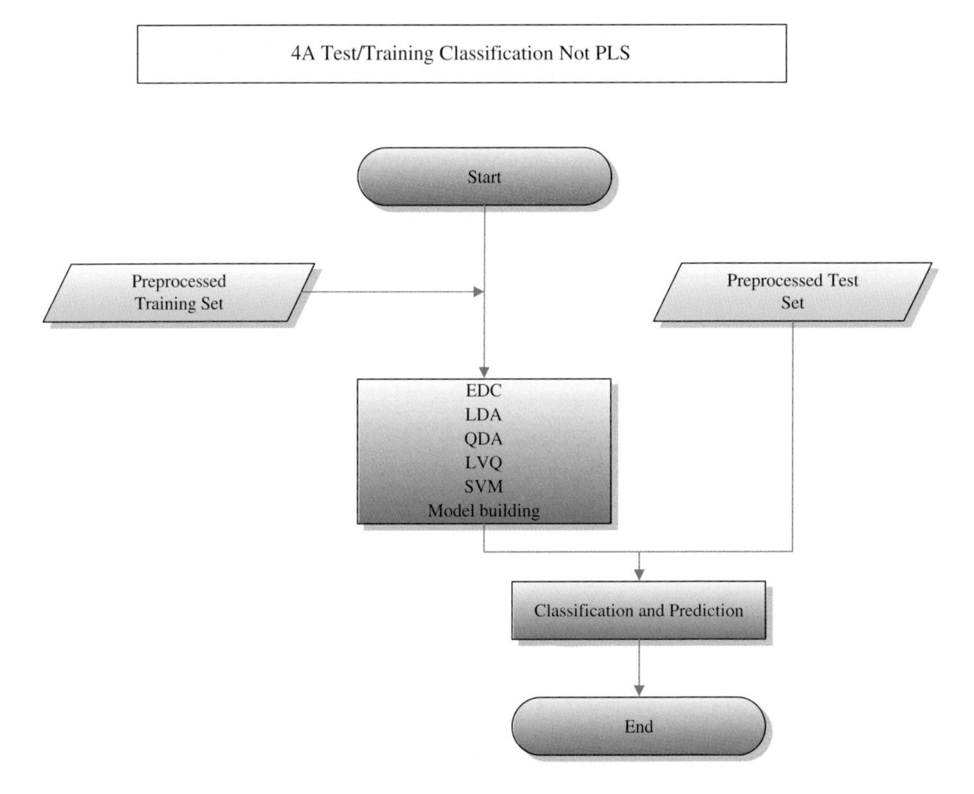

Figure 4.19 Flow charts for level 4 processes

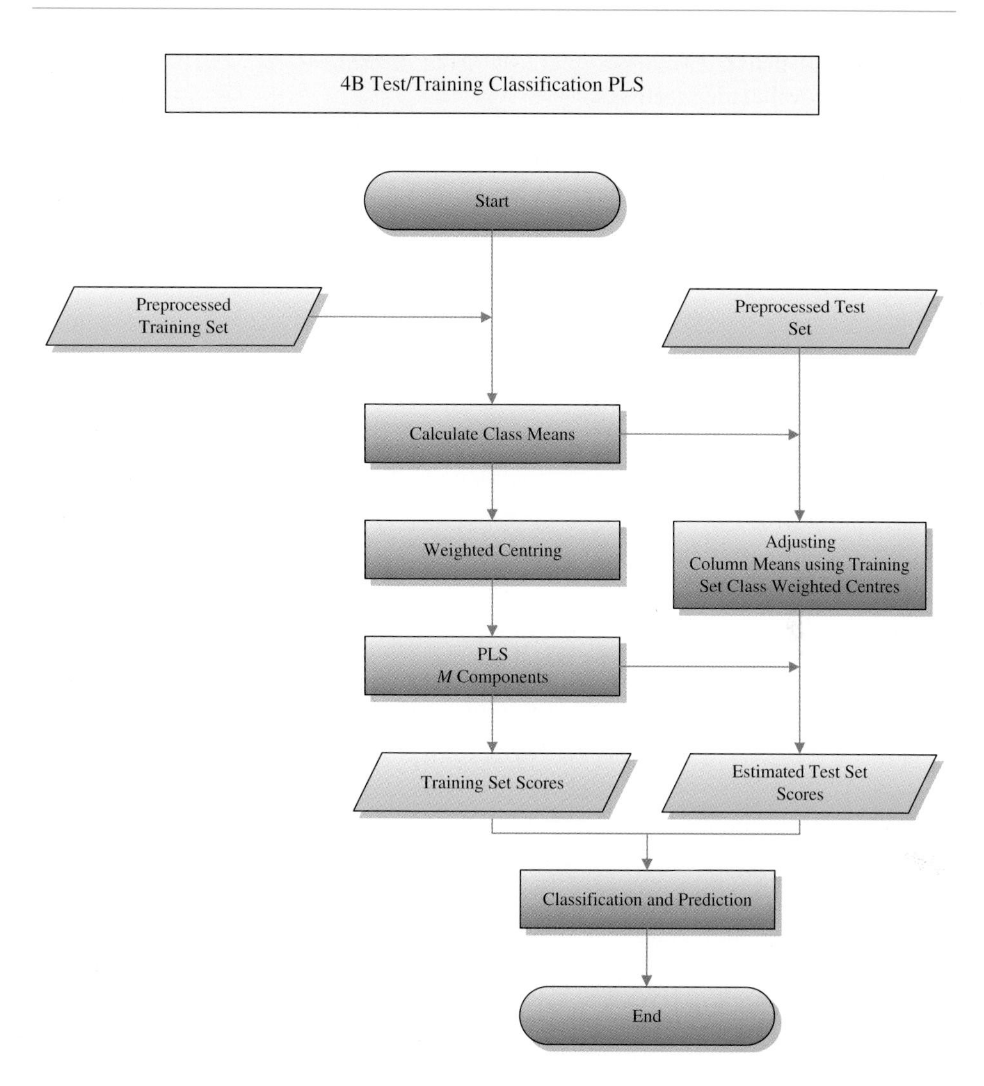

Figure 4.19 (continued)

Flow charts 3D to 3F are the corresponding ones when LOO cross-validation is performed for I samples.

Note that level 4 is never called from level 3, which always returns back to level 2. The reason why levels 3 and 4 are distinguished is that they perform quite different functions. Level 3 processes, however are always performed before level 4.

4.4.5 Level 4

The flow charts for processes at the fourth and final level that we discuss are illustrated in Figure 4.19 and involve direct input to the classifier, to obtain predictions about class membership of individual samples and also about determining how well the classifier performs when validating methods.

In flow chart 4B if, as is usual, there are equal representatives of each class in the training set, column centring is sufficient and it is only necessary to use the weighted class means to readjust the column centres if there are unequal numbers of samples.

Further levels could be represented that are called by the processes represented by flow charts 4A and 4B; however, the main aim of the flow charts in this chapter is to describe how the data are scaled, primarily down the columns, and the remaining algorithms can be adequately described as a series of steps, without the need for flow charts. It is important not to overcomplicate the use of flow charts for representing computational procedures where they are easiest described as a linear set of steps, but data scaling is quite complex and best described in the form of flow charts.

Bibliography

Bootstrap and Cross-Validation

S. Wold, Cross-Validatory Estimation of Number of Components in Factor and Principal Components Models, *Technometrics*, **20**, 397–405 (1978).

M. Stone, Cross-Validatory Choice and Assessment of Statistical Predictions, *J. R. Statistical Soc., Ser. B – Statistical Methodology*, **36**, 111–147 (1974).

B. Efron, R.J. Tibshirani, *An Introduction to the Bootstrap*, Chapman & Hall, New York, NY, USA, 1993.

J.S.U. Hjorth, *Computer Intensive Statistical Methods: Validation, Model Selection and Bootstrap*, Chapman & Hall, New York, NY, USA, 1994.

R. Wehrens, H. Putter, L.M.C. Buydens, The Bootstrap: a Tutorial. *Chemometrics Intell. Lab. Syst.*, **54**, 35–52 (2000).

5

Two Class Classifiers

5.1 Introduction

Many problems in chemometrics can be reformulated as classification problems, which involve trying to use measurements on a series of samples first to determine whether there is a relationship between interesting properties of the samples and the analytical data, and then if there is so, to determine what this relationship is, in the form of a mathematical model. Once the model has been set up, we try to determine the provenance of an unknown sample. In the case of classification this involves determining whether a sample belongs to one or more predetermined groups. Such methods are often called supervised methods as they require some form of information about the origins of the samples used to form a model, in advance, unlike exploratory methods such as PCA, discussed in Chapter 3. Classification differs from calibration (which we will not discuss in this text), in that for classification we are looking at whether samples can be related to groupings whereas in calibration we are looking at whether samples can be related to one or more continuous numerical properties.

There are numerous approaches in the literature to classification. To describe everyone's favourite approach would involve many volumes as there are several hundreds of named classification algorithms described in the literature. One problem arises is that to obtain a grant application or a PhD it is often helpful to propose a new named method – this allows a paper to be written or a conference presentation to be given high profile. Often the developers of these methods then try to prove their approach is better (than the hundreds else out there), and find some dataset and quote some statistic that shows its superiority to others, as a justification for what might be a considerable amount of funding, a sponsor hungry for results, or a PhD student wanting to graduate. However in a text this size it would be impossible to describe every approach: to do this would be more to write an encyclopeadia or historical record of the development of pattern recognition algorithms in chemometrics, and so this chapter focuses more on generic principles and some of the main groups of approaches towards classification. In fact

Chemometrics for Pattern Recognition Richard G. Brereton

most named methods are based on a small number of principles, and differ primarily in the way data are preprocessed and the precise nature of the steps used to obtain the output from the classifier.

Classifiers can be divided into two types, two class classifiers, in which samples are assigned to one of two (or more – the extension to multiclass situations is discussed in Chapter 7) groups, and one class classifiers (Chapter 6) in which each group is modelled separately. The former type of classifier is also sometimes called a hard model, because the data space is divided into regions, each of which corresponds to one class of sample, and the latter soft models in which each class is modelled separately and so there can be overlapping or empty regions of the data space. A few multiclass hard models can define void portions of the dataspace (Chapter 7) but this is rare; however hard models never involve overlap between classes.

5.1.1 Two Class Classifiers

In this chapter we will focus on two class or binary classifiers. These are used to decide whether:

1. There is sufficient evidence that the analytical data obtained can be used to determine whether a sample is a member of one of two predefined groups.
2. If so, which group a sample belongs to.

In Chapter 8 we will be looking at ways to determine how good the classification is and in Chapter 7 look at ways of handling data when more than one group is of interest. In this chapter we will introduce generalized methods for developing classification models but will illustrate their application graphically to the case studies introduced in Chapter 2.

Many of the case studies discussed in this book can be treated as two class problems. Case Studies 1 (forensic), 4 (pollution) 5 (sweat) , 8a (mouse urine – diet study), 9 (NMR spectroscopy), 10 (simulations) and 11 (null) can all be formulated as binary classification problems, for example, are the mice on a diet or not (Case Study 8a) or are the subjects male or female (Case Study 5)? Case Studies 3 (polymers), and 7 (clinical) could be expressed as binary classification problems, for example, whether a polymer is amorphous or semi-crystalline (Case Study 3) but there is also some further structure within some or all of the main groups, e.g. the two main types of polymer are divided in nine subgroups. The remaining case studies cannot be described by binary classifiers. The datasets that can be employed as examples of binary classification problems are summarized in Table 5.1. Note that there are often other factors that influence the data in addition to the binary class structure. For example, the data of Case Study 9 (NMR spectroscopy) also involve individuals and time, but these factors are balanced in that there are the same number of individuals and the same time series for samples in both groups; in this text we will primarily treat this case study as a binary classification problem, but there are other ways of handling these data. For Case Study 7, we could also study the structure of the diseased group. However in this chapter we will ignore this secondary structure (which may be of additional interest and require other approaches) and focus on trying to answer whether we can distinguish between the two groups of samples as specified.

Table 5.1 Main binary classification problems discussed in this book, together with the two classes and the numbers of samples in each class

Case study	Classes
1 Forensic	A (+1): Contaminated (46) B (−1): Background (59)
3 Polymers	A (+1): Amorphous (92) B (−1): Semi-crystalline (201)
4 Pollution	A (+1): Polluted (179) B (−1): Unpolluted (34)
5 Human sweat by GCMS	A (+1): Male (415) B (−1): Female (495)
7 Clinical	A (+1): Diseased (94) B (−1): Control (446)
8 (a) Mouse urine by GCMS	A (+1): On diet (59) B (−1): Control (60)
9 NMR	A (+1): Mouthwash (48) B (−1): Control (48)
10 (a) (b) (c) Simulations	A (+1): (100) B (−1): (100)
11 Null dataset	A (+1): (200) B (−1): (200)

For two class problems we divide the data into class A and B:

- General guidance is that class A is the one consisting of samples whose properties we are primarily interested in detecting, for example, in Case Study 1 (forensic) we are primarily interested in whether a series of banknotes are contaminated with drugs and the background could be a heterogeneous group which is of less direct interest; in Case Study 4 (pollution) our main aim is to decide whether a sample is polluted and so this group is class A; in Case Study 7 (clinical) the interest is detecting whether a person has a disease; in Case Study 9 (NMR spectroscopy) class A represents the saliva samples treated with mouthwash; in Case Study 8a (mouse urine – diet study) we are interested primarily in whether we can determine whether a mouse has been on a high fat diet. For biomedical applications the controls are usually class B. A member of class A can be called 'positive' whereas class B are called 'negative', rather like diagnosing a disease – is the patient positive for the disease? When we discuss concepts such as likelihood ratios (Section 8.2.1) there is a difference as to which class is labelled positive.
- If both classes are of equal interest, e.g. in Case Study 3 (polymers), then if the numbers in each group are very different, we call the class with less samples class A. This is because one often wants to detect a sample with a specific characteristic from a large background, e.g. determining whether a person has a disease against a large number of healthy subjects, and so the group of 'positives' is often considerably smaller than the background population.

- For Case Studies 5 (sweat), 10 (simulations) and 11 (null) there is no real distinction between the groups, both being approximately equal in size and of similar significance. In Case Study 5 we choose, arbitrarily, to assign males to class A and females to class B because we will adopt a convention that class A is represented by blue symbols (usually indicative of males) and class B red symbols.

5.1.2 Preprocessing

In Chapter 4 we discussed in detail methods for preparing data prior to classification.

For visualization of the class models in this chapter we will first calculate PC scores on overall datasets (as discussed in Section 3.2), prior to which we employ default dataset specific preprocessing as described in Chapter 2 for each case study. When column centring and standardizing are defaults we use methods described in Sections 4.2 and 4.4 for scaling the data prior to PCA. The data are visualized using 2D PC scores plots; in this chapter we will employ autopredictive two class models as outlined using flowchart 1A.

Although it is possible to apply many of the classifiers to raw data without first performing PCA, the reason for illustrating the performance of classification techniques on the first two PCs in this chapter are threefold, as follows:

1. Some of the classification techniques such as those based on the Mahalanobis distance (Linear Discriminant Analysis (LDA) (Section 5.3) and Quadratic Discriminant Analysis (QDA) (Section 5.4)) fail if the number of variables is greater than the number of samples, and so reduction by PCA is employed for all techniques as a method of data reduction for direct comparison between methods.
2. We want to visualize the classifiers in two dimensions and so select just two variables, being the first two PCs in this case. Of course classifiers can be visualized in three dimensions and computed in any number of dimensions, and in Sections 4.3.1 and 4.3.2 we discuss how to choose the number of variables or PCs as input to the classifier.
3. PCs have no direct relationship to the classifier and mainly are used to maintain the overall data structure, and so are unprejudiced methods for simplifying the data.

We do not, however, advocate that it is always (or even most usually) appropriate to perform classification just using the scores of the first two PCs; the reason for representing the methods in this chapter using the scores of the first two PCs is for illustrative purposes, although all methods are described in general terms allowing any number and type of variables (except in the case of Mahalanobis distance type measures) as input to the classifier, as discussed below. The purpose of this chapter is primarily to describe and illustrate the main classes of algorithm employed in this text.

5.1.3 Notation

In this chapter and elsewhere we denote the preprocessed data that are input to the classifier by a matrix $\boldsymbol{X}$ whose rows represent the samples and whose columns represent the variables. In the case where only 2 PC scores characterize the variables, this matrix will consist of two columns. We will assume there are I samples and J variables and so the dimensions of $\boldsymbol{X}$ are $I \times J$. Each sample is characterized by a row vector of measurements $\boldsymbol{x}_i$. In class A there are I_A samples and in class B I_B samples, with g denoting a general class

(= A or B in this context when there are only two classes), and so I_g equals the number of samples in class g. For Case Study 3 (polymers), for example, $I_A = 92$. For further discussion of the general aspects of notation in this text, see Section 2.2.

A separate vector $\boldsymbol{c}$ contains a label for each sample, either +1 or −1 according to which class it belongs to. In some circumstances, class A is denoted the 'in group' which has a label of +1 and class B the 'out group' which has a label of −1. The vector $\boldsymbol{c}$ will not be mean centred unless there are equal numbers of samples in each class. There are special techniques that can be used to handle unequal class sizes if desired, as discussed in Chapter 10.

5.1.4 Autoprediction and Class Boundaries

In Chapter 8 we will discuss ways of validating class models, that is, determining whether there is truly an underlying relationship between the analytical data and the classifier (or $\boldsymbol{X}$ and $\boldsymbol{c}$) or not. In this chapter we will introduce the concept of classifiers using autopredictive models, but it is important to understand the difference between these types of models and what they mean.

For two class classifiers, all models can be represented visually as boundaries between classes, that is, the result of a model will be to draw a border between two groups of samples, each side of the border consisting primarily of samples from classes A and B. Sometimes a perfect boundary cannot be drawn but an important aim of all two class classification algorithms is to be able to find a boundary (of given complexity) that can separate two classes as well as possible within the certain limitations placed on the type of boundary.

As an example to illustrate autopredictive approaches, Figure 5.1 is of a seating plan of men and women – perhaps they are listening to a lecture? Can we draw boundaries between where the men and women sit and so classify the person into male or female

Figure 5.1 Seating plan of men and women with boundaries drawn between seats where clusters of males and females are sitting

Figure 5.2 Dividing the seating area up into male and female regions

according to their position in the seating plan? If we know where everyone is sitting, of course we can, and this is illustrated in Figure 5.2, we can then determine a person's gender according to where they sit. However this boundary is quite complicated and probably suggests that there is no underlying system in where people sit. But what is to stop us producing such as boundary? Fundamentally nothing, we could say that we are able to model exactly where a person sits and as such perfectly predict the gender of a person according to their position in the room. This is called an autopredictive model, because it uses known information to provide a distinction between two groups.

Why does this argument fall down? It depends a little on why one wants to establish which regions of the seating plan are occupied by males and females. In some situations a perfect autopredictive model is fine, as the model is a 'one-off'. A cartographer wanting to divide up a map according to geographical features is probably entirely happy by such models even if they are extremely complicated; one can take a region and divide it up into hilly and low lying areas; then after this is done all one does is enter a map reference and one can tell perfectly whether the area is hilly or not, which may be useful for pilots or military exercises. An advertiser who wishes to place on seats, in the interval of a performance, advertising material that is firmly aimed at females (for example, a new type of cosmetic) or males (for example, a new type of shaver) might like this information, in order not to waste leaflets. But can we predict the gender of a person likely to sit in any particular place without knowing all the information in advance? This is a different question and one that relates to the major topic of this book. In fact it is quite a difficult question and might depend on why people are sitting in the room. In some socially conservative countries of the world in large public meetings there often is a tendency for there to be segregation of sexes, for traditional reasons, and so we would expect big groups of females and of males in different parts of the room; these groups might initially be seeded randomly, for example, if three or four women sit in a corner they are more likely to be joined by other women than by men, forming a cluster of women in one part of the room. We may then be able to predict that whether, for example, someone sitting in the right-hand corner is likely to be female, or indeed whether a seat in the right-hand corner is more likely to be occupied by a women even if it is empty at the time, just by observing a few people, maybe at the interval of the performance. To see whether the place a person sits at is a good way of predicting a person's gender could be interesting, as we may want to study cultural segregation in communities across the world, and so we pose a hypothesis, that in certain cultures in public meetings there are likely to be regions in the seating plan where there are mainly men and regions where there are mainly women. In different circumstances this model may fall down, for example, in family entertainment, such as a play or a circus, most often there is an almost random mix of sexes sitting next to each other in the audience as couples and family groups are usually in consecutive seats, according to the ticketing agency.

It is still possible to produce a perfect autopredictive model in both cases, the public meeting in a traditional society and the family entertainment, by drawing borders around groups of men and women, and in both cases claim perfect classification using autopredictive models. The case of Figure 5.2 represents an autopredictive model that perfectly predicts a person's gender from where they sit.

However, there are two principle differences between the model for each of the two situations discussed above (the traditional society and the family entertainment):

1. For the family entertainment situation the model becomes very complicated as it is not always possible to draw simple boundaries between groups of people of different gender. In the traditional society situation, however, where there may be large groups of people of the same gender, often a simpler border can be drawn between the male and female groups.
2. The models for the seating plan for the family entertainment may have low predictive ability, as will be discussed in Chapter 8. That is, although it is possible to determine boundaries in the seating plan that perfectly matches each person's gender, if we were to remove some of the information, for example, leave an unknown who is sitting in a few of the seats, and use the remaining people to predict the gender of people assigned to sit in these seats, we may get it very wrong. For the traditional society situation we are more likely to get it right.

In this chapter we will look at a variety of ways of producing autopredictive models, starting at simple models and ending up with very complex ones. These methods will be represented graphically by boundaries between two groups of samples. Because of the difficulty of visualizing more than two dimensions these will all be represented in two dimensional space using plots of the scores of the first 2 PCs as the axes. It is important to remember that we do not know in advance how complex the boundary (or class structure – in more technical terms) is likely to be. In traditional analytical chemistry often simple linear boundaries are adequate as the data are quite linear, whereas in biology, for example, there is no reason why the data should be linearly separable and we might expect more complex boundaries. It is also important to realize that boundaries may not always perfectly separate two (or more) groups of samples, and it is a mistake to always search for a boundary (however complex) that perfectly divides the data space into regions that are occupied by each group. Sometimes there can be mislabelled samples or outliers; sometimes there is noise and measurement error. Sometimes, although we might wish to find a perfect boundary there is actually no evidence in the data that it is separable, for example, we might be interested in whether we can diagnose an early onset of a disease from a patient's urine sample: this may be possible occasionally but in other circumstances, it is not always possible – we can still try to draw as good a boundary as possible but this may not classify unknown samples correctly. In traditional analytical chemistry it is often quite certain that there must be an underlying model and so the aim of computational algorithms is to search for a way of dividing the data that comes up with a boundary that is to all intense and purposes perfect – if this cannot be done either the analytical method or the algorithm is at fault and the lower the misclassification, the better the model. In modern chemometrics, which interfaces with a large number of disciplines such as medicine and biology, there is often no certain answer in advance that the experimental data can come to the conclusion we may wish about the data, which may depend on unproven hypotheses. In Chapter 8 we will discuss how to cope with these situations, especially in hypothesis driven science, but below we focus simply on producing a boundary between two classes using a variety of different methods as a way of understanding how different algorithms work.

In the sections below we will describe several common approaches for finding these boundaries, illustrated by autopredictive models using the scores of the first two PCs of Case Studies 1, 3, 4, 5, 7, 8a, 9, 10 and 11. The PC projections have been presented earlier in Figures 3.6 to 3.17, and we will focus on how to obtain boundaries between two groups of samples when projected onto the first 2 PCs, using the 2D score plot of PC2 versus PC1. By default, the boundaries will be drawn at the point where the classifier suggests an equal numerical value of the classification statistic of belonging to each of the two classes.

5.2 Euclidean Distance to Centroids

A very simple classifier is the Euclidean distance to centroids. The principle is illustrated in Figure 5.3. We assume that data has already been preprocessed by a suitable method as discussed in Chapters 2 and 4. For each class g, and each of the variables, the mean is calculated over all samples in that class to give a centroid $\bar{\boldsymbol{x}}_g$. No other class distribution information apart from the mean is employed, and it is implicitly assumed that the distribution of samples around the centroid is symmetrical and similar in the original variable space for each class. The Euclidean distance for sample i (represented by a row vector $\boldsymbol{x}_i$) is calculated to the centroid of each class g by:

$$d_{ig}^2 = \left(\boldsymbol{x}_i - \bar{\boldsymbol{x}}_g\right)\left(\boldsymbol{x}_i - \bar{\boldsymbol{x}}_g\right)'$$

where d_{ig}^2 is the squared Euclidean distance between sample i and the centroid of class g. The sample is assigned to the class with the lowest distance. A linear boundary can be drawn between each class representing the line where the distance to each class centroid is equal. If the data are linearly separable a perfect cut-off can be found, otherwise there will be a few misclassified samples each side of the line.

The Euclidean boundaries for the datasets listed in Table 5.1 are illustrated in Figure 5.4 (10(b) is omitted for brevity). Misclassified samples are indicated with closed symbols. Most of the boundaries are straightforward to interpret. We see that Case Study 5 (sweat) is

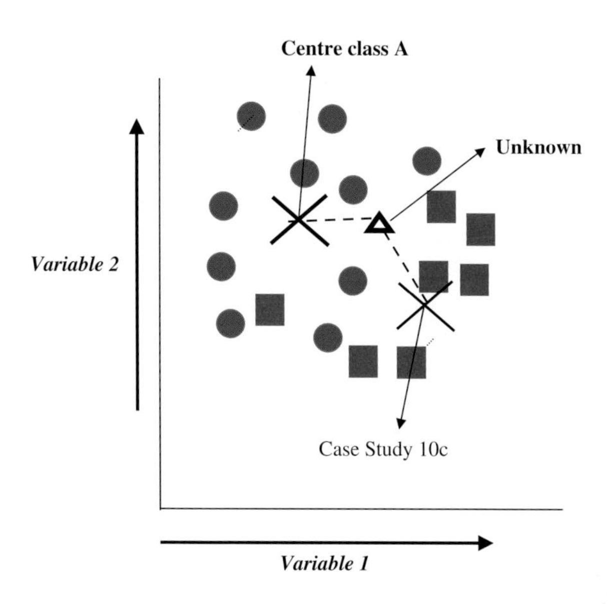

Figure 5.3 Principle of Euclidean distance measure, for two variables and two classes. The distance of an unknown calculated to the distance to the centroid of each of the classes is assigned to the class whose centroid it is nearest to

not separable in the first two PCs (in fact this is because there are other factors that influence it, and we can get good class models but using more PCs). For Case Study 11 the position of the line is quite arbitrary and we will show (Chapter 8) that if we remove one third of the samples (the remaining samples forming a training set), the position of this line is quite unstable, according to which samples have been removed. Case Study 7 (clinical) is the only perfectly linearly separable dataset using projections onto the first 2 PCs. Case Study 4 (pollution) is an interesting one. Whereas three of the samples appear to be outliers and fall within the unpolluted group (these are all from soil 4) we might have expected the boundary to move to the left to best discriminate the two groups. In fact the variance of class A is greater than class B in both PCs but EDC assumes equal variance in each dimension, and as such is not the method of choice if each class has a different level of variability.

5.3 Linear Discriminant Analysis

One weakness of the Euclidean distance is that it takes no notice of the different spread or importance of each variable. Sometimes, for example, variables could be measured on entirely different scales. In chemistry we may be looking at the amount of an element (e.g. a heavy metal) present in a series of samples. Some may be detected on entirely different ranges, for example, one might be in ppb and another ppm, and yet the more abundant element may not be very diagnostic of the variation between the samples we are interested in. We say that these variables have different spreads or standard deviations, in absolute terms. Figure 5.5 is such a case in

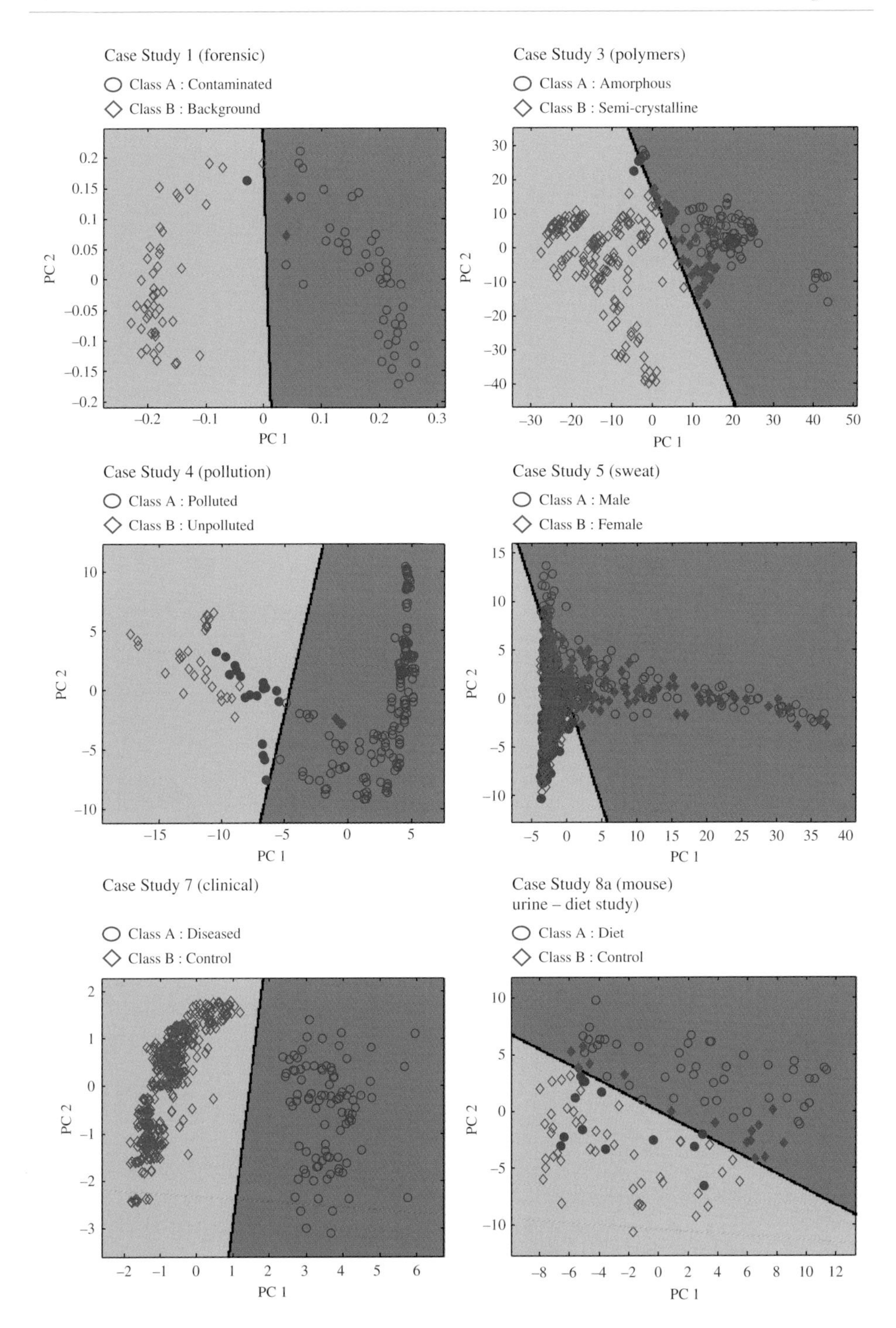

Figure 5.4 Visualization of boundaries obtained by EDC for the datasets discussed in this chapter, using default data preprocessing, and data represented by the scores of the first two PCs. Misclassified samples are represented by filled symbols, and the respective regions belonging to each class are coloured appropriately

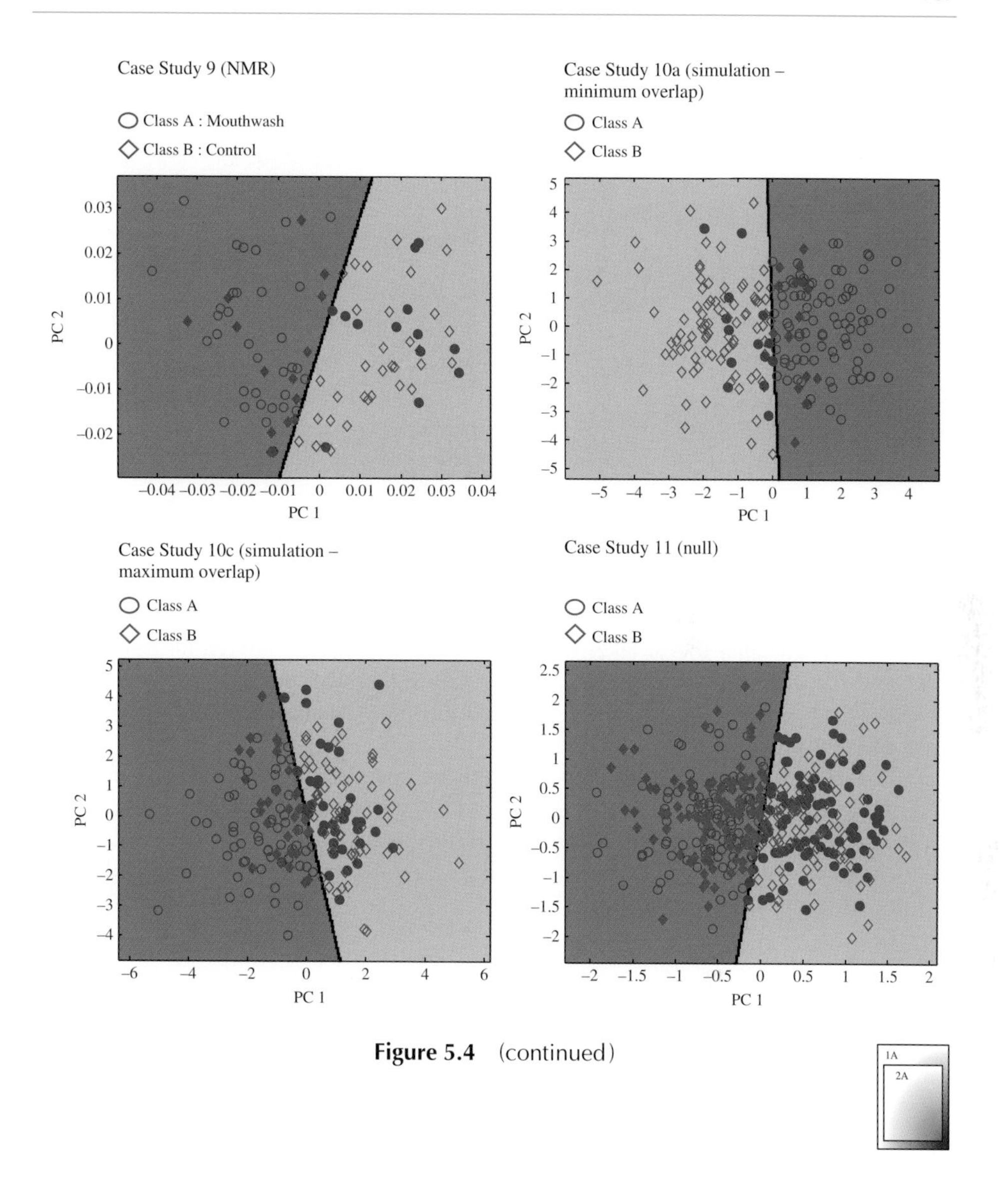

Figure 5.4 (continued)

question: the 'xs' mark the Euclidean centroids of each class, and an unknown is marked by a triangle – by visual inspection it should belong to class B. However if we calculate the Euclidean distance to centroids this is not actually what emerges. If the distance to the centroid of class A is 0.040 and that for class B 0.670, this means that it appears very much nearer the centroid of class A. The reason for this is simple – the range of variable 2 is very much smaller in absolute terms than that from variable 1 (this can be checked by looking at the scales of the two axes), and so the measure used to classify an unknown is overwhelmingly influenced by the horizontal distances and the unknown lies very much closer to the mean of variable 1. LDA (Linear Discriminant Analysis) takes the different variances of each variable into

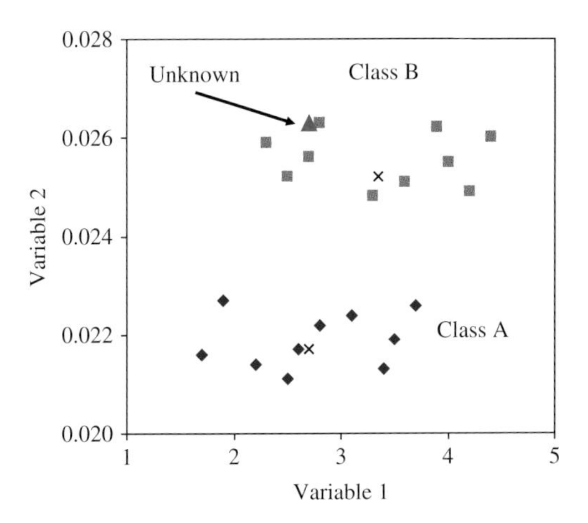

Figure 5.5 Example for LDA in Section 5.3

account (which could also be accomplished by standardizing the variables), but has an additional property that it also takes into account the correlation between the variables: if there are 20 variables, for example, and 19 convey similar information they will be correlated and as such should not be weighted too much against variables that measure quite different properties.

LDA, as the name suggests, also produces a linear boundary between two classes just as the Euclidean distance. The acronym LDA is actually used to cover a range of techniques, but the most common is Fisher Discriminant Analysis which we will use below. In this book, the LDA distance to each class centroid is calculated by using a measured distance, called the Mahalanobis distance as follows:

$$d_{ig}^2 = (\boldsymbol{x}_i - \bar{\boldsymbol{x}}_g)\boldsymbol{S}_p^{-1}(\boldsymbol{x}_i - \bar{\boldsymbol{x}}_g)'$$

where $\boldsymbol{S}_p$ is the pooled variance-covariance matrix, calculated for two classes as follows:

$$\boldsymbol{S}_p = \frac{(I_{\mathrm{A}} - 1)\boldsymbol{S}_{\mathrm{A}} + (I_{\mathrm{B}} - 1)\boldsymbol{S}_{\mathrm{B}}}{(I_{\mathrm{A}} + I_{\mathrm{B}} - 2)}$$

and where $\boldsymbol{S}_g$ is the symmetric variance–covariance matrix for class g of dimensions $J \times J$ whose diagonal elements correspond to the variance of each variable and whose off-diagonal elements are the corresponding covariance between each variable. It is important to realize that this approach uses the Mahalanobis distance based on a variance-covariance matrix obtained on the entire dataset, rather than for each class independently. There is some confusion in chemometrics based papers where authors state that they use the Mahalanobis distance but it is not clear whether they are referring to LDA as in this section or QDA (Section 5.4). As with the Euclidean distance the LDA boundary is where the distance measure to the centroids of each class is equal.

The LDA boundaries for the case studies discussed in this chapter are illustrated in Figure 5.6. In most cases, LDA does not represent a dramatic improvement over EDC,

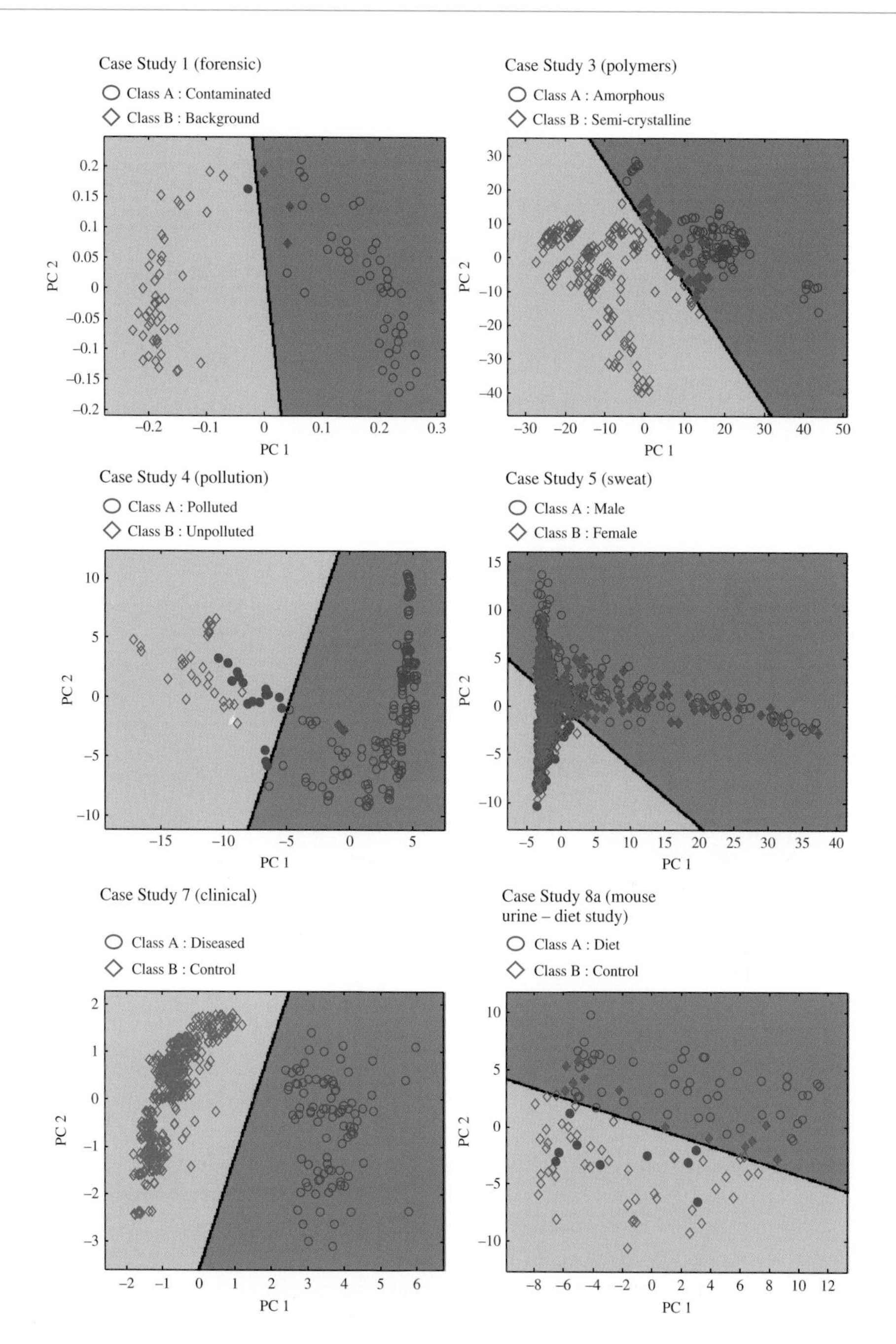

Figure 5.6 Vizualisation of boundaries obtained by LDA for the datasets discussed in this chapter, using default data preprocessing, and data represented by the scores of the first two PCs. Misclassified samples are represented by closed symbols, and the respective regions belonging to each class are coloured appropriately. Boundaries for PLS-DA using the algorithm described in this chapter are identical to those for LDA

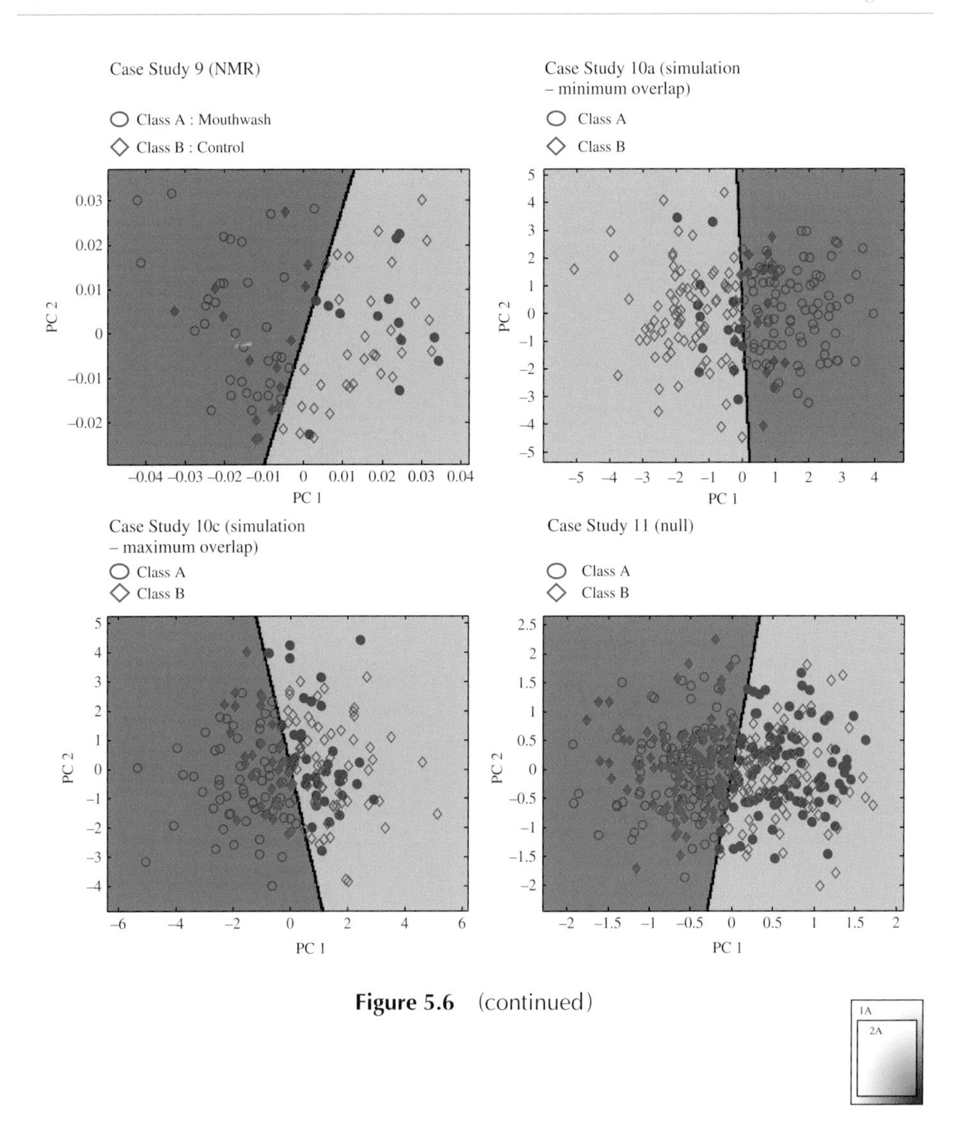

Figure 5.6 (continued)

as it does not take into account the difference in spread for each class, just whether the variables have different standard deviations. There is some improvement for Case Study 3 (polymers) in the classification boundary over that for EDC as PC2 has a substantially smaller variance to PC1. For Case Study 7 (clinical), the two PCs have very different variances and the boundary line changes in orientation compared to EDC, but since both classes are already separable this makes no practical difference when the problem is viewed as one of binary classification. There is some improvement in the classification for Case Study 8a (mouse urine – diet study) compared to EDC with 14 diet and 12 control samples misclassified for EDC compared to 12 diet and 9 control samples using LDA. Interestingly enough, Case Study 11 (null) shows almost identical results for LDA and EDC: this is because the first two PCs are very similar in size and variance (see Table 4.3). If two PCs are similar in variance there will not be a significant difference between the boundaries when using EDC and LDA. The differences between LDA and EDC boundaries for several of the case studies in this text are illustrated in Figure 5.7.

As we will comment below, the boundaries obtained using LDA are identical to those obtained using PLS-DA (Section 5.5) providing the data is preprocessed as discussed below, although there are many other diagnostic advantages in PLS-DA which is a method we employ widely in this book.

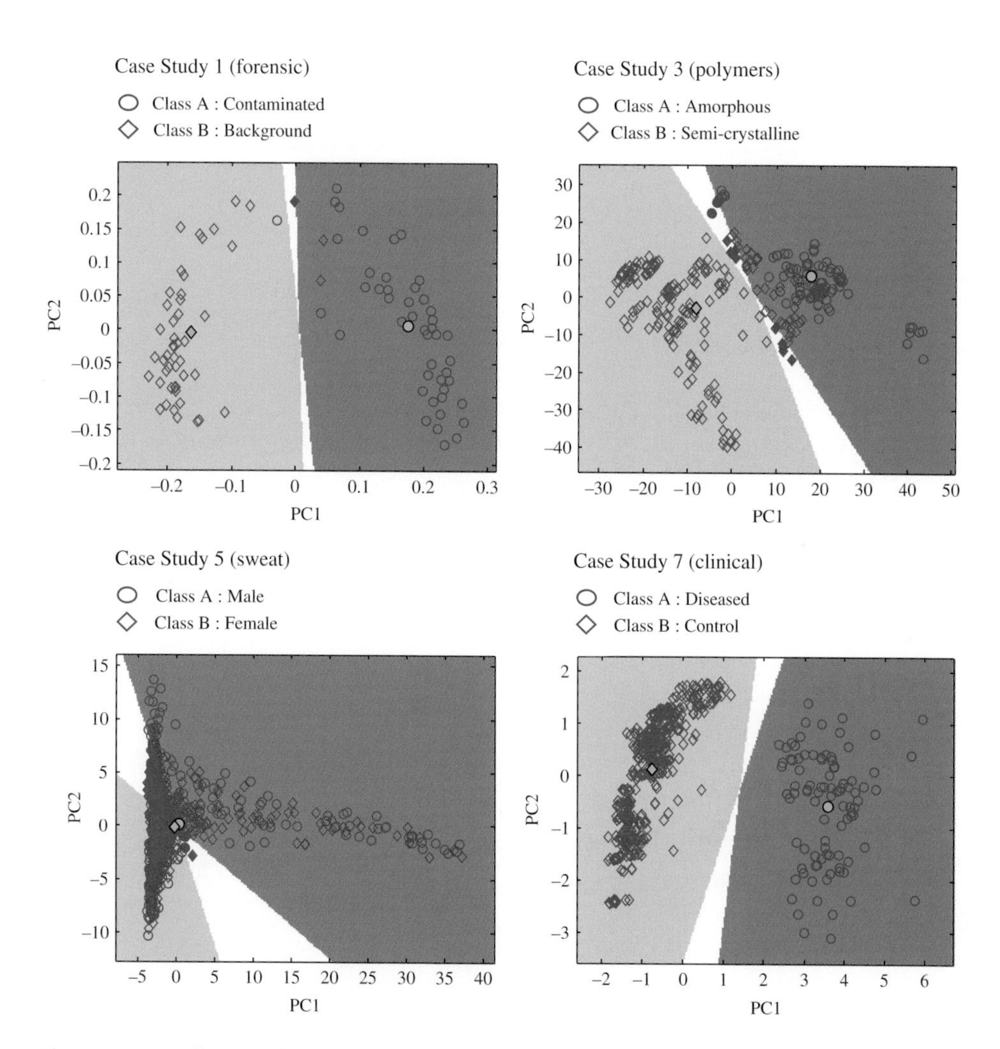

Figure 5.7 Difference between EDC and LDA boundaries for several of the case studies in this text; white regions indicate the differences between the boundaries, class centroids represented in green and all samples between boundaries are represented by filled symbols

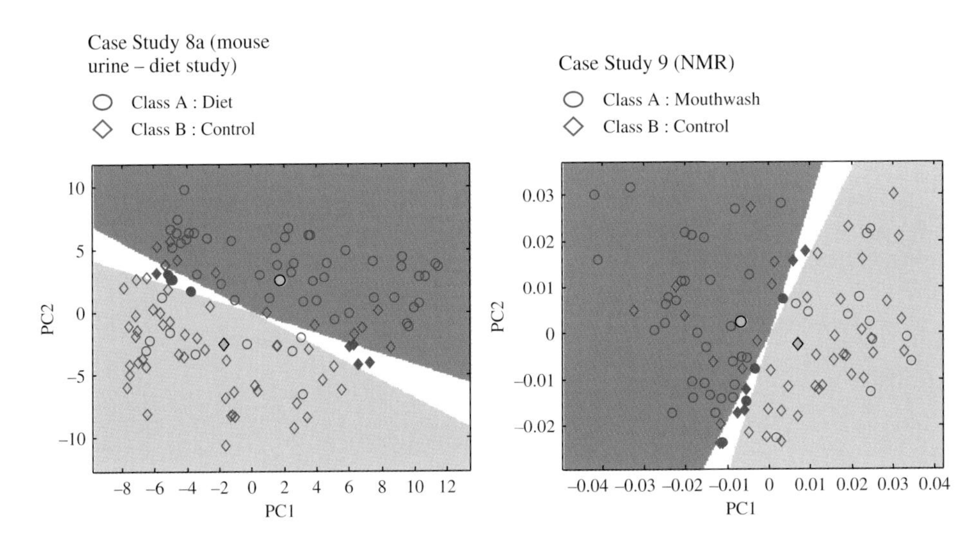

Figure 5.7 (continued)

5.4 Quadratic Discriminant Analysis

LDA overcomes two limitations of the Euclidean distance measure, namely that variables may have different variances and that there can be correlation between the variables. However sometimes each class can in itself have very difference variance structure. Consider Figure 5.8 where we are trying to discriminate mice from elephants. We could set up a class consisting of mice and of elephants and then ask which class the baby

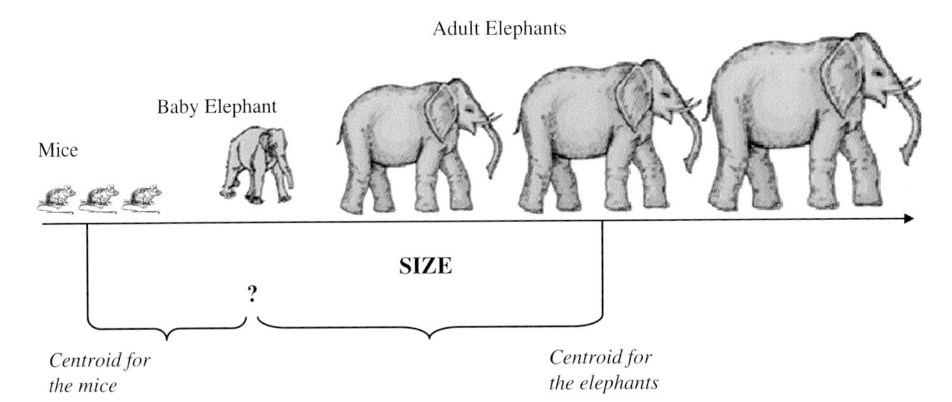

Figure 5.8 Different classes may have different variances

elephant best fits into. In fact the size of the baby elephant is closer to the centroid of the mice than the centroid of the elephants using almost any measurement we can think of (weight, height, length, etc.). Only when we consider that the mice and elephants have separate distributions can we see that the baby elephant resembles more closely the adult elephants than the mice, and so an added elaboration is to calculate the variance of each group separately and calculate the Mahalanobis distance to each class centroid using different variances.

Quadratic Discriminant Analysis (QDA) is very similar to the LDA metric: however, the distance to each class centroid is calculated using the sample variance – covariance matrix of each class rather than the overall pooled matrix:

$$d_{ig}^2 = (\boldsymbol{x}_i - \bar{\boldsymbol{x}}_g)\boldsymbol{S}_g^{-1}(\boldsymbol{x}_i - \bar{\boldsymbol{x}}_g)'$$

Some authors include an extra expression of the logarithm of the determinant of the variance–covariance matrix. However, this additional term relates to the Bayesian formulation which will be discussed in Chapter 10 and we restrict ourselves to the simple equation above in this chapter and elsewhere.

The shape of the boundary using QDA is no longer linear, and indeed can divide the data space into several regions, or enclose one class in a circular region, and so results in much more sophisticated boundaries compared to LDA. In some ways this additional level of sophistication is a logical extension but in other ways it could be regarded as introducing extra complexity that is not necessarily justified by the existing data. However, as chemometrics moves more out of mainstream analytical chemistry and into application based domains such as biology, such linear models may no longer be appropriate. We will see in Chapter 6 that the SIMCA method is based on QDA – but for single classes.

The QDA models for the ten datasets discussed in this text are illustrated in Figure 5.9 and the boundaries are drawn where the quadratic distance to each centroid is equal. Two of the case studies (1 – forensic; 8a – mouse diet) show very little change compared to either EDC or LDA. The optimal separation is linear and each class has similar variances for each PC. For Case Study 7 (clinical) although the classes are still perfectly linearly separable, the boundary moves closer to that of class B (control) which reflects the larger relative variability in PC1 of class A. Case Studies 3 (polymers) and 4 (pollution) are better modelled using QDA, reflecting the distribution of the two groups. Case Study 11 (null) is particularly interesting – we will see later that the QDA boundaries are extremely sensitive in this case to the removal of a few samples (Chapter 8). QDA has latched onto the slightly higher densities of the samples from class A in the bottom left-hand and top right-hand corners, and tries to take this into account. In fact for the QDA models the classification accuracy of each group may be quite different although the overall classification rate should normally be close to 50 %. Although the difference in distribution between the classes is quite small, all random samples contain some clumping which acts as a handle for QDA. Interestingly Case Study 5 (sweat) incorrectly suggests that most samples belong to class A (males); there is a very small female region in the centre that is hard to see as it is obliterated by the datapoints. This is because the centroids are almost identical but the class A variance is greater than that of class B; for two classes with identical centroids, samples will always be classified into the class with the

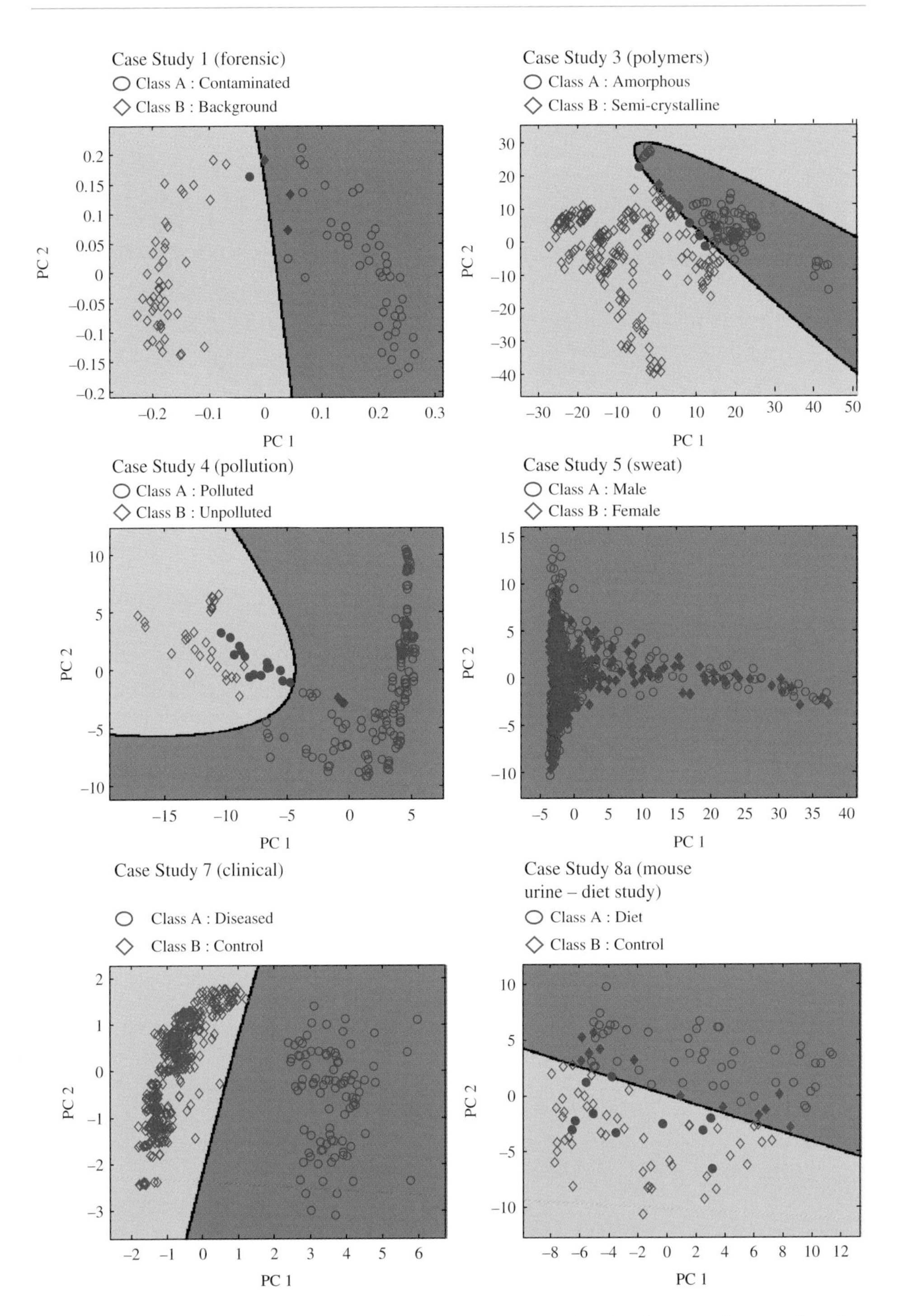

Figure 5.9 Visualization of boundaries obtained by QDA for the datasets discussed in this chapter, using default data preprocessing, and data represented by the scores of the first two PCs. Misclassified samples are represented by filled symbols, and the respective regions belonging to each class are coloured appropriately

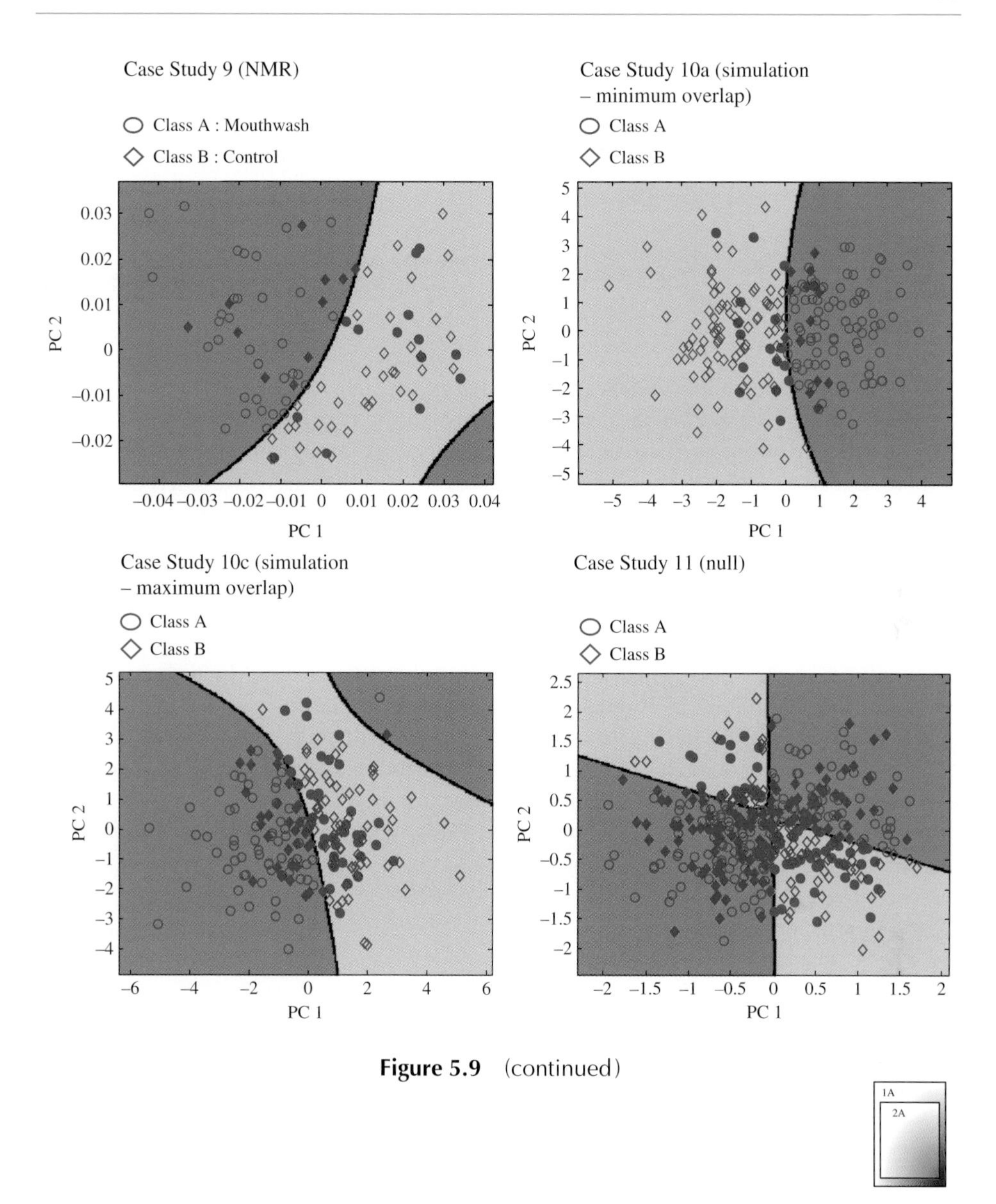

Figure 5.9 (continued)

greater variance providing that this variance is greater in both dimensions. Of course using more PCs to represent this data does permit discrimination but cannot be represented graphically. Case Study 9 (NMR spectroscopy) is divided into three regions. This is because the variance of class A (mouthwash) is greater (along the diagonal) than class B (control), and so although there are no samples in the bottom left-hand corner, the variance scaled distance to the centroid of class A is less than that for class B. A similar type of situation is found for Case Study 10c (simulation – maximum overlap).

5.5 Partial Least Squares Discriminant Analysis

The PLS (Partial Least Squares) method has a long vintage in chemometrics circles. One could almost divide chemometricians into two groups, those that use PLS for nearly everything and those that rarely use PLS. It is however an important approach for classification, and one that will be employed in frequent places in this text. Like LDA, PLS in its basic form is a linear method, and we will show that under certain circumstances it gives identical results to LDA. Although it is possible to include squared terms, this is rarely done, and in this book we will restrict ourselves to linear models.

The major advantage in PLS-DA comes from its flexibility, especially when coping with situations, unforeseen in traditional statistics, where the number of variables far outnumber the samples. Traditional pattern recognition studies are primarily concerned with whether two groups can be separated rather than why. We may be interested if we can separate two groups 'of face' that are visually distinct using a computational algorithm as that may be important for image processing: we are not primarily interested in whether people can be distinguished from the size of a pimple on the left of their face, but more whether we can produce an automated model to determine the differences between two or more groups. In chemometrics we are often interested in further diagnostics, for example, we measure a large number of variables, which may be related to chemicals, i.e. which ones are most influential on the classifier. In addition to classification abilities similar to LDA (any differences are primarily relating to data scaling) there is a wealth of diagnostics and there are a large number of statistics that have been historically developed in the chemometrics literature especially for determining the importance of the variables for discrimination, as discussed in Chapter 9. For some other classifiers, discriminatory power can also be measured but sometimes not so easily, for example, for LDA and QDA there cannot be more variables than samples and in the case of SVMs (Section 5.7) the main aim is to classify the samples rather than determine the influence of variables. Many historic classifiers, for example, in traditional biology, were developed in an era when the number of samples usually exceeded variables, which is not always the case in chemometrics; hence the need for different ways of handling the data.

In fact if one performs PLS on data retaining all non-zero PLS components and then performs PLS-DA as described below, and summarized in flowchart 4B, PLS-DA gives identical results to LDA for the classifier. This is summarized in Figure 5.10.

5.5.1 PLS Method

Although there are several algorithms, the main ones due to Wold and to Martens, the overall principles are quite straightforward. PLS tries to relate two types of variables, called the $\boldsymbol{X}$ block (in our case the experimental measurements) and the $\boldsymbol{c}$ block (in our case the classifier). The algorithm is applied to a $\boldsymbol{c}$ block that consists of a single vector or variables, which in our case consists of class labels $+1$ and -1 according to whether a sample is a member of class A or class B respectively, and two sets of equations are obtained as follows:

$$\boldsymbol{X} = \boldsymbol{T}\boldsymbol{P} + \boldsymbol{E}$$
$$\boldsymbol{c} = \boldsymbol{T}\boldsymbol{q} + \boldsymbol{f}$$

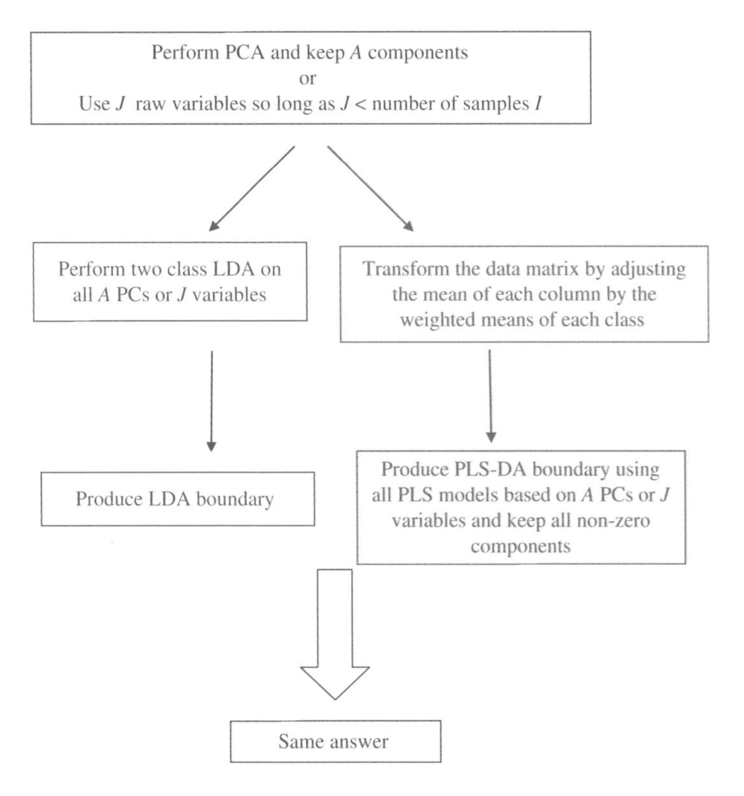

Figure 5.10 Equivalence of PLS-DA and LDA

where $\boldsymbol{q}$ has analogies to a loadings vector (see Section 3.2.2 on PCA), although is not normalized using the equations above. The dimensions of $\boldsymbol{T}$, $\boldsymbol{P}$ and $\boldsymbol{q}$ are $I \times A$, $A \times J$ and $J \times 1$ where A is the number of PLS components used in the model. The product of $\boldsymbol{T}$ and $\boldsymbol{P}$ approximates to the experimental data and the product of $\boldsymbol{T}$ and $\boldsymbol{q}$ to the $\boldsymbol{c}$ vector, the common link being $\boldsymbol{T}$. Note that $\boldsymbol{T}$ and $\boldsymbol{P}$ for PLS are different to $\boldsymbol{T}$ and $\boldsymbol{P}$ obtained in PCA. It is important to recognize that there are several algorithms for PLS available in the literature, and although the predictions of $\boldsymbol{c}$ should be the same in each case, providing the data have been suitably preprocessed in the same way and the same decision function is employed, the scores and loadings are not. In this book the algorithm given in Section 5.5.2 is employed. Although the scores are orthogonal (as in PCA), the $\boldsymbol{P}$ loadings are not (which is an important difference to PCA), and, furthermore, the loadings are not normalized, and so the sum of the squares of each $\boldsymbol{p}$ vector does not equal 1. In addition to scores and loadings, a further matrix $\boldsymbol{W}$ of dimensions $J \times A$, also called a PLS weight matrix, is obtained as a consequence of the algorithm.

There are various ways of determining A, the optimal number of PLS components, which we will discuss in Section 8.5.1 in more detail. In this chapter we illustrate using two variables and use two PLS components in the model for illustrative purposes. However when there are several variables (as in most situations) it is important to choose the optimal number of components correctly.

Additionally, the size of each PLS component can be calculated, in analogy to the eigenvalue of a PC, by multiplying the sum of squares of both $\boldsymbol{t}_a$ and $\boldsymbol{p}_a$ together, and so we define the magnitude of a PLS component as:

$$\kappa_a = \left(\sum_{i=1}^{I} t_{ia}^2\right)\left(\sum_{j=1}^{J} p_{aj}^2\right)$$

This will have the property that the sum of values of κ_a for all non zero components add up to the sum of squares of the original (preprocessed) data. Notice that in contrast to PCA, the size of successive values of κ_a does not necessarily decrease as each component is calculated. This is because PLS does not only model the $\boldsymbol{X}$ data, and is a compromise between $\boldsymbol{X}$ and $\boldsymbol{c}$ block regression, and the equation above does not include terms for q. There is of course no universal agreement as to how to determine the size of each successive PLS component, but the method described above is a common and well accepted one.

When there is more than one variable in $\boldsymbol{C}$ an alternative method called PLS2 can be employed . In fact there is little conceptual difference to PLS1 (which we will abbreviate as PLS in this book), except that the latter allows the use of a matrix, $\boldsymbol{C}$, rather than vectors for each individual group, and the algorithm is iterative. The equations above alter slightly in that $\boldsymbol{Q}$ is now a matrix not a vector, and so that:

$$\boldsymbol{X} = \boldsymbol{TP} + \boldsymbol{E}$$
$$\boldsymbol{C} = \boldsymbol{TQ} + \boldsymbol{F}$$

The number of columns in $\boldsymbol{C}$ and $\boldsymbol{Q}$ are equal to the number of groups of interest. In PLS1 one class is modelled at a time, whereas in PLS2 all known classes can be included in the model simultaneously; the algorithm is described in detail in Section 7.4.1. PLS2 can be used when there is more than one class in a dataset, but there are alternatives using PLS1, as discussed in Chapter 7.

5.5.2 PLS Algorithm

Model Building

The first step involves model building, that is, forming a relationship between both types of variables ($\boldsymbol{X}$ and $\boldsymbol{c}$). There are several algorithms in the literature for PLS1, but a simple one is as follows: it is assumed that the data have been preprocessed (e.g. row scaling, PCA, standardization, centring, as discussed in Chapter 4 previously). In this text we do not centre $\boldsymbol{c}$, but there are many variants on a theme.

1. Calculate the PLS weight vector $\boldsymbol{w}$:

$$\boldsymbol{w} = \boldsymbol{X}'\boldsymbol{c}$$

   where $\boldsymbol{X}'$ and $\boldsymbol{c}$ are the preprocessed experimental data and classifier.
2. Calculate the scores which are given by:

$$\boldsymbol{t} = \boldsymbol{X}\,\boldsymbol{w} \Big/ \sqrt{\sum w^2}$$

3. Calculate the x loadings by:

$$\boldsymbol{p} = \boldsymbol{t}'\boldsymbol{X} \Big/ \sum t^2$$

4. Calculate the c loading (a scalar) by:

$$q = \boldsymbol{c}'\boldsymbol{t} \Big/ \sum t^2$$

5. Subtract the effect of the new PLS component from the datamatrix to get a residual datamatrix:

$$^{resid}\boldsymbol{X} = \boldsymbol{X} - \boldsymbol{tp}$$

6. Calculate the residual value of $\boldsymbol{c}$:

$$^{resid}\boldsymbol{c} = \boldsymbol{c} - \boldsymbol{t}q$$

7. If further components are required, replace both $\boldsymbol{X}$ and $\boldsymbol{c}$ by their residuals and return to step 1. We discuss criteria for choosing the number of PLS components in Section 8.5.1 in the context of classification.

A weights matrix $\boldsymbol{W}$ can be obtained, with each successive column corresponding to a successive PLS component.

Prediction

Once a model is built it is then possible to predict the value of $\boldsymbol{c}$ both in the original data (autoprediction) and for future samples of unknown origins as follows, or for test sest samples.

The relationship between $\boldsymbol{X}$ and $\boldsymbol{c}$ can be expressed by:

$$\boldsymbol{c} = \boldsymbol{Xb} + \boldsymbol{f} = \boldsymbol{Tq} + \boldsymbol{f}$$

where $\boldsymbol{b}$ is a regression coefficient vector of dimensions $J \times 1$: hence an unknown sample value of c can be predicted by:

$$\hat{c} = \boldsymbol{xb}$$

The estimation of $\boldsymbol{b}$ can be obtained through the pseudo-inverse of $\boldsymbol{X}$ (denoted $\boldsymbol{X}^+$ – Section 2.2) as follows:

$$\boldsymbol{b} = \boldsymbol{X}^+\boldsymbol{c} = \boldsymbol{Wq}$$

where $\boldsymbol{W}$ contains the weight vectors of the PLS components. It is important to rescale $\boldsymbol{c}$ if necessary to return to the original scale.

5.5.3 PLS-DA

PLS-DA (Discriminant Analysis) uses the PLS algorithm as above, but the vector $\boldsymbol{c}$ contains a number which is a label for the class an object belongs to. There are various strategies for how to indicate numerically which class a sample belongs to but in this text we use a value of +1 if a sample is a member of the ‘in class’ and −1 otherwise. If there are equal numbers of samples in each class the $\boldsymbol{c}$ vector will be centred but it is not necessary to centre $\boldsymbol{c}$ even if the numbers of samples in each class are different. Of course there is no general agreement on this but there are other ways of correcting for unequal class sizes, as discussed in Chapter 8, one of which is through adjusting the acceptance threshold which can be visualized using ROC (Receiver Operator Characteristic) curves (Section 8.5.2). An additional recommendation is to adjust the column means of $\boldsymbol{X}$ according to the number of samples in each group represented in $\boldsymbol{c}$ (see Section 4.2.3) or $\bar{\bar{\boldsymbol{x}}} = (\bar{\boldsymbol{x}}_A + \bar{\boldsymbol{x}}_B)/2$, subtracting this from each row in the matrix (weighted centring). Whereas this is not mandatory for the PLS-DA algorithm, we will employ this by default. If there are equal numbers of samples in each class, this, of course, is the same as the overall mean. The reason for adjusting the $\boldsymbol{X}$ block column means is illustrated in Figure 5.11, which represents the

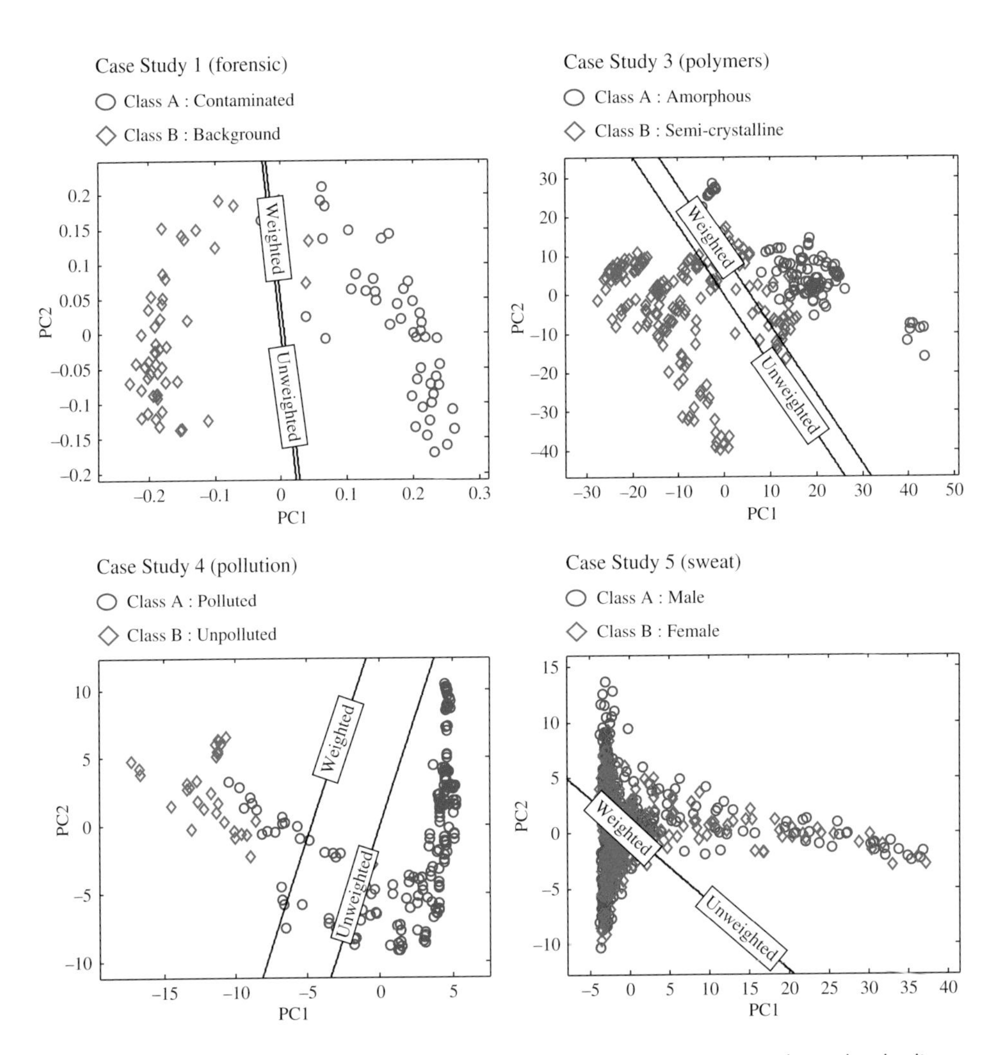

Figure 5.11 PLS-DA boundaries using both unweighted mean centring and weighted adjustment of means for Case Studies 1, 3, 4, 5, 7 and 8a; weighted methods are the default for this book

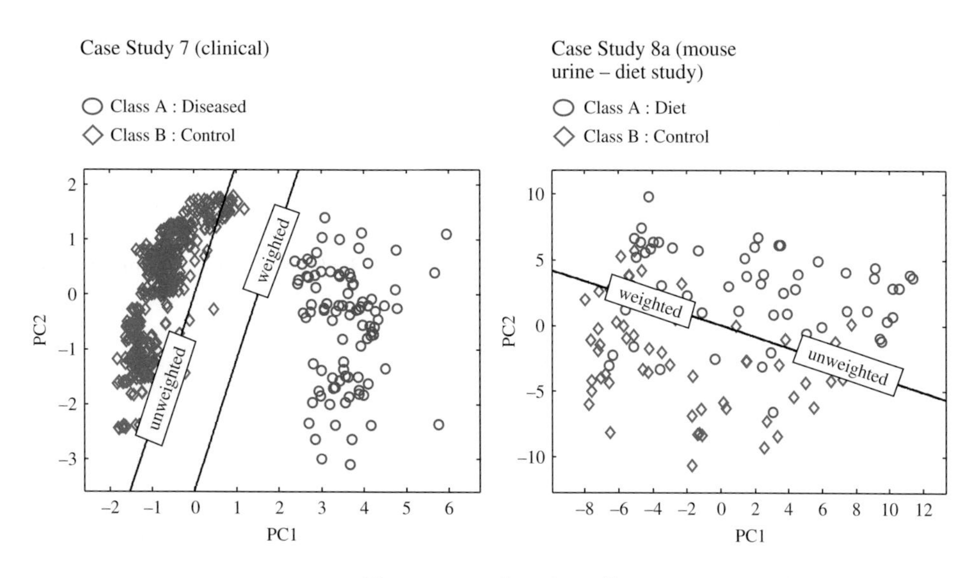

Figure 5.11 (continued)

PLS-DA boundaries for case studies where the two classes are unequal in size. If columns are mean centred (unweighted), the boundary moves towards the larger class: see, for example, Case Studies 7 and 4 where the numbers of samples in each class are very different. This introduces classification errors, for example, in Case Study 7, whereas the two classes are linearly separable, unweighted centring in PLS-DA places the boundary within the larger class (B). Adjusting the column mean according to the average mean of both classes places the boundary between each class, in an identical position to that obtained from LDA.

For any sample the value of $\boldsymbol{c}$ can be predicted according to the PLS-DA model as discussed in Section 5.5.2. There are a number of criteria for deciding which class a sample belongs to but the simplest for a two class model is to set up a cut-off threshold at $c = 0$: if positive it belongs to the 'in group' of class A, while if negative to the other class, and this determines the boundaries between the two classes. We will look at how this criterion can be changed in Section 8.5.2 and what the consequences are. Using the algorithmic description in this chapter and flowchart 4B (Chapter 4), LDA and PLS-DA results are identical to within computer precision and are presented in Figure 5.6. The advantage of PLS-DA lies in its greater interpretability, especially in terms of the original variables, which we will discuss in detail in Chapter 9.

5.6 Learning Vector Quantization

All the methods above treat samples from a single class as ideally belonging to one homogeneous group, and look for a single linear or quadratic boundary between classes. In traditional statistics most populations are homogeneous and can be described often in the form of a normal or log-normal distribution. However, in many cases in chemistry and biology this is not necessarily so. Let us say that we want to distinguish amorphous from semi-crystalline polymers (Case Study 3). The amorphous polymers consists of several different types or subgroups. Each subgroup in itself may be homogeneous but the overall polymer class consists of a set of these subgroups that may not necessarily form a nice 'neat' homogeneous distribution. An example is illustrated in Figure 5.12. One of the easiest ways to separate these groups is via a boundary consisting of series of straight lines.

Figure 5.12 Example of two groups, with one consisting of four subgroups represented by open symbols and one consisting of three subgroups represented by closed symbols, where these are best separated by a series of straight lines

Such situations can be quite common: a good example is in biological pattern recognition, where two populations may not be homogeneous, for example, even if diseased patients all have the same disease they might best be divided into subcategories, e.g. according to genetic origin. A traditional statistical argument may be that if this is so then the best way to approach this is to identify these subgroups and model these independently. However this is not always possible as first of all one has to identify these subgroups which is sometimes quite difficult, and second to model each subgroup adequately on its own there should ideally be sufficient samples in each such group. We will discuss issues of sample sizes in Chapter 8, but increasing the number of samples may be expensive and take time whilst one is waiting for results on the existing dataset, and so it is often useful to model smaller datasets first often as a proof of principle or to determine whether it is worth investing more resources in developing the methodology.

LVQ (Learning Vector Quantization) forms boundaries between groups based on a series of straight lines. It resembles the simpler kNN (k Nearest Neighbour) method; however in contrast to kNN, instead of using the original samples to form a model, it uses codebooks which are representative vectors of each group. The more the codebooks, the more complex the boundary. An advantage of LVQ over kNN is that the models are easier to validate, as discussed in Chapter 8. One reason is that for kNN and training set or autopredictive models, if training samples are left in the model there will always perfect prediction for $k = 1$; if we want to optimize k for autoprediction then there is no competition, and so for an autopredictive (or training set) model it is important to remove each sample from the model when testing how well it is predicted unlike the other approaches discussed in this text; the problem here is that different kNN models will be formed when testing each sample even when doing autoprediction. Many of the methods of validation and optimization fall down in this situation, as discussed in Chapter 9, if the kNN results are not directly comparable to those of the other methods discussed in this text and we cannot therefore directly compare the performance of kNN against any of the other classifiers; LVQ overcomes this problem by creating representative codebooks for each class which do not correspond to specific samples. A second problem relates to situations where class sizes are different: if there are more samples representative of one group than another, this biases the probability that an unknown is assigned to a particular class. Whereas this could be overcome by some form of weighted 'voting', for small values of k (e.g. $k = 1$) this is not very practical and so the algorithms have to be enhanced making them quite complicated. A disadvantage of LVQ is that the codebooks are obtained using a neural network and so will not be reproducible. As an example of variability we illustrate the boundaries obtained using three iterations of LVQ and boundaries for models based on two and four codebooks (the more the codebooks, the more complex the boundary), for Case Study 9 (NMR spectroscopy) in Figure 5.13 using the LVQ1+3 algorithm discussed below. Figure 5.14 represents another way for four of the case studies in this text, showing how the boundaries change (using autopredictive models) repeated 100 times with different random starting points and three codebooks per class. The position of the boundaries are primarily inflenced by which samples are included as the random 'seeds' representative of each class for each of the 100 models. For Case Study 11 (null) the boundaries are basically random as expected, and very unstable. For Case Study 1 (forensic) there is a stable boundary for negative PC2 scores, but some instability at positive values of PC2 (top of diagram). This is because there are a few samples that overlap from each class and so the boundary is quite uncertain in this region of the data space (there are some misclassifications also); however most datapoints are characterized by more negative scores in PC2 and as such are represented by a much more stable boundary. Case Study 4 (pollution) is interesting as there are three samples of class B (unpolluted – soil 4) buried within the main grouping represented by class A (polluted); if one of these samples is chosen as a 'seed' the boundary extends to the right; however the chance that one of these samples are chosen is relatively small (each sample has a 3 out of 34 chance of being chosen so the chance that at least one of these 3 samples being chosen is around 25 %, if there are three codebooks in the model) . For Case Study 3 (polymers) there are sufficient samples close to the border of each class to make the boundaries quite stable and not unduly influenced by choice of initial guesses for the codebooks. For autopredictive models, in this text, we use only one iteration, and show only one possible boundary as an illustration of the method. However, in this book we advocate

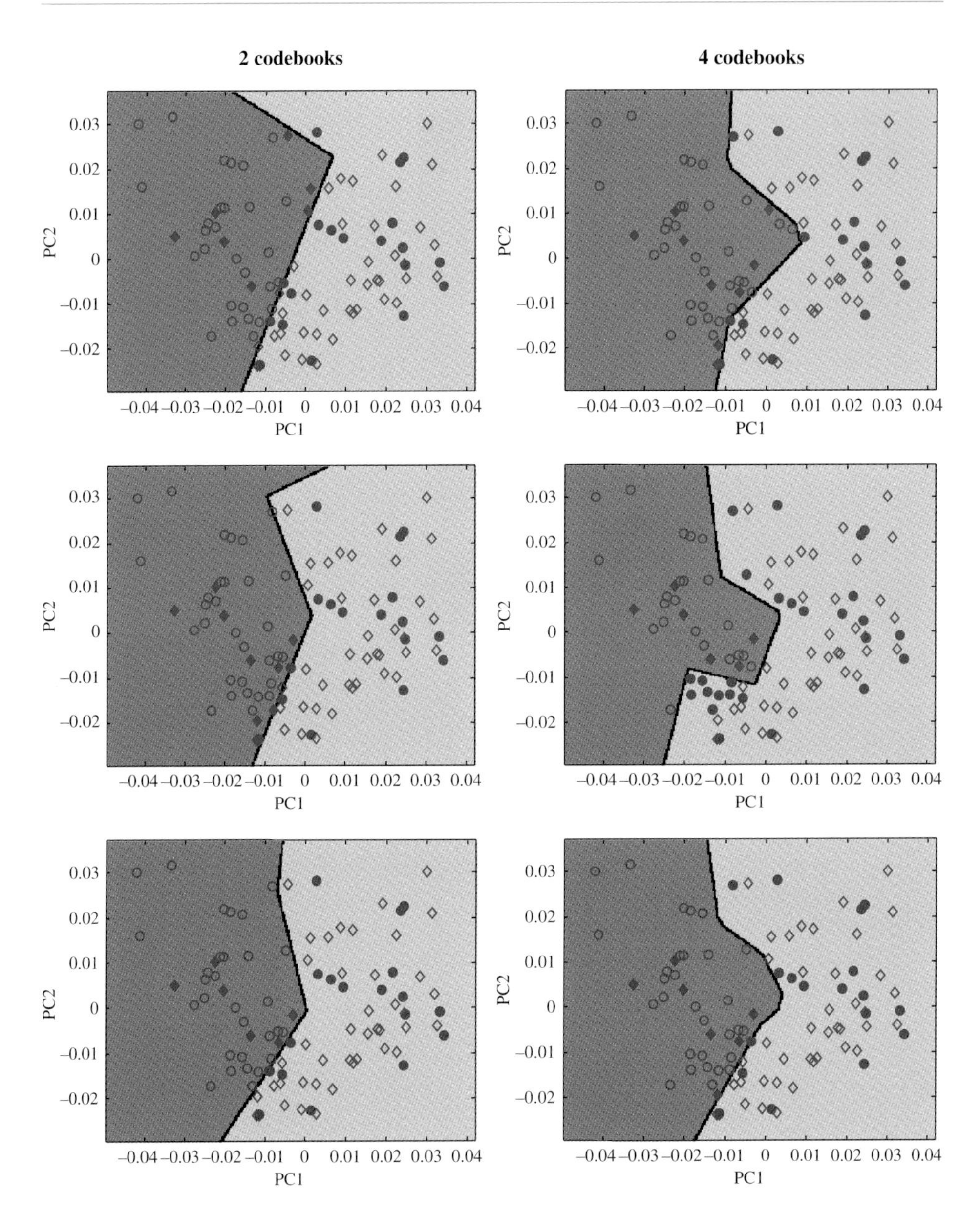

Figure 5.13 Three repeat iterations of LVQ on Case Study 9, using different numbers of codebooks, to demonstrate variability of the method

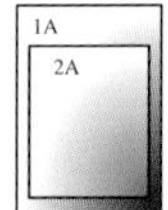

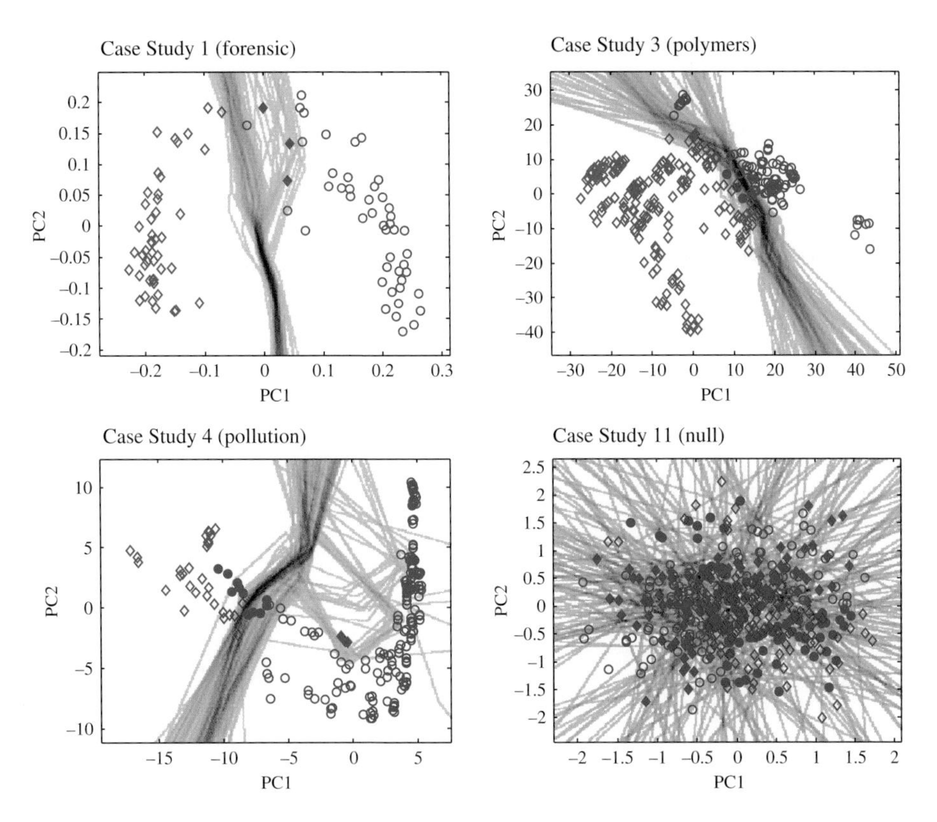

Figure 5.14 Variability of LVQ boundaries when the algorithm is performed 100 times using different starting points (three codebooks per class)

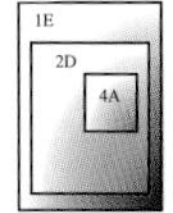

many iterative methods for testing and optimizing models (Chapter 8) and the classification results are typically based on a consensus of 100 iterations and as such are acceptably stable; even though there may occasionally be an outlying boundary, it is therefore recommended to perform the LVQ several times if one wishes to obtain useful statistics about the performance of this classifier (in this chapter we will illustrate the method using a single iteration primarily for the purpose of visualization). LVQ is implemented in an iterative manner and although computationally intensive, by using the modern generation of PCs is quite feasible to be performed on a powerful desktop computer.

For readers more familiar with kNN, many of the principles of LVQ are directly applicable, the main difference is that instead of using actual samples for classification we use codebooks that are representative vectors of each class. For brevity we employ only LVQ in this text, as the method is safer in situations where there are unequal class sizes and misclassification errors are directly comparable to the other approaches in this text.

5.6.1 Voronoi Tesselation and Codebooks

A useful concept for illustrating LVQs is Voronoi tesselation. Figure 5.15 represents a two dimensional space in which each point represents a codebook vector. The space is divided into regions so that each region contains one codebook. Each region is characterized by the area that has a specific codebook as its nearest neighbour. It is these lines and regions that make up the Voronoi tessellation. By removing lines that are between codebook vectors of the same class, the Voronoi regions can be merged so that each class region is represented. Any point falling into the defined class region will have a codebook from that class as its nearest neighbour. The lines remaining between codebook vectors from different classes therefore represent the boundaries between the classes.

Figure 5.16(a) represents 22 samples from to two classes that have been recorded using two measurements. If a linear border is used to separate the two classes then some misclassifications will occur. If LVQ is used, however, a piecewise linear border can be defined by adjusting the positions of a number of codebook vectors so that the

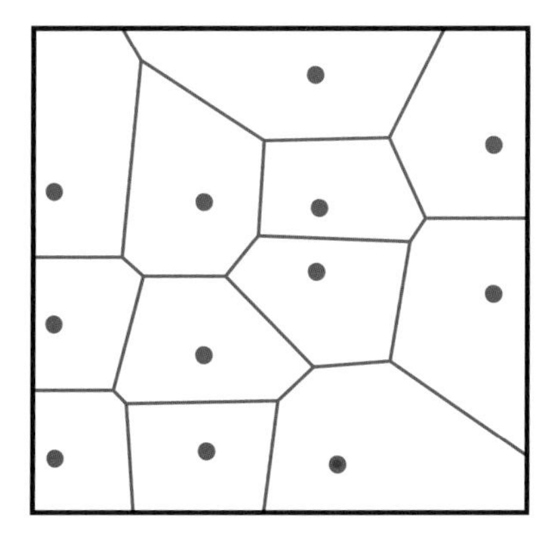

Figure 5.15 Voronoi tesselation of 12 points

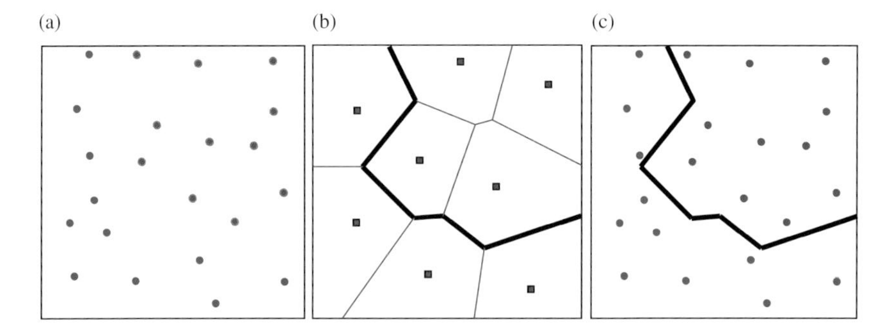

Figure 5.16 (a) Sample data (two classes, 22 samples). (b) Finalized codebook vector positions (points) and their Voronoi tessellation (lines) using four codebooks per class. The class border is defined by bold lines. (c) Class border (lines) with respect to the classification of the sample data

misclassification rate is approximately minimized. Figure 5.16(b) shows the finalized positions of eight generated codebook vectors (four for each class) and the Voronoi tessellation for these points. Since the codebook vectors have been assigned to class regions some of the lines (bold) can be used to distinguish between the codebook vectors from different classes. The border defined by the codebook vectors can be also be used to classify the samples (Figure 5.16(c)) as they were used to optimize the final positions of the codebook vectors so that there is a minimum chance of misclassifying a sample. The border between classes therefore approximates the border without requiring information about the distributions of each class.

The number of codebook vectors for each class can have a significant effect on the classification ability of the LVQ algorithm. Using multiple codebook vectors per class can improve the description of the classification borders, but too many may cause overfitting and result in poor classification of samples not included in the training set. The number of codebook vectors can be determined empirically or by modifying the LVQ algorithm to add and/or remove codebook vectors where needed. In this chapter and elsewhere we use a default of three codebooks per class, but one can use the normal approaches of cross validation or the bootstrap to change the number of codebooks in a model.

The original (LVQ1) algorithm was first described by Teuvo Kohonen, who also suggested further improvements in the forms of LVQ2 , LVQ2.1 and LVQ3. Often, the LVQ1 algorithm is sufficient to obtain satisfactory classification. Accuracy can generally be increased however by following the LVQ1 algorithm by one of the improved algorithms using a low initial learning rate in an attempt to fine tune the location of the decision borders. In this text we recommend that LVQ1 is followed sequentially by LVQ3 and so further discussion will be limited to these cases for the sake of brevity. LVQ2, although reported in the literature as a formal algorithm, is rarely used and as such has been superseded by LVQ3, and so it is not necessary to describe this in a general text, although of interest to specialists in the field. The sequential combination of the two algorithms will be referred to as LVQ1+3.

Once codebook vectors are defined it is easy to determine the predicted class membership of a sample; it is just necessary to measure the distance between the sample and all the codebooks and assign it to the class of the one it is nearest. Note that even if there are several codebooks per class LVQ just looks at the single nearest codebook for class membership information. Boundaries between classes are the positions in variable space which are equidistant from two codebooks of different classes. The more the codebooks, the more complex the boundary.

5.6.2 LVQ1

The LVQ1 algorithm (Figure 5.17) is the simplest algorithm because the position of only one of the codebook vectors is updated at a time. There are several enhancements to this basic algorithm which attempt to optimize the number of codebook vectors per class, or make changes to the way in which the codebook vectors are updated, but the basic algorithm however, remains almost unchanged.

The first step is to generate the codebook vectors for each class that are then combined into a $B \times J$ matrix $\boldsymbol{\Phi}$, where B is the total number of codebook vectors and J the number

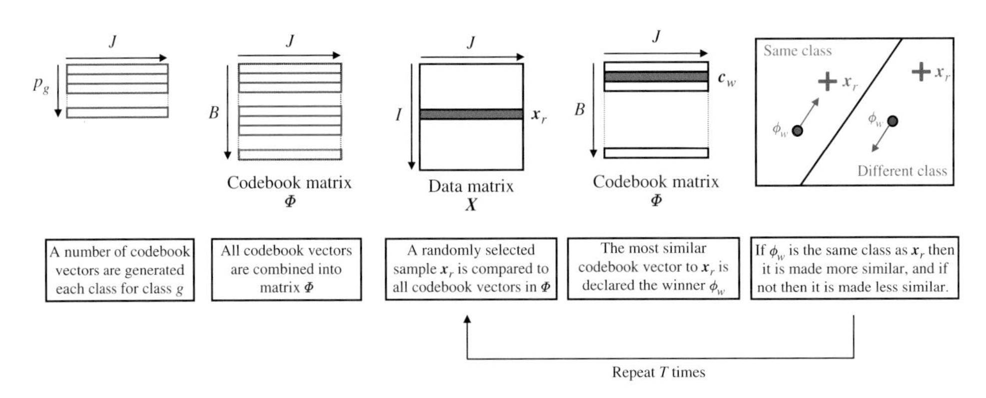

Figure 5.17 Schematic of the LVQ1 algorithm

of variables. This is to ensure that the codebook stays within its class region: if the codebook were to stray into the wrong class region it may be that it is pushed away from the class in every iteration causing it to never converge on an optimal position. Since the class regions are not yet known, one way to ensure the codebook vectors start within the correct class region is to use a sample that can already be correctly classified using another method. The codebook vectors should initially be placed within their class region of the variable space and usually are generated randomly, although there exist algorithms that are available that provide more systematic approaches, which we will not discuss for brevity. Note that if one splits the data into test and training sets (Chapter 8), it is of course necessary to choose these initial guessed codebooks just from the training set.

For each iteration, t, which takes integer values from 1 to the maximum number of iterations, T, there is a learning rate α_t for which a simple definition which we will use in this text is:

$$\alpha_t = \alpha_0 \left(\frac{T - t}{T} \right)$$

where α_0 is the initial learning rate. The learning rate determines how much the codebook vectors are moved and hence how much the decision boundaries are adjusted in each iteration. The exact form of this function is not critical provided that it reaches 0 over a suitable number of iterations, and there are several variants in the literature such as OLVQ1 (Optimized LVQ1) which result in somewhat faster convergence. Recommended values are for α_0 to be no larger than 0.1, and for T to be 50–200 times the number of codebook vectors. Although it is possible to choose different numbers of codebooks for each class it is easier (and computationally more efficient) to set the same number of vectors for each class. This allows classes of different sizes to be modelled and classified without bias. The number of codebooks should not be too large to avoid over-fitting. For the default examples in this book the number of codebooks is set to three for each class, unless otherwise stated.

For any iteration after choosing the number of codebook vectors, their positions are then iteratively updated for each value of t according to the following steps:

1. A sample vector $\boldsymbol{x}_r$ is selected from $\boldsymbol{X}$, where r is a randomly selected integer in the interval 1 to I which is generated for each iteration.
2. The Euclidean distance between each of the codebook vectors and the selected sample is calculated as follows:

$$d_{rb} = \sqrt{\sum_{j=1}^{J} (x_{rj} - \phi_{bj})^2}$$

3. The codebook with the smallest distance to the sample is declared the winner, $\boldsymbol{\phi}_w$ where w is the index of the winning codebook in matrix $\boldsymbol{\Phi}$ for each iteration. Depending on the class of the winner and the sample, the winner will be adjusted to try and improve the classification rate.
4. If the winning codebook is from the same class as the chosen sample, it is moved an amount proportional to the learning rate towards the selected sample:

$$\boldsymbol{\phi}_w = \boldsymbol{\phi}_w + \alpha_t(\boldsymbol{x}_r - \boldsymbol{\phi}_w)$$

If the winning codebook is associated with a different class to the sample then it is moved a proportional amount away from the sample:

$$\boldsymbol{\phi}_w = \boldsymbol{\phi}_w - \alpha_t(\boldsymbol{x}_r - \boldsymbol{\phi}_w)$$

All other codebook vectors remain unchanged:

$$\boldsymbol{\phi}_b = \boldsymbol{\phi}_b \text{ for } b \neq w$$

When training of the codebook vectors is complete ($t = T$), the class borders are approximately located so that samples can be classified by assigning them to the same class as their nearest neighbouring codebook vector with the minimum misclassification rate.

5.6.3 LVQ3

The basic LVQ1 algorithm can be modified to better approximate the class borders by updating two codebook vectors at a time, rather than just one. Additionally, the codebook vectors are only updated if the sample $\boldsymbol{x}_r$ falls in a window defined around the midpoint of its two nearest codebooks.

Figure 5.18 shows how the window is defined between two codebook vectors in a 2-dimensional example. The window is included to ensure that the class borders are adjusted such that the misclassification rate of samples on the edge of a class is at a minimum. If d_{rk} is the Euclidean distance of the sample $\boldsymbol{x}_r$ from its kth nearest codebook, then the sample is considered to be inside window of width s relative to the sample dimensions if:

$$\frac{d_{r1}}{d_{r2}} > \frac{1-s}{1+s}$$

This means that only codebook vectors with a sample near to the midplane of the two codebook vectors will be updated, thus ensuring that the class border is defined most accurately where samples are close to the border. If the value of s is too small then the number of times a codebook vector is updated will be small, even for a large number of

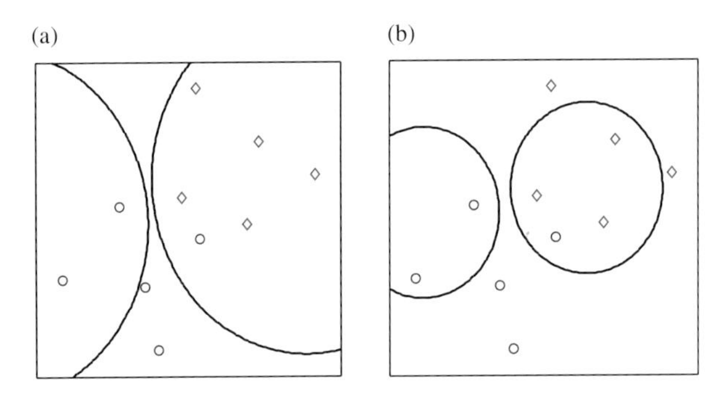

Figure 5.18 Illustration of the window for LVQ3. Each class is represented by five codebooks, coloured according to class. (a) $s = 0.1$ (b) $s = 0.2$

iterations. This means the codebook vectors have not been sufficiently trained which results in poorer accuracy. A relative window width of $s = 0.2$ to 0.3 is generally recommended, depending on the number of samples available for the learning process.

In LVQ3 there are two winning vectors, $\boldsymbol{\phi}_v$ and $\boldsymbol{\phi}_w$. Step 4 of the LVQ1 algorithm therefore needs to be updated for the codebook vectors to be adjusted appropriately as there are now four combinations for the class of the codebook vectors and the class of the sample that need to be considered (Table 5.2). If only one of the neighbouring codebook vectors is from the same class as $\boldsymbol{x}_r$, then the rules applied are similar to those used in LVQ1. If $\boldsymbol{\phi}_v$ is the winning vector of the same class as $\boldsymbol{x}_r$ then it is moved towards the sample:

$$\boldsymbol{\phi}_v = \boldsymbol{\phi}_v + \alpha_t(\boldsymbol{x}_r - \boldsymbol{\phi}_v)$$

whilst $\boldsymbol{\phi}_w$, which is from a different class to $\boldsymbol{x}_r$, is moved away:

$$\boldsymbol{\phi}_w = \boldsymbol{\phi}_w - \alpha_t(\boldsymbol{x}_r - \boldsymbol{\phi}_w)$$

The above rules may cause the codebook vectors to converge on a non-optimal solution. To overcome this, an additional learning rule is implemented for when $\boldsymbol{\phi}_v$, $\boldsymbol{\phi}_w$ and $\boldsymbol{x}_r$ are all

Table 5.2 *Rules for updating codebook vectors in LVQ3 when $\mathbf{x}_r$ falls within the defined window about the midplane of $\boldsymbol{\phi}_v$ and $\boldsymbol{\phi}_w$*

x_r class	c_v class	c_w class	Update rule
A	A	A	Move positions of $\boldsymbol{\phi}_v$ and $\boldsymbol{\phi}_w$ towards $\boldsymbol{x}_r$
A	A	B	Move position of $\boldsymbol{\phi}_v$ towards $\boldsymbol{x}_r$ Move position of $\boldsymbol{\phi}_w$ away from $\boldsymbol{x}_r$
A	B	A	Move position of $\boldsymbol{\phi}_v$ away from $\boldsymbol{x}_r$ Move position of $\boldsymbol{\phi}_w$ towards $\boldsymbol{x}_r$
A	B	B	Do not move any codebook vectors

from the same class. The additional rule includes an extra learning factor ε to ensure that the codebook vectors converge on their optimal positions:

$$\boldsymbol{\phi}_u = \boldsymbol{\phi}_u + \varepsilon\alpha_t(\boldsymbol{x}_r - \boldsymbol{\phi}_u)$$

where $\boldsymbol{\phi}_u \in \{v, w\}$. The exact value of ε is dependent on window size, but generally it is advisable for it to have a value between 0.1 and 0.5. LVQ3 is generally used after the LVQ1 algorithm to fine tune the location of the codebook vectors. For this reason the initial learning rate α_0 for LVQ3 should start no higher than 0.01.

An unknown sample is then assigned to the class of the nearest codebook. Note that even if each class is characterized by several codebooks only the single nearest one is used for classification (differing from kNN where several samples may be used for the test if $k > 1$).

5.6.4 LVQ Illustration and Summary of Parameters

The recommended range of parameters and the defaults used in this text are summarized in Table 5.3. In practice these parameters have very little effect on the resultant boundaries; more crucial considerations are how many codebooks are used to define each class and the initial guesses of codebooks.

Table 5.3 *Parameters used for training LVQ: (a) recommended range; (b) defaults used in this text*

(a)

Parameter	Description	Recommended values
α_0 (LVQ1)	Initial learning rate	≤0.1
α_0 (LVQ3)	Initial learning rate	≤0.01
s (LVQ3)	Window width	0.2–0.3
ε (LVQ3)	Second learning rate	0.1–0.5
T	Number of iterations	50–200 × Number of codebook vectors

(b)

Parameter	Description	Default value
α_0 (LVQ1)	Initial learning rate	0.1
α_0 (LVQ3)	Initial learning rate	0.01
s (LVQ3)	Window width	0.2
ε (LVQ3)	Second learning rate	0.2
T	Number of iterations	10 000

The LVQ boundaries for the case studies discussed in this chapter are illustrated in Figure 5.19, using three codebooks per class and a single set of iterations. In some situations such as Case Study 4 (pollution) LVQ more accurately represents the boundary between the classes, resulting in less misclassifications than any of the other methods discussed so far. For Case Study 11 (null) there is quite significant flexibility to adjust the

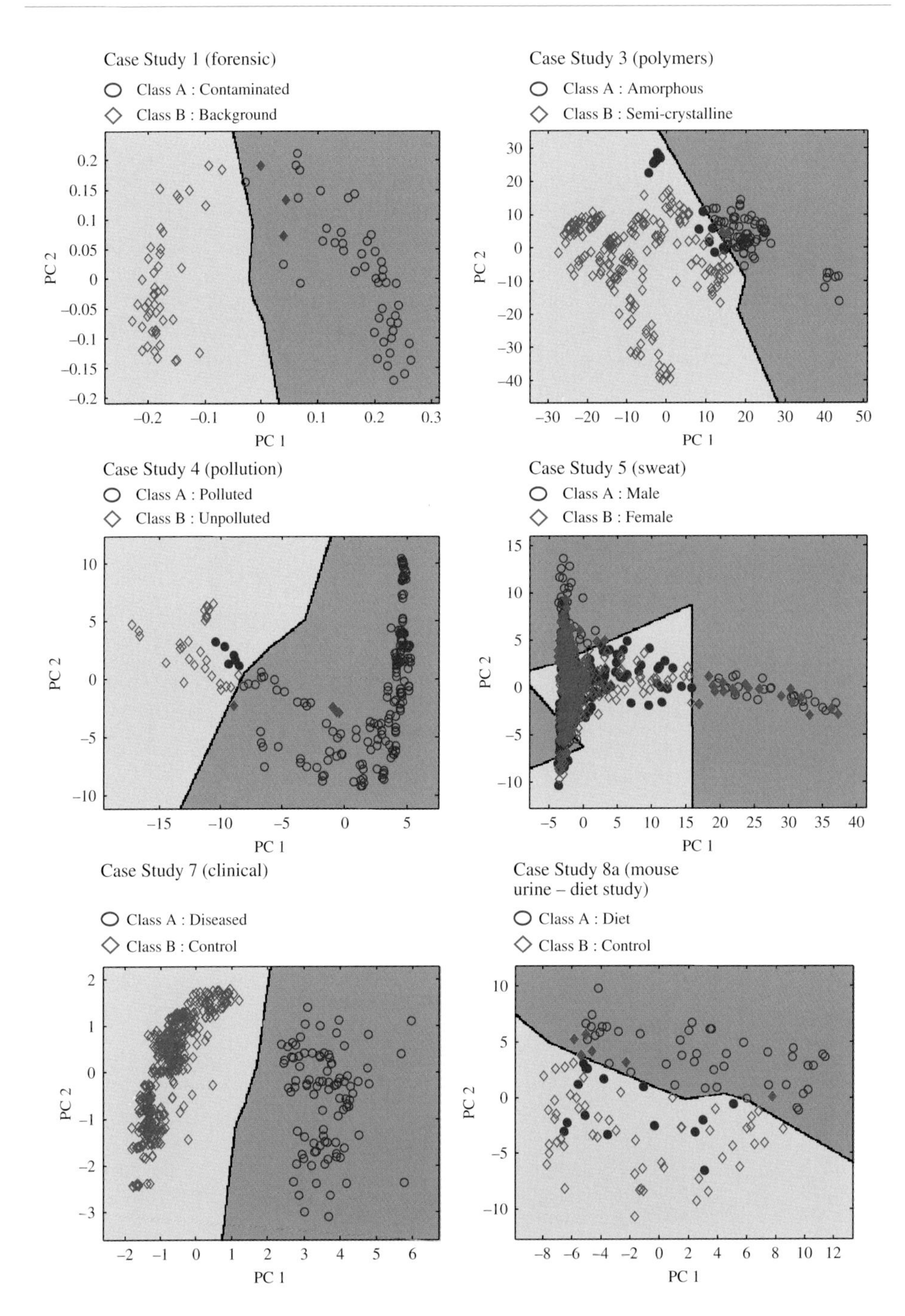

Figure 5.19 Visualization of boundaries obtained by LVQ for the datasets discussed in this chapter, using default data preprocessing, and data represented by the scores of the first two PCs. Misclassified samples are represented by closed symbols, and the respective regions belonging to each class are coloured appropriately. Boundaries are formed using three codebooks per class

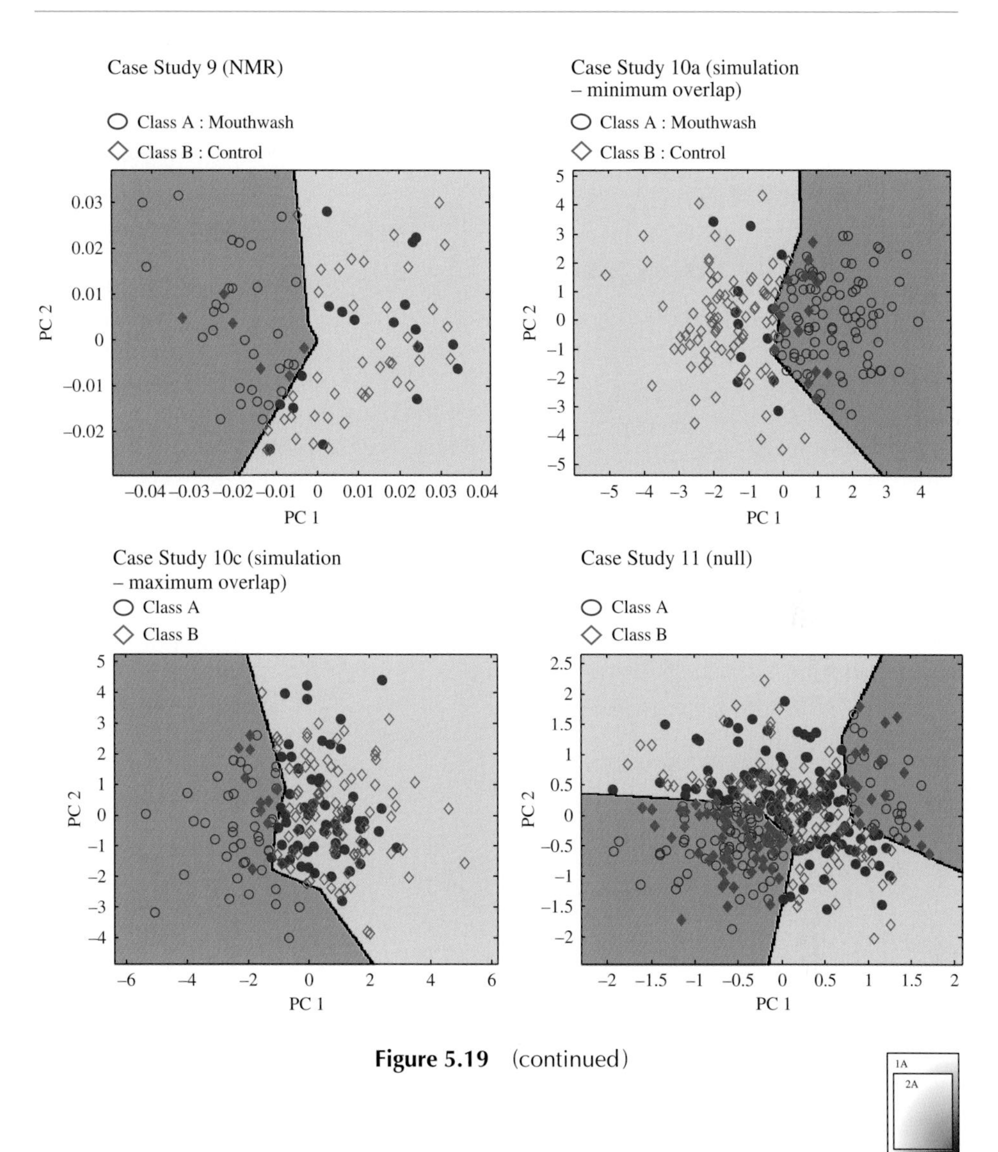

Figure 5.19 (continued)

boundaries compared to other approaches. Remember that LVQ can produce very complex boundaries if the number of codebooks is increased; however this comes with the risk of overfitting, which will be discussed in Chapter 8 and so careful validation approaches are necessary.

5.7 Support Vector Machines

Whereas LVQ results in boundaries that are a combination of straight lines (or hyperplanes when more than two variables are used for the model), SVMs (Support Vector Machines) can produce complex curved boundaries. In fact SVM models can involve boundaries of indefinite complexity, as is sometimes necessary when studying non-linear processes such as in biology. The more complex the boundary, the better it is likely to perfectly separate objects in the training set. However such a boundary can overfit the data, that is, it may work well on a training (or autopredictive) set of samples but poorly once unknowns or test set samples are introduced. How far to go down the route of complexity is difficult to judge as each problem is unique. We will discuss issues of validation of models in Chapter 8 in more depth.

The description of the SVM algorithm below is in three parts. First, the basic mathematical derivation for linearly separable classes, second, the extension to the non-linearly separable case with the use of kernel functions and third the generalized solution with the incorporation of a trade-off parameter to control complexity. The description in this section will be restricted to two classes, while extensions will be described in Chapters 6 and 7. SVM differ from other models discussed in this chapter in that the boundaries are determined just by a small number of samples from the original dataset, called support vectors.

Enthusiasts of SVM will advocate that almost any boundary or classification problem can be performed using SVM, even simple ones where the classes are linearly separable. However in a text of this nature it is important to provide a broad overview of the methods available and many chemometrics problems are well solved using much simpler and less computationally intense approaches. The trick of the chemometrician is to know that these alternative approaches are available and decide when it is necessary to move towards a more flexible and computationally intense method such as SVM or when to stick to an easily understood and computationally cheap approach. This depends on the type of problem and nature of the data.

5.7.1 Linear Learning Machines

The simplest type of classifier is a linear classifier. However SVM treat linear clsssification problems somewhat differently to the other methods discussed in this chapter as described below. For a simple linear problem the method is probably overkill but if first understood this is an initial conceptual building block for understanding SVM.

Consider a binary classification problem where samples $\boldsymbol{x}$ have been obtained that have membership of two classes g (=A or B) with labels $c = +1$ for class A and -1 for class B and are perfectly linearly separable. These samples can be used to determine a decision function to separate two classes, which in its simplest form can be expressed by a linear boundary:

$$g(\boldsymbol{x}_i) = \mathrm{sgn}(\boldsymbol{w}\ \boldsymbol{x}_i' + b) = \mathrm{sgn}\left(b + \sum_{j=1}^{J} w_j x_{ij}\right)$$

where $\boldsymbol{w}$ and b are often called weight and bias parameters that are determined from the training set. Note the similarity between this basic equation and those for linear classifiers e.g. PLS-DA which may could be expressed as:

$$g(\boldsymbol{x}_i) = \mathrm{sgn}(\boldsymbol{x}_i \boldsymbol{b})$$

where $\boldsymbol{b}$ in the case of PLS-DA is the vector of regression coefficients, equivalent to $\boldsymbol{w}$ (the weights for SVM) and $\boldsymbol{x}$ has already been preprocessed (we recommend adjusting column means to take into account the average mean of each class in the dataset). Although there is some incompatibility between notation (for example 'b' is commonly employed as a bias term involving changing the intercept, in SVM, but used for regression coefficients in PLS-DA, and the meaning of a 'weight' is different using both techniques), we will adopt commonly agreed notation for SVM in this text so as to be compatible with existing literature, with the exception that $\boldsymbol{x}$ for an individual sample is a row vector, to be compatible with this text. However it is important to understand the relationship between different approaches, and that SVM when expressed in its simplest linear form has analogies to PLS-DA which also is related to LDA, and advocates of SVM would argue that SVM is a more generalized and universal approach, although one which carries risks of overfitting and unnecessary complexity if one is not careful, and which forms a boundary using a different criterion to PLS-DA (or other approaches discussed above). However, this simple classification function corresponds to representing the border between two classes as a hyperplane, or a line if $\boldsymbol{x}$ is 2-dimensional. The sign of g determines which class a sample is assigned to, if positive class A and if negative class B. A generic hyperplane ($\boldsymbol{w}$, b) can be defined by coordinates $\boldsymbol{x}$ satisfying the condition $\boldsymbol{wx}' + b = 0$ which divides the data space into two regions opposite in sign. If the two classes are separable we can define a 'margin' between the two classes, such that:

$$\begin{aligned} \boldsymbol{wx}' + b \geq 1, \quad c = +1 \\ \boldsymbol{wx}' + b \leq -1, \quad c = -1 \end{aligned}$$

since no samples will be precisely on the boundary. The value ±1 can be obtained by by scaling $\boldsymbol{w}$ and b appropriately. The hyperplane should be equidistant from the two extreme samples in each class. The hyperplane separates the classes with no error if any sample $\boldsymbol{x_i}$ is projected in the region of the data space with a sign equal to the respective class membership c_i or at least on the boundary, i.e. the hyperplane must satisfy the condition:

$$c_i(\boldsymbol{wx}_i' + b) \geq 1$$

for all the samples providing they are perfectly linearly separable. However there are an infinite number of possible hyperplanes (w, b) satisfying this condition, as illustrated in Figure 5.20(a), for two variables, where the border is a line and so there needs to be a further rule to determine which of these hyperplanes is best.

The optimal separating hyperplane, as chosen using SVM, is defined by the parameters $\boldsymbol{w}$ while b is one for which the margin between the most similar samples in each group is largest. It can be shown that this hyperplane is the one that minimizes $0.5\,(\boldsymbol{ww}')$, subject to the constraint $c_i(\boldsymbol{wx}_i' + b) \geq 1$ for all samples. This is illustrated in Figure 5.20(b). The samples on the margins are called support vectors (SVs) as illustrated in the figure.

Algebraically, this optimization task can be expressed by the structure error function:

$$\varphi(\boldsymbol{w}, b, \boldsymbol{\alpha}) = 0.5(\boldsymbol{ww}') - \sum_{i \in sv} \alpha_i [c_i(\boldsymbol{wx}_i' + b) - 1]$$

where N_{sv} is the number of samples for which both $c_i\left(wx_i' + b\right) \geq 1$ and in addition $\alpha_i > 0$, which are a subset of the original samples called the Support Vectors (SVs). The samples

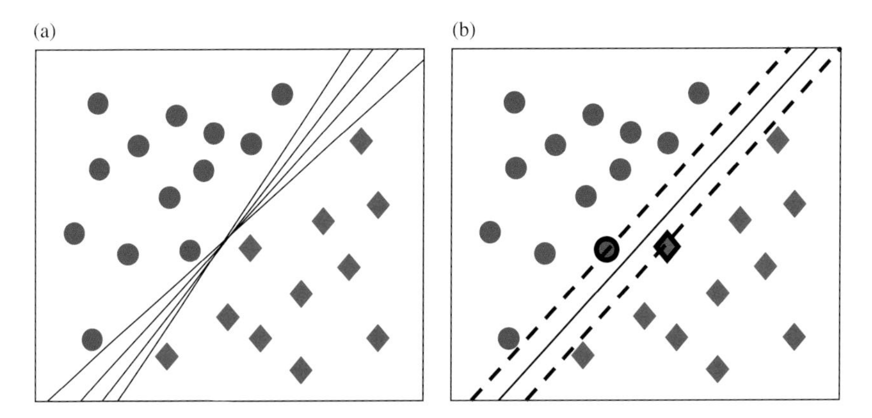

Figure 5.20 (a) Possible separating hyperplanes and (b) optimal separating hyperplane (right) maximising the margin between two classes, the closest samples being indicated as support vectors

that have $\alpha_i > 0$ are those that are closest to the boundary. Hence the SV solution depends only on samples close to the boundary between two (or more) classes. In this way SVM models differ from other models described in this chapter in that they are only determined by a small number of the samples in the original dataset.

The parameter α is called a Lagrange multiplier and is common in calculus. For readers not familiar with Lagrange multipliers, they are used to optimize a function subject to one or more constraints. A simple example involves finding the dimensions of a rectangle that gives the maximum area (optimization) and whose sum of the length of the sides is 16 (constraint). The constraint can be algebraically expressed as:

$$g(x) = x_1 + x_2 = 16$$

and the function we want to optimize is:

$$f(x) = x_1 x_2$$

where x_n is the length of the nth dimension of the rectangle. The Langrange multipliers are defined by the value of α that is obtained from the following equation, subject to the constraint:

$$\nabla f = \alpha \nabla g$$

where ∇ represents the partial derivative over each of the variables. In the case of the box, we have two variables over which we can differentiate and so there are three equations we need to solve, the latter representing the constraint:

$$x_2 = \alpha x_2$$
$$x_1 = \alpha x_1$$
$$x_1 + x_2 = 16$$

We can see from this analysis that $\alpha = 1$ and the rectangle must have all sides equal and of length 4, given the condition imposed by the 3rd equation.

In the context of SVMs, the value of φ has to be minimized with respect to $\boldsymbol{w}$ and b and maximised with respect to α_i. The minimum of φ with respect to $\boldsymbol{w}$ and b is given by:

$$\frac{\partial \varphi}{\partial b} = 0 \Rightarrow \sum_{i \in sv} \alpha_i c_i = 0 \text{ and } \frac{\partial \varphi}{\partial \boldsymbol{w}} = \boldsymbol{0} \Rightarrow \boldsymbol{w} - \sum_{i \in sv} \alpha_i c_i \boldsymbol{x}_i' = \boldsymbol{0}$$

Hence:

$$\varphi(\boldsymbol{\alpha}) = \frac{1}{2} \sum_{i \in sv} \sum_{l \in sv} \alpha_i c_i (\boldsymbol{x}_i \boldsymbol{x}_l') c_l \alpha_i - \sum_{i \in sv}^{l} \alpha_i$$

The optimization task is that of minimizing $\varphi(\boldsymbol{\alpha})$ with respect to $\boldsymbol{\alpha}$, a vector consisting of Lagrange multipliers, satisfying the constraints:

$$\alpha_i \geq 0 \text{ and } \sum_{i \in sv} \alpha_i c_i = 0$$

Finally, the optimal $\boldsymbol{\alpha} = (\alpha_1, \alpha_2, \ldots, \alpha_{Nsv})$ allows determination of the weight vector $\boldsymbol{w}$ of the optimal separating hyperplane:

$$\boldsymbol{w} = \sum_{i \in sv} \alpha_i c_i \boldsymbol{x}_i$$

while the offset b can be calculated from any pair of samples of opposite classes satisfying the conditions that their values of α are greater than 0. The optimization of φ is a quadratic programming problem, which can be generally written in the form:

$$\min_{a} \left(\frac{1}{2} \alpha \boldsymbol{H} \boldsymbol{a}' + z\alpha' \right)$$

where $\boldsymbol{H}$ has elements $h_{i,l} = c_i(\boldsymbol{x}_i' \boldsymbol{x}_l) c_l$ (for samples i and l) and z is a row vector of '-1s'. This is a well known type of an optimization problem that is straightforward to solve because it has only one global minimum, thus making the learning procedure reproducible. We will not discuss the details of this optimization method in this text as it can be found in many general references on numerical programming. The expression for φ contains a scalar product of vectors and explains why the approach is particularly fast and suitable when dealing with samples having many variables. Last but not least, this opens the way to treat some of the more complicated non-linearly separable cases using kernels as discussed in Section 5.7.2.

The classifier can be directly expressed as a decision function in terms of the SVs' $\boldsymbol{s}_i$ (those samples with values of $\alpha > 0$) as follows:

$$g(x_l) = \text{sgn}\left(\sum_{i \in sv} \alpha_i c_i \boldsymbol{s}_i \boldsymbol{x}_l' + b_c \right)$$

If positive the samples are assigned to class A, otherwise to class B.

SVs are often visualzised as being on the margins of each class, with a hyperplane representing the decision boundary. Which side of the hyperplane a sample lies relates to its class membership, whilst the SVs and margins are the extremes of each class. In an ideal situation there will be an empty space between the margins, providing that classes are completely separable.

5.7.2 Kernels

Determining a class boundary in the form of a separating hyperplane is adequate for simpler cases where the classes are nearly or completely linearly separable. However, this is a situation where arguably many other methods as discussed above would return satisfactory results and SVMs would not appeal very much due to their relatively complex formulation, and so are most useful where classes are not linearly separable. As an example we will look at how to produce a boundary between the two classes represented in Figure 5.21, which are not linearly separable.

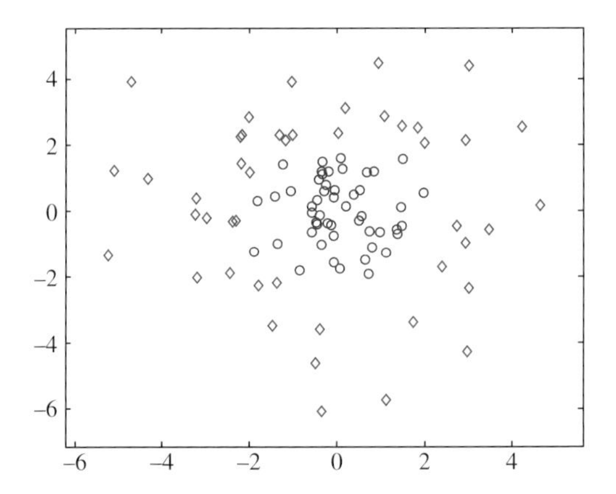

Figure 5.21 Example of the use of kernels with SVM

SVMs handle this by adding an extra step to the procedure described above. Instead of forming a boundary in the original variable space, where the two classes are not separable, a new higher dimensional (feature) space, where the samples are projected by means of a feature function $\Phi(\boldsymbol{x})$, is defined. The back-projection of the optimal separating boundary (in the form of a hyperplane) from this new feature space to the original variable space will then result in a non-linear boundary of given complexity that better suits the distribution in the original variable space, providing that the feature space is correctly defined, as illustrated in Figure 5.22. This new dataspace is often of high dimensionality with one dimension per SV. Their mappings by means of $\Phi(\boldsymbol{x})$ allows the determination of a hyperplane that separates them. A feature function is found that makes separation easier in higher dimensions. Finally the back-projection of this plane into the original dataspace generates a non-linear boundary of given complexity.

In most situations, the set of functions $\Phi(\boldsymbol{x})$ that is used to map the data is of very high dimensionality, which means that many more dimensions are generally added rather than only one, but it is consequently possible to find boundaries to suit a variety of complex distributions. Mathematically, this is done by reformulating the optimization task by replacing the scalar product of input vectors $(x_i x_l')$ with the scalar product of the respective feature functions defined by $\langle \boldsymbol{\Phi}(x_i), \boldsymbol{\Phi}(x_l) \rangle$ as follows:

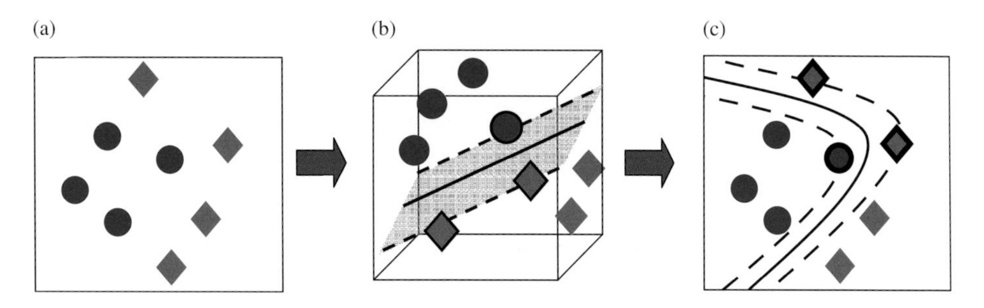

Figure 5.22 Creation of the boundary for a non separable case. (a) Two linearly inseparable classes in two dimensions. (b) Projection onto a higher dimensional space where it is possible to separate the classes using a plane, with three support vectors indicated. (c) Projection back into two dimensions

$$\varphi(\boldsymbol{\alpha}) = \frac{1}{2}\sum_{i\in sv}\sum_{l\in sv}\alpha_i c_i \langle \boldsymbol{\Phi}(\boldsymbol{x}_i),\ \boldsymbol{\Phi}(\boldsymbol{x}_l)\rangle\, c_l\alpha_l - \sum_{i\in sv}^{l}\alpha_i$$

$$\text{replacing}\ \ \varphi(\alpha) = \frac{1}{2}\sum_{i\in sv}\sum_{l\in sv}\alpha_i c_i (\boldsymbol{x}_i\boldsymbol{x}_l{}')\, c_l\alpha_l - \sum_{i\in sv}^{l}\alpha_i$$

and so it is mainly necessary to find these functions to develop an SVM model using kernels.

An important concept in SVMs is that there exist kernel functions K in the original variable space that corresponds to the dot product of functions in the new feature space:

$$K(\boldsymbol{x}_i, \boldsymbol{x}_l) = \langle \boldsymbol{\Phi}(\boldsymbol{x}_i), \boldsymbol{\Phi}(\boldsymbol{x}_l)\rangle$$

The optimization task can therefore be re-written as follows:

$$\varphi(\boldsymbol{\alpha}) = \frac{1}{2}\sum_{i\in sv}\sum_{l\in sv}\alpha_i c_i K(\boldsymbol{x}_i, \boldsymbol{x}_l) c_l\alpha_l - \sum_{i\in sv}\alpha_i$$

The optimization task still involves a quadratic convex programming problem, which is particularly easy to handle, but most importantly, by means of $K(\boldsymbol{x}_i, \boldsymbol{x}_l)$ rather than $\boldsymbol{\Phi}(\boldsymbol{x})$, it is possible to proceed with the transformation, omitting the intermediate step of creating the feature space and working only in the original dataspace where $K(\boldsymbol{x}_i, \boldsymbol{x}_l)$ is defined (which can be added as an extra dimension). This powerful attribute is known as the kernel trick and it is what makes SVMs effective in addressing complex tasks. The classification decision function can be re-written as:

$$g(\boldsymbol{x}) = \text{sgn}\left[\sum_{i\in sv}\alpha_i c_i K(\boldsymbol{s}_i, \boldsymbol{x}) + b\right]$$

and is still explicitly expressed in a dependence on the SVs.

Only certain kernels can be employed (as they also must satisfy some additional conditions). Some of the most common are as follows:

1. Radial basis function (RBF) defined as $K(\boldsymbol{x}_i, \boldsymbol{x}_j) = \exp\left(-\gamma \|\boldsymbol{x}_i - \boldsymbol{x}_j\|^2\right)$ or $K(\boldsymbol{x}_i, \boldsymbol{x}_j) = \exp - \|\boldsymbol{x}_i - \boldsymbol{x}_j\|^2 / 2\sigma^2$ where $\gamma = 1/2\sigma^2$.
2. Polynomial function (PF) $K(\boldsymbol{x}_i, \boldsymbol{x}_j) = \left(a\ \boldsymbol{x}_i \boldsymbol{x}_j' + b\right)^c$.
3. Sigmoidal function (SF) $K(\boldsymbol{x}_i, \boldsymbol{x}_j) = \tanh\left(a\ \boldsymbol{x}_i \boldsymbol{x}_j' + b\right)$.

These kernel functions can be visualized as creating an extra dimension, involving a sum of functions centred on each sample that is assigned as a support vector. The creation of this kernel (in this example an RBF) is exemplified in Figure 5.23, in which a third dimension, representing the decision function multiplied by the class membership label of each sample and by its Lagrange multiplier, is added for which the two classes are separable. The mesh is the value of the distance of each sample from the centre in the RBF higher dimensional space that cannot be visualized (the additional vertical axis should not be confused with the additional higher dimension feature space). Note that values of the mesh are given by $\Sigma_{i \in sv} \alpha_i c_i K(\boldsymbol{s}_i, \boldsymbol{x})$. Samples that are not SVs are projected onto this space. The procedure of forming boundaries and backpropagation are represented in Figure 5.24. The value of b is a decision threshold and can be visualized as creating a plane that separates the surface into two parts, those above the plane (assigned to class A) and those below (class B). In Figure 5.25, we illustrate how the model depends on the SVs. Each of the samples that are identified as SVs are the centre of a Gaussian RBF, the sign being positive for members of class A ($c_i = 1$) and negative for members of class B ($c_i = -1$). 'Non-SVs' are projected onto the surface, but are not used to form this surface. We can rotate the surface onto the original dataplane to see the distribution of the SVs, or at

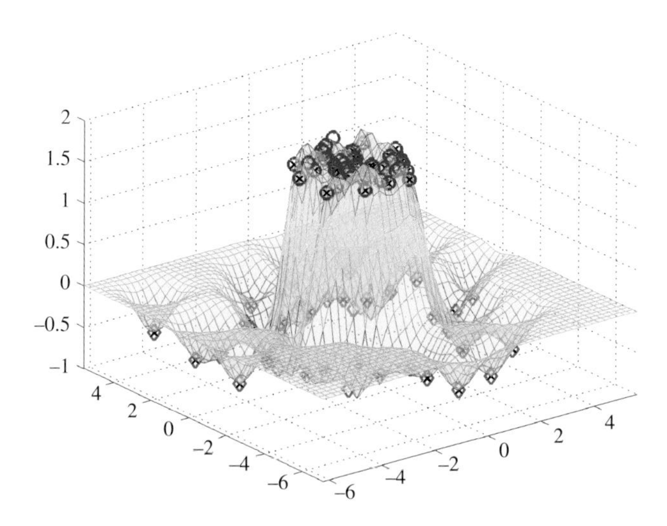

Figure 5.23 Creating a kernel function for separating two classes in the dataset of 5.21. The vertical axis relates to the decision function multiplied by the class label and the Lagrange multiplier for each SV

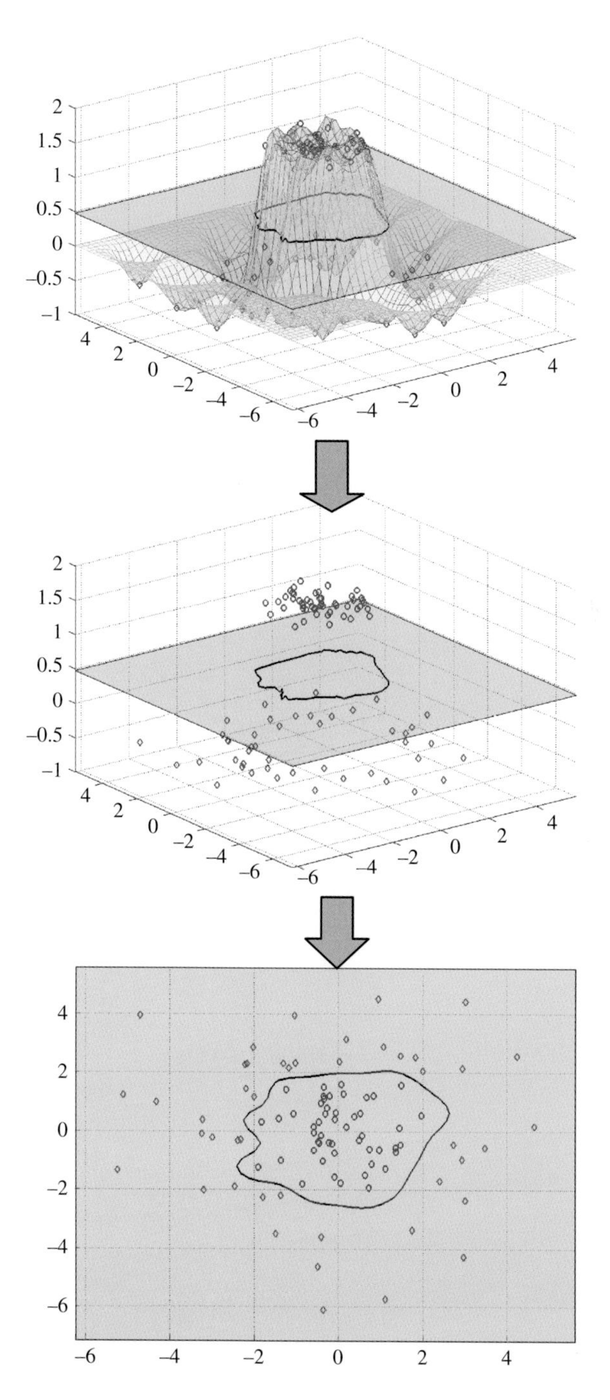

Figure 5.24 Developing an SVM model

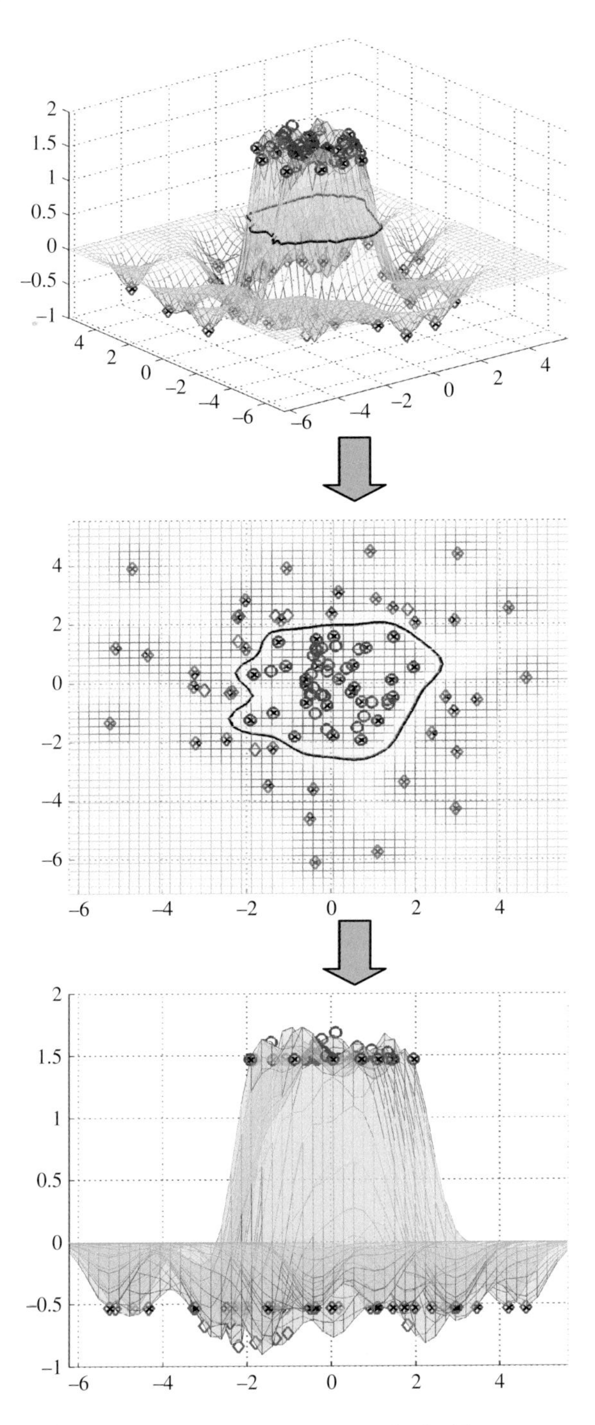

Figure 5.25 Support vectors – marked with a cross

right angles to this to see where these are distributed and so the empty margins between the SVs for each class.

Each kernel has a set of parameters that must be optimized for a particular problem. The RBF is particularly popular because it requires only one parameter to be optimized (the value of γ or σ). This has many advantages in optimization as SVMs are computationally intensive procedures and so optimizing several parameters using procedures such as cross-validation can be time consuming; if there are several parameters for optimization and if incorrectly performed can lead to risks such as overfitting. In this text we will restrict the illustration to RBFs, which should cover the vast majority of situations encountered in chemometrics. The interested reader may want to look at some of the source literature if it is felt that this type of function is inadequate. There is usually a limit to the level of complexity that can reasonably be modelled, especially when datasets are limited in size and contain experimental errors, and RBFs result in some quite complex boundaries and so are probably at the upper limit of what a chemometrician might encounter; biologists mining large databases (e.g. in genetics) may have problems that justify going farther. We will discuss the influence of different RBF parameters on the performance of SVM in Section 8.7.

5.7.3 Controlling Complexity and Soft Margin SVMs

Intuitively, because the kernel trick allows SVMs to define complex boundaries, the risk of overfitting is particularly high, that is, it is possible to define almost any boundary around training set samples even if there is no particular significance to these complex boundaries, and so as complexity increases there is also a risk that the overcomplicated boundaries have no real predictive power. To this end a concept called Structural Risk Minimization has been developed. SVMs are equipped with an additional parameter (slack variable ξ) that allows a control on complexity. To introduce this parameter it is easiest to recall the example of the simplest case where the optimal separating hyperplane is determined in the original dataspace, without projecting the samples into a higher dimensional feature space. If these cannot be perfectly separated by a hyperplane, one may also allow deviations defined by $\xi_i > 0$ for individual samples $\boldsymbol{x}_i$. Those samples for which $\xi_i = 0$ are on the margin of its correct class, those with $\xi_i = 1$ on the separating plane, and those with $\xi_i > 1$ the wrong side of the dividing line, or misclassified samples. This is illustrated in Figure 5.26, for which there are five SVs, where three are exactly on the margins, and two between the margins in what would be empty space if the classes were perfectly separable. Of the two between the margins, one is misclassified and so has $\xi_i > 1$. This now allows a number of samples to be misclassified, and also for samples to be between the margins. Mathematically, the optimization task of Section 5.7.1 requires simultaneously maximizing the margin $0.5(\boldsymbol{ww}')$ and minimizing the empirical error, given by the sum of the allowed deviations $\Sigma_{i=1}^{l} \xi_i$, hence becoming: $\varphi(\boldsymbol{w}, b, \xi) = 0.5(\boldsymbol{ww}') + C\,\Sigma_{i \in sv}\, \xi_i^p$ subject to the constraint $c_i\ (\boldsymbol{wx}_i{}' + b) \geq 1 - \xi_i$ C is called the penalty error; the higher it is the more significant misclassifications are but the more complex the boundary (see below). It should be noted that the margin errors ξ_i become training errors only when $\xi_i > 1$. The SVs are now no longer all exactly on the margins but are somewhere between the two extreme margins. Every sample on or between the margins is an SV. When $p = 1$ the SVM is called a Level 1 (L1) Support Vector Machine, and when $p = 2$ a Level 2 (L2) Support Vector

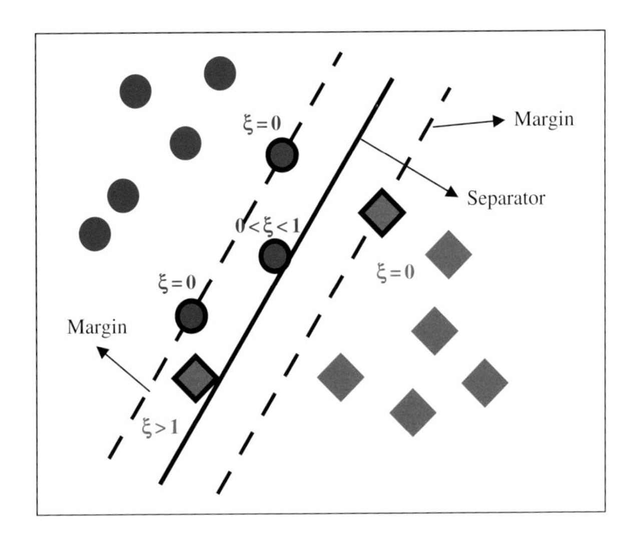

Figure 5.26 Illustration of slack variables. The support vectors for two classes are illustrated with closed symbols. Samples with $\xi = 0$ are on the margins, with $\xi > 0$ between the margins and with $\xi > 1$ are misclassified. The five SVs are indicated with borders around the symbols

Machine. In this text we illustrate SVMs using L1 methods, for simplicity, but when using packages or comparing results be sure to check whether the method is L1 or L2.

SVMs can be divided into two categories, hard and soft margin SVMs. Hard margin SVMs require two classes to be perfectly separable, and aim to find the optimal boundary that exactly separates the classes, with the maximum possible margin between the two classes, and in practice involve setting an infinite value of C: this was employed for the examples of Figure 5.21 and Figure 5.23 to Figure 5.25. However, it is always possible to find a feature space in which the two classes are perfectly separable when using a kernel function such as RBF, and forcing the algorithm to search for this feature space may lead to overfitting of the model. To avoid this, most people use soft margin SVMs which tolerate a degree of misclassification, and are designed to balance the classification error against the complexity of the model; in this text we will illustrate our examples using soft margin SVMs which are the most common available. The parameter C is set to determine the level of tolerance the model has, with larger C values reflected in lower tolerance of misclassification and more complex boundaries. C is included as an upper bound on the Lagrange multipliers, so that:

$$0 \leq \alpha_i \leq C$$

This additional parameter determines which one of the two criteria is emphasized during the optimization (either $0.5(\boldsymbol{ww}')$ or $\Sigma_{i=1}^{l}\xi_i$). Lower penalty error values emphasize the first term, allowing higher deviations from the margin ξ_i; hence the emphasis will be on margin maximization rather than minimizing the distance of misclassified samples from the boundary. In contrast, higher penalty error values will emphasize the second term, hence allowing smaller deviations across the boundary ξ_i and minimizing the training error. C offers the opportunity to pursue a trade-off between complexity of the boundary and the importance attached to misclassified samples or samples near the boundary. Note that a very high value of C tends towards a hard margin SVM, as this occurs when there is a very large penalty error for misclassification, i.e. one tries to construct boundaries that perfectly model the training set.

As an example we look at the simulated dataset of Section 5.7.2 but introduce an outlier, a member of class A that happens to fall within the region of class B. We see the effect of changing C for a constant value of σ in Figure 5.27. When C is reduced, more samples become SVs. For a constant value of σ this is the only difference. The more samples that are support vectors, the 'broader' the surface becomes since the neighbouring Gaussians overlap more; although the change in number of SVs may seem small (there are only three rather than seven 'red' class B samples that are not SVs when C is reduced); this is sufficient to smooth the surface. This means that the outlying sample is not longer a spike and brought below the boundary and so is correctly classified within the 'red group'.

As an example of the influence that the value of C has on the projected boundaries of the case studies described in this text, we examine its effect on Case Study 4 (pollution) using a default value of $\sigma = 1 \times$ average standard deviation of variables in the dataset in Figure 5.28.

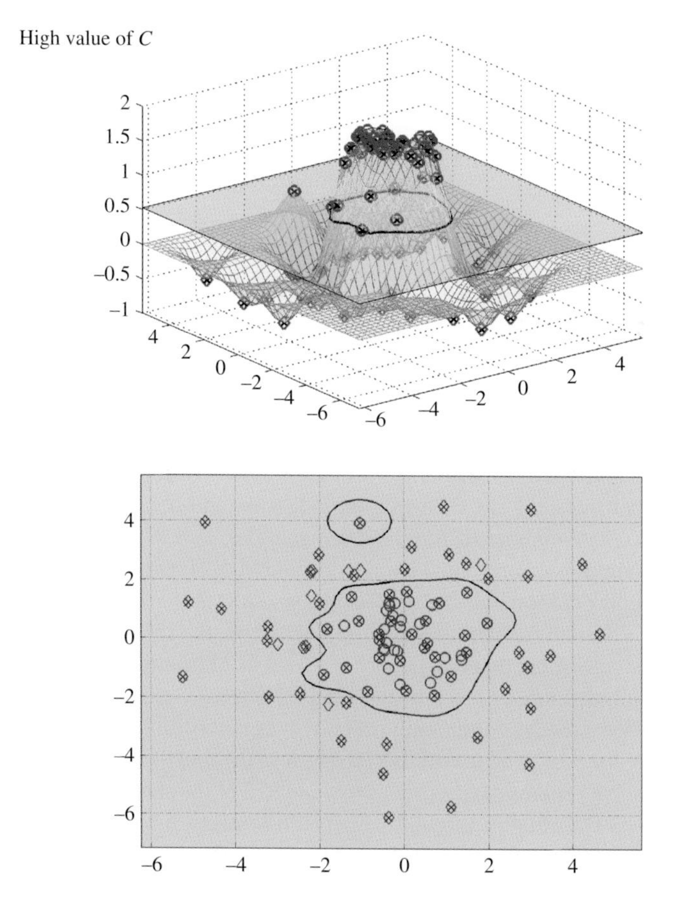

Figure 5.27 Illustration of effect of C on the SVM boundaries, in the case of an outlying sample

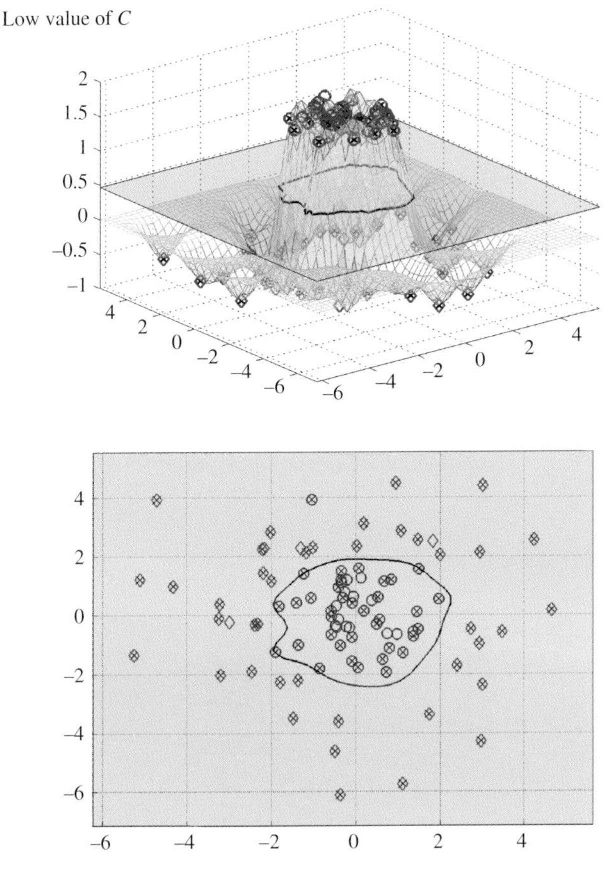

Figure 5.27 (continued)

More details are discussed in Chapter 8. At first glance the result for when $C = 0.01$ appears strange, in that all samples are classified into class A, and none into class B. Clearly this represents the highest number of misclassified samples, and as C increases the number of such samples decreases (represented by closed symbols). However if we are constructing a series of overlapping Gaussians, they will form a smooth surface when added together, and if there are a very different number of samples in one class, the surface will appear to have the sign of this class: remember that only 34 out of a total of 213 samples are of class B, and so by misclassifying all samples of class B and none of class A, over 84 % of the samples are correctly classified, which is quite high and involves a very simple model. This value of C though is clearly too low to be useful. Increasing it to 0.1 makes a major difference and now we can see two clear regions of the graph. Now only 12 of the samples are misclassified. The boundary for class B is roughly circular which is still quite a simple shape. Once $C = 1$, this boundary is defined rather more precisely, with only 9 samples misclassified. Increasing C to 10 refines the shape even further, particularly adjusting the fine shape of the boundary between the two main groups, to reduce misclassification to 5 samples. For $C = 100$, a

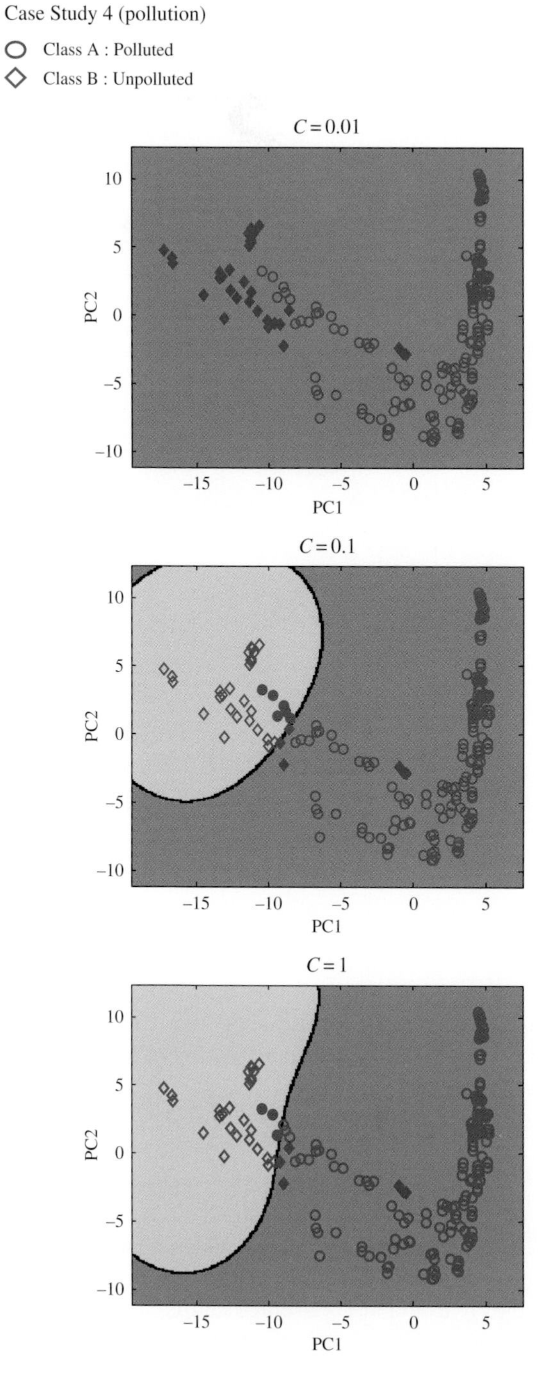

Figure 5.28 Illustration of the influence of penalty error C on SVM boundaries, using an RBF with $\sigma = 1 \times$ average standard deviation of the variables in the dataset, for Case Study 4 (pollution), and an L1 SVM. Misclassified samples are represented by filled symbols

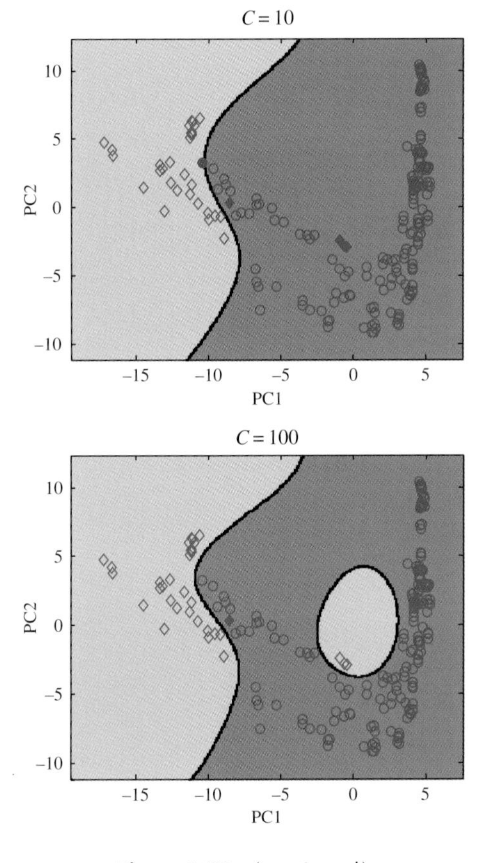

Figure 5.28 (continued)

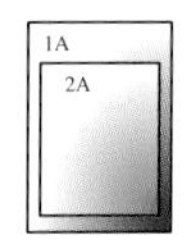

second region is defined for class B, consisting of the 3 samples from soil 4: although these now can be correctly classified it is debatable whether these samples actually have been assigned to the wrong group, and so we may be overinterpreting the evidence and overfitting the data. Only one sample remains misclassified.

5.7.4 SVM Parameters

In this text we will use RBF SVMs by default. In chemometrics it is rarely necessary to apply more sophisticated models. We will use as a default a value of $C = 1$. For the RBF radius σ we will employ as a default $1 \times$ average sample standard deviation of the variables in the data after initial preprocessing. The effect of changing these parameters and methods for optimizing SVM will be discussed in Section 8.7.

The illustration of SVM boundaries among the case studies discussed in this chapter are presented in Figure 5.29 using autoprediction and default parameters. We can see that

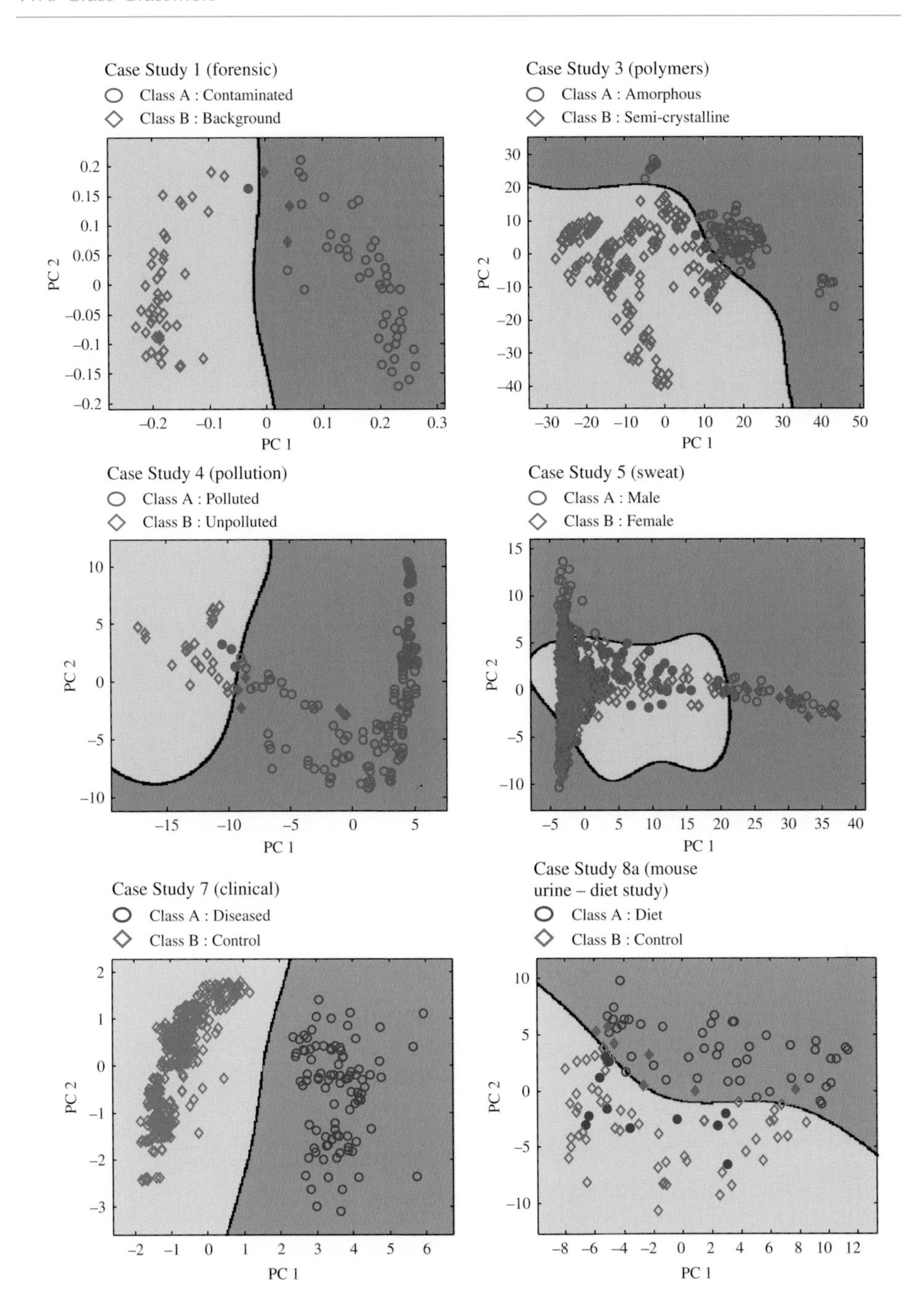

Figure 5.29 Visualization of boundaries obtained by SVM for the datasets discussed in this chapter, using default data preprocessing, and data represented by the scores of the first two PCs. Misclassified samples are represented by filled symbols, and the respective regions belonging to each class are coloured appropriately. A default of $C=1$ and $\sigma = 1 \times$ average sample standard deviation of the variables in the dataset is used, and an L1 SVM is employed by default

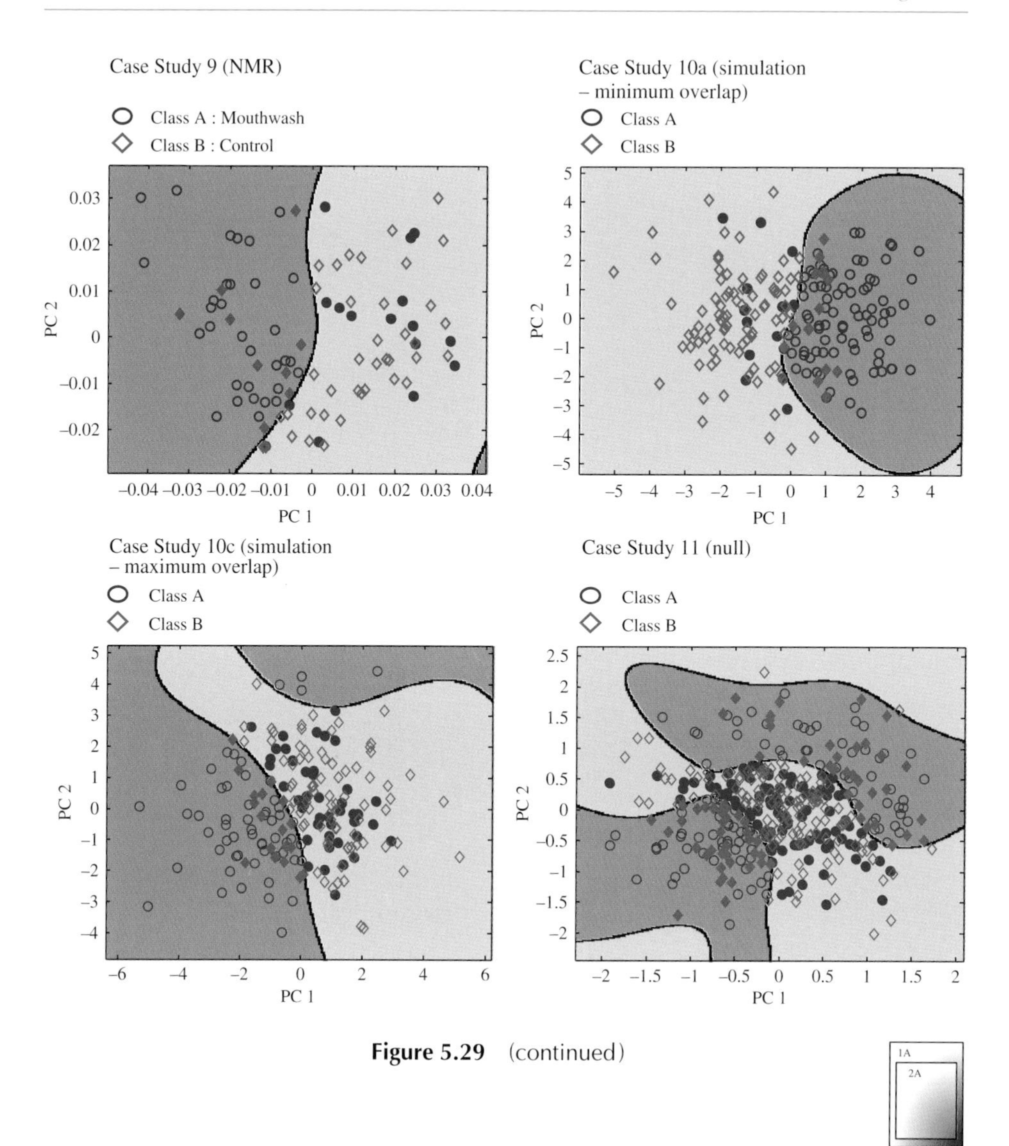

Figure 5.29 (continued)

when classes are well separated such as in Case Studies 1 (forensic) and 7 (clinical) there is very limited difference between these and the boundaries obtained by other methods. For Case Studies 3 (polymers) and 4 (pollution) SVM defines these boundaries somewhat better, especially in the interface. For Case Study 11 (null) the dataset SVM homes in on areas of clumping for each class, splitting class A into two regions. We will see in Chapter 8 that answers of indefinite complexity can be obtained if we adjust both σ and C, but can, of course, result in overfitting. Interestingly, SVM appear to define a definable region in the centre for Case Study 5 (sweat) for class B (female), which only QDA is capable of doing for the other methods. On the whole the choice of parameters used in

Figure 5.29 for SVM seem quite sensible, allow for better definition of curved boundaries without overfitting, and is recommended as a default in the absence of optimization.

Bibliography

Discriminant Analysis

R.A. Johnson, D.W. Wishern, *Applied Multivariate Statistical Analysis*, Prentice Hall, London, UK, 1988.

R.A. Fisher, The Use of Multiple Measurements in Taxonomic Problems, *Ann. Eugen.*, **7**, 179–188 (1936).

I.E. Frank, J.H. Friedman, Classification: Oldtimers and Newcomers, *J. Chemometrics*, **3**, 463–475 (1989).

J.H. Friedman, Regularized Discriminant-Analysis, *J. Am. Stat. Assoc.*, **84**, 165–175 (1989).

R. De Maesschalck, D. Jouan-Rimbaud, D.L. Massart, The Mahalanobis distance, *Chemometrics Intell. Lab. Systems*, **50**, 1–18 (2000).

Partial Least Squares Discriminant Analysis

S.J. Dixon, Y. Xu, R.G. Brereton, A. Soini, M.V. Novotny, E. Oberzaucher, K. Grammer, D.J. Penn, Pattern Recognition of Gas Chromatography Mass Spectrometry of Human Volatiles in Sweat to Distinguish the Sex of Subjects and Determine Potential Discriminatory Marker Peaks, *Chemometrics Intell. Lab. Systems*, **87**, 161–172 (2007).

L. Ståhle and S. Wold, Partial least squares analysis with cross-validation for the two-class problem: A Monte Carlo study, *J. Chemometrics*, **1**, 185–196 (1987).

P. Geladi, B.R. Kowalski, Partial Least Squares: a Tutorial, *Anal. Chim. Acta*, **185**, 1 (1986).

M. Baker and W. Rayens, Partial least squares for discrimination, *J. Chemometrics*, **17**,166–173 (2003).

Learning Vector Quantization

G. Lloyd, R. Faria, R.G. Brereton, J.C. Duncan, Learning vector quantization for multi-class classification: application to characterization of plastics, *J. Chem. Inform. Model.*, **47**, 1553–1563 (2007).

T. Kohonen, J. Kangas, J. Laaksonen, K. Torkkola, In *LVQPAK: A Software Package for the Correct Application of Learning Vector Quantization Algorithms*, IJCNN., International Joint Conference on Neural Networks, 1992, pp. 725–730.

T. Kohonen, *Self-Organizing Maps*, Third Edition, Springer-Verlag, Berlin, Germany, 2001.

Support Vector Machines

S. Abe, *Support Vector Machines for Pattern Classification*, Springer-Verlag, London, UK, 2005.

V.N. Vapnik. *The Nature of Statistical Learning Theory*, Second Edition, Springer-Verlag, New York, NY, USA, 2000.

C.J.C. Burges, A Tutorial on Support Vector Machines for Pattern Recognition, Data Mining and Knowledge, *Discovery*, **2**, 121–167 (1998).

N. Cristianini, J. Shawe-Taylor, *An Introduction to Support Vector Machines and Other Kernel-based Learning Methods*, Cambridge University Press, Cambridge, UK, 2000.

Y. Xu, S. Zomer, R.G. Brereton, Support Vector Machines: a Recent Method for Classification in Chemometrics, *Crit. Rev. Anal. Chem.*, **36**, 177–188 (2006).

B. Schölkopf, A.J. Smola, *Learning with Kernels*, MIT Press, Cambridge, MA, USA, 2002.

6

One Class Classifiers

6.1 Introduction

In Chapter 5 we introduced a number of classifiers that aim to divide the data space up into two separate regions, each of which corresponds to one class of samples. The classification of a sample is given by which region of the data space it falls into. These classifiers are often sometimes hard called models, in that they divide space up into sections using one, or a series of, boundaries, which principle is the basis of the most widespread classification methods. However when there are several classes it is often hard to re-express these classifiers in a multiclass form, and the boundaries become quite complicated, as discussed in Chapter 7. Furthermore, hard models find it difficult to deal with outliers, that is, samples that belong to none of the predefined groups – the inherent assumption of such classifiers is that all samples must belong to a predefined group – and cannot deal well with samples that are genuinely ambiguous. A final weakness of hard models is that they have to be reformed if new groups are introduced, unless these groups are subsets of the existing groups.

In order to overcome these limitations a set of modelling techniques which we will call one class classifiers have been developed. The approaches are often sometimes called soft models, and in the area of chemometrics, SIMCA is the best known, although by no means unique. Although there are a large number of potential approaches, only some have become widespread and we will restrict the discussion in this chapter to a few of the most common one class classifiers, whilst allowing the reader to apply the general principles to others if appropriate. The main job is to recognize when a classifier is a one class classifier and when it is a two class classifier and to understand that these two approaches are based on quite different principles.

A one class classifier models each group independently. There is no limit to the number of groups that can be modelled, and a decision is made whether a sample is a member of a predefined group or not. The difference between one class and two class classifiers is illustrated in Figure 6.1. For the two class classifiers a line (or more complex boundary) is drawn between the two classes and a sample is assigned according to which side of the line

Chemometrics for Pattern Recognition Richard G. Brereton

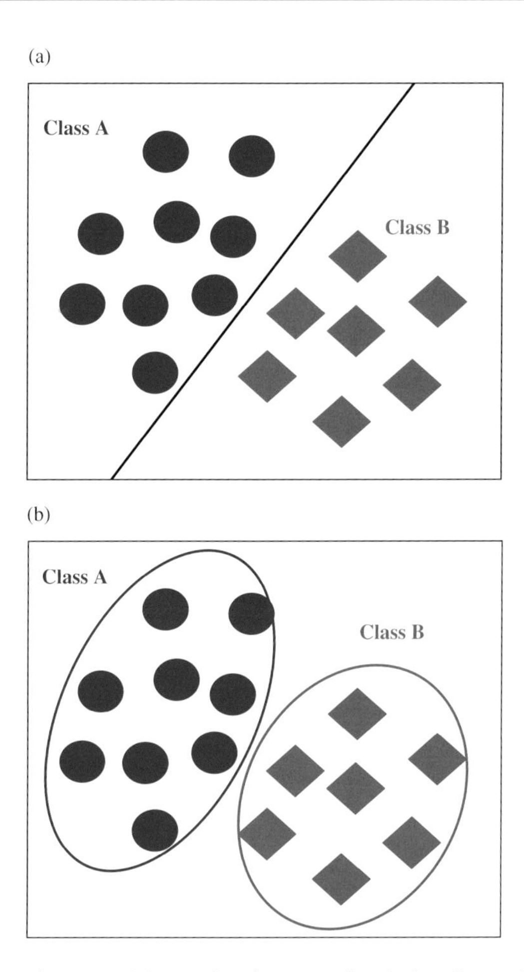

Figure 6.1 Difference between (a) two class binary or hard classifier and (b) one class or soft classifier

it falls. For the data illustrated, there are two possible one class classifiers, and these can be represented by boundaries that in Figure 6.1 are ellipsoidal. Samples outside these boundaries would be assigned as outliers belonging to neither known class.

Figure 6.2 extends this theme. In this case there are three groups: although classes A and B are separate, class C overlaps with both of them; in addition there is an outlier that belongs to none of the three classes. A one class classifier establishes a model for each class separately and is able to conclude that samples belong to 'no class', or to more than one class simultaneously. Although the latter verdict may defy commonsense in many situations (how can a person be simultaneously male and female, for example – unless they are an hermaphrodite of course) within chemometrics a key consideration is whether the analytical data are of sufficient quality to allow a sample to be unambiguously assigned to one class or another. We may be monitoring patients for a disease and the tests on some patients may be ambiguous; indeed in some cases we expect this – consider, for example,

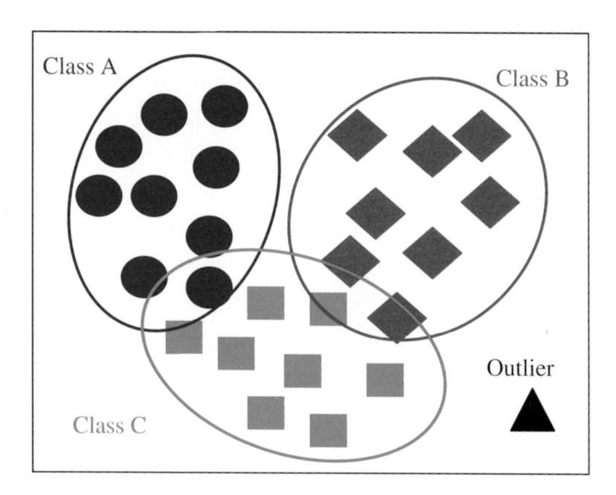

Figure 6.2 Example of three classes with some overlap and an outlier

testing for early stages of a disease by measuring chemicals in plasma: a particular set of chemical markers may be indicative of development of a disease but be dependent on the patient's subsequent lifestyle or environment and as such whether a person develops a disease or not is ambiguous – not all people with these marker compounds will become diseased and in some cases there may be other unknown factors that mean a presence of these markers do not lead to the disease. So a verdict that we cannot decide which group a sample belongs to is often entirely acceptable. The advocates of one class classifiers such as SIMCA regard this as a major advantage, but it is important to recognize that two class classifiers also can provide this sort of information, for example, how close a sample is to a boundary or what is a sample's Model Stability (as discussed in Section 8.4.1): if low then the sample's classification is ambiguous. However one class classifiers are a valuable alternative to two class classifiers and this chapter describes the main features of such approaches.

There are various incompatible terminologies in the literature but we will call the class that is to be modelled the 'in group' and all other samples the 'out group'. Where there are several classes this terminology can be made more distinct, e.g. 'in group A' and 'out group A'.

There is a very wide range of named one class classifiers in the chemometrics literature ranging from SIMCA to Gaussian density estimators to UNEQ to name just a few of the most well known; however they are mainly based on a few generic principles which we outline below.

6.2 Distance Based Classifiers

The basis of most one class classifiers is to determine the centre of each class and distance to this centre. There are only two fundamental groups of method, one based on the Euclidean distance, analogous to the EDC two class classifier in which the Euclidean distance to the centroid of each class is measured, $d_{ig}^2 = (\boldsymbol{x}_i - \overline{\boldsymbol{x}}_g)\ (\boldsymbol{x}_i - \overline{\boldsymbol{x}}_g)'$, while the other fundamental and more widespread group of methods is based on the Mahalanobis distance to the class centroid, defined by $d_{ig}^2 = (\boldsymbol{x}_i - \overline{\boldsymbol{x}}_g)\boldsymbol{S}_g^{-1}(\boldsymbol{x}_i - \overline{\boldsymbol{x}}_g)'$, analogous to the

QDA two class classifier, as discussed in Section 5.4. Note that there is no direct analogy to LDA because the latter method requires the variance-covariance matrix to be computed over all classes, but one class classifiers consider each class in turn. In most practical situations QDA will provide a more realistic distance measure to EDC because the variance of each variable often is quite different.

The basis of these classifiers is that the further an object is from the centroid of a class, the less likely it is to be a member of the class (see Figure 6.3) where we can contour the distance from the centre of class A; if EDC is used these contours are circular, whereas for QDA they are ellipsoidal. The contours for Case Studies 1 (forensic) and 3 (polymers – 2 groups) are illustrated in Figure 6.4 for autopredictive models after the data have been first reduced using PCA to the first two PCs (for visualization). Usually, statistically based criteria are then used to convert the distance into a probability, as discussed below, although distances themselves can be employed if preferred, or else if there are sufficient samples the cut-off can be performed empirically by determining the region within which a specific proportion of samples from a specific group lie. Note that it is not usual to compute probabilities for EDC based models as distances are in the original space and as such are not relative, for example, to a standard deviation, and so the majority of statistical indicators are applied to QDA based models which we will focus on below. However the Q statistic, which is used for a different purpose in chemometrics, could conceivably be employed to convert a Euclidean distance to a probability when performing class modelling, but since this is not common we will not report this below for brevity, and Section 6.4 will primarily relate to QDA based models.

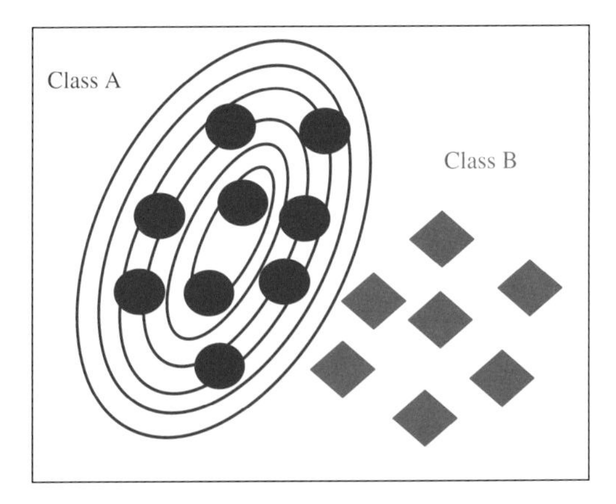

Figure 6.3 Principle of using distance from a centroid to determine the probability of a sample being a member of class A

6.3 PC Based Models and SIMCA

After the raw data are preprocessed (element transformation and row scaling) as discussed in Chapter 4 it is common to further transform the data using PCA. When using QDA based

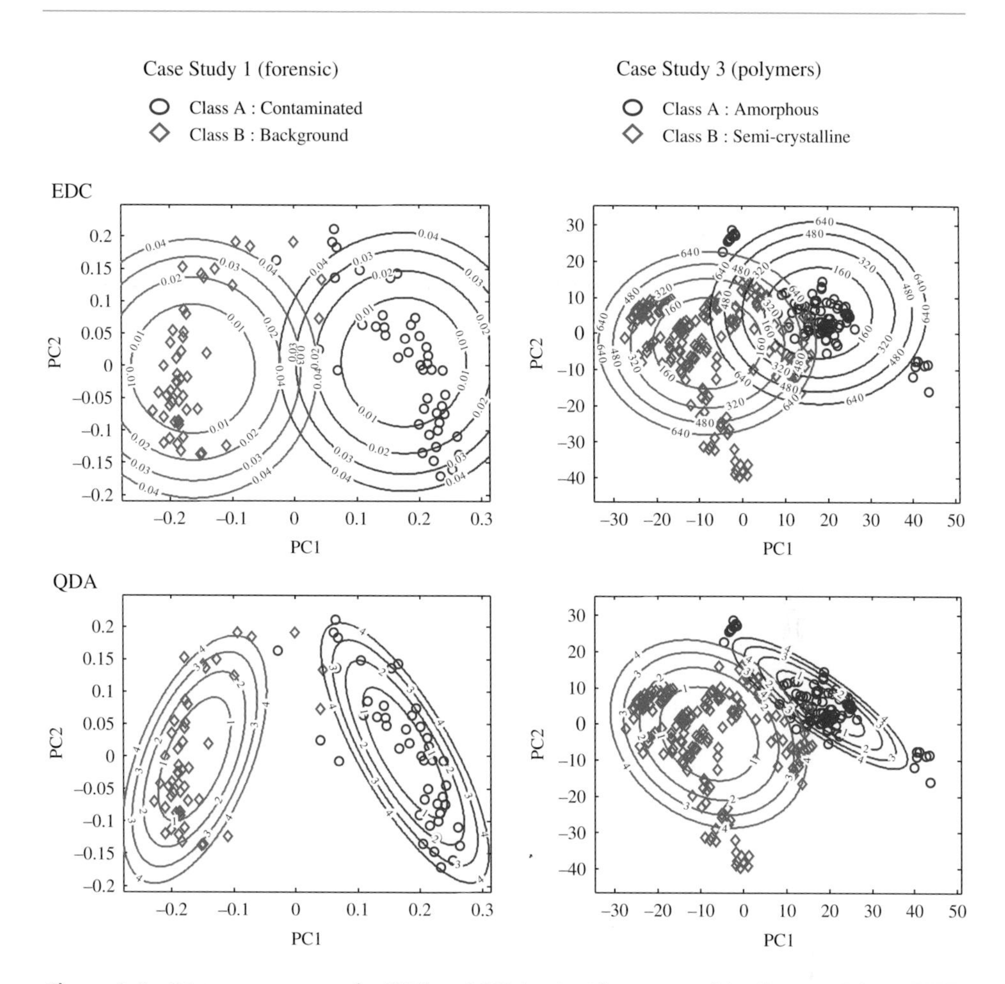

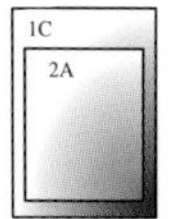

Figure 6.4 Distance contours for EDC and QDA using the scores of the first two PCs as the variables, for Case Studies 1 and 3 and conjoint PC models. The contours are equidistance for the class centroids, labelled with the squared Euclidean (EDC) or Mahalanobis (QDA) distances as appropriate

distances models this is essential if the number of variables exceeds the number of samples. For visualization in this chapter we always first reduce the data using PCA prior to representing the class boundaries. Prior reduction of the data by PCA is an essential first step of SIMCA as originally advocated by Wold and co-workers, and bears some resemblance to PC-DD (PC data description) of Tax.

The reduction of data using PCA involves somewhat more complex decisions for one class classifiers than for two or higher class classifiers. In the latter situation the only decision is how many and which components to retain for the entire dataset consisting of all samples in the autopredictive or training set model. For one class classifiers it is possible to perform PCA on each individual class in addition to, or instead of, the overall dataset. There are two options as follows:

1. For visualization of data in a reduced data space it is possible to reduce the overall dataset to 2 PCs, and then perform PCA again on each individual class, retaining 2 PCs on the re-centred class model. This method can be extended either to first reduce the overall dataset to A PCs or retain all J original variables and then perform PCA separately on each dataset but retain all non-zero PCs. Although the class distances from each class centroid are modelled independently, PCA on each individual class has the effect of rotating the scores rather than projecting onto a subspace. When visualizing the boundaries in two dimensions we will follow flowchart 1C. We will call these conjoint PC models, which are used by default elsewhere in the text. Conjoint PC models can also be obtained when performing PCA on the entire dataset without doing this separately again for each class by any number of PCs so long as the number is no more than the smaller of the number of variables or samples.
2. An alternative is to either use all the original J variables in the raw data or reduce the data first to A PCs. Then for each class g separately, PCA is performed independently and A_g PCs are retained where $A_g < A$ or $A_g < J$ so long as some non-zero PCs are discarded, even if the number of PCs retained for each class is the same. The choice of A_g can be made in a variety of ways, either using a predetermined cut-off value, or by cross-validation or the bootstrap as discussed in Section 4.3.1. Figure 6.5 illustrates this principle, whereby one of the two classes (represented by blue symbols) is modelled using 2 PCs, projecting onto a plane; the other (represented by red symbols) falls outside this plane. So long as the number of PCs is less than the number of variables in the original data space this additional test as to whether the model of samples from one class are described well by another class, can be performed and is an additional advantage of soft models. In this text we will independently scale each class using its own statistics and follow flowchart 1D under such circumstances. We will call these disjoint PC models. They are only applicable to one class classifiers.

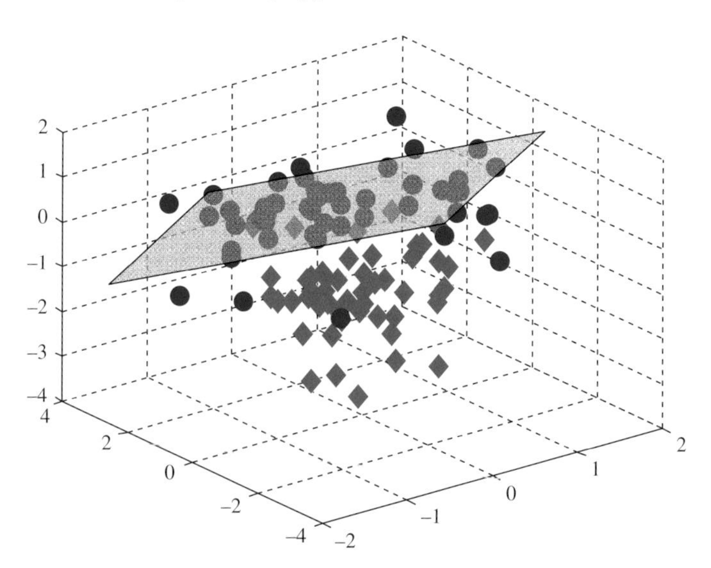

Figure 6.5 Principle of disjoint PC models. Three variables are measured for each class. The blue class is projected onto a plane represented by 2 PCs, which does not fit the red class well

The advocates of SIMCA came for a background heavily influenced by statistics and so unlike some of the approaches such as SVM or LVQ, where it is hard to convert numbers into probabilities, many users of SIMCA convert geometric distances into probabilities. The basis is that we want to determine the confidence that a sample is a member of a specific class, e.g. 95 % confidence, and this is done in a variety of ways, a common one involves the assumption that the distribution of samples in each class is multinormal, allowing for the computation of various statistics as discussed in Section 6.4; in addition for disjoint PC models a second assessment is how closely samples fall into the PC space of a predefined group. These statistics have mainly been employed in the context of Multivariate Statistical Process Control where the 'in group' are samples that appear to be behaving in a normal way (the NOC or Normal Operating Conditions), and the 'out group' are all other samples, but can easily be adapted for most standard classification problems, and are by no means restricted to process chemistry. In Process Control often only one group can be well defined so although there is considerable interest in whether a sample falls into a given class or not (if not, there are potential problems with the process), only the NOC samples are adequate in number and distribution for class modelling.

6.4 Indicators of Significance

6.4.1 Gaussian Density Estimators and Chi-Squared

Strangely in chemometrics there misleadingly appear to be quite different types of classifier, so called density methods and approaches such as SIMCA. In fact the assessment of whether a sample is a member of a class is based on very similar principles for both groups of methods, of which the simplest involves using Gaussian density estimates for one class classifiers. It is assumed that the distribution of samples is more dense close to the centroid of a class, forming by default a Gaussian distribution.

Figure 6.6 illustrates a typical example of a Gaussian density estimator, for 100 data points, with confidence limits 'marked-in'. It can be seen that 23 of these points are within the 25% or $\alpha = 0.25$ region, and 51 points within the $\alpha = 0.5$ region. These approximate to the expected values, although the Gaussian estimate is not exact even if the underlying population comes from a multinormal distribution. To fit a Gaussian model it is usual to determine the Mahalanobis distance of each point from the centroid of the class given by:

$$d_{ig}^2 = (\boldsymbol{x}_i - \overline{\boldsymbol{x}}_g)\boldsymbol{S}_g^{-1}(\boldsymbol{x}_i - \overline{\boldsymbol{x}}_g)'$$

where $\overline{x}_g$ is the mean of class $\boldsymbol{g}$ and $\boldsymbol{S_g}$ its variance–covariance matrix (for the training samples). If there is only one measurement this simplifies to:

$$d_{ig}^2 = (x_i - \overline{x}_g)^2 / s_g^2$$

or the distance to the mean divided by the standard deviation. If we assume that d is normally distributed, that is, the probability of obtaining a specific values of d can be approximated by a normal distribution function, there is an important relationship between the normal distribution (z) and the chi squared (χ^2) distribution. If one takes a critical value

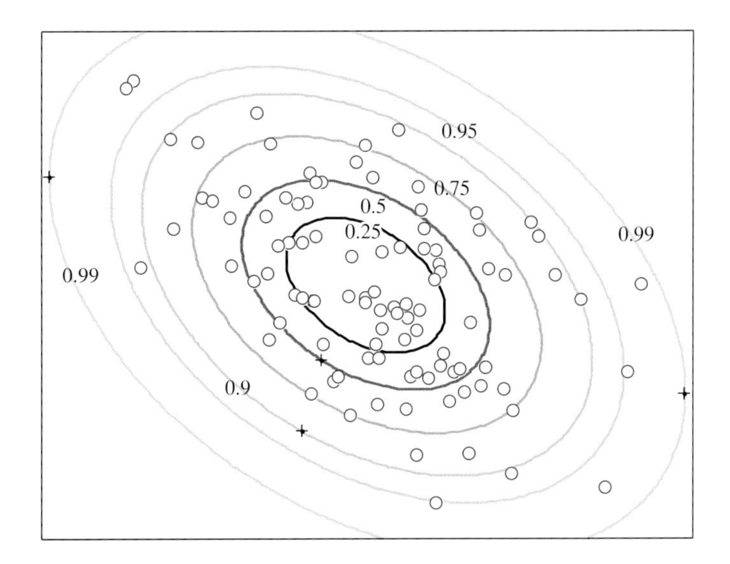

Figure 6.6 Typical Gaussian density estimator for a dataset characterized by two variables, with contour lines at different levels of certainty indicated

of the normal distribution and squares it, one will get the corresponding chi-square value with one degree of freedom, but twice the area in the tails. Hence if we take, for example, the critical value z at 0.10, representing the area of the last 10 % of the tail of the normal distribution (Figure 6.7), we obtain a value of 1.282, implying that 10 % of values obeying a normal distribution will have a value at least 1.282 standard deviations above the mean.

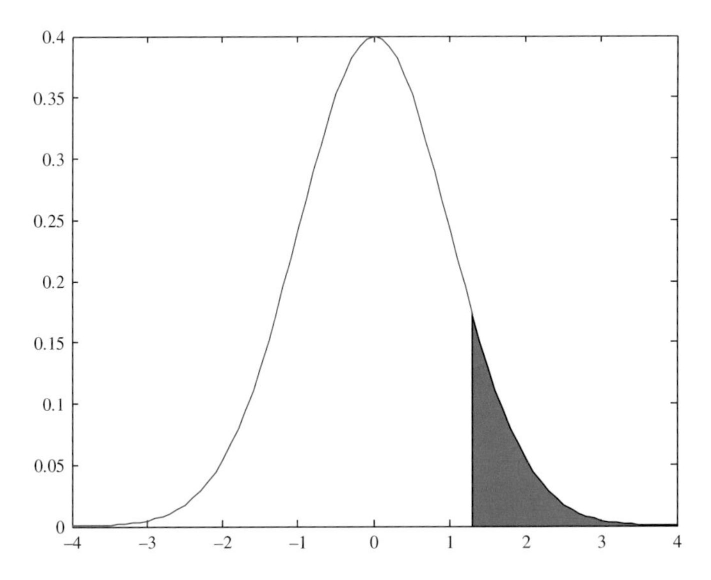

Figure 6.7 Illustration of $z = 0.1$; the filled area of the normal distribution equals 10 % of the overall area under the curve

If we are interested in the squared distance from the mean (d^2) we would expect this area of be twice this since both positive and negative values are included or 20 %; hence the area under the curve for the χ^2 distribution greater than 1.642 (= 1.282^2) is 20 % of the total area, and we expect 20 % of samples to have a value of χ^2greater than 1.642 if there is only one variable.

The Mahalanobis distance in more than one dimension can be in any direction from the centre; hence the need to use d^2 rather than d and the χ^2 statistic with J degrees of freedom where J equals the number of variables. The Mahalanobis distance can only be computed if the number of variables is less than the number of samples; if this is not so it is necessary to reduce the dimensionality by PCA or remove some of the original variables. If the variables are PCs calculated on the training samples and each column is centred, so that PCs are centred on the centroid of the class, the variance covariance matrix will be diagonal and equal to the eigenvalues of each PC divided by the number of samples in the training set.

Because more people use Hotelling's T^2 (Section 6.4.2) and because the difference with the confidence boundaries using χ^2 is quite small unless the number of samples is small (see below), we will not illustrate these confidence bounds in this section. However confidence contours can be obtained for any level of α, for example, a contour at $\alpha = 0.95$ should envelope approximately 95 % of the training set samples. We can then establish our boundaries at any given probability level, and any sample at a Mahalanobis distance from the centroid corresponding to confidence level α has a $1 - \alpha$ chance of being a member of the class.

One small trick if the covariance matrix does not have an inverse (or is nearly singular), which can happen if there are a large number of variables, some of which contain correlations, is to approximate the inverse by adding a small regularization term ρ to $\boldsymbol{S}$, $\boldsymbol{S} = (1 - \rho) + \rho\boldsymbol{U}$ where $\boldsymbol{U}$ is the identity matrix.

Using Gaussian density estimators and the χ^2 statistics assumes that samples in classes are approximately multinormally distributed. In many cases, individual variables are not, as we will see in Chapter 9, but this is often a reasonable assumption for samples, providing suitable data preprocessing has been performed and there are not too many outliers that could be unduly influential on the model, as a sum of symmetric distributions tends to a normal distribution. If in doubt, simple graphical representation of the samples in PC space will give an idea whether such models are suitable. We will briefly discuss tests for normality in Section 6.4.6.

There are a variety of other approaches to density estimation involving more elaborate functions, often composed of several different overlapping functions, for example, instead of one Gaussian, the class distribution can be modelled by a series of Gaussians, centred on each point. It is beyond the scope of this book to list all such possible classifiers.

6.4.2 Hotelling's T^2

Hotelling's T^2 is another common statistic for determining the significance of multivariate distances. In order to understand this consider calculating the t-statistic to determine whether a sample i characterized by a univariate measured variable x_i belongs to a population of samples from group g (the training set), each characterized by a single measured variable. This is given by the following:

$$t_{ig} = \frac{x_i - \bar{x}_g}{s_g/\sqrt{I_g}}$$

where $\bar{x}_g$ is the mean value of group g, s_g the corresponding standard deviation and I_g the number of samples in the group (note that there are slightly different equations for t according to whether it is assumed that both groups being compared have equal variances – in this case we are comparing a group of I_g samples to a group of 1 sample – the test sample, and have no information about the standard deviation of the test sample – see Section 9.2.2 for an alternative definition where there are two groups with different variances). Thus, the squared t statistic is given by the following:

$$t_{ig}^2 = I_g(x_i - \bar{x}_g)(s_g^2)^{-1}(x_i - \bar{x}_g)$$

This is then easy to extend to the multivariate situations:

$$T_{ig}^2 = I_g(\boldsymbol{x}_i - \bar{\boldsymbol{x}}_g)\boldsymbol{S}_g^{-1}(\boldsymbol{x}_i - \bar{\boldsymbol{x}}_g)'$$

which is called the Hotelling's T^2 statistic.

There are two common equations for the distribution when J variables are being measured:

$$T_{ig}^2 \approx \frac{J(I_g - 1)(I_g + 1)}{I_g(I_g - J)} F(J, I_g - J)$$

and a second one which is used if it is assumed that the mean is known and the uncertainty is primarily in the variance:

$$T_{ig}^2 \approx \frac{J(I_g - 1)}{(I_g - J)} F(J, I_g - J)$$

When I_g is large, these two equations converge and in practice distributions do not obey normality so well that it is possible to accurately distinguish them. In this text we will use the former equation. The confidence limits for any desired value of α can be established using the equation above and the F distribution with J and I_g–J degrees of freedom (see Tables 9.12 and 9.13 for numerical values which can alternatively easily be computed); to determine what is the probability a sample belongs to a specific group, compare the value of T_{ig}^2 (which is the squared Mahalanobis distance multiplied by the number of samples in the training set for group g) to the critical value above.

The Hotelling's T^2 statistic is closely related to χ^2. The F and χ^2 statistics are very similar in that, after a normalization, χ^2 is the limiting distribution of F as the denominator degrees of freedom goes to infinity. The normalization is χ^2 = (numerator degrees of freedom) × F. Since the formula for T^2 is $T_{ig}^2 \approx J(I_g - 1)/(I_g - J)F(J, I_g - J)$, if I_g is large, $T_{ig}^2 \approx JF(J,\infty)$ or the χ^2 distribution. The T^2 statistic is more commonly used than χ^2 as it allows for small sample sizes, although when the number of samples in the training set is large this does not make much difference.

6.4.3 *D*-Statistic

The *D*-statistic is basically an extension of Hotelling's T^2 for use after PCA – other than that, it is very little different; however, the terminology was introduced to distinguish from application to the raw data. Note that when using one class classifiers it is usual to centre the data over each class separately prior to PCA. Using the loadings $\boldsymbol{P}_g$ obtained from group g samples, for each sample represented by vector $\boldsymbol{x}_i$, a predicted scores vector $\hat{\boldsymbol{t}}_{gi}$ of dimensions $1 \times A_g$ can be calculated for the class g model:

$$x_i = \hat{\boldsymbol{t}}_{gi}\boldsymbol{P}_g + \boldsymbol{e}_i \text{ so } \hat{\boldsymbol{t}}_{gi} = x_i\boldsymbol{P}'_g$$

remembering that the loadings vectors are orthonormal, and so $(\boldsymbol{P}\,\boldsymbol{P}')^{-1}$ is a unit matrix.

Assuming that T_g is centred, the *D*-statistic for sample i to group g can then be calculated by:

$$\boldsymbol{D}_{ig,A_g} = \hat{\boldsymbol{t}}_{gi}\left(\frac{\boldsymbol{T}'_g\boldsymbol{T}_g}{I_g - 1}\right)^{-1}\hat{\boldsymbol{t}}'_{gi}$$

$$\approx \frac{A_g(I_g - 1)(I_g + 1)}{I_g(I_g - A_g)} F(A_g, I_g - A_g)$$

It is assumed that the *D*-statistic follows an *F*-distribution. A_g is the number of PCs employed in the model and I_g the number of samples in class g. A confidence level of α can be used for the *F*-distribution that allows the computation of the confidence that a sample is a member of a class, on the assumption of a multinormal distribution of samples (note that even if the variables are not normally distributed, in many cases the samples may fall approximately into a normal distribution). Thus, a value of D exceeding a limit given by α indicates that the corresponding sample does not fall within the class boundaries using the set confidence limits, even though it may maintain the correlation structure between the variables in the model. Hence it is possible to determine the Mahalanobis distance from the centre of a class that gives a specific confidence that a sample is a member of the class. Typically 99 % or 95 % confidence limits are computed. A sample that is within these limits is assigned to the specific class.

Examples of confidence ellipsoids using the *D*-statistic and PC models based on conjoint models are presented in Figure 6.8 for Case Studies 1, 2, 3, 5, 8a, 8b and 11. Note that for purpose of visualization we model each class independently using 2 PCs which is the same as the number of variables used to visualize the data in this case – in practice the number of PCs used would be less than the number of PCs or variables employed to describe the entire dataset. Since we are using one class classifiers we are no longer restricted to two class models, and can add or remove classes as desired, although it is important to realize that the conjoint PC model will change if samples are removed (it is still possible to produce a conjoint PC model of samples belonging to several groups including outliers but then only develop a classification model on a subset of these samples and it is also possible to introduce new samples to the existing PC space and model these using QDA without recalculating the PCs). A sample that belongs to a group that has not been modelled is considered an outlier. We can clearly see that there are regions of overlap between the classes, and we are no longer required to decide whether a sample is assigned one side of a boundary or not. We can see clearly for Case Study 11 (null) that most samples equally well fit both class models, as anticipated, whereas for Case Study 2 (NIR spectroscopy of food) there are clear separations between the classes. For Case Study 8a (mouse urine – diet study) we can see that the two classes are quite overlapping and so it might be difficult to model a third class, whereas for Case Study 2 it is likely that new

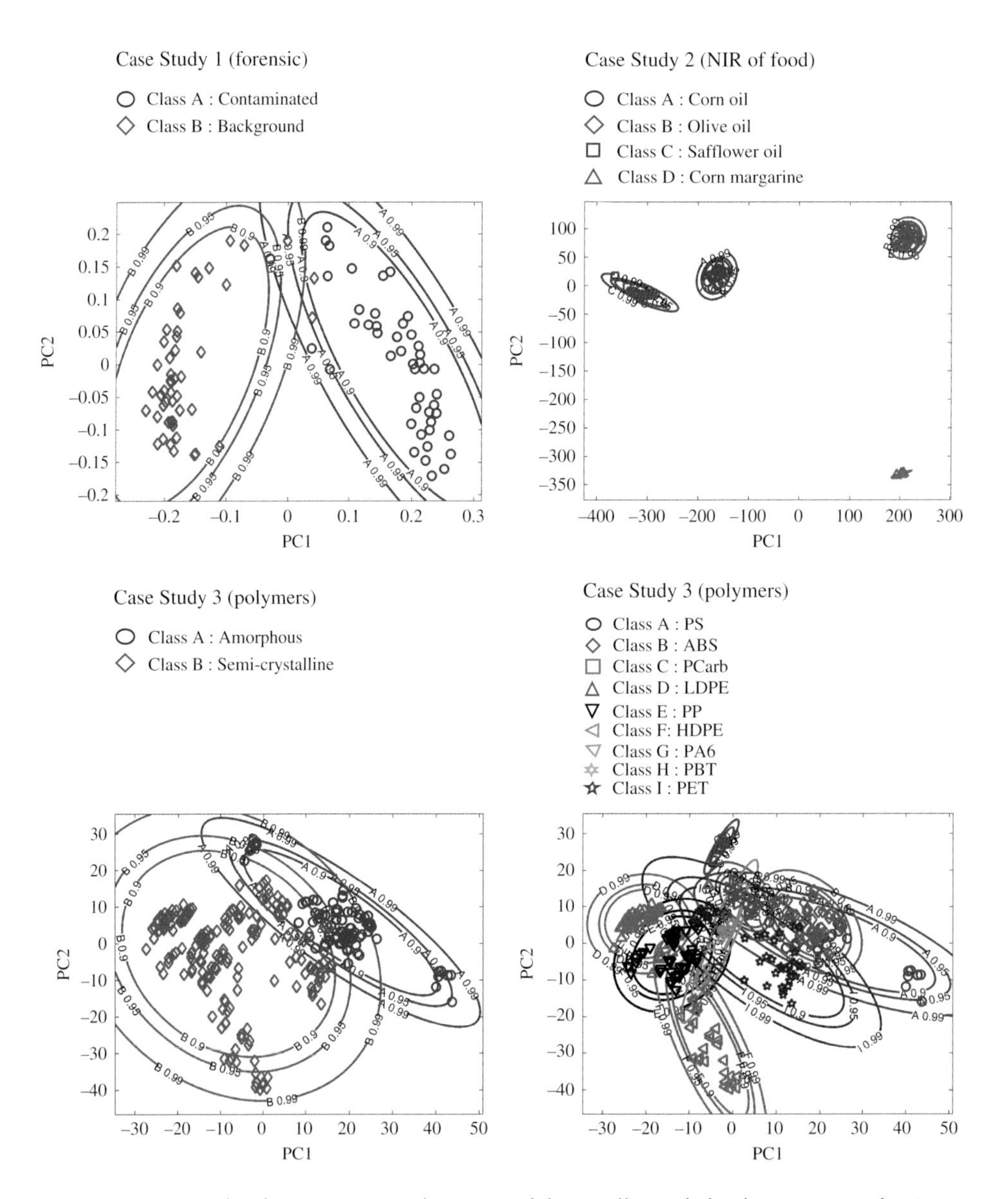

Figure 6.8 Example of 90 %, 95 % and 99 % confidence ellipsoids for the *D*-statistic for Case Studies 1, 2, 3, 5, 8a, 8b and 11 using conjoint PC models

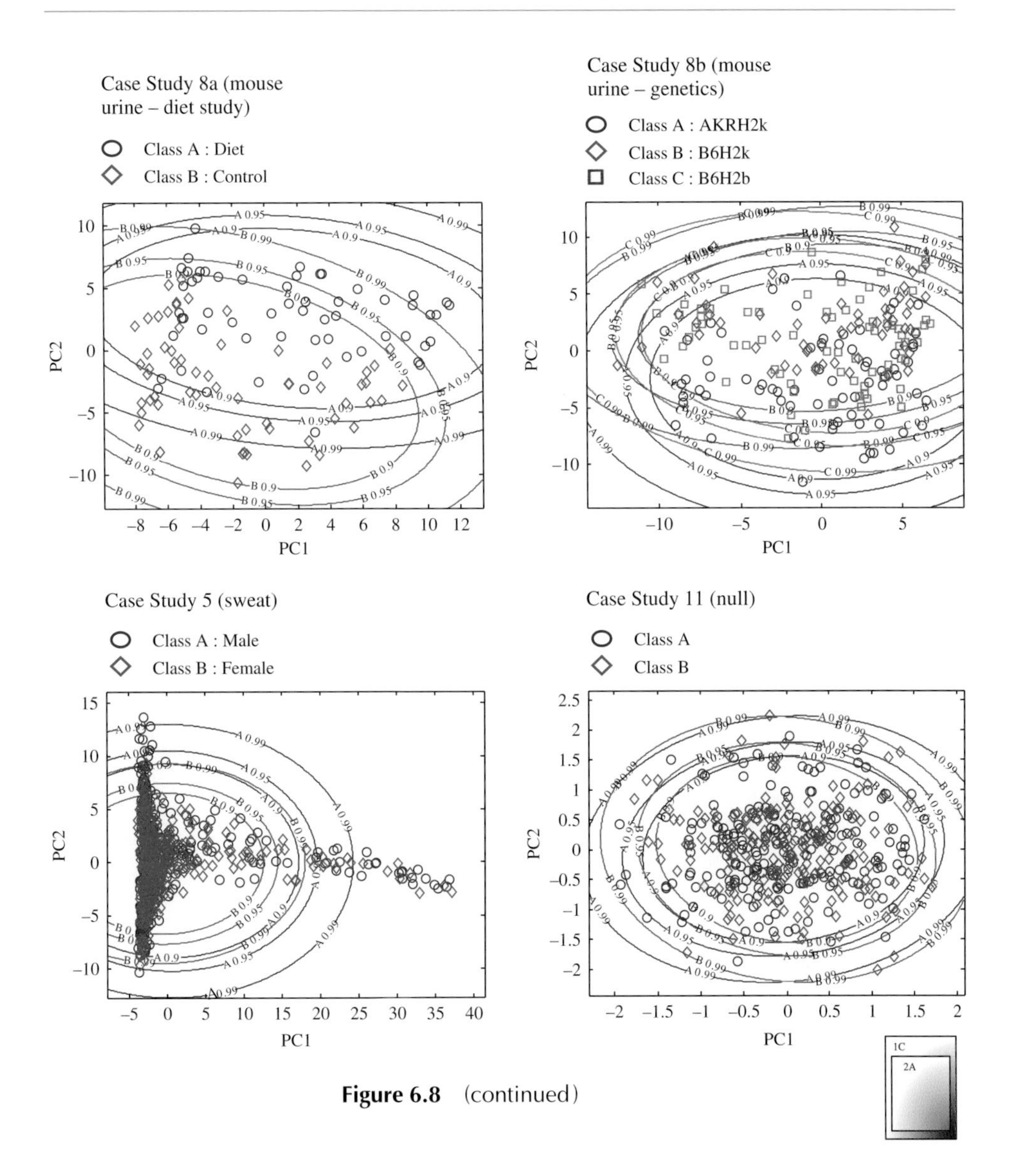

Figure 6.8 (continued)

groups could be easily distinguishable. In Case Study 3 (polymers) we can distinguish most of the nine groups but there is still overlap: in fact more than 2 PCs are necessary to obtain good discrimination between all nine groups – however they are in fact well distinguishable if more PCs are employed in the model. The first 2 PCs can distinguish the main types (amorphous and semi-crystalline) with 90 % confidence with only a few in the boundary region.

The *D*-statistic can also be presented as a bar chart with any desired confidence limit; Figure 6.9 shows the 95 % confidence limits for Case Studies 1, 2 and 8a. Case Study 2 (NIR spectroscopy of food) is all highly separable with very large *D*-statistics for the ‘out

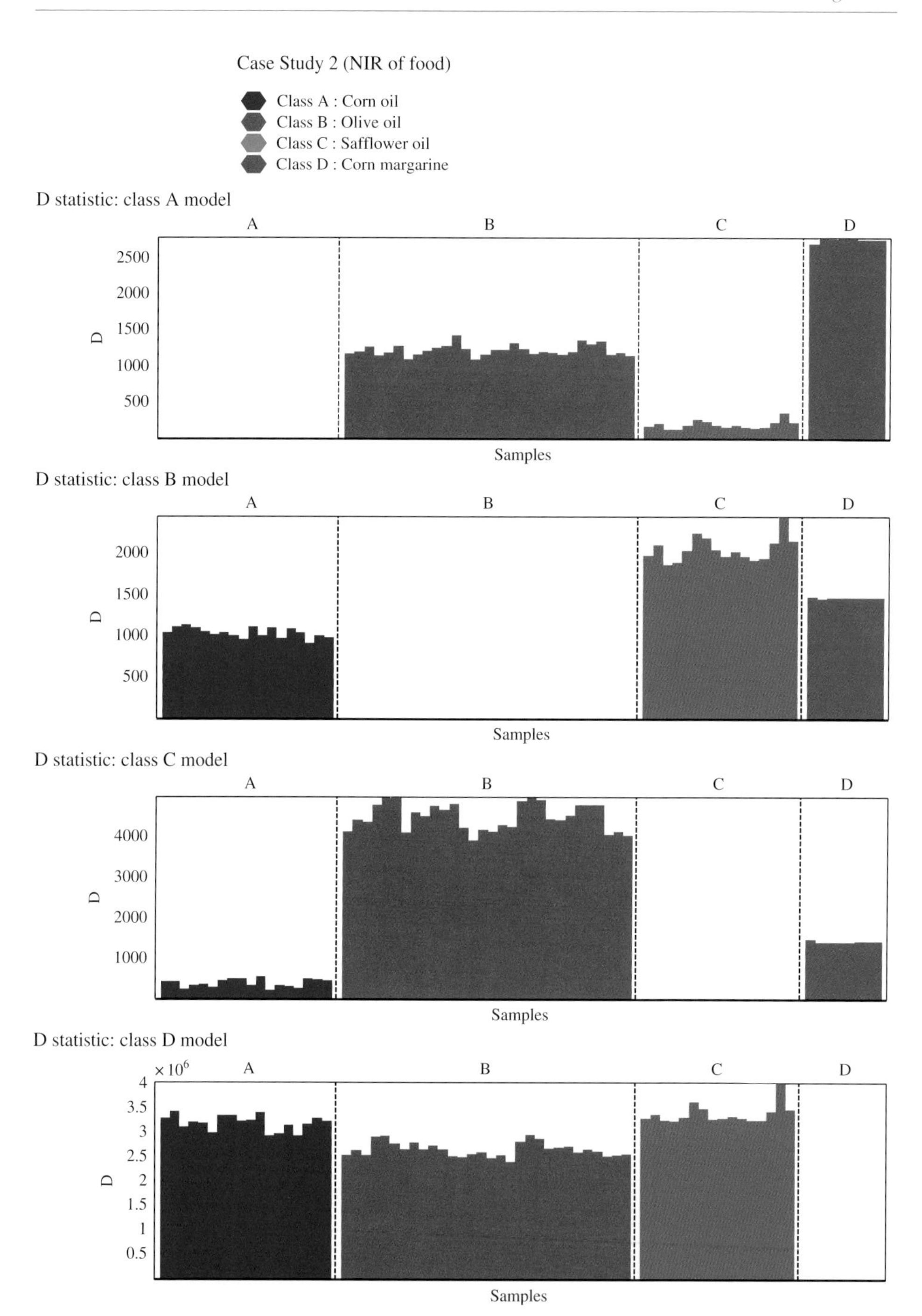

Figure 6.9 D and Q statistics for Case Studies 1, 2 and 8a, using conjoint PC models. Misclassified samples using each statistic individually are represented in light colours with the 95 % confidence levels being indicated, for case study 2 they are very low

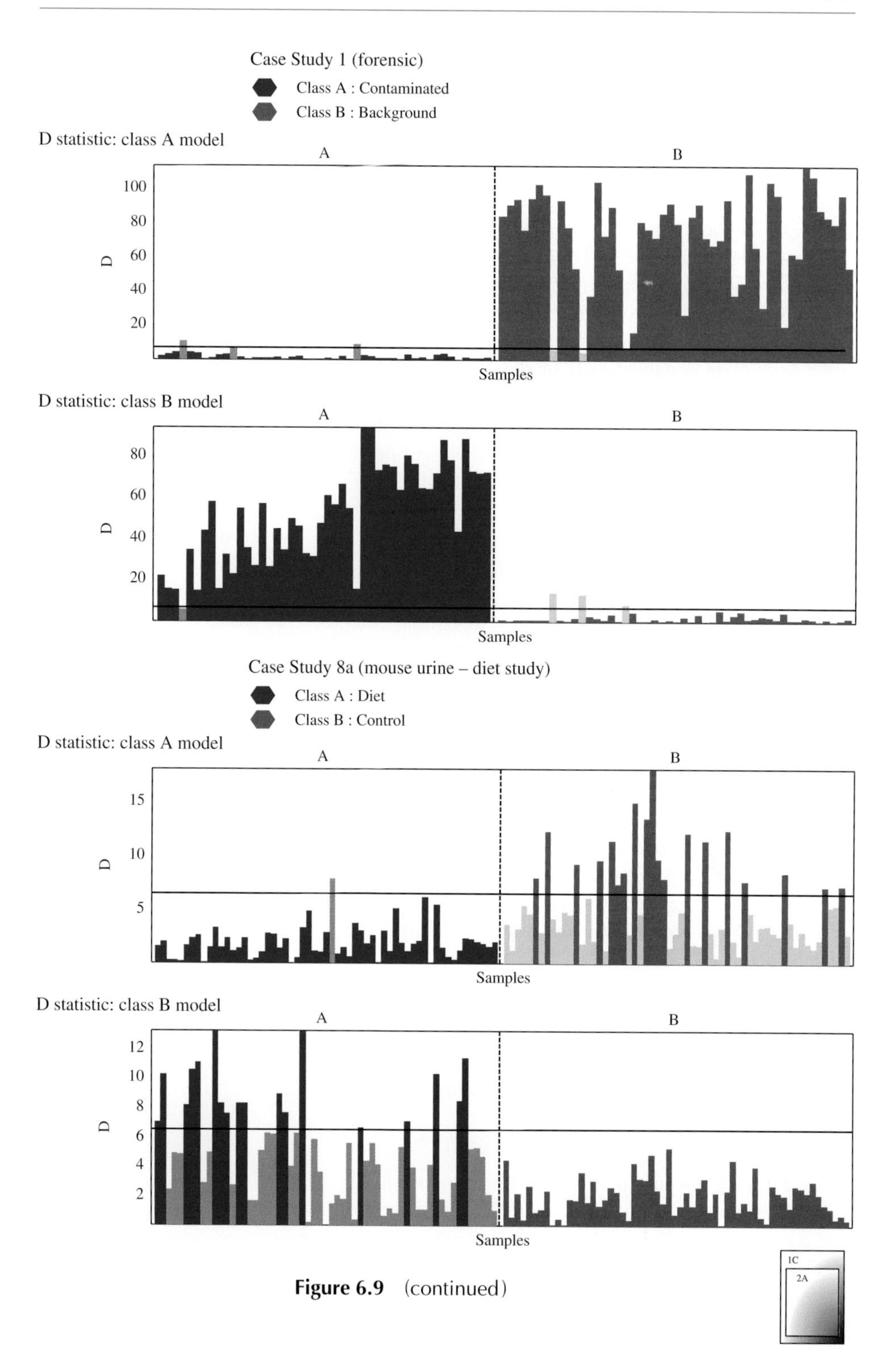

Figure 6.9 (continued)

groups' in all models. However what is interesting is that we can see which groups of samples are most similar, for example, groups A and C seem closely related, as group C samples have a low D-statistic when modelling group A and vice versa. Much classical chemometrics have been developed based on NIR spectroscopic measurements in areas such as food and process chemistry and as we see in this text, the NIR spectroscopic case study, if correctly preprocessed, gives excellent predictions, and so is less of a challenge, but classical chemometrics allows visualization of these data even though conventional spectroscopic methods could probably obtain the same information but less well presented: presentation is important for 'selling' a subject and to allow people who are not expert spectroscopists to understand trends in the data and bring methods to a wider audience. For Case Study 1 (forensic) we find excellent separation between the classes, using both class A and class B models. A small number of samples for each 'in group' model are above the 95 % limit, but this is to be expected as 1 in 20 samples are likely to exceed this level. The 'out groups' are also well distinguished, with only two samples using the class A model misclassified at 95 % confidence, and one sample using the class B model. Case Study 8a (mouse urine – diet study) is less easy to distinguish, with around half the 'out group' within the 'in group' 95 % confidence limits. However the 'in group' samples mainly have much lower D-statistics than the 'out group' especially when modelling the dieting mice. A reason for this problem is that the number of PCs is probably inadequate to model the groups, and more PCs would be desirable. Dataset 8a, from metabolomic profiling, is typical of the sort of data that are available for the new generation of chemometricians and is much harder to interpret using straightforward approaches, requiring a more sophisticated application of chemometrics. In Section 6.4.5 we will discuss the visualization of D- and Q-statistics for disjoint PC models.

6.4.4 Q-Statistic or Squared Prediction Error

The Q-statistic or SPE (Squared Prediction Error) is used to see how well samples fit into a PC model based on one class. For one class models we can perform PCA separately for each group, centring and standardizing the data in the group independently of the remaining samples, and then see how well samples from this and other groups fit the PC model. It is defined as the sum of squares of the residuals obtained from the PCA model of the training set as follows:

$$Q_{gi}^2 = \sum_{j=1}^{J} (e_{gij})^2$$

where e_{gij} is the error for reconstruction of the ith sample with the jth variable given by:

$$\boldsymbol{e}_{gi} = \boldsymbol{x}_i - \boldsymbol{x}_i \boldsymbol{P}_g' \boldsymbol{P}_g$$

where $\boldsymbol{P}_g$ is the loadings matrix for the training set for group g and Q_{gi} is the Q-statistic for sample i and group g. Note that although $\boldsymbol{PP}'$ is a unit matrix (using our terminology of $\boldsymbol{X} = \boldsymbol{TP}$), $\boldsymbol{P}'\boldsymbol{P}$ is not providing the number of PCs and is less than the number of original variables.

The Q-statistic geometrically represents the Euclidean distance between the new sample vector and its projection onto the class model. The Q-statistic has mainly been used in the area of quality control of manufacturing processes, the 'in-group' consisting of samples of known quality, but can be extended to use in all classification problems. When the Q-statistic of a sample is outside the limits of confidence obtained from a model using the 'in-group', this means that the constructed model is not valid for that sample and that a new type of variation is present in that sample which is not present in the samples for the 'in group': there is a change in the correlation structure of the model since this new variation was not taken in account and the orthogonal distance of this new sample to the hyperplane defined by the 'in-control' model will increase. This type of sample is sometimes denominated as an alien outlier. Therefore, the D-statistic of this sample has little meaning under such circumstances since the PC model is not valid for that sample.

Control limits for the Q-chart can be calculated for a specific significance level (α). The most widely used equation for computing these limits is given by:

$$Q_{lim,\alpha} = \theta_1 \left[1 + \theta_2 h_0 \left(\frac{h_0 - 1}{\theta_1^2} \right) + \frac{z_\alpha \sqrt{(2\theta_2 h_0^2)}}{\theta_1} \right]^{1/h_0}$$

where $\boldsymbol{V}$ is the covariance matrix of the residual matrix $\boldsymbol{E}$ (after performing PCA on the training set), θ_1 is the trace of $\boldsymbol{V}$, i.e. the sum of the variances of the J variables, θ_2 is the trace of $\boldsymbol{V}^2$, θ_3 is the trace of $\boldsymbol{V}^3$, $h_0 = 1 - (2\theta_1\theta_3/3\theta_2^2)$ and z_α is the standardized normal variable with a $(1-\alpha)$ confidence level. Hence the Q-statistic can also be computed for different confidence levels. An alternative equation was proposed by Box in the 1950s but provides very similar results, and we will stick with the equation above in this text.

6.4.5 Visualization of D- and Q-Statistics for Disjoint PC Models

A major feature of one class classifiers is that not only is it possible to produce separate models for how distant a sample is from a class centroid, but it is also possible to project samples onto disjoint PC models for each class. Up to now we have visualized models in the PC space of all the samples together. Figure 6.10 illustrates these for Case Studies 1, 2, 3, 5, 8a, 8b and 11 with 95 % limits indicated; although other confidence limits could be visualized, further discussion of additional statistics would be too complex if we were to introduce too many confidence bands. There are several important observations, as follows:

1. A separate PC model is obtained for each class, and so there are many more possible model spaces, requiring individual visualization for each group separately.
2. The 2 PC projections obtained using each class are often quite different to those for all the data together. At first they look as if this procedure has worsened separation but an aim is to spread out the class of interest to maximize its variability over the projection. Hence we can take tightly defined groups, e.g. in Case Study 2 (NIR spectroscopy of food) and find they are now much more spread out; another example is class C of Case Study 3 (polymers – Pcarb) which appears as a very tight group in the conjoint PC plot (Figure 6.8) but much more spread out in the disjoint PC space.
3. In many cases, which at first may appear surprising, the separation between groups that are clearly distinct when conjoint PCA is performed, is often lost. A good example is Case Study 3 (polymers – amorphous and semi-crystalline) where the two groups are now mixed, particularly if we model amorphous polymers, the semicrystalline group is densely packed in the middle. Even for Case Study 1 (forensic) which appears quite

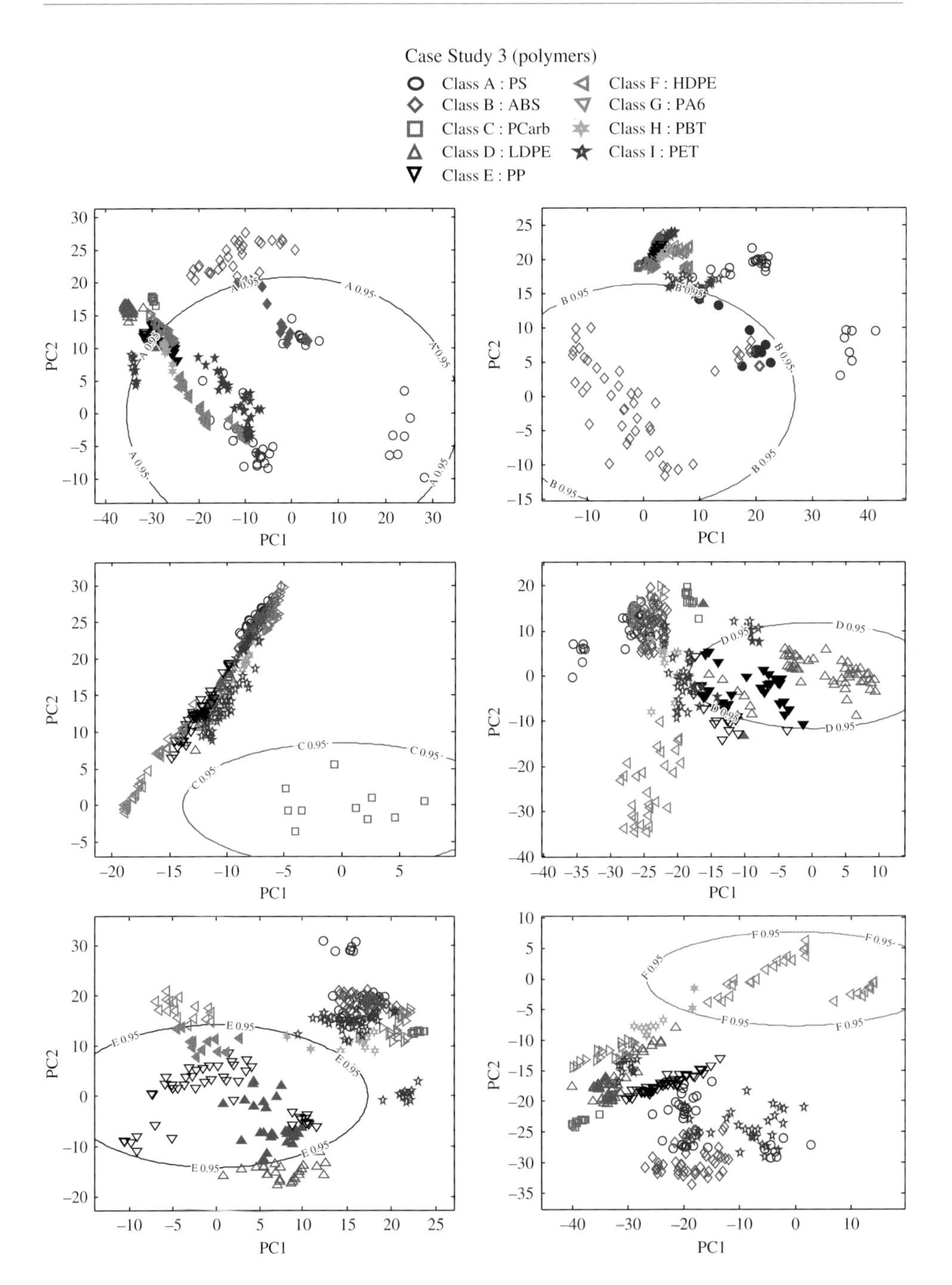

Figure 6.10 95 % confidence limits for the *D*-statistic for Case Studies 1, 2, 3, 5, 8a, 8b and 11, for each class using projections onto 2 PCs for each class separately (disjoint PC models). The colour of the contour band relates to the class being modelled, both via PCA and the *D*-statistic. Misclassified samples are indicated with filled symbols

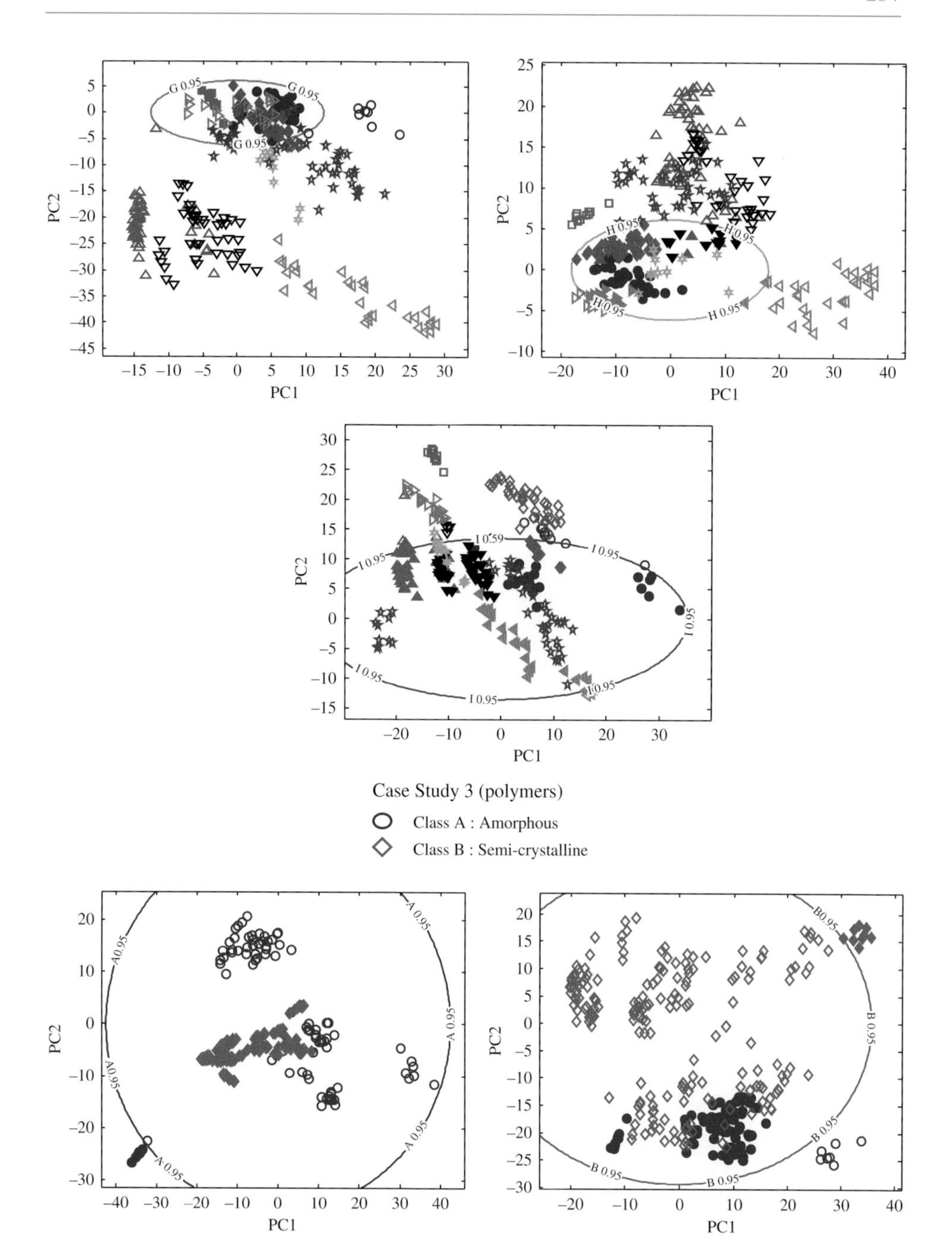

Figure 6.10 (continued)

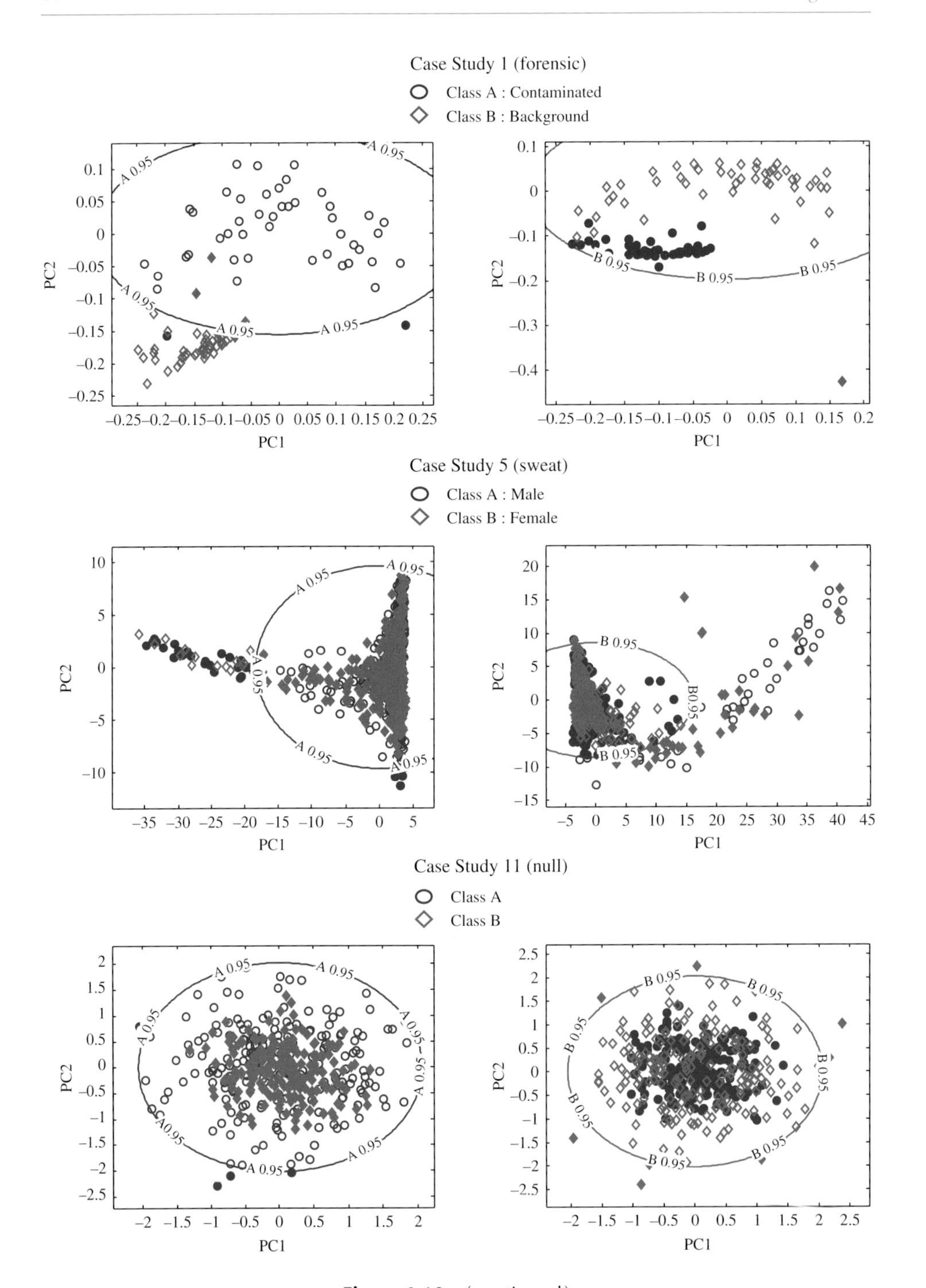

Figure 6.10 (continued)

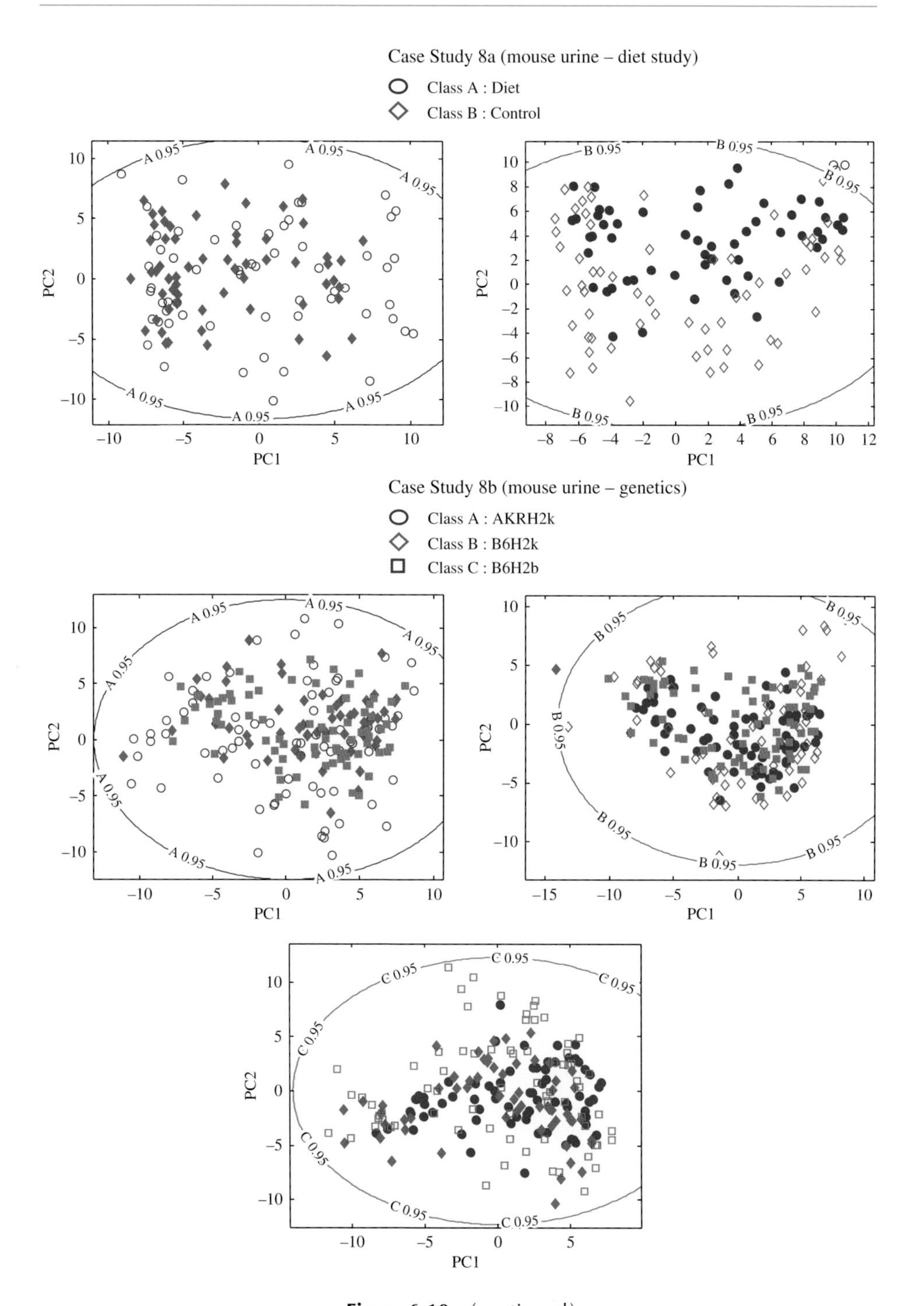

Figure 6.10 (continued)

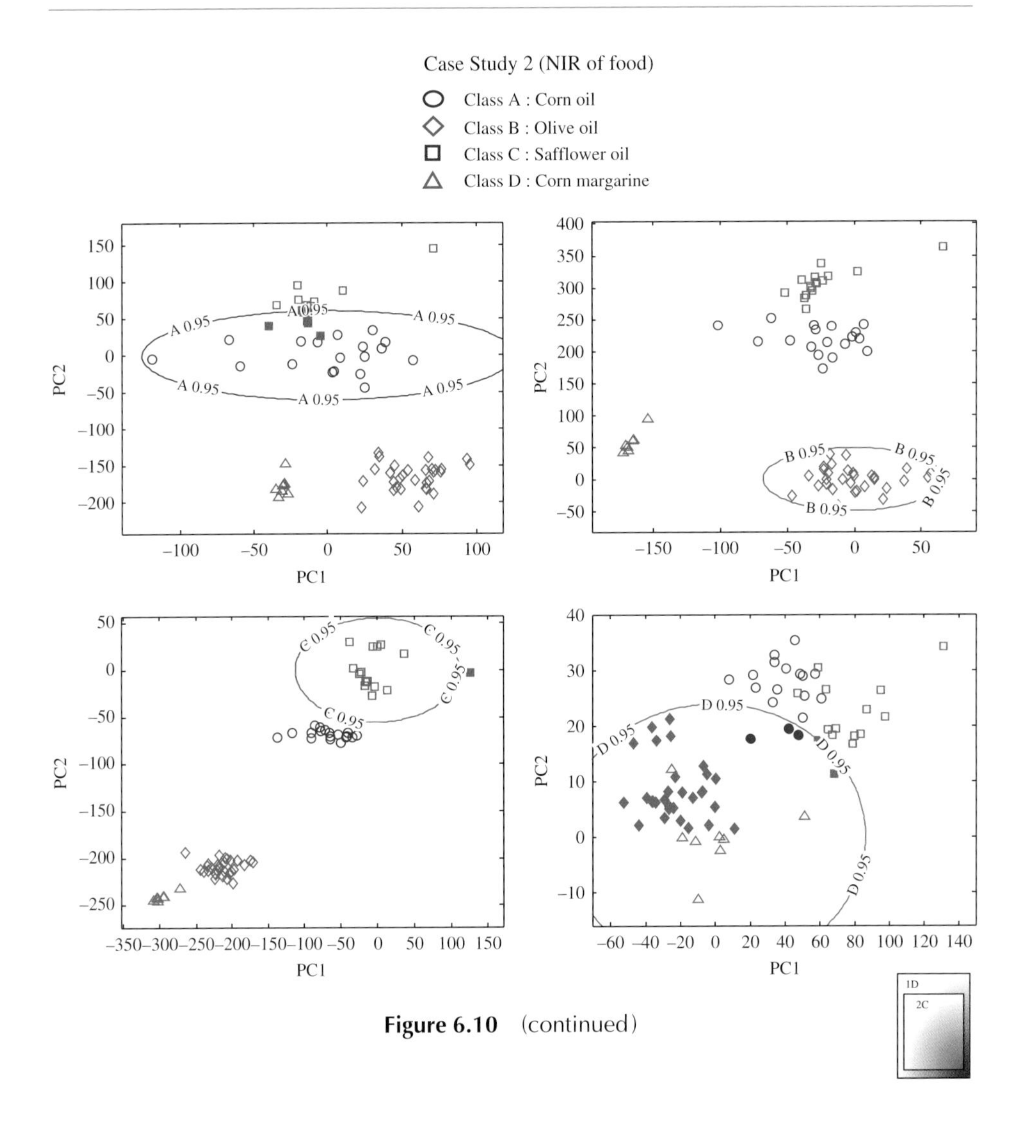

Figure 6.10 (continued)

separable by all other indicators the two groups are now hard to distinguish, for example, using a class B model (background), class A samples all fall within the 95 % confidence limit of class B. For Case Study 11 (null), the 'out group' appears to be better modelled by the 'in group' than the 'in group' is modelled by itself: this is probably because the 'out group' appears to be modelled by random deviation from the centre and there is some small structure in the 'in group' which is seized upon but has no relevance to the 'out group'. The reason for this apparent problem is that when PCA is performed on the entire dataset, often the most important factor is the separation between groups. If it is performed just on one class, this factor is lost and so the difference between the two classes may be lost. This is illustrated in Figure 6.11 for two classes in 2D projected onto 1D. The top projection is onto the first PC of both classes together and we can see that the

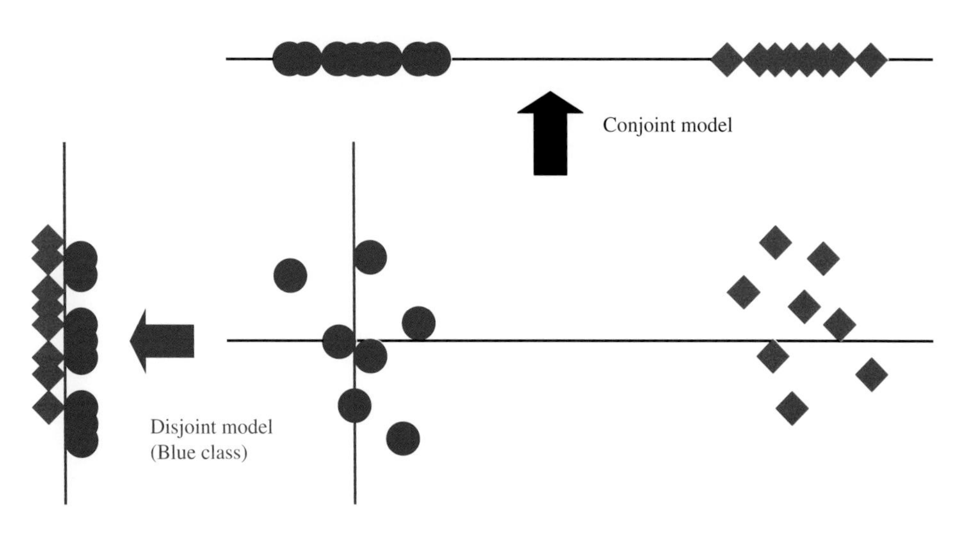

Figure 6.11 Illustration of projection of a 2D dataset consisting of two classes, onto 1D models: top for a conjoint PC model and left for a disjoint PC model for one class

class separation is maintained well, as this is the dominant factor. If we model just one of the classes (left projection), we see that the two classes are no longer distinct.

Hence if PCA is performed independently on each class, often the results of the D-statistic are worse than if PCA is performed on the overall dataset. Early chemometrics studies often focused on quite easily soluble problems such as NIR spectroscopy where spectroscopic differences are quite distinct (e.g. Case Study 2) and so this problem was less noticeable. But if each class is modelled separately, the D-statistic is rarely sufficient. Note that when using disjoint PC models the classes appear much less easy to separate.

Fortunately the Q-statistic comes to our rescue. Although the distinction between the classes is often blurred when samples of an 'out group' are projected onto a model formed by the 'in group' they do often do not fit the 'in group' PC model well. In order to demonstrate this, we illustrate both 95 % confidence limits for Case Studies 1 (forensic), 3 (polymers) and 11 (null) and for class A of Case Studies 2 (NIR spectroscopy of food) and 8b (mouse urine – genetics), using a projection into 2 PCs based on each of the classes in Figure 6.12. The vertical axis represents the value of the Q-statistic and the two axes of the plane correspond to the axes of Figure 6.10 or the disjoint PC scores. The 95 % cut-off for the Q-statistic is illustrated by a plane coloured according to the class model, and the 95 % D-statistic limits by an ellipsoid. Note that in these diagrams we illustrate samples that are outside the Q-statistic limit with filled symbols, and those within with open symbols. For Case Studies 1 and 3 we can now immediately see that although the 'out group' samples are mainly within the 95 % D-confidence limit, as illustrated also in Figure 6.10, they are in fact all outside the 95 % Q-confidence limit. A few samples from the 'in group' are also outside their own 95 % confidence limits, as we would expect 1 out of 20 to be out of this region. For Case Study 11 (null) all the samples are within each other's 95 % confidence limits thus suggesting that the PC model of class A fits class B well and vice versa, which is as anticipated. We illustrate also the class A models for Case Study 2. When we look at the *D*-statistic alone using a PC projection onto class A rather than all classes together, several samples from class C appear to be within the 95 % limits for class A, but it is clear from the *Q*-statistic that they are above the limit. For Case Study 8b (mouse – genetics) whereas there appears no strong distinction between the classes using the *D*-statistic alone, we now see that the *Q*-statistic can differentiate some of the class C samples from the rest.

The *D*- and *Q*-statistics can be visualized for each sample and each model. Figure 6.13 shows these for Case Studies 1, 2 and 8a and should be compared to Figure 6.9. For Case Study 1 (forensic), we see that using the *D*-statistic and a class B model, class A is

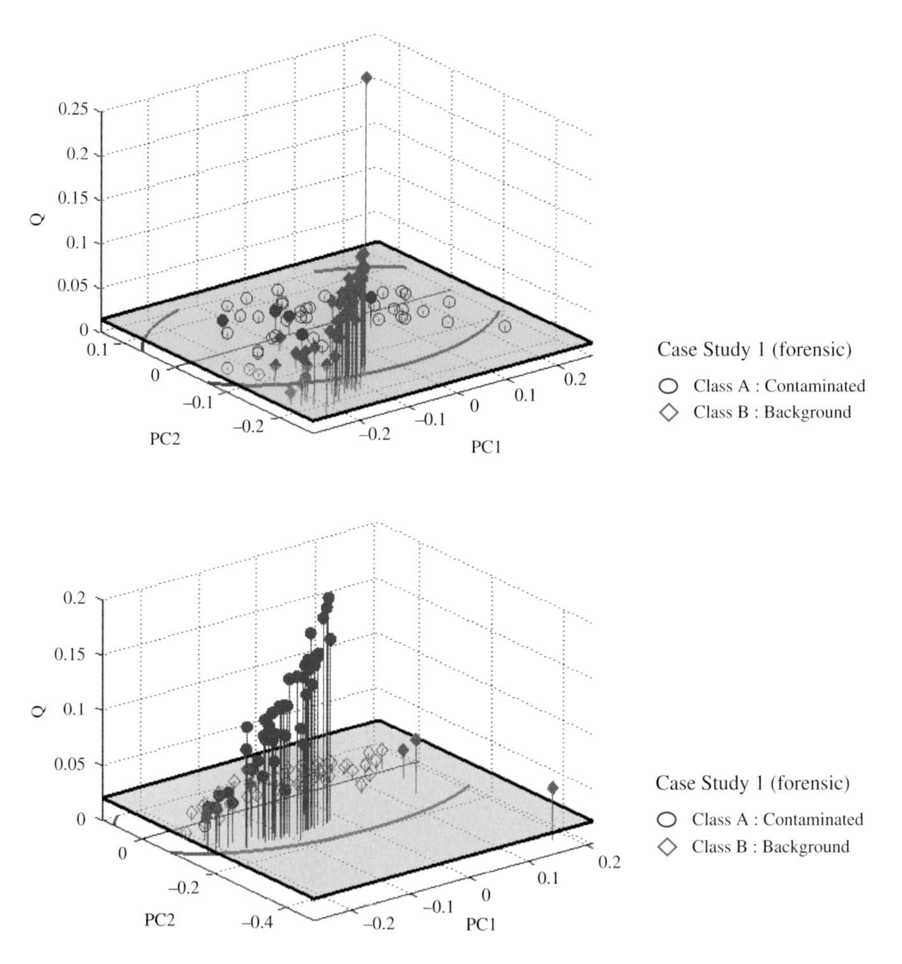

Figure 6.12 Superimposed D and Q 95 % confidence limits for Case Studies 1, 3 and 11 and Case Studies 2 and 8b (class A) using disjoint PC models. Q limits are represented by a plane and D limits and ellipsoidal, coloured according to class. Samples outside the Q limit are indicated by filled symbols

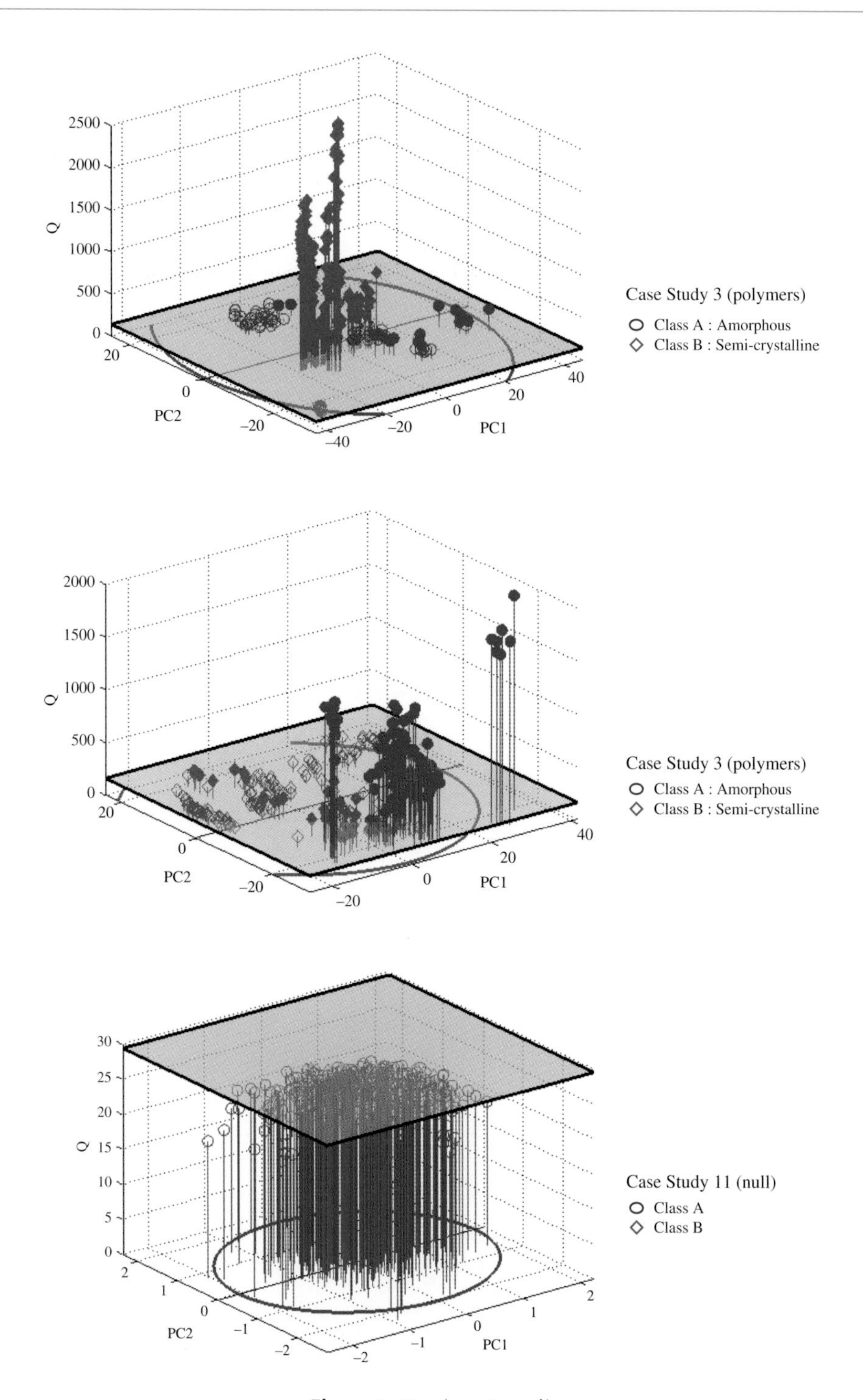

Figure 6.12 (continued)

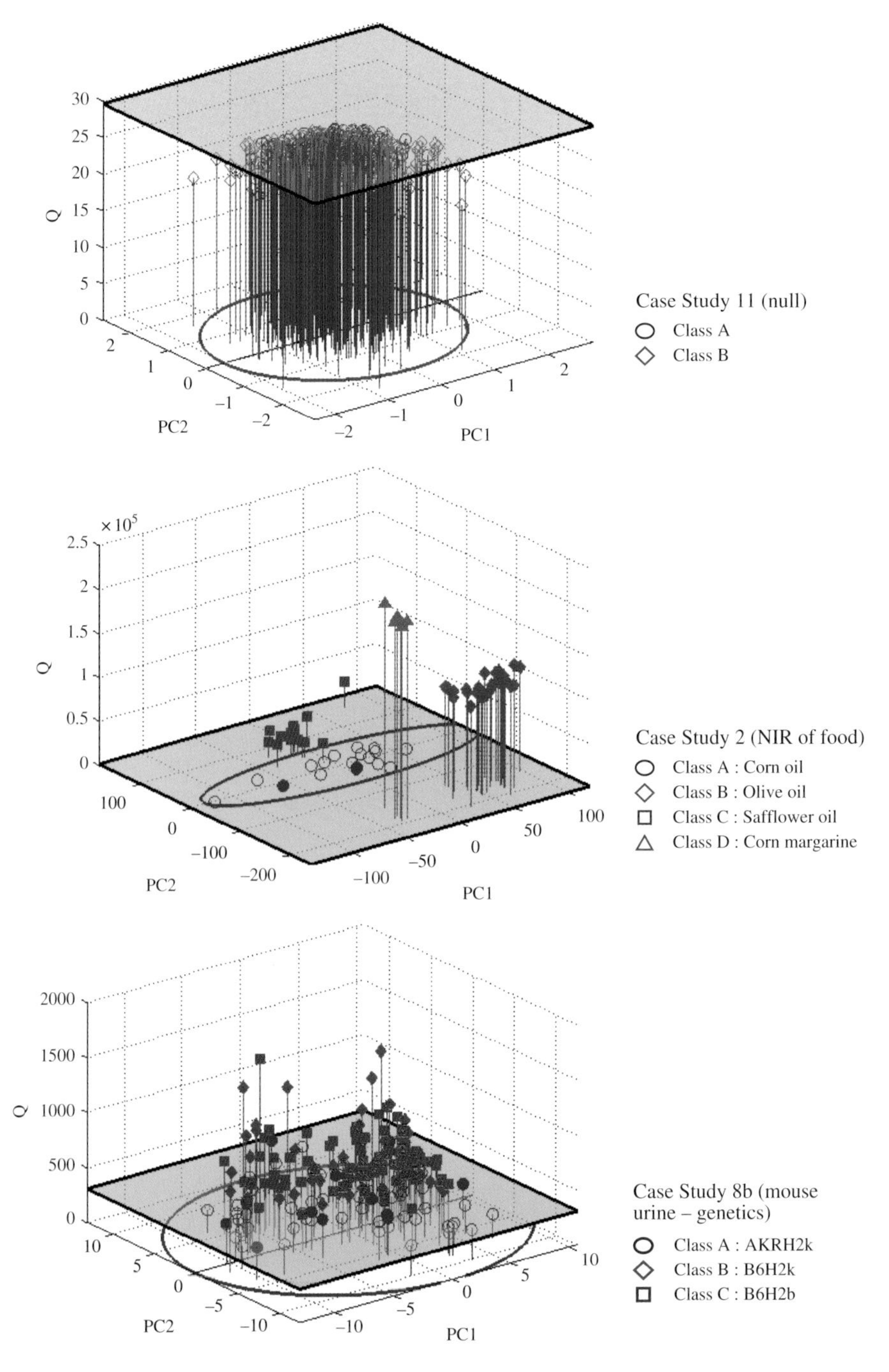

Figure 6.12 (continued)

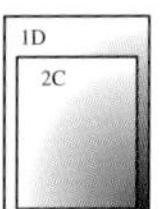

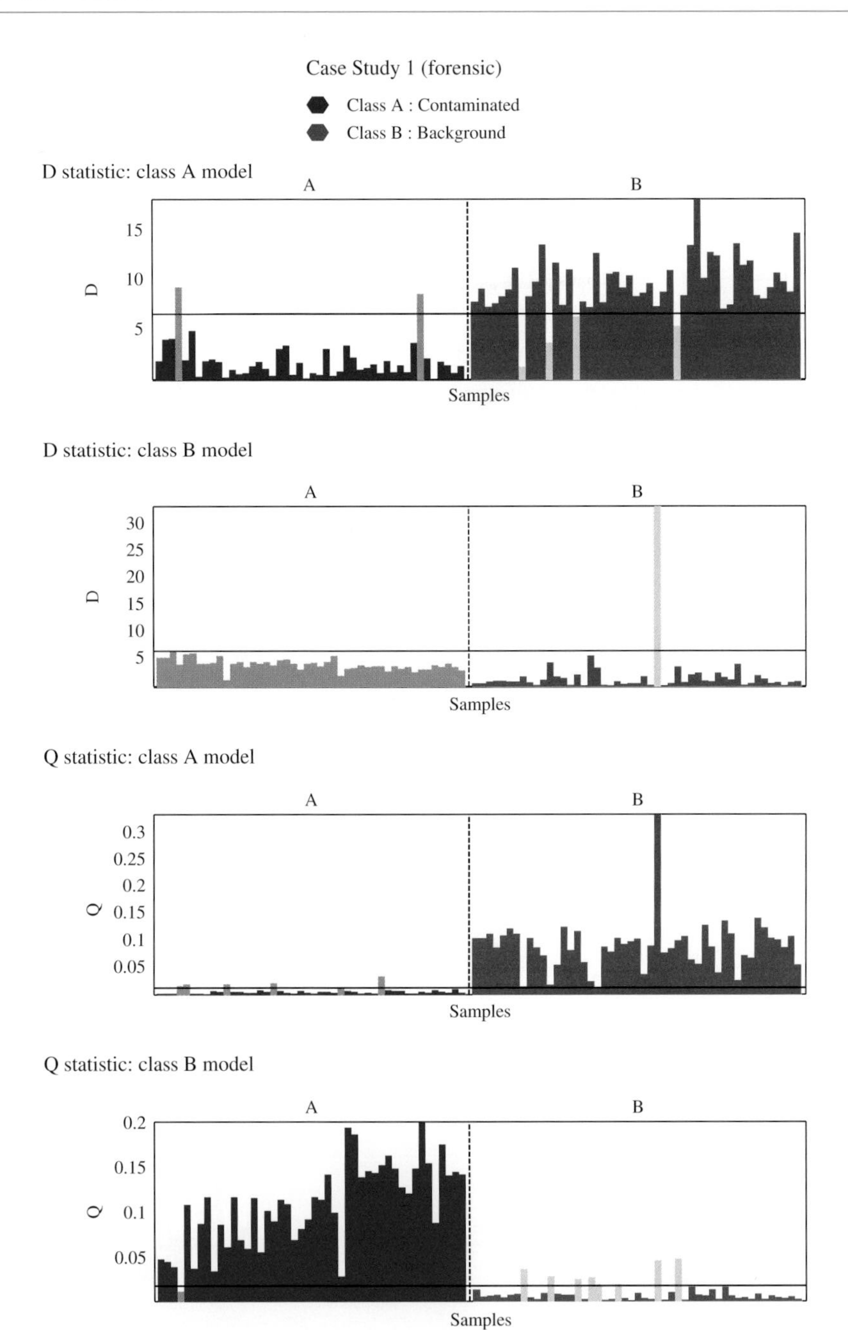

Figure 6.13 *D* and *Q* statistics for Case Studies 1, 2 and 8a, using disjoint PC models. Misclassified samples using each statistic individually are represented in light colours. The 95 % confidence levels are indicated

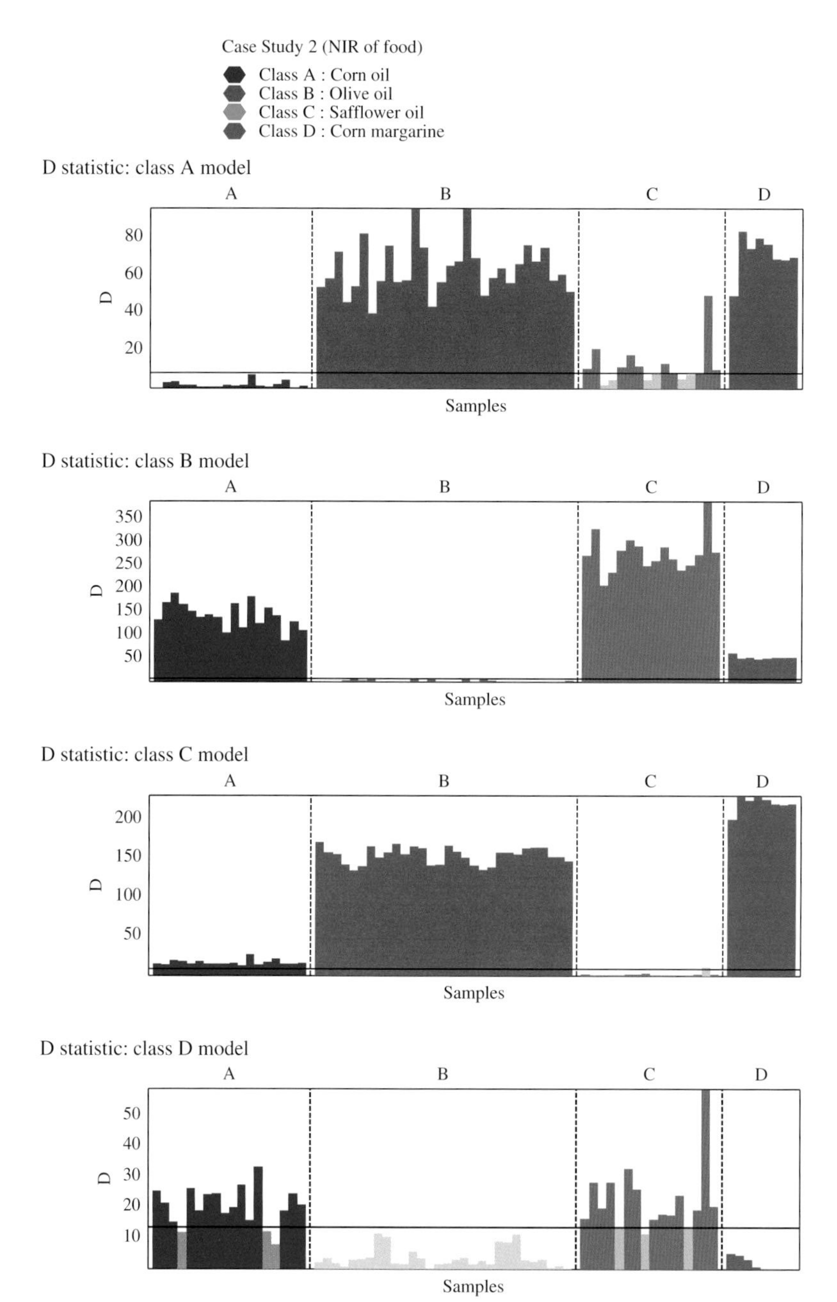

Figure 6.13 (continued)

Q statistic: class A model

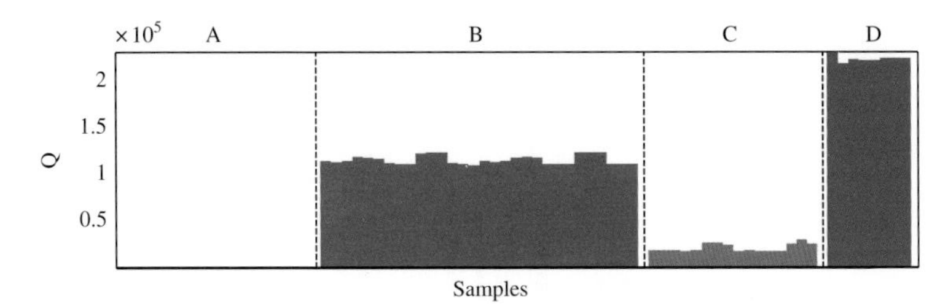

Q statistic: class B model

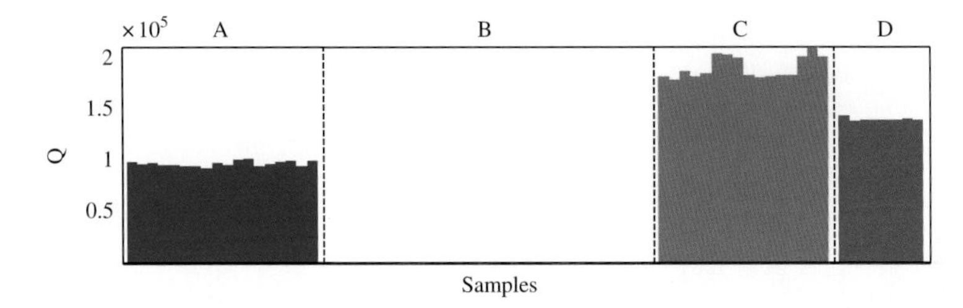

Q statistic: class C model

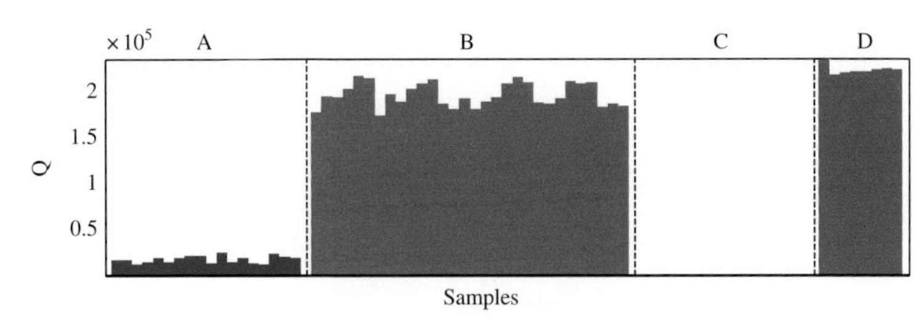

Q statistic: class D model

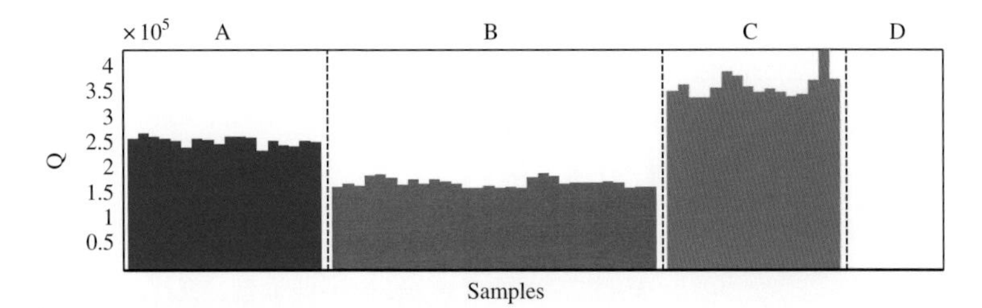

Figure 6.13 (continued)

Case Study 8a (mouse urine – diet study)
Class A : Diet
Class B : Control

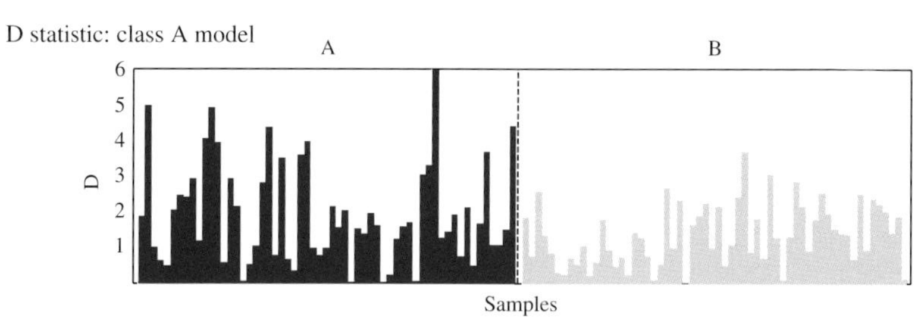

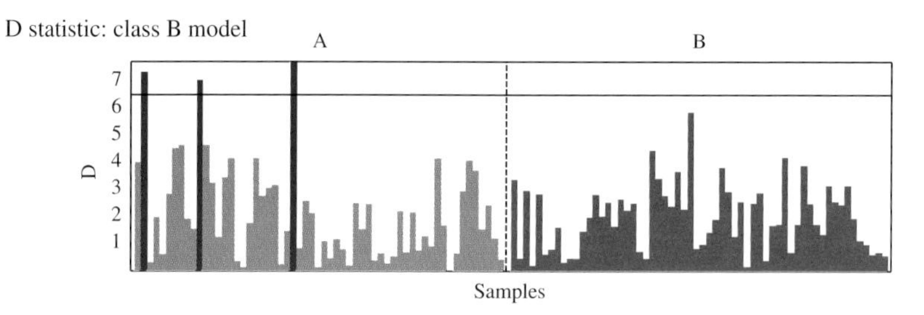

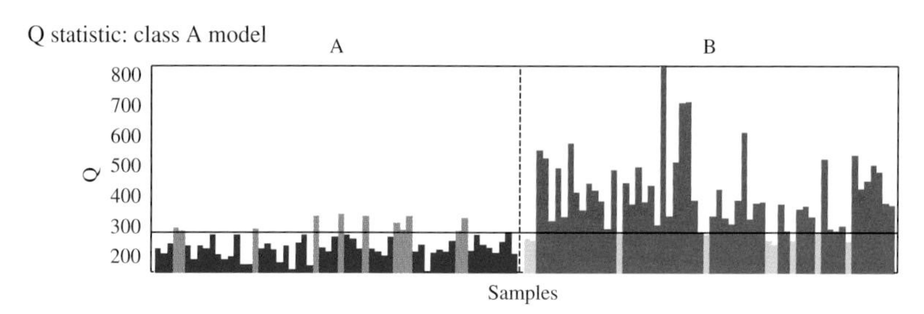

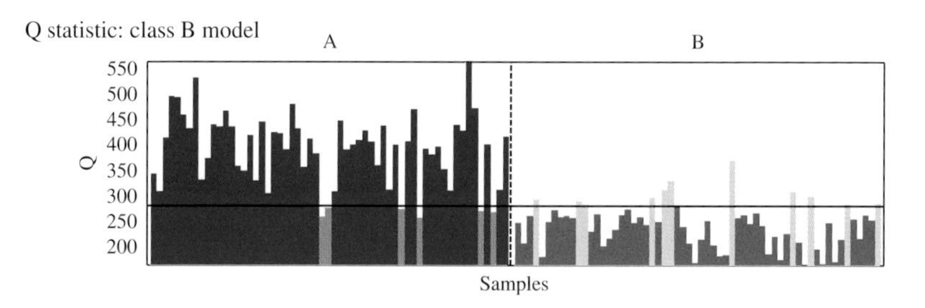

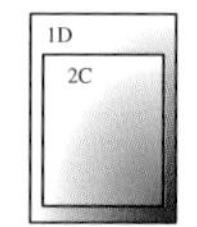

Figure 6.13 (continued)

consistently incorrectly predicted within the 95 % confidence level using disjoint PC models, in contrast to conjoint models and only if we first use the Q-statistic can we obtain a true picture. For Case Study 2 (NIR spectroscopy of food), in most cases the D-statistic alone is sufficient, except when using the class D model, where class B is confused with class D. The reverse is not true though, and this probably represents the fact that class D is small (8 samples) and so perhaps not really large enough for robust models. The Q-statistic though comes to our rescue and is quite unambiguous. However, it is important to recognize that many early chemometrics papers using Q- and D-statistics often found that the D-statistic was adequate (e.g. in process monitoring) even for disjoint PC models, probably reflecting the fact that NIR spectroscopic measurements are easily distinguishable. Case Study 8a (mouse urine – diet study) is interesting. We see from the D-statistic of the disjoint PC model that there is very little distinction between the groups, with the 'out class' falling mainly within the 'in class' limits. However, the Q-statistic does offer quite good distinction between the groups, which is not found in the conjoint PC model. Naturally the analysis can be extended to more that 2 PCs, but the main aim of this chapter is to illustrate generic principles.

These graphs can be used to demonstrate the advantages and disadvantages of using disjoint PC models in contrast to conjoint models. First of all, for many simpler problems the conjoint models are adequate, and often clearer. Second, the Q-statistic is usually essential for disjoint models, except in the simplest cases; because of an early chemometrics focus on NIR spectroscopy this perhaps was not so obvious. Third, under certain circumstances (Case Study 8a) the Q-statistic combined with disjoint PC models does offer an advantage over conjoint models. Finally, whereas for many chemometrics problems conjoint models are often adequate, in areas such as process monitoring where only a single class (usually the NOC or Normal Operating Conditions) can be modelled to any degree of accuracy, independent models may be essential.

6.4.6 Multivariate Normality and What to do if it Fails

Most of the probabilities above assume that the samples are approximated by a multinormal distribution. As we discuss in Section 9.2.2, most variables are not, but this does not mean that the samples are not so distributed, as the sum of distributions often approximates well to a multinormal distribution. If in doubt it is possible to perform tests for multivariate normality, but these should be in the data space for which the statistics are computed. There are a large number of such tests and it is not the job of this text to enumerate these in detail, simply to advise that it is often a useful idea to run these tests if in doubt. One test is the Doornik–Hansen Omnibus Test for normality (which we will not describe in detail here for brevity), and the results for the data sets in this text using 2PC conjoint and disjoint models are presented in Table 6.1. The higher the p value the more likely it is to be multinormal (some people use $1 - p$ but this simply involves subtracting p from 1, and the same conclusions are obtained). At first glance the results appear somewhat discouraging. However many tests of normality are critically dependent on the presence of outliers, and often fail if the number of samples is quite small. For example, for Case Study 1 (forensic) both class A and B fail for the conjoint model. This is probably due to a few outliers (see Figure 6.8 for conjoint PC plots). For Case Study 3 (polymers) we find that the main groups A and B fail the normality test for both conjoint and disjoint models, probably because they are composed of several

Table 6.1 Results of Doornik–Hansen Omnibus Test for multivariate normality for each class in this text using 2 PC models; values of $p > 0.1$ are highlighted (bold). Note that the higher the value of p, the more likely the classes are normal

Case study	Class	Conjoint model	Disjoint model
1 (forensic)	A	0.000	**0.517**
	B	0.000	0.000
2 (NIR spectroscopy of food)	A	**0.714**	**0.131**
	B	**0.227**	**0.409**
	C	0.005	0.000
	D	0.059	0.014
3 (polymers – 2 types)	A	0.004	0.004
	B	0.000	0.000
3 (polymers – 9 groups)	A	**0.775**	0.000
	B	**0.908**	0.001
	C	0.032	**0.411**
	D	0.002	0.000
	E	0.016	0.008
	F	**0.230**	0.028
	G	**0.917**	0.079
	H	**0.627**	**0.189**
	I	**0.598**	0.000
4 (pollution)	A	0.000	0.000
	B	0.018	0.001
5 (sweat)	A	0.000	0.000
	B	0.000	0.000
6 (LCMS of tablets)	A	**0.227**	0.004
– origin	B	0.008	**0.504**
	C	**0.410**	**0.302**
– batch	D	0.044	0.006
	E	**0.883**	**0.542**
	F	0.054	0.011
	G	**0.526**	0.001
	H	**0.877**	0.000
7 (clinical)	A	0.001	**0.187**
	B	0.000	0.000
	C	0.019	0.100
	D	0.055	**0.444**
	E	**0.366**	**0.134**
	F	**0.231**	**0.496**
8a (mouse urine – diet study)	A	0.035	0.069
	B	0.005	0.001
8b (mouse urine – genetics)	A	0.006	**0.118**
	B	0.001	0.087
	C	0.001	0.000
9 (NMR spectroscopy)	A	0.062	**0.625**
	B	0.055	0.003
10(a) (simulation)	A	**0.633**	**0.929**
	B	**0.876**	**0.842**
10(b) (simulation)	A	**0.957**	**0.908**
	B	**0.500**	**0.685**

Table 6.1 (*continued*)

Case study	Class	Conjoint model	Disjoint model
10(c) (simulation)	A	**0.168**	**0.453**
	B	**0.422**	**0.197**
11 (null)	A	**0.493**	**0.306**
	B	**0.454**	**0.397**
12 (GCMS)	A + B (S1)	0.083	0.020
	C + D (S2)	**0.279**	**0.478**
12 (microbiology)	A + B (S1)	**0.703**	0.002
	C + D (S2)	**0.001**	0.000

subgroups, and so would appear bimodal or clumped, but many of the subgroups (or polymer types) are normally distributed. The simulations as perhaps anticipated all pass normality tests: this raises a warning that it is often hard to mimic reality and although the simulations can provide useful information about the performance of algorithms, it is quite rare that they are realistic – Case Study 11 (null) is good for guarding against overfitting and Case Study 10 (simulations – varying overlap) for studying the differences when the overlap between classes changes, but they may not necessarily be realistic models for NIR spectroscopic or GCMS or NMR spectroscopic data, often involving many more correlations, and so should be viewed on context. In addition, small groups often fail normality tests, and so, for example, groups C and D of Case Study 2 (NIR pectroscopy of food) that consist of 16 and 8 samples respectively fail in both the disjoint and conjoint space whereas the larger groups pass.

Despite this apparently pessimistic view, in fact in most cases this is because the middle of the distribution is modelled poorly by a model based on a normal distribution. Once we get to high values of α such as 95 % or 99 %, the predictions are often fairly accurate, for example, for Case Study 3 (polymers – 2 types) class B using the conjoint model, 15 out of 201 samples are outside the 90 % limit or 7.5 % compared to an expected 10 %, 8 are out of the 95 % limit (4 % compared to an expected 5 %) and none outside the 99 % limit. Visual inspection of the 95 % limits show that in most cases they do enclose the vast majority of samples which are within these limits in both the conjoint and disjoint PC model space.

Hence it is usually not unreasonable to take the 95 % or 99 % limit as far from the mean and a good 'data independent' indicator of whether a sample a member of the 'in group' or not. If data do fail a normality test badly, it is worth looking empirically at it first and although it may not be easy to accurately model 50 % or 75 % confidence limits very well, but we are mainly interested as to whether an unknown sample is far from the centre of a class. If we take the conjoint PC models for Case Study 2 (NIR spectroscopy of food), we see that even though two of the groups appear not to be multinormal in distribution, groups are well distinguished using high confidence limits. The limits can be regarded as a way of scaling the distance from the centroid of a class so that it is comparable for different groups using different numbers of variables, rather than using raw distances which may be on different scales because the raw data are collected and recorded in different ways. Hence unless classes are very irregularly shaped, usually setting up 95 % or 99 % limits for rejecting samples that are outside these as being members of the 'in group' is satisfactory providing these values are not interpreted too strictly in terms of probabilities.

There are alternatives – one is just to use a distance and not assign a probability to it. Another approach is simply to count how many samples are at a given (scaled – usually Mahalanobis) distance from the centre and use these for empirical probability contours; if, for example, there are 95 out of 100 samples within a Mahalanobis distance of 3 from the centre this becomes the empirical 95 % limit. If one wants a finer gradation it is possible to interpolate. Additionally there are methods for iterative computation of confidence limits involving a number of approaches, such as removing some samples, calculating the limits, then using a different subset and so on, rather like the test/training validation methods discussed in Chapter 8.

If however there is a strong reason for developing methods with confidence boundaries that are not ellipsoidal, other approaches can be employed of which SVDD discussed in Section 6.5 is one alternative.

6.5 Support Vector Data Description

SVMs are usually introduced as a form of binary classifiers (Chapter 5) but one class modifications are available. Rather than being used to separate two or more classes, one class SVMs are built on a single class. For more discussion of the basic concepts of SVM, see Section 5.7. There are two main one class SVM algorithms, one is called 'Support Vector Data Description' (SVDD) and another is called 'ν-Support Vector Classifier' (ν-SVC). We will restrict discussion to the SVDD method in this text, whilst reminding readers that there are, as always, several alternatives available. We will assume that we are using a RB kernel function.

For SVDD, we fit our kernel function to only one class at a time, rather than both or all classes in the data set. We use a soft margin at L1 (level 1) to be consistent with our description of SVM for two class problems. Of course other variations are possible but the aim of this section is to describe one widespread alternative to two class SVMs. The structure error function (see Section 5.7.3 for more details) for the support vectors is defined as:

$$\varphi(R, \boldsymbol{\mu}, \xi) = R^2 + C \sum_{i \in sv} \xi_i$$

with μ being the centre of the data in higher dimensional space, R an adjustable parameter, which can be visualized as a radius in this space, ξ_i is the slack variable modelling the training error and the tolerance of the training error is controlled by the penalty term C which controls the size of $\boldsymbol{\xi}$ and therefore controls the fraction of training samples lying outside the boundary as we will see below. SVs are either on the boundary ($\xi_i = 0$) or outside the boundary ($\xi_i > 0$). It is important to realize that R changes the appearance of the boundary but is controlled by C. The more rigid the boundary, the larger the value of R^2. The mathematics is rather complicated but the RBF in kernel space is the same as for two class SVMs (Section 5.7.2) except that the SVs come from only one class, and so the kernel function is always positive. This means that the position of the separating hyperplane (or decision function) must be changed, and the decision function is given by $0.5(b - R^2)$ where b is defined slightly differently (and is negative in value) to b for two class SVMs. For more details it is suggested to reference the original literature. The principles are illustrated in Figure 6.14.

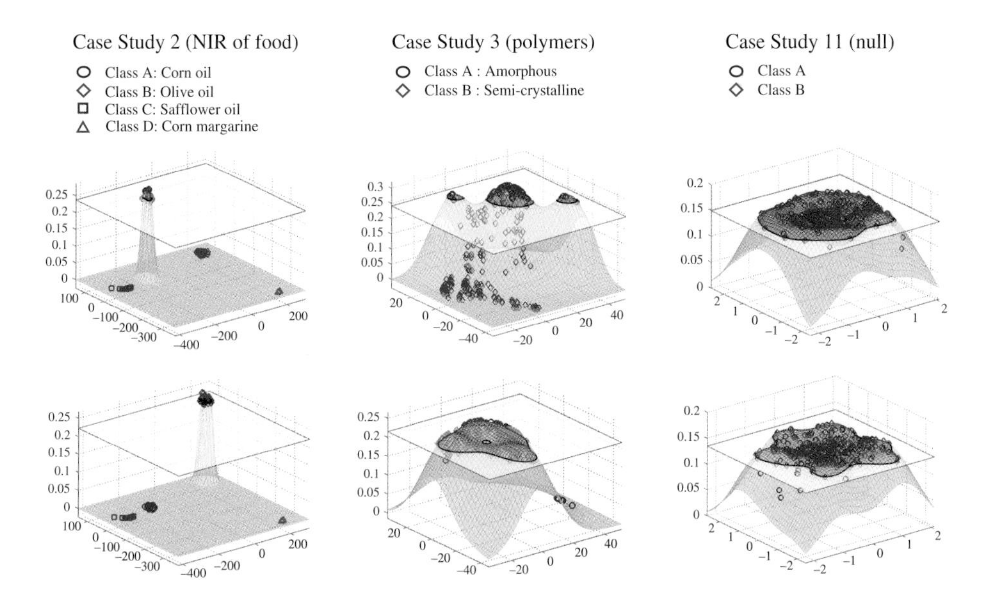

Figure 6.14 Principle of the one class SVM method. The vertical scale relates to an RBF kernel function, formed using just one class, and so all positive, with a decision plane. In this case, hard margin models ($D = 0$) are employed for several case studies (only classes A and B are represented for Case Study 2 – NIR spectroscopy of food), where avalue of $\sigma = 1 \times$ average standard deviation of the variables is employed

The interpretation of C differs somewhat to that used in two class SVMs and is not strictly analogous. Because there is only one class to be modelled, the boundary can either be set to include all samples or to misclassify a certain proportion of samples. The larger the number samples misclassified, the simpler the boundary becomes but the smaller the region in space. However unlike two class SVMs, where the boundary can become very complex, one class models tend not to, as they do not use information about the 'out group' and as such are not trying to avoid samples of different classes. Instead of C we will introduce a parameter D which equals the proportion of samples outside the boundary, defined by:

$$D = \frac{1}{I_g C}$$

to emphasize that C no longer has a similar meaning to that in two class SVMs, and so that D is directly comparable to a confidence band. A $D = 0.2$ is comparable to an 80 % confidence band within which 80 % of the training set samples are included; note that this number is approximately what is aimed for in the model and may not provide an exact fit. The change in appearance of the boundary with D is illustrated in Figure 6.15 for Case Study 8a (mouse urine – diet study) and an RBF using a value of $\sigma = 1 \times$ average sample standard deviation for the variables. Unlike the methods based on QDA the boundaries are no longer ellipsoidal. However they may be regarded as hard boundaries for a given confidence limit, but if superimposed they would not fall into concentric regions, and so the different levels of D cannot easily be represented as contours, unlike in the models of Section 6.4. For example, when modelling class A when D changes from 0.0 (all samples perfectly classified – effectively an infinite value of C equivalent to a hard

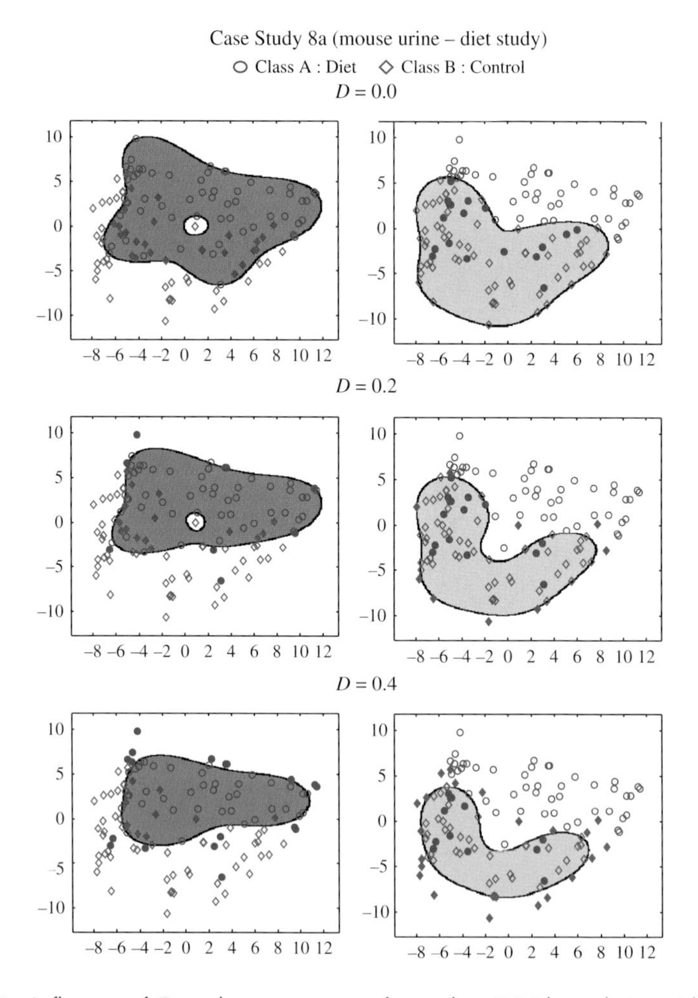

Figure 6.15 Influence of D on the appearance of one class SVM boundaries with an RBF of $\sigma = 1 \times$ average standard deviation of the variables in the dataset, using Case Study 8a, an L1 classifier and conjoint PC models. Misclassified samples are indicated using closed symbols

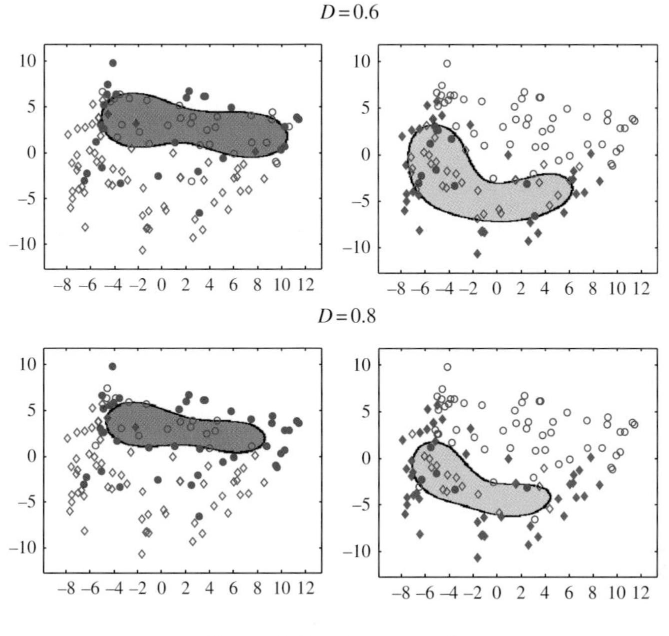

Figure 6.15 (continued)

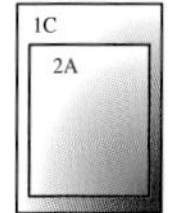

model SVM) to 0.4, the number of regions changes as the 'hole' in the middle of the class disappears. As D increases the boundaries become smoother but the region is smaller.

The conjoint boundaries for Case Studies 1 (forensic), 3 (polymers), 8b (mouse urine – genetics) and 11 (null) are presented in Figure 6.16, using a value of $D = 0.2$. We deliberately do not use a $D = 0.95$ as this value is not exactly analogous to a confidence limit so cannot be directly compared to the D-statistic for QDA, but in Chapter 8 discuss how this value can be optimized. It is important to remember that the boundaries represented are autopredictive and as such can be prone to overfitting. There are several important observations. First of all, for the parameters chosen most of the boundaries are very compact for Case Study 3 (polymers) and appear to fit groups (and subgroups) very tightly. Compared to the two class SVMs (Figure 5.29) the boundaries for dataset 11 (null) seem much smoother and approximately central. The classifier has no need to consider the 'out group' and so is not influenced by clumps of the samples from different classes: the data are most densely clustered around the centre and as such the best model to include most samples is around the centre, even if the 'out group' is then misclassified. Interestingly for the chosen values of D and σ there is a tendency for there to be a 'hole' in the middle of several of the classifiers – possibly a space where there are not many sample from the 'in group'; this can be rectified by changing the values of the adjustable parameters.

One class SVDD can also be applied to disjoint PC models – see Figure 6.17 for some examples. Similar considerations apply as to the conjoint models. There are many

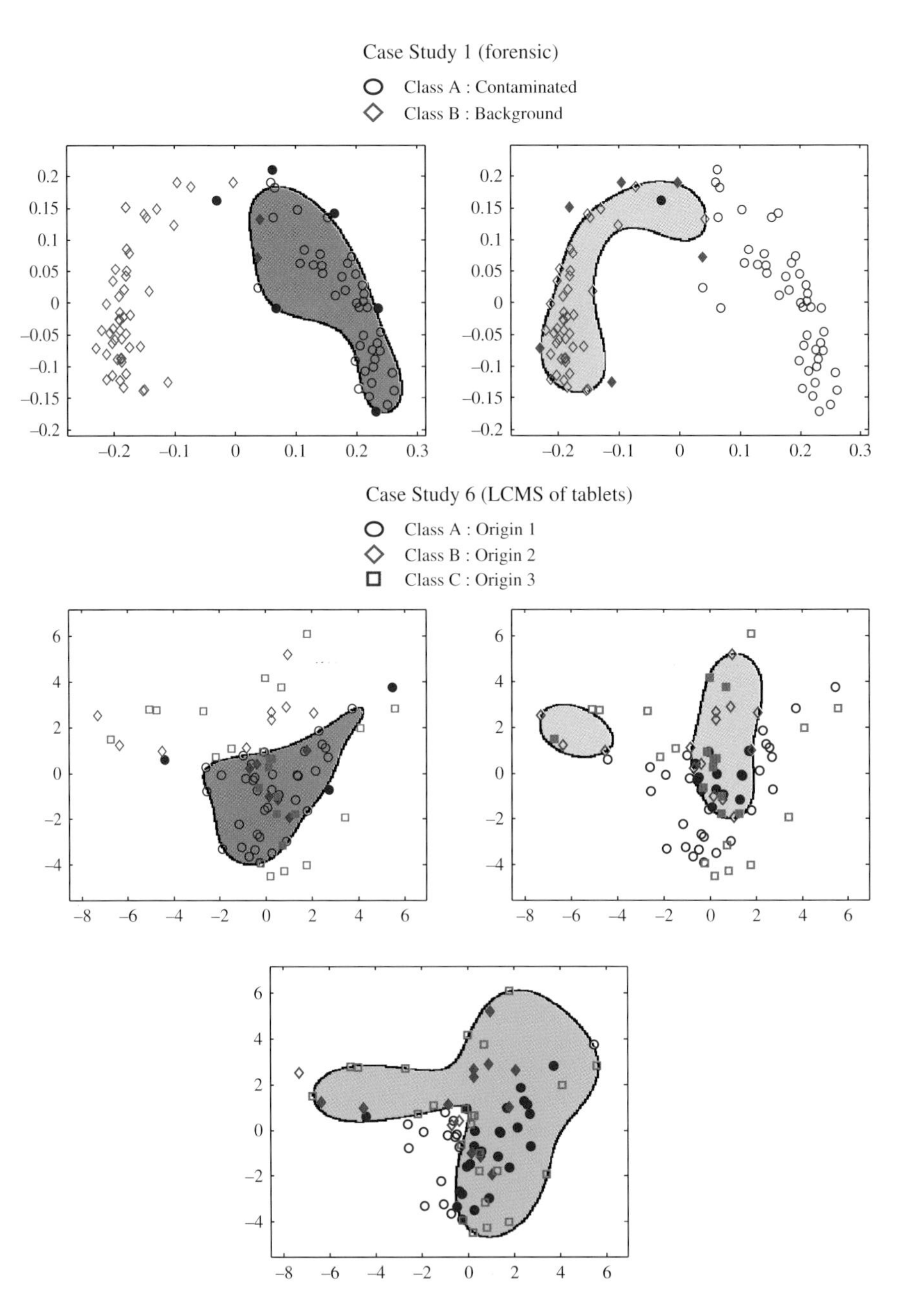

Figure 6.16 Illustration of SVDD boundaries for Case Studies 3, 1, 6 (origins), 8b and 11, using autoprediction, an RBF of $\sigma = 1 \times$ average standard deviation of the variables in the data, $D = 0.2$, an L1 classifier and conjoint PC models. Misclassified samples are indicated using closed symbols and arrows are used to indicate very compact regions

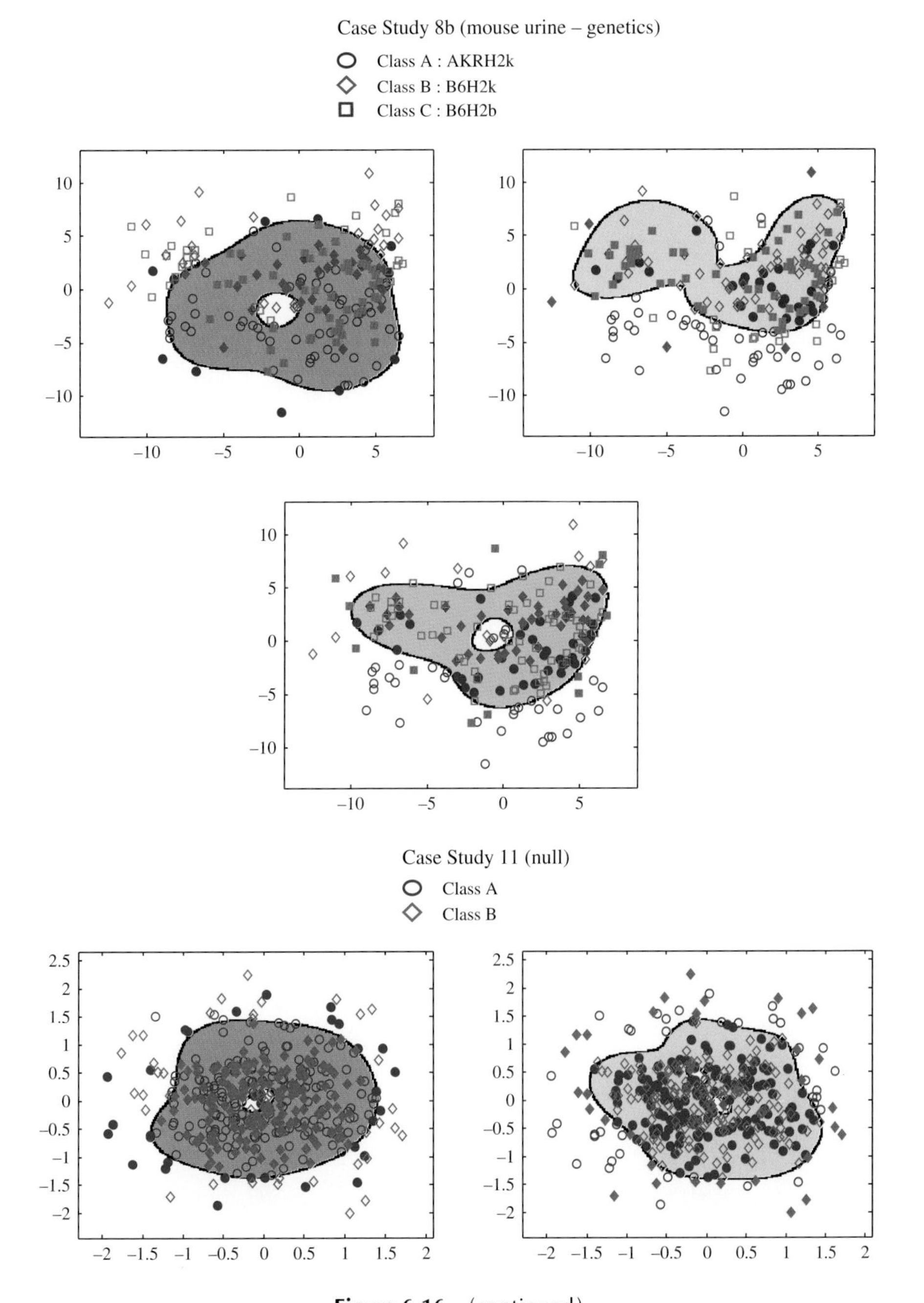

Figure 6.16 (continued)

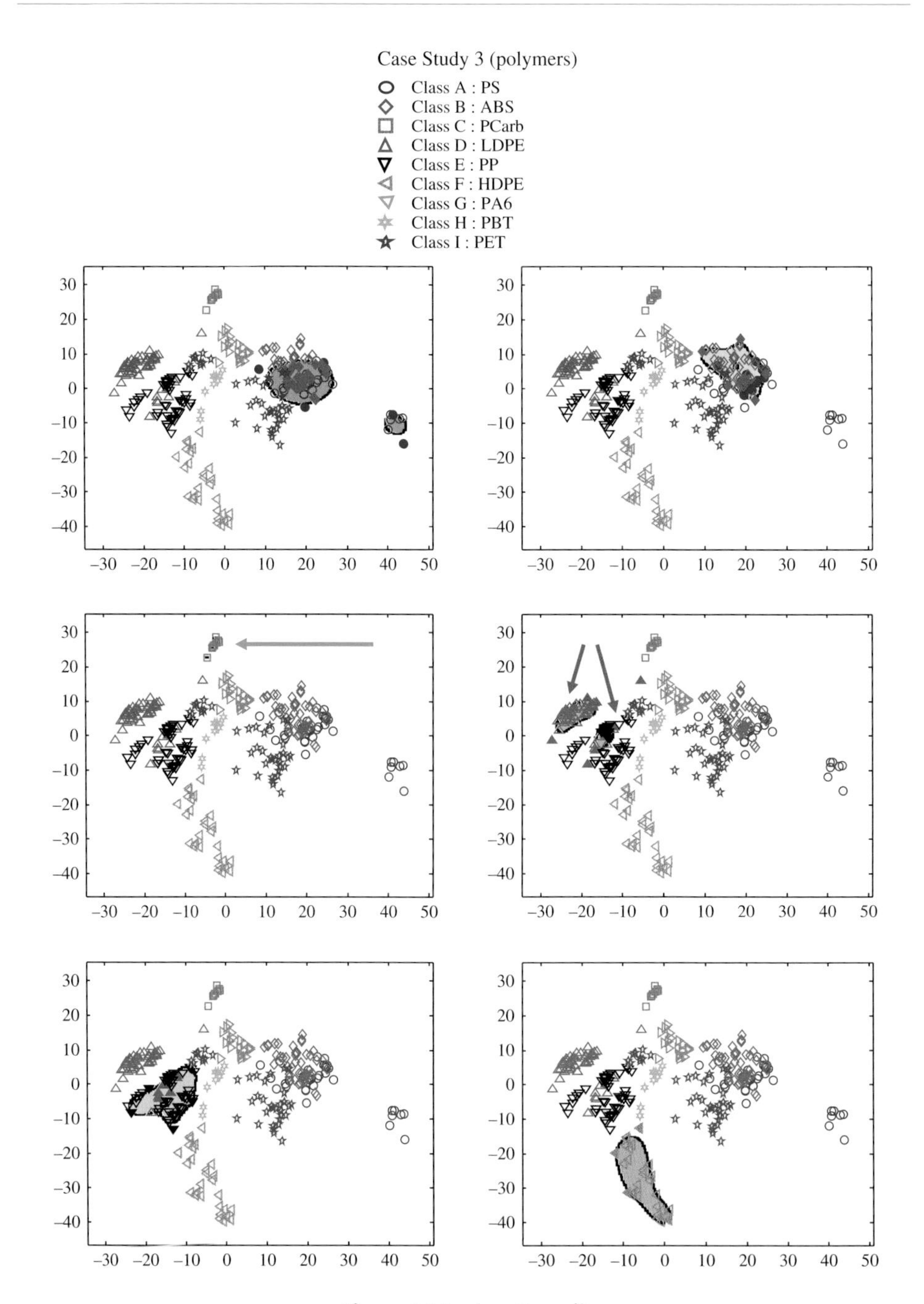

Figure 6.16 (continued)

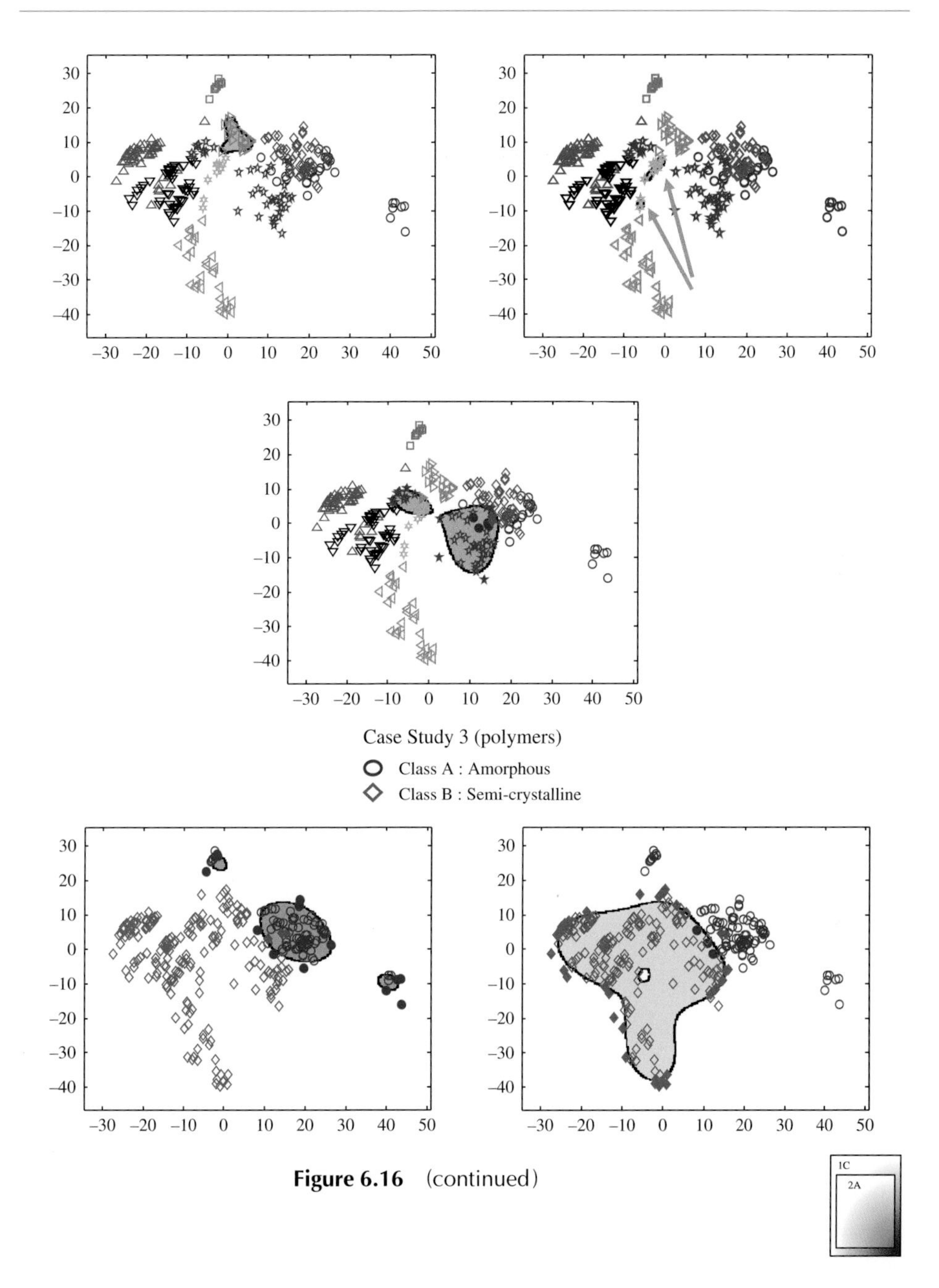

Figure 6.16 (continued)

advantages over the straight use of QDA as there is no need for the data to be multinormal, and so there are significant improvements for Case Study 1 (forensic) – compare to Figure 6.10 – although if the Q-statistic is first used for sifting outliers, this advantage may be lost in many cases. However the flexibility of SVM to model more complex boundaries can on some occasions be used to advantage.

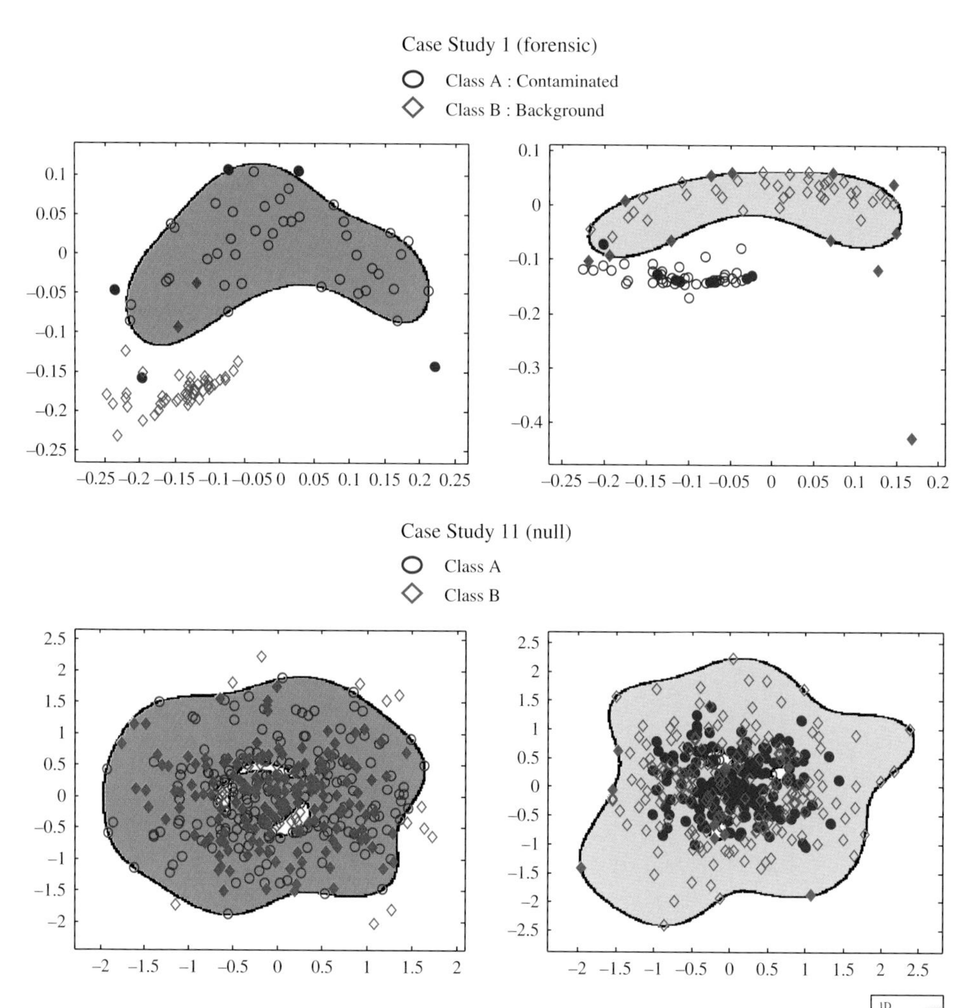

Figure 6.17 SVDD on disjoint PC models of Case Studies 1 and 11, using similar parameters to Figure 6.16

SVM is probably the most well developed boundary method and represents an alternative to the more established statistical methods. There are several other methods for one class classifiers, and most methods including LVQ and SOMs can be formulated as one class classifiers. However often chemometrics data are not so complex to require such approaches, unlike for example, face recognition or genetic databases where data structure may be more complex, and there is little point using an overly complicated method if the problem is soluble by more straightforward approaches. Within the chemometrics community much emphasis is placed on multivariate models, reflecting the large number of variables that are often able to be acquired per dataset, and so the way PCA is employed prior to classification is also an important decision. However, one class SVMs do represent an interesting alternative in some cases and is an example of a one class classifier that has been developed primarily by the machine learning community rather than the statistics community.

6.6 Summarizing One Class Classifiers

There are several techniques for summarizing the results of one class classifiers, which are somewhat different in nature to binary classifiers. We will illustrate these using QDA based classifiers and conjoint PC models, for brevity, but these principles can be extended to most other classification protocols discussed in this chapter.

6.6.1 Class Membership Plots

For most one class classifiers each sample can be given a numerical value of how well it fits a class model, for example, the distance to a centroid or the probability it fits the class or a statistic such as the Q- and D-statistic, which can be plotted for each class. This type of plot is also sometimes called a Coomans plot, and the principle is illustrated in Figure 6.18, where each axis represents how close a sample is to the centre of a class. The axes are often in the form of confidence or probability, and the limits (dashed lines) are given by a value of α, e.g. 99 % or 95 %. Autopredictive class membership plots for Case Studies 1 (forensic), 8a (mouse urine – diet) and 11 (null) are illustrated in Figure 6.19, all involving QDA based models. The axes represent the statistics (D and Q) as appropriate, with 95 % confidence bands indicated. Note that we can now analyse the data now using more than 2 PCs if required, as we do not need to visualise the data using a scores plot; and that we can also use a slightly different method of scaling for the 2 PC conjoint models to previously, as it is no

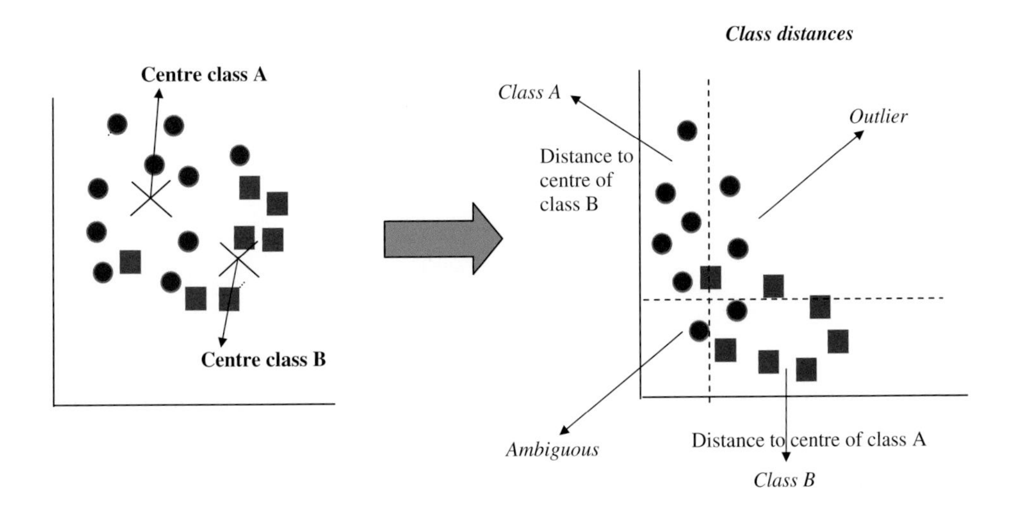

Figure 6.18 Class membership plot

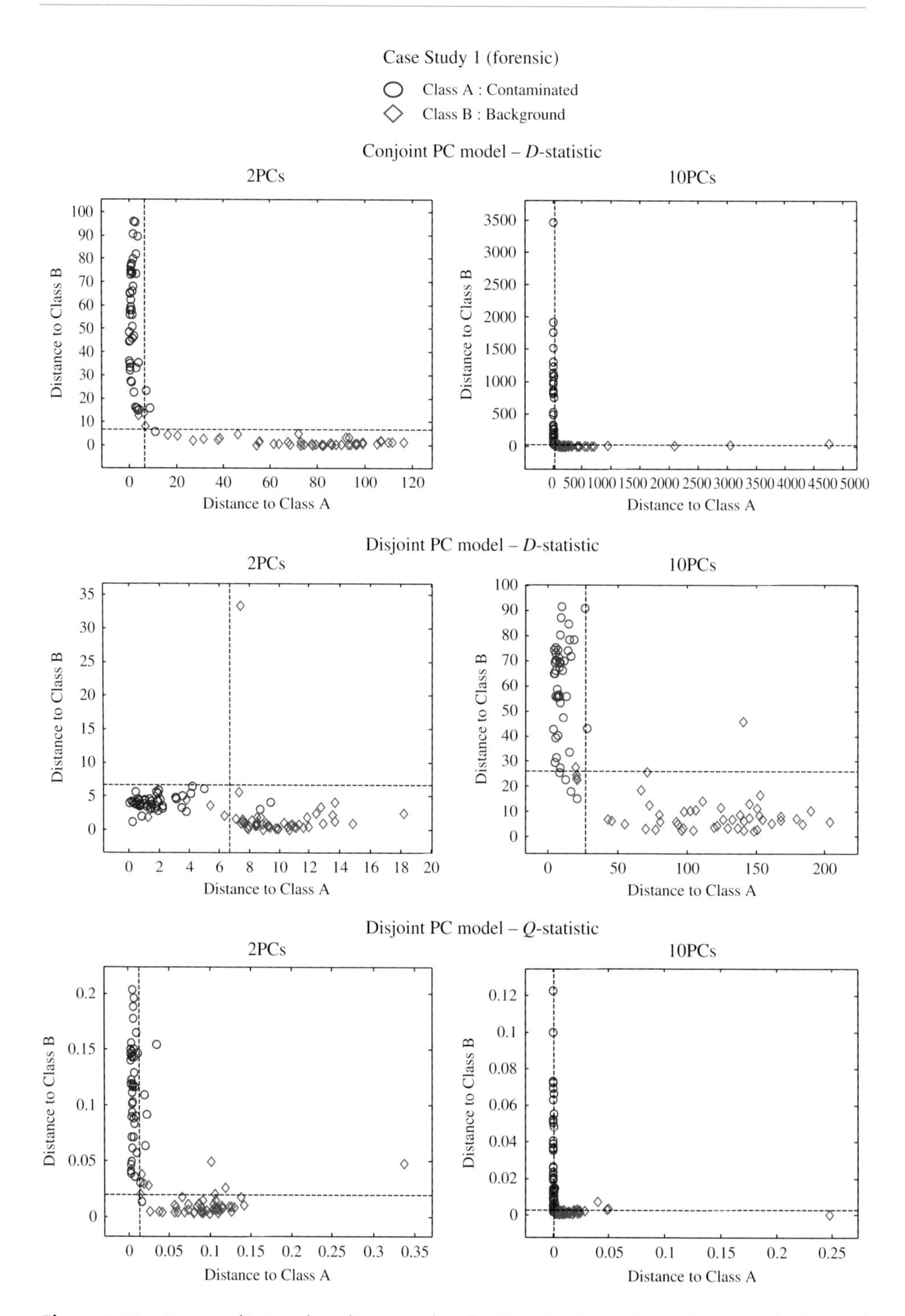

Figure 6.19 Autopredictive class distance plots for Case Studies 1, 8a and 11 using both 2 and 10 PC models, with 95 % confidence limits indicated

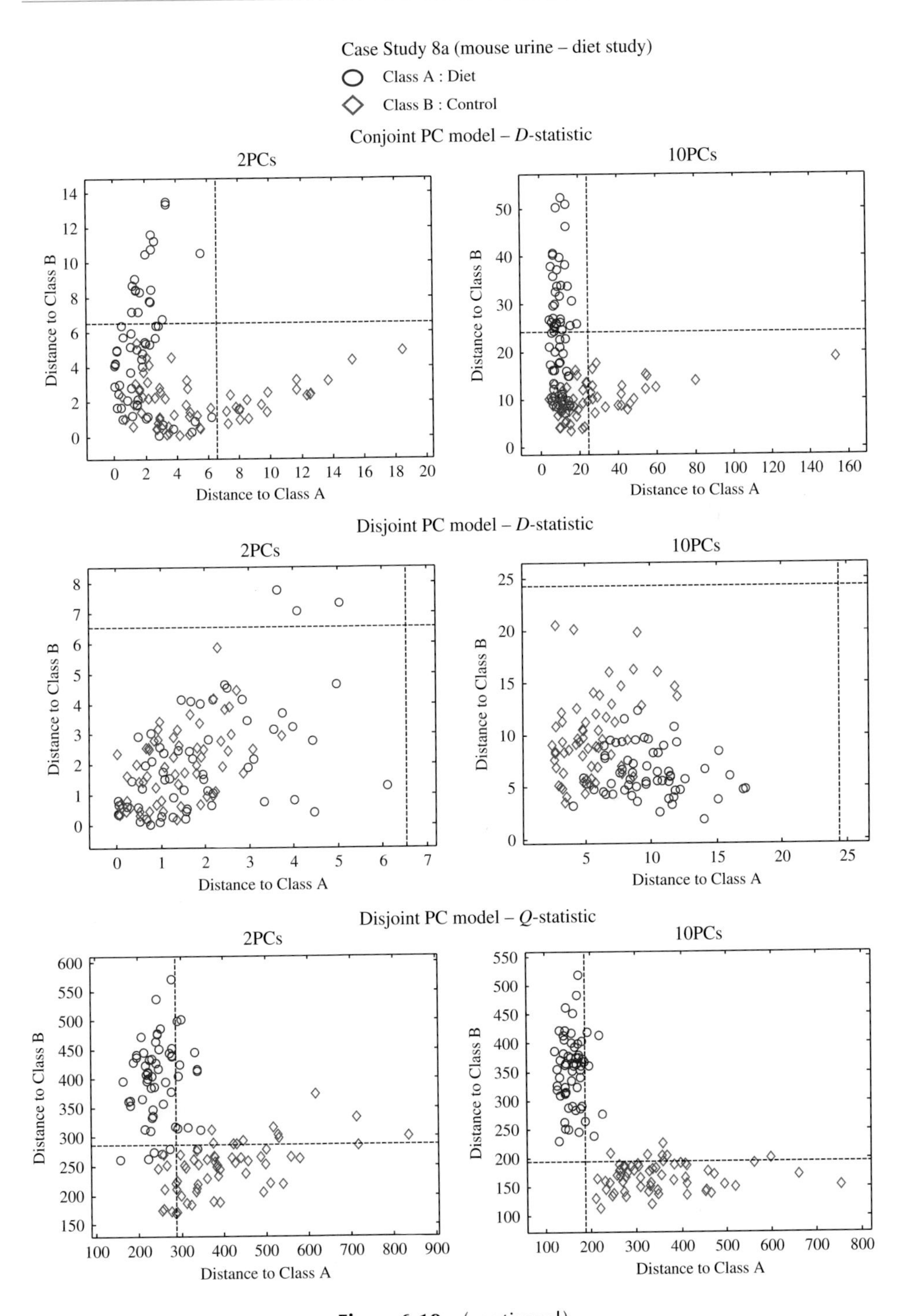

Figure 6.19 (continued)

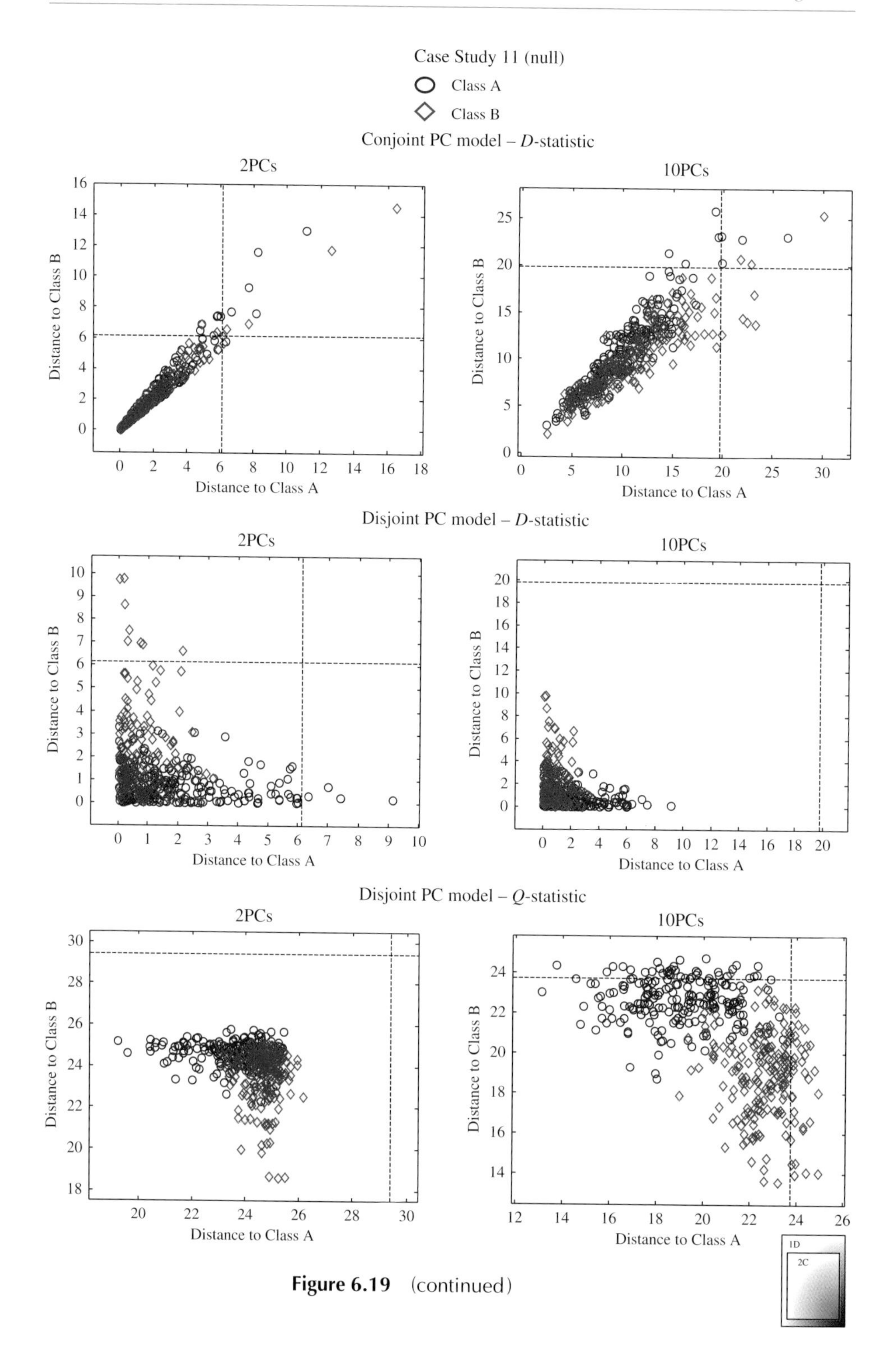

Figure 6.19 (continued)

longer necessary to visualize the data in a PC scores plot, but this make very limited difference in this case.

1. As expected, Case Study 1 (forensic) is readily separable using a conjoint PC model and the *D*-statistic, and even with 2 PCs there are just a few ambiguous samples. For the disjoint PC plot and the *D*-statistic, using 2 PCs, we find that whereas group B (background) can be distinguished well, the contaminated samples are ambiguous. This can be understood by looking at the 2D PC projection in Figure 6.10. Whereas for the most part class B is outside the 95 % confidence bands of class A, apart from four samples (these four samples fall into the bottom left-hand corner of the *D*-statistic for the disjoint model using 2 PCs), all the class A samples fall within the class B boundary. We can see in the PC projections that two class A samples lie outside the class A boundary and are equivalently in the bottom left of the disjoint 2 PC plot. An outlier, easily identified in the PC plot, is also identifiable in the *D*-statistic class distance plot. Once 10 PCs have been calculated, the two groups can now be distinguished using a disjoint projection, but the outlier is still clear. However, the *Q*-statistic is even more diagnostic, and we can clearly see that the two classes are separable and the outlier, this time, is assigned to the correct class.
2. Case Study 8a (mouse urine – diet) exhibits interesting behaviour with most samples being ambiguous when using 2 PCs and a conjoint method, although considerably better separation is obtained for class A once 10 PCs are employed for the model. This suggests that 2 PCs are inadequate. For the disjoint models we cannot obtain any reasonable distinction using either 2 or 10 PCs via the *D*-statistic. However there is excellent separation using 10 PCs and the *Q*-statistic. This conclusion that the classes are separable should, however, be treated with caution. There should normally be a stage in which autopredictive disjoint PC models give a successful result, as the more the PCs in the model, the more closely the training set is fitted, but the real test involves validating the model which will be discussed in Chapter 8, and it is possible also to obtain test set class distance plots.
3. Encouragingly for Case Study 11 (null) almost all models come up with an ambiguous result. There will appear to be a slight separation when performing disjoint models if several PCs are employed using autopredictive models, but graphically it is obvious that samples cannot readily be separated into groups.

An advantage of class distance plots is that they allow visualization for any number of PCs unlike scores plots that are limited to 3 PCs as a maximum, and are most informative when projected onto two dimensions. They can be extended to test and training sets and are a powerful way of looking at data.

6.6.2 ROC Curves

A final topic is ROC (Receiver Operator Characteristic) curves. If we develop a classification model for an 'in group', we may decide to assign a sample to this group if its distance to a centroid is below a given threshold. If it is closer to the class centroid than this threshold but is in fact a member of the 'out group' this is called a false positive (FP). If it is correctly assigned to the 'in group' it is a called a true positive (TP). For any threshold it is possible to calculate the following numbers:

- *TP* (the number of members of the 'in group' that have been correctly classified).
- *FP* (the number of members of the 'out group' that have been incorrectly predicted to be members of the 'in group').

As the threshold distance is decreased, the number of samples predicted to be members of the 'in group' increases correspondingly, and so the number of FPs and TPs increases. Dependent on the type of application it may be more important to find a threshold for which the number of TPs is maximized or one for which the number of FPs is minimized. If evidence is to be presented to a court of law it is often most important to minimize the risk of FPs (if a positive would lead to conviction) to be sure that the evidence is really sound and that innocent people are not falsely convicted. In preliminary medical diagnosis of a disease, where however, the opposite is normally so, it is better to scan all patients that possibly show symptoms of a disease and so reduce the number of FNs, even if this means a few FPs are included. Figure 6.20 represents a set of samples (from an 'in group' and an

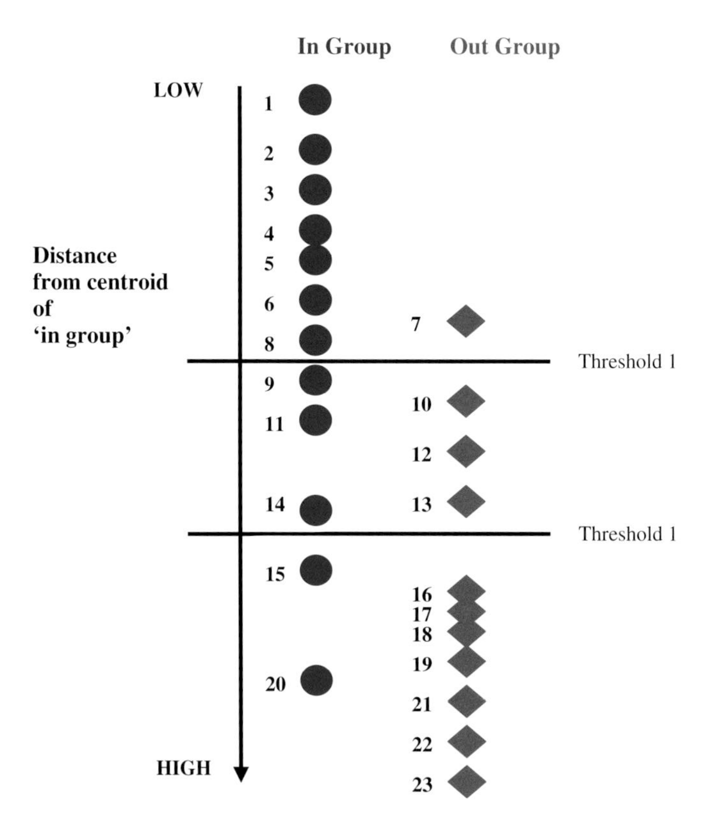

Figure 6.20 Illustration of thresholds. The distance from the centroid on an 'in group' (or a statistic such as the *D*- or *Q*-statistic), is plotted for samples in two groups, plotted along with the samples for each class, and the samples are ordered in rank of their distance. Two possible thresholds are indicated

'out group' consisting of all other samples) together with the values of a distance or statistic as calculated by the classifier. There is no sharp dividing line and so there will always be some FPs or FNs whatever decision threshold is employed. The choice of threshold corresponds to the choice of dividing line between the two classes and the choice of which type of misclassification error we are most comfortable with. For the example of Figure 6.20 we indicated two possible thresholds, each of which results in a different number of TPs and FPs. This is presented numerically in Table 6.2.

Table 6.2 *The number of TPs and FPs for the two thresholds of Figure 6.20*

	Threshold 1	Threshold 2
TP	7	10
FP	1	4

An ROC curve involves plotting the proportion of TPs (vertical axis) against the proportion of FPs (horizontal axis), as the threshold changes. This allows visualization of whether there are optimal conditions that minimize each type of error, and can provide guidance as to the setting of a threshold. These curves have important characteristics according to how disjoint or otherwise the two classes are. The ROC curve for the example of Figure 6.20 is presented in Figure 6.21. Models that randomly assign samples into two groups tend to have ROC curves that are along the diagonal axis and are poor classifiers without any discrimination, whereas ROC curves that consist of a vertical line followed by a horizontal line represent highly discriminating classifiers.

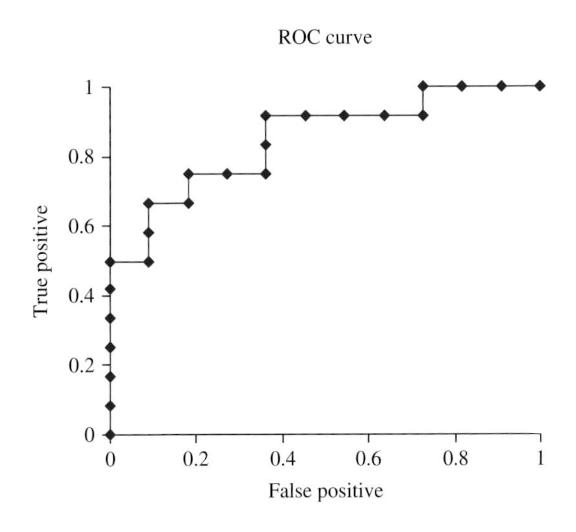

Figure 6.21 ROC curve for data of Figure 6.20

The quality of a model can be assessed by the AUC (Area under the Curve) which can be between 0 and 1, but in practice is unlikely to be less than 0.5, and is simply the area underneath the ROC curve. The higher the AUC, the better the model. For the example of Figure 6.21 this value is 0.849. ROC curves are useful for one class classifiers as they allow visualization of what happens for each class model separately when confidence levels or class distances are changed. Whereas a one class classifier will only model the 'in group', and separate classifiers can be determined for each class in a dataset, an effective boundary, distance or cut off threshold, can be determined according to the distribution of FPs and TPs, taking all samples into account.

The autopredictive ROC curves for 2 PC models are presented in Figure 6.22 for Case Studies 1, 8a and 11 for both conjoint and disjoint models. Because probability is monotonically related to distance from the centroid (in scaled units), there is no difference as to whether the distance or probability is employed for constructing these graphs. We can see various trends. First of all Case Study 1 (forensic) is mainly very well modelled. Previously there were some problems with the class B disjoint model using the D-statistic, as samples from class A appeared to be within the 95 % boundary for class B, and so were ambiguous. We see that although that particular model is still not as good as the others, there is a threshold where good discrimination is possible, with nearly 90 % TP and less than 5 % FP (top left of the graph), but this involves setting a low distance from the centroid. If this threshold is missed, the difference between TP and FP gets much worse. Examining the scores plot of Figure 6.10 (disjoint class B model) suggests that the 95 % margin is probably too wide, and reducing this limit, although it may miss out some of class B, dramatically reduces misclassification of class A. Case Study 8a (mouse urine – diet) is interesting – we can see that for 2 PC disjoint models the Q-statistic is quite effective, whereas the D-statistic shows almost no discrimination. Case Study 11 (null) is important, as this is a benchmark against the others. For the conjoint PC models the ROC curve is almost linear with an AUC of around 0.5, suggesting no discrimination as anticipated. However the disjoint autopredictive PC model shows some apparent discrimination when using the Q-statistic as a distance measure. Given that this is a randomly generated dataset this suggests overfitting and that too many PCs are used in the model (a '0' PC model which just involves distance from the mean may be more appropriate). This problem can be detected using a proper form of validation. Ultimately as more PCs are added the model appears to predict that there is a distinction between the two classes when performing autopredictive disjoint models – since each class is modelled separately, ultimately a full PC model (using all non-zero PCs) will exactly fit the 'in group' but not the 'out group' and so deviations from this model will make it appear that the 'out group' is predicted worse than the 'in group'. Using a conjoint PC model reduces this danger as it tries to predict all samples together. Hence performing PCA separately on each class carries a risk of overfitting, which can be avoided by proper use of validation schemes as discussed in Chapter 8. The corresponding AUCs are presented in Table 6.3, where the closer to 1, the better the model classifies samples.

ROC curves are important graphic tools, allowing models with greater than three components to be visualized. For Case Study 11 (null) we can see that we are in danger of overfitting using a disjoint model, whereas for Case Study 1 (forensic) we

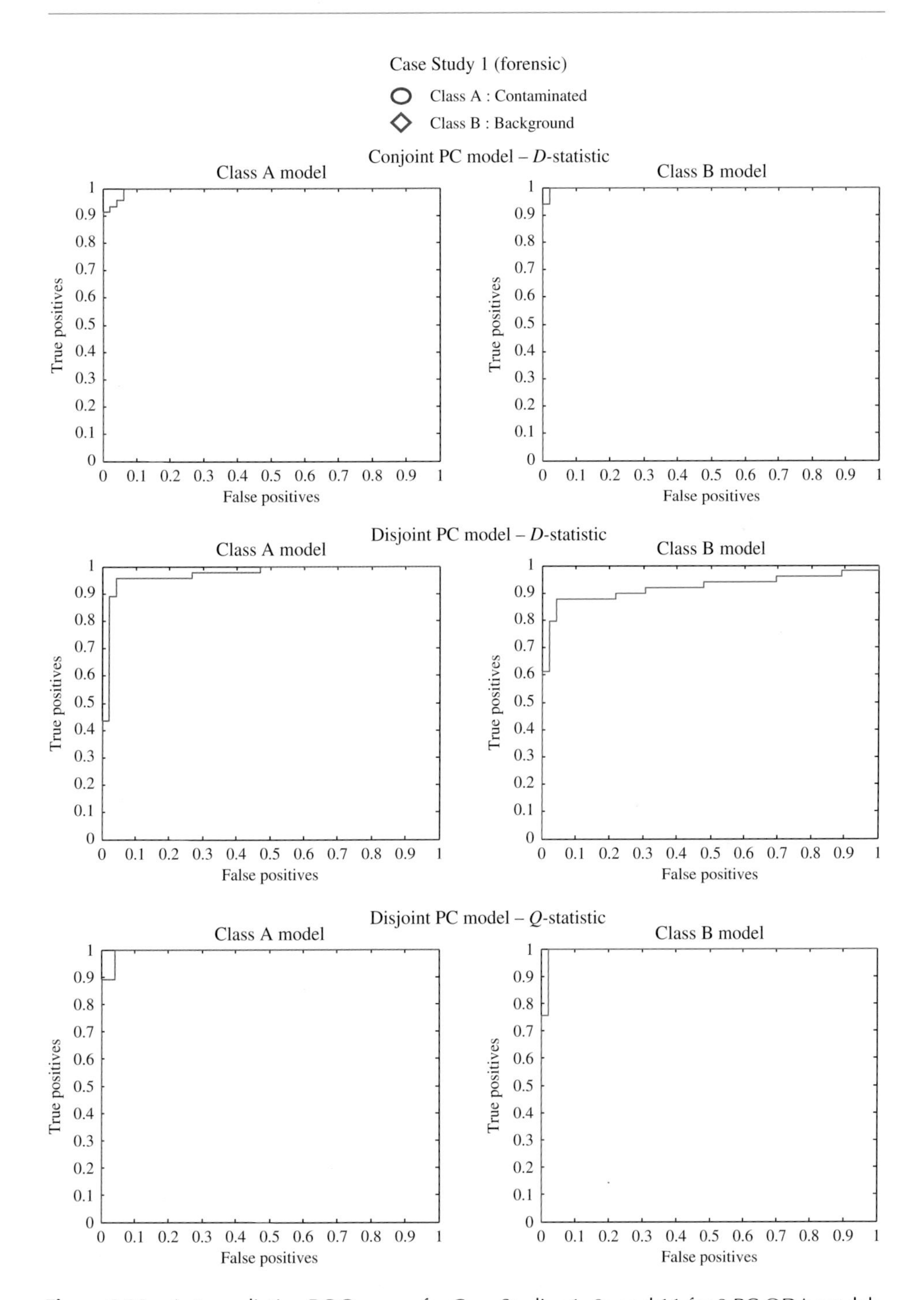

Figure 6.22 Autopredictive ROC curves for Case Studies 1, 8a and 11 for 2 PC QDA models

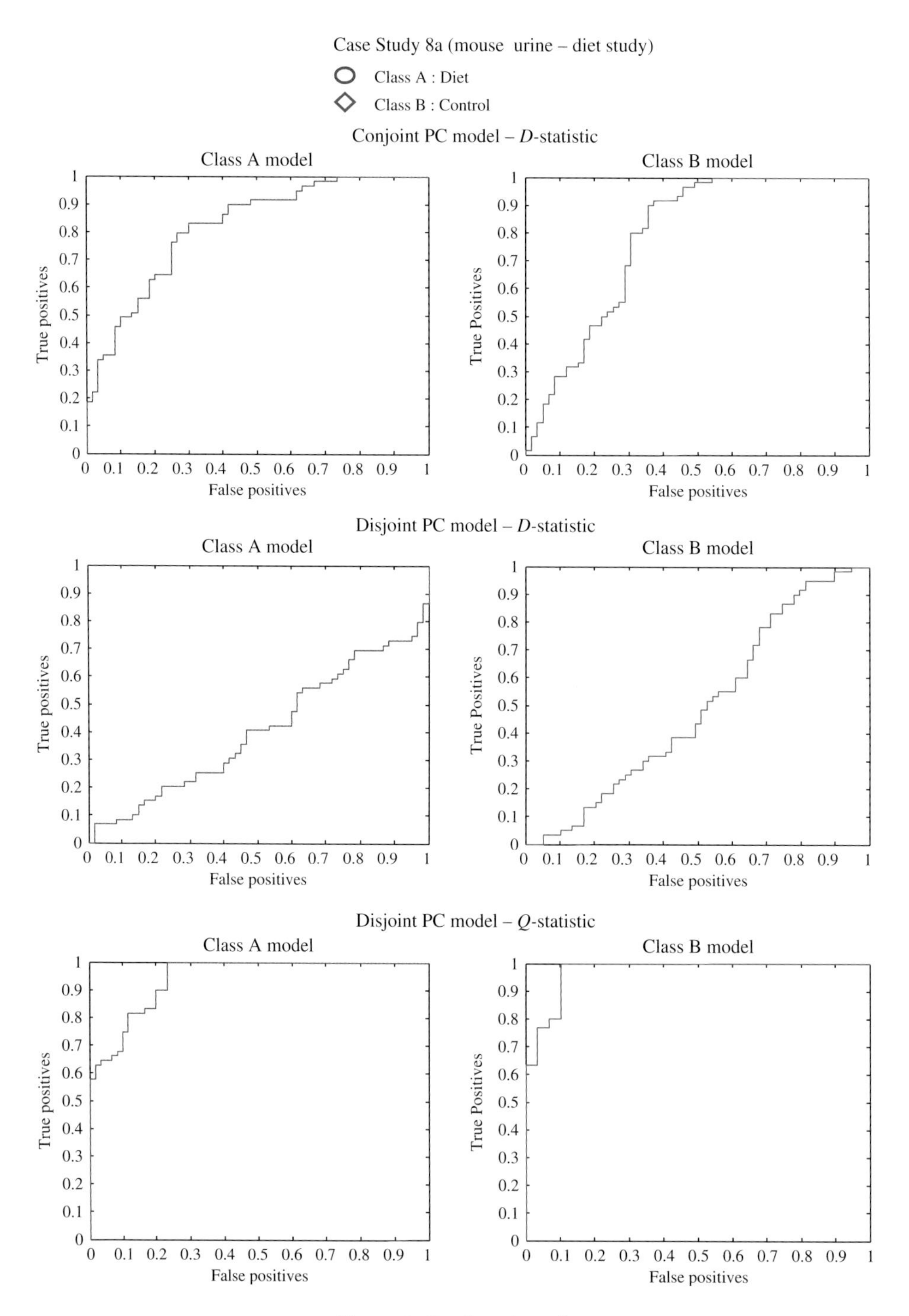

Figure 6.22 (continued)

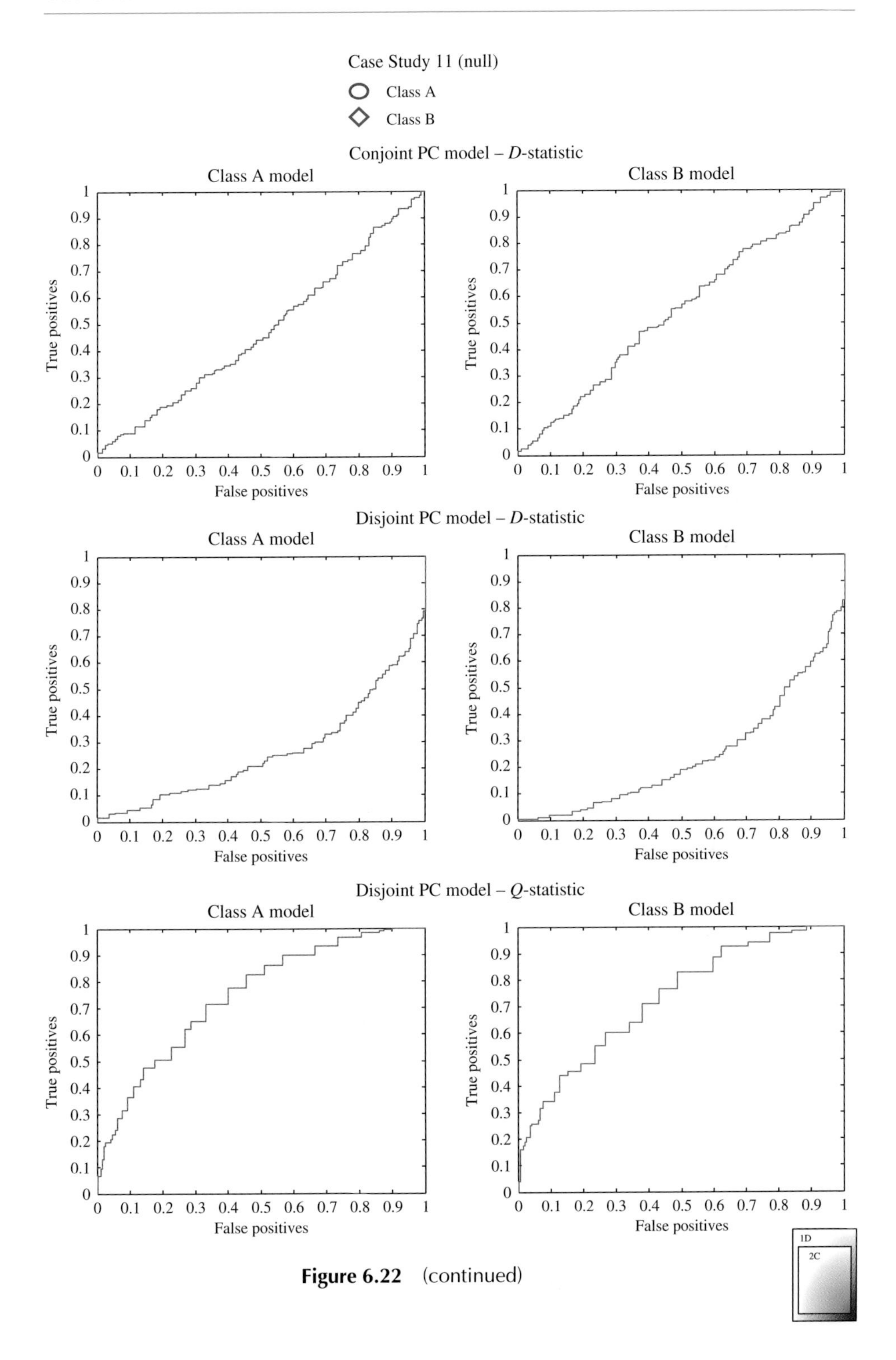

Figure 6.22 (continued)

Table 6.3 AUCs for the ROC curves of Figure 6.22

Conjoint models D statistic		
Case study	A	B
Case Study 1	0.985	0.979
Case Study 8a	0.813	0.777
Case Study 11	0.449	0.555
Disjoint models D statistic		
Case study	A	B
Case Study 1	0.968	0.917
Case Study 8a	0.397	0.489
Case Study 11	0.257	0.238
Disjoint models Q statistic		
Case study	A	B
Case Study 1	0.954	0.935
Case Study 8a	0.901	0.935
Case Study 11	0.802	0.811

see that by using a different threshold we can much get better discrimination in the disjoint model.

Note that although this chapter focuses on autopredictive models, test and training set ROC curves can also be produced. These concepts together with the related ideas of contingency tables will be introduced in Chapter 8.

Bibliography

SIMCA, *Q*- and *D*-Statistics

J.E. Jackson, G.S. Mudholkar, Control Procedures for Residuals Associated with Principal Components Analysis, *Technometrics*, **21**, 341–349 (1979).

P. Nomikos, J.F. MacGregor, Multivariate SPC Charts for Monitoring Batch Processes, *Technometrics*, **37**, 41–59 (1995).

S.J. Qin, Statistical process monitoring : basics and beyond, *J Chemometrics*, **17**, 480–502 (2003).

S. Wold, Pattern-Recognition by Means of Disjoint Principal Components Models, *Pattern Recognition*, **8**, 127–139 (1976).

T. Kourti, J.F. MacGregor, Process analysis, monitoring and diagnosis, using multivariate projection methods, *Chemometrics Intell. Lab. Systems*, **28**, 3–21 (1995).

One Class Classifiers and SVDD

D.M.J. Tax, One-class classification, Concept-learning in the absence of counter-examples, PhD Thesis, University of Delft, 2001 [http://www-ict.et.tudelft.nl/~davidt/papers/thesis.pdf].

D.M.J. Tax, R.P.W. Duin, Support Vector Data Description, *Machine Learning*, **54**, 45–56 (2004).

ROC Curves

C.D. Brown, H.T. Davis, Receiver operating characteristics curves and related decision measures: A tutorial, *Chemometrics Intell. Lab. Systems*, **80**, 24–38 (2006).

A. Swets, Measuring the accuracy of diagnostic systems, *Science*, **240**, 1285–1293 (1988).

Multinormality test

J.A. Doornik, H. Hansen. An Omnibus Test for Univariate and Multivariate Normality, *Oxford Bulletin of Economics and Statistics*, **70**, 927–939 (2008).

7

Multiclass Classifiers

7.1 Introduction

In Chapters 5 and 6 we introduced two class and one class classifiers. When it is desired to classify more than two groups, there are two approaches. The first is to perform independent one class classification on each group as discussed in Chapter 6. The second is to extend two class (or hard) classifiers to situations where there are more than two groups. This requires dividing the data space up into more than two types of region, creating a different region for each class, which will be the focus of this chapter. Figure 7.1 illustrates the principle when there are three groups. Unlike one class classifiers, every sample is uniquely classified into a single group and there is no provision for outliers. However it might be known that there are only a certain number of groups in a dataset and as such the main aim is to try to determine whether an unknown is a member of one of these predefined groups. Or it may be desirable to ask whether a set of samples can be successfully modelled using a number of predefined groups or not, using methods described in this chapter together with approaches for validation of models (Chapter 8); if not whether there are sufficient classes to describe a dataset or more are required, or even that some classes that we feel are distinct are not so different. For all these reasons extension of two class classifiers to more than two classes serves an important purpose, but is not always entirely straightforward.

There are several case studies in this text where there are more than two groups; this includes Case Study 2 – NIR spectroscopy of food (4 groups), Case Study 3 – polymers (which can be represented by 9 groups), Case Study 4 – LCMS of pharmaceuticals (which can be represented either by 3 groups or by 5 groups), Case Study 7 – disease study by atomic spectroscopy (which can be represented by 5 groups – the controls and 4 types of disease) and Case Study 8b – mouse genetics (3 groups). These are summarized in Table 7.1. In this chapter, we will primarily illustrate the methods using the scores of the first two PCs after suitable data preprocessing. It is important to remember that using two PCs may not be the most suitable for classification and more PCs are often needed for a good classification model: choosing the number of PCs is discussed in Section 4.3,

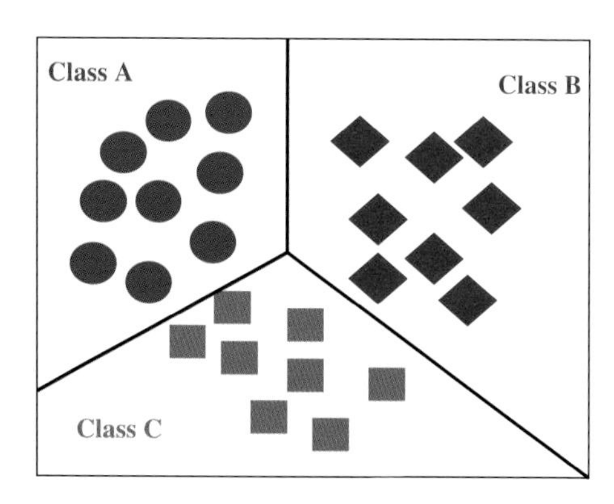

Figure 7.1 Multiclass classifier when there are three groups

Table 7.1 *Muliclass datasets described in this text*

Case study	Classes
2 (NIR spectroscopy of food)	A: Corn oils (18) B: Olive oils (30) C: Safflower oils (16) D: Corn margarines (8)
3 (Polymers)	A: PS (35) B: ABS (47) C: PCarb (10) D: LDPE (56) E: PP (45) F: HDPE (30) G: PA6 (20) H: PBT (10) I: PET (40)
5 (LCMS of tablets)	A: Origin 1 (39) B: Origin 2 (15) C: Origin 3 (25)
	A: Batch 1 (16) B: Batch 2 (15) C: Batch 3 (16) D: Batch 4 (16) E: Batch 5 (16)
7 (Clinical)	B: Control (446) C: Diseased – CV (31) D: Diseased – CA (19) E: Diseased – RH (21) F: Diseased – MH (23)
8(b) (Mouse urine by GCMS – genetics)	A: AKRH2k (71) B: B6H2k (59) C: B6H2b (75)

and methods for determining what is the most appropriate model in Chapter 8. However projections onto the two largest PCs do offer insight into the method employed.

In Chapter 5 we focused on describing classification algorithms in the framework of binary classification problems, that is, assigning a sample to one of two classes. When there are more than two classes, there are two strategies that can be employed. The first is to break down the approach into a series of two class comparisons, and the second is to employ a single multiclass classifier. Some algorithms, such as LVQ, can naturally be extended to a situation where there is more than one class, whereas others, such as SVM, find it hard. We will focus on the 6 common classifiers introduced in Chapter 5, although most of the principles can be extended in general to other classification methods. For PLS and SVM, however, extensions are necessarily specific to the algorithm.

7.2 EDC, LDA and QDA

These three distance methods can easily be adapted to a situation where there are more than two classes. In the simplest form the distance to the centroids of each of the classes is measured and the one to which a sample is closest is selected as the class it is a member of. The class boundaries for Case Studies 2 (NIR spectroscopy of food), 3 (polymers – 9 groups), 7 (clinical) and 8b (mouse – genetics) for autopredictive 2 PC models are presented in Figures 7.2–7.4. The EDC and LDA boundaries being linear are quite simple to visualize but the QDA boundaries somewhat more complex, especially when there are several classes and each class has a different distribution, as in Case Study 3.

For Case Study 2 (NIR spectroscopy of food), as expected, the EDC and LDA classifiers divide the space up without any ambiguities or misclassified samples. Each region is of approximately similar size, even though the distribution of samples in class D is much more compact; this is because neither approach takes into account the variance of each class and the relative difference between classes relates primarily to the difference in position of their centroids. The slight differences between the two sets of plots primarily relate to the different variability along each of the two PC axes. The QDA boundaries, whilst perfectly separating each class, appear very different. Class D has a very small variance in the direction of PC2 compared to the other groups and so is represented by a long a thin region. The class C's region is approximately diagonal as its main variability is down the diagonal. Classes A and B have relatively large variance in PC2 compared to the other groups, and so they are represented by several distinct regions. The reason why a group can be represented by more than one contiguous region is illustrated in Figure 7.5, which represents two classes in a one dimensional dataset; one, the blue class with a relatively large standard deviation (4) centred at position 5, and the other (red class) with a small standard deviation (1) centred at position 10. The Mahalanobis distance (or QDA) classifies a region of data space according to how close a sample is to the centre of a class in units of standard deviations, the class with least standard deviations being the winner. We can see that there are two regions belonging to the blue class and one relatively small region in the middle belonging to the red class. In the context of QDA where there are several dimensions and several groups with different variance structure the resultant regions of space can be quite complex.

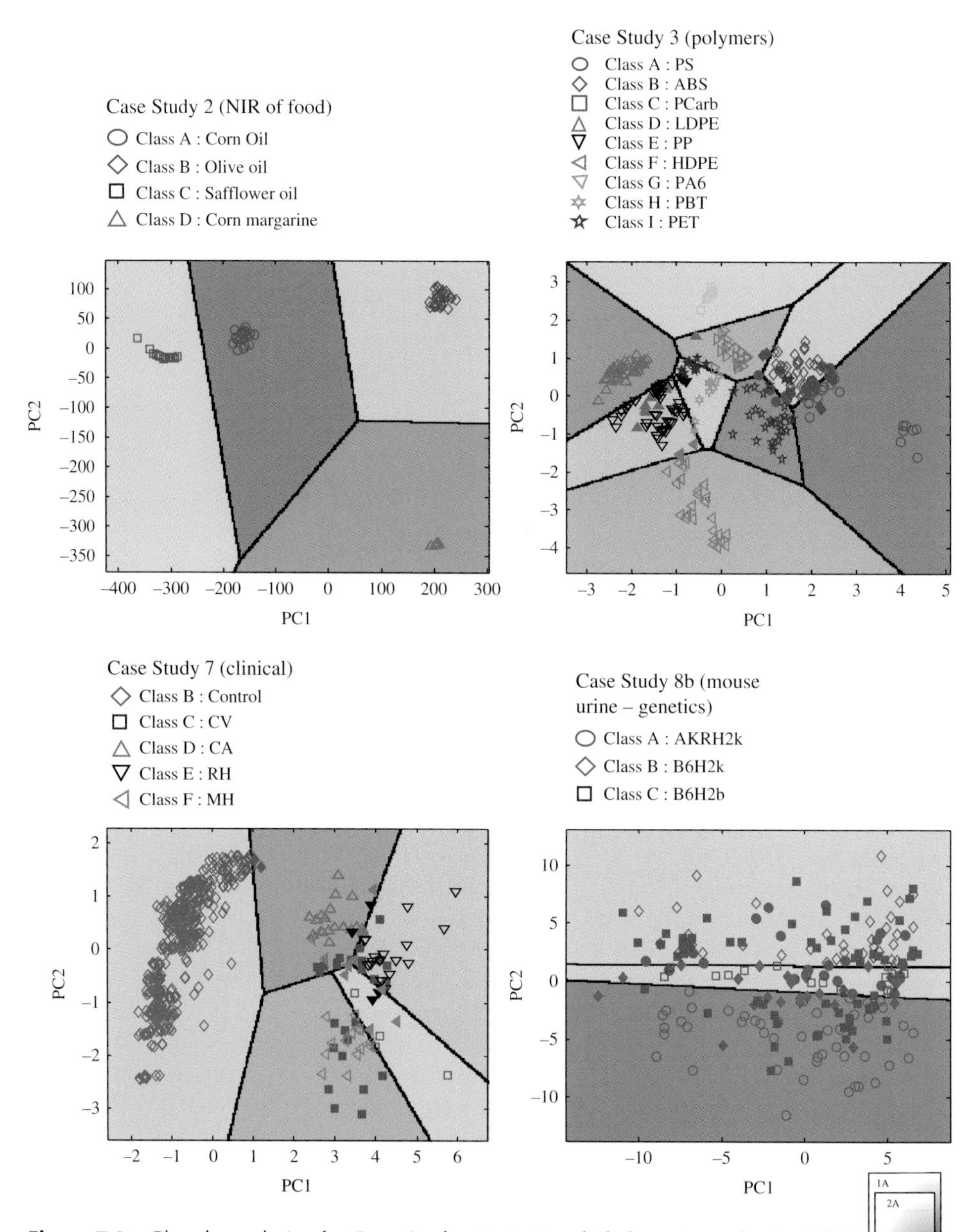

Figure 7.2 Class boundaries for Case Studies 2, 3, 7 and 8b for autopredictive 2 PC models using multiclass EDC

For Case Study 3 (polymers) we see that both EDC and LDA manage to divide the graph into nine regions, one for each group. Because for some classes, samples are characterized by subgroups (for example D and I), linear methods will never perfectly separate the groups unless more PCs are used in the model (and even then this may still not be possible).

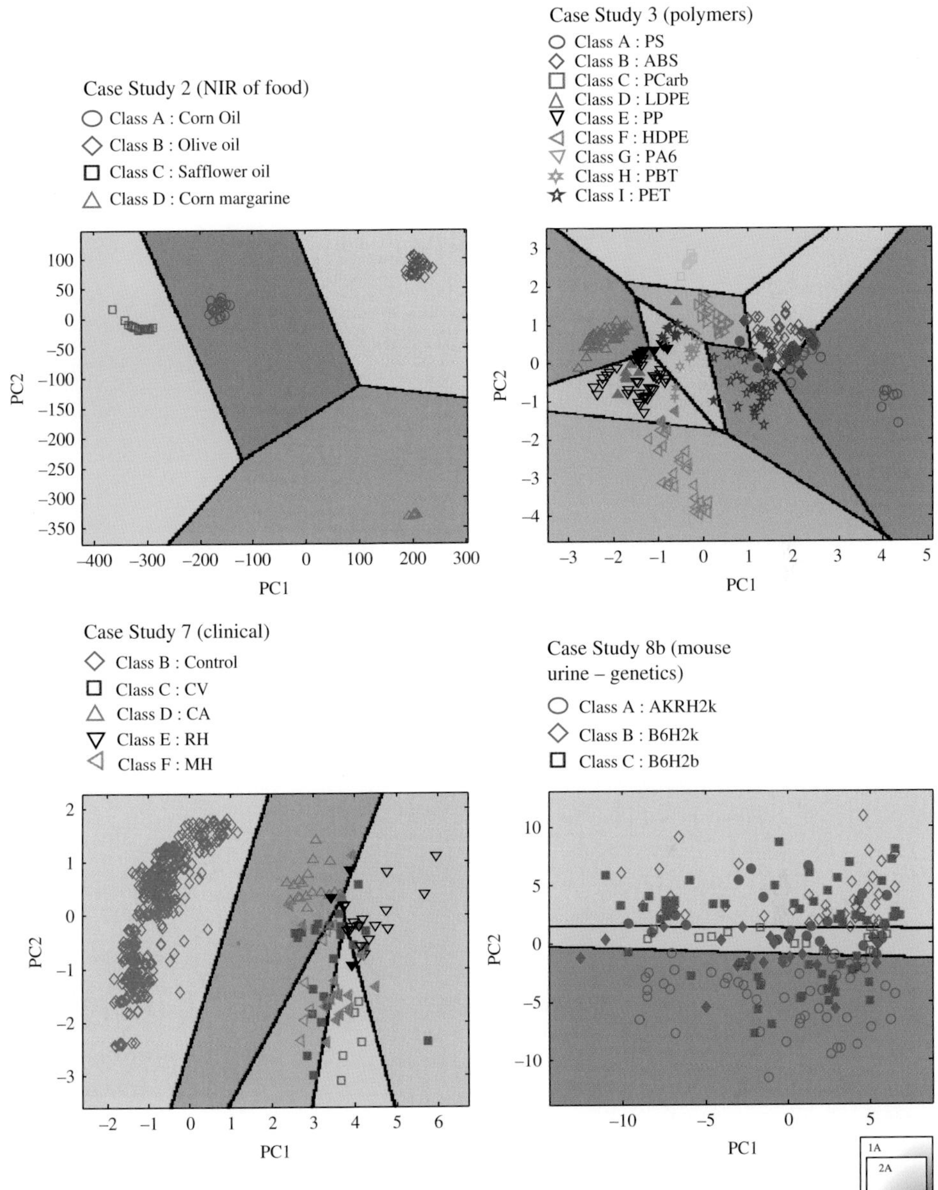

Figure 7.3 Class boundaries for Case Studies 2, 3, 7 and 8b for autopredictive 2 PC models using multiclass LDA

QDA however can take this into account. Class H is represented by a small area reflecting the somewhat compact group, which now allows both parts of class I to be correctly classified. Not much can be done about class D as some of the samples almost totally overlap with class E using 2 PC models. Class C which is small and compact similarly is represented by a small region. Clearly linear models are inadequate in this situation, but the danger with

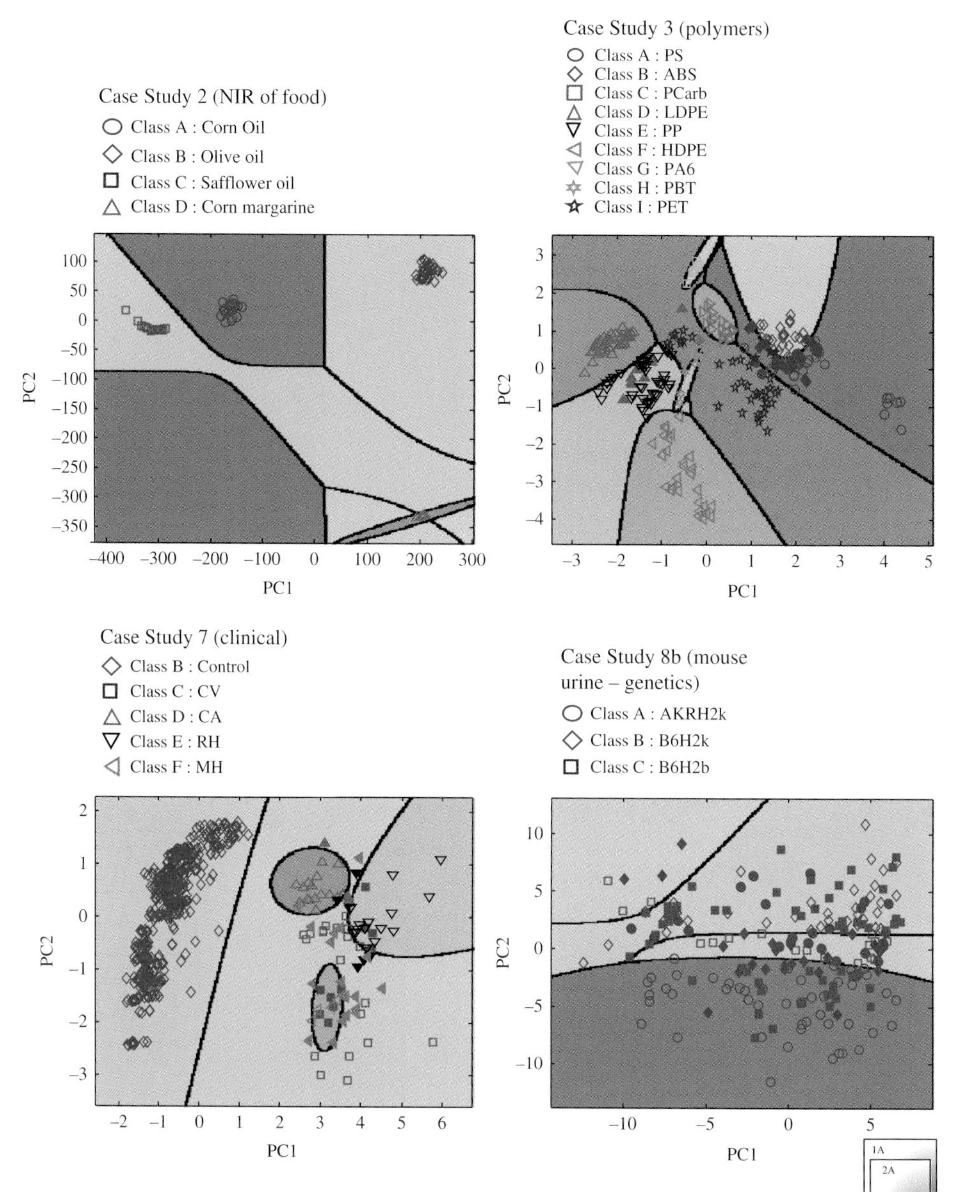

Figure 7.4 Class boundaries for Case Studies 2, 3, 7 and 8b for autopredictive 2 PC models using multiclass QDA

quadratic models can be overfitting as we will see in Chapter 8. However QDA has improved the distinction between classes somewhat, and the polymer graphs are not linearly separable using the first two PC scores.

There is not much that can be done to improve the classification of objects for Case Study 8b (mouse urine – genetics) using just the first 2 PCs; however with further PCs QDA does in fact produce good models.

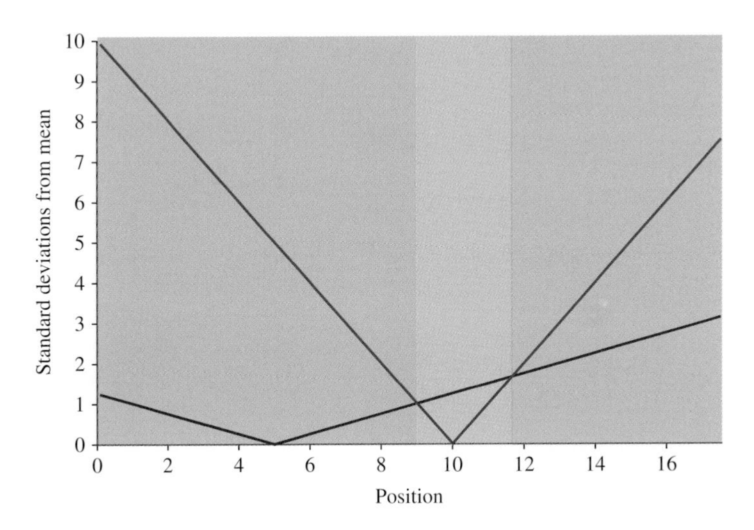

Figure 7.5 Illustration of how QDA can divide data space into several regions for one group

7.3 LVQ

The LVQ algorithm can also easily be generalized for several classes, and indeed is well suited for such problems. The complexity of the boundaries depends on the number of codebooks used for each class. Class boundaries for Case Studies 2 (NIR spectroscopy of food), 3 (polymers – 9 groups), 7 (clinical) and 8b (mouse urine – genetics) for autopredictive 2 PC models are presented in Figure 7.6 using models based on three codebooks per class. Note that the LVQ models change according to the initial guess of the codebooks and so may differ a little each time. However if classes are well defined and cover the majority of the data space, the instability is often not so great as the areas are much better defined, in contrast, for example to Case Study 1 (forensic) where there is a large region where no samples are found which could result in unstable boundaries.

For Case Study 2 (NIR spectroscopy of food), LVQ offers little advantage over simpler linear methods as the groups are already very well separated. LVQ is unlikely to result in the complex structures obtained using QDA, as the models are seeded by codebooks within each class which will normally try to carve out approximately equal regions. For Case Study 3 (polymers), LVQ tends to find approximately equal sized regions in the centre, as the codebooks are spread out and attempt to occupy approximately similarly sized region for each class. Interestingly class I is characterized by two separate regions, which since a three codebook model per class was employed, will be characterised by two and one codebooks located in each region respectively. If LVQ finds two separate regions for a class, at least one of the initial codebook 'guesses' must be in each region, for three codebooks per class the chances that this is so are 3 in 4; when more codebooks are included the chances that at least one codebook is seeded within each region increases.

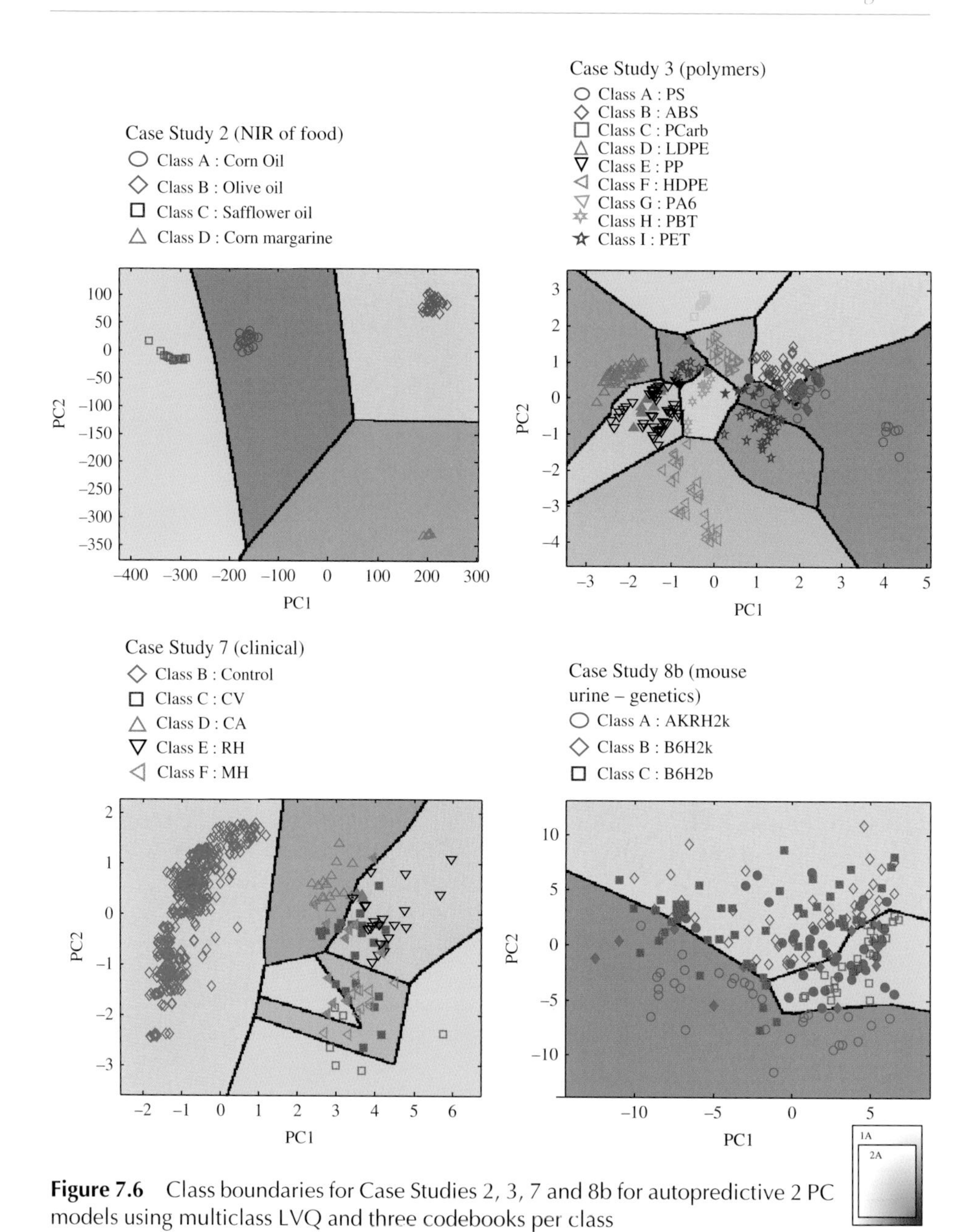

Figure 7.6 Class boundaries for Case Studies 2, 3, 7 and 8b for autopredictive 2 PC models using multiclass LVQ and three codebooks per class

Figure 7.7 illustrates this principle for models based on three and on five codebooks per class. All the five codebook models manage to model the two separate regions for class I, but one of the three codebook models fails. Interestingly, despite being a neural network, LVQ is quite effective and reproducible when there are several classes. For the polymers, simple linear models such as LDA and EDA are inadequate but LVQ results in a good

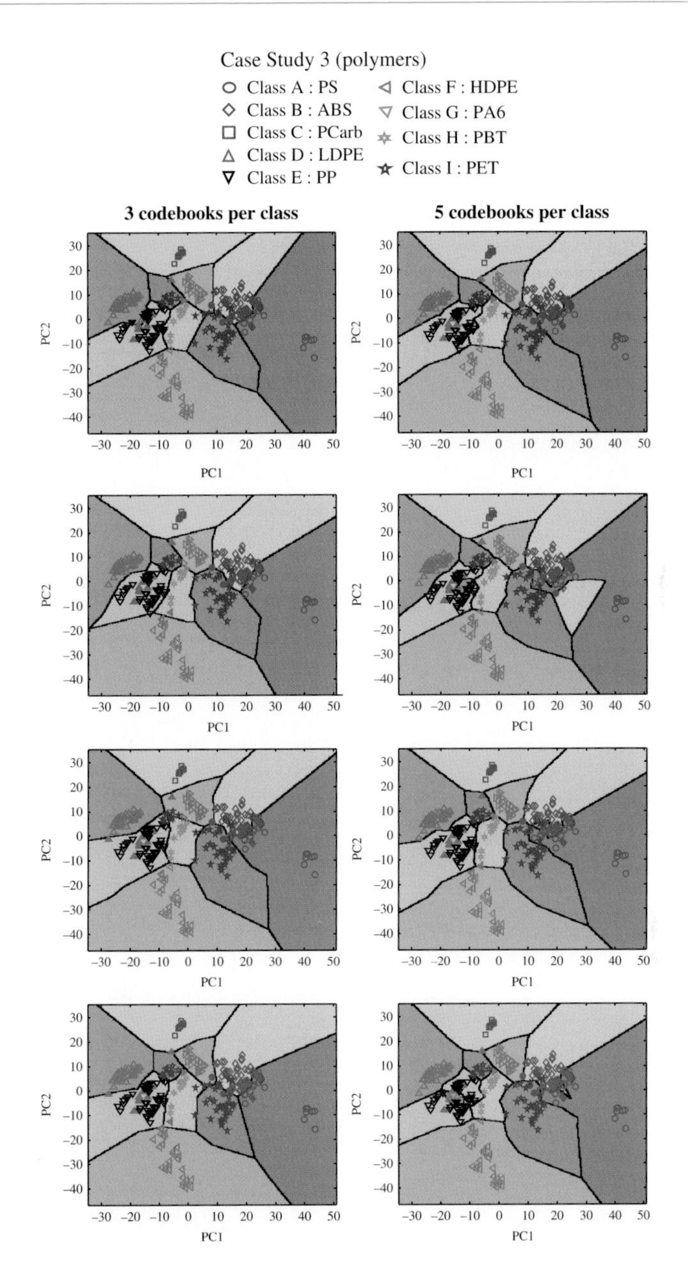

Figure 7.7 LVQ boundaries for Case Study 3 (polymers), using three and five codebooks per class, and four separate iterations

representation because it can cope with separate regions for each group – providing the number of codebooks per class is adequate, and at least one of the codebooks is seeded within each contiguous region of each class. LVQ tries hard to separate the classes in Case Study 8b (mouse urine – genetics) but is not very successful because of overlap.

7.4 PLS

PLS-DA is best employed as a two class modelling technique and its application when there is more than one class is somewhat awkward. There are several ways around the problem.

7.4.1 PLS2

The classical approach is to perform PLS2, which allows a matrix $\boldsymbol{C}$ of dimensions $I \times G$, each of whose columns consist of a class membership function (+1 or −1) for each of the G groups. Then instead of using the PLS1 algorithm, PLS2 can be employed which decomposes the matrix as follows:

$$\boldsymbol{X} = \boldsymbol{TP} + \boldsymbol{E}$$

$$\boldsymbol{C} = \boldsymbol{TQ} + \boldsymbol{f}$$

This is a straightforward, iterative, extension of PLS1 (Section 5.5). Only small variations in the algorithm are required. Instead of $\boldsymbol{c}$ being a vector it is now a matrix $\boldsymbol{C}$ and instead of q being a scalar it is now a vector $\boldsymbol{q}$:

1. An extra step is required to identify a vector $\boldsymbol{u}$ which can be a guess (as in PCA), but can also be chosen as one of the columns in the initial matrix, $\boldsymbol{C}$.
2. Calculate the vector:

$$\boldsymbol{w} = \boldsymbol{Xu}$$

3. Calculate the estimated scores by:

$${}^{new}\hat{\boldsymbol{t}} = \boldsymbol{Xw} \Big/ \sqrt{\sum w^2}$$

4. Calculate the estimated x loadings by:

$$\hat{\boldsymbol{p}} = \frac{\hat{\boldsymbol{t}}'\boldsymbol{X}}{\sum \hat{t}^2}$$

5. Calculate the estimated c loadings (a vector rather than scalar in PLS2) by:

$$\hat{\boldsymbol{q}} = \frac{\boldsymbol{C}'\hat{\boldsymbol{t}}}{\sum \hat{t}^2}$$

6. If this is the first iteration, remember the scores, and call them ${}^{initial}\boldsymbol{t}$ and then produce a new vector $\boldsymbol{u}$ by:

$$\boldsymbol{u} = \frac{\boldsymbol{C}\hat{\boldsymbol{q}}}{\sum q^2}$$

   and return to step 2.
7. If this is the second iteration, compare the new and old scores vectors for example, by looking at the size of the sum of square difference in the old and new scores, i.e. $\Sigma({}^{initial}\hat{t} - {}^{new}\hat{t})^2$. If this is small the PLS component has been adequately modelled,

set the PLS scores ($\boldsymbol{t}$) and both types of loadings ($\boldsymbol{p}$ and $\boldsymbol{q}$) for the current PC to $\hat{\boldsymbol{t}}$, $\hat{\boldsymbol{p}}$ and $\hat{\boldsymbol{q}}$. Otherwise calculate a new value of $\boldsymbol{u}$ as in step 7 and return to step 2.

8. Subtract the effect of the new PLS component from the datamatrix to get a residual datamatrix:

$$^{resid}\boldsymbol{X} = \boldsymbol{X} - \boldsymbol{tp}$$

9. Determine the new estimate by:

$$^{new}\hat{\boldsymbol{C}} = {}^{initial}\hat{\boldsymbol{C}} + \boldsymbol{tq}$$

and sum the contribution of all components calculated to give an estimated $\hat{\boldsymbol{C}}$. Calculate:

$$^{resid}\boldsymbol{C} = {}^{true}\boldsymbol{C} - \hat{\boldsymbol{C}}$$

10. If further components are to be computed, replace both $\boldsymbol{X}$ and $\boldsymbol{C}$ by their residuals and return to step 1.

Whereas PLS2 can always be used to provide estimates of c for each class there are several decisions that need to be made. The most important decision is how to preprocess $\boldsymbol{X}$. As we saw for PLS1 (Section 5.5) we use a form of column adjustment by the mean of each group (for unequal groups). For PLS2 the decision rules could become quite complex, and so we will use a default involving centring the $\boldsymbol{X}$ matrix. The decision of which group a sample belongs to also has many options, a simple one being to take the class with the highest value of predicted c. The results, on Case Studies 2 and 3, are presented in Figure 7.8 and are

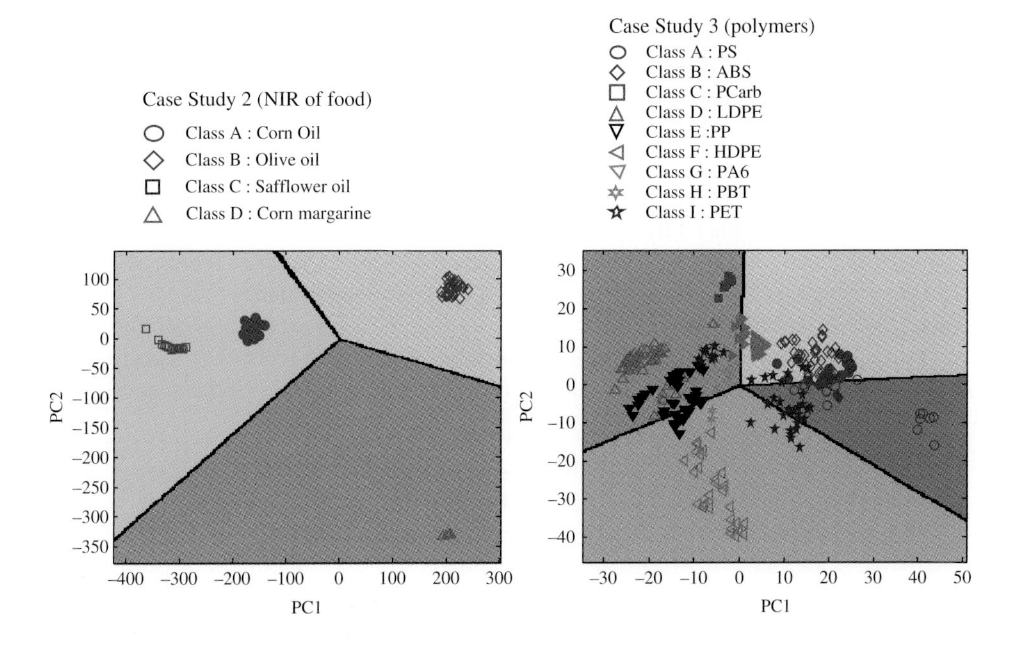

Figure 7.8 Application of PLS2 as discussed in Section 7.4.1 to Case Studies 2 and 3

rather disappointing. Several of the groups are missed altogether. The boundaries meet at the centre (this point is a consequence of centring the $\boldsymbol{X}$ matrix). Whereas it is possible to change the decision rule for PLS2 and also the column means in the $\boldsymbol{X}$ matrix, on the whole, this method is not usually very satisfactory, and involves including elaborations in the PLS2 algorithm moving it from one that is quite simple to a complex and elaborate one. PLS2 was originally developed for computational convenience, especially when computers were limited in memory, since if there are several variables in the 'c' block it is time consuming to repeat calculations for each column of the 'c' block using PLS1 each time. However PLS1 offers more flexibility and we discuss alternatives below.

7.4.2 PLS1

PLS1 can be applied in a variety of ways, as a series of binary decisions. We will denote the class that is being modelled as the 'in group' labelled by +1 and the class that it is compared against the 'out group' labelled by –1 for G groups:

1. *One v all.* The simplest approaches involves performing G comparisons of one versus the rest of samples. So if we are testing a model for class A, the 'in group' is class A and the 'out group' the remaining samples. G comparisons can be made. The binary decision boundaries are presented in the top row of Figure 7.9 for Case Studies 2 (NIR spectroscopy of food – 4 groups) and 3 (polymers – 9 groups). However it would be hard to determine the multiclass boundaries from these binary ones, which in some cases can be somewhat arbitrary and unstable. A decision rule is needed and we illustrate the decision rule to assign a sample to the class whose value of c is largest. The position of the decision boundary is influenced by how the means of the columns of the $\boldsymbol{X}$ matrix are adjusted. Using the protocol of Section 5.5.3, we subtracted $\bar{\bar{\boldsymbol{x}}} = (\bar{\bar{\boldsymbol{x}}}_{in} + \bar{\bar{\boldsymbol{x}}}_{out})/2$ from each column in the matrix, where $\bar{\boldsymbol{x}}_{in}$ is the mean of the 'in group' and $\bar{\boldsymbol{x}}_{out}$ of the 'out group' for each of the PLS1 class models for the decision boundaries of Figure 7.9. An alternative (Figure 7.10) involves subtracting $\bar{\bar{\boldsymbol{x}}} = (\bar{\boldsymbol{x}}_{in} + (\sum_{h \neq in} \bar{\boldsymbol{x}}_{h})/(G-1))/2$. We can see that, in contrast to PLS2, regions for all classes are identified. However some regions are misassigned, for example, Class C of Case Study 3 is not correctly positioned on the map. This relates to our decision function that assigns a sample to a class for which it has a highest c value, rather than the value that is closest to 1. Changing the decision function can make a difference to the predictions but on the whole elaborate changes in decision rules result in quite complex models that are not very general and likely to be valid only in certain specialist cases and possibly prone to overfitting as the decision rules are adjusted to fit a specific application. It can be seen however that changing the way the column means are adjusted does influence the appearance of the map.

2. *One v one.* This method involves performing a series of two class classifications between each of the possible G groups, involving $G(G-1)/2$ pairwise comparisons; for example, if there are four classes, there are six possible pairwise comparisons, namely between classes A and B, A and C, A and D, B and C, B and D, and C and D. The comparisons are performed using just the samples from two groups to form the PLS model; the class membership of other samples or regions of the data space are predicted, using the approaches discussed in Section 5.5.3. So for Case Study 2 (NIR spectroscopy of food) there will be 6 comparisons, and for Case Study 3 (polymers)

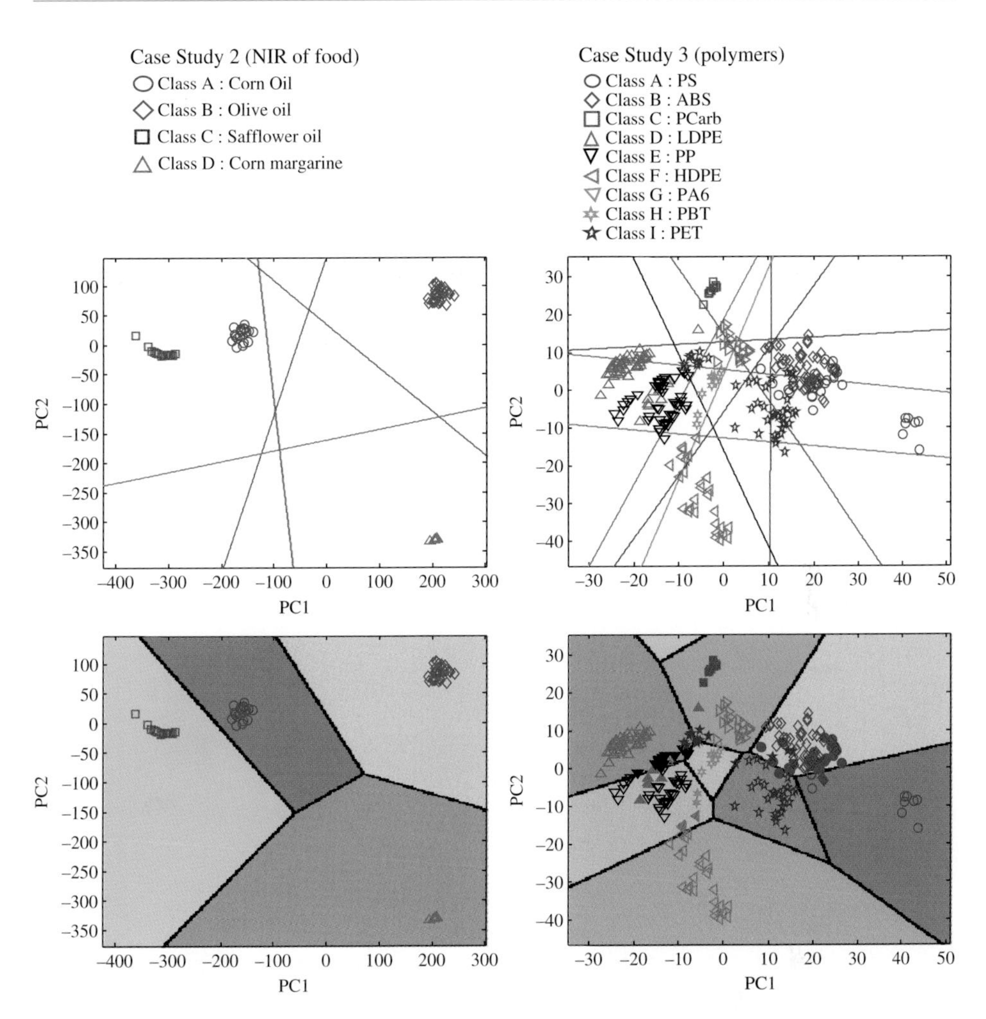

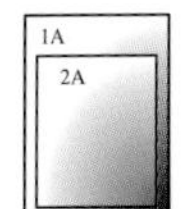

Figure 7.9 Application of PLS1 as a series of binary 1 v all decisions to Case Studies 2 and 3, adjusting the mean according to the weighted mean of the 'in group' and 'out group'. Individual boundaries are given in the top diagrams, and classification according to the maximum value of c below

36 comparisons. A classification rule is needed, which can be by 'majority vote', that is, the class receiving most votes (i.e. the one to which samples are most frequently classified) is predicted to be a member of the corresponding class: regions where there are ties are void and coloured in white (see Figure 7.11). Void regions can be eliminated by using fuzzy decision rules as follows:

- For each point in dataspace, perform all comparisons 1 v 2, 1 v 3 to 1 v G and predict c for each model.
- If the predicted value of c is greater than 1 for any comparison, set it to 1.

- Keep the minimum value of c for this set of comparisons – call it m_1.
- Perform the full set of comparisons for all classes G, giving m_g for each class g.
- Assign the sample or region of data space to the class for which m_g is a maximum.

The result of using fuzzy rules is presented in the bottom row of Figure 7.11. It can be seen that the classes are now quite well represented apart from classes D and I of Case Study 3 (polymers – 9 groups), which overlap with other groups and cannot be easily modelled using two PCs and linear models.

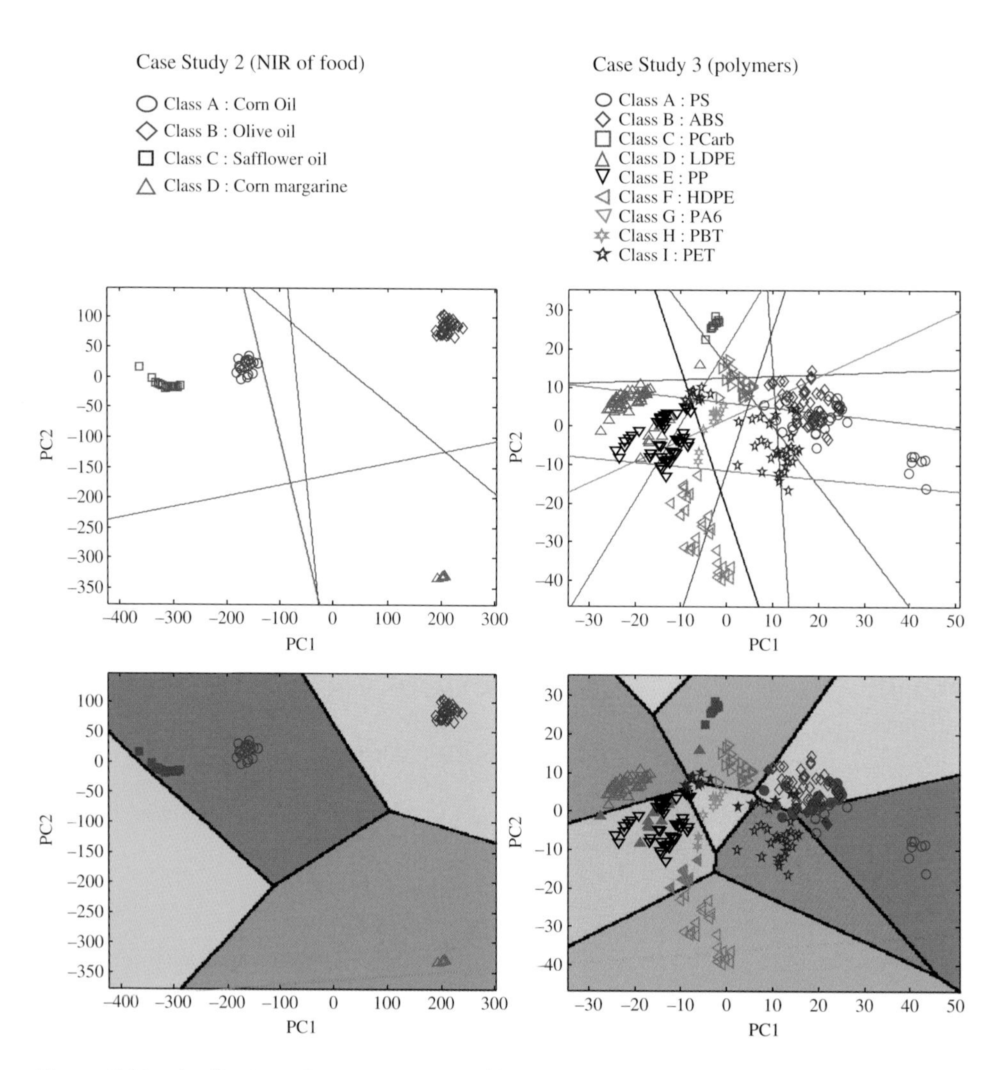

Figure 7.10 Application of PLS1 as a series of binary 1 v all decisions to Case Studies 2 and 3, adjusting the mean according to the weighted mean of the 'in group' and 'out group', but weighting the 'out group' mean according to the average of the out group means. Individual boundaries are given in the top diagrams, and classification according to the maximum value of c below

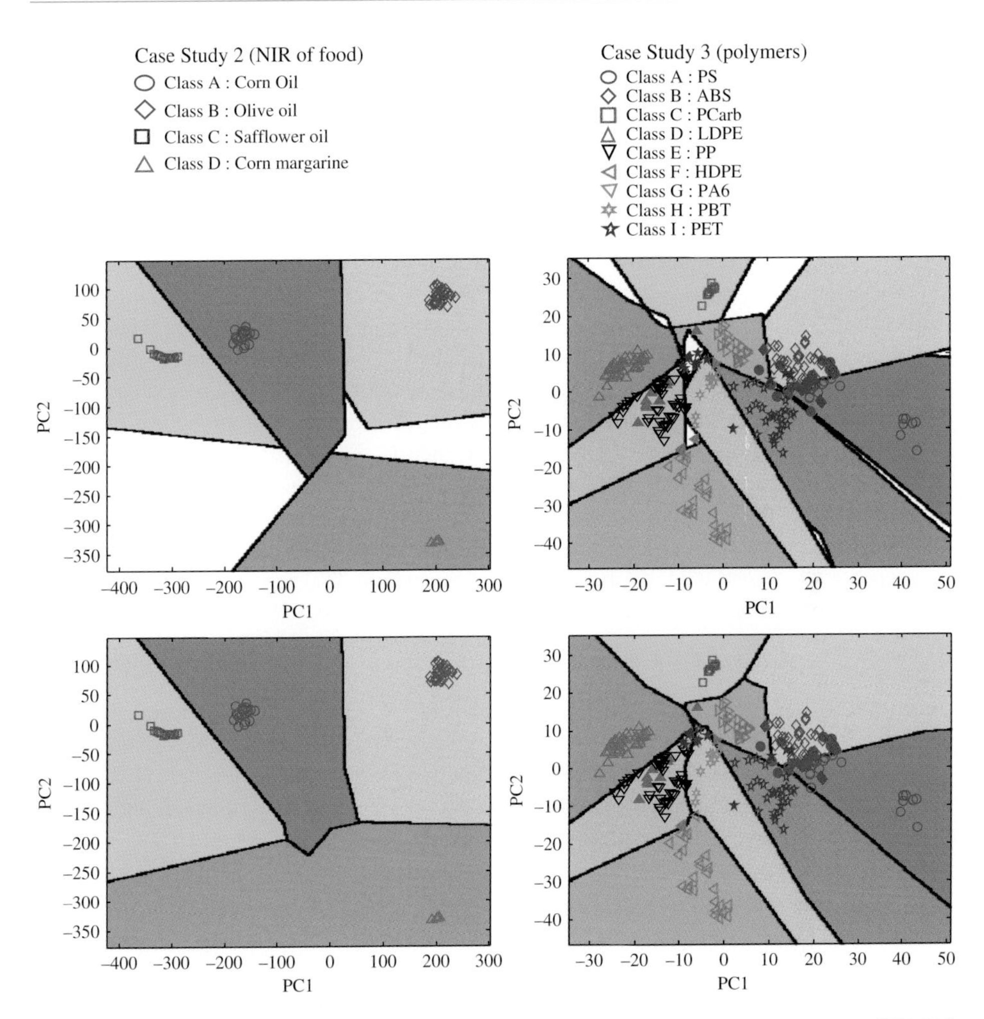

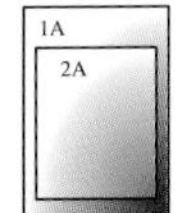

Figure 7.11 Application of PLS1 as a series of binary 1 v 1 to Case Studies 2 and 3, adjusting the mean according to the weighted mean of the 'in group' and 'out group'. Top, using 'majority vote' to decide class membership (white regions being unclassified) and bottom, using fuzzy rules to classify all regions

More elaborate rules both in decision making (how to decide whether a sample belongs to a class using predicted values of c) or column scaling (primarily adjusting of means) can be incorporated but PLS which is fundamentally a statistically based approach, becomes closer to machine learning where there is a set of elaborate decision rules, and loses its original simplicity. Whereas PLS-DA is an excellent two class classifier, and as we will see in Chapter 9 is especially useful when we want to look at which variables are most diagnostic of a particular separation, its extension to several classes can be rather difficult.

7.5 SVM

Support Vector Machines (SVMs) are not naturally a multiclass method, but there are extensions that are available:

1. *One v all* is the earliest method reported for extending binary SVMs to solve the multiclass problem and involves determining how well a sample is modelled by each class individually and choosing the class it is modelled by best analogy to the multi-distance approaches in Section 7.2 where a sample is assigned to the centroid it is closest to. Given G classes under consideration, G binary SVM models are constructed, samples either being considered as part of the class or outside it. The gth ($g = 1, \ldots, G$) SVM model is trained with all of the samples in the gth class being labelled by $+1$ and all other samples being labelled by –1 (note that the alternative approach of one class SVDD has been discussed in Chapter 6). Hence G SVM decision functions can be obtained, for each model. Instead of using a sign function to determine the class membership, the numerical outputs of the decision functions for each SVM model are compared, as described below. The membership $g(\boldsymbol{x})$ of an unknown sample $\boldsymbol{x}$ is determined by finding the class which for which the corresponding decision function is a maximum, as follows: $g(\boldsymbol{x})=\max\limits_{g=1,G}((\sum\limits_{i\in sv_g}\alpha_i c_i K(\boldsymbol{s}_{ig},\boldsymbol{x}))+b_g)$ where $\sum\limits_{i\in sv_g}\alpha_i c_i K(\boldsymbol{s}_{ig},\boldsymbol{x}))+b_g$ is the decision function for class model g. Figure 7.12 illustrates these boundaries for Case Studies 2 (NIR spectroscopy of food), 3 (polymers), 7 (clinical) and 8b (mouse urine – genetics) and default values of σ and C. Interestingly for Case Study 3, the method failed to find any region for class H, although the shape of the regions will be changed if different values of the RBF width and penalty error are employed. Interestingly, also, class D (Case Study 2) and class F (Case Study 7) are represented by very small regions.
2. *One v one* is also possible using SVMs, which are illustrated using both 'majority vote' and fuzzy rules (Section 7.4.2) in Figures 7.13 and 7.14. Note that for Case Studies 2 and 7, because the entire data space is mapped onto one of the classes using the majority vote, there is no difference between the fuzzy rules decision and the 'majority vote' decision. For Case Study 3, although a region for class H has now been found, the region chosen is quite a poor representation. Case Study 2 (NIR spectroscopy of food) is quite well represented, as the structure is very simple and so there is no reason for complex boundaries and as such the model is actually much simpler than QDA, and more successful that PLS-DA. However Case Study 8b (mouse urine – genetics) results in the formation of a very complex model possibly unjustified by the data. However setting a larger value of the RBF value σ could result in less complex models.

7.6 One against One Decisions

In Sections 7.4.2 and 7.5 we discussed the use of a series of one v one classifiers to determine the origin of a sample or classification of a region of data space in the context of SVM and PLS-DA. However these principles can be extended to other

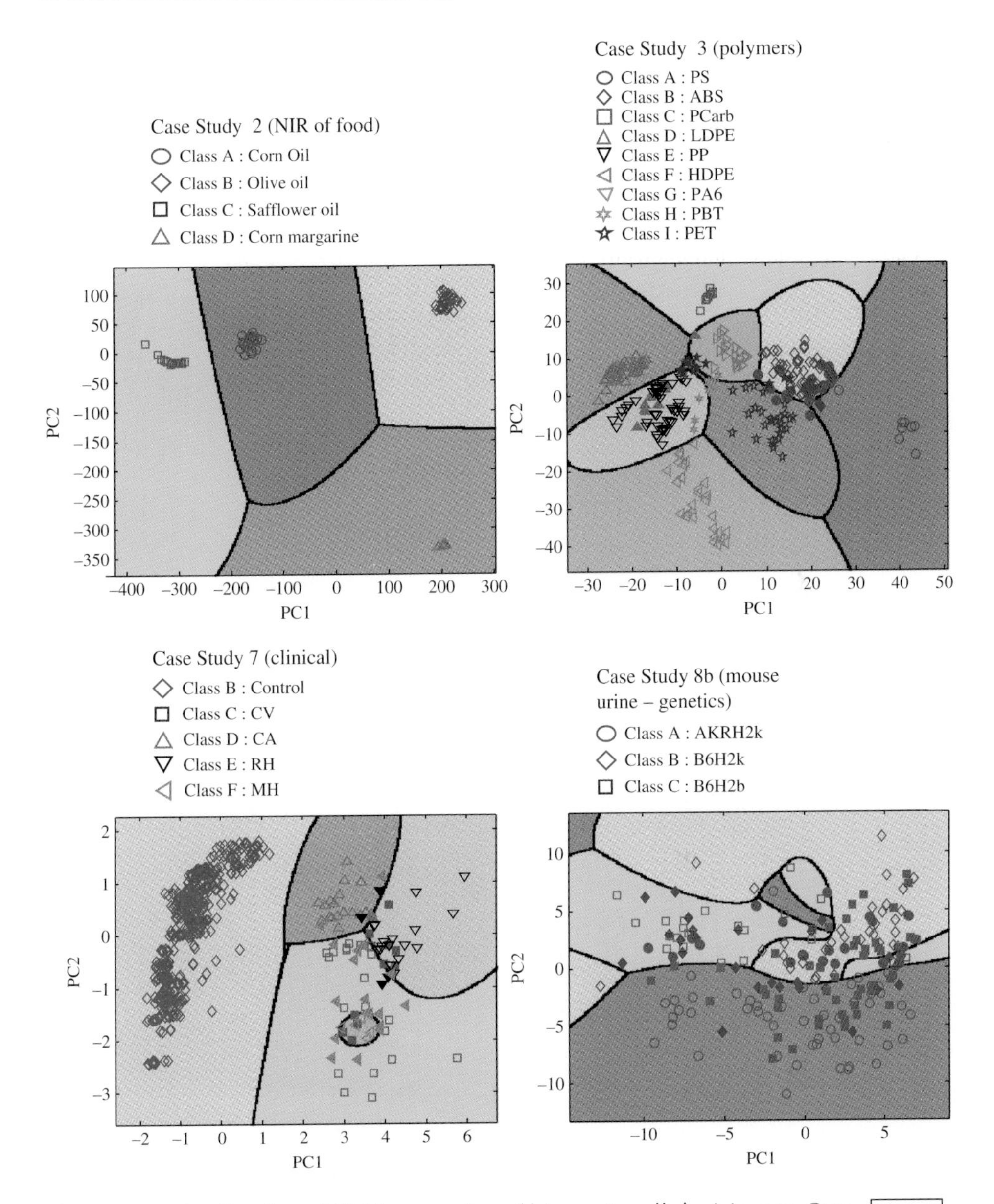

Figure 7.12 Application of SVM as a series of binary 1 v all decisions to Case Studies 2, 3, 7 and 8b. A default of $C=1$ and $\sigma = 1 \times$ average standard deviation of the variables in the dataset is used, and an L1 SVM is employed by default

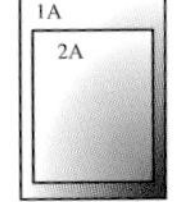

classifiers including all those described in this chapter. For reasons of brevity we do not illustrate these calculations, and for methods such as EDC, LDA and QDA offer no advantages.

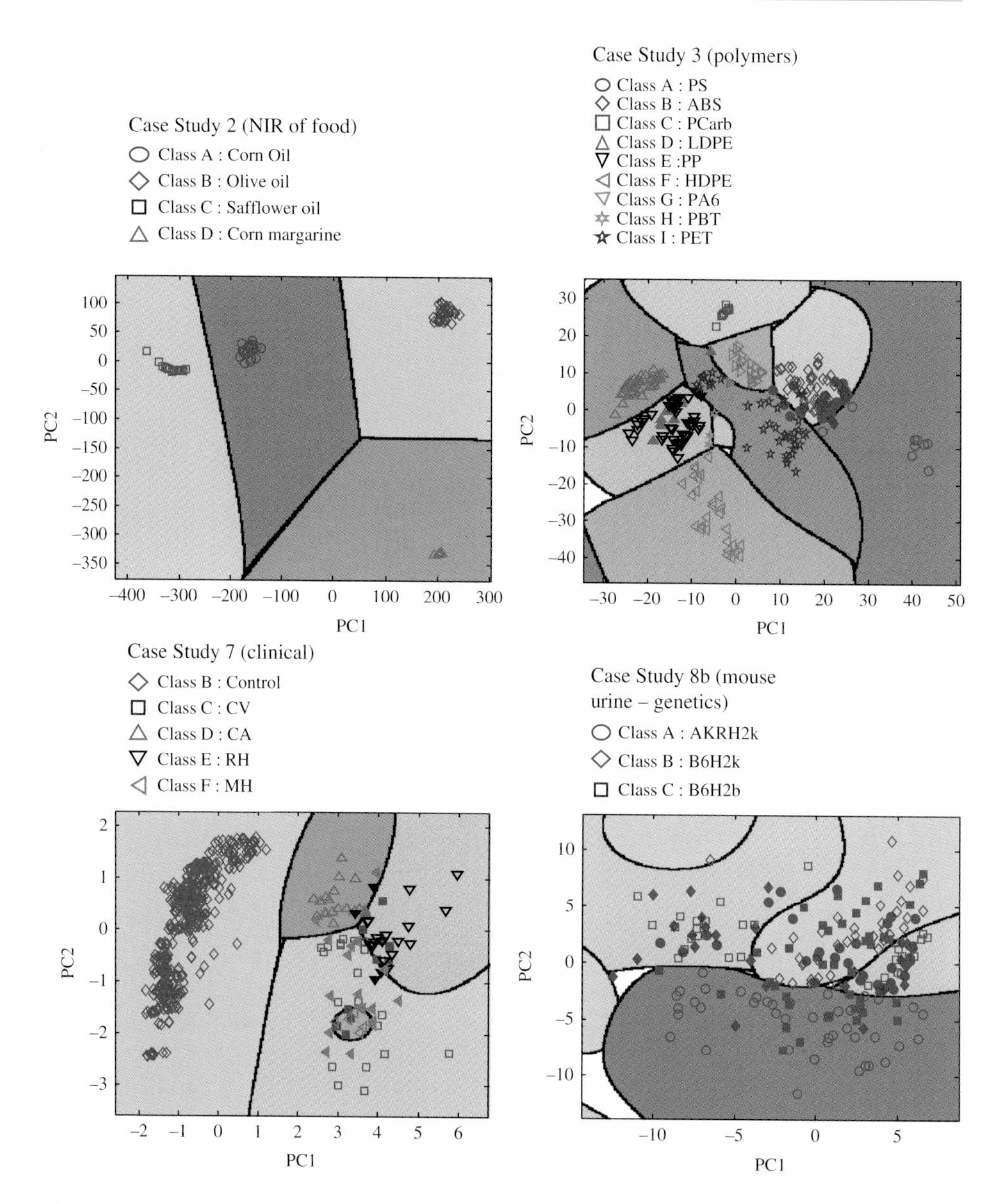

Figure 7.13 Application of SVM as a series of binary 1 v 1 decisions to Case Studies 2, 3, 7 and 8b. A default of $S = 1$ and $\sigma = 1 \times$ average standard deviation of the variables in the dataset is used, and an L1 SVM is employed by default. Classification is via 'majority vote'

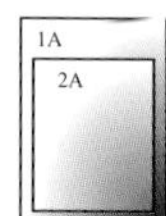

However when there are a large number of classes, the number of comparisons can be quite substantial, especially if there are also test and training set comparisons (Chapter 8) and the model is formed iteratively (e.g. LVQ) and so computationally more efficient methods are often desirable. A DAG (Directed Acyclic Graph) tree is an alternative and more

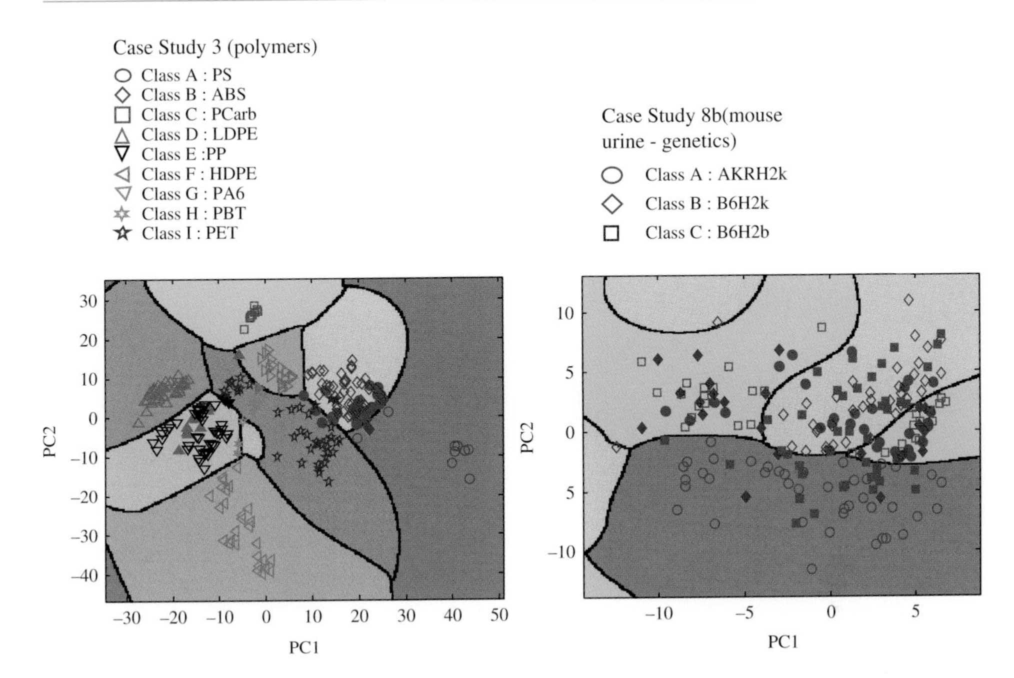

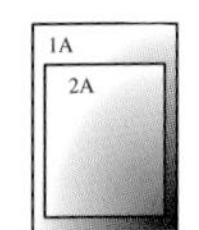

Figure 7.14 Application of SVM as a series of binary 1 v 1 decisions to Case Studies 3 and 8b. A default of S = 1 and $\sigma = 1 \times$ average standard deviation of the variables in the dataset is used, and an L1 SVM is employed by default. Classification is via fuzzy rules. Note that there is no difference for Case Studies 2 and 7 compared with Figure 7.13

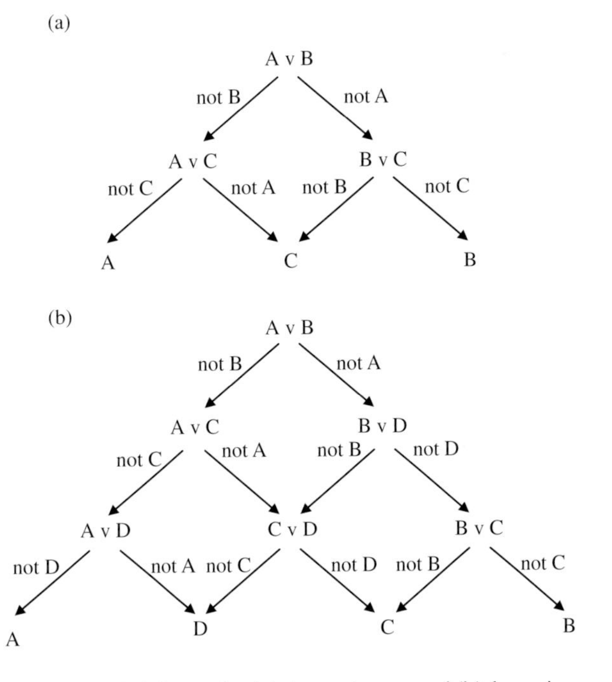

Figure 7.15 DAG tree for (a) three classes and (b) four classes

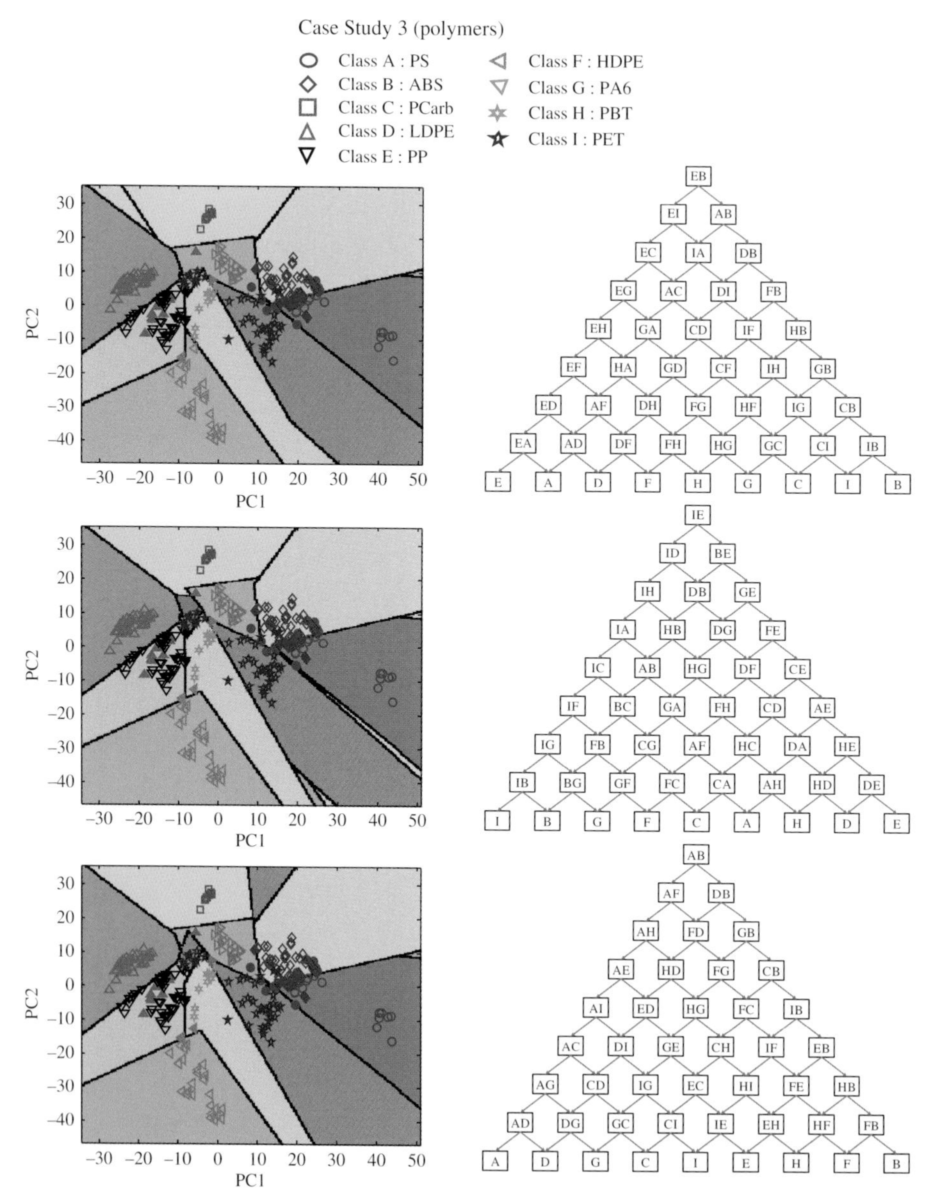

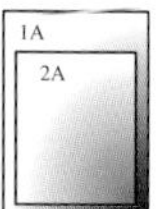

Figure 7.16 DAG decision trees and resultant class boundaries using PLS1 DA, for three DAG trees and Case Study 3 (9 polymer types). Each binary decision is represented by a box with the left hand arrow indicating the next comparison if the sample is grouped into the left hand class and vice versa

computationally efficient approach to the one against one method described, although it starts on the same basis of forming all possible one against one models on the training set. But when predicting the origins of a sample, it uses a rooted binary DAG tree with $G(G-1)/2$ internal nodes (or decisions) and G leaves (or outcomes). Each node involves a binary one against one model. A test sample starts from the root node and the binary decision function is evaluated depending on the output value of the decision function. Examples of DAG trees for 3 and 4 class problems are shown in Figure 7.15. The computational time for a DAG tree is shorter than 'one-against-one' because as it is not necessary to go through every binary classifier to test an unknown sample, only G comparisons are necessary, and so this could shorten the time considerably for predictive models, for example, if there are 9 classes, instead of forming 36 models and taking the majority vote, only 9 comparisons are required. There is no universal guidance as to which two classes are tested in each node, but for most problems the method is not strongly influenced by the arrangement of the nodes. A possible strategy, for example, would be to choose the two most different classes for the first decision in the tree and so on. The improvement in computational efficiently will be significant when predicting unknowns especially if algorithms are quite slow, but does not benefit much the model building or validation stages as these still require all possible binary models to be formed. As an illustration of the use of DAG trees for PLS-DA we present the results of three possible solutions in Figure 7.16. There are slight differences according to which decision tree is employed. However when there are a large number of classes, this can speed up the model building substantially, and although there is some variability between solutions, this is usually no more than the variability due to other decisions such as data preprocessing or choice of decision function.

Bibliography

Topics are covered in the main references to Chapter 5.

8

Validation and Optimization

8.1 Introduction

In this chapter we will be discussing approaches for validation and optimization of pattern recognition techniques. The main methods for pattern recognition were introduced in Chapters 5 to 7 where we discussed autopredictive models. In this chapter we will be primarily illustrating methods for validating models, that is, determining how well they are obeyed avoiding the pitfalls of overfitting, and allow us to decide whether the data are of sufficient quality for underlying hypotheses that there are groups in the data. We will in addition look at how to optimize all the main classifiers, in particular SVM and PLS-DA. Optimization and validation are often confused in the chemometrics literature, and as such their decision is combined in one chapter, but have clearly different aims which are described below, although the approaches for determining whether a model is a suitable one and for optimizing a classifier are often quite similar in nature and based on common principles, for example, investigating the effect on classification accuracy of leaving one or a group of samples out from the modelling process.

This chapter will be based on several principles:

1. The performance of methods should be benchmarked against a null dataset (Case Study 11).
2. Computationally intensive methods will be preferred to less intensive methods where they show a clear advantage.
3. Optimization should be clearly separated from validation.
4. The number of samples correctly classified reflects both the method used and the underlying data structure.

8.1.1 Validation

In traditional analytical chemistry although it is certainly important to determine the accuracy and precision of most techniques, and there is a large literature that has developed

Chemometrics for Pattern Recognition Richard G. Brereton

around this, validation as in the context of pattern recognition was not so strongly emphasized. In quantitative analytical chemistry the main aim is to determine how much of a compound is in a mixture, for example, how much of a pollutant can be measured in a sample of soil. We want to measure this well, and want to ensure that sampling is representative and that our sample is not an outlier, but we usually do not use statistical methods to determine whether we can actually measure this pollutant or not: this is done by building instruments, from chemical knowledge and by engineering. Once we have an instrument that we know can measure the specific pollutant (it may be an element and we may use atomic spectroscopy or ICP to measure its concentration), then other types of statistical data analysis come into play, often based around the uncertainty or precision of a measurement.

Much of early chemometrics was based around this world of certainties and some of the biggest advances in chemometrics in the 1980s were in the area of multivariate calibration where there is often a high degree of certainty in the existence, and usually, amount, of the calibrant. The difficulties were to determine how well we can perform a measurement, and the more accurate the determination (lower percent error) the better, and data analytical techniques were not used to determine whether the measurement is possible – that is an engineering and chemical problem. The analytical chemist strives to produce instruments, sampling techniques and data analysis approaches that reduce this error, and the lower it is the better. Chemometrics had its origins primarily in analytical chemistry, and this world of certainty definitely influenced the direction that many of the earlier pioneers took in developing their algorithms. Some problems in traditional chemometrics, though, were classification problems, but with a certain outcome. An example may be how we can use a spectrum to determine the underlying structural class of a compound – is it a ketone or an aldehyde? There is a certain answer and we are sure that spectroscopy can help solve it; we may like to use spectroscopy combined with chemometrics to determine the structural class in an automated way, and the higher the number of samples correctly classified, the better. Chemometricians, like analytical chemists, would strive to increase the proportion of correctly classified samples to as high a level as possible. An algorithm, combined with an instrumental method, that correctly classifies 95 % of the samples is better than one that correctly classifies 90 %. As chemometrics moved into more applied fields, noticeably the area of food, similar philosophies took hold. Can we use chemometrics to distinguish cheeses from one area and another combined with spectroscopic measurements? We are certain that the cheeses are different, and probably the traditional cheese tester could tell by smell, feel and taste very accurately, and we know there is a certain answer out there, but the main aim of data analysis is to produce some rapid automated method that replaces the traditional tester, and the greater the proportion correctly classified, the better.

But chemometrics has strayed from the comforting certainties of analytical chemistry into new territory. Because many of the current building blocks of chemometrics were developed in the 1970s and 1980s, the traditionally accepted conceptual framework finds it hard to cope with this new world of uncertainties. Biology and medicine in particular are full of uncertainties. Biology, unlike analytical chemistry, is a very hypothesis driven subject. Many of the best biologists work on a ‘hunch’; they are not sure of the answer in advance. So once we come to analysing biological, or medical, data we may not necessarily be sure whether there is an answer. For example, can we tell by analysing a person's urine, that they have early stage diabetes? Can we classify subjects into a group of people who will not develop diabetes and a group that will (say within a five year period), by looking at the chemical signal? We are not sure, and so pattern recognition is not just about calibrating to a known answer, but to an uncertain one.

This hypothesis driven world is familiar to biologists but difficult for many in the analytical chemistry community. Now no longer is an answer that we can predict whether a person has a disease at a probability of 100 % or 95 % necessarily the most appropriate answer. No longer is it always possible, or sensible, to develop a method that maximizes the number of samples correctly classified. We may well be able to obtain groups of people, those that developed diabetes within five years and those that did not, and may have their samples stored. We may know very well that there the urine samples were taken at the same time and stored in the same way. We may be able to perform numerous analyses on these samples. But can we be sure that we can use chemometrics combined with analytical chemistry to successfully classify someone with early stage diabetes from someone without, in other words can we be sure that there is a sufficiently distinct chemical signature in the urine samples that characterizes people with early stage diabetes from those without? The answer is no, and so chemometrics methods have to answer an additional question, does the data support the hypothesis?

It is, as will be shown in this chapter, relatively easy to come up with high classification success rates even starting with random data. If as a researcher the aim is to impress one's boss and get a promotion, or to obtain a grant, or write a paper, it is usually quite possible to put almost any data through a chemometrics 'package' and get a high classification rate, and in addition there have been developed many methods for improving graphical display to the extent that even data originating from random sources can look as if they are separated into groups if so desired: many modern algorithms seize on any correlations between, for example, class membership and analytical data, and force the solution so that the graphs look good. If the method for obtaining this high classification rate is accurately described (not just 'I put it through a package') probably there is no great harm, but if the results are used for real decisions which may have an economic consequence (e.g. in process monitoring to decide whether to destroy a batch) or a medical consequence (e.g. whether to amputate a person's leg), then there can be serious problems. Actually, surprisingly, chemometrics methods are not used very much for real world decision-based classification, as most of the reports in the literature are exploratory: it is an excellent tool for scoping and laboratory based investigations. Many chemometrics methods can be likened to a sniffer dog in customs in an airport when detecting drugs: the sniffer dog will direct the customs officer to a suitcase with reasonable accuracy, but the proof is when the suitcase is opened – a court will not prosecute someone just on the evidence of a sniffer dog. Chemometrics has had a much bigger real world role in calibration and in visualization, simplifying large datasets and allowing groupings to be determined, for example, by PC plots.

In order that chemometrics techniques are used in hypothesis driven as well as predictive science it is essential to understand the importance of model validation. It is also essential to distinguish model validation from optimization. Many early (and even current) applications of chemometrics fail to make this distinction – for example, cross-validation can be employed to determine the number of PCs or PLS components in a model – very legitimately – but this is an optimization problem – however, the cross-validation error then is sometimes used to determine how well the model performs, which should be assessed on an independent test set. Practitioners are all very aware of this, and the principles are not too difficult to appreciate, but the majority of applications of chemometrics come from use of packaged software and it is entirely possible to miss this distinction.

In addition to the increased need for pattern recognition tools for exploratory data analysis, there has been a huge increase in cost effective computer power over the past two decades; an observation often called Moore's law, in its various incarnations, suggests that computer power (per unit cost) doubles every one to two years. Using the intermediate value of 18 months, this effectively means 10 fold in 5 years, 1000 fold in 15 years and an incredible 1 million fold in 30 years. This may seem exaggerated but is not. Many people can remember swapping Mbyte for equivalent Gbyte memory over a period of 15 years – the number of Mbytes of memory in a typical PC in the early 1990s equals the number of Gbytes now. Disk space has similarly increased apace, a 40 Mbyte disc drive being typical in 1990, whereas 100 Gbyte or more is routine in 2008. Thirty years ago (1978) Apple produced some of the first commercial PCs with 4 kbyte memory, around one million times less than now.

Yet many of the pioneering chemometrics methods such as cross-validation or NIPALS started to become well accepted around thirty years ago, involving ingenious algorithms that could function effectively in the very modern (at the time) environment of microprocessors – a revolution to do multivariate analysis on one's desktop rather than by a remote mainframe, doing away with paper tape, punchcards or teleprinters. These algorithms are still deeply engrained in the mind of many in chemometrics and were indeed suitable and feasible for the type of problem available then. Software companies were formed in this area in the early 1980s, many devoted to marketing and developing the approaches of these pioneers. But computing power has changed radically since then, and methods inconceivable to all but the very largest establishments at the time (mainly in the defence industry who do not usually report results in public) are now routinely available for use on a desktop PC. The trouble is that it can be time consuming to set up new companies around new philosophies especially when there are already software leaders who have well established sales networks and contacts, and customers (unlike twenty years ago) now demand very user friendly packages or they will not buy (especially large corporations). So many new methods that can routinely take advantage of more accessible computing power, whereas they are available to experts (especially Matlab users in this field) and are well regarded in the literature, are as yet, not readily available to the laboratory based user, causing a major dilemma in chemometrics. In areas such as biology often specialist bioinformaticians or biometricians handle data and they are trained in the development and application of computer algorithms, but in chemistry there is a demand often from laboratory based chemists to handle data hands-on without a specialist appreciation of the mathematical background to the methods they are employing, hence a strong reliance on packages – and on choosing software that is user friendly and displays data in the most optimistic way.

A major topic of this chapter is guidance as to the correct procedures for validation of models, that is, determining exactly how well models fit the data. It is well known that certain models can overfit data – common examples are as follows:

1. Choosing too many PLS components in PLS-DA.
2. Choosing a model that is very complex in SVM.
3. Preselecting variables that have already been shown to have high classification ability and then forming models using these variables.

Test sets form an important role in determining whether a model really can predict classes with a high degree of confidence. If not, this may be that the model is not suitable or it may simply be that the data are not of adequate quality for a good classification (e.g. the null Case Study 11). We would hope that poor classification results are obtained when the null dataset is used, but cannot guarantee good results on all real datasets, if there is no underlying trend. Test sets can be used both for model validation and for optimization as discussed in this chapter. Validation of models using test sets is described in Sections 8.3 and 8.4.

8.1.2 Optimization

Optimization of the classifier is often an important step. Many people confuse this with validation, for example they may use cross-validation to determine both the optimum model and the performance of the model. This is incorrect as both are separate procedures. The %CC (or percent of samples correctly classified – see Section 8.2) obtained whilst optimizing a model can provide a falsely high estimate of the model performance, and so should be a separate step.

Optimization can be performed both on the overall dataset (for autoprediction) or on the training set (for validation). If we want to determine whether there is an underlying relationship between classifier and analytical signal, the training set model should be optimized independently of the test set. If we are sure that our hypothesis is correct and want to obtain the highest possible %CC, optimisation can be performed in autopredictive mode.

A second procedure often confused with optimization of a classifier involves optimizing the choice of variables as input to the classifier. These may, for example, be the most discriminatory PCs or the individual variables that show the most different distribution between two groups. This procedure is often called variable selection, feature selection, variable reduction or feature reduction and is discussed in Chapters 4 and 9. In this chapter we will assume that a choice has already been made of the most appropriate variables as input to the classifier.

There are different approaches for optimizing each classifier, some of the main ones are described in Sections 8.5 to 8.7.

8.2 Classification Abilities, Contingency Tables and Related Concepts

8.2.1 Two Class Classifiers

Fundamental to understanding how classification methods can be validated is to determine a measure for success. Table 8.1 represents the result of a two class classification algorithm. Each sample is either correctly assigned to its correct group or not. In analogy to the

Table 8.1 Typical results of a classification algorithm presented as a contingency table

	Group A (True class)	Group B (True class)
Group A (Predicted class)	60	10
Group B (Predicted class)	20	90

methods discussed in Chapter 5, each sample is either the correct side of the line or not. Along the columns we place the samples that we know (in advance) to belong to each class, and along the rows the samples that are predicted to belong to each class. This is commonly called a 2×2 contingency table or, when there are more than two groups, a confusion matrix, and is the basic building block of our methods.

From this contingency table we can compute several numbers:

1. The simplest and most basic is the overall percentage correctly classified (%CC), the higher this is, the more samples have been assigned to their correct groups. In the example of Table 8.1, there are 180 samples of which 150 (= 60 + 90) have been correctly assigned to their group, or an overall %CC of $100 \times (150/180)$ or 83.33 %. The higher this number, the more the samples have been assigned to their correct group. Note that a higher %CC does not always imply a more suitable algorithm – this we will discuss under the context of validation.
2. If we denote class A as the 'in group' e.g. mice on a diet (Case Study 8a) or polluted samples (Case Study 4) and class B as the 'out group', then if a sample assigned to the 'in group' is regarded as positive we can define the following:
   - TP: True Positives, samples that are genuinely in class A and are classified as such.
   - TN: True Negatives, samples that are genuinely in class B and are classified as such.
   - FP: False Positives, samples that are genuinely in class B but are falsely classified as class A.
   - FN: False Negatives, samples that are genuinely in class A but are falsely classified as class B.

   Even if it is not clear which group should be regarded as the positive group (e.g. a test for females and males) we can still set up one class as an 'in group' and the other as an 'out group' and use this terminology. The numbers calculated from the data of Table 8.1 are presented in Table 8.2. They can be cited in percentage format as follows:
   - $\%TP = 100 \times TP/(TP + FN)$; the sensitivity of a method is defined by $TP/(TP + FN)$.
   - $\%TN = 100 \times TN/(FP + TN)$; the specificity of a method is defined by $TN/(FP + TN)$
   - $\%FN = 100 \times FN/(TP + FN)$.
   - $\%FP = 100 \times FP/(FP + TN)$.
3. It is possible to determine %CC for each class, as follows:
   - $\%CC_A = \%TP$.
   - $\%CC_B = \%TN$.

   Note that the overall %CC (83.33 %) defined by $100 \times (TP + TN)/(TP + TN + FP + FN)$ is not, in this case, the average of the %CCs for each class (75 % and 90 %), as the numbers in each class are different. These are presented in Table 8.2.

Table 8.2 Some definitions associated with a contingency table, calculated for the data of Table 8.1

	'In group' (True class)	'Out group' (True class)	Likelihood ratio	Probability
'In group' (Predicted class)	TP (60 = 75 %)	FP (10 = 10 %)	LR^+ (75/ 10 = 7.5)	$P^+ = 0.88$
'Out group' (Predicted class)	FN (20 = 25 %)	TN (90 = 90 %)	LR^- (25/90 = 0.277)	$P^- = 0.12$
%CC for classes	75 % Sensitivity 0.75	90 % Specificity 0.90		
Likelihood ratio	$LR^+ = 7.5$	$LR^- = 0.28$		

4. Likelihood ratios are sometimes employed, especially in clinical and forensic studies. We can define two likelihood ratios as follows:
   - LR^+, or in our case, LR^A, is the odds that a sample is a member of the 'in group' if the classifier assigns it to that group, to the sample not being a member of that group, and can be defined in a number of ways as follows:

     $LR^+ = \text{sensitivity}/(1 - \text{specificity}) = [TP/(TP + FN)] / [FP / (FP + TN)] = \%TP/\%FP.$

     It can be converted to a probability that a sample is a member of the 'in group' if the sample is assigned to this group:

     $$P^+ = LR^+/(LR^+ + 1) = \%TP/(\%TP + \%FP)$$

     Note importantly that this probability is not equal to the ratio of True Positives to False Positives, in our example. The reason for this is that the group sizes are different. For example, we may have 50 subjects in one group (e.g. diseased) and 1000 in another control group as they are easier to find and study, and the likelihood ratio corrects for this imbalance in experimental subjects. We can of course change the weighting between the two groups in any way we like, but this is best expressed using Bayesian statistics, which are discussed in Chapter 10.
   - LR^-, or in our case, LR^B, is the odds that a sample is a member of the 'in group' if the classifier assigns it to the 'out group' (a negative result) and the sample is not a member of that group:

     $LR^- = (1 - \text{sensitivity}) / \text{specificity} = [FP/(FP + TN)] / [TP / (TP + FN)] = \%FP/\%TP.$

     $$P^- = LR^-/(LR^- + 1) = \%FP/(\%FP + \%TN)$$

   The likelihood ratios and associated probabilities are calculated in Table 8.2 for our example. The historic origin of these terms arises particularly from clinical diagnosis, where we might be asking for example whether a patient symptoms of a specific disease, and so class A, the 'in-group', has particular significance compared to the 'out group' unlike in most classification studies where we regard each class as of equal importance. These values are often cited in the literature and so are introduced here for completion to allow the methods in this text to be translated into language that is used in parallel literature.

In this and subsequent chapters we will mainly use %CC as our indicator of how an algorithm has performed and also whether there is sufficient information within a dataset to support a particular underlying hypothesis.

8.2.2 Multiclass Classifiers

When there are more than two classes, a similar type of contingency table or confusion matrix can be obtained but with dimensions $G \times G$ where G is the number of classes. The diagonals represent the number of samples correctly classified. A typical confusion matrix is illustrated in Table 8.3(a). Note that the number of samples in each class may be different and so the numbers in a confusion matrix are often presented as percentages, with the columns adding up to 100 %. The confusion matrix can tell us which classes are more often confused with each other, for example, classes A and B are more often mistaken for each other than they are for classes C and D, in the example given, as they may be more closely related or even can be better viewed as one single group. Confusion matrices (or contingency tables) can also be calculated for each class separately, for example, class A in Table 8.3 (b), which shows how well one group of samples is classified against all the rest. Confusion matrices, can, of course be calculated for autoprediction, for the training set, for the test set and any other options such as the bootstrap or cross-validation as discussed below.

Table 8.3 *(a) A typical confusion matrix for samples originating from four classes, presented as numbers of samples classed into each class (left) and percentages (right). (b) Confusion matrix for class A*

(a)

		True numbers in each class								
		A	B	C	D		A	B	C	D
Predicted numbers in each class	A	60	11	0	7	A	80.00 %	18.33 %	0.00 %	7.37 %
	B	10	45	3	0	B	13.33 %	75.00 %	4.29 %	0.00 %
	C	2	4	58	3	C	2.67 %	6.67 %	82.86 %	3.16 %
	D	3	0	9	85	D	4.00 %	0.00 %	12.86 %	89.47 %
	Total	75	60	70	95					

(b)

		True				
		A	Not A		A	Not A
Predicted	A	60	18	A	80 %	8 %
	Not A	15	207	Not A	20 %	92 %

8.2.3 One Class Classifiers

For classical binary (or multiclass) classifiers, as the concept of confusion matrices is well established, every sample is unambiguously assigned to one class or the other. For one class classifiers, this is not necessarily so, and some adaptations are required if these approaches are to be useful.

Table 8.4 *Example of how contingency tables can be formed for one class classifiers*

	True class		Predictions class A model	Predictions class B model
1	A	Correct	A	Not B
2	A	Correct	A	Not B
3	A	Correct	A	Not B
4	A	Ambiguous	A	B
5	A	Misclassified	Not A	B
6	A	Correct	A	Not B
7	B	Correct	Not A	B
8	B	Correct	Not A	B
9	B	Correct	Not A	B
10	B	Correct	Not A	B
11	B	Ambiguous	A	B
12	B	Outlier	Not A	Not B
13	B	Correct	Not A	B
14	B	Correct	Not A	B

Contingency table
Class A model

	A	B
A	5	1
Not A	1	7

Contingency table
Class B model

	B	Not B
B	7	2
Not B	1	4

This is illustrated by Table 8.4. It can be seen that in contrast to two class classifiers there are now four verdicts:

1. Correctly classified (10 out of 14 or 71.4 % of samples)
2. Misclassified (1 out of 14 or 7.1 % of samples)
3. Ambiguous (2 out of 14 or 14.3 % of samples)
4. Outlier (1 out of 14 or 7.1 % of samples) rather than two verdicts, correctly or incorrectly classified.

A separate contingency table can be drawn up for each of the two one class classifiers if required; when using hard model (binary or multiclass) classifiers these contingency tables will be identical to each other except possibly with the rows and columns swapped round. Note that in many practical situations, e.g. in medicine or forensics, where there may be a simple and important binary decision (diseased or not; guilty or not) a one class classifier may be adequate, and it may only be necessary to model the group of interest, and not the 'out group'. In fact in some cases this is more appropriate than a two class classification model. Imagine trying to determine whether a set of banknotes are forged. The legitimate banknotes will probably be all of very similar standard and so form a tight group. The forged banknotes may come from many different sources (which are hard to control) and so form a very inhomogeneous group which is hard to model and hard to find representative samples from, and so it would be difficult to produce an effective model of forged against legitimate banknotes. However so long as the legitimate banknotes are modelled well, all we are interested in is whether a specimen falls within this region, and so only one type of model and one type of contingency table is needed.

Contingency tables or confusion matrices based on a consensus model from all known classes in the data are much harder to envisage for one class classifiers, and depend in what assumptions are made about each sample. They would also assume, inherently, that there are no outliers and no ambiguous samples; whereas ambiguous samples could be assigned equally to the relevant classes, it would be incorrect to assign outliers to either of the specific classes.

8.3 Validation

8.3.1 Testing Models

Fundamental to understanding how to correctly apply classification procedures is to understand the distinction between autopredictive and set models.

In Chapters 5 to 7 we discussed autopredictive models. For such models the entire dataset is used to model itself. Autopredictive models can often be overoptimistic and provide results that overfit the data, that is, appear good but once applied to unknowns give poor predictions; however these are useful for understanding the performance of classifiers.

In order to illustrate the principle consider again the seating plan of Figure 5.1. In the original figure we found we could draw perfect boundaries around groups of males and females, although these boundaries were rather complex. However if we accept that these complex boundaries are allowed we can get perfect autoprediction (%CC = 100) by the criterion that each person is correctly assigned within the rather complex boundary to their appropriate class.

Consider now a situation in which we remove one in three of the people – see Figure 8.1. We now try to predict the gender of the people that have been removed. This can be done by trying to establish new boundaries between the Male and Female groups. To do this we need some rules. Normally this involves trying, in the absence of further evidence, to

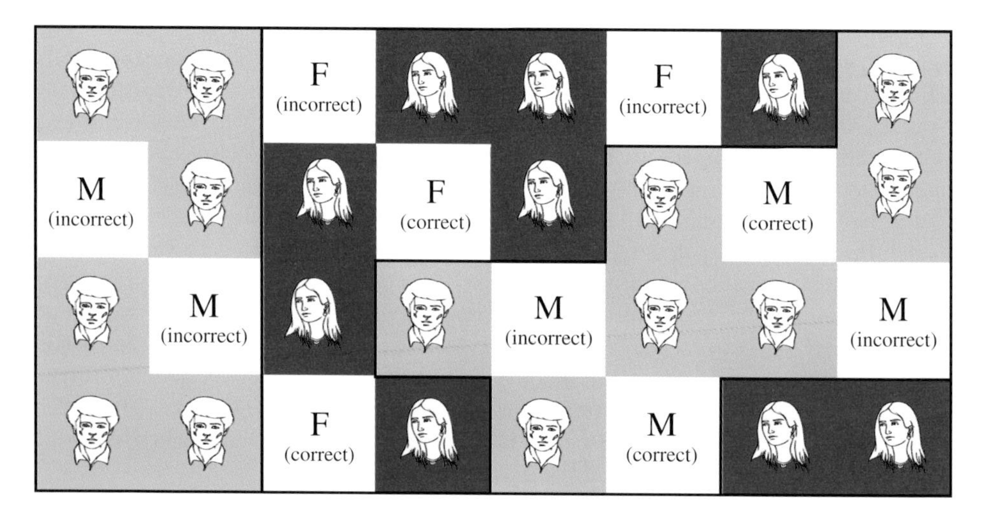

Figure 8.1 Male/female seating plan as of Figure 5.1 with one in three people removed and a simple nearest neighbour approach for predicting the gender of the missing people

produce simple boundaries unless there is overwhelming evidence of complexity. A simple set of rules could be as follows:

1. For each unknown place look at the gender of the nearest neighbours both horizontally and vertically. For places in the middle of the seating plan this will be four, for places at the edges three, and in the corners just two.
2. Look at the gender of the nearest neighbours and assign the empty seat to the majority gender. We would need a rule about ties if there is an even number of neighbours, but fortunately in our example this does not happen.

Figure 8.1 presents our predictions of the genders of the missing people using this simple system. Referring back to Figure 5.1, we find that only four out of ten of our predictions are correct, representing a % CC or 40 %. This suggests that our classification rules are not so good after all and basically randomly estimate the gender of missing people. The nearest neighbour rule we used tried to produce smoother and simpler boundaries and was also based on a social assumption that in certain circumstances men tend to sit next to men and women to women. We could adopt an inverse rule, that opposite sexes tend to like sitting next to each other, but that rule, whilst creating more complex boundaries would only give us 60 % CC – not much better.

The reason why we cannot predict the people sitting in the missing seats, is because the data are more or less randomly organized, it is not really possible (unless we had more information for example, about family groupings, if this was family entertainment) to predict where people of different genders sit, and so no real evidence that our boundaries have much significance.

Consider now a different situation, that of Figure 8.2, in which there appears more grouping between the sexes, although still no rigid boundaries. Using the same rules as above we can now predict 8 out of 10 or 80 % of the people left out correctly. This suggests our methods are suitable for the data, and we get better %CC. The rules have not changed, nor has the size of the dataset, nor assessment of the quality of our predictions.

The people removed from the seating plan are analogous to a test set. It is only via trying to fill in these holes that we are able to assess whether the model is appropriate or not. In one case we obtain 40 %CC and in the other 80 %. The difference is because one dataset corresponds to a fairly random mix of sexes sitting next to each other, and in the other there is considerable clustering. Rather than trying to increase our %CC to be as high as possible, we may be more interested in testing a hypothesis, is there an underlying pattern in where people sit? We could be studying cultural phenomena, looking at social patterns in different societies or countries or types of meetings. The value of the %CC for the test set can tell us something about the society we are studying, and so help us develop a hypothesis. A low %CC is not necessarily a bad result, and simply tells us about whether the underlying hypothesis is suitable for the data or not, which could be valuable information.

8.3.2 Test and Training Sets

The overall dataset is called an autopredictive dataset, consisting of I samples, but can be split into the following:

1. a training set consisting of I_{train} samples upon which a model is built;
2. a test set consisting of I_{test} samples which tests the quality of the model and are predicted by the training set model.

(a)

(b)

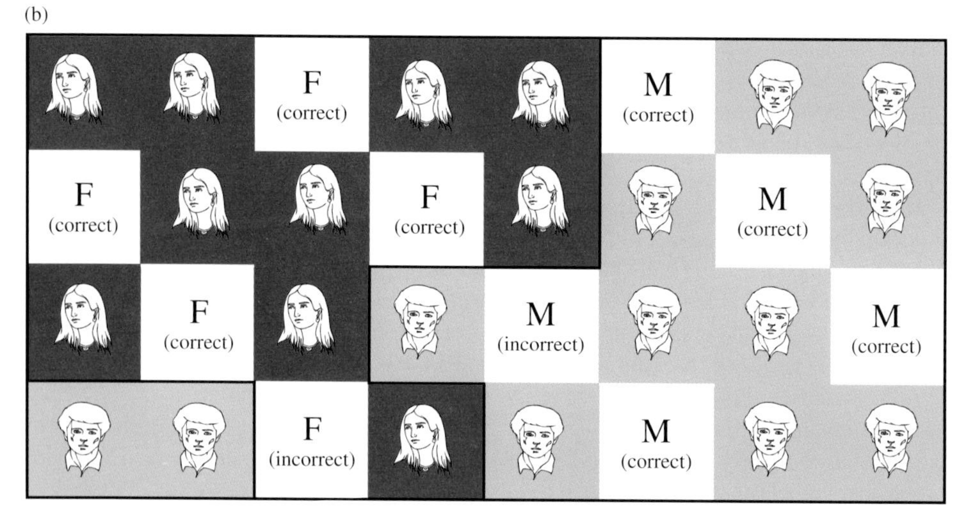

Figure 8.2 (a) A different male/female seating plan to Figure 8.1. (b) Predictions of missing people using the rules of Figure 8.1

In the above, $I_{train} + I_{test} = I$, as illustrated in Figure 8.3. There is no generally accepted rule as to what proportion of samples is assigned to the test set, but a good guide is that 1/3 of the samples are test set samples and 2/3 are training set samples, so that $I_{train} = 2I_{test}$, and although this is not a hard and fast rule, we will adopt this in this book. Some authors call the test set a validation set, and there are no agreed rules but we will use the terminology test set in this text.

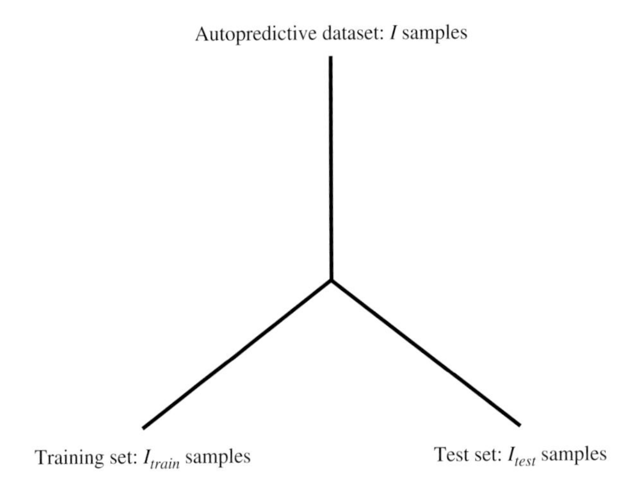

Figure 8.3 Split of a dataset into training and test sets

The training set is used to form a model between the experimental data and the classifier; for example, if we use QDA, the centres and variances of each class are computed using the 2/3 of samples in the training set. The classification of each test set sample is then predicted as follows:

- For statistically based one class models, if a sample falls within a given specified confidence limit for a class obtained from the training set, it is assigned to that class (e.g. using the D statistic at 95 % confidence).
- For two class EDC, LDA and QDA, the centres and where appropriate, variances, of the training sets are computed and the test set samples assigned to whichever centre they are closest to using an appropriate distance measure.
- For LVQ the codebooks are obtained for each class in the training set, and then test set samples are assigned according to which codebook they are nearest to.
- For SVM, an unknown is assigned to a class if it is within the boundary of that class established using just training set samples, for both one class and two (or multiclass) models.
- For PLS-DA the value of $\hat{c} = \boldsymbol{xb}$ is calculated for an unknown, ideally using the data scaling obtained from the training set.

Note that in Chapters 5 to 7 we visualize the class boundaries. When doing this, in order to place all data on the same picture, we have to use the autopredictive data scaling even for test and training sets, because the training set is likely to have a slightly different mean and standard deviation to the overall dataset, dependent on which samples are selected, which would mean that the positions in the data space would be distorted if we superimposed a model obtained, for example, by centring a training set, on the full dataset, leading to distortions in the boundaries and positions of the samples. For this reason when we visualize classification boundaries we use the data scaling methods of flowcharts 1A, 1C, 1E and 1G, but when we produce graphical or tabular data without corresponding boundary graphs we use flowcharts 1B, 1D, 1F and 1H as appropriate (Chapter 4) unless otherwise stated. In practice this difference in scaling usually makes very little difference unless there are strong outliers in the data, and other key decisions like the choice of samples to be included in the training set, for example, are often as important, but to be strictly correct the test set should not influence the scaling of the training set. Where it helps, the method of scaling used to obtain results is indicated with a flowchart symbol in the relevant figure or table.

An additional complication arises when there are unequal numbers of samples in each class. We will discuss further developing the models when class sizes are unequal in Chapter 10, but one approach is to use equal sample sizes for each class in the training set, so that $I_{trainA} = I_{trainB}$, if there are two classes (with a similar approach if there are more classes in the data) but the numbers of samples in the test set for each class will then be different. For example, in Case Study 8a (mouse urine – diet study), there are 59 samples in class A and 60 in class B; we use 39 samples from each class in the training set and 20 samples from class A and 21 from class B in the test set; for Case Study 3 (polymers – 2 types) there are 92 samples in class A and 201 in class B, and we use 61 samples from both classes for the training set, but 31 for class A and 140 for class B in the test set. This approach is effective so long as the size of the training set is not too small. For one class classifiers, in contrast, it is of course irrelevant how many samples are in the 'out group', and the training set will consist just of 2/3 of the 'in group' samples.

The %CC can be estimated for:

1. Autoprediction, i.e. forming a model on all I samples.
2. Training set, i.e. forming a model on just the I_{train} samples.
3. Test set, i.e. forming a model on the training set and using it to predict the I_{test} test set samples.

As we will see below, the test and training set %CCs are often best expressed as the average of several iterations which we will discuss below.

8.3.3 Predictions

In order to illustrate the difference between the three types of prediction we show the results of using two class LDA (and PLS-DA) and QDA models on Case Studies 1 (forensic), 3 (polymers), 8a (mouse urine – diet study), 9 (NMR spectroscopy) and 11 (null) in Table 8.5 and Figure 8.4 for 2 PC projections (using 2 PLS-DA components where appropriate for which the models are the same as for LDA) as illustrative purposes. Autopredictive boundaries have been illustrated in Chapter 5. These are illustrated for one random split of samples into test and training sets for which there are equal representatives of each class; we will discuss in Section 8.4 the differences in the models as different numbers of iterations are performed. Each sample is represented as either a training or test set sample, and so appears in either the left-hand (training set) or right-hand (test set) diagram.

Table 8.5 *Contingency tables and %CCs for Case Studies 1, 3, 8a, 9 and 11, using both LDA/PLS-DA and QDA for training and test set models using 2 PCs for the input to the classifier and a single test/training split*

		LDA and PLS-DA				QDA		
Case Study 1 (forensic)								
Autoprediction		Class A	Class B	Overall		Class A	Class B	Overall
	Class A	97.83	6.12	**95.79**	Class A	97.83	6.12	**95.79**
	Class B	2.17	93.88		Class B	2.17	93.88	
Training set								
	Class A	100	3.33	**98.34**	Class A	100	3.33	**98.34**
	Class B	0	96.67		Class B	0	96.67	
Test set								
	Class A	93.75	10.53	**91.43**	Class A	93.75	10.53	**91.43**
	Class B	6.25	89.47		Class B	6.25	89.47	
Case Study 3 (polymers – 2 types)								
		LDA and PLS-DA				QDA		
Autoprediction		Class A	Class B	Overall		Class A	Class B	Overall
	Class A	100	19.40	**86.69**	Class A	95.65	3.98	**95.90**
	Class B	0	80.60		Class B	4.35	96.02	
Training set								
	Class A	100	16.39	**91.80**	Class A	98.36	4.92	**96.72**
	Class B	0	83.61		Class B	1.64	95.08	
Test set								
	Class A	100	20.71	**83.04**	Class A	93.55	5	**94.74**
	Class B	0	79.29		Class B	6.45	95	
Case Study 8a (mouse urine – diet study)								
		LDA and PLS-DA				QDA		
Autoprediction		Class A	Class B	Overall		Class A	Class B	Overall
	Class A	84.75	21.67	**81.51**	Class A	83.05	21.67	**80.67**
	Class B	15.25	78.33		Class B	16.95	78.33	
Training set								
	Class A	87.18	23.08	**82.05**	Class A	87.18	23.08	**82.05**
	Class B	12.82	76.92		Class B	12.82	76.92	
Test set								

(continued overleaf)

Table 8.5 (*continued*)

	Class A	80	23.81	**78.05**	Class A	80	23.81	**78.05**
	Class B	20	76.19		Class B	20	76.19	
Case Study 9 (NMR spectroscopy)								
		LDA and PLS-DA				QDA		
Autoprediction		Class A	Class B	Overall		Class A	Class B	Overall
	Class A	68.75	25	**71.88**	Class A	70.83	22.92	**73.96**
	Class B	31.25	75		Class B	29.17	77.08	
Training set								
	Class A	71.88	18.75	**76.56**	Class A	75	15.63	**79.68**
	Class B	28.12	81.25		Class B	25	84.37	
Test set								
	Class A	56.25	31.25	**62.50**	Class A	62.5	31.25	**65.62**
	Class B	43.75	68.75		Class B	37.5	68.75	
Case Study 11 (null)								
		LDA and PLS-DA				QDA		
Autoprediction		Class A	Class B	Overall		Class A	Class B	Overall
	Class A	53	50.5	**51.25**	Class A	71	63	**54.00**
	Class B	47	49.5		Class B	29	37	
Training set								
	Class A	54.48	44.78	**54.85**	Class A	76.87	70.15	**53.36**
	Class B	45.52	55.22		Class B	23.13	29.85	
Test set								
	Class A	45.45	37.88	**53.78**	Class A	78.79	72.73	**53.03**
	Class B	54.55	62.12		Class B	21.21	27.27	

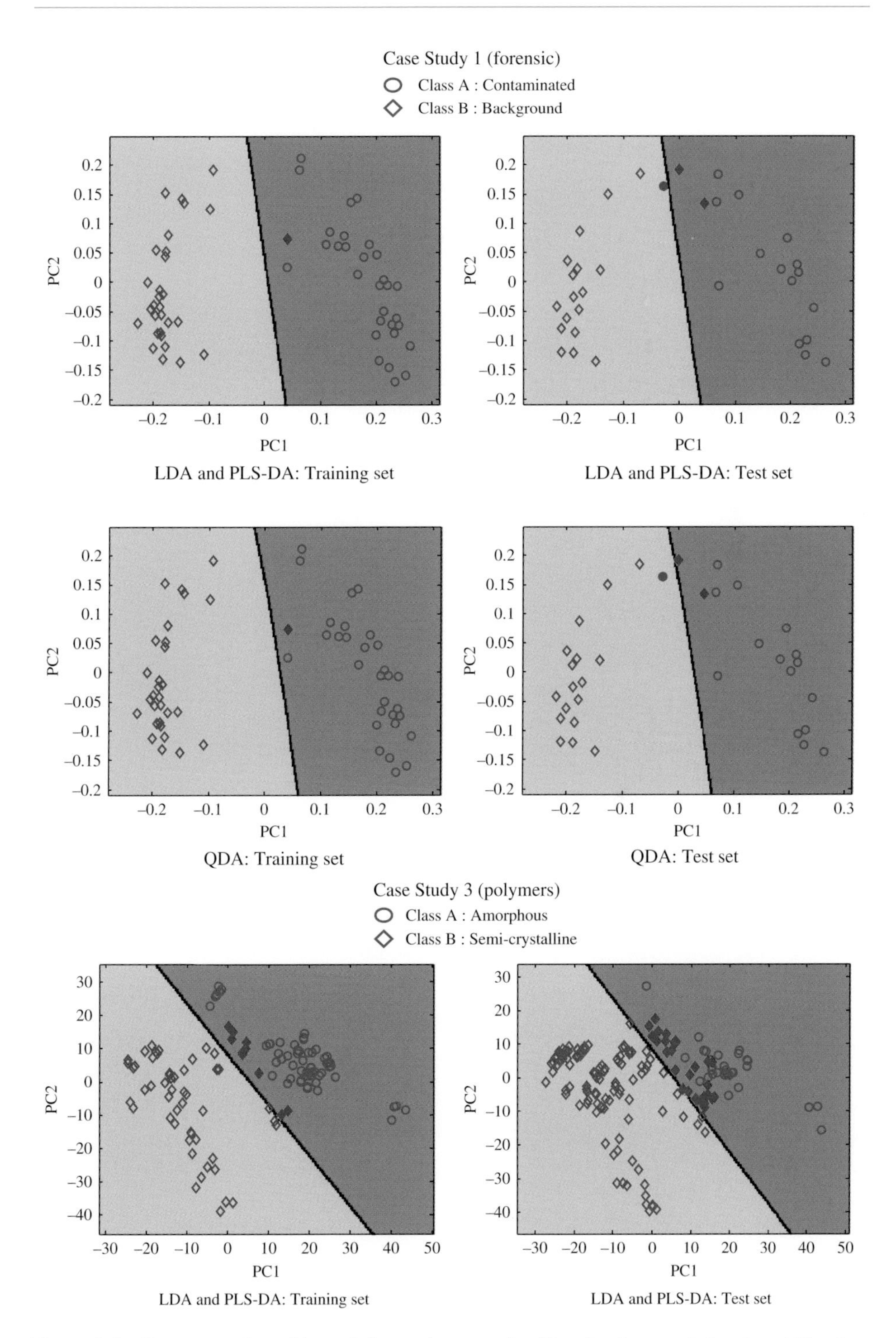

Figure 8.4 Representation of the training and test set classifiers for datasets 3, 8a , 9 and 11 using the methods of Table 8.5. Misclassified samples are indicated by filled symbols

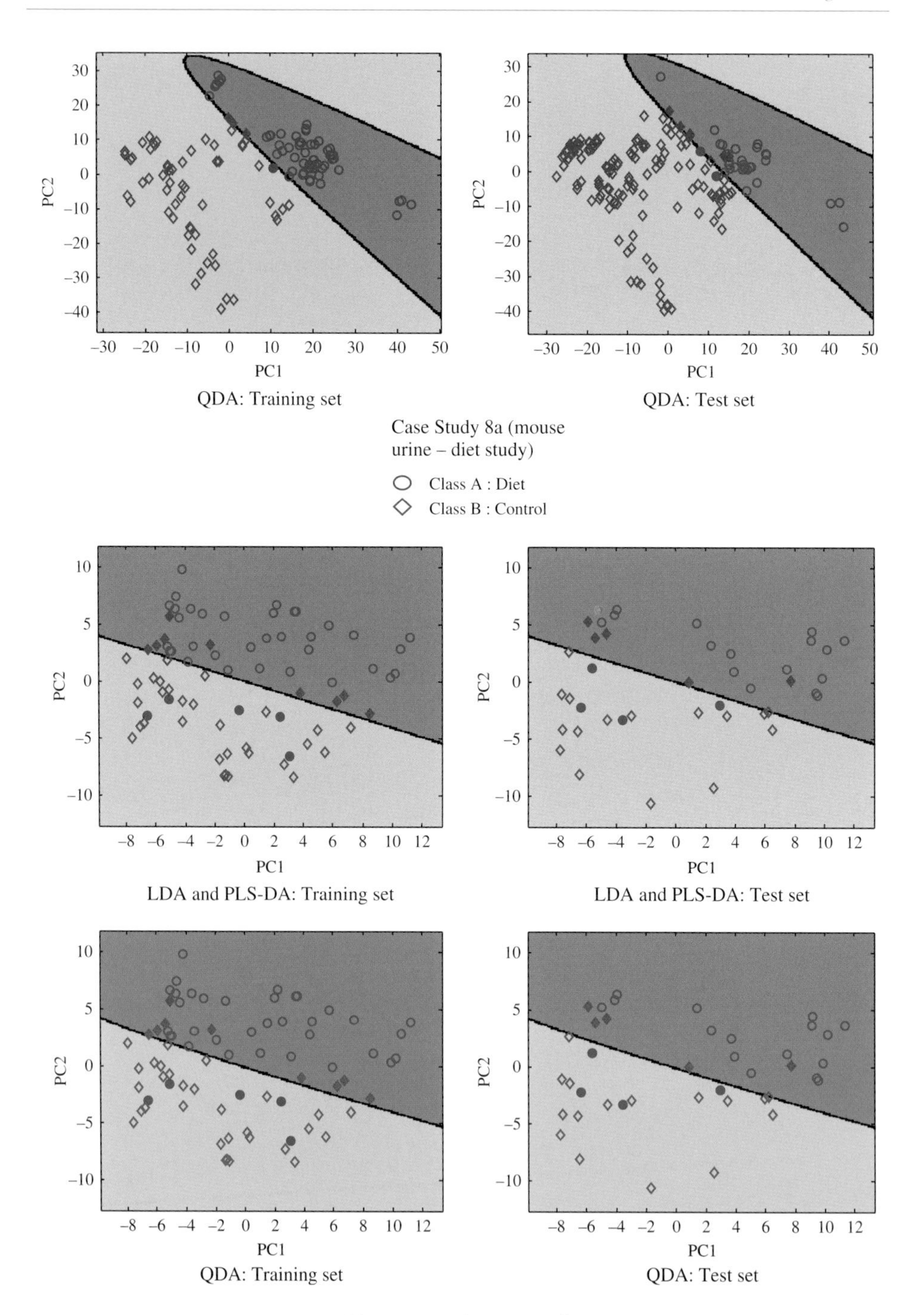

Figure 8.4 (continued)

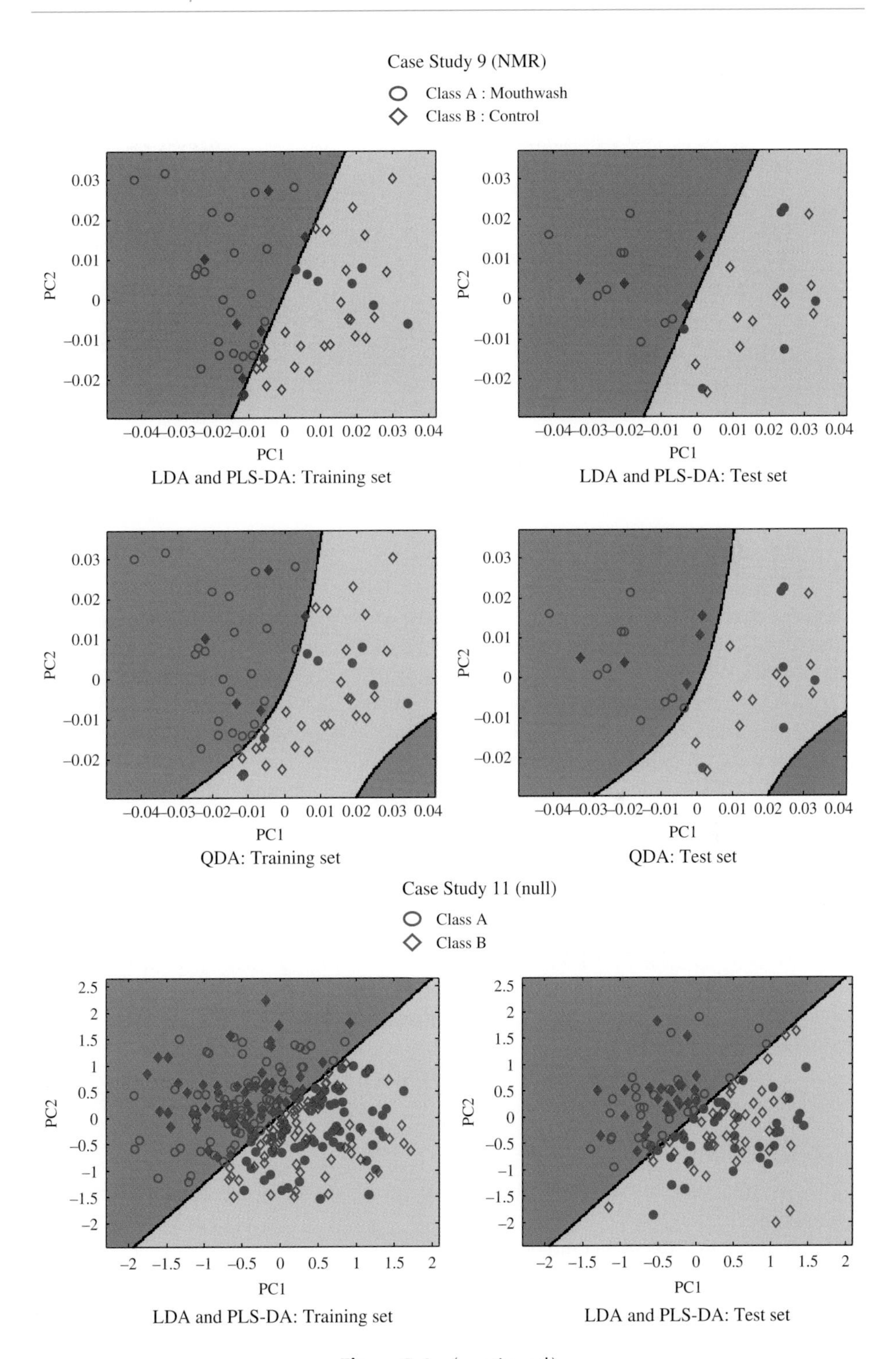

Figure 8.4 (continued)

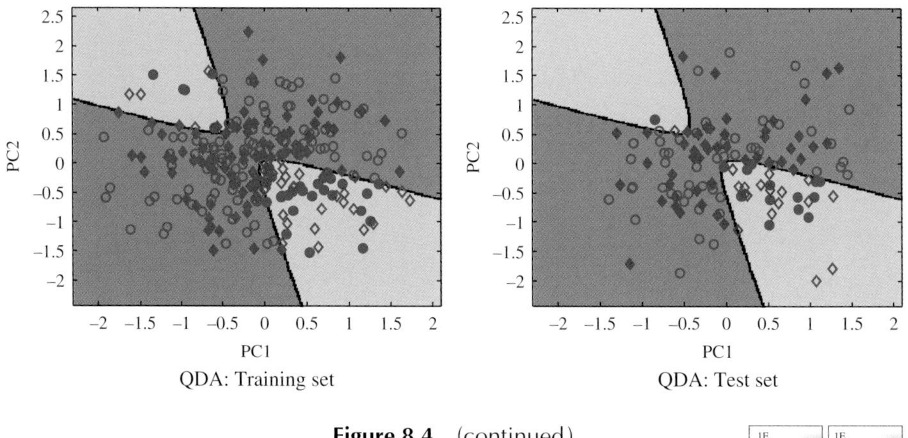

Figure 8.4 (continued)

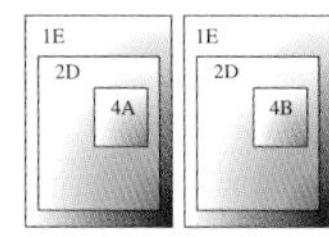

From the tables we can see as follows:

- Autopredictive and training set %CCs are on the whole higher than test set %CCs. This is clear for Case Studies 8a and 9. For Case Study 11 (null) this is not so clear, because almost all methods give a %CC close to 50 % as anticipated; however, as we will see in Section 8.4.2 the exact value is dependent on which samples are included in the training set, and so small differences of a few percent are not important.
- There is often an imbalance as to which class is predicted better; this is very clear from QDA on the Case Study 11 (null) – see the autopredictive results. However the overall %CC is close to that anticipated.
- The overall test set %CC can be used as an indication of how well it is possible to separate classes; one close to 50 % suggests there is no particular relationship between the data and class membership. The higher it is, the more confident we can be in the underlying hypothesis that the classes are separable.
- Surprisingly, for Case Study 3 it appears that the training set %CC is much higher than the autopredictive %CC, especially for LDA/PLS-DA. This is a consequence, in part, that the two original classes are very different in size. In the training set an equal number of representatives of each class are used. Since class A is the smallest class but the one that is easiest to model, it has a larger influence on the overall %CC than in the autopredictive and test set results.

The graphs should be compared to the figures, as follows:

- The test and training set boundaries are identical, as these are obtained using the training set.
- The numbers in Table 8.5 could be ascertained from close inspection of Figure 8.4. For example, for QDA and the training set for Case Study 3 (polymers) we see that only 1.64 % of class A is misclassified: there are 61 training set samples for this class (out of 92 in total), and so this represents 1 sample. There are in fact 2 misclassified samples in class A in the test set (note that the symbols are overlapping), or 2/31 or 6.45 %. For Case

Study 8a (mouse urine – diet study), using LDA / PLS-DA we see that in the test set, 4 out of 20 samples are misclassified for class A, as can be verified by inspecting the corresponding graph, or 20 %.

- For Case Study 1 (forensic), it is clear that the test and training set %CCs are critically dependent on which of the borderline samples are chosen for the test and training sets.

8.3.4 Increasing the Number of Variables for the Classifier

For 2 PC models there is often not much challenge and usually many more than two variables are necessary to describe a dataset adequately. We illustrate most of the methods graphically using 2 PCs but do not necessarily advocate that this is the most appropriate choice of number of significant components. We will be discussing the optimum choice of variables in more detail below, which should always be a consideration prior to pattern recognition. In Section 4.3.1 we discussed the use of cross-validation and the bootstrap for determining how many PCs should be retained; however, for pattern recognition purposes it is not always necessary to fully model a dataset. Most pattern recognition problems involve describing samples using very many variables, sometimes hundreds or thousands of variables. Instead of the boundaries being lines or curves, they become hyperplanes or hypersurfaces, if represented in many dimensions, and an important aim of classification algorithms is to search for the optimal hyperplanes or hypersurfaces dividing multidimensional variable space into regions. The same principles as have already been discussed are used to develop models and predictions and so we use the two variable examples for illustrative purposes.

However we cannot visualize boundaries in more than 3 dimensions and there are problems with representing multidimensional boundaries in 2 dimensions. This problem is illustrated in Figure 8.5 where a boundary represented in 3 dimensions cannot easily be projected onto 2 dimensions; equivalently a boundary in 10 or 100 dimensions cannot be

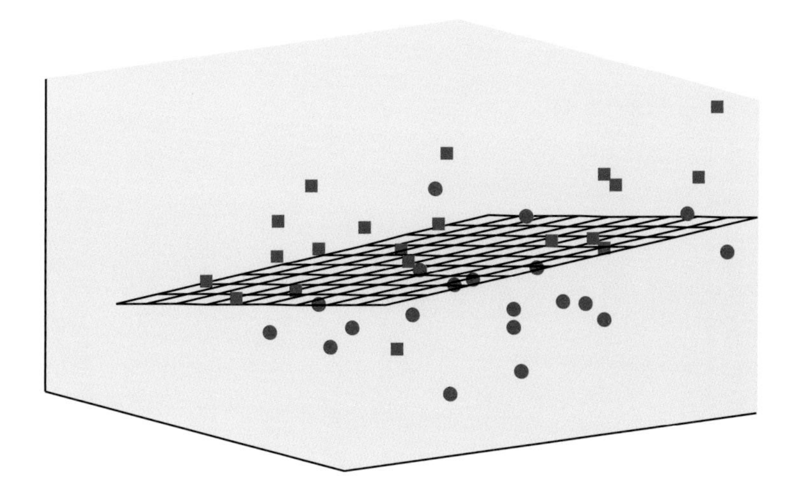

Figure 8.5 A boundary in three dimensions which cannot effectively be projected into two dimensions

Table 8.6 *Contingency tables and %CCs for Case Studies 1, 3, 8a, 9 and 11, using both PLS-DA (2 components) and QDA for training and test set models using 10 PCs for the input to the classifier and a single test training split*

		PLS-DA				QDA		
Case Study 1 (forensic)								
Autoprediction		Class A	Class B	Overall		Class A	Class B	Overall
	Class A	97.83	6.12	**95.79**	Class A	97.83	2.04	**97.90**
	Class B	2.17	93.88		Class B	2.17	97.96	
Training set								
	Class A	100	3.33	**98.34**	Class A	100	0	**100**
	Class B	0	96.67		Class B	0	100	
Test set								
	Class A	93.75	10.53	**91.43**	Class A	87.5	5.26	**91.43**
	Class B	6.25	89.47		Class B	12.5	94.74	
Case Study 3 (polymers – 2 types)		PLS-DA				QDA		
Autoprediction		Class A	Class B	Overall		Class A	Class B	Overall
	Class A	100	2.99	**97.95**	Class A	100	0	**100**
	Class B	0	97.01		Class B	0	100	
Training set								
	Class A	100	8.20	**95.90**	Class A	100	0	**100**
	Class B	0	91.80		Class B	0	100	
Test set								
	Class A	100	11.43	**90.64**	Class A	100	0	**100**
	Class B	0	88.57		Class B	0	100	
Case Study 8a (mouse urine – diet study)		PLS-DA				QDA		
Autoprediction		Class A	Class B	Overall		Class A	Class B	Overall
	Class A	84.75	6.67	**89.08**	Class A	89.83	11.67	**89.07**
	Class B	15.25	93.33		Class B	10.17	88.33	

Training set								
	Class A	87.18	10.26	**88.46**	Class A	97.44	10.26	**93.59**
	Class B	12.82	89.74		Class B	2.56	89.74	
Test set								
	Class A	85	4.77	**90.24**	Class A	90	19.05	**85.36**
	Class B	15	95.23		Class B	10	80.95	
Case Study 9 (NMR spectroscopy)								
		PLS-DA				QDA		
Autoprediction		Class A	Class B	Overall		Class A	Class B	Overall
	Class A	100	6.25	**96.88**	Class A	100	4.17	**97.92**
	Class B	0	93.75		Class B	0	95.83	
Training set								
	Class A	100	6.25	**96.88**	Class A	96.88	0	**98.44**
	Class B	0	93.75		Class B	3.12	100	
Test set								
	Class A	87.5	12.5	**87.50**	Class A	93.75	12.5	**90.62**
	Class B	12.5	87.5		Class B	6.25	87.5	
Case Study 11 (null)								
		PLS-DA				QDA		
Autoprediction		Class A	Class B	Overall		Class A	Class B	Overall
	Class A	57	45	**56.00**	Class A	69	35.5	**66.75**
	Class B	43	55		Class B	31	64.5	
Training Set								
	Class A	55.22	44.78	**55.22**	Class A	57.46	24.63	**66.42**
	Class B	44.78	55.22		Class B	42.54	75.37	
Test Set								
	Class A	53.03	53.03	**50.00**	Class A	51.52	42.42	**54.55**
	Class B	46.97	46.97		Class B	48.48	57.58	

projected onto 2D space even though we can visualize the data, for example, by a scores plot in 2D. So we cannot use graphical 2 dimensional plots to represent multidimensional boundaries although there are other approaches for the representation of such results.

The results for autoprediction, training and test set predictions for five case studies are presented in Table 8.6, using both PLS-DA and QDA. In this case, instead of 2 PCs as input to the classifier we use the first 10 PCs (but for PLS-DA retain 2 PLS components) so the results now differ from those that would be obtained using LDA. Ways to select the number of variables input to the classifier are covered in Chapters 4 and 9. Note that this is entirely different from optimizing the classifier (e.g. determining the optimum number of PLS components as discussed in Sections 8.5 to 8.7). The results should be compared to those of Table 8.5. Note that we use 2 PLS components for PLS-DA but 10 PCs for data reduction. Since the number of PLS components is less than the number of PCs, PLS-DA is different to LDA for these models, and we will not discuss LDA for brevity:

- We see that in all cases the autopredictive %CC either stays the same, or in most cases increases, sometimes quite dramatically. For Case Study 9 (NMR spectroscopy) the increase in autopredictive ability is 25 % for PLS-DA compared to the 2 PC model. Using more PCs in autopredictive models will always give a better fit to the data and so result in a higher %CC unless already close to the maximum.
- For the test set results there are different trends. For Case Study 1 (forensic) we find no change in the PLS-DA model, suggesting that extra PCs do not add any further information, and very limited change to the QDA model. For Case Study 11 (null) the PLS-DA %CC actually reduces to 50 %: we will see later that this value varies according to the number of iterations (Sections 8.4). For the Case Study 9 (NMR spectroscopy), however, the increase is dramatic from 65.6 % to 90.6 %, suggesting that the extra PCs add a lot of useful information. The polymers are perfectly identified in the test set using QDA.

The samples that are correctly classified can be visualized for all types of models, plotted in the space of the first two PCs, as illustrated in Figure 8.6 for Case Study 8a (mouse urine – diet study) using a single PLS-DA model and 10 components for PCA but 2 for PLS. We see that the misclassified samples no longer fall along a clearly defined boundary as the boundary is in 10 dimensions, and so we cannot cleanly divide the graph into two regions.

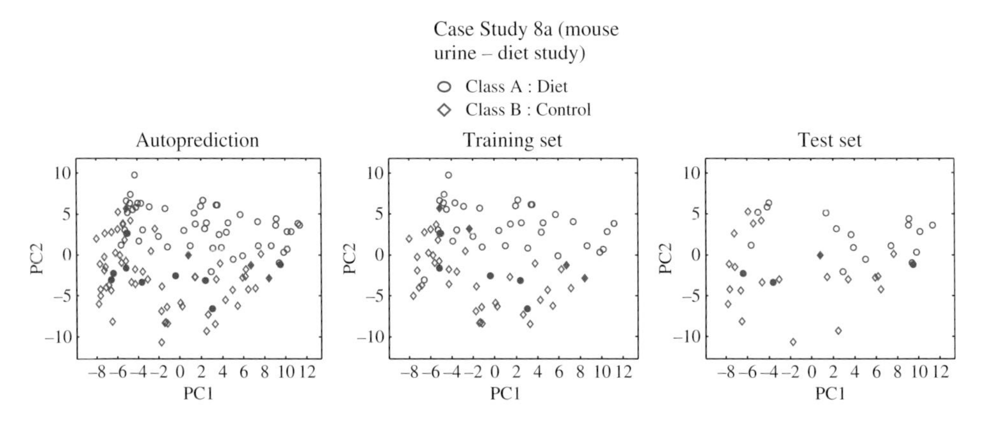

Figure 8.6 Autopredictive, training set and test set samples for Case Study 8a, with misclassified samples indicated by filled symbols, using 10 PCs and a single PLS-DA model formed using 2 PLS components

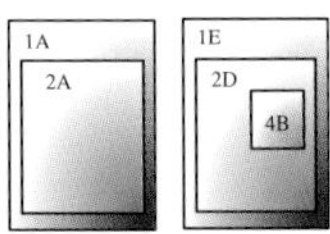

8.4 Iterative Approaches for Validation

An important consideration is that in the methods described above we choose a random set of samples to form the test set. This is the traditionally accepted method for validation of algorithms. However this approach has been around for more than 20 years, during which time, if we accept one of the formulations of Moore's law, that computing power (in all its aspects) doubles every 18 months per unit cost, then our PCs are around 1000 times more powerful than two decades ago, and so it is not necessary to be so limited. A difficulty is that the pioneering work in chemometrics in the late 1970s and early 1980s laid a conceptual framework that was the foundation for most commercial package software and has become established practice.

However there is no reason why we should not form more test sets. A single division of data into test and training sets is essentially arbitrary, and the %CC from this may not necessary be representative and is critically dependent on choice of samples in the training set. If there are outliers, the inclusion of an outlier in the training set could have a significant influence on the model; if the number of samples is small for any group then models could be unstable; finally, for complex boundary methods such as SVMs that tightly fit groups using non-linear models, the training set can be subject to overfitting. The way to overcome this limitation is to repeatedly reform the model by iteratively dividing the dataset into test and training sets many times over, using a random number generator to select those samples that are to be assigned to the training set.

8.4.1 Predictive Ability, Model Stability, Classification by Majority Vote and Cross Classification Rate

In order to interpret the results of iterative methods for validating models we need several new definitions. Consider a situation in which a test set is formed 100 times in an iterative fashion. In each iteration every sample, whether in the training or test set, is assigned to a specific class (A or B for two class classifiers, the 'in group' for one class classifiers and so on).

1. The %Predictive Ability (%PA) is the percentage of times a sample is assigned to its correct class. Consider the following:
   - A sample belongs to class A.
   - It is chosen 36 out of 100 times to be part of the test set, and 64 out of 100 times to be part of the training set.
   - In 13 out of 36 times it is chosen to be part of the test set, it is correctly assigned to class A, whereas in 55 out of 64 times in the training set.
   - The %PA is defined for each of the samples as the percentage of times it is correctly classified over all iterations, or:

$$100 \times (13/36) = 36.1\,\% \text{ for the test set}$$
$$100 \times (55/64) = 85.9\,\% \text{ for the training set}$$

The mean %PA can then be obtained for all samples, which is different to the %CC (see below), for both test and training sets.

2. The %Model Stability is a measure of how stable the classification of a sample is, and is defined by:

$$\%MS = 2|(\%PA - 50)|$$

Hence a sample with a %PA of 50 % has a model stability of 0 % as it is assigned half the time to class A and half the time to class B.

The relationship between PA and MS is represented numerically in Table 8.7:

- A sample with a high PA and MS is likely to be classified well and be far from the boundary.
- A sample with a low MS and consequently a PA around 50 % is likely to be close to the boundary and its classification may be in doubt.
- A sample with a low PA and high MS is consistently misclassified and may be an outlier or a sample that was mislabelled (e.g. a misdiagnosed patient) and will usually be far from the boundary.

Table 8.7 *Relationship between %predictive ability (PA) and %model stability (MA) for a sample that is known to be a member of class A*

% of time predicted as		%PA	%MS
Class A	Not class A		
100	0	100	100
75	25	75	50
50	50	50	0
25	75	25	50
0	100	0	100

The model stability for the sample discussed in 1 above is 28 % for the test set and 72 % for the training set (it is coincidence that these numbers add up to 100 % in this case). Whereas the %PA can only be calculated for known samples as part of a testing procedure, the %MS can be calculated for any unknown, and the lower it is, the more likely the sample is close to a boundary. Hence even for two class classifiers we can compute a statistic analogous to confidence in model predictions in a one class classifier. The mean model stability can also be calculated for all samples. In addition, the MS can be calculated separately for samples that are correctly and incorrectly classified or for specific classes separately. An advantage of this measure is that it can be calculated also for any unknown sample. On the whole, misclassified samples and samples near the border will have a low MS and this is indicative that confidence in predictions on the samples is quite low.

3. In this text the %CC is obtained via majority vote. That is, for a two class classifier, if over a number of iterations a sample is classified more frequently into class A it is assigned into that class (i.e. has a %PA > 50 %), otherwise into class B. If the class it is assigned to is the correct class then the sample is considered correctly classified, or otherwise it is a misclassification. Ties should be extremely rare but if so we recommend using a random assignment; for the purpose of computing %CC, the sample is half

assigned into each class. It will only happen if the sample was selected an even number of times and was equally assigned to each class in half the iterations: providing the dataset is large enough this is likely to be a very rare occurrence. Note that if there are several classes the assignment is into the class the sample is most often classified over all iterations, which may be less than 50 % of the time. It is important to note that because samples are randomly chosen to be included in test and training sets, the number of times each sample is selected in either category over all iterations may differ slightly. In addition when class sizes are unequal, samples from the larger class will be selected more frequently to be part of a test set than those of the smaller class, providing there are equal representatives of all modelled classes in the training set. However the %CC results in a single assignment for each sample, no matter how many or few times it is chosen.

4. For multiclass classifiers (Chapter 7) we can also calculate the cross class classification rate (%CR) as the percentage of times a sample is classified into each class. If, for example, a sample *i* (which happens to be in group A) is selected 30 times in a training set, and it is assigned 20 times to its own group (A), 6 times to group B, and 4 times to group C, then we define $\%CR_A = 67\ \% = 100 \times 20/30$, $\%CR_B = 20\ \% = 100 \times 6/30$ and $\%CR_C = 13\ \% = 100 \times 4/30$ for sample *i*. The %CR for a sample's own group is the same as its %PA and so the %PA for sample *i* is 67 % in this case. For a two class classifier it is only necessary to know the %PA as all misclassified samples from class A are assigned to class B and vice versa, but for multiclass situations there are more than two decisions about class membership.

These calculations of %PA, %MS and %CC are exemplified using a two class classifier in Table 8.8 for a small example dataset where there are:

1. 10 samples, 5 of which are members of class A and 5 of class B.
2. The model is formed 5 times.
3. Each time, 4 samples (2 from each class) are removed to be the test set and the remaining 6 samples are the training set.

Some observations can be made:

- The number of times each sample is picked to be in a test and training set may vary slightly.
- The %CC and average %PA are different measures, and do not necessarily provide the same results.
- In this small example there are risks of ties, that is, samples that are equally assigned to both classes and have a %PA of 0; however, this is because the number of iterations is very small and in practice we recommend many more iterations for a stable model.

The %PA and %MS for both training and test set samples for Case Studies 1, 3, 5, 8a and 11 for two class models both using 2PCs and 10 PCs are presented in Table 8.9 using the default preprocessing of Chapter 2 and default parameters for SVM and LVQ as described in Chapter 5. Note that 2 PLS components are employed for both types of models, and so PLS-DA is identical to LDA only for the 2 PC model. Note also that since it is no longer necessary to visualize the scores plots, the data scaling is slightly different to previously, resulting in small differences in classification ability to those in previous sections. We recommend a default of 100 test/training splits, and so the numbers in the table are averaged from 100 such iterations.

Table 8.8 *Example of calculation of %CC, %PA and %MS for a small dataset consisting of 10 samples, 5 originating from class A and 5 from class B. 5 iterations are performed and 4 out of 10 samples (2 from each class) are assigned in each iteration to the test set. Samples picked in iterations 1 to 5 are indicated together with their assignments. Incorrect assignments are indicated in italics, and the majority decision of the classifier for each sample in bold. The results of the assignments for each sample and each iteration are presented together with the statistics on the models built*

	Classification of samples in each iteration										Training set						Test set					
	Iteration 1		Iteration 2		Iteration 3		Iteration 4		Iteration 5		Train number	Assignment			%PA	%MS	Test number	Assignment			%PA	%MS
	Train	Test	Train	Test	Train	Test	Train	Test	Train	Test												
Class A												A	B					A	B			
1	A		A			A		A		*B*	2	2	0	**A**	100	100	3	2	1	**A**	67	33
2		A	A		A		A		*B*		4	3	1	**A**	75	50	1	1	0	**A**	100	100
3	A			A	A		A			A	3	3	0	**A**	100	100	2	2	0	**A**	100	100
4	*B*			*B*		*B*	A		A		3	2	1	**A**	67	33	2	0	2	***B***	0	100
5		*B*	*B*		*B*			*B*	A		3	1	2	***B***	33	33	2	0	2	***B***	0	100
Class B																						
6	*A*		B			B	*A*		*A*		4	3	1	***A***	25	50	1	0	1	**B**	100	100
7		B	B		B			B		*A*	2	0	2	**B**	100	100	3	1	2	**B**	67	33
8	B		B			B	B		B		4	0	4	**B**	100	100	1	0	1	**B**	100	100
9		B		B	B		B			B	2	0	2	**B**	100	100	3	0	2	**B**	67	33
10	B			*A*	B			*A*	B		3	0	3	**B**	100	100	2	2	0	***A***	0	100
											Mean PA and MS				80	77	Mean PA and MS				60	80
											Correct			8			Correct			7		
											Incorrect			2			Incorrect			3		
											%CC			80			%CC			60		

Table 8.9 *Predictive ability in the form %CC (average PA%, average MS%) for two class classifiers and Case Studies 1 (forensic), 3 (polymers), 8a (mouse urine – diet study) and 11 (null), using default parameters (3 codebooks per class for LVQ and an L1 SVM with $C = 1$ and $\sigma = 1 \times$ average standard deviation of the variables in the dataset are used). (a) 2 PCs; (b) 10 PCs using 2 PLS components for PLS-DA and 100 test/training set splits. Note that PLS-DA gives identical results to LDA using 2 PC models (apart from the effect of random splits into test and training sets)*

(a) 2 PCs

Case Study 1 (forensic)

	Training A	Training B	Training overall	Test A	Test B	Test overall
EDC	97.83,(97.86,99.93)	95.92,(95,98.17)	96.84,(96.39,99.02)	97.83,(97.83,100)	93.88,(94.56,98.64)	95.79,(96.14,99.3)
LDA/PLS-DA	97.83,(98.39,98.86)	93.88,(94.27,99.21)	95.79,(96.27,99.04)	97.83,(97.95,99.75)	93.88,(94.19,99.38)	95.79,(96.01,99.56)
QDA	97.83,(98.03,98.8)	93.88,(94.96,97.84)	95.79,(96.45,98.3)	97.83,(97.57,99.24)	93.88,(94.43,98.89)	95.79,(95.95,99.06)
LVQ	97.83,(98.39,98.86)	93.88,(94.27,99.21)	95.79,(96.27,99.04)	97.83,(97.95,99.75)	93.88,(94.19,99.38)	95.79,(96.01,99.56)
SVM	97.83,(98.6,97.53)	95.92,(95.99,97.66)	96.84,(97.25,97.6)	97.83,(97.08,97.76)	93.88,(94.09,98.73)	95.79,(95.54,98.26)

Case Study 3 (polymers)

	Training A	Training B	Training overall	Test A	Test B	Test overall
EDC	97.83,(95.75,93.17)	82.09,(83.07,94.47)	87.03,(87.05,94.06)	93.48,(93.39,93.06)	82.59,(82.64,94.46)	86.01,(86.01,94.02)
LDA/PLS-DA	100,(99.75,99.5)	82.09,(83.19,96.16)	87.71,(88.39,97.21)	100,(99.55,99.1)	82.09,(82.73,96.63)	87.71,(88.01,97.41)
QDA	96.74,(96.7,97.47)	96.02,(95.6,94.78)	96.25,(95.94,95.62)	96.74,(95.7,96.28)	95.02,(94.96,94.52)	95.56,(95.19,95.07)
LVQ	100,(99.75,99.5)	82.09,(83.19,96.16)	87.71,(88.39,97.21)	100,(99.55,99.1)	82.09,(82.73,96.63)	87.71,(88.01,97.41)
SVM	98.91,(97.98,96.06)	97.01,(95.27,93.67)	97.61,(96.12,94.42)	96.74,(97.21,94.83)	96.02,(94.14,92.27)	96.25,(95.1,93.07)

Case Study 5 (sweat)

	Training A	Training B	Training overall	Test A	Test B	Test overall
EDC	29.28,(32.65,72.39)	78.38,(72.58,75.5)	55.99,(54.37,74.08)	25.66,(31.13,72.71)	77.37,(71.71,76.17)	53.79,(53.2,74.59)
LDA/PLS-DA	44.1,(43.55,60.92)	60.3,(61.22,60.51)	52.91,(53.16,60.7)	42.77,(41.69,60.17)	60.71,(60.25,60.69)	52.53,(51.79,60.45)
QDA	95.78,(92.54,87.76)	7.68,(11.74,82.29)	47.86,(48.59,84.78)	95.78,(92.59,87.94)	8.28,(11.67,82.16)	48.19,(48.57,84.8)
LVQ	41.33,(41.8,54.62)	77.68,(70.45,59.55)	61.1,(57.38,57.3)	41.69,(40.59,55.16)	75.56,(70.19,59.25)	60.11,(56.69,57.39)
SVM	44.58,(44.42,84.31)	75.66,(74.57,86.18)	61.48,(60.82,85.33)	41.33,(41.26,84.22)	73.74,(73.66,85.74)	58.96,(58.89,85.05)

(continued overleaf)

Table 8.9 (continued)

Case Study 8a (mouse urine – diet study)

	Training A	Training B	Training overall	Test A	Test B	Test overall
EDC	79.66,(78.99,90.65)	76.67,(77.33,84.68)	78.15,(78.16,87.64)	77.97,(77.07,90.84)	76.67,(75.58,85.07)	77.31,(76.32,87.93)
LDA/PLS-DA	83.05,(82.33,91.88)	77.5,(77.2,87.09)	80.25,(79.75,89.47)	83.05,(80.8,89.79)	75,(74.59,88.07)	78.99,(77.67,88.92)
QDA	83.05,(82.07,91.96)	76.67,(77.87,85.95)	79.83,(79.95,88.93)	81.36,(80.48,89.23)	75.83,(74.88,86.14)	78.57,(77.66,87.67)
LVQ	83.05,(82.33,91.88)	77.5,(77.2,87.09)	80.25,(79.75,89.47)	83.05,(80.8,89.79)	75,(74.59,88.07)	78.99,(77.67,88.92)
SVM	77.97,(78.07,84.01)	90,(88.1,82.29)	84.03,(83.13,83.14)	74.58,(71.56,79.44)	86.67,(81.48,77.62)	80.67,(76.56,78.52)

Case Study 11 (null)

	Training A	Training B	Training overall	Test A	Test B	Test overall
EDC	59,(53.33,23.84)	57,(52.98,23.44)	58,(53.16,23.64)	49.5,(49.6,22.45)	53.5,(50.35,22.88)	51.5,(49.98,22.67)
LDA/PLS-DA	59,(53.36,23.7)	56.75,(52.99,23.36)	57.88,(53.17,23.53)	49.25,(49.5,22.45)	53.25,(50.33,23.07)	51.25,(49.91,22.76)
QDA	69.5,(57.22,22.65)	53.75,(50.97,20.08)	61.63,(54.1,21.37)	55.75,(52.38,20.7)	44,(48.01,21.15)	49.88,(50.2,20.92)
LVQ	59,(53.36,23.7)	56.75,(52.99,23.36)	57.88,(53.17,23.53)	49.25,(49.5,22.45)	53.25,(50.33,23.07)	51.25,(49.91,22.76)
SVM	57,(52.18,12.71)	86,(57.15,16.27)	71.5,(54.66,14.49)	32.25,(46.52,15.61)	63,(53.12,15.59)	47.63,(49.82,15.6)

(b) 10 PCs

Case Study 1 (forensic)

	Training A	Training B	Training overall	Test A	Test B	Test overall
EDC	97.83,(97.83,100)	95.92,(95.12,98.41)	96.84,(96.43,99.18)	97.83,(97.83,100)	93.88,(94.73,98.3)	95.79,(96.23,99.12)
LDA	97.83,(98.62,98.42)	95.92,(96.76,97.42)	96.84,(97.66,97.9)	97.83,(97.83,100)	93.88,(93.03,97.85)	95.79,(95.35,98.89)
QDA	100,(99.34,98.68)	100,(99.82,99.64)	100,(99.59,99.17)	93.48,(94.87,96.24)	93.88,(93.77,96.88)	93.68,(94.3,96.57)
PLS-DA	97.83,(98.42,98.81)	93.88,(94.7,98.35)	95.79,(96.5,98.58)	97.83,(98.02,99.62)	93.88,(93.88,99.55)	95.79,(95.88,99.58)
LVQ	97.83,(98.52,98.23)	97.96,(96.81,96.79)	97.89,(97.64,97.48)	97.83,(97.35,98.78)	93.88,(93.8,96.79)	95.79,(95.52,97.76)
SVM	100,(100,100)	100,(100,100)	100,(100,100)	97.83,(96.48,96.8)	91.84,(91.83,97.5)	94.74,(94.08,97.16)

Case Study 3 (polymers)

	Training A	Training B	Training overall	Test A	Test B	Test overall
EDC	98.91,(96.16,93.64)	88.06,(87.3,94.17)	91.47,(90.08,94)	94.57,(94.42,92.98)	86.07,(85.97,93.51)	88.74,(88.62,93.34)
LDA	100,(100,100)	100,(99.87,99.73)	100,(99.91,99.82)	100,(100,100)	100,(99.62,99.25)	100,(99.74,99.48)
QDA	100,(100,100)	100,(100,100)	100,(100,100)	100,(100,100)	100,(99.86,99.73)	100,(99.91,99.81)

	Training A	Training B	Training overall	Test A	Test B	Test overall
PLS-DA	100,(99.92,99.84)	91.79,(91.82,91.68)	94.37,(94.36,94.24)	100,(99.57,99.14)	89.55,(89.98,91.41)	92.83,(92.99,93.83)
LVQ	100,(99.02,98.04)	100,(99.39,98.77)	100,(99.27,98.54)	100,(98.01,96.02)	99.5,(98.88,97.97)	99.66,(98.61,97.35)
SVM	100,(100,100)	100,(100,100)	100,(100,100)	100,(100,100)	100,(99.97,99.95)	100,(99.98,99.96)

Case Study 5 (sweat)

	Training A	Training B	Training overall	Test A	Test B	Test overall
EDC	66.51,(65.68,88.1)	71.31,(72.18,88.04)	69.12,(69.21,88.06)	64.7,(63.85,87.99)	70.81,(70.81,88.33)	68.02,(67.64,88.17)
LDA	71.33,(70.45,87.4)	68.38,(68.93,86.83)	69.73,(69.63,87.09)	69.16,(68.76,87.83)	67.58,(67.64,88.55)	68.3,(68.15,88.22)
QDA	66.51,(66.15,77.03)	73.74,(72,79.78)	70.44,(69.33,78.53)	63.13,(63.46,77.95)	71.62,(69.59,79.29)	67.75,(66.79,78.68)
PLS-DA	69.64,(68.74,91.4)	70.3,(70.25,91.04)	70,(69.56,91.2)	67.47,(67.1,91.12)	69.09,(69.04,91.76)	68.35,(68.16,91.47)
LVQ	62.41,(58.44,57.34)	75.76,(71.61,64.67)	69.67,(65.6,61.33)	60.12,(55.96,56.73)	73.43,(69.97,64.16)	67.36,(63.58,60.77)
SVM	88.67,(85.71,86.11)	85.96,(84.39,83.85)	87.2,(84.99,84.88)	68.19,(67.17,73.69)	69.19,(66.45,76.09)	68.74,(66.78,75)

Case Study 8a (mouse urine – diet study)

	Training A	Training B	Training overall	Test A	Test B	Test overall
EDC	81.36,(82.2,91.71)	90,(88.29,94.09)	85.71,(85.27,92.91)	77.97,(78.54,92.49)	86.67,(85.06,91.34)	82.35,(81.83,91.91)
LDA	88.14,(86.83,86.55)	95,(92.22,90.09)	91.6,(89.55,88.33)	77.97,(79.58,89.73)	90,(87.78,86.5)	84.03,(83.72,88.1)
QDA	91.53,(92.23,88.8)	95,(91.52,87.62)	93.28,(91.88,88.21)	83.05,(82.11,87.69)	85,(82.25,83.79)	84.03,(82.18,85.72)
PLS-DA	88.14,(86.19,90.38)	95,(92.31,90.88)	91.6,(89.27,90.63)	83.05,(80.27,91.19)	89.17,(87.31,86.51)	86.13,(83.82,88.83)
LVQ	89.83,(85.78,83.12)	98.33,(93.25,88.09)	94.12,(89.54,85.63)	76.27,(75.58,83.61)	92.5,(86.16,79.97)	84.45,(80.91,81.77)
SVM	100,(99.92,99.84)	100,(99.88,99.76)	100,(99.9,99.8)	71.19,(69.36,79.32)	90.83,(87.87,85.35)	81.09,(78.69,82.36)

Case Study 11 (null)

	Training A	Training B	Training overall	Test A	Test B	Test overall
EDC	60.5,(57.98,47.13)	59.75,(57,46.13)	60.12,(57.49,46.63)	51.25,(50.57,40.35)	46,(49.73,41.41)	48.63,(50.15,40.88)
LDA	60,(57.91,46.59)	61.25,(57.3,45.86)	60.63,(57.61,46.23)	51.75,(50.6,39.88)	47.25,(49.63,41.75)	49.5,(50.12,40.81)
QDA	77,(64.96,42.6)	81.5,(70.45,50.01)	79.25,(67.7,46.3)	45,(48.47,33.04)	54.75,(51.96,37.28)	49.88,(50.22,35.16)
PLS-DA	60.25,(57.86,46.56)	61.25,(57.33,45.85)	60.75,(57.6,46.2)	51.75,(50.54,39.94)	47,(49.67,41.85)	49.38,(50.11,40.9)
LVQ	74.25,(59.52,27.31)	70.5,(57.42,24.99)	72.38,(58.47,26.15)	55.25,(51.6,21.8)	48.5,(49.21,22.71)	51.88,(50.4,22.26)
SVM	100,(99.98,99.96)	100,(99.97,99.94)	100,(99.97,99.95)	53,(51.06,24.54)	48.25,(48.85,24.19)	50.63,(49.96,24.37)

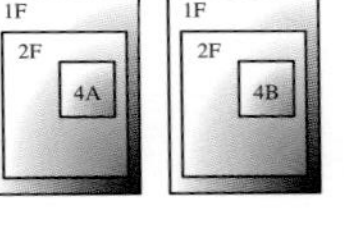

There are many deductions that can be made from this table:

- We can see that the MS is an important parameter. For Case Study 11 (null) there is a low model stability for all 2 PC models of around 15–25 % as anticipated. This suggests that the classification of samples is highly dependent on which are chosen for the test and training sets. It also implies that a single division into test and training sets is insufficient to provide an adequate view of the data. On the whole, model stability is greatest for Case Study 1 (forensic) and least for Case Study 11 (null) and relates to how easy it is to find a stable boundary and how influenced this is by choice of samples.
- SVM tend to find very complex training set boundaries (dependent on the parameters chosen) and when there are opportunities results in very high %PA and %MS in the training set, an example being the 10 PC SVM model for Case Study 11 (null) where the training set %CC, %PA and %MS are extremely high: basically every point is modelled correctly, however the %MS for the test set is low and the %CC and %PA around 50 % as anticipated, suggesting that whilst the training set model has overfit the data, using the test set is quite safe as an assessment of the model. This emphasizes the importance of proper validation.
- Some methods model one group very much better than the other group, examples being LVQ and SVM on the 2 PC model of Case Study 11 (null), but this comes at the cost of a %PA of around 50 % and low %MS which is a warning sign that the classification is finely balanced.
- The %PA provides additional 'fine structure' to the %CC. An example is the QDA 10 PC model on Case Study 3 (polymers) for which all %CCs are 100 % but the test set %PA is slightly lower. A perfect model would have 100 % CC and PA which is not quite achieved.
- For Case Study 11 (null) the test set %CC is very close to 50 % for all test set models. This contrasts a little to Table 8.5 and Table 8.6 where the difference is somewhat larger in most cases. This is because models on a single randomly generated dataset are rather unstable and as such each can give different results, but when 100 models are performed the average approximates to 50 %.
- For Case Study 5 (sweat) we see now that we can discriminate between the two classes, using 10 PCs, based on the test set %CC, even though the PC scores plots looks as if this is not easy. This is because several PCs are needed for an adequate model and visualization in 2 PCs is insufficient. It suggests that gender is not the most important factor for distinction between the two groups, but nevertheless is there, if more components are included in the model.

These results show that if we iteratively form test and training sets many times we can obtain additional information about the model stability, and also need several such splits in order to provide a robust model.

Furthermore, we can divide samples into those that are correctly classified and those that are misclassified and separate out the contributions to %MS as in Table 8.10 for 2 PC models and Case Studies 1, 3 and 8a. We can see that the misclassified samples on average have lower a MS than the correctly classified samples, as anticipated, because most (apart from some mislabelled samples or outliers), are close to boundaries: this technique can also be used for unknowns as we can form many different training set models on these samples if required to see the assignment of test set samples are unstable, and provides a measure of confidence in our classification ability of a sample.

Table 8.10 *Average model stability for overall, correctly and incorrectly classified samples for Case Studies 1, 3, and 8a, in the form %MS overall (%MS correctly classified, %MS incorrectly classified), using the 2 PC models of Table 8.9*

	Training A	Training B	Training overall	Test A	Test B	Test overall
Case Study 1 (forensic)						
EDC	99.93,(100,96.92)	98.17,(98.09,100)	99.02,(99.03,98.97)	100,(100,100)	98.64,(100,77.78)	99.3,(100,83.33)
LDA/PLS-DA	98.86,(100,47.69)	99.21,(100,87.06)	99.04,(100,77.22)	99.75,(100,88.57)	99.38,(100,89.9)	99.56,(100,89.57)
QDA	98.8,(99.6,63.08)	97.84,(100,64.65)	98.3,(99.8,64.26)	99.24,(99.35,94.29)	98.89,(100,81.82)	99.06,(99.68,84.94)
LVQ	97.53,(99.52,7.69)	97.66,(98.86,69.55)	97.6,(99.18,48.93)	97.76,(98.09,82.86)	98.73,(99.55,86.12)	98.26,(98.83,85.31)
SVM	98.86,(100,47.69)	98.6,(100,77.11)	98.73,(100,69.76)	99.75,(100,88.57)	99.03,(99.89,85.86)	99.38,(99.94,86.54)
Case Study 3 (pollution)						
EDC	93.17,(94.39,38.16)	94.47,(97.83,79.07)	94.06,(96.62,76.92)	93.06,(96.2,48.09)	94.46,(96.71,83.81)	94.02,(96.53,78.58)
LDA/PLS-DA	99.5,(99.5,NaN)	96.16,(99.01,83.13)	97.21,(99.18,83.13)	99.1,(99.1,NaN)	96.63,(98.73,87.01)	97.41,(98.86,87.01)
QDA	97.47,(98.65,62.51)	94.78,(96.84,44.98)	95.62,(97.41,49.76)	96.28,(97,74.89)	94.52,(97.05,46.25)	95.07,(97.03,52.86)
LVQ	96.06,(97.07,4.76)	93.67,(94.94,52.38)	94.42,(95.61,45.57)	94.83,(97.81,6.34)	92.27,(94.01,50.18)	93.07,(95.21,38.22)
SVM	97.05,(99.52,23.61)	97.68,(98.21,71.72)	97.48,(98.62,51.1)	97.86,(99.47,50.15)	97.43,(98.43,65)	97.56,(98.75,60.05)
Case Study 8a (mouse urine – diet study)						
EDC	90.65,(93.29,80.29)	84.68,(90.87,64.32)	87.64,(92.1,71.69)	90.84,(92.98,83.28)	85.07,(88.84,72.68)	87.93,(90.91,77.78)
LDA/PLS-DA	91.88,(94.25,80.3)	87.09,(92.28,70.03)	89.47,(93.29,74.31)	89.79,(91.14,83.16)	88.07,(91.51,77.78)	88.92,(91.31,79.93)
QDA	91.96,(93.98,82.04)	85.95,(92.41,64.74)	88.93,(93.22,71.95)	89.23,(92.3,75.83)	86.14,(90.6,72.75)	87.67,(91.48,74.06)
LVQ	84.01,(89.88,63.26)	82.29,(88.05,30.44)	83.14,(88.89,52.9)	79.44,(82.18,71.42)	77.62,(81.1,54.95)	78.52,(81.6,65.69)
SVM	89.91,(94.9,72.26)	86.43,(89.78,61.05)	88.15,(92.16,68.34)	88.36,(90.18,81.92)	85.18,(91.72,56.02)	86.75,(90.98,70.05)

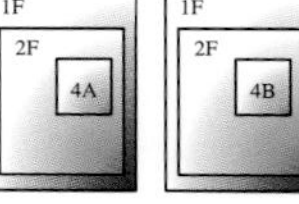

We can also look at the %PA and %MS for individual samples to potentially identify those that are close to the border between classes.

Figure 8.7 represents graphs of %PA and %MS for 2 PC models and LDA / PLS-DA for Case Studies 1 (forensic), 8a (mouse urine – diet study) and 11 (null). For Case

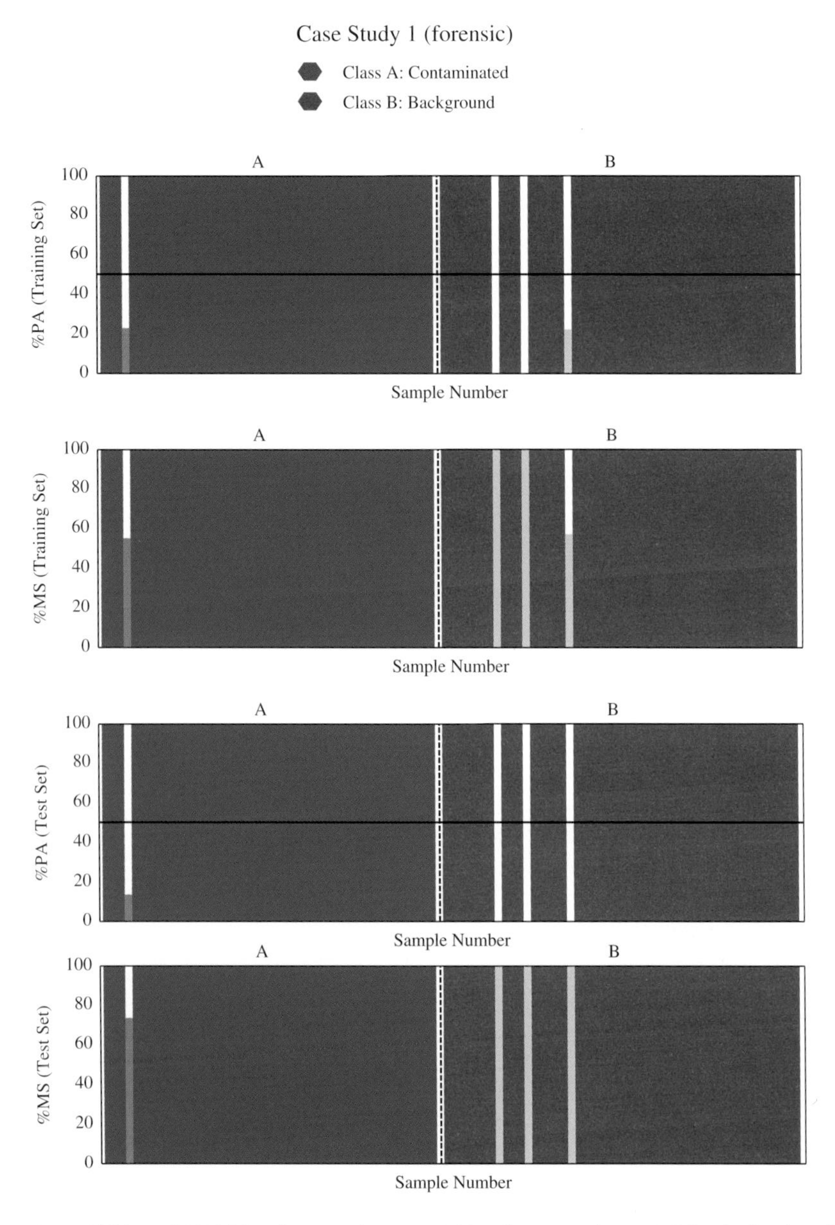

Figure 8.7 %PA and %MS for datasets 1, 8a and 11 using as an input 2 PCs, for both training and test sets and for LDA/PLS-DA. Misclassified samples are indicated in light colours, and 50 %PA is indicated by a horizontal line

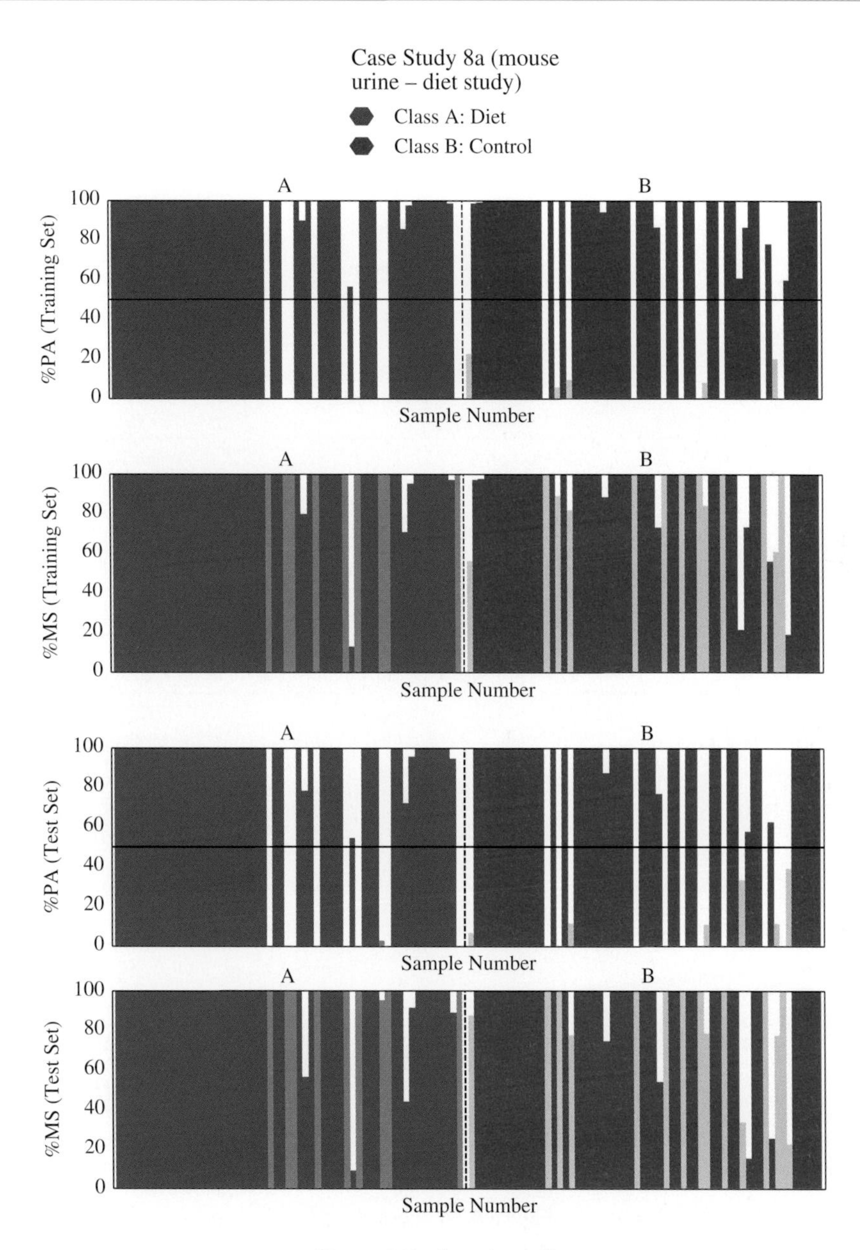

Figure 8.7 (continued)

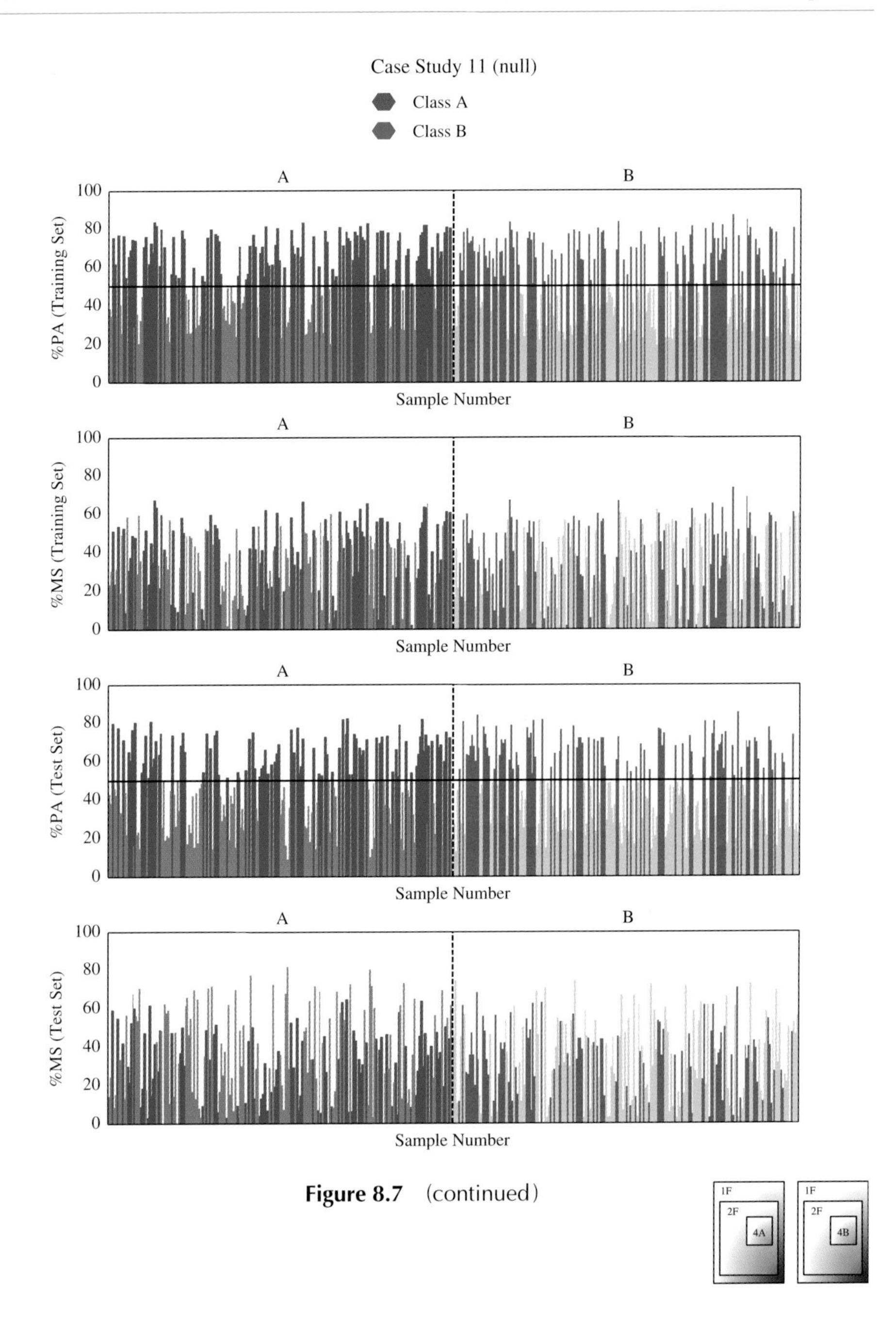

Figure 8.7 (continued)

Study 1 (forensic), we clearly see that there are two four samples that are misclassified, two of which are class B samples consistently misclassified into class A and the other two (one from class A and one from class B) which are often misclassified, but not always. Whereas this information could be obtained from scores plots when using

2 PCs, for models with more than 2 PCs this information may be harder to ascertain by visual inspection of the PCs. Interestingly for Case Study 8a (mouse urine – diet study) there are several samples with low %PA and high %MS, and only a few samples with %PA close to 50 %, suggesting that the overall model %CC is influenced by a few consistently misclassified samples. For Case Study 11 (null) we see quite a different pattern, with most samples having a %PA around 50 % and none with very high %PA, but the majority with low %MS. It is important to distinguish situations where poor predictive ability is due to all samples having low model stability (such as the null dataset) in which cases none of the samples are well modelled or most samples having high model stability (such as the mouse urine – diet study) but a few samples having very poor predictive ability possibly because they are outliers or mislabelled, which is best done by visualizing the %PA and %MS of individual samples. In Table 8.9, we see that for the 2 PC model and LDA / PLS-DA the average %MS is considerably higher than the %PA for Case Study 8a, but considerably lower for Case Study 11.

When there are more than two classes we can extend the use of these tables. As an example we will look at the predictions for Case Study 8b (mouse urine – genetics), the confusion matrix for QDA models based on 10 PCs is presented in Table 8.11. We can see that autopredictive and training set models give high predictive abilities as anticipated, but for the test set, although class A is clearly distinct, classes B and C are often confused with each other. This is anticipated as class A is a different strain to classes B and C, which in turn differ just in one background genes. This can also be represented by a graph of %PA which indicates clearly that samples in class A are mainly well modelled (Figure 8.8). The cross classification rate can also be visualized (Figure 8.9): the vertical axis of the %CR indicates the percentage of times a sample is assigned to a given group; it is clear that classes B and C are more regularly confused with each other than with class A, which is anticipated as classes B and C are from the same strain.

Table 8.11 *Confusion matrices for Case Study 8b (mouse urine – genetics), using QDA models based on 10 PCs, with %CCs indicated. Class A = AKRH2k, Class B = B6H2k and Class C = B6H2b*

Autoprediction		A	B	C
	A	98.59	0.00	0.00
	B	1.41	89.83	16.00
	C	0.00	10.17	84.00
Training set		A	B	C
	A	98.59	0.00	0.00
	B	1.41	98.31	12.00
	C	0.00	1.69	88.00
Test set		A	B	C
	A	92.96	3.39	0.00
	B	7.04	76.27	34.67
	C	0.00	20.34	65.33

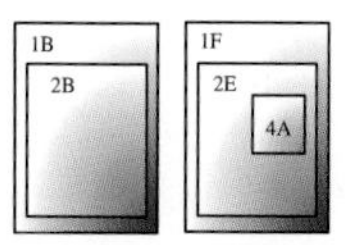

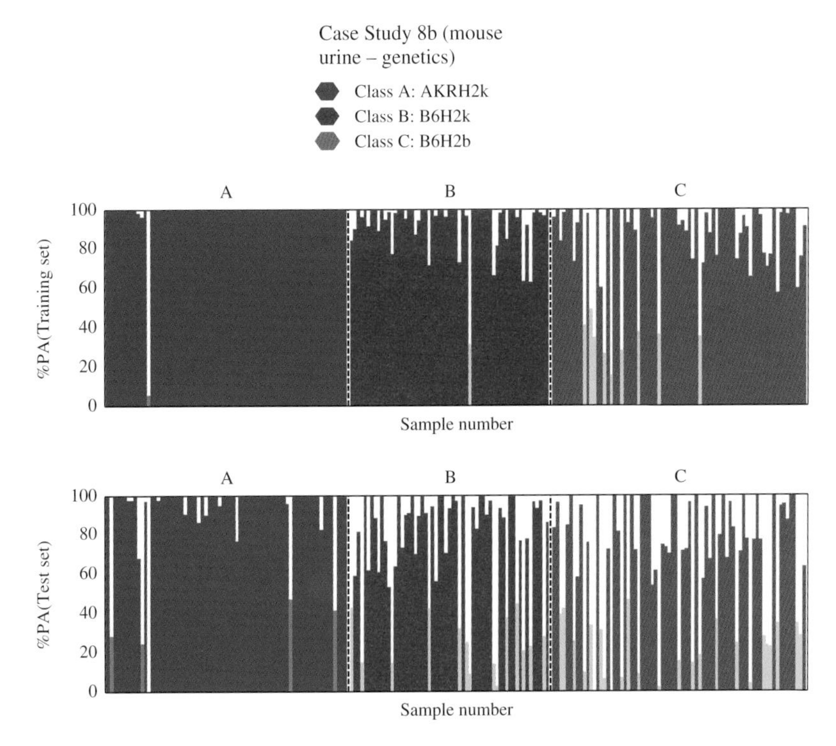

Figure 8.8 Graph of %PA for individual samples for Case Study 8b (mouse urine – genetics), using the model of Table 8.11. Samples that are incorrectly classified are indicated in a light colour

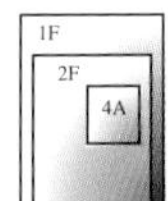

8.4.2 Number of Iterations

How many iterations are required to obtain a stable or useful model? A sensible choice of number of iterations may be limited by computing power especially if datasets are very large and it is necessary to form (and optimize) complex models such as SVMs. However it is useful to get an insight into how many iterations might be necessary to obtain a stable model for any specific application.

Ensuring Samples are Chosen for both Training and Test Sets

In order to determine both test and training set predictions for individual samples there must be sufficient iterations that each sample is included at least once, ideally very many times.

The example of Table 8.8 is very small but there is a slight risk that a sample may never be picked for the test (or training set). The probability a sample is never chosen to be part of the test set is $(1 - I_{test}/I)^N$ where N is the number of iterations with an equivalent equation for the training set. If only a few iterations are performed there is a small chance that specific samples will never be part of the test set and so a PA and MS cannot be calculated, and so clearly increasing the number of iterations allows prediction of all the samples in both the training and test set models. However, even if some samples are

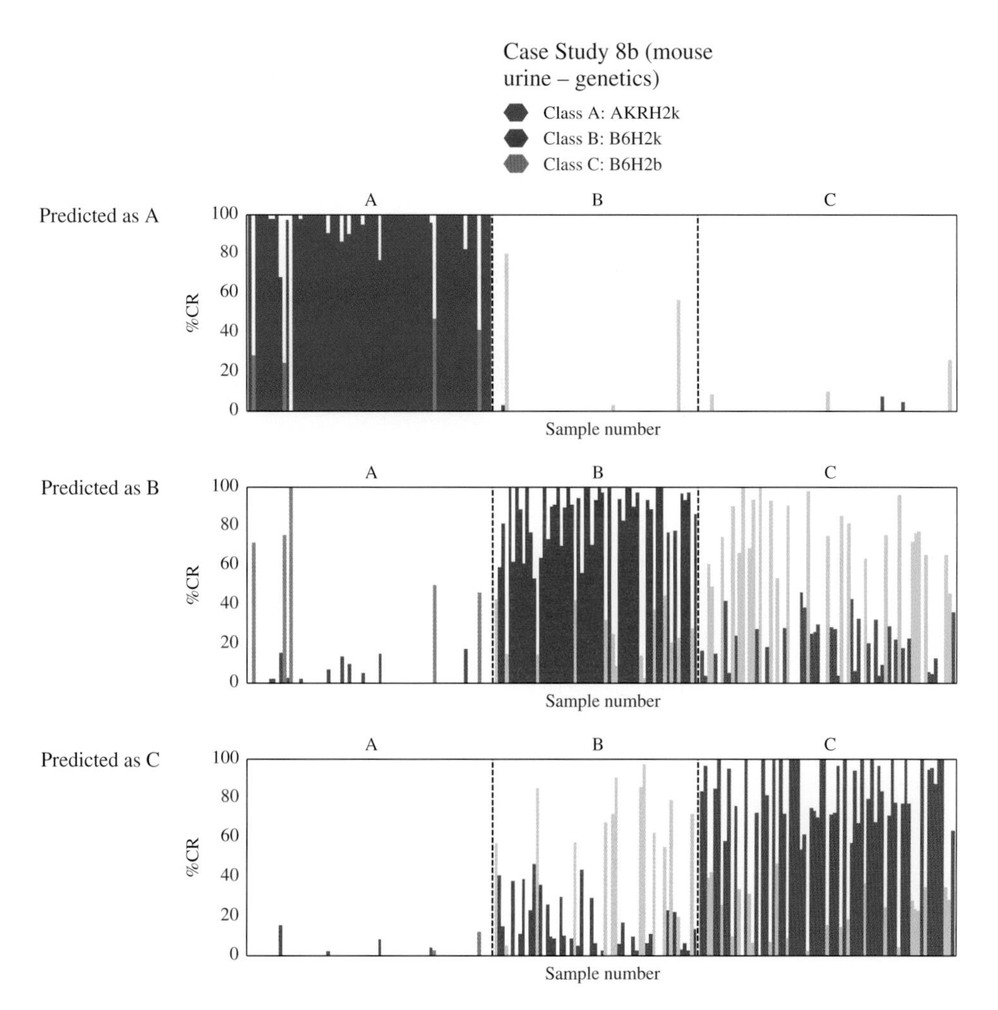

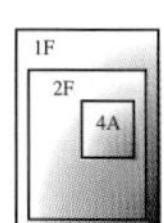

Figure 8.9 Graph of test set %CR for samples in Case Study 8b using the model of Table 8.11, with incorrectly classified samples indicated in a light colour

missed out from the test or training set does not prevent the overall %CC being calculated, of course.

In cases where the class sizes are unequal, in order to obtain equal representatives for each group in the training set, the chance an individual sample is selected is different for each group. If there are unequal class sizes there may be a significantly higher probability of picking samples from the larger compared to the smaller group in the test set, dependent on the ratio of samples in each group, whereas the training set probability diminishes. If the ratio of samples is I_A / I_B and $(M-1)/M$ of the samples are chosen from the smaller group for the training set (typically $M = 3$) then the probability of choosing a sample from the larger group (A in this case) becomes $I_{trainA}/I_A = [(M-1)/M)]/(I_B/I_A)$, or, in other words, if

2/3 of the samples from the smaller group are chosen for the training set and there are 5 times more samples in the larger group, this becomes 2/3 × 1/5 or 2/15. After N iterations the chance a sample is never chosen for the training set is $(1 - I_{trainA}/I_A)^N$. Similar equations can be set up for the test set. The probabilities that a sample is never chosen are given in Table 8.12.

Table 8.12 *Probability that a sample will never be chosen for either the training set or test set of the largest group when the size of two groups are unequal, for varying number of iterations when 1/3 of the samples are randomly selected for the test set. Note that for the smaller group, or when both group sizes are equal, the probabilities in the columns for equal group sizes represent the chances a sample will never be chosen, as given in the columns headed 1 (all values in percentages)*

Iterations	Ratio of samples in largest to smallest groups					
	1	2	5	1	2	5
1	33.333%	66.667%	86.667%	66.667%	33.333%	13.333%
5	0.412%	13.169%	48.895%	13.169%	0.412%	0.004%
10	0.002%	1.734%	23.907%	1.734%	0.002%	0.000%
25	0.000%	0.004%	2.794%	0.004%	0.000%	0.000%
50	0.000%	0.000%	0.078%	0.000%	0.000%	0.000%
100	0.000%	0.000%	0.000%	0.000%	0.000%	0.000%
	Training set			Test set		

Stability of Classifier

A second reason for using several iterations is to determine a stable estimate of %CC. Figure 8.10 illustrates the distribution of individual values of %CC for Case Study 11 (null) using a PLS-DA model, and 200 different test/training set splits. It can be seen that although the average is 48.43 %, the standard deviation is 3.83 % meaning that there is considerable difference each time a new test and training set is formed, dependent on which samples are included in the training set.

The more the iterations, the more stable this value. Figure 8.11 illustrates the standard deviation in %CC for test and training sets when a set number of iterations are repeated 20 times, using Case Study 11 (null) and a 2 PC QDA two class model. Hence the 50 iteration model involves (a) performing 50 iterations and finding the 'majority vote' %CC over all 50 iterations, (b) repeating this 20 times and (c) finding the standard deviation of the %CC from each of these 20 repeats of 50 iterations. It can be seen that the standard deviation of the estimate of the %CC for each individual class is quite high, around 400 iterations are needed to reduce this to 3%, whereas for the overall %CC (averaged over both classes) a standard deviation of 1.5 % is achieved after about 100 iterations for both test and training sets. Note that this standard deviation is different to the one presented in Figure 8.10 which is of the individual %CCs obtained in each of 200 iterations. However it is quite clear that unless a dataset is very easy to separate into groups, it is advisable to perform several test and training set splits, and statistics obtained on a single test set could be atypical and provide misleading results. It is especially important to recognize that comparisons of the

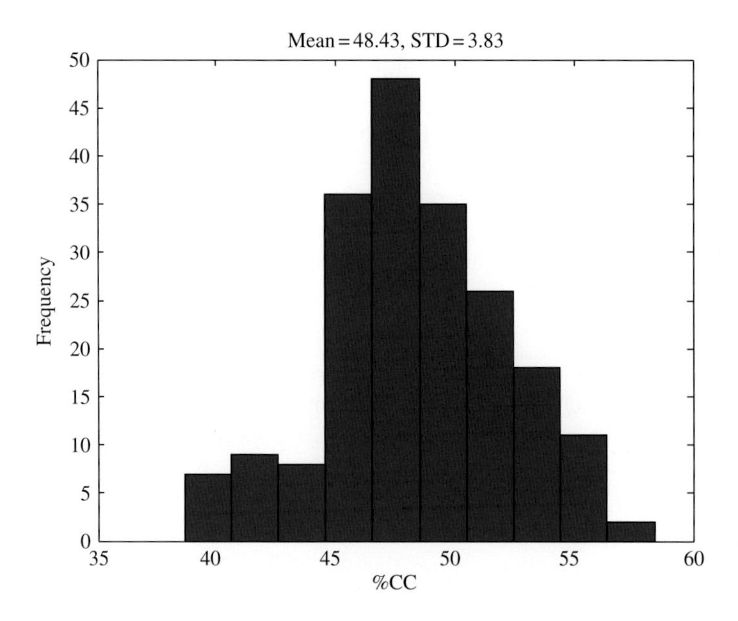

Figure 8.10 Distribution of values of %CC for PLS-DA predictions using 200 random test/training set splits for Case Study 11 (null), a 2PC model and test set

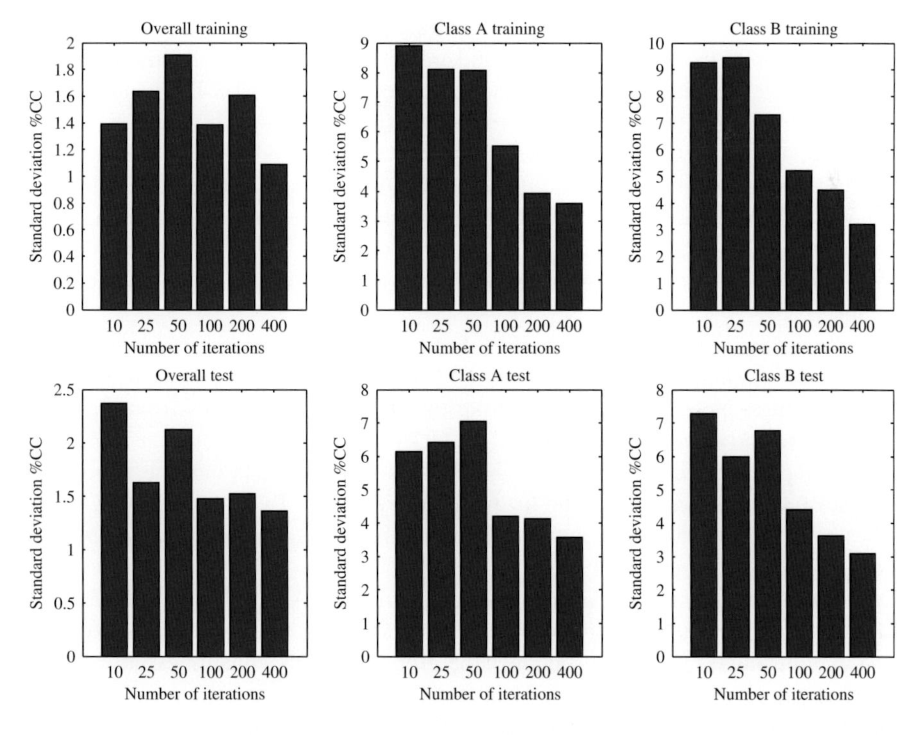

Figure 8.11 The standard deviations in 20 estimates of %CC when averaged over 10, 25, 50, 100, 200 and 400 iterations for QDA, a 2PC model and Case Study 11 (null)

classification ability of methods on the basis of, for example, a single test/training set split are unlikely to be safe, especially when comparing the performance of different methods whose classification abilities differ by just a few percent. With the power of modern computers it is normally quite easy to perform several iterative splits of test and training sets in order to obtain a stable estimate of %CC and other statistics such as %PA and %MS.

We recommend as a default 100 iterations. The number can be adjusted according to the dataset structure and type of model, but this compromise usually leads to a stable %CC and sufficient samples in the training and test set to accurately estimate %PA and %MS.

8.4.3 Test and Training Set Boundaries

If we project the data onto two dimensions and form our models in two dimensions we can visualize boundaries. Each time a dataset is split into test and training sets, a new boundary is obtained which is dependent on the samples selected for the training set. If 100 iterations are performed, each split of samples leads to a different boundary, and so 100 boundaries can be superimposed, each representing one iteration. The boundaries can be represented as a shaded area; the broader this area, the less well defined the boundary, the density of shading relating to the number of lines that cross the specific region of space, with a maximum black colour representing the maximum density of lines crossing a given pixel.

The 100 test/training set boundaries for two class classifiers and Case Studies 1 (forensic), 3 (polymers), 4 (pollution), 9 (NMR spectroscopy) and 11 (null) are presented in Figure 8.12:

- For Case Study 1 we see that the boundaries for EDC and LDA are quite tightly defined, with two samples that have a low %MS and are misclassified in both the training and test sets within the boundary region. Note that, in these figures, the PLS-DA boundaries only differs from LDA on the basis of random choice of samples for the training set, which are not reproducible. LVQ has quite a well defined boundary at negative values of PC2, but because there are a small number of misclassified samples at positive values of PC2, the boundary is much broader, being influenced by whether these are in the training set, and in turn whether they are chosen as initial guesses of codebooks. SVM has a surprisingly well defined boundary. QDA is somewhat broad, dependent on a few influential samples.
- In Case Study 3, EDC and LDA (PLS-DA differing only in the random choice of samples for training and test set splits) both have a well defined central region with broader wings. Note that the small group of misclassified samples from class A are within this poorly defined region for EDC. QDA, LVQ and SVM manage to define the more complex boundary quite well and what is remarkable is that the misclassified samples mainly belong within the fuzzy boundary region. Even though QDA misclassifies more class A samples than SVM or LVQ, the small group of class A samples are within the region of uncertainty.
- Case Study 4 is interesting. All methods (apart from LDA and PLS-DA which are the same except for the iterative split of samples) behave quite differently. QDA attempts to find an ellipsoidal region that separates the groups but only sometimes finds the three soil 4 samples from class B that are within the region of class A. LVQ occasionally finds these, and so there are some models that enclose this small group and others that neglect it. SVM tries to enclose class B in an almost circular grip, but for the default penalty error

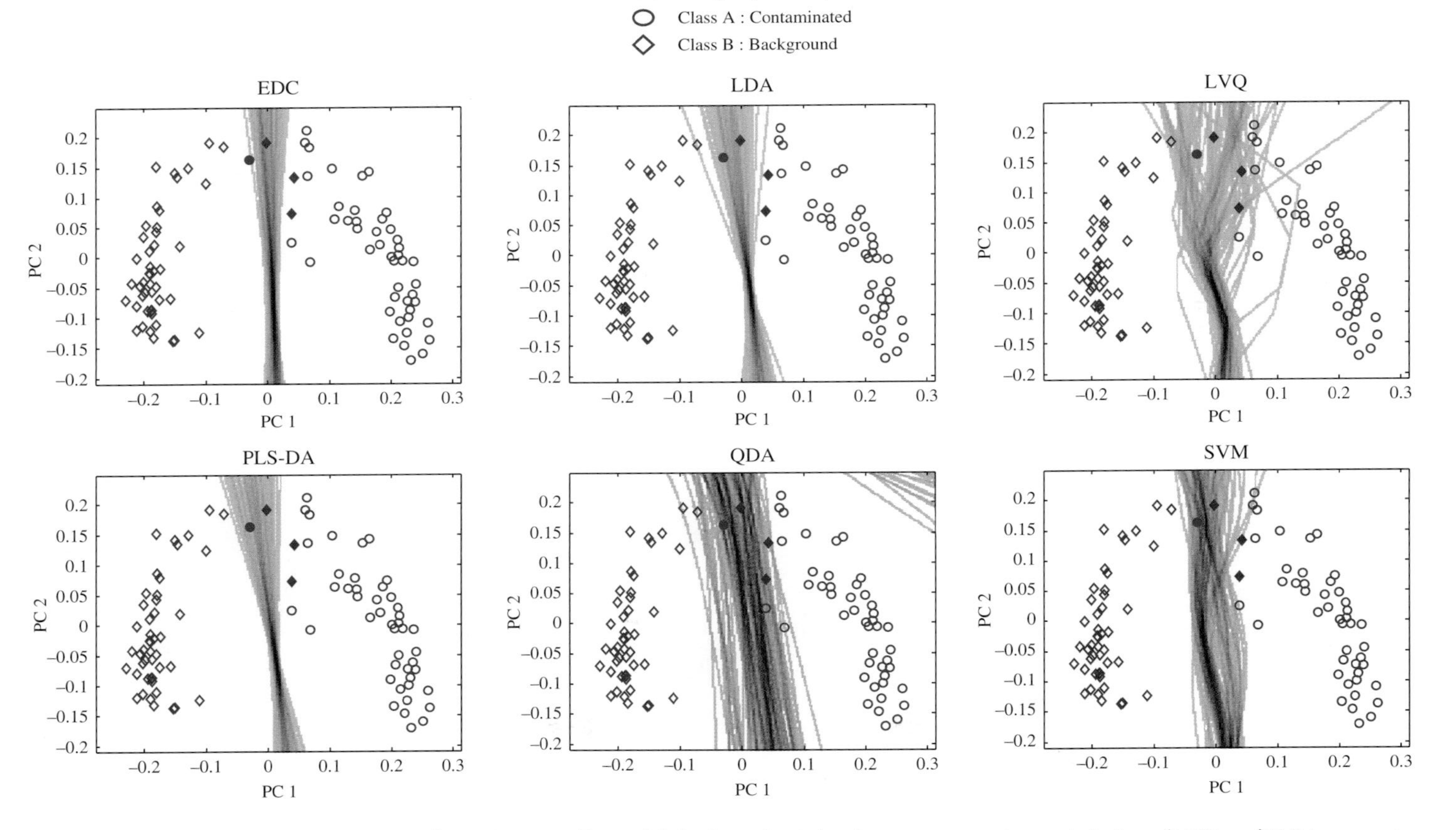

Figure 8.12 Boundaries using 100 test and training set splits and default methods for data preprocessing and choice of LVQ and SVM parameters for two class classifiers as applied to the first 2 PCs of Case Studies 1, 3, 4, 9 and 11, Misclassified samples indicated using filled symbols

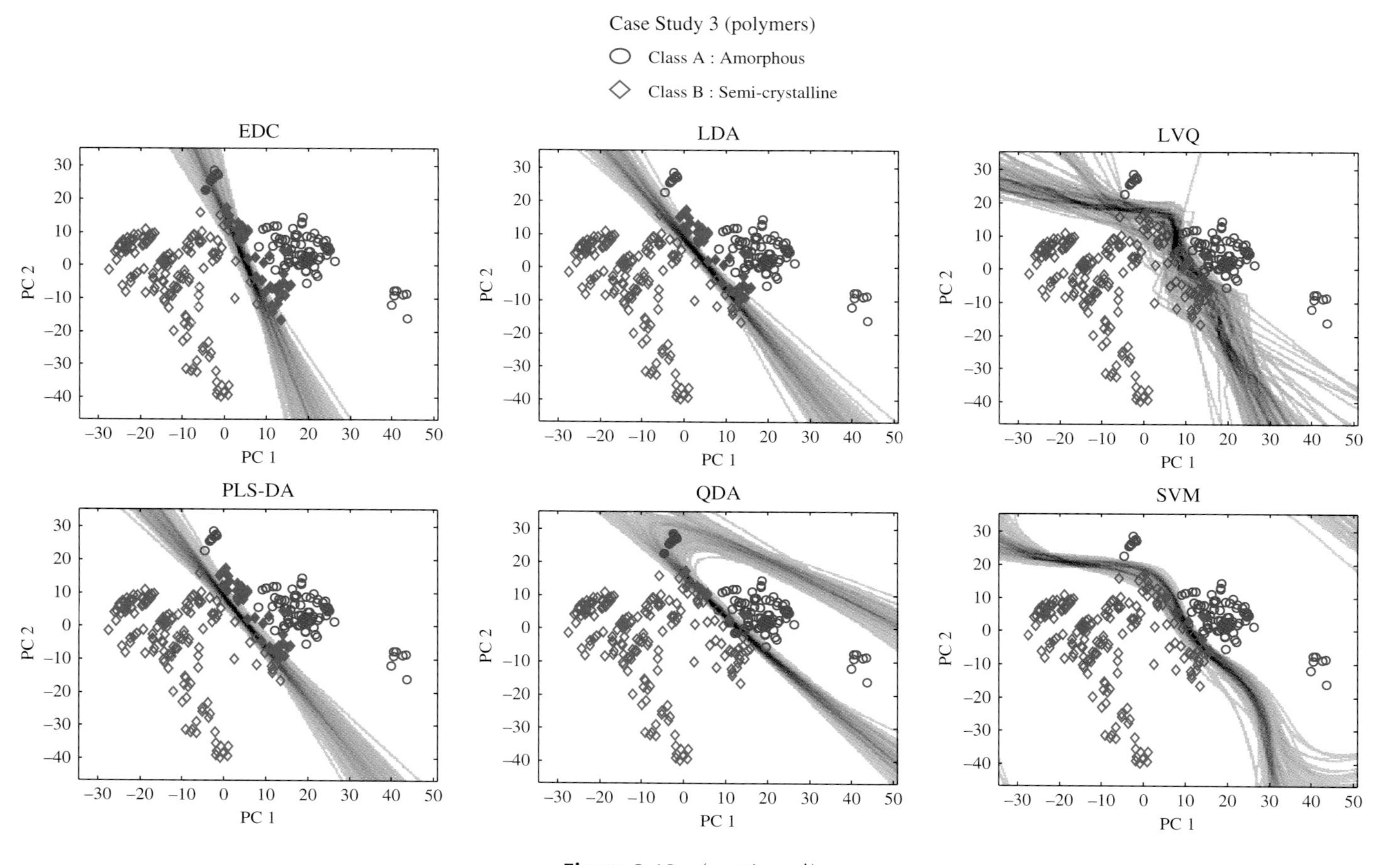

Figure 8.12 (continued)

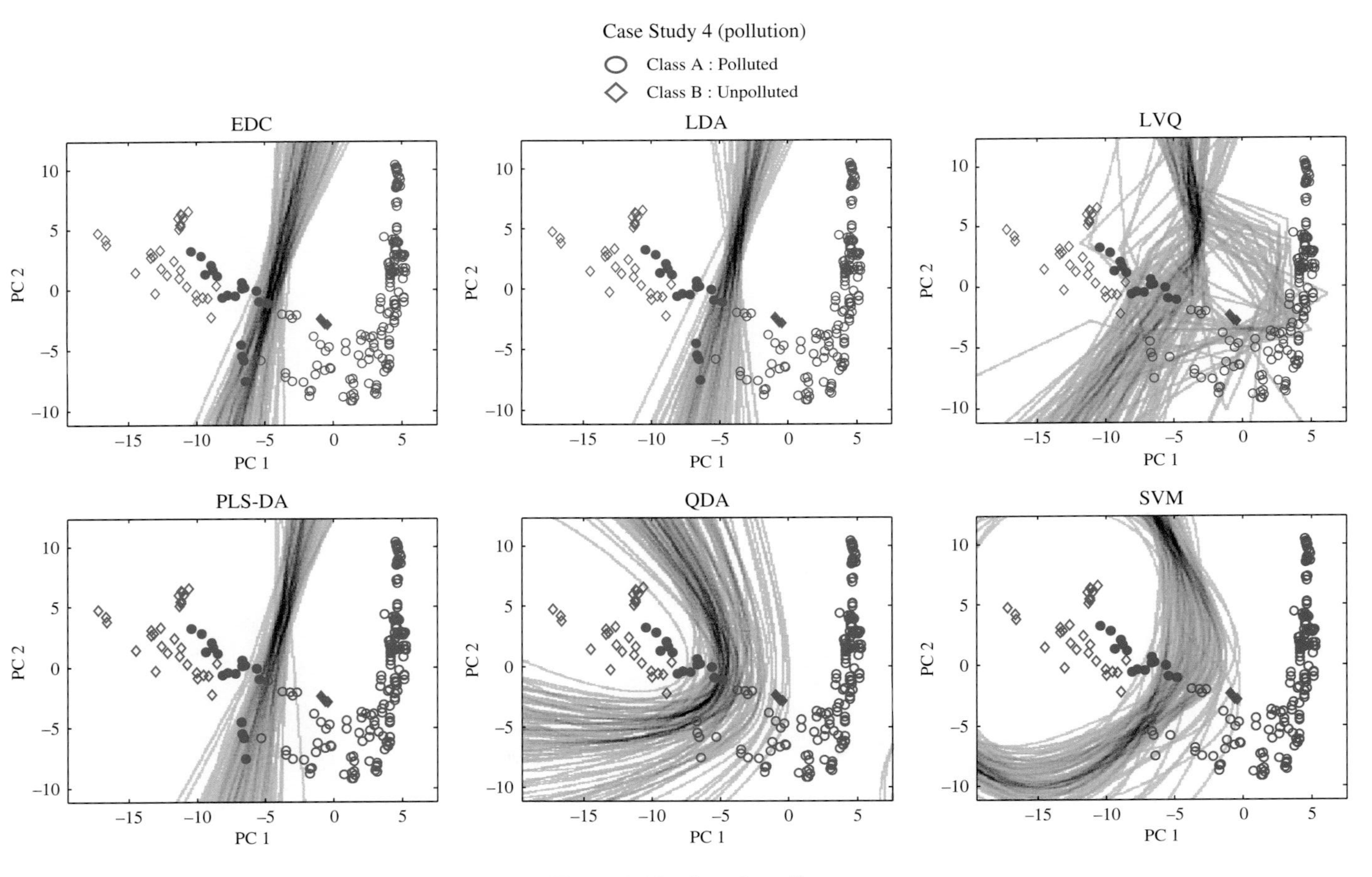

Figure 8.12 (continued)

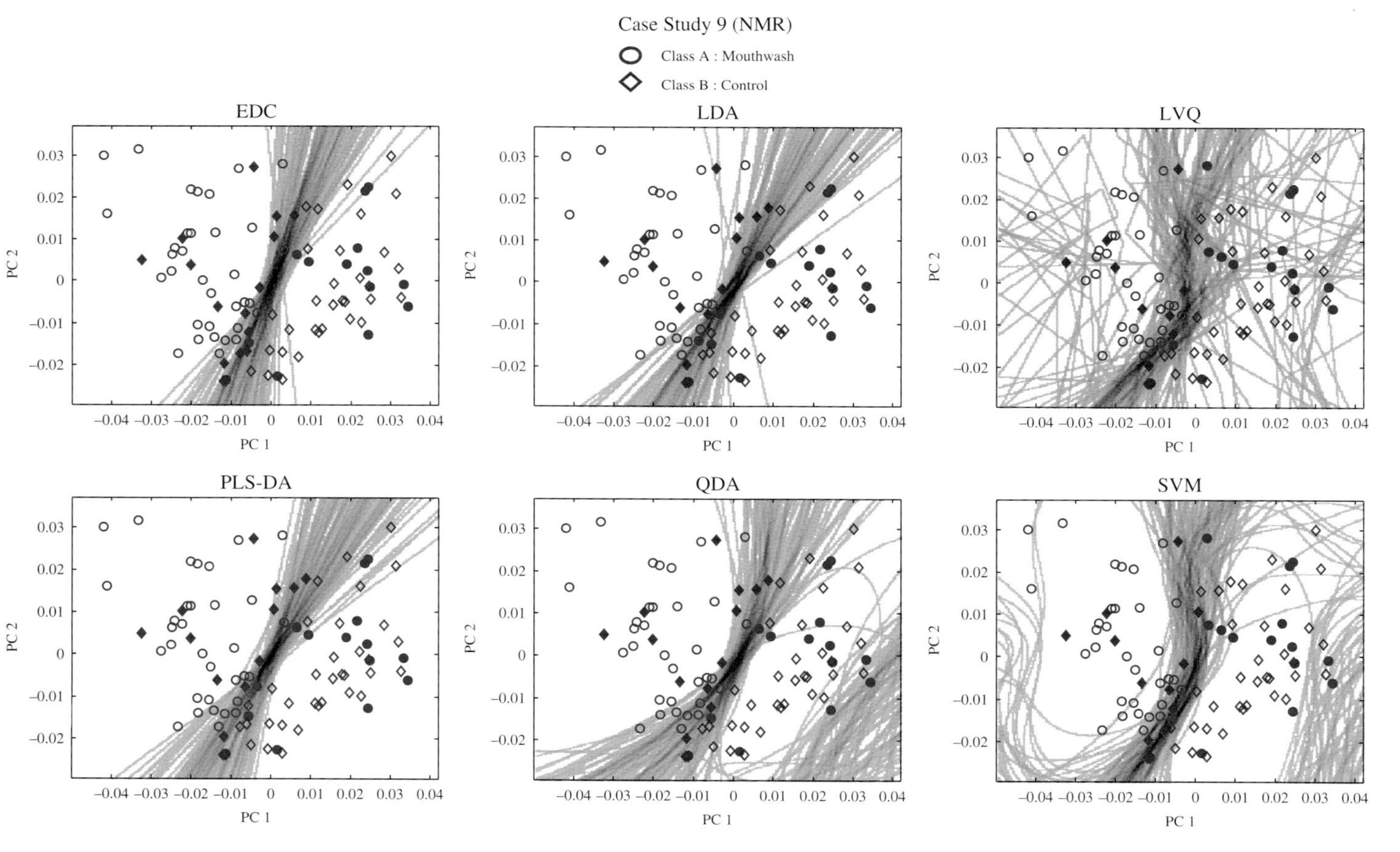

Figure 8.12 (continued)

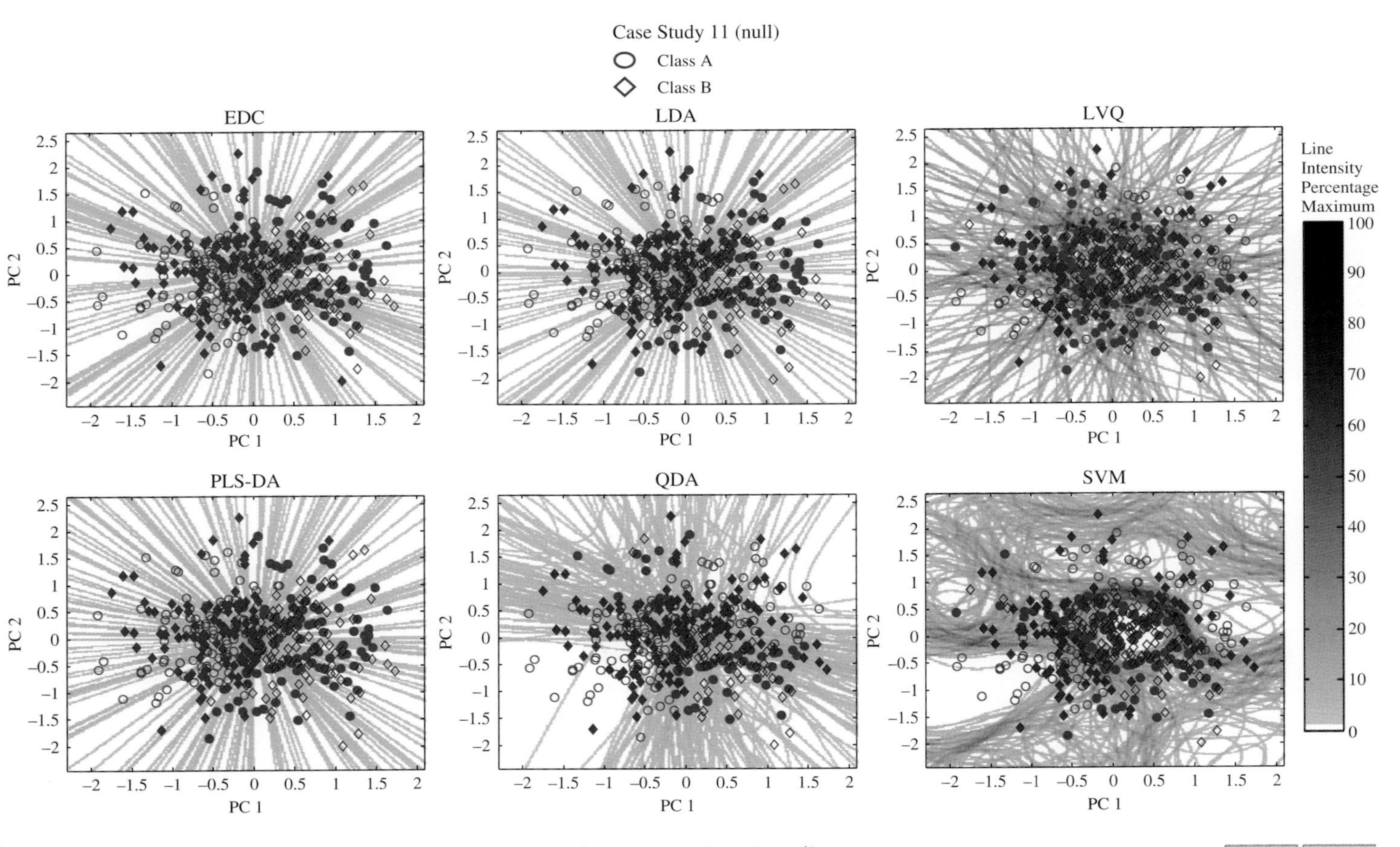

Figure 8.12 (continued)

($C = 1$) and the RBF parameters used for these calculations are not unduly influenced by the three soil 4 samples unlike LVQ.

- Case Study 9 is typical of the complexity that is often found in metabolic profiling. The linear methods EDC, LDA and PLS-DA all behave quite similarly. LVQ is highly influenced by a few influential samples. QDA and SVM behave in quite a similar way and are able to represent non-linear boundaries quite well.
- For Case Study 11 we see that the three linear methods (EDC, LDA and PLS-DA) result mainly in a series of radial lines meeting close to the centre. The orientation of these lines is essentially random, suggesting that the choice of training set samples has a large influence on the boundaries. There is not much pattern, although certain regions, e.g. the bottom left-hand corner, where there are more class A samples than class B, tends to be more free of boundaries. LVQ is similarly influenced with no real pattern. QDA manages to find some clusters of samples, for example, in the bottom left and so looks for small pockets of structure and thus has a little more flexibility than the linear methods. SVM using the default penalty error and RBF can identify more complex regions of the data space where there are clusters of samples, for example in the bottom left (more class A than class B) and in the middle (more class B than class A) or mid right (more class A than class B). The SVM backdrop is thickest in regions where there are not many samples. Although Case Study 11 has been generated randomly, as of all random patterns there are still small clusters of samples from a particular group which provides a handle for the SVM classifier.

Figure 8.13 represents multiclass classifiers. For PLS-DA we illustrate the boundaries using PLS2, although there are other possibilities as discussed in Chapter 7:

- For Case Study 2 (NIR spectroscopy of food) although at first this may seem a very straightforward dataset, we can see very different performances from several of the classifiers. Remarkably, EDC, LVQ and SVM all perform quite similarly. Three of the four main boundary lines are very well defined. The boundary where the regions defined by classes A and D meet on the bottom left of the figures is somewhat less well defined; this is probably because it is quite far away from any of the groups, so not so easy to pinpoint. SVM, although it often searches for complex boundaries, if they are not there, it is equally capable of producing well defined linear boundaries. LDA appears to behave somewhat differently to EDC which is slightly unexpected. However the generation of the training set involves removing one third of the original samples. The relative variance along the two axes is critically dependent on which samples are retained and because the centroids are so far apart, this is a major factor that will influence the appearance of the boundaries.
- Case Study 3 (polymers – 9 groups) is quite interesting. Some groups, F, G and C, seem to be defined relatively well (apart from when PLS2 is employed). Class I, interestingly, can be split into two subgroups. Whereas linear methods might not be able to model these well, when many iterations are performed, it can be seen that all methods apart from PLS place the top left subgroup into a separate pocket which is much less densely populated by boundary lines. Hence building up boundaries using iterative methods actually allows linear classifiers to define disjoint regions for each group. The SVM boundaries are characterised by several 'holes' which correspond roughly to a single group. It is also interesting that class C (top of diagram) seems to be enclosed in a thick

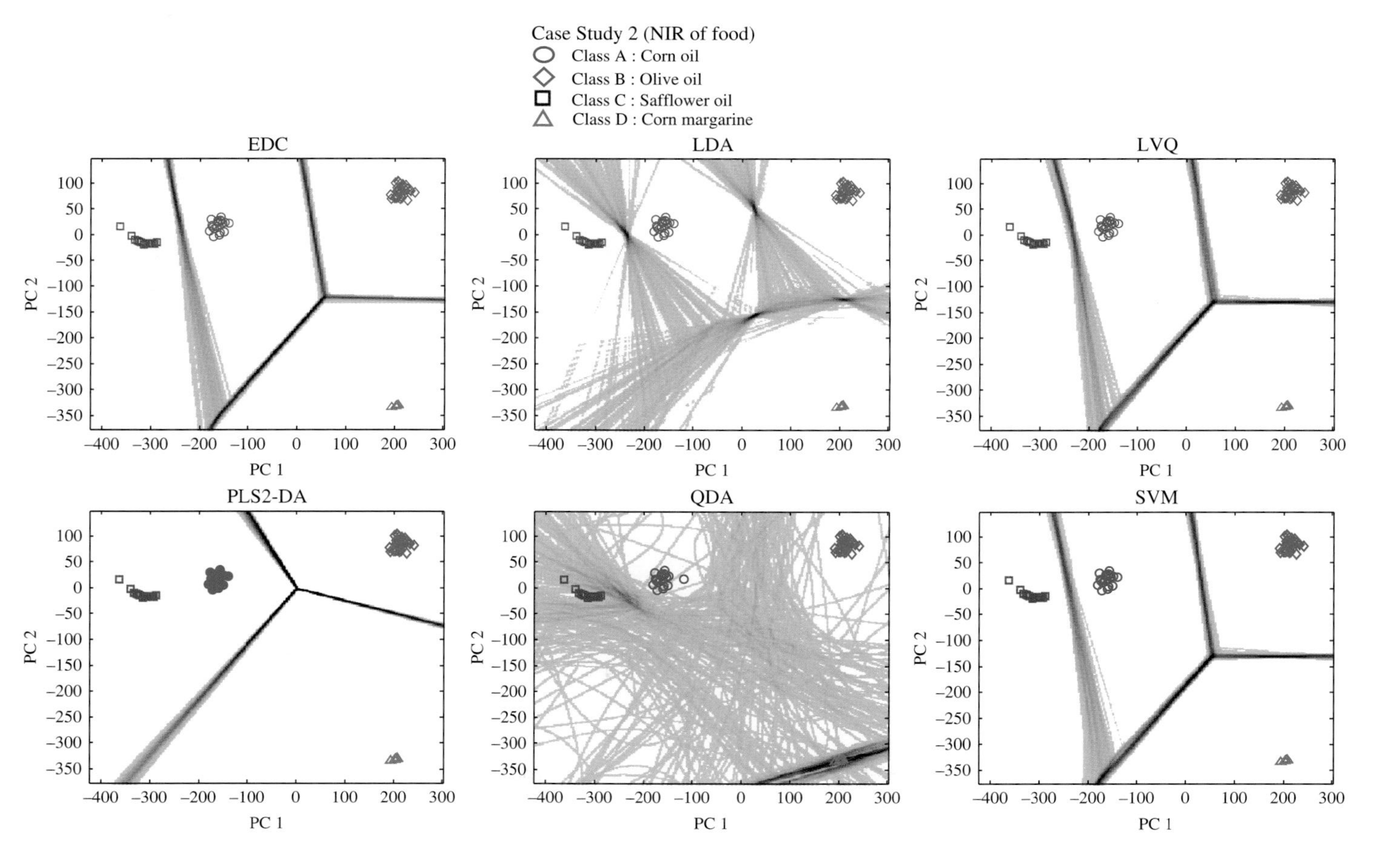

Figure 8.13 Boundaries using 100 test and training set splits and default methods for data preprocessing and choice of LVQ and SVM parameters for multiclass classifiers as applied to the first 2 PCs of Case Studies 2 and 3. PLS2 is used for multiclass PLS, and 1 v 1 'majority vote' SVM, Misclassified samples indicated using filled symbols

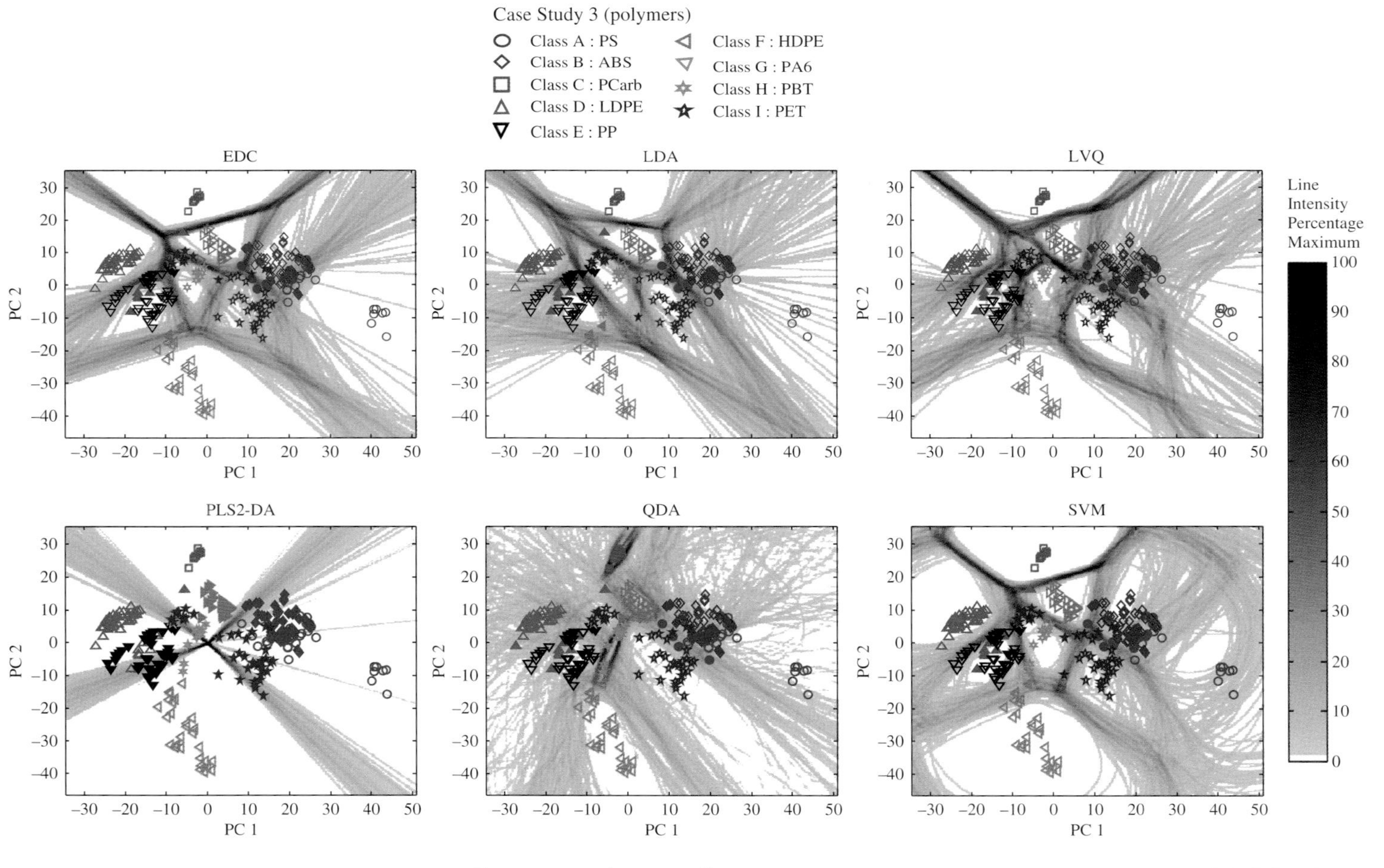

Figure 8.13 (continued)

and distinct boundary using EDC, LDA, LVQ and QDA, and class G, for example is enclosed in a very distinct pocket. Interestingly, using QDA, these groups are much more compact but are enclosed by very narrow and well defined borders.

Finally, Figure 8.14 represents boundaries for one class classifiers using Case Studies 3 (polymers – 2 types), 8a (mouse urine – diet) and 11 (null). QDA (95 %) and SVDD ($D = 0.2$) boundaries are drawn using conjoint PC models:

- For Case Study 3, the boundaries are more or less as expected, and quite stable in appearance. For SVDD there is a considerable variation in the nature of the boundaries, but what is interesting for class B, is that there are small empty regions within the group where there are no samples where the boundaries cross (represented in grey) suggesting that SVDD is strongly influenced by the substructure of this class.
- For Case Study 8a, the QDA boundaries are similarly tight. For SVDD there is a strongly defined region for each class.
- For Case Study 11, both sets of QDA boundaries almost coincide, suggesting that the samples chosen for the training set are more influential than the class membership of a sample. SVDD almost gets this circular region but is somewhat influenced by outlying groups of samples. It is interesting to compare these boundaries with those for the comparable two class classifier, which are much less clean (Figure 8.12) because for one class classifiers the boundaries are only influenced by the structure of a single group.

We can see from this section that a great deal of insight into the behaviour of the classifier can be obtained by visualizing training set boundaries, which also demonstrate the importance of performing enough test/training set splits to obtain a stable model.

8.5 Optimizing PLS Models

8.5.1 Number of Components: Cross-Validation and Bootstrap

In PLS, the main task of optimization is to determine the number of PLS components that are most appropriate to the model. There are numerous ways of choosing the number of PLS components which is a large topic in the chemometrics literature probably stretching over many hundreds of papers over three decades. However, the original applications of PLS were in regression, so called PLS-R (where R stands for regression), and aimed primarily to obtain a quantitative answer, e.g. the concentration of an analyte rather than a decision as to whether a sample belongs to one group or another, and a great deal of importance is attached to the precise value of this resultant numerical value.

In classification we usually assign a label such as an integer to each class, and do not ultimately care whether the answer from PLS-DA is, for example, 1.234 or 0.897; we care primarily whether it belongs to class A or class B (in a two class problem). If class A is labelled as +1, then it is likely that a sample belongs to this class if the result of PLS-DA is a numerical value close to 1. What we want to do is optimize the classification ability rather than the accuracy of a numerical prediction of a value associated with a sample. This means that some of the more numerically intense methods for obtaining the optimal

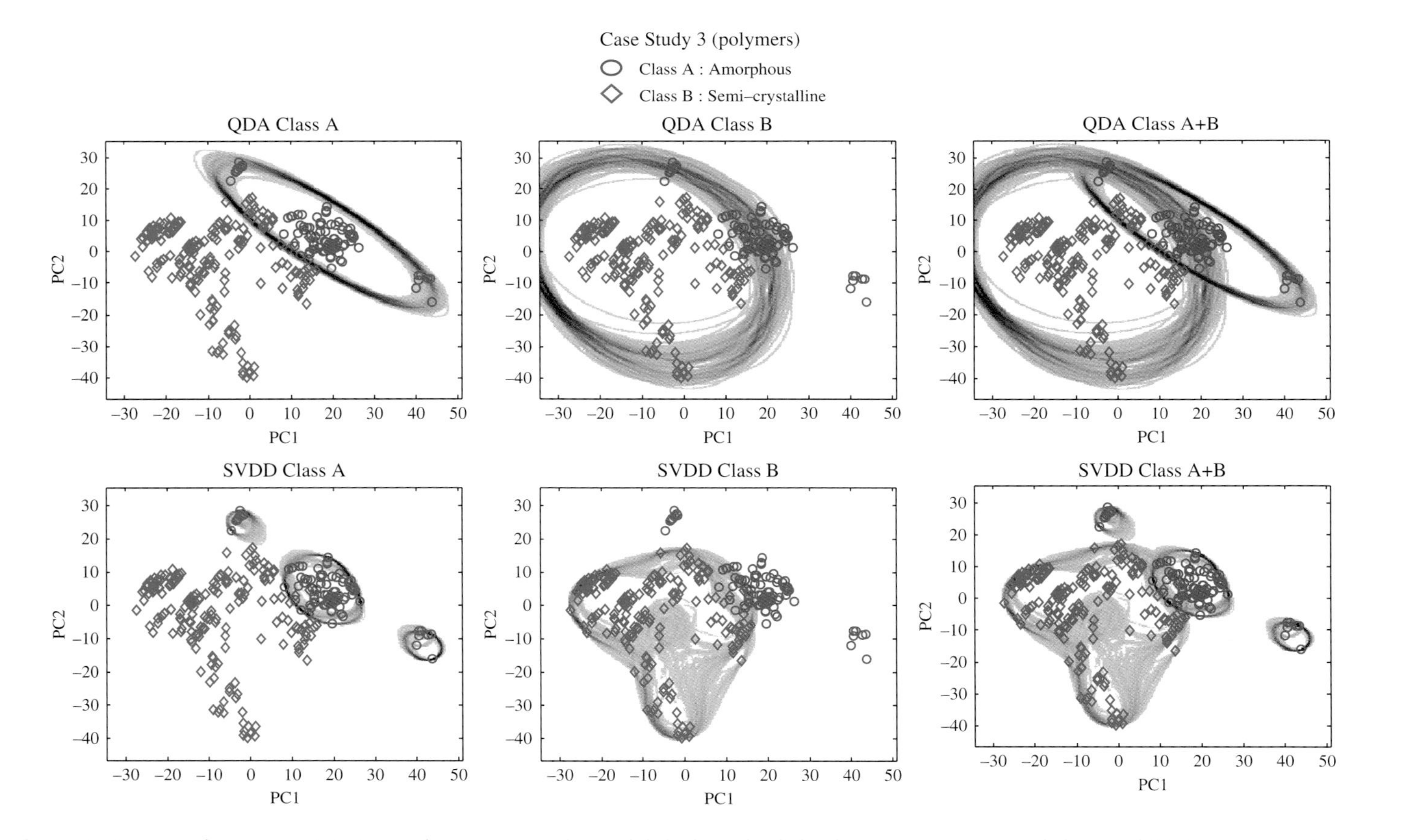

Figure 8.14 Boundaries using 100 test and training set splits and default methods for data preprocessing and choice of SVM parameters for one class classifiers as applied to the first 2 PCs of Case Studies 3, 8a and 11. For QDA, 95 % confidence limits are visualized

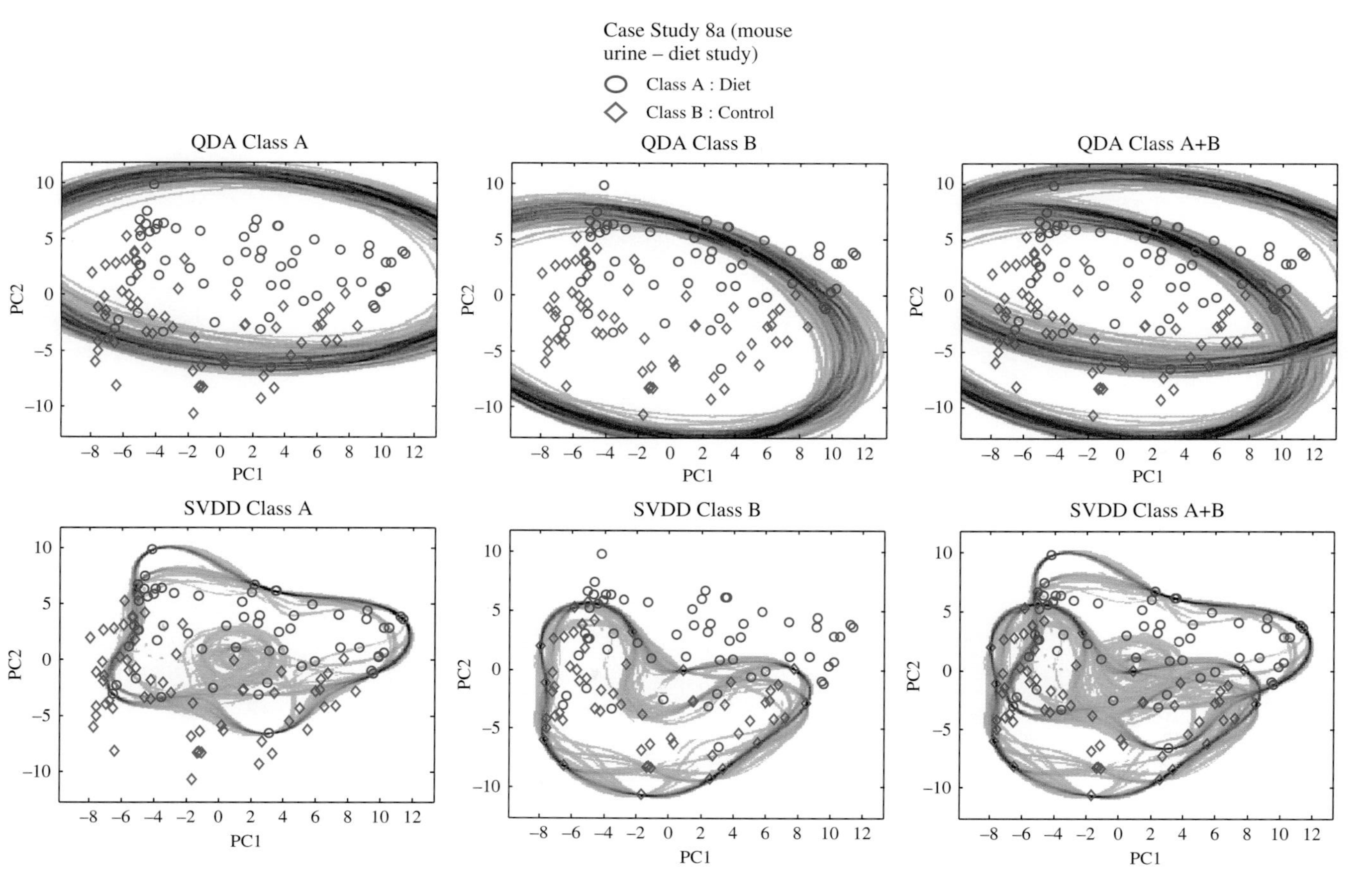

Figure 8.14 (continued)

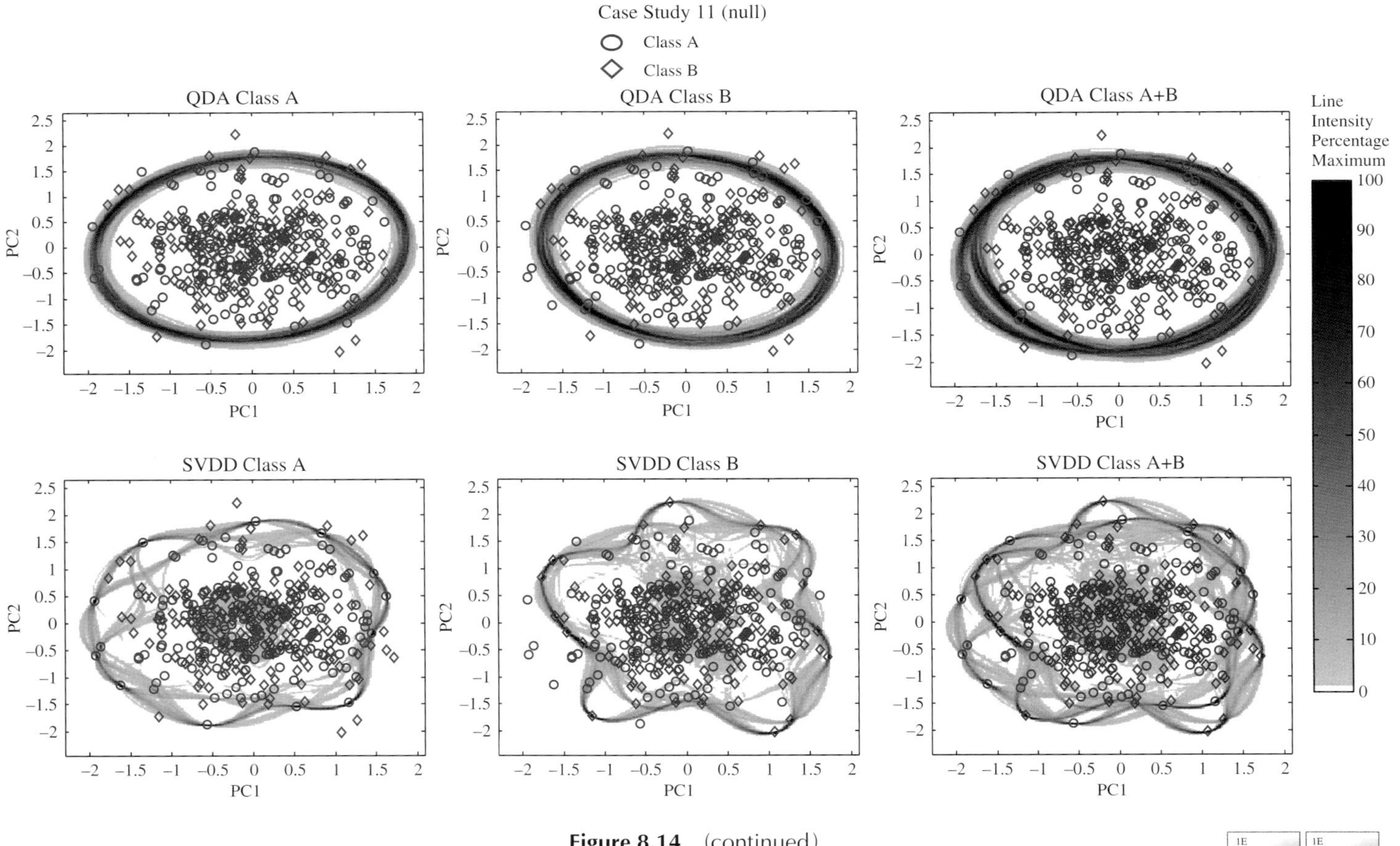

Figure 8.14 (continued)

number of components may not be so directly relevant to classification studies. In pattern recognition, the prime interest is to find the most appropriate number of PLS components that will optimize the %CC of the training set. One of the conceptual difficulties in chemometrics is that this is done using methods quite similar to those used for the testing and validation of models. Most approaches involve removing a fraction of the data, performing PLS on the remaining data and then seeing how well the samples that have been left out are predicted. This of course involves a similar principle to testing a model but in order to be effective should be strictly separated or otherwise the result of validation may be over optimistic. In traditional analytical chemistry where we often know there must be a certain underlying answer (we may know what analyte we added to a series of standard reference samples and the concentration it was added) it is just a question of improving its accuracy; in hypothesis based research this is not so.

It is important to recognize that it is not inherently necessary to optimize the number of PLS components, and models can be formed using a fixed number of components – which sometimes has advantages that different datasets and analysis protocols can be compared directly; for example in Chapter 5 we have set 2 PLS components as the default for PLS models and in Sections 8.3 and 8.4 use models based on both 2 and 10 PLS components for comparison. In fact when testing whether a model is suitable or not there is no requirement it is optimized, although an overly pessimistic assessment of its performance might be obtained but this errs on the side of caution and so the result of the assessment might be quite safe, although if a method is to be used for real life predictions it is preferable to optimize it eventually, sometimes as a separate later stage.

Cross-validation

The classic approach for optimizing the number of PLS components is to use cross-validation. There are various methods for cross-validation but the most widespread in chemometrics is the LOO (Leave One Out) method in which one sample is removed at a time and the model is formed on the remaining samples. The property of the sample left out is predicted from a model obtained using the remaining samples and so on until every sample is removed. In our case, if we have, for example, 100 samples in a dataset, sample 1 is first left out and then a model is formed from samples 2 to 100 to predict the class membership of sample 1, then sample 2 is left out and a model formed from samples 1 and 3 to 100 to predict sample 2 and so on. The sample left out can be considered as a mini-test set of size 'one'. Cross-validation ensures that each sample is selected once (and once only) as part of this unique test set. For classification studies, the %CC can then be obtained via this procedure, being the percentage of samples (left out using LOO) that are correctly classified.

In the context of PLS-DA, the aim of cross-validation (or model optimization) is to determine the optimal number of PLS components that gives the best classification for the training set. The conceptual framework is that later components model noise rather than structure in the data. PLS components can be divided into those that model the structure (the first A components) and those that model the noise. In order to find how many components are optimal, the value of %CC for cross-validation can be calculated as an increasing number of components is used in the model. Once the maximum value of %CC has been obtained then this is the correct number of PLS components for the model as

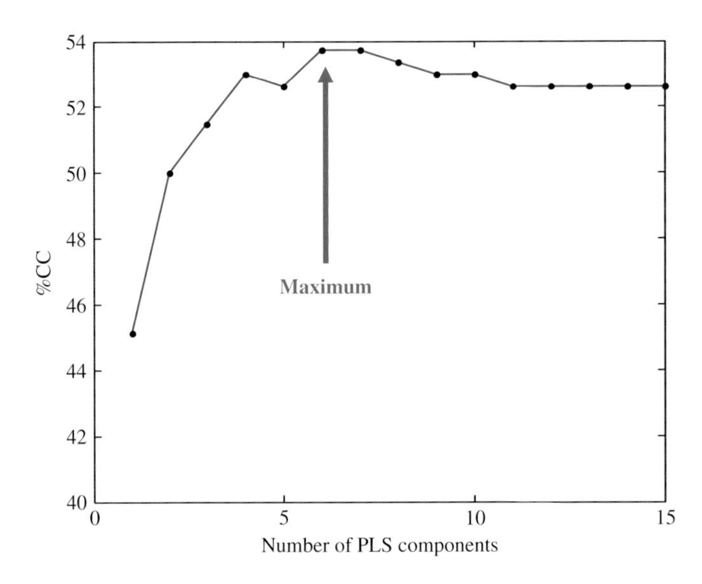

Figure 8.15 LOO Cross-validation of a single training set (from Case Study 11 – null) in order to determine the optimum number of components

illustrated in Figure 8.15. It is important to note that using full autopredictive datasets the %CC will always increase as more components are added to the model, resulting in a perfect %CC of 100 % if as many components (as samples or variables) are included. Cross-validation in the context of PLS-DA serves a different purpose compared to PCA where we are trying to optimize a numerical value which can be assessed, for example, by RMSEP (Section 4.3.1) and will always change slightly according to the number of components: it we are looking at classification accuracy, one model may classify 96 and the other 97 out of 100 samples – there is not a continuous scale, and so statistics such as RMSEP and RSS are not so easy to translate into PLS-DA although they can be used in PLS-R.

Although cross-validation is often very effective for many numeric calculations there is a problem when using it for classification. If, for example, there are 100 samples, and only 10 or so around the borderline, often the cross-validated models will be very flat near the optimum as the %CC can be easily influenced by the assignment of one or two samples. However a more serious problem relates to validation. If we want to validate a model we need to split it into test and training sets. The model must be formed (and optimized) on the training set. If we repeatedly form the training set, then there will be different models which may have a different optimum number of components for each training/test set split or iteration. The problem here is that the boundary and so the model is very critically dependent on which samples are included in the training set and so different solutions are obtained according to the iteration or training/test set split. This is illustrated in Figure 8.16, for Case Studies 11 (null), 8a (mouse urine – diet study), 3 (polymers – 2 types) and 9 (NMR spectroscopy), using flowchart 3E (Figure 4.19) for computing column means and standard deviations as appropriate. Note that these data have not been reduced

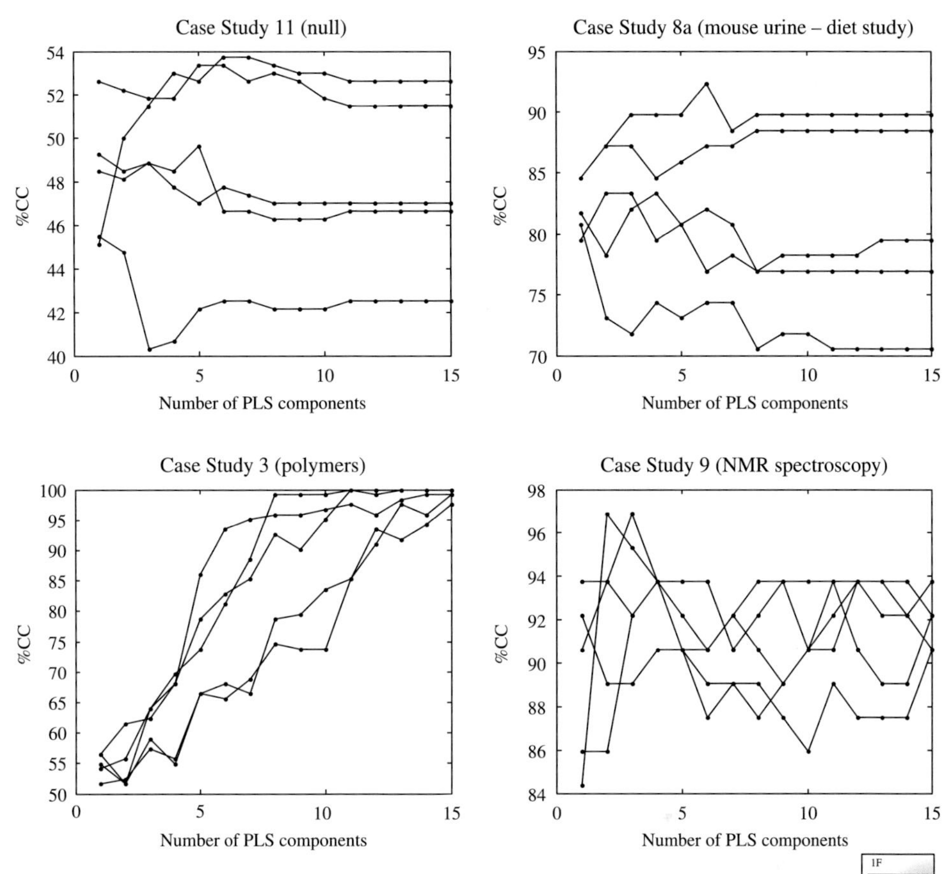

Figure 8.16 Results of 5 LOO cross-validation runs on Case Studies 11, 8a, 3 and 9 using a training set that consists of a random selection of 2/3 of the samples (of the smaller class, balanced by equal numbers from the larger class)

first by PCA; in fact, using PLS, this is usually unnecessary and an advantage of PLS is that it can be applied to the raw data direct (unlike QDA or LDA, for example – when there are more variables than samples), although we have used PC reduction prior to performing PLS previously so that different methods can be compared directly. It can be seen that each time a training set is formed different results are obtained, and it is hard to determine what is a stable solution. Therefore cross-validation cannot necessarily find a stable maximum. This is primarily because we are dealing often with small changes in the numbers of samples being misclassified, and in fact the nature of the split between training and test sets can be more significant than the number of components selected. Were we optimizing a numerical value, such as in a calibration, this would not be such a serious drawback, but in classification each sample results in an integral difference in the number of correctly classified samples. However cross-validation is not computationally intensive and was developed at a time when computing power was more limited, and for more computationally intensive algorithms (e.g. SVMs) still has a major role.

Bootstrap

An alternative is the bootstrap, first introduced in Section 4.3.1, but discussed in more detail below, which, in the context of classification, can be implemented as follows:

1. For optimizing models the training set is split into two subsets: a bootstrap-training set and a bootstrap-test set. Note that in the implementation described in this text we recommend that there are equal numbers of samples from each class in the training set, which should result (see below) in approximately equal numbers of samples from each class in the bootstrap-training set. Of course, other criteria could be applied. This division is illustrated in Figure 8.17.
2. The number of samples in the bootstrap-training set of the class is the same as the number of samples in the training set of the class. Samples for the training-bootstrap set are chosen randomly with repetitions. This means that samples are selected I_{train} times

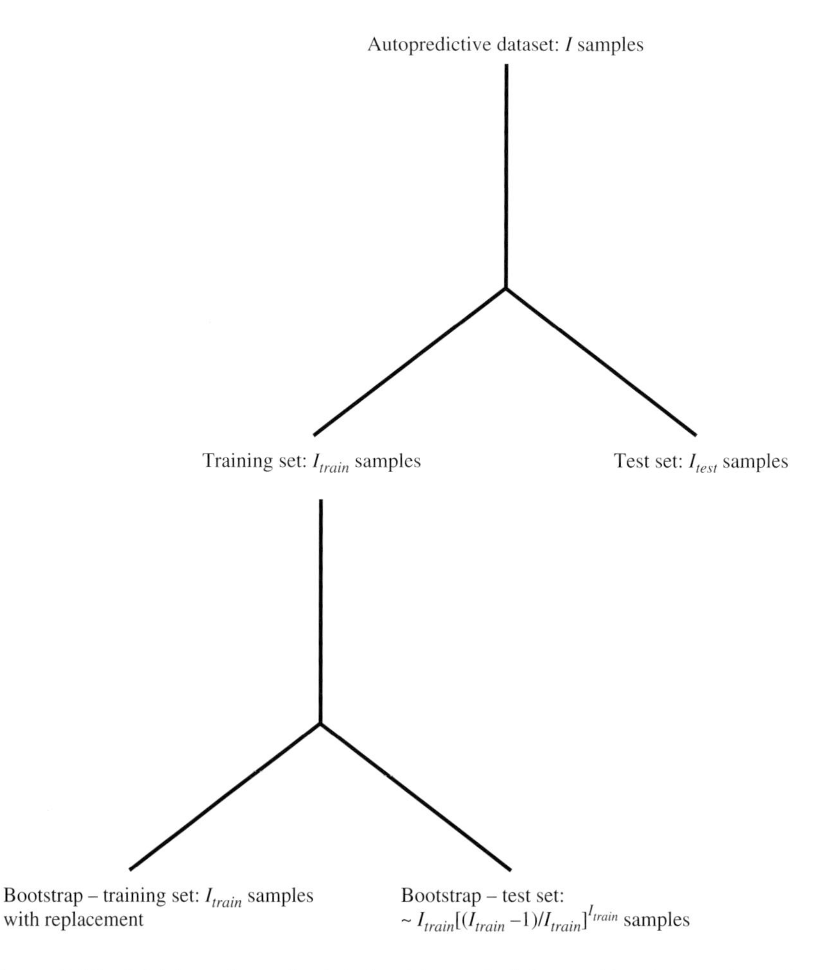

Figure 8.17 Division of samples into bootstrap – training and bootstrap – test sets

randomly, some of which are selected more than once. Note that if a sample is represented several times in the bootstrap-training set, we recommend that these repeated occurrences are used for calculations of the mean and standard deviation of the bootstrap-training set – see flowchart 3B (Figure 4.19) – so if a sample is present three times, it is represented three times when calculating the column parameters.

3. The remaining samples, never chosen for the bootstrap-training set, belong to the bootstrap-test set. It can be shown that if I_{btest} is the number of samples in the bootstrap-test set and I_{train} is the number of samples in the overall training set, then $(I_{train}-1)/I_{train}$ is the probability of not being chosen for the bootstrap-training set for one sample (= the probability of being chosen for the bootstrap-test set). This procedure is repeated I_{train} times, on all I_{train} samples so that $I_{btest} \approx I_{train}[(I_{train}-1)/I_{train}]^{Itrain}$. As guidance if there are 50 samples in the training set, the approximate number in the bootstrap-test set is 18, for 100 samples 37 and for 200 samples 73. The number will vary each time the bootstrap is performed. Table 8.13 illustrates this procedure for a small set of 10 samples. In the first column, 10 samples are chosen randomly while some samples are chosen more than once, and so appear more the once in the bootstrap-training set. Three samples are never chosen, in this Case Samples 3, 5 and 6, and are assigned to the bootstrap-test set. According to our equation we expect on average 3.48 samples in the bootstrap-test set, and so the number of samples in this illustration is well within expected limits.

Table 8.13 *Illustration of the choice of samples for the bootstrap using a dataset of 10 samples numbered from 1 to 10. Samples 1, 7 and 9 are each chosen twice for the bootstrap training set*

Bootstrap training samples	Bootstrap test samples
10	3
7	5
1	6
9	
8	
7	
4	
1	
2	
9	

4. The number of components that give the best model (in our case highest %CC) for the bootstrap-test set is chosen as the optimum number of PLS components. Note that the bootstrap-test set is not really a validation set but one that is used for optimization.
5. Generally the bootstrap procedure is repeated a number of times, typically 200 times, each repetition involving repeating the random split into bootstrap-training and bootstrap-test sets. Each of these repetitions may result in different answers for how many PLS components are optimal for a specific problem, although in many cases this will not differ dramatically. The number of repetitions can be adjusted if need be for any specific dataset by looking at the standard deviation in %CC as the number of repetitions is increased, just as for the test and training set split discussed in Section 8.4.2.

However, unless there is a serious problem in computing power we recommend a default of 200 times whilst not being prescriptive, which we use in this text

6. There are various approaches to choosing the number of PLS components that are optimal for any training set, but a common one, which we will use, is to choose the number of PLS components that are most frequently obtained, for example, if over 200 bootstraps 2 components are chosen 40 times, 3 components 90 times, 4 components 40 times and 5 components 30 times, the optimal number of PLS components for the training set is selected to be 3.

The bootstrap is inherently more stable than cross-validation in that on the whole it results in sharper optima and is more reproducible for each training test set iteration as illustrated in Figure 8.18 the average bootstrap models based on five training sets equivalent to those of Figure 8.16, meaning that it is easier to find an optimum. For Case Study 11 (null) it may at first appear that the bootstrap is less reproducible, but in fact the vertical scale represents

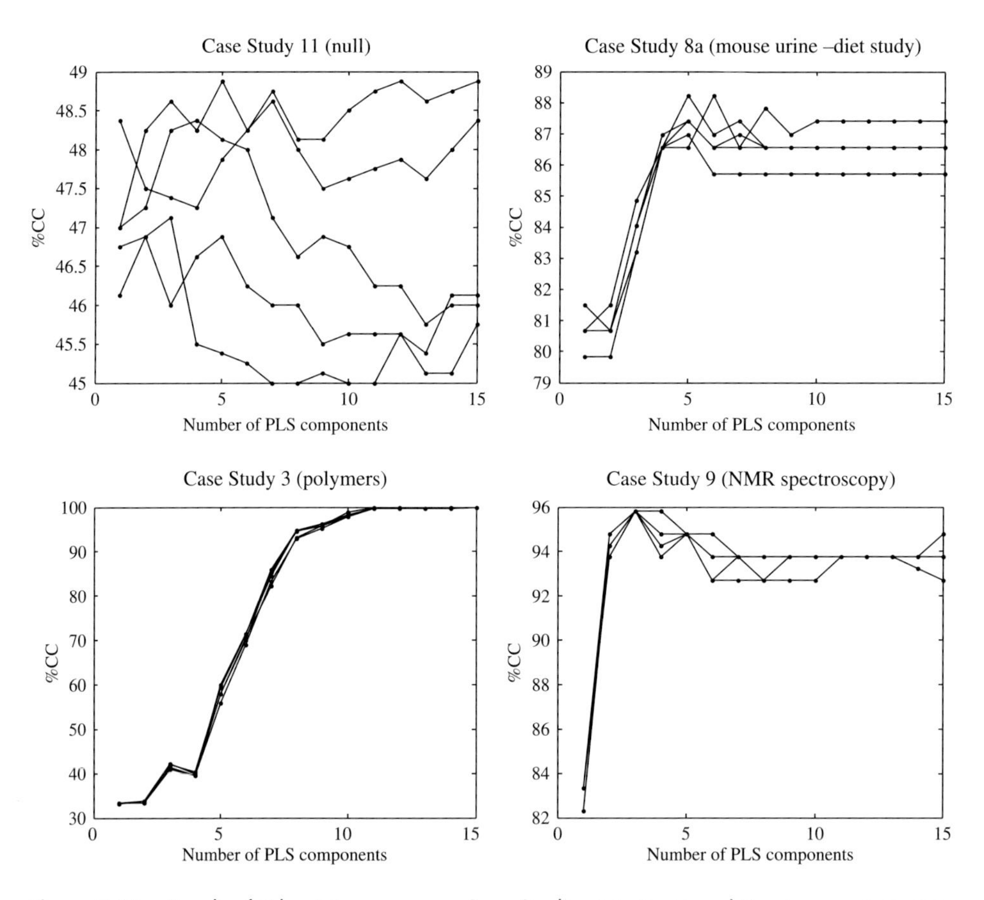

Figure 8.18 Result of 5 bootstrap runs on Case Studies 11, 8a, 3 and 9 using a training set that consists of a random selection of 2/3 of the samples (of the smaller class, balanced by equal numbers from the larger class). Each bootstrap run consists of 200 bootstrap test sets

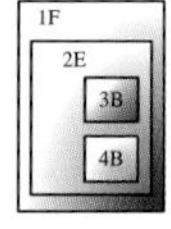

less than 5 %, compared to 15 % for cross-validation (Figure 8.16) and varies very little within a very narrow band.

A typical combined validation and optimization strategy may involve 200 bootstrap repetitions and 100 training/test set splits, or forming 20 000 models if feasible. For each iteration:

1. There is a random split into training and test sets.
2. The bootstrap is used to determine the optimum number of PLS components for the training set.
3. Then this number of components is used to build a model on the training set (which incorporates the bootstrap training and test sets).
4. Finally the model is applied to the test set (that has been left out of the optimization procedure) to determine the test set %CC which is usually taken as an indicator of the suitability of the model.
5. For each new iterative split of test and training there may be a different number of PLS components that best describes the training set. Hence each of these 100 splits may involve models formed using different numbers of PLS components.

This procedure, using 100 test/training set splits and 200 bootstrap-training sets for each split, is illustrated in Figure 8.19.

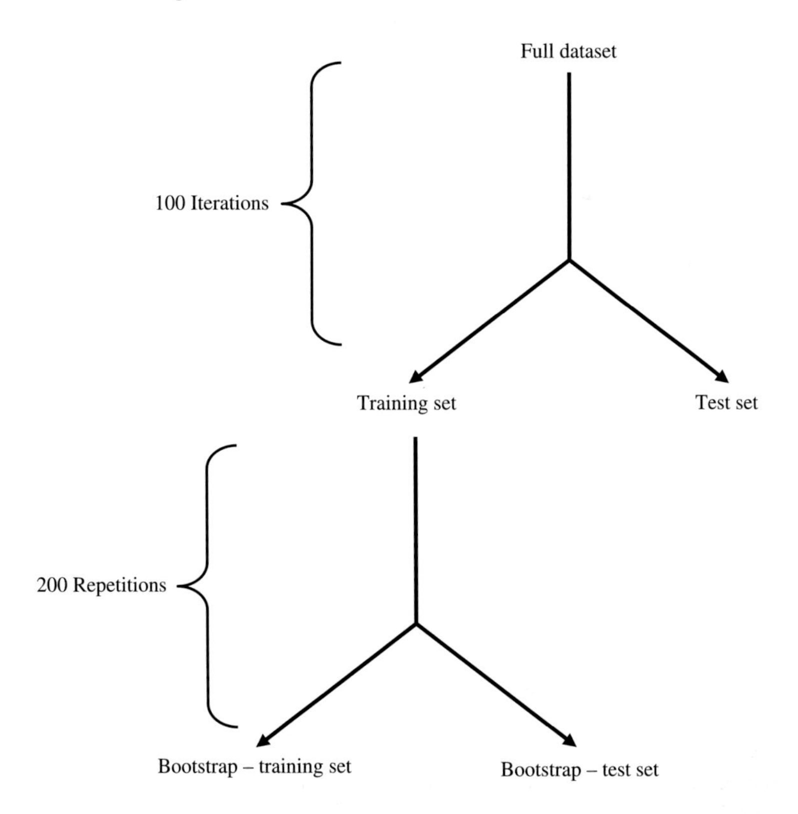

Figure 8.19 Typical iterative approach for splitting the data 100 times into test and training sets and 200 times for bootstrap

Once the bootstrap has been performed there are now many different %CCs that can be computed.

We define the following:

1. Autopredictive %CC is simply the proportion of samples that are correctly classified using an autopredictive model. If using the bootstrap we simply perform this on the entire dataset to obtain the optimal number of PLS components, and then perform autopredictive PLS using the number chosen.
2. Training set %CC. For each test/training set split an optimal number of PLS components is chosen using the bootstrap on the training set, and then a model is performed on the training set as a whole using this number of PLS components. For each iteration p (typically $P = 100$ in total) a different optimum number of PLS components is obtained and the samples selected for the training set are classified into groups. The overall %CC is obtained using 'majority vote', i.e. the class a sample is assigned to most times.
3. The test set %CC is similar in conception to the training set %CC.
4. Bootstrap-test set %CC. There are a variety of ways of defining this. We recommend the following:
   - For each of the test/training set splits, p, and (typically) 200 bootstrap repetitions r, samples will be associated with the bootstrap-test set several times.
   - Using the optimum number of PLS components in each of the R (typically 200) bootstrap-test sets for iteration p, classify those samples left in the test set.
   - For each of the P test/training set splits, obtain a bootstrap test set %CC by 'majority vote' for each sample, for example, if a sample occurs 60 times in the bootstrap-test set and if assigned correctly to class A 40 times and class B 20 times, it is considered to best belong to class A, if this is known to be a member of class A, correctly classified.
   - There will be a bootstrap-test set %CC for each of the P test/training set iterations: since different samples will be in the training set each time, we recommend averaging this to get an overall bootstrap-test set %CC.
5. Bootstrap-training set %CC. We define the %CC similarly to the bootstrap-test set %CC; however if a sample is included more than once in a training set it contributes more than once to the majority vote, and so if a sample is included three times in the bootstrap-training set and correctly classified three times, its classification will be corresponding weighted three times. Of course there could be other possible definitions.

In Table 8.14 we illustrate the calculation of these %CCs for Case Studies 11 (null), 8a (mouse urine – diet) and 9 (NMR spectroscopy). For the autopredictive model three PLS components were found as an optimum for Case Study 11, four for Case Study 8a and three for Case Study 9. It can be seen that encouragingly, the test set and bootstrap-test set %CCs are all around 50 % for Case Study 11 and between 90 % and 95 % for the other case studies (remember that PLS has been performed on the entire dataset and not the PC reduced dataset in contrast to Table 8.5 and Table 8.6 which previous results were performed on the PCs for comparison with LDA and QDA models among others), and so we anticipate slightly higher %CCs. The autopredictive errors all suffer from overfitting with a particularly high training set %CC for Case Study 11, and 100 % predictive ability for Case Study 8a. We have discussed this problem in Section 4.3.3 in which we show that PLS-DA can be prone to overfitting for autopredictive models; however if performed correctly is safe to

Table 8.14 *%CCs associated with testing and optimization using PLS-DA*

	Case Study 11 (null)			Case Study 8a (mouse urine – diet study)			Case Study 9 (NMR spectroscopy)		
	Class A	Class B	%CC overall	Class A	Class B	%CC overall	Class A	Class B	%CC overall
Autoprediction									
Class A	72	30	**71**	100	0	**100**	95.83	2.08	**96.88**
Class B	28	70		0	100		4.17	97.92	
Training set									
Class A	84	17.25	**83.375**	100	0	**100**	100	0	**100**
Class B	16	82.75		0	100		0	100	
Test set									
Class A	45.5	50.75	**47.375**	91.52	6.67	**92.44**	95.83	6.25	**94.79**
Class B	54.5	49.25		8.48	93.33		4.17	93.75	
Bootstrap-training set									
Class A	98.29	1.57	**98.36**	100	0	**100**	100	0	**100**
Class B	1.71	98.43		0	100		0	100	
Bootstrap-test set									
Class A	50.55	49.06	**50.74**	88.26	7.27	**90.49**	98.25	4.37	**96.94**
Class B	49.45	50.94		11.74	92.73		1.75	95.63	

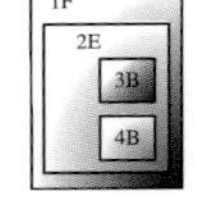

use on test sets. Note that the bootstrap-training set %CCs are very high for Case Study 11 (null) – near to 100 % suggesting heavy overfitting but the test set results are all very safe. In this text we will recommend using the bootstrap to determine the optimal number of PLS components for classification if desired, although this is by no means the only approach available.

8.5.2 Thresholds and ROC Curves

One limitation of the implementation of PLS-DA so far described is that it uses a decision threshold of 0 (for two class classifiers). Above this, a sample is classified as originating from class A (the 'in group') and below from class B (the 'out group'). However, the result of a classifier may be used in different ways. Consider performing separate tests for each group. A 'class A' test involves determining whether a sample has a predicted value of the classifier above a specified threshold. If above this threshold, but in fact a member of class B, this is a false positive (see Section 8.2 for definitions). A false negative results in a sample having a value below this threshold but being in fact a member of class B. For any threshold it is possible to calculate a contingency table as discussed in Section 8.2. Note that both a class A test (selecting samples above a threshold as being members of class A or the 'in group') and a class B test (selecting samples below a threshold as being members of class B or the 'out group') could be performed; however both will provide the same information when using PLS-DA and a two class classifier, and so we will restrict the discussion to a class A test. Given any threshold, four types of statistics can be obtained as follows:

TP – the number of members of class A that have been correctly classified
FP – the number of members of class B that have been incorrectly predicted as class A
TN – the number of members of class B that have been correctly classified
FN – the number of members of class A that have been incorrectly predicted as class B

as discussed in Sections 6.6.2 and 8.2.1.

As the threshold is reduced, the number of samples predicted to be members of class A increases correspondingly, and so the number of FPs and TPs increase, and at each threshold we could produce a different contingency tables. In the context of PLS-DA different contingency tables can be produced for any chosen threshold of c. These can be produced for all types of model and assessment including (1) the autopredictive model, (2) the average test set model after optimization, (3) the average training set model after optimization, (4) the average bootstrap-test set model and (5) the average bootstrap-training set model. The contingency tables for Case Study 8a (mouse urine – diet study) showing the threshold changes are presented in Table 8.15, the %CCs for a threshold of 0 being identical to those of Table 8.14. The effect of changing the decision threshold can also be viewed visually as illustrated for the autopredictive model based on 2 PCs for Case Study 3 (polymers) in Figure 8.20. If c is less than 0 then the boundaries move towards class A, as more samples are rejected from class B (label –1) and accepted in class A (label +1). Changing the decision threshold has a similar influence on predictive abilities as changing the column mean as illustrated in Figure 5.10 or by using Bayesian priors as discussed in Chapter 10.

Table 8.15 *Contingency table for Case Study 8a (mouse urine – diet study) for PLS-DA using different thresholds*

	Threshold = –0.5			Threshold = 0			Threshold = 0.5		
	Prediction (%CC)			Prediction (%CC)			Prediction (%CC)		
	Class A	Class B	CC overall	Class A	Class B	CC overall	Class A	Class B	CC overall
Autoprediction									
Class A	100	10	**94.96**	100	0	**100**	89.83	0	**94.96**
Class B	0	90		0	100	0	10.17	100	
Training set									
Class A	100	1.67	**99.17**	100	0	**100**	98.31	0	**99.16**
Class B	0	98.33		0	100		1.69	100	
Test set									
Class A	93.22	39.17	**76.89**	91.52	6.67	**92.44**	66.95	0	**83.61**
Class B	6.78	60.83		8.48	93.33		33.05	100	
Bootstrap-training set									
Class A	100	0.46	**99.77**	100	0	**100**	98.82	0	**99.41**
Class B	0	99.54		0	100		1.18	100	
Bootstrap-test set									
Class A	96	53.9	**71.05**	88.26	7.27	**90.49**	51.24	0.15	**75.55**
Class B	4	46.1		11.74	92.73		48.76	99.85	

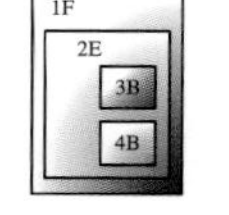

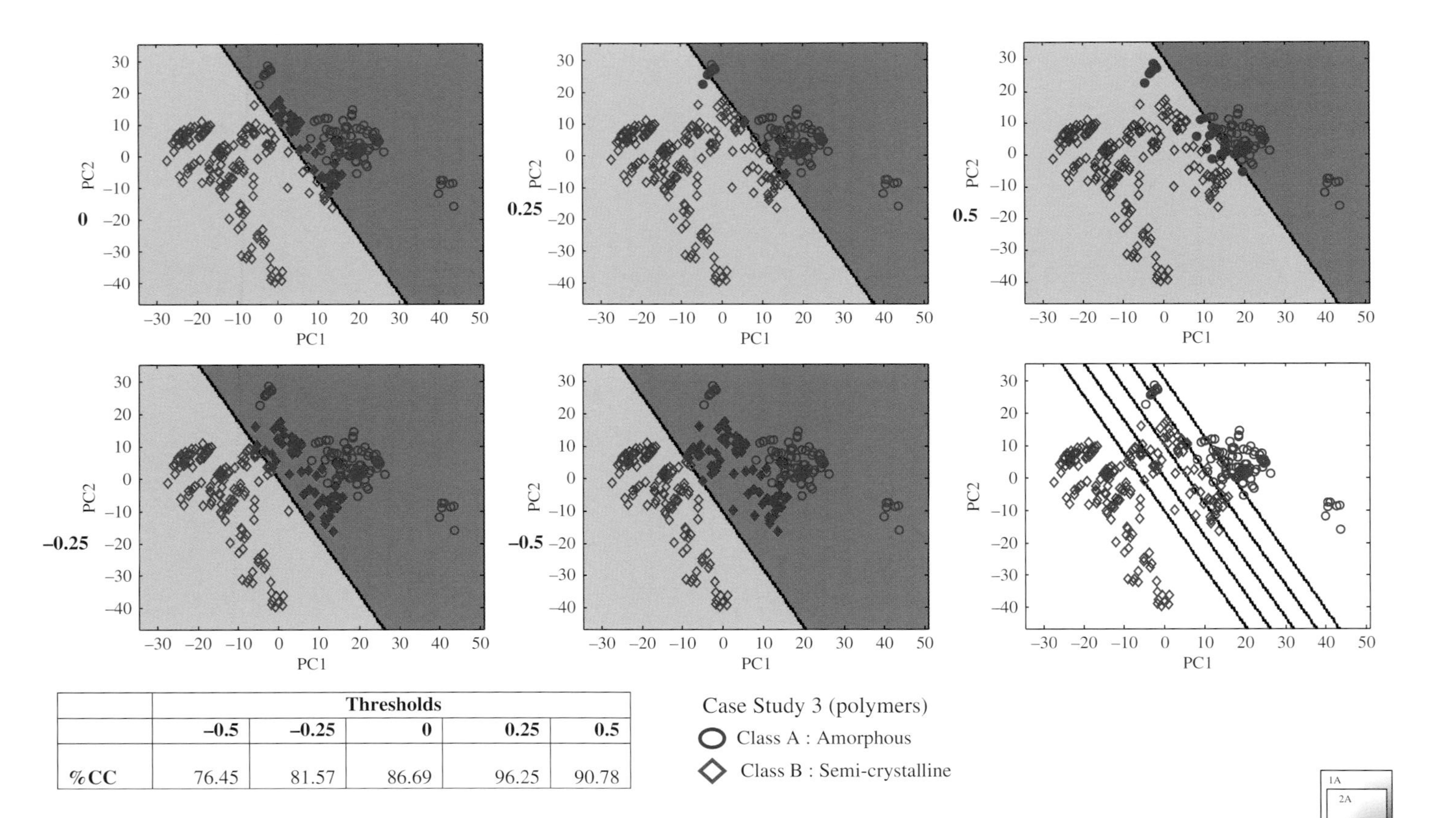

	Thresholds				
	−0.5	**−0.25**	**0**	**0.25**	**0.5**
%CC	76.45	81.57	86.69	96.25	90.78

Figure 8.20 Illustration of how changing PLS-DA decision thresholds influences an autopredictive contingency table, by changing the position of the decision boundary

Another way of visualizing the model performance is via an ROC (Receiver Operator Characteristic) which involves plotting the proportion of TP (vertical axis) against the proportion of FP (horizontal axis), as the threshold changes, as discussed in Section 6.6.2 in the context of one class classifiers. This allows a visualization of whether there are optimal conditions that minimize errors due to FPs or due to FNs as desired, and can provide guidance as to optimal choice of threshold. These curves have important characteristics according to how disjoint or otherwise the two classes are. Models that randomly assign samples into two groups tend to have ROC curves that are along the diagonal axis and are poor classifiers without any discrimination, whereas ROC curves that consist of a vertical and horizontal line represent highly discriminating classifiers.

The quality of a model can also be assessed by the AUC (Area Under the Curve) which can be between 0 and 1, but in practice is unlikely to be much less than 0.5. In PLS-DA, ROC curves can be obtained for the autopredictive, training set, test set, bootstrap-test set and bootstrap-training set as appropriate. In Figure 8.21 the ROC curves are illustrated for the autopredictive, test and training sets for Case Studies 11 (null), 8a (mouse urine – diet study) and 5 (sweat), together with the AUCs, for PLS-DA models as the threshold is changed using the optimum number of PLS components obtained via the bootstrap. The most important are the test set ROC curves where we can see that there is no particular trend for Case Study 11 as anticipated for the null dataset, whereas Case Study 8a shows quite good discrimination with an AUC of 0.932. Note that the test set AUCs are normally considerably lower than the autopredictive and training set AUCs. Figure 8.22 represents the superimposed test set ROC curves for 100 test/training set splits as an indicator of the variability as different test sets are formed. For the training sets we calculate an average ROC curve of all 100 ROC curves obtained at each iterative split.

8.6 Optimizing Learning Vector Quantization Models

As in many methods originating from Machine Learning there are huge variations in how LVQ can be implemented, and each can lead to a different solution. Because of the large number of parameters, it is not possible to use such formalized methods for optimization as in PLS-DA. However the methods are often quite robust to many of the choices of parameters. The following describes the main groups of parameters and decisions. The reader is referred back to Section 5.6 for several of the diagrams and tables which will not be repeated for brevity:

1. Parameters used during development of the network. There are several such parameters, as discussed in Section 5.6. Recommended values have been presented in Table 5.3. In practice they can be varied within quite a broad range without seriously influencing the model.
2. Choice of initial codebooks. There are several approaches for this. One problem is if a codebook that seeds a specific class is an outlier, e.g. is within the region of another class by accident; this can sometimes happen. There are several ways of preventing this; one is to use a method such as *k*. Nearest Neighbour or *k* means clustering to choose these initial vectors so that LVQ has a starting point. The reproducibility of boundaries

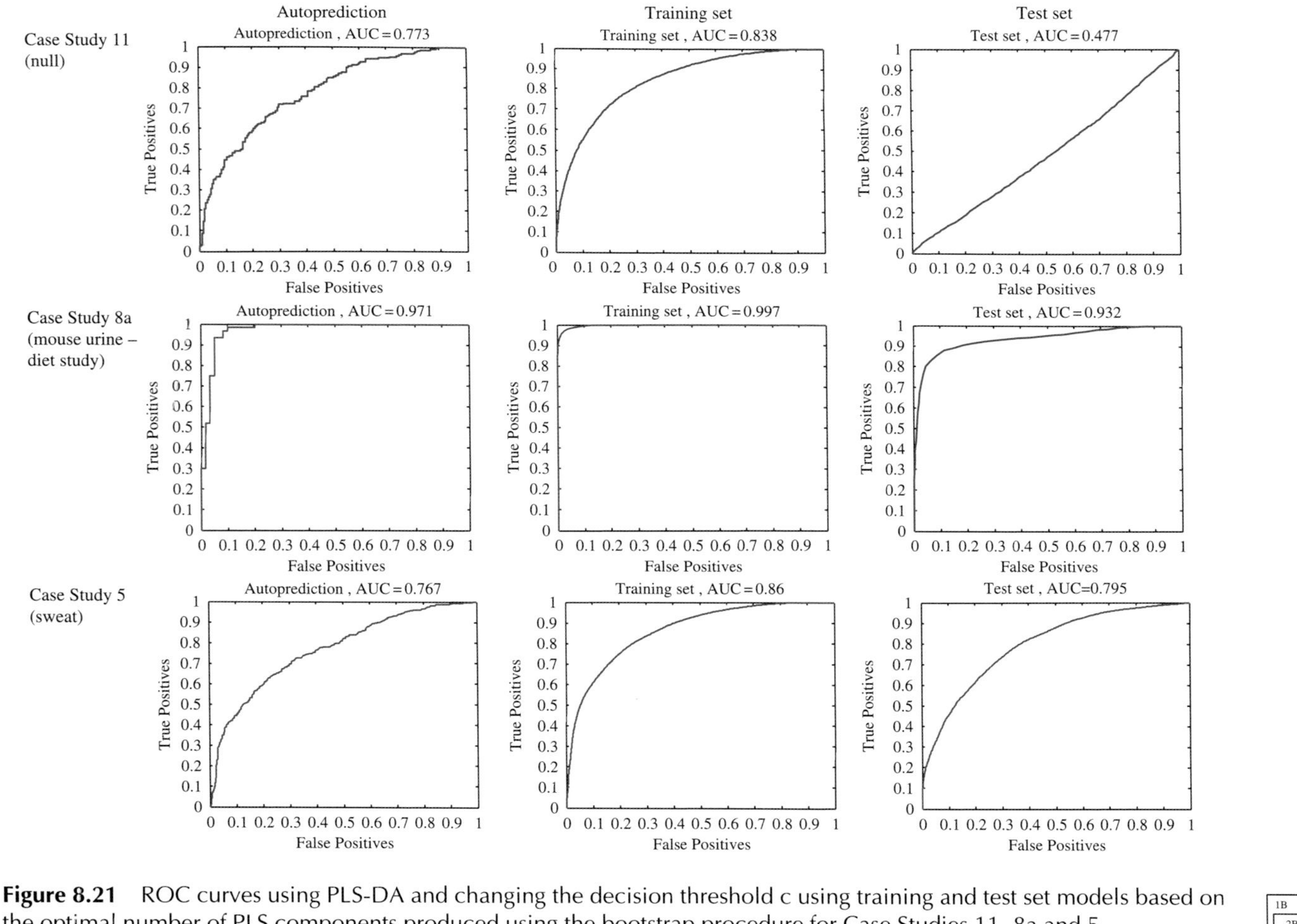

Figure 8.21 ROC curves using PLS-DA and changing the decision threshold c using training and test set models based on the optimal number of PLS components produced using the bootstrap procedure for Case Studies 11, 8a and 5

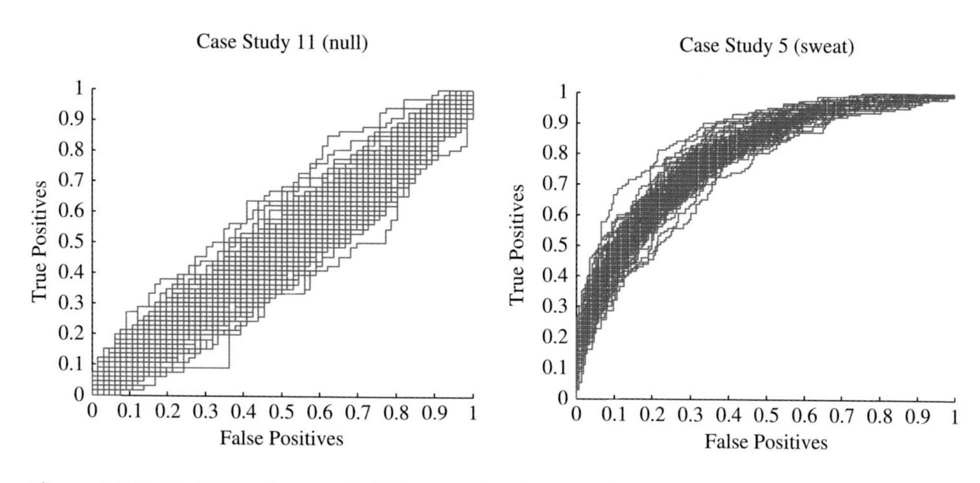

Figure 8.22 Stability of test set ROC curves for Case Studies 11 and 5 over 100 iterations (PLS-DA using an optimal number of PLS components chosen by the bootstrap)

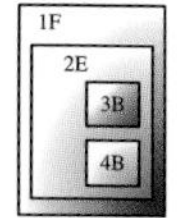

using different initial guesses of codebooks is illustrated in Figures 5.13 and 5.14. If iterative methods are used to assess LVQ models, e.g. using 100 test and training sets, this irreproducibility of the boundary will be so that after several iterations a good consensus will be reached overcoming the problems of just running LVQ once. If a single LVQ model is to be formed, and each class is characterized just by a small number of codebooks, where classes are disjoint, e.g. consist of several distinct regions, there can be problems and so careful choice of initial guesses could be important.

3. Choice of number of codebooks. This is the most critical decision, and relates to the complexity of the boundary between classes. Figure 5.13 illustrates change in appearance of the boundaries as the number of codebooks per class is increased from 2 to 4 in a 2 PC autopredictive model: the %CC increases as the number of codebooks is increased. Of course datasets will normally contain far more than two variables and so this example is a simple one for illustrative purposes, but a complex model with too many codebooks may overfit the data. However in terms of optimization, providing that the only parameter is the number of codebooks per class (related to the model complexity) approaches such as the bootstrap can be employed, since only one parameter needs to be adjusted. The number of codebooks in LVQ has some relationship to the penalty error C in SVMs, and relates to model complexity.

Of course there are other potential sophistications and there is a large literature in LVQ algorithms. The main elaboration in the context of chemometrics is to be able to model different classes using different numbers of codebooks, for example, one class could be modelled by 3 codebooks and the other 6. This however makes optimization slow and should only be done if there a specific need, as it can lead to training set boundaries that are unjustifiably complex. In this book we will settle on basic methods which are probably sufficient as a first step in the area of pattern recognition.

8.7 Optimizing Support Vector Machine Models

There are a large number of approaches for optimizing SVMs, and, unlike PLS-DA, less general agreement. We have discussed how the model is developed in Section 5.7, and will not repeat this here, but will primarily be concerned with how the adjustable parameters influence the appearance of SVM boundaries and hence the %CC. This aspect of SVM is normally routinely implemented in most SVM toolboxes.

The main issue involves optimizing what chemometricians call tuneable parameters. A recommended default approach is to use a soft margin L1 RBF (Sections 3.7.2 and 3.7.3). For such models there are two common parameters, as follows:

1. For RBFs, the radial width σ, where $K(\boldsymbol{x}_i, \boldsymbol{x}) = \exp(\|\boldsymbol{x}_i - \boldsymbol{x}_j\|^2)/2\sigma^2$
2. In addition, the value of C or penalty error: the higher this is, the more significant a misclassification in the training set is and the more complex the boundary.

It is important to realize that SVMs using appropriate parameters could result in perfectly classified training sets, but this may result in overfitting and the test set results may appear more or less random. In order to understand how these two parameters influence the appearance of the boundaries our datasets, we look at the RBF kernel function and separating plane for the PC projections of Case Studies 4 (pollution) and 11 (null) in Figure 8.23. The left-hand side graph is the projection onto 1D with the value of SVM RBF decision function as the vertical axis, and the horizontal axis corresponding to scores of PC2, showing the decision boundary plane; the central graph illustrates the RBF decision function (mesh), together with the original data points, and the left-hand side graph the projections into data space. SVs are indicated with a cross. Class A is indicated in blue and class B in red. Non-SVs are projected onto this grid. The plane cuts through the grid. Class A regions are where the samples lie above the plane and class B below, as we have labelled class A as +1 and class B as –1.

For Case Study 4 (pollution) we can see the effect of penalty error by seeing how the boundaries change for $C = 1$ from 0.1 to 100. At low values of C there are more SVs, and higher misclassification rates. This is because there is a greater tolerance of misclassification. We see for $C = 0.1$ that all samples from class B and roughly equal numbers for class A (which consists in total of 179 samples compared to only 34 in class B), are SVs. The surface is fairly smooth – the more the SVs, the smoother the surface – and class B is represented by a fairly well defined 'well' in the surface. As C increases, the number of SVs decreases and the surfaces becomes less smooth, allowing for bumps, and so by the time $C = 100$, the surface now has two dips: there are very few SVs and these are close to the rather more complex boundary. Only one sample is misclassified (from class A) and that is close to the boundary. The influence of the RBF can be seen by comparing graphs using $C = 1$ and $\sigma = 0.1$, 1 and 10. A σ of 0.1 results in a number of sharp spikes which well separate the two groups in this case. Class B samples 'puncture' the separating plane. Almost all samples are support vectors, which then result in the grid formed from the RBF kernel being essentially planar with spikes on it representing SVs as the RBF is very narrow, and so each SV is characterized by a narrow Gaussian. The boundaries are very well defined around class B samples: because the majority of samples come from class A the 'background' is class A; almost all samples are SVs. Using a σ of 10 has the effect of spreading out the Gaussians so they are effectively flat and the kernel function becomes

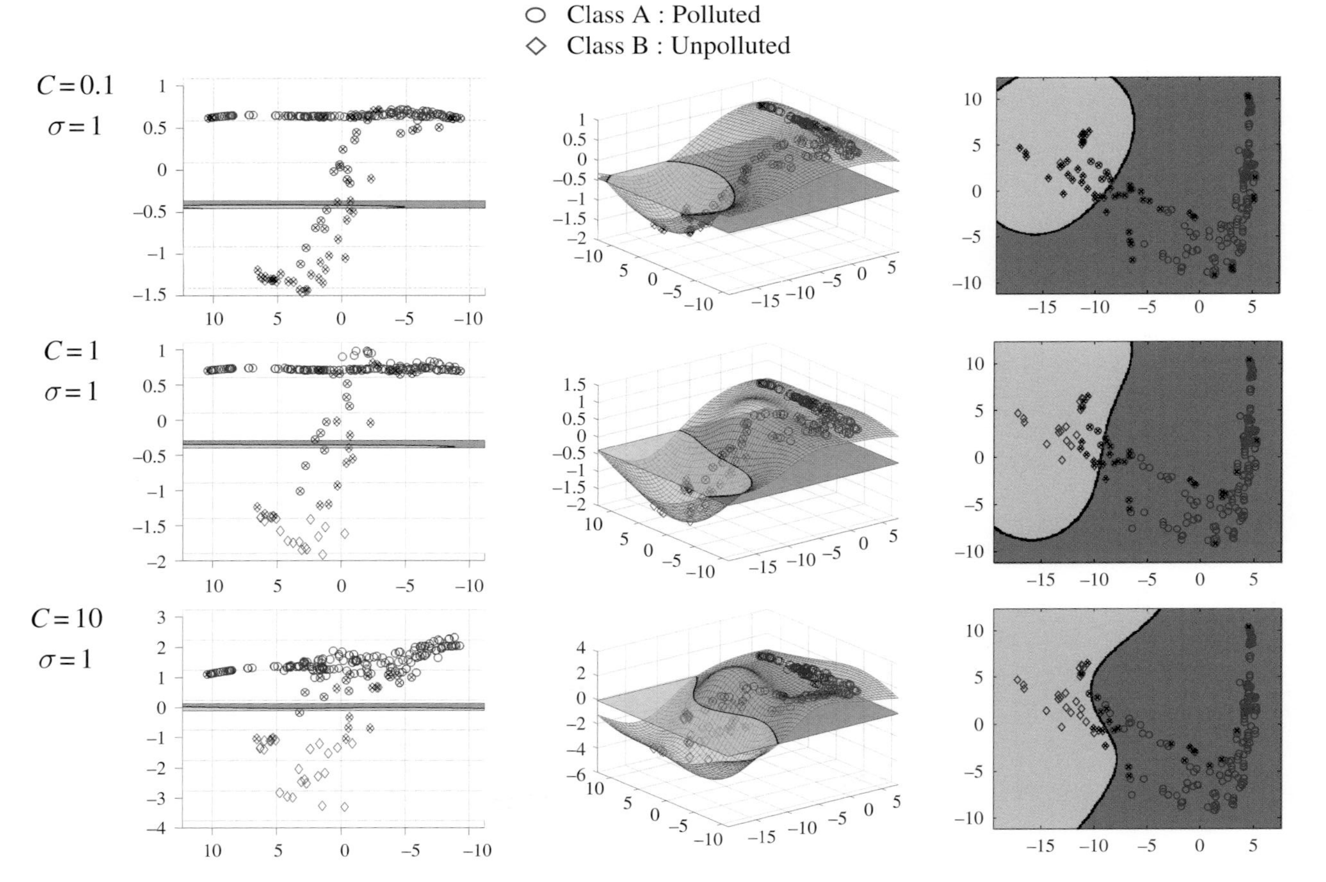

Figure 8.23 Influence of penalty error (C) and RBF σ (in units of mean standard deviation of training set variables) on SVM boundaries. For details see text

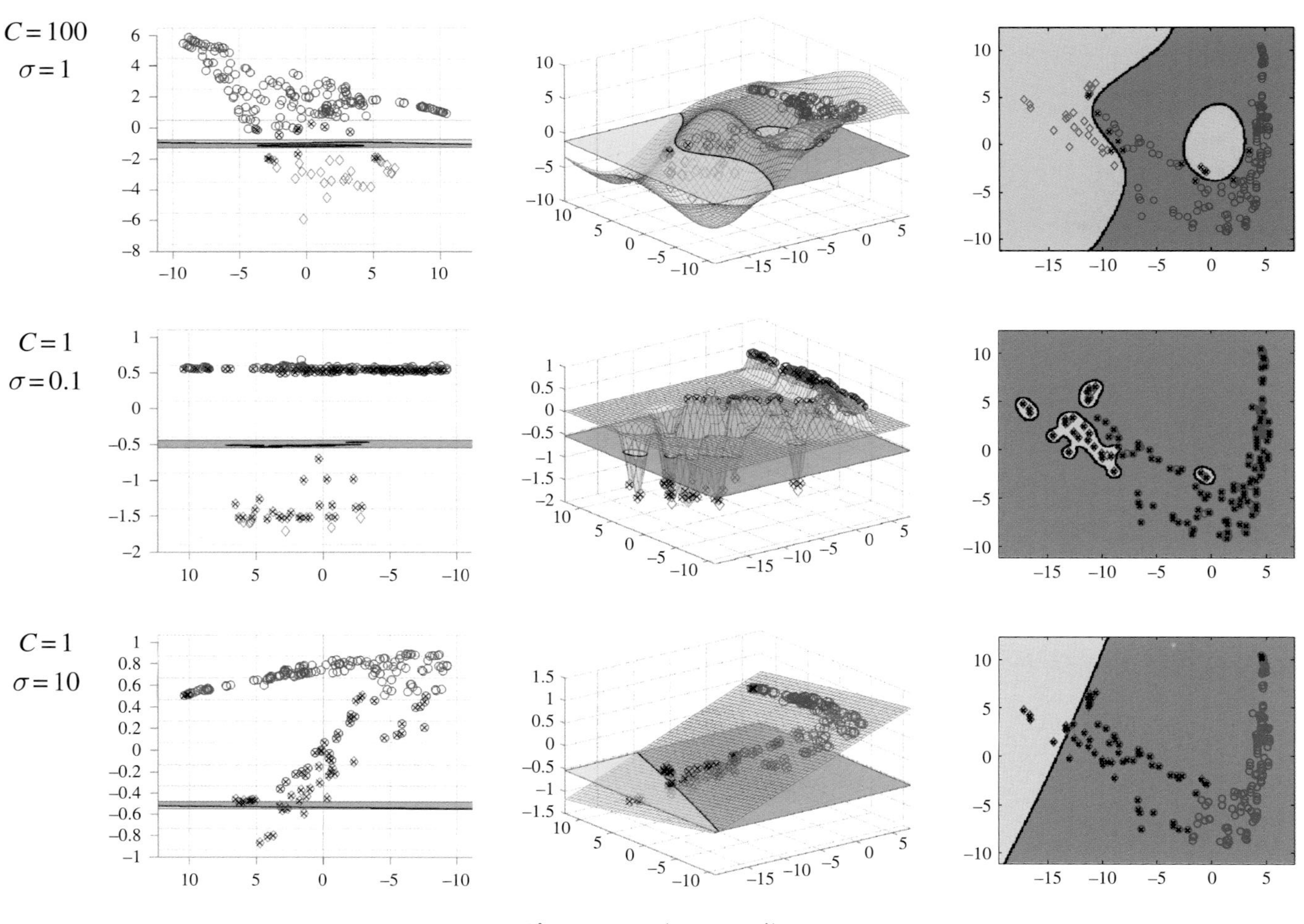

Figure 8.23 (continued)

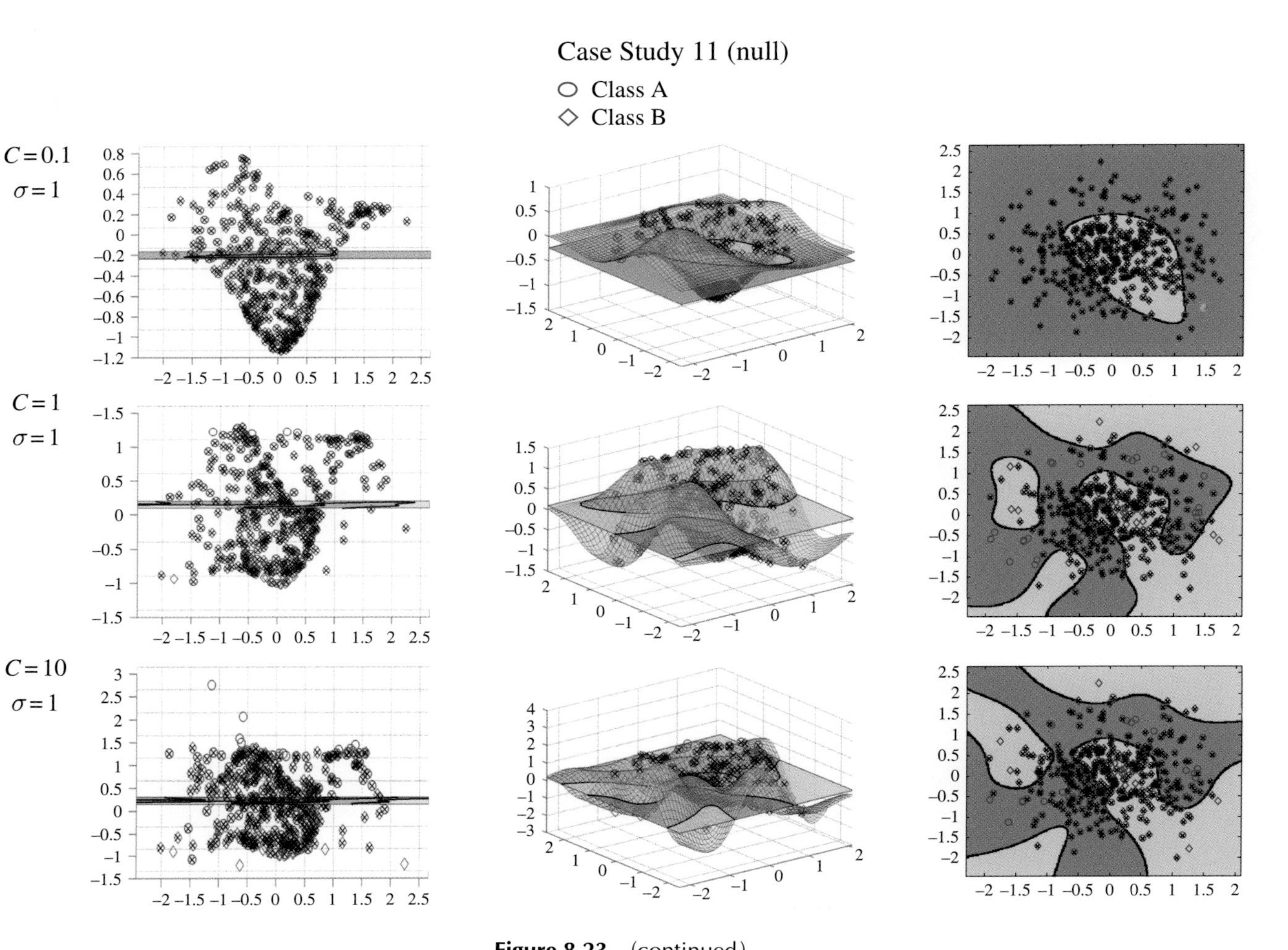

Figure 8.23 (continued)

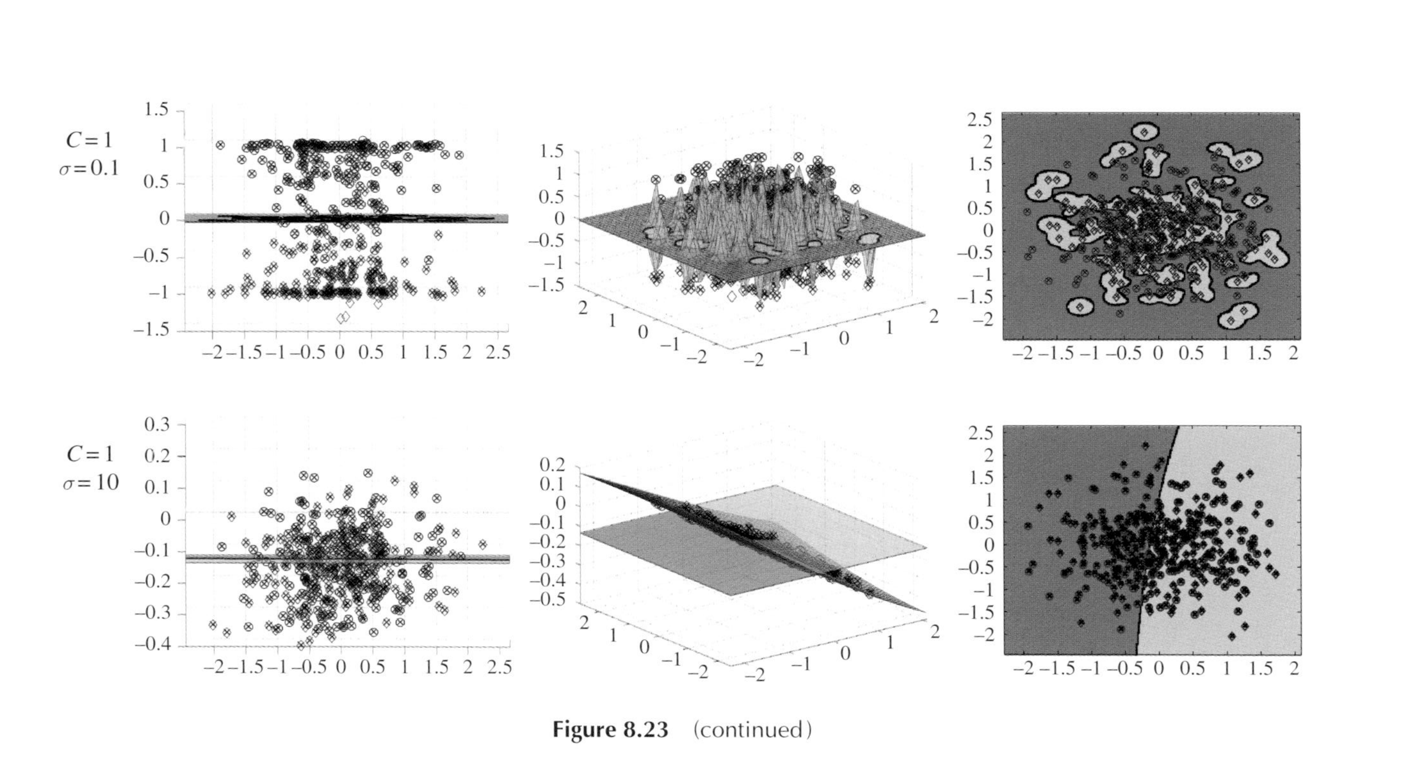

Figure 8.23 (continued)

almost planar, resulting in a linear boundary between the classes. The higher the σ, the broader the Gaussians and hence the smoother the surface formed from the kernel function and so the more likely the boundary is to be linear: a low σ results in sharp peaks and so potentially boundaries that can surround samples; however the number of samples that are used for forming the SVM model relate to C and so if C is high this number reduces.

Case Study 11 confirms these trends, with a small value of σ of 0.1 resulting in individual groups of samples from class B being surrounded by tight boundaries. Note that because the separating plane is slightly below 0, the background belongs to class A, but this will change in a very sensitive way. A value of $\sigma = 10$ results in a boundary that is nearly linear due to the smooth nature of the RBF surface. One can also observe the effect of C. A high value reduces the number of SVs, and so the surface becomes more bumpy and the regions in data space somewhat more complex. For $C = 0.1$, in practice all samples are SVs and as such all lie on a fairly smooth surface. Since slightly more class B samples lie in the centre, there is a dip in the centre which defines a class B region.

In order to optimize these types of models, it is necessary to adjust two parameters, σ and C, and determine their influence on the test set predicted %CC. Unlike the more established approach of PLS there is less agreement how to optimize SVMs. In some ways this does not matter too much providing validation is performed correctly, if a model is slightly less than optimal; so long as the validation results can be trusted then the model gives a reliable picture – and a good rule of thumb might be to run it on the null dataset first to see what happens. In areas such as bioinformatics, the problems tend to be much larger and sometimes quite fundamental, and so spending days or weeks optimizing a one-off calculation on a large cluster of computers is quite acceptable, but in chemometrics, whereas we can certainly entertain much more computationally intensive approaches than a decade ago, nevertheless having a solution that can be obtained within a few hours, or a day or so, on perhaps a quad core computer is about the typical maximum size of the problem. If a comparable text were written a decade later, more computationally intense solutions would probably be more routinely accessible. Visual examples of changing C and σ are presented in Figure 8.24 and Figure 8.25 for Case Studies 8a (mouse urine – diet study), 4 (pollution), 11 (null) and 1 (forensic).on 100 training set models. Filled symbols are misclassified samples via the 'majority vote' test set. The influence of σ is quite dramatic; for very low values, it usually surrounds individual samples. However note that we are classifying samples from the perspective of a 'majority vote'. For Case Study 11 (null), we see most individual samples or small clusters of samples surrounded by boundaries, but some are correctly classified and others not. This is because a training set boundary may form, for example, 49 times or 51 times around a sample – if 49 times it will be misclassified and the solution is quite unstable. In Figure 8.23 we see that the SV autopredictive solution forms boundaries around class B (red group) samples. Whether the compact boundaries are formed around class A or class B samples is very finely balanced and depends on the exact position of the separating plane, and so we would expect this plane to change 'colour' dependent on which samples are part of the training set, and individual representations would oscillate between those with a blue background and tight boundaries around the red samples and vice versa. The overall training/test set boundaries are a combination and as such it appears all samples are

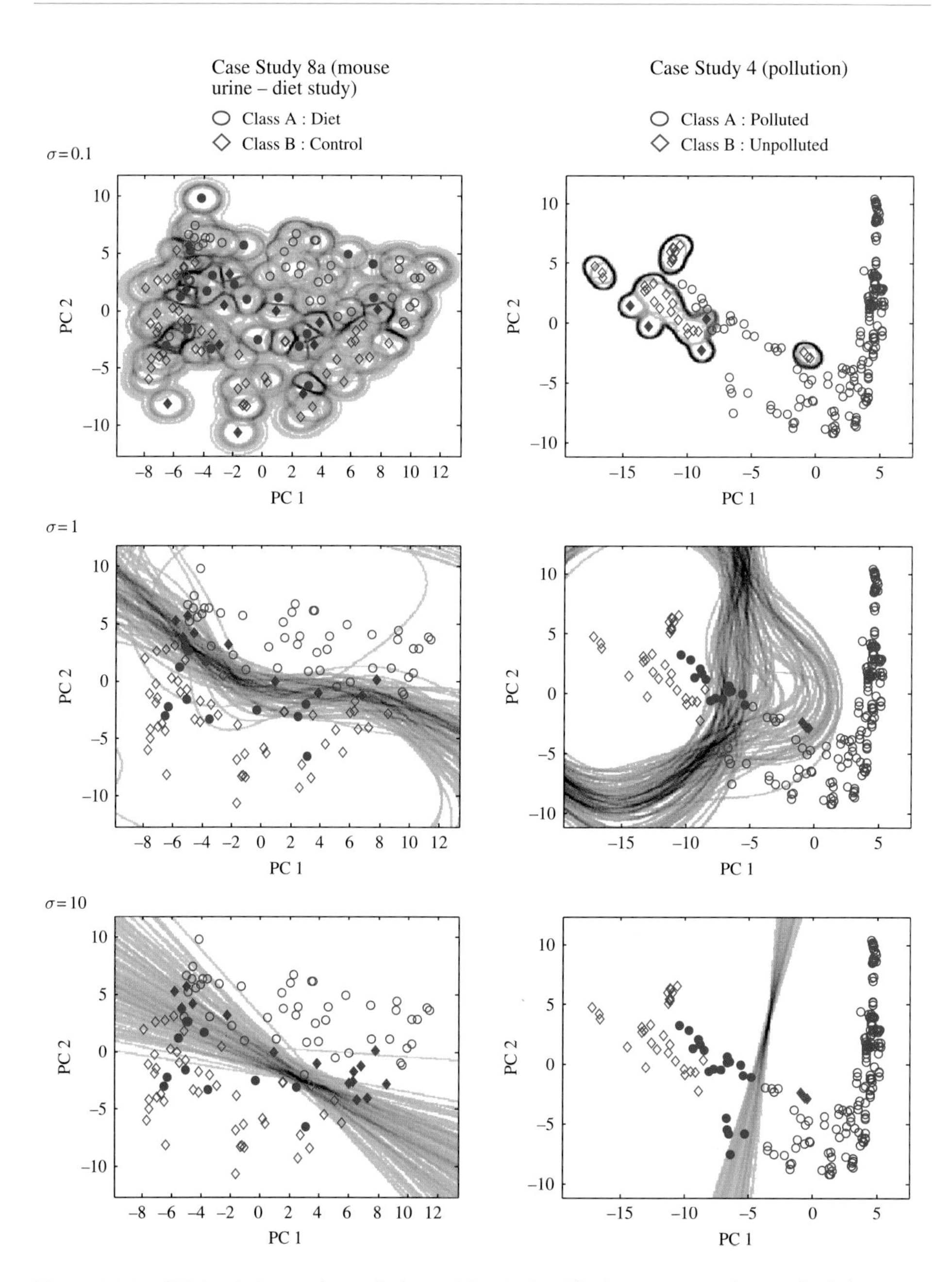

Figure 8.24 SVM training set boundaries, with misclassified test set samples marked, for Case Studies 8a, 4, 11 and 1, using 2 PC models , an RBF with $\sigma \times$ average standard deviation of the variables in the dataset, C = 1 and an L1 SVM

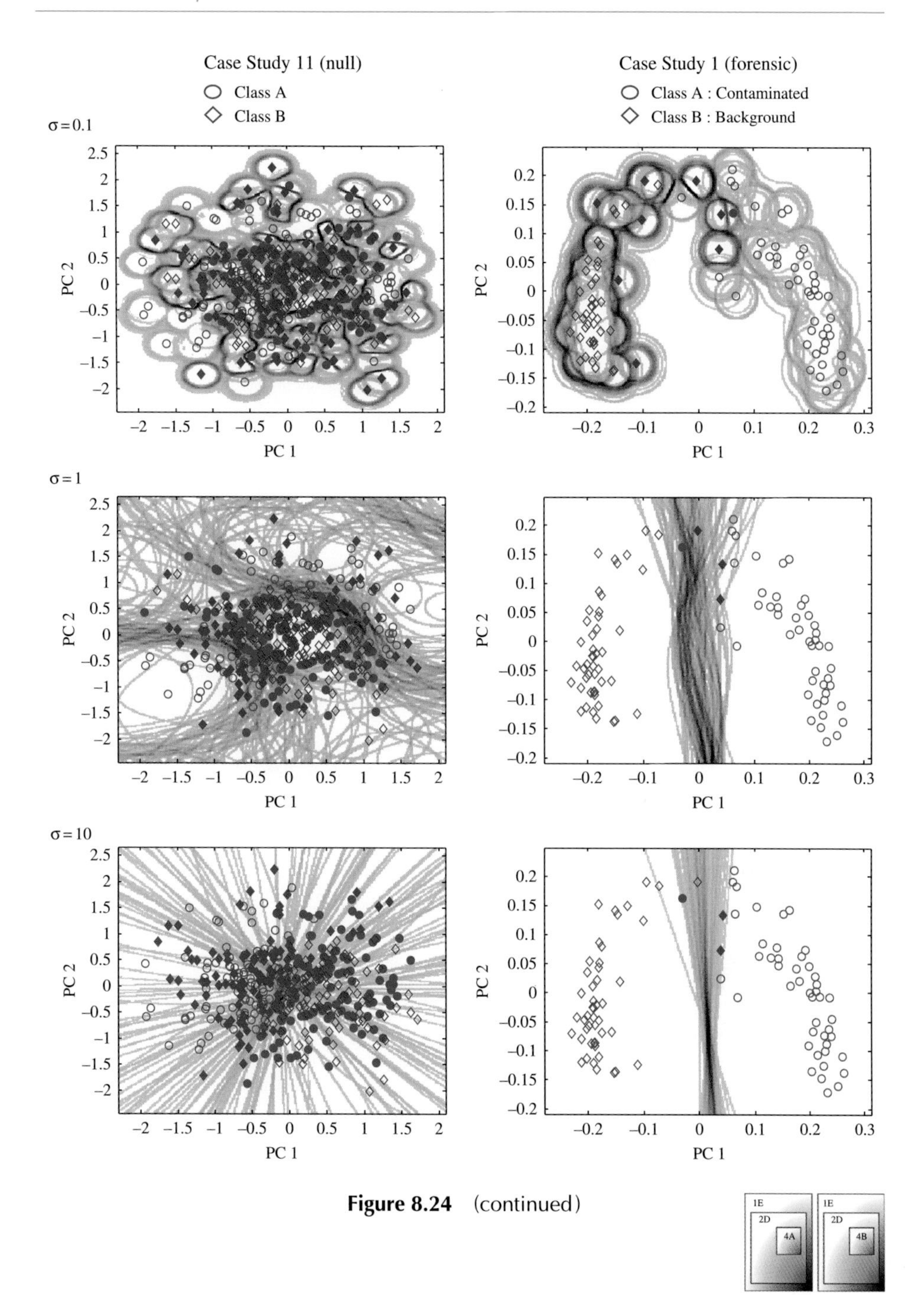

Figure 8.24 (continued)

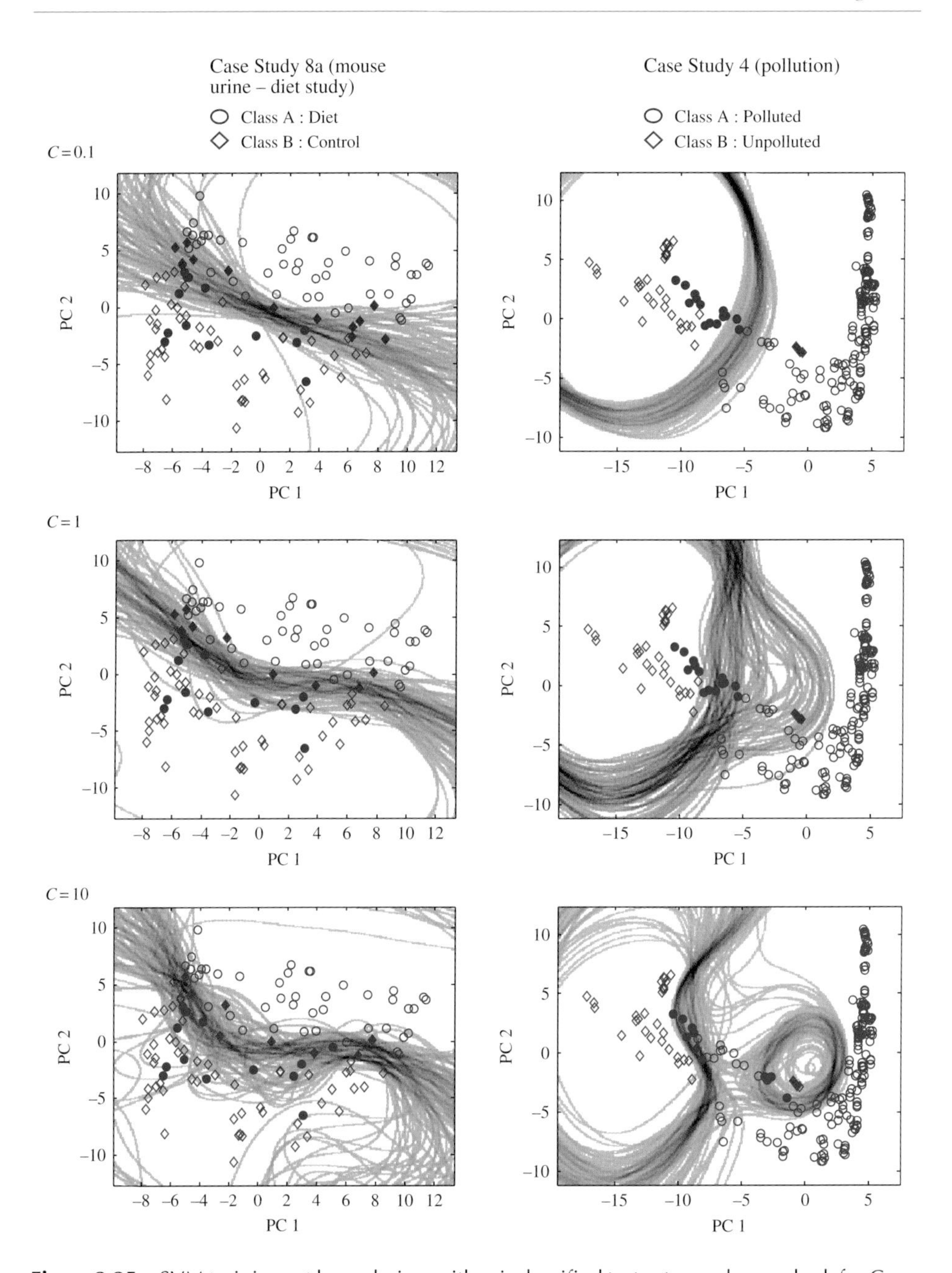

Figure 8.25 SVM training set boundaries, with misclassified test set samples marked, for Case Studies 8a, 4, 11 and 1, using 2 PC models , an RBF with $\sigma = 1 \times$ average standard deviation of the variables in the dataset, and an L1 SVM

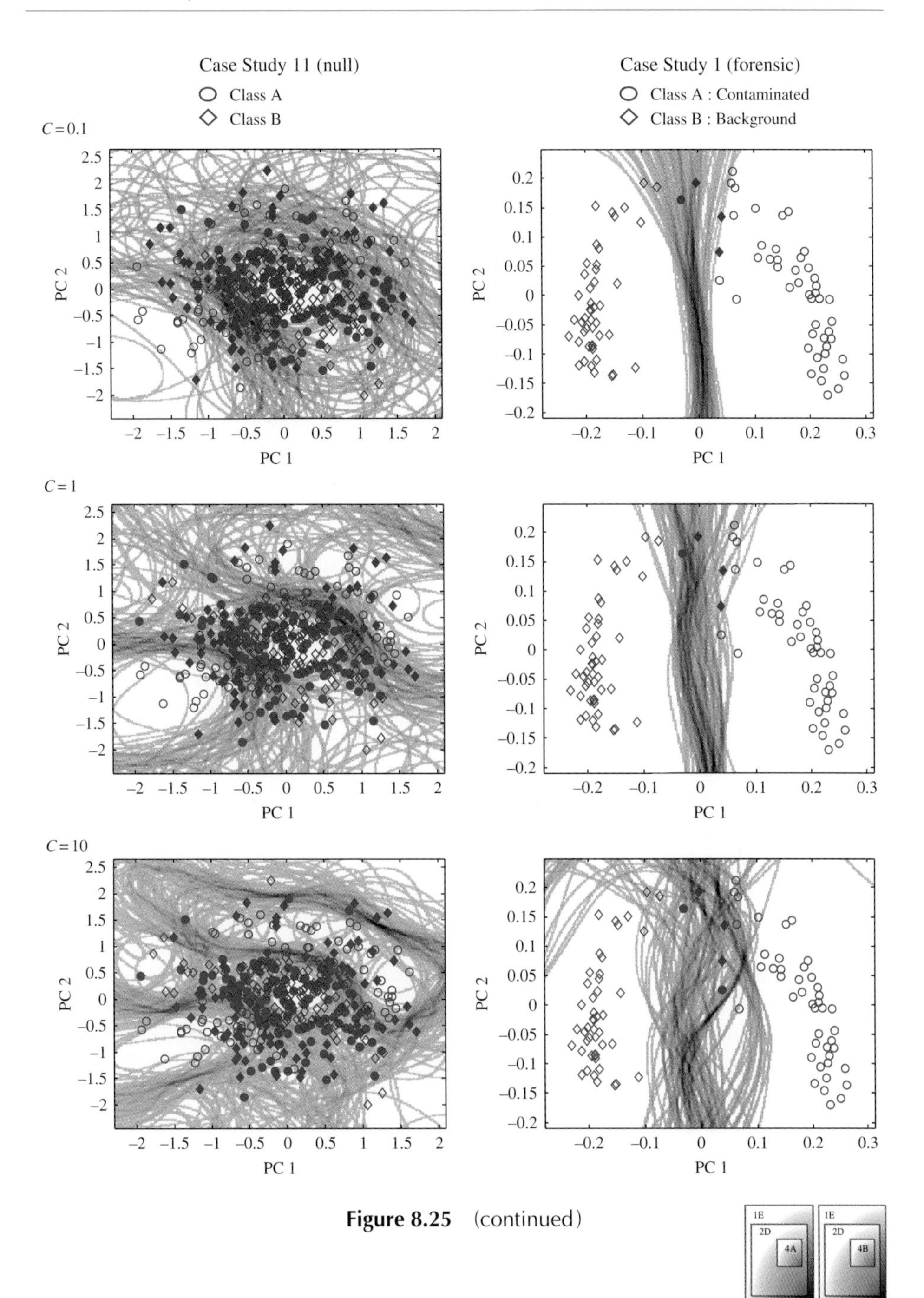

Figure 8.25 (continued)

surrounded by a compact boundary, whilst some are correctly classified and some not. The change in value of C appears to have less effect, but that is because we have chosen a value of $\sigma = 1$, and in fact, in cases where there is possibly some substructure (e.g. Case Study 4 – pollution) we can see that having a higher penalty error does allow for class B to be modelled as two separate regions. But interestingly the small group of three samples from soil 4, although in some iterations encompassed within the boundary of its own class, this happens only less than 50 % of the time, and so the three samples are not assigned to the class B once a consensus, 'majority vote', view is taken into account.

Formal computational optimization of SVMs in analogy to PC or PLS based models can be quite hard, and many people use visual or empirical ways of establishing the parameters. It is possible, for example, to perform a grid search involving 3 or 4 values of C and 3 or 4 values of σ – so 9 to 16 calculations – and then on each combination of parameters, perform cross-validation or the bootstrap to determine the optimal model on the samples left out. This can be very computationally intense, and there are several drawbacks. The first is that it is likely that there will be only a quite narrow range of σ that is suitable. For example, in practice a value of $\sigma < 1$ is unlikely to be suitable; and within this range the optimum might be quite flat. The second and quite important issue is that unlike in the cases of PLS or PCA, only certain samples (the SVs) are required to form a model. Hence removing samples that are not SVs from the dataset and forming a model without these samples, makes no difference to the solution, and so cross-validation and other methods should in practice be modified to take this into account – but of course as the SVM model parameters change, so do the numbers of SVs. There are in fact no hard and fast rules for SVMs unlike PLS and there is much less emphasis on optimization of tuneable parameters in the SVM literature compared to the PLS literature. This may come as a surprise to chemometricians. However SVMs have many more parameters that can be adjusted unlike PLS where the only parameter (apart from preprocessing which is important in all cases including the application of SVM models) normally of interest is the number of PLS components, and so optimization is a much harder problem. However most chemometrics datasets are rarely of sufficient complexity to merit the more elaborate SVM kernel functions, and so although SVMs definitely should be considered for chemometric pattern recognition, as datasets are generally somewhat less complex than in many other areas of applied science simpler methods such as PLS-DA or QDA have a proportionately larger role to play. In addition it is essential to understand that providing validation has been done correctly one is safe from overfitting, and problems occur if validation and model optimization are not clearly distinguished. SVMs however are an increasingly popular and powerful alternative especially with numerous public domain toolboxes, but the basic principles need to be understood prior to the correct application.

Bibliography

General

R.G. Brereton, Consequences of sample sizes, variable selection, model validation and optimisation for predicting classification ability from analytical data, *Trends in Anal Chem,* **25,** 1103–1111 (2006).

Definitions

S.J. Dixon, R.G. Brereton, Comparison of performance of five common classifiers represented as boundary methods: Euclidean Distance to Centroids, Linear Discriminant Analysis, Quadratic Discriminant Analysis, Learning Vector Quantization and Support Vector Machines, as dependent on data structure, *Chemometrics Intell. Lab. Systems,* **95**, 1–17 (2009).

M.S. Pepe, *The Statistical Evaluation of Medical Tests for Classification and Prediction,* Oxford University Press, New York, NY (2003).

C.D. Brown, H.T. Davis, Receiver operating characteristics curves and related decision measures: A tutorial, *Chemometrics Intell. Lab. Systems,* **80**, 24–38 (2006).

Optimisation

See Bootstrap and Cross-validation (Chapter 4), Learning Vector Quantisation (Chapter 5) and Support Vector Machines (Chapter 5).

9

Determining Potential Discriminatory Variables

9.1 Introduction

One of the most important aspects of pattern recognition, especially in biological and medical studies, is to determine which variables are best discriminators. The process of identifying these variables and then forming models on a reduced set of variables is sometimes called feature selection, variable selection, feature reduction or variable reduction according to author, although each word has a slightly different connotation. The process of just identifying these variables and not using this information further for modelling is sometimes called biomarker discovery when applied to biomedical applications. We will call this process 'determining potential discriminatory variables', but not focus on how these can be used for modelling in this chapter, although we will examine a few fallacies in the incorrect application of feature reduction for discrimination and show how it can be performed safely. Finding out which compounds, for example, can distinguish one or more groups, in itself is often an important exercise.

In this chapter we will deal with methods for determining the following:

1. Which of the variables in a dataset show most difference in distribution between two or more classes?
2. Which of these variables shows a significant difference in distribution (which involves setting a decision threshold)?

We will only deal with raw variables that can be interpreted directly in terms of the original data, for example, corresponding to specific potential biomarkers as identified by chromatographic peaks, or wavelengths in a spectrum that can be interpreted as originating from specific compounds or groups of compounds. We will not discuss variables such

Chemometrics for Pattern Recognition Richard G. Brereton

as principal components in this chapter, as these usually cannot so easily be directly interpreted in terms of specific compounds or chemical signals, although PCA for variable reduction has been covered in Sections 4.3.1 and 4.3.2.

9.1.1 Two Class Distributions

The simplest situation is where there are two classes and the interest is to determine whether there are variables whose distributions differ significantly between these classes. Figure 9.1 shows the distribution of intensity of a variable between two classes. Is this distribution sufficiently different between the two classes that the difference is considered significant? This is sometimes quite a hard question to answer. Some of the considerations and strategy are as follows:

1. The difference between classes can either be in the form of intensities or presence/absence. In the latter case a variable is considered significant if it is detected many more times in one class than in another.
2. Using statistical criteria several variables can be compared and ranked according to their difference; the more different, the higher the rank (we will use rank = 1 as the highest or most significant).
3. Often the difference in distributions is best modelled by multivariate models, because variables may interact. This is typical in biology, where the presence of compounds is often interrelated. An example might be an enzymatic reaction. Compound B may be formed from compound A using enzyme C. The presence of B may be highly diagnostic of a specific group but is only detected both if A is present (which may come from food eaten by the subject), and enzyme C is also present (which may be a consequence of genetics). So the presence of A, B and C are related, but if we only knew about A and B we would find that B is only sometimes found in the presence of A, and the correlation between these two compounds will be less than 1: for such a simple example it is easy to deduce the relationship but in real life

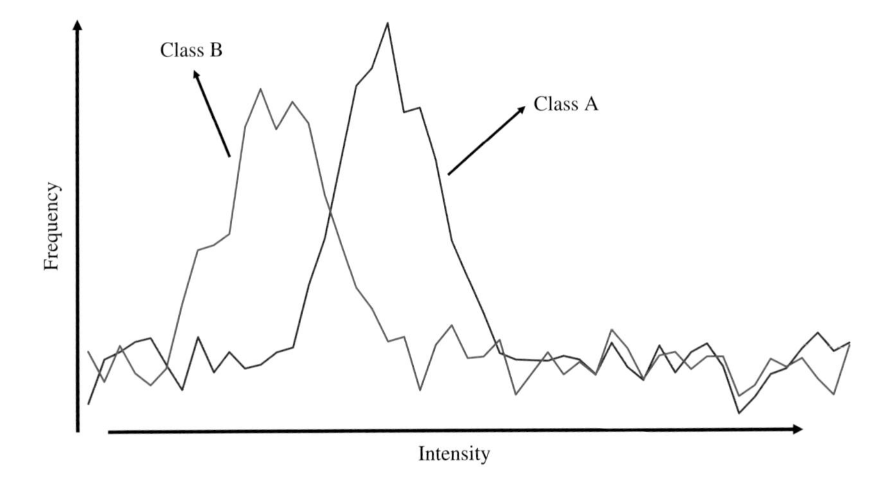

Figure 9.1 The distribution of the intensity of a variable between two classes

when there may be thousands of enzymes and metabolites simple pathway models break down and bulk statistical models are often necessary for an understanding of which compounds are potential markers.

4. Some people like to place a probability on whether a marker is discriminatory; however, for many tests this assumes normal distributions. In most chemometrics this assumption cannot be taken for granted, especially where data have been preprocessed prior to pattern recognition. In this text we usually avoid citing probabilities.

Many chemometricians prefer multivariate models under these circumstances although univariate statistics are also available and in some cases provide comparable results to multivariate tests as discussed below.

9.1.2 Multiclass Distributions

A somewhat harder problem arises when there are several different classes, for example, we may be interested in which compounds are markers to distinguish three different strains of microbe.

Many of the classic chemometric approaches such as PLS-DA are best suited for two class models, and can answer questions such as whether we can safely say that one compound is more prevalent or more intense in one group out of the three or more groups by creating a two class model – the first class being the 'in group' and the second all other groups. In many cases this is all that is of interest, for example, can we find markers that are characteristic for a specific species. All the normal two class multivariate methods are available for this type of assessment.

However, sometimes we are more interested in whether there are variables or markers that vary between classes rather than are characteristic for specific classes. An example may involve studying 10 people, each of whom has donated 20 samples over a period of time. We may be interested in whether there are compounds that are markers for specific groups of people, without in advance knowing which ones. Consider the situation of Figure 9.2 in which the distribution of a particular variable (e.g. a marker compound) is

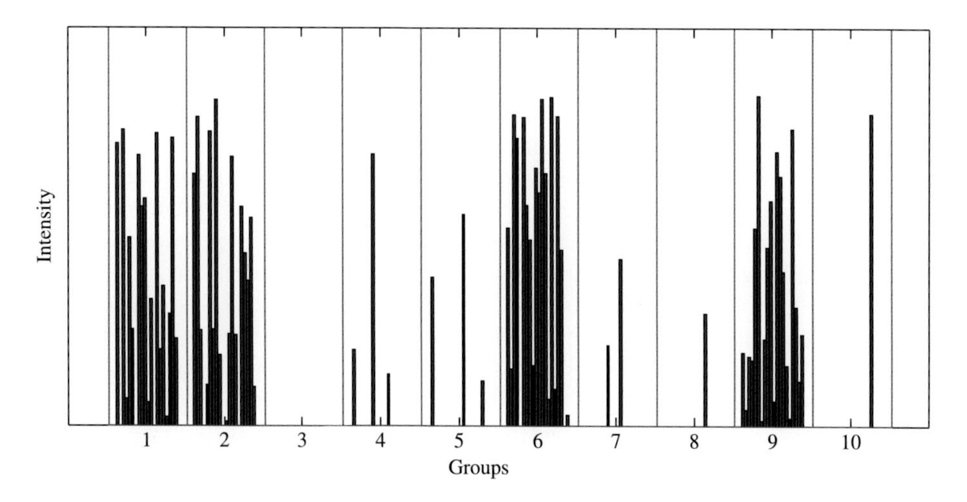

Figure 9.2 Intensity of a variable in 10 groups each consisting of 20 samples

studied between 10 groups, each consisting of 20 samples. It appears to be a good marker for groups 1, 2, 6 and 9, but not for the others. If we do not know in advance that these four groups form a subset or are clustered together, we have no prior information to link these together. If we were performing a one against all comparison we may actually miss some of these potential markers, as these are not good markers for any specific individual group, only for a combination of groups.

If the groups represent individual humans, there may be many factors that influence the distribution of markers, such as genetics, age, gender, etc. and it may happen that groups 1, 2, 6 and 9 correspond to repeat samples from individuals of the same gender, and that quite different distributions are found for different compounds. If all we are interested in at this stage is whether there are markers that can be associated strongly with individuals or groups of individuals, rather than characteristic of a specific individual (rather like characteristics such as eye or hair colour) we are mainly concerned whether the marker clumps in a non-random way according to each group. To determine this, there are a variety of univariate approaches such as the Fisher weight, which can extend two group concepts to a situation where there are many groups.

Multivariate approaches are rather awkward as this is quite a complex situation, and the most effective approaches might be to determine first whether there are any strong groupings in the data, e.g. by exploratory approaches first, and then by seeing if there is good evidence via proper use of validation that these groups actually do form classes, and then look for discriminators between these new groups.

9.1.3 Multilevel and Multiway Distributions

A third and more complex situation is where each sample can be grouped in more than one way, but each way is of equal interest. An example is where we are characterizing mice by their age and also by their genetic characteristics. Each sample could be grouped in one of two ways, and there are a plethora of approaches to dealing with such types of data. In this text we will only discuss this briefly as these situations, whereas quite common, for example, in clinical medicine, are less well established in chemometrics, although with the emergence of biomedical applications it is anticipated that such situations will occur more frequently. A key to whether more sophisticated approaches are suitable depends in part on how well the underlying experiments are designed, and formally trained statisticians spend much time on such considerations. There is a big culture gap between statisticians and most chemometricians, as the former group often will not accept and will not attempt to analyse data unless the experiments have been designed very carefully. However, in many practical situations these perfect designs are not possible, and although the data are perhaps of less value something often can be obtained from it. Figure 9.3 illustrates a typical experimental design where the interest is to determine the effect of age and genetics on characteristics of mice. In this design, there are two genetic and three age groups, and each combination of age $\times$ genetics is replicated nine times.

The simplest approach is to ignore the link between the two factors, e.g. age and genetics. Each type of effect is modelled separately, and so we try to find markers for age and for genetics as separate exercises. There are 27 samples in each of the two genetics groups and 18 in each of the three age groups, and these are modelled separately as

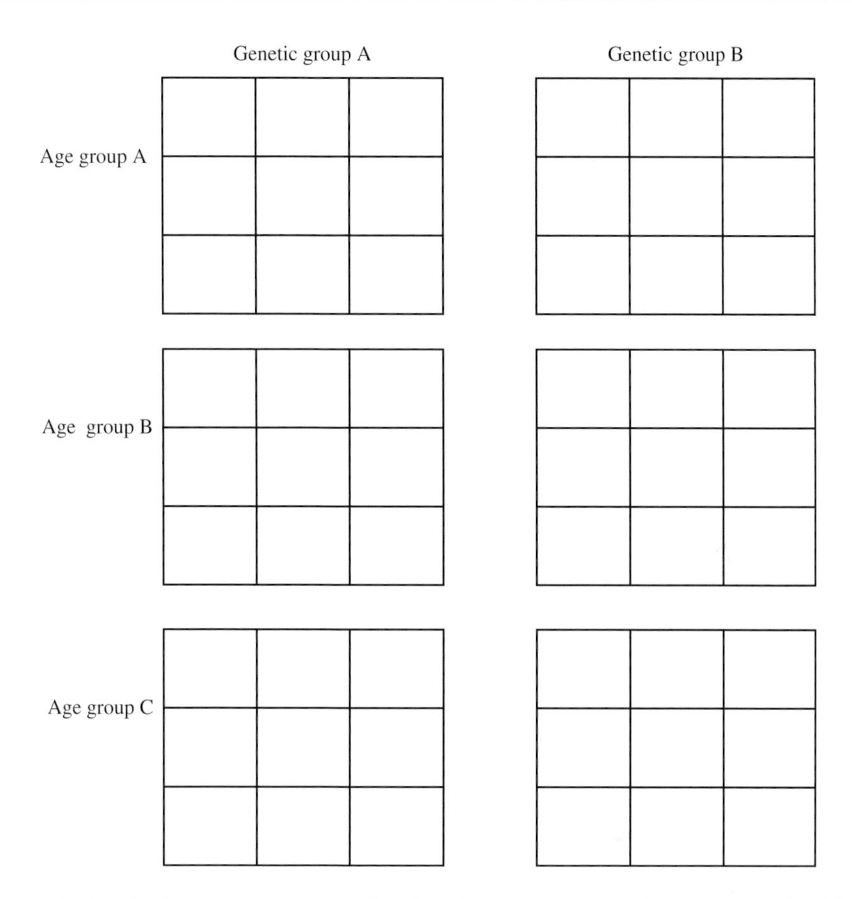

Figure 9.3 Typical design where multilevel and multiway methods may be applicable for determination of significant variables. Each square represents a sample, and each of the nine possible combinations of age $\times$ genetics is replicated 9 times, making 81 experiments

independent classes: the experimental has the advantage it is balanced. If experiments are not formally designed (and in some situations it may be expensive and impracticable) then this is often the only way to go about it, and the variable selection methods discussed in this chapter can be applied independently to each class. Even for formally designed experiments it is possible to analyse data this way, although it will miss some trends.

However when datasets are the result of formally statistically designed experiments, we can take other considerations into account. Many statisticians are very keen on what are called interactions. These imply that variables are not independent of each other. An example may be that young mice of one specific genetic group have certain specific habits. These habits are not common to all young mice and not common to all members of the specific genetic group and so there is an interaction between age and genetics, but still there are markers that can be found. Often methods based on MLR (Multiple Linear Regression) and ANOVA (Analysis of Variance) can be applied to analyse these types of experiment, and many traditional texts on statistics are full of such methods; if there are several factors ANOVA based methods are also called multilevel methods. The strength of ANOVA is that these methods can take interactions into account, but the weakness is that as commonly applied they are univariate: we will discuss a number of univariate methods in Section 9.2.3 in the context of variable selection when there is more than one factor. Within the chemometrics community there have been a few recent papers about multivariate extensions to ANOVA, although they are not widespread and common as yet, which take into consideration interactions between variables and factors at the same time.

Other extensions can involve treating the data as multiway datasets; this depends on the following:

1. There being some form of sequential meaning to the data, and so each type of grouping becomes a sequential value that can be expressed numerically.
2. The dataset being fairly complete without too many missing values.

If the measurements (e.g. HPLC or NIR spectroscopy) are taken into account, the data can then be rearranged as a box or tensor as illustrated in Figure 9.4, allowing multiway methods to be applied. This unleashes many opportunities, but is dependent on high quality data. Many chemometrics experts get interested in these prospects although in reality only a small number of datasets can be organized in this way, and a special set of techniques are required to make best use of this information. Each dimension corresponds to a specific factor and the data can be decomposed to determine the influence of each factor of the data.

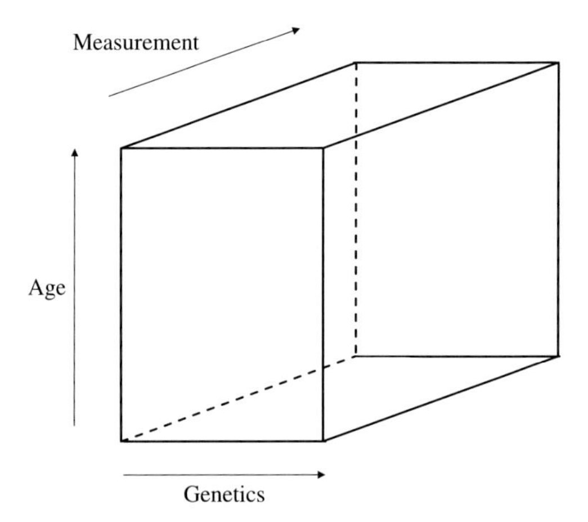

Figure 9.4 Organization of data of Figure 9.3 in the form of a box, taking into account the spectral dimension (measurement)

Trained statisticians will probably be more torn towards multilevel methods which in many cases ignore the multivariate nature of the data and treat each variable independently whereas trained chemometricians will be more drawn towards multiway approaches.

9.1.4 Sample Sizes

Before describing approaches for the determination of which variables are potentially discriminatory it is important to understand what is meant by a significant variable. In many areas of science, especially biology, it is common to use methods described in this chapter to determine which variables represent markers, e.g. for a disease or a genetic group. These are the variables (usually identified compounds) whose distributions are most distinct between one or more groups, for example, diseased patients and controls.

A fundamental problem with interpretation of such information results in a dilemma fairly characteristic of chemometrics, that often the sample to variable ratio is very low. A typical small experiment might involve 50 test subjects but measuring several thousand variables, e.g. unique compounds identified by GCMS. In order to understand this, consider a situation in which we toss a coin several times, and want to know whether it is biased. If an unbiased coin, having an underlying probability of 0.5 that it turns up 'H' (Heads) and 0.5 that it turns up 'T' (Tails), is tossed N times the chances of obtaining M heads is given by $N!/[M!(N-M)!]\ 0.5^N$. This distribution, called the binomial distribution, is illustrated in Figure 9.5 for situations where an unbiased coin is tossed 10 and 20 times, respectively. It can be shown that if an unbiased coin is tossed 10 times, around 0.1 % of the time it will come up 'H' 10 times, and around 1 % of the time 'H' 9 times, even though the coin is unbiased. If the number of times the coin is tossed is increased to 20, then the chance that an unbiased coin will turn up 'H' 18 out of 20 times (equivalent to 9 out of 10 for the smaller experiments) is reduced to 0.02 % – a fifty fold decrease for the same proportion of 'Hs' compared to 10 tosses. Many people get quite confused by these concepts: a coin turning 'H' 9 out of 10 times could well be an unbiased coin with an underlying 50 % chance of turning up 'Hs', we are not certain, but it is essential to hold this at the back of one's mind, and interpretation depends on the number of times the coin is tossed.

How does this aid our understanding of the problem of identifying potential markers or discriminatory variables? Let us consider an experiment where we are trying to find potential markers that can distinguish two groups of subjects. Let us say that in a series of chromatograms coming equally from the two groups (e.g. 10 chromatograms from group A and 10 from group B), a marker is detected in precisely half. The ideal discriminator would always be detected in one group but never in the other, and so such a compound is a candidate for a perfect marker. However, whether it could be a marker depends on how it is distributed between the two groups. If it is found equally in both groups it is not likely to be a good marker. Determining whether a candidate is a good marker is analogous to determining whether a coin is biased. Can an apparently uneven distribution between the groups arise from a non-discriminatory marker? This is equivalent to asking whether an unbiased coin can turn up 'H' many more times than 'T' or vice versa. In the analysis above we commented that if an unbiased coin is tossed 10 times, it will turn up 'H' 9 times around 1 % of the time. This is equivalent to taking 10 subjects in group A and 10 in group B and finding that a marker that is detected in half the samples, is detected 9 times in group A and only once in group B (or vice versa), or is 9 times more predominant in group A. Extending the analogy to 8 'Hs', the probability is almost 5 %; this is equivalent to a marker being detected in half the samples overall but detected 4 times more frequently in group A to group B as [8 ('H' or A)]/[2 ('T' or B)] = 4. So a compound whose underlying distribution is unbiassed detected in half the samples will be found 4 times more

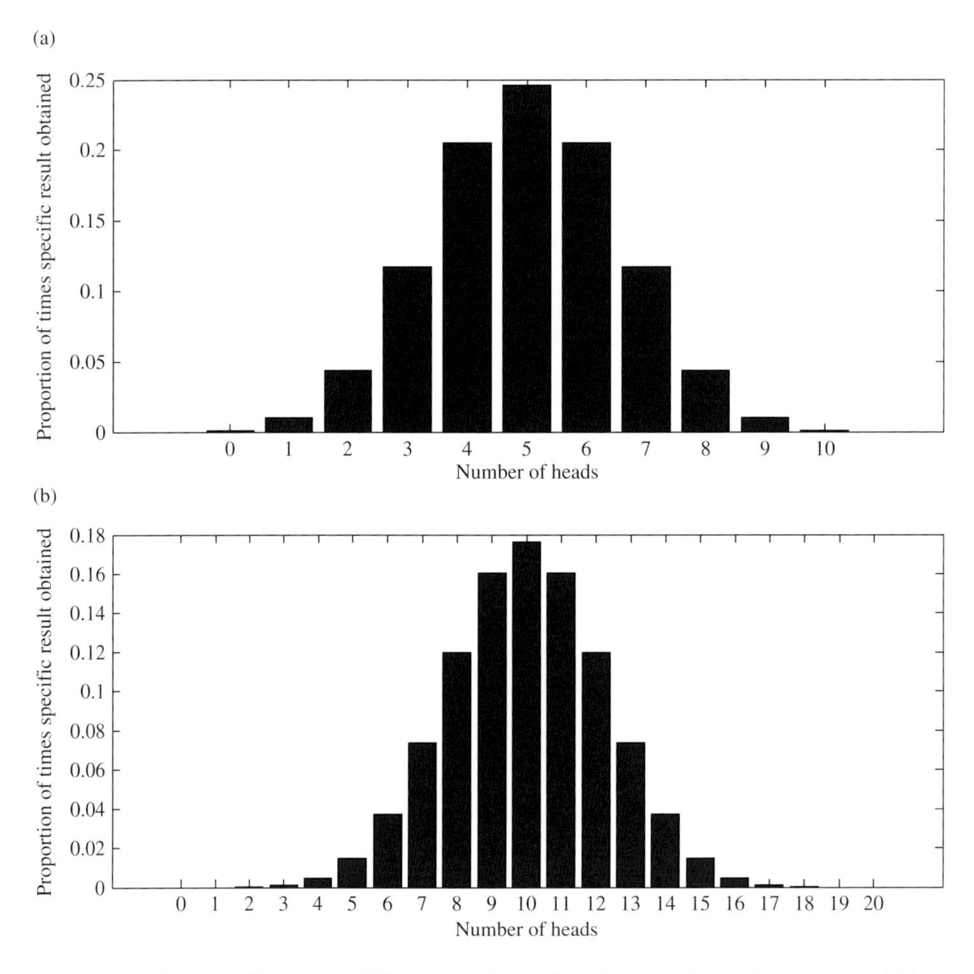

Figure 9.5 Chances of obtaining different numbers of heads if an unbiased coin is tossed (a) 10 times and (b) 20 times

frequently in one of the two groups 10 % of the time (dependent which is the most frequent group), in a sample of 20 subjects equally drawn from two groups. Typically a method such as GCMS identifies a large number of peaks – let us say 1000. Investigating 1000 peaks as potential markers is analogous to repeating 1000 times the experiment of 20 tosses of a coin. By the logic above (which is somewhat simplified and assumes each peak is detected in exactly half the samples), 100 of the 1000 peaks will be detected in 4 times more samples in one group that the other, even if they have an underlying probability of detection that is equal in each group. Even a 9 fold difference in detection frequency will be observed for 20 of the 1000 peaks (assuming each peak is detected in half the samples – of course, there are different calculations according to the proportion of samples a compound is detected in but this discussion is primarily to illustrate general principles). So it may appear that there are many exciting marker compounds, but in fact an observed unequal distribution has just occured by chance, and we call these peaks False Positives. In areas such as bioarrays where thousands or tens of thousands of genes can be tested, this problem becomes even more extreme. The converse, of course, is if not many samples are analysed, markers that in fact do have a strong underlying discriminatory power may not appear as such, resulting in False Negatives.

How can this problem be overcome? The obvious statistical approach is to analyse more samples. But then additional problems may start to emerge. In many experimental situations such as in biology, as more samples are obtained, more unique compounds are identified and so the number of potential markers increases, rather like repeating the toss of a coin more and more times providing more opportunities for FPs. Another problem relates to finding homogeneous populations, if, for example, we want to study 100 people with a disease and 100 controls, there may be other factors such as age, ethnicity, gender, personal habits, etc. that will influence the chromatographic or spectroscopic signal so these additional hidden factors may come into play: often finding a suitably larger number of subjects (usually volunteers) which are matched to a control group is difficult, expensive and time consuming. Finally analytical problems may come into play, for example, there can be problems about storage – if samples are obtained at different times (often necessary for large experiments), they may need to be stored for different periods of time, which could influence the peaks detected in the chromatograms, and there are usually analytical problems about long term stability and instrumental performance, for example, in chromatography about changing, cleaning and chopping (for GC) columns, and so the analytical system will introduce an additional factor into the analysis. Hence increasing sample sizes whilst sometimes feasible, in other studies can be very difficult, and certainly requires careful attention to the experimental design. Often factors, such as instrumental drift over time, in chromatography changing columns, extraction procedures and storage are as important – sometimes more important – than the factors of direct interest, and so increasing the number of experiments manyfold can involve additional problems creeping in. We will not look in depth at this issue, which could involve writing another volume in its own right, except to point out that this is a real difficulty that is often encountered and means that sometimes chemometricians have to compromise.

9.1.5 Modelling after Variable Reduction

There are a number of serious misconceptions about using potential discriminatory variables for subsequent modelling. A typical strategy might involve the following:

1. Determine the most discriminatory variables
2. Form a model with these variables.

Whether this is a sensible strategy or not depends very much on what the aim of the analysis is. In traditional predictive analytical chemistry we know in advance there must be an answer and the aim is primarily to obtain this answer with as high degree of accuracy as possible, and the strategy above may be suitable. However in hypothesis driven science this is not such a good approach. We may not know in advance whether the experimental method or data are adequate to support our hypothesis, for example, can we determine from a sample of plasma whether a person has the onset of a disease or can we determine from analysis of saplings whether a tree will be productive or not in many years time? There may be no certain answer. As discussed below, it is possible to determine variables that are potential discriminators, and we can list them in order of significance according to the value of a statistic, as discussed in Section 9.2. We will illustrate this for dataset 11 (null).

In Table 9.1 (a) we present the %CC for Case Study 11 (the null dataset) when the top 10 and 20 variables that appear to be best discriminators over the entire dataset, using the *t*-statistic (Section 9.2.2), are used for the model (PLS-DA model with 2 PLS components). The classification accuracy for the test set is close to 60 % (using 100 iterations as discussed in Section 8.4), the reason being that the variables are selected over the entire dataset, which includes the test set, hence resulting in an over-optimistic %CC for the test set, which may inadvertently suggest that there is some discrimination between the two classes. We can see from the plots of the first two PCs using these 10 or 20 variables that there does indeed appear to be some separation (Figure 9.6) as these PCs have been formed on the variables that happen to appear most discriminatory (out of the 100 original variables in total). In analogy, one might toss an unbiased coin several times and record the amount of 'Hs'. Then this experiment is repeated 100 times. Then one chooses the 10 experiments (out of the 100) for which the proportion of 'Hs' is most, and ignores the rest. One would probably come to a false conclusion that the coin is biased, as one is throwing away the majority of evidence coming from the less conclusive experiments. Surprisingly this fallacy is very common in chemometrics data analysis, often because there are preconceived conceptions and because after reducing the variables graphs such as PC scores plots 'look better'.

An alternative and safer strategy is as follows:

1. Perform 100 test and training set alone splits.
2. For each split select the best discriminatory variables for the training set.
3. Apply this model to the test set, in each case, using discriminatory variables obtained from the training set.

The latter procedure is a safe one because the variables in the test set are not influenced by the training set and give the results of Table 9.1(b). We can see that although the training set %CC is approximately 75 % for both examples using the top 10 and top 20 variables, the test set results are within a safe region (46 %) as anticipated for the null dataset. The training set %CC is higher than previously because variable selection is performed only on the training set rather than over both training and test sets. Interestingly the gap between the training and test set %CCs widens from 6 % (top 10 variables) to 30 % when performing variable selection correctly. It is (falsely) accepted wisdom in many circles that the smaller the gap between test and training set %CC, the less the model is overfitting. In fact we see that the correct model of Table 9.1(b) exhibits a wider gap between the training and test set %CCs. It is important to understand that only the test set model (if correctly formed) can answer whether there is a genuine underlying difference between two or more sets of samples. However once this has been established, it is, of course, then possible to optimize the model subsequently using autopredictive approaches.

One problem with the latter procedure is that, although the assessment of %CC on the test set has a valid meaning the variables selected in each iteration will differ. There is often a

Table 9.1 *(a) %CC for autoprediction, test and training set models for dataset 11 (null dataset) when respectively best discriminatory 10 and 20 variables are included in the model, taken from the overall dataset. (b) %CC for test and training set models for dataset 11 (null dataset) when respectively best discriminatory 10 and 20 variables are included in the model, taken from the training set in each of 100 iterations. PLS-DA with 2 PLS components is used for all models*

(a)

Top 10 variables			
Training set			
Class A	63.75	32.25	65.75
Class B	36.25	67.75	
Test set			
Class A	55.75	37.75	59.00
Class B	44.25	62.25	
Top 20 variables			
Training set			
Class A	67.25	32.50	67.38
Class B	32.75	67.50	
Test set			
Class A	57.00	41.25	57.88
Class B	43.00	58.75	

(b)

Top 10 variables			
Training set			
Class A	76	23.25	76.38
Class B	24	76.75	
Test set			
Class A	46.5	54.50	46.00
Class B	53.5	45.50	
Top 20 variables			
Training set			
Class A	75.75	22.75	76.50
Class B	24.25	77.25	
Test set			
Class A	45.50	53.50	46.00
Class B	54.50	46.50	

case for reducing the number of variables; for example, in a typical biological study, there may be several thousands, many coming from analytical procedures, environment, unrelated (or confounding) factors we are not interested in, and so we ultimately would like to form a model using a reduced number of variables; in fact forming a model using all the original variables often makes no sense as the vast majority are irrelevant to the study in hand (for

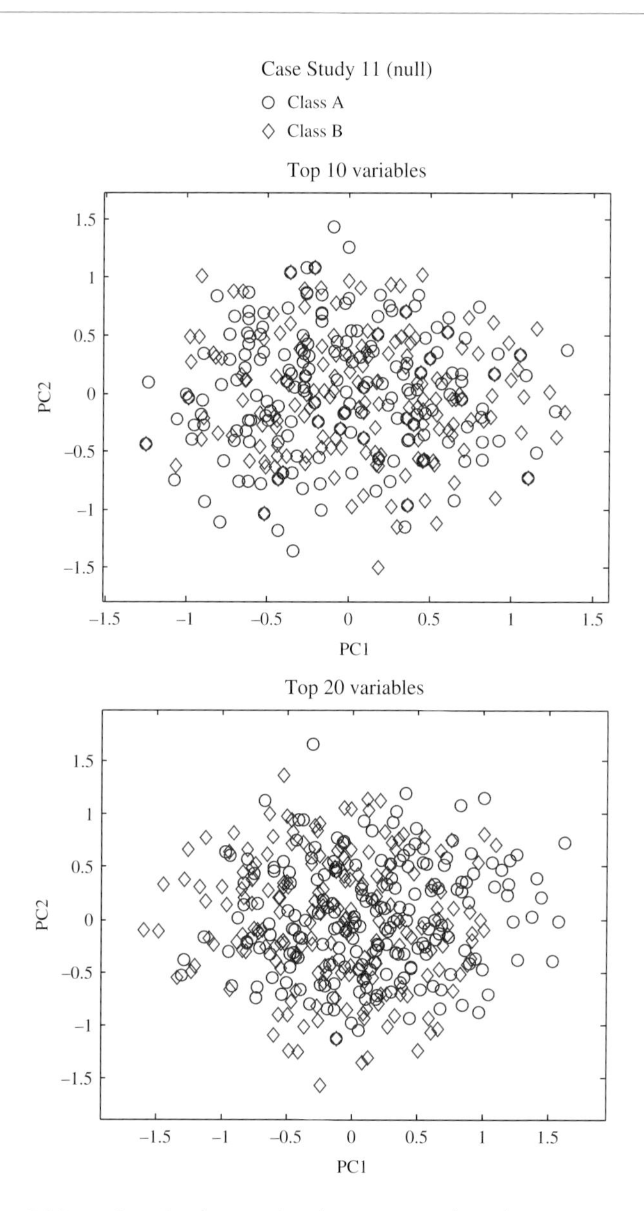

Figure 9.6 PCA on Case Study 11 using the top 10 and 20 discriminatory variables

example, we may be interested in predicting whether a person has a disease by analysing their urine – of 500 or so detectable compounds in the urine only 20 or 30 may be related to the disease), and so we do want to form models on a reduced set of variables, but the problem is that our models could be over-optimistic if the variables are selected from the entire dataset and lead to dangerously false conclusions. In Section 9.3.3 we will discuss a strategy to overcome this problem.

9.1.6 Preliminary Variable Reduction

Although it is inadvisable to perform classification models after reducing the dataset to only the most discriminatory variables unless great care is taken, it is still legitimate to reduce the number of variables prior to discrimination, even for hypothesis testing, providing the criteria do not relate to the main objective of the classification procedure. Typical reasons for removing variables are as follows:

1. Removing variables that are known artefacts of the analytical procedure, an example may be siloxanes that are related to sample preparation, often detected e.g. by spectroscopic properties. Variables that relate to the instrument a sample has been recorded on can also be removed; a typical example is Case Study 6 (LCMS of pharmaceuticals) where one factor is when the sample was analysed and on what instrument – removing the known chromatographic peaks that are believed to be due to instrumental variation should not influence the ability to assess the quality of classification of samples according to their process origins.
2. Removing variables that are rarely detected. In metabolomic studies, for an example, a compound that is detected only in a few samples is unlikely to be a marker for a specific class. So, for example, a compound detected in 5 % of a set of samples is unlikely to be a marker for a group, providing the group size represents a reasonable proportion of the samples in the overall dataset. If variables are to be removed for this reason (and sometimes tens of thousands of variables can be reduced to a few hundred) the criterion of rarity must be over all samples and not within a specific class to avoid bias, although the threshold of number of times the samples is detected can be related to minimum class size.

Some prefiltering of variables is desirable in certain situations simply because there may be many irrelevant variables that will introduce noise into the classification model, dependent on the nature of the dataset. In the case studies discussed in this book we have already prefiltered variables, in situations where this is desirable (e.g. Case Studies 5 and 8).

9.2 Which Variables are most Significant?

9.2.1 Basic Concepts: Statistical Indicators and Rank

Most methods attach a statistical indicator to each variable according to its power to discriminate between two or more groups. We will discuss a number of common indicators in Sections 9.2.2 to 9.2.4. The greater the magnitude of this indicator, the more the variable contributes towards the discrimination between groups.

In most cases, variables can then be sorted according to the size of the statistical indicator function. Each variable j is assigned a rank, r_j: if equal to 1 the variable appears to have the highest discriminatory power, while if equal to J (total number of variables), the lowest. It is important to realize that this rank relates to the particular property or feature that is being studied and the particular experiment being performed. Variables can then be listed in order from most to least discriminatory. Many indicator functions are based on different principles and the lists obtained from different indicators can be compared, although as we show (Section 9.2.5) there is some relationship between several of the common methods.

Some indicator functions have a sign, for a two class discriminatory problem, positive values being variables that are of high intensity or frequency in class A and negative values in class B (so long as we label class A +1 and class B –1): we can say that a variable that has the highest positive value is the best marker for class A and vice versa. Under such circumstances it is possible either to convert the values of the indicator functions to their absolute values and so rank as above, or else pick the most significant markers for each group, creating two rank tables, one for positive and one for negative value markers. The difference between the two methods, both aiming to select 6 out of a possible 16 candidate variables as markers, is presented in Table 9.2 (also compare Figures 3.1 and 9.6 for the top 20 variables). For the sign method, the 3 most positive markers for each group are selected, whereas for the magnitude method, the 6 markers with highest magnitude are selected. These two methods give different answers, with the magnitude method selecting 4 markers for group A but only 2 for group B. Usually selecting markers by the magnitude of their statistical indicator is satisfactory, and for some statistics there is no sign; however in a few situations it may be important to find positive markers for each group: consider trying to distinguish a set of diseased subjects from controls, if most of the significant markers are found to be at a high level in the control group, it might be useful to have a more balanced set of marker compounds. By default in this text we will select markers by magnitude even if there is a sign attached.

One key conceptual problem that taxes many people is that the statistics calculated even if apparently very definitive are for the samples actually studied (or in formal

Table 9.2 *Example where 16 variables have values of an indicator function, where variables assigned according to two different criteria are indicated in grey as discussed in the text*

Signs			Magnitudes		
Marker	Indicator	Class	Marker	Indicator	Class
16	1.690	A	10	1.915	B
11	1.625	A	16	1.690	A
4	1.423	A	11	1.625	A
8	1.099	A	1	1.477	B
12	0.862	A	4	1.423	A
3	0.760	A	8	1.099	A
9	0.631	A	12	0.862	A
2	0.399	A	5	0.766	B
13	0.229	A	3	0.760	A
7	0.127	A	14	0.704	B
15	−0.368	B	9	0.631	A
6	−0.594	B	6	0.594	B
14	−0.704	B	2	0.399	A
5	−0.766	B	15	0.368	B
1	−1.477	B	13	0.229	A
10	−1.915	B	7	0.127	A

terms the experimental population). Let us say we are building up a database of chemicals that are found in sweat that can distinguish males from females, and we take a sample of individuals, and find those features that differ significantly between the two groups. It might be that in our population we had more blue eyed females than males but more dark haired males than females. Unless our population is very large and representative there is bound to be imbalance between some features of both groups just by chance. Now let us say we find some characteristics (e.g. chemicals) that can distinguish both groups. We may find that in our sample population these can unambiguously allow us to discriminate between males and females, but how do we know these are not a consequence of having blue eyes or dark hair? For fairly straightforward traits this may be easy to identify but for more subtle traits (for example, more females may possess an enzyme that allows the digestion of fat even though this has nothing to do with 'femaleness' or 'maleness' but is simply inherent in the population by chance), it is often hard or impossible to identify such hidden confounding factors. There will be thousands of metabolic pathways and without a lot of effort we cannot necessarily be sure of the origins of each chemical found in a person's sweat (if that were so, the problem would have been solved already). Yet by building up a statistical model, using these chemicals, we can probably predict whether an unknown person in our sample population is male or female, but the statistics says nothing about the origins of the chemicals in the sweat. We may be able to identify markers with great confidence and be sure that for our population that the model is really excellent. It may be we are sampling for a specific town or village where men and women have very specific habits, for example, drinking a particular local beer (men) or using a specific local perfume (women) that influences the chemical signal. The markers that are identified whilst probably statistically sound for the sampled population may not necessarily be appropriate in a different population. Whereas a solution might be to include a much wider range of people in the analysis, there is a limit beyond which one cannot go, and so there are likely always to be population dependent factors in play. The methods described in this chapter, whilst able to identify potentially significant variables with high confidence, are entirely dependent on the nature of the experimental population and it is up to the scientist to decide whether these variables can be then used for predicting the properties of a wider range of samples. In some situations, for example, the characterization of polymers (Case Study 3) the answer is probably yes, as this is a traditional analytical chemistry problem whereas in other situation such as human sweat (Case Study 5) the answer is less sure.

Whereas there are a very large number of potential indicator functions and the general principles of obtaining numerical indicators of the significance of variables can be applied to any such statistic, we will restrict the detailed discussion to three main groups of methods.

9.2.2 *T*-Statistic and Fisher Weights

The t-statistic is a common univariate method to determine which variables differ most between two groups, and looks at the ratio between the difference of the means and the pooled standard deviation of the intensities of each variable j. This statistic takes no account of interactions between variables, and aims to interpret each variable separately. Although there are several variations on this theme, in the context of a simple two class problem where the aim is to look at the difference in distribution of a variable between these two groups, it is defined as follows:

1. The mean ($\bar{x}_{jg}$) and standard deviation s_{jg} of each variable is found for each class g. In this chapter we employ the population standard deviation of the variable to be consistent with Chapter 4, rather than the sample standard deviation. If sample sizes are equal in each class this makes no difference; if they are unequal there is only a major difference if one of the classes is quite small, but since we do not recommend small sample numbers this probably is not very important. It has the advantage that some of the relationships with ranks obtained using PLS-DA weights are exactly linear if the data matrix is standardized prior to PLS-DA (Section 9.2.5).
2. The t-statistic for each variable $t_j = \left(\bar{x}_{jA} - \bar{x}_{jB}\right) / s_{jpool}\sqrt{1/I_A + 1/I_B}$ is calculated, with s_{jpool} corresponding to the pooled standard deviation over the two groups defined by:
$$s_{jpool} = \sqrt{\left[(I_A - 1)s_{jA}^2 + (I_B - 1)s_{jB}^2\right] / (I_A + I_B - 2)}$$
where there are I_g samples in group g. It is important to note that s_{jpool} is not the overall standard deviation over all samples but is a weighted mean of the standard deviation within each class.
3. The variables with the highest absolute values of t are considered as candidates for the most significant markers, their signs indicating which group they are markers for, the variable with the highest value being ranked 1 and the lowest ranked J.

The 20 variables with the highest absolute value of the t-statistic for Case Studies 8a, 4, 10b and 10c are listed in Table 9.3. For Case Study 10b (simulation – intermediate overlap) we know which variables we intended to be discriminatory, which are listed in Table 2.6. We find that all the 17 top ranked variables also appear in the list of discriminatory variables that were included in the simulation; however, variables numbered 34, 71 and 15 are not within the known discriminators and variables 45, 59 and 79 are missed out. For Case Study 10c (simulation – maximum overlap), variables 16, 60, 17 and 40 are not discriminators, but appear within the top 20, whereas variables 42, 64, 65 and 68 are missed out. Furthermore variables 16 and 60 (which should not be discriminators), are ranked within the top 10. Hence the greater the overlap, the more the possibility that some of the highly ranked variables will be False Positives. Nevertheless there is still quite a good correlation even for Case Study 10c, where 16 out of the top 18 variables are indeed within the 20 known discriminators. Note that different results might occur if the structures of the simulation were different. This is quite a complex area and simulation is a large field, and the simulation used was quite straightforward, and although there are some correlations between the discriminatory variables, these have not been deliberately introduced. Different simulations might result in univariate methods performing less effectively; however, the t-statistic does illustrate how one can pull out relevant discriminators and that it is more effective if classes are more separate.

For Case Study 8a (mouse urine – diet study) we do not of course know for certain which variables should be discriminatory. However we can see that the top 8 discriminatory variables all have a positive value for the t-statistic, suggesting they are more intense in class A, which involve mice on a special diet. Case Study 4 (pollution) also shows disproportionate number of mass spectrometry marker peaks for class A (polluted) as

Table 9.3 *20 Variables with the highest rank of* t*-statistic in Case Studies 8a, 4, 10b and 10c together with the value of the statistic and its sign*

Case Study 8a (mouse urine – diet study)			Case Study 4 (pollution)		
1	197	9.76	1	35	21.43
2	55	7.83	2	23	20.43
3	110	6.68	3	37	20.41
4	263	6.56	4	9	18.87
5	124	6.22	5	3	−18.14
6	238	5.89	6	7	17.89
7	175	5.44	7	4	−16.95
8	149	5.41	8	26	−16.83
9	262	−5.31	9	49	16.72
10	270	5.20	10	21	16.65
11	201	5.14	11	63	−15.54
12	271	−5.13	12	27	15.52
13	185	4.96	13	51	13.80
14	218	−4.90	14	34	13.40
15	4	4.90	15	77	13.01
16	285	−4.90	16	65	12.91
17	77	−4.62	17	15	−12.83
18	69	4.58	18	18	−12.79
19	212	4.58	19	13	−12.79
20	209	4.31	20	50	12.71
Case Study 10b (simulation – intermediate overlap)			**Case Study 10c (simulation – maximum overlap)**		
1	78	−4.97	1	59	−4.26
2	100	−4.69	2	18	−4.04
3	9	−4.65	3	11	−3.63
4	77	−4.49	4	81	3.13
5	16	4.28	5	16	3.04
6	26	4.21	6	10	2.98
7	89	4.19	7	54	2.97
8	27	3.70	8	61	2.95
9	24	−3.56	9	63	2.85
10	58	3.55	10	60	−2.81
11	33	−3.53	11	20	2.80
12	13	−3.50	12	22	2.69
13	92	3.35	13	52	2.66
14	10	3.27	14	43	−2.54
15	30	3.19	15	44	−2.53
16	17	−3.14	16	12	2.45
17	67	2.74	17	7	−2.43
18	34	2.56	18	79	−2.19
19	71	2.47	19	17	−2.09
20	15	−2.36	20	40	−2.08

might be anticipated, with only 4 from the top 16 markers for class B (unpolluted). The distribution of intensities of the 20 marker variables for these case studies is presented in Figure 9.7. We can see that for Case Study 8a for some variables the difference in distribution is due to a differing frequency of detection in additional to intensity (e.g. variables 197 and 262), which suggests that on the whole some compounds are only detected frequently in one for the two groups. Variable 55, in contrast, shows a difference in distribution due to intensity. There are also some strong differences of distributions for variables between classes for Case Study 4 (pollution). At first glance it may appear slightly strange that there are marker variables for the unpolluted samples: we might anticipate that the background mass spectrum is the same for both groups of samples, with additional markers for the polluted class. However it is important to realize that the

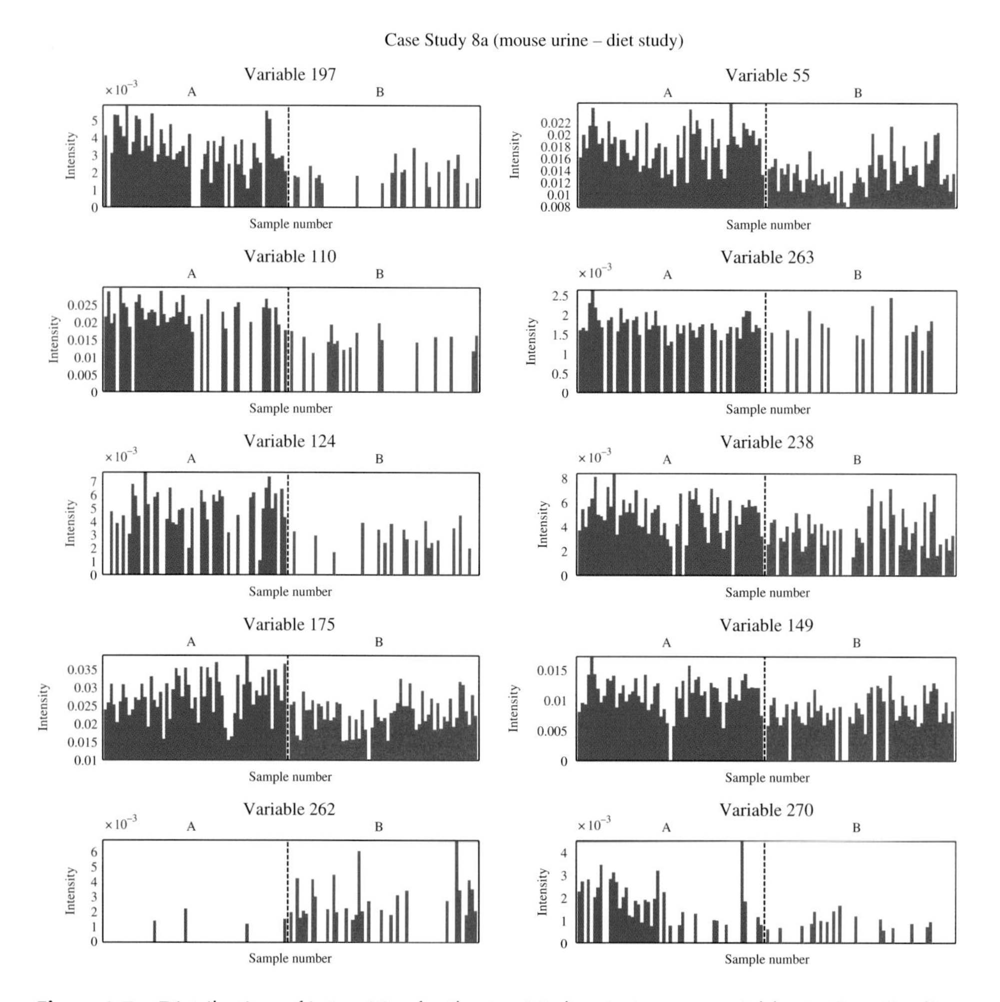

Figure 9.7 Distribution of intensities for the top 20 discriminatory variables in Case Studies 8a and 4, as assessed using the *t*-statistic with each class A and B indicated – row scaled to constant total

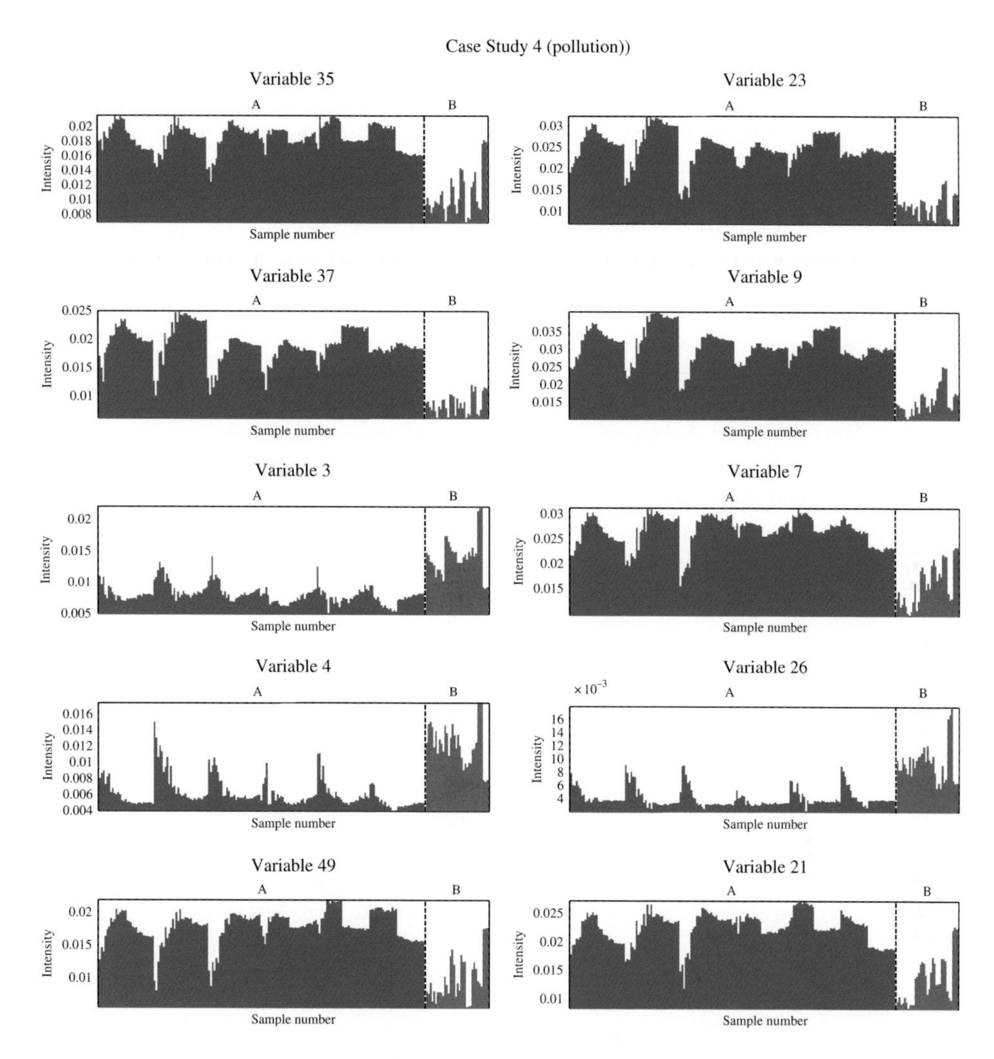

Figure 9.7 (continued)

data have been row scaled to constant total beforehand. This means that the background peaks present in class B (unpolluted) might be consistently higher in proportion than in class A (polluted) and so appear as markers. This dilemma can only be resolved by using data that are not row scaled. However in many cases this is not easy because the amount of sample cannot be controlled, and so for many of the case studies described in this text we have had to row scale the data prior to pattern recognition, resulting in the possibility that some markers are found as a consequence of this transformation called 'closure'. This does not mean the variables selected are invalid as statistical indicators that allow one to model a distinction between groups; what the statistics say is that they are found in greater proportion in one class as opposed to another, but this is all we can state for sure if the absolute intensities are not available. It is important to understand this concept because sometimes the identities of markers are interpreted in terms of metabolic pathways or specific chemical processes but yet they may simply be a byproduct of the method of data scaling, although are legitimate features that can be employed for statistical discrimination between classes.

An alternative method for looking at significance of variables is the Fisher weight, defined by $f_j = \left[\sum_{g=1}^{G} I_g(\bar{x}_{jg} - \bar{x}_j)^2\right] \Big/ \left[s^2_{jpool} \sum_{g=1}^{G} (I_g - 1)\right]$, when there are G classes, where $\bar{x}_j$ is the mean of variable j of all classes which uses the ratio of within class variance to between class variance. When simplified for a two class model ($G = 2$) the rank of the variables is the same as the t-statistic although there is no sign, the square root of the Fisher weight being linearly related to the t-statistic. The t-statistic, however, has advantages in that a sign is attached to each variable and so it can be directly assigned as a marker for a specific class, and when there are two classes, we will use only the t-statistic for variable selection.

The two methods only provide different results for the ranks of the variables when there are more than two classes in the dataset. As the t-statistic can only be calculated between two classes, for more classes it is usually applied as a series of 'one vs. all' comparisons, whereas the Fisher weight can be calculated for as many classes as required. The t-statistic is still valuable under such circumstances; if we were interested in, for example, 5 classes, the t-statistic might identify markers that are characteristically present (or absent) uniquely in one class, whereas the Fisher weight might identify markers that were characteristic for several classes. As an example, consider the distribution of three variables over 10 groups each consisting of 20 samples as illustrated in Figure 9.8. The Fisher weights and t-statistics for classes A, B and H against all other samples are calculated in Table 9.4. We see that variables are ranked differently according to which test is employed. The most extreme difference in statistics is for the t-statistic class B test, variable 2 being overwhelming more important than the others. Both variables 1 and 2 have positive t values for this test as they are both detected in class B, but variable 3 has a negative t value being never found in class B. A similar conclusion is obtained when examining the t-statistic for the class H test, variable 3 being the most significant and then variable 1. The t-statistic is not as large as that for variable 2 in the class B test, as variable 3 is found only in some samples from class H and hence, whereas this is a good marker for this class, not so overwhelmingly predominant. Class A has no variables unique to it, but variables 2 and 3 are never found in class A and as such have a negative t-statistic. At first it may be strange that variable 2 appears a more significant marker than variables 1 and 3 over the three t values, but if one looks at the relative proportions of samples this variable is detected in, in the 'in group' (class A) and 'out group' (the remainder), this exhibits the most extreme difference (of course, the intensity also has some influence on the statistic as well). The Fisher weight gives an overview. Variable 3, whilst only found in class H, is not always found in class H and as such is not as heavily diagnostic as variable 2. Variable 1 has some discriminatory ability, and so is intermediate in the list.

The Fisher weight can be used to rank variables in order just as the t-statistic, and is useful if there are more than two groups. As a practical example we compare the t-statistic ranks for Case Study 8b (mouse urine – genetics) to the Fisher weight ranks in Table 9.5. We see that of the 20 most significant variables according to Fisher weights, 16 are within

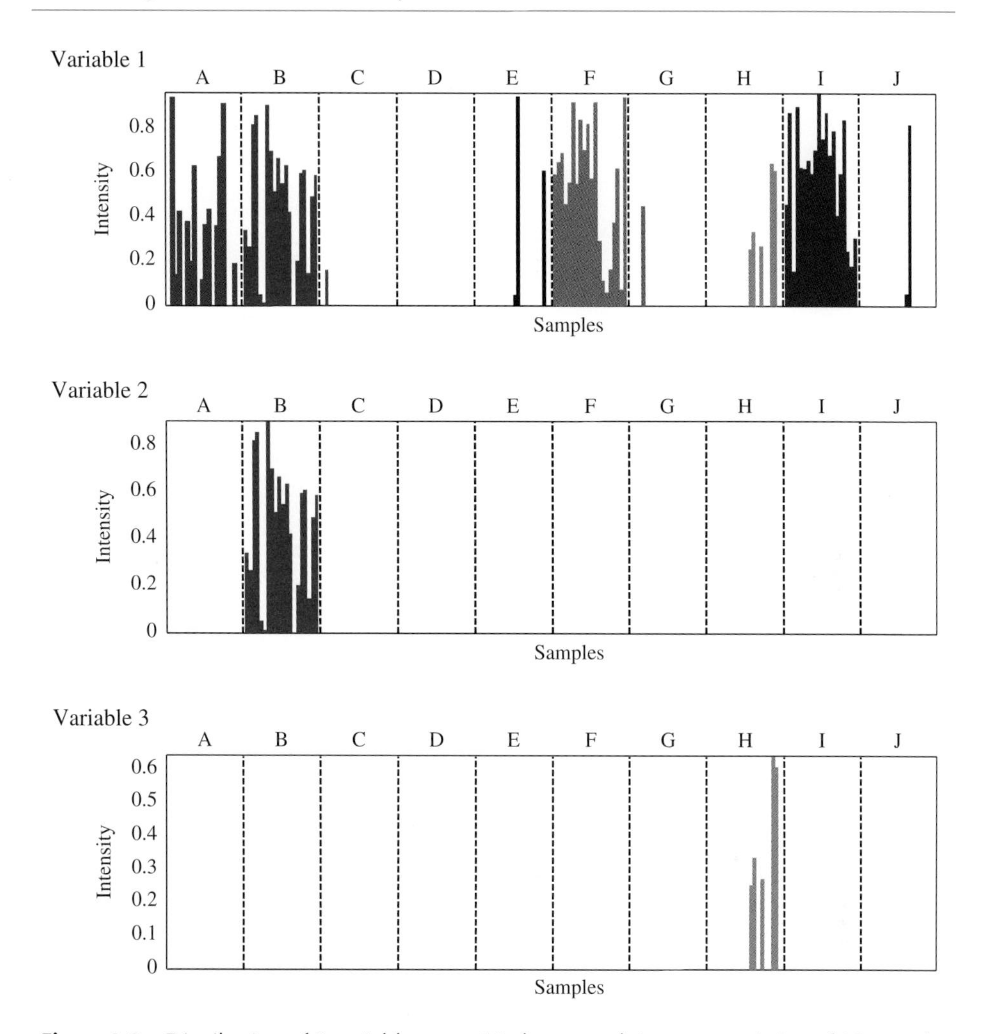

Figure 9.8 Distribution of 3 variables over 10 classes each in turn consisting of 20 samples

the top 2 for class A, 8 for class B and 10 for class C. Class A interestingly contains most of the discriminatory variables detected using the *t*-statistic, as it is more different from the other two classes because it represents a different strain (Section 2.10).

It is possible to attach probabilities to the *t*-statistic and Fisher weight; however these depend on underlying assumptions such as normality, and this often is not obeyed in practice in chemometrics, both because data preprocessing often distorts the underlying structure (e.g. row scaling or square root scaling – see Chapter 4) and also because of the unrepresentativeness of many experimental populations (see discussion of Section 9.2.1). In many cases it is unlikely that the underlying distribution of variables falls into a normal distribution; remember that to compare probabilities obtained from several different variables all should fall approximately in a normal distribution (for each class separately).

Table 9.4 *Fisher weights and t-statistics (classes A, B and H) for the variables of Figure 9.8*

Fisher weight	Rank	Variables	Fisher weight
	1	2	2.66
	2	1	1.19
	3	3	0.25
t-*statistic, class A versus the rest*	Rank	Variables	*t*-statistic
	1	2	−1.34
	2	1	1.05
	3	3	−0.70
t-*statistic, class B versus the rest*	Rank	Variables	*t*-statistic
	1	2	22.94
	2	1	3.954
	3	3	−0.70
t-*statistic, class H versus the rest*	Rank	Variables	*t*-statistic
	1	3	7.00
	2	1	−1.70
	3	2	−1.34

Table 9.5 *Rank of variables for Case Study 8b (mouse urine – diet study), using Fisher weights and* t-*statistics. Variables found to be within the top 20 most significant via the Fisher weight are highlighted (bold) in the* t-*statistics*

Fisher weight			*t*-statistic: class A v others			*t*-statistic: class B v others			*t*-statistic: class C v others		
Rank	Var	Fisher weight	Rank	Var	*t*-Statistic	Rank	Var	*t*-statistic	Rank	Var	*t*-statistic
1	151	0.78	1	**151**	−11.69	1	**151**	7.35	1	**266**	8.53
2	209	0.57	2	**93**	10.24	2	**209**	6.50	2	**74**	−7.94
3	93	0.52	3	**209**	−10.14	3	134	6.38	3	**313**	6.72
4	74	0.48	4	**254**	9.77	4	**277**	6.20	4	**271**	−6.40
5	254	0.47	5	**3**	9.53	5	200	5.94	5	**158**	−6.26
6	158	0.47	6	**169**	−9.42	6	83	5.07	6	276	5.84
7	169	0.45	7	**158**	9.40	7	304	−4.74	7	**169**	5.77
8	3	0.45	8	74	8.57	8	**59**	−4.67	8	214	5.76
9	266	0.39	9	**232**	7.78	9	**3**	−4.39	9	**232**	−5.50
10	313	0.33	10	**33**	7.52	10	142	4.32	10	155	5.46
11	232	0.32	11	**142**	−7.34	11	24	4.23	11	230	5.25
12	33	0.28	12	313	−7.12	12	**271**	4.17	12	260	5.23
13	142	0.28	13	**234**	−6.92	13	210	4.15	13	93	−5.16
14	234	0.25	14	**219**	−6.91	14	**254**	−4.13	14	137	5.15
15	219	0.24	15	**54**	6.63	15	31	4.09	15	**234**	4.73
16	59	0.24	16	**59**	6.53	16	**54**	−4.04	16	**33**	−4.50
17	54	0.23	17	**201**	6.51	17	86	−3.98	17	**254**	−4.43
18	271	0.22	18	213	−6.32	18	222	3.98	18	39	4.36
19	201	0.21	19	**266**	−6.09	19	178	3.89	19	193	4.36
20	277	0.21	20	71	−6.04	20	308	−3.86	20	293	4.28

If it is desired to use the *t*-test (or test based on the Fisher weight) to obtain a probability that a variable is significant it is important first to perform a test of normality. There are many possible ones, but a well established one is the Kolmorogov–Smirnov test, which gives a numerical value of how the experimental distribution differs from a normal distribution and should be performed as follows for each class g separately, and not on the overall dataset: if there is a strong difference between classes, we would not expect the combined data to obey a normal distribution, but we might hope that the distribution of each variable within each group separately does the following:

1. Order the values of variable j from the lowest (1) to the highest (I_g), with e_{ij} being the value of the rank of variable j in sample i.
2. Determine the cdf (cumulative distribution function) for these variables $d_{ij} = e_{ij}/I_g$ (so if there are 100 samples in class g, the value of d for the 25th lowest is 0.25).
3. Standardize the values of variable j over all samples in the class of interest, and plot the value of d_{ij} against these values (having first ordered the sample by rank).
4. Compare this to the cdf for the standard normal distribution.
5. Compute the maximum difference between the cdf and the standard normal distribution. The larger this is, the less likely the distribution of variables fall onto a normal distribution. This value can be interpreted graphically (just by looking at the two curves), or by using standard critical tables available in books or on the Internet.

Because we discussed the Kolmorogov–Smirnov test in a different context in Section 3.3.3, so we will not dwell on this in more detail in the context of this chapter except to note as follows:

1. If the results of the *t*-statistic or Fisher weight calculations are to be converted directly to probabilities it is important that the distribution of the variables in each class are approximately normal.
2. To do this a test of normality should be applied to the distribution of variables, scaled in the way they will be for input to the classifier.
3. In most situations it is rare that variables are normally distributed; however there may be some datasets for which approximate normality is obeyed.

The results of performing this test on the top 10 most significant variables according to their *t*-statistic are presented in Table 9.6, with an indication as to whether they pass (P) or fail (F) at 95 % level of significance, for each class and Case Studies 8a (mouse urine –diet) and 4 (pollution). It can be seen, especially for Case Study 8a, that the majority of values fail. Hence converting to the *t*-statistic directly to a probability has little meaning. Despite this problem, which often occurs particularly in biological applications of pattern recognition, biologists in particular like citing so called ‘p values’ even though these can be quite misleading. This sometimes causes a major dilemma for the chemometrician presenting results to biological audiences.

Nevertheless the numerical value of the *t*-statistic (and other indicators discussed below) is still useful for ranking the significance of variables. If the values of the *t*-statistic are to be converted to probabilities, use a two tailed *t*-test, the aim being to see whether the means of variable j in each group of samples are significantly different from each other; see Table 9.7 for critical values, with the ‘df’ being the number of degrees of freedom or total number of samples minus 2.

Table 9.6 *Kolmorogov–Smirnov test for normality at 95 % for the 10 variables with the highest values of the t-statistic for Case Studies 8a and 4. P= Pass and F= Fail for classes A and B*

Case Study 8a (mouse urine – diet study)				Case Study 4 (pollution)			
		A	B			A	B
1	197	F	F	1	35	P	P
2	55	P	P	2	23	P	P
3	110	F	F	3	37	F	P
4	263	P	F	4	9	P	P
5	124	F	F	5	3	F	P
6	238	P	F	6	7	F	P
7	175	P	P	7	4	F	P
8	149	P	P	8	26	F	P
9	262	F	F	9	49	P	P
10	270	F	F	10	21	F	P

Table 9.7 *Two tailed t-test: columns represent probabilities and rows represent degrees of freedom*

	0.10	0.02	0.01
1	6.314	31.821	63.656
2	2.920	6.965	9.925
3	2.353	4.541	5.841
4	2.132	3.747	4.604
5	2.015	3.365	4.032
6	1.943	3.143	3.707
7	1.895	2.998	3.499
8	1.860	2.896	3.355
9	1.833	2.821	3.250
10	1.812	2.764	3.169
11	1.796	2.718	3.106
12	1.782	2.681	3.055
13	1.771	2.650	3.012
14	1.761	2.624	2.977
15	1.753	2.602	2.947
16	1.746	2.583	2.921
17	1.740	2.567	2.898
18	1.734	2.552	2.878
19	1.729	2.539	2.861
20	1.725	2.528	2.845
25	1.708	2.485	2.787
30	1.697	2.457	2.750
35	1.690	2.438	2.724
40	1.684	2.423	2.704
45	1.679	2.412	2.690
50	1.676	2.403	2.678
55	1.673	2.396	2.668
60	1.671	2.390	2.660
65	1.669	2.385	2.654
70	1.667	2.381	2.648
80	1.664	2.374	2.639
90	1.662	2.368	2.632
100	1.660	2.364	2.626
∞	1.645	2.327	2.576

In this text we will primarily employ the *t*-statistic and Fisher weight as numerical indicators of rank rather than for the purpose of computing probabilities, but will describe an approach for obtaining empirical probabilities in Section 9.3.2 that does not depend on assumptions of normality.

9.2.3 Multiple Linear Regression, ANOVA and the *F*-Ratio

Multiple Linear Regression

MLR (Multiple Linear Regression) can be employed in a variety of ways to determine the significance of variables and as such is very flexible, but is normally applied in a univariate manner to assess the significance of each variable. In order to illustrate the method, we will assume all variables have been standardized prior to employing MLR for variable selection, using population standard deviations (the difference between population and sample standard deviations is small if there are many samples but is done so that all indicators are on a comparable scale). Note that this standardization is over both (or all) classes and not each class separately.

For a two class classifier, using MLR we can set up an equation that relates the observed values of $\boldsymbol{x}_j$ to the classifier $\boldsymbol{c}_i$ in various ways, the simplest being:

$$x_{ij} \approx b_{0j} + b_{1j}c_i$$

or in matrix form:

$$\boldsymbol{x}_j \approx \boldsymbol{C}\boldsymbol{b}_j$$

with $\boldsymbol{x}_j$ being a column vector containing the standardized values of measurement j, $\boldsymbol{b}_j$ being a 2×1 vector containing the value of the two regression coefficients for variable j, and $\boldsymbol{C}$ being an $I \times 2$ matrix whose first column contains '1s' and whose second the value of the classifier (+ or – 1 in our case), so that the estimates are:

$$\boldsymbol{b}_j = \boldsymbol{C}^+\boldsymbol{x}_j$$

where '+' denotes the pseudoinverse (Section 2.2). The magnitude of b_{1j} is an indicator of how strongly related the value of variable j is to the classifier, and so how significant it is, and the sign to which class it is a marker for. If c and $\boldsymbol{x}$ are first mean centred then the equation becomes:

$$b_j = {}^{\mathrm{mc}}\boldsymbol{c}^+\boldsymbol{x}_j = {}^{\mathrm{mc}}\boldsymbol{c}'\boldsymbol{x}_j/I$$

where we will use ${}^{\mathrm{mc}}\boldsymbol{c}$ to denote the mean centred $\boldsymbol{c}$ vector (only necessary if the number of samples in each class is unequal), b_j to denote the regression coefficient and I is the total number of samples (the pseudoinverse of $\boldsymbol{c}$ being equal to its transpose divided by the number of samples providing the classifier is given a value of ± 1 as can be verified by the interested reader). If there are equal class sizes and the data have been centred (standardized) beforehand the equation for b becomes:

$$b = \delta/2s$$

where s is the standard deviation of the data and δ the difference in means (in the original units prior to standardization).

It can be shown that the value of b obtained from MLR is related to t obtained from the t-statistic. For simplicity we will assume that there are two groups of equal size ($I/2$ samples) and each group has equal variance. The results follow for unequal variance but the derivations are beyond the scope of this text. The t-statistic becomes:

$$t = \delta/(p\sqrt{2/I + 2/I}) = \delta/(p\sqrt{4/I})$$

since there are $I/2$ samples in each group where p is the pooled standard deviation. It can be shown that if the standard deviation for each group is the same:

$$p^2 = s^2 - \delta^2/4$$

providing we use the population definitions throughout. But:

$$\frac{b}{t} = \frac{\sqrt{4/I}p}{2s}$$

so that:

$$\frac{b}{t} = \frac{\sqrt{4/I}\sqrt{s^2 - \delta^2/4}}{2s} = \sqrt{\frac{1}{I}}\sqrt{1 - \delta^2/(4s^2)}$$

But since $b = \delta/2s$ we have:

$$\frac{b}{t} = \sqrt{\frac{1}{I}}\sqrt{1 - b^2}$$

or:

$$t = b\sqrt{\frac{I}{(1 - b^2)}}$$

The implies that there is a direct relationship between t and b (providing the data are standardized – which is usually essential if variables are on different scales), and we label the classes +1 and –1. This relationship is illustrated in Figure 9.9 for a simple dataset presented in Table 9.8. Providing data have been standardized, the higher the value the b_1, the more significant the variable is considered to be, and variables can be ranked in order of their regression coefficients, just as for the other approaches discussed in this chapter.

This relationship implies that if suitably scaled, the rank of variables using the t-statistic and the MLR b coefficient will always be the same and so these both measure equivalent quantities.

There are a number of important points to note:

1. MLR methods as described in this section are univariate and do not take into account interactions between the variables.
2. The inverse equation for MLR, $c_i \approx a_0 + a_{1j}\,x_{ij}$, can be solved for individual variables; however in such a case the larger is a, then the less significant is the variable, and so we prefer the reverse equation.

The discussion above would suggest that MLR regression coefficients have no advantage over using the t-statistic for ranking variables for two class models. However MLR

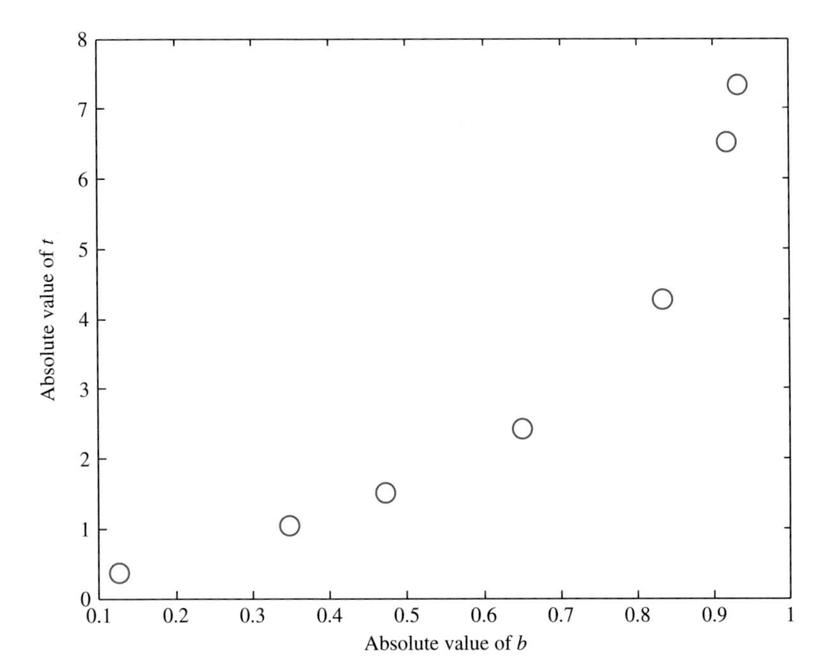

Figure 9.9 Relationship between the *t*-statistic and the value of *b* obtained on the standardized data using MLR for the small dataset of Table 9.8

Table 9.8 *Small datamatrix consisting of 8 samples in two classes characterized by 7 variables, to illustrate the relationship between the* t*-statistic and the* b *coefficient from MLR*

X							**c**
Var 1	Var 2	Var 3	Var 4	Var 5	Var 6	Var 7	
7	11	8	3	9	4	13	1
3	4	7	6	4	3	16	1
4	1	3	1	8	2	11	1
2	6	4	2	7	4	12	1
1	3	12	9	3	5	5	−1
0	7	10	9	6	3	6	−1
2	1	9	10	1	4	4	−1
5	2	11	9	5	2	2	−1

methods do have a use if it is desired to include additional terms in a model, for example, we might want to include squared terms such as:

$$x_{ij} \approx b_0 + b_{1j}\, c_i + b_{11j}\, c_i^2$$

if we suspect the behaviour of the classifier is non linear.

Another important extension is when samples can be characterized by two types of classifier, for example, a sample may be classifier into two or more groups, e.g. we may want to divide patients into groups of (1) male and female and (2) diseased or not diseased. If we set up a classifier for each class, say c_1 for gender and c_2 for disease, we can then model $\boldsymbol{x}_j$ by:

$$x_{ij} \approx b_0 + b_{1j}c_{1i} + b_{2j}c_{2i} + b_{12j}c_{1i}c_{2i}$$

where c_{1i} is a label for the class membership of sample i according to the first grouping with an equivalent definition for c_{2i}; this can easily be re-expressed as matrix terms, for example as:

$$\boldsymbol{X} \approx \boldsymbol{CB}$$

Note the term b_{12j} in the equation above: this is called the interaction term and if important can imply that the two classifiers are not completely independent of each other. For example, the effect of a disease on female patients may be different to male patients, and so we cannot model each factor (or classifier) separately.

An advantage of MLR methods compared to most multivariate methods is that they can take into account the interactions between the classifiers, however not between the variables measured, when expressed as above. MLR is one of the main tools in traditional statistics; most statisticians are interested in interactions and responses that can be characterized by several factors, but are less used to multivariate approaches, and so historically within mainstream statistics univariate approaches have been developed, whereas in traditional chemometrics there has been less emphasis on interactions between factors and more on multivariate approaches. PLS however is a good alternative that can take into account the interactions between the variables but not, in the basic PLS1 algorithm, the interactions between the classifiers.

MLR methods as discussed in this section are only applicable if there are only two groups in each class, or if more than two groups that they can be related sequentially, e.g. different age groups, so that each group can be assigned a sequential number. If there is no sequential relationship between groups, interpretation and analysis can become quite tricky and although possible is not recommended, but the extension into ANOVA based approaches is more usual.

ANOVA

If this were a book aimed primarily at statisticians, ANOVA (Analysis of Variance) based methods for determining the relative significance of variables would play a much larger role. If this were a book for the Machine Learning community it is unlikely to be mentioned. There are many different cultures and expectations. The types of problems encountered by chemometricians and statisticians tend to be somewhat different in nature. Statisticians tend primarily to be concerned with analysing designed experiments as are possible often in very controlled environments, but although there is certainly a well established field of multivariate statistics, many applied statisticians encounter multivariate problems much more rarely. Chemometricians, however, often have to deal with less well controlled systems, a typical example being in industrial process monitoring where it is not easy to perform planned experiments, but usually encounter data that are highly multivariate. Hence a chemometrician would approach the analysis of a problem primarily by using multivariate tools; in the case of variable selection, PLS weights and regression coefficients (Section 9.2.4) are typical of the approaches commonly applied.

ANOVA and associated F-ratios are however an essential tool for the professional statistician and can be applied to selected chemometrics problems for the determination of which variables are most significant for distinguishing groups. In this text we will limit our discussion because such types of dataset are less common in chemometrics or analytical chemistry, often because they cannot be obtained or the expense would be prohibitative or the problem is a simple one in terms of the number of factors being studied, whilst the complexity is usually in the multivariate nature of the data. Nevertheless, a review of the main approaches in the context of determining which variables are important for distinguishing between groups is valuable. Usually statisticians use ANOVA for hypothesis testing. They may be concerned with whether the value of a discrete factor (which we may call a class) has a significant influence on one or more observed properties (which we call variables). Usually a statistic, often an F-ratio, is computed, and the higher its value, the more significant the factor is: turning the problem around in terminology related to this text, we are interested in whether a variable can significantly distinguish between one or more groups and wish to compute a statistic whose value relates to its significance. Statisticians often then look at the value of the statistic and associate a probability with it that gives the confidence that there is a significant relationship between the factor (or class) and the property (or variable). Chemometricians are primarily interested in the inverse problem, but since we are usually not attempting to determine causality (for example, we are not trying to determine whether a chemical in a person's blood causes a disease or whether it is a consequence of a disease, just whether there is a link between the presence of a chemical in a person's blood and a disease) the same significance indicators can be calculated.

One way ANOVA is the simplest approach, and is very similar in concept to both the t-statistic and Fisher weight (Section 9.2.2). Consider variable j which is measured over I samples, and g classes. ANOVA tries to divide the total variance into two parts (a) between classes and (b) within classes, as illustrated in Figure 9.10. Algebraically the equations are

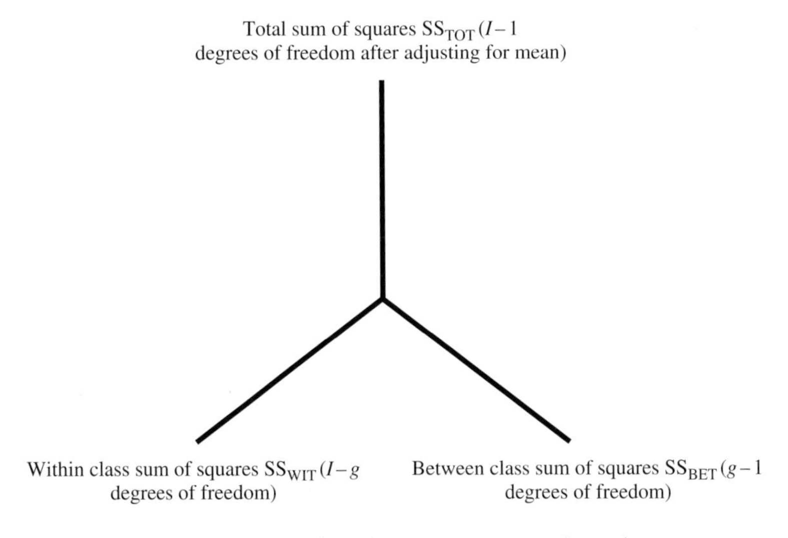

Figure 9.10 Principles of one way ANOVA for g classes

quite simple; the total variance is often called the sum of squares after mean centring and is given by the following equations:

1. $SS_{TOT} = \sum_{i=1}^{I} (x_{ij} - \bar{x}_j)^2$ where $\bar{x}_j$ is the mean of variable j (= 0 if standardized) over the entire dataset. Note that SS_{TOT} will be I for standardized data using a population standard deviation for the scaling.
2. SS_{BET} or 'the between class' sum of squares is given by $SS_{BET} = \sum_{g=1}^{G} I_g (x_{gj} - \bar{x}_j)^2$ where $\bar{x}_{ig} = \sum_{i \in g} x_{ij} / I_g$ is the mean of variable j over class g.
3. SS_{WIT} or the within class sum of squares is given by $SS_{WIT} = \sum_{g=1}^{G} \sum_{i \in g} (x_{ij} - \bar{x}_{jg})^2$. Note that $SS_{BET} + SS_{WIT} = SS_{TOT}$.

Each sum of squares has a number of degrees of freedom (df) associated with it as given in Figure 9.11. The mean sum of squares is given by:

$$MS = SS/df$$

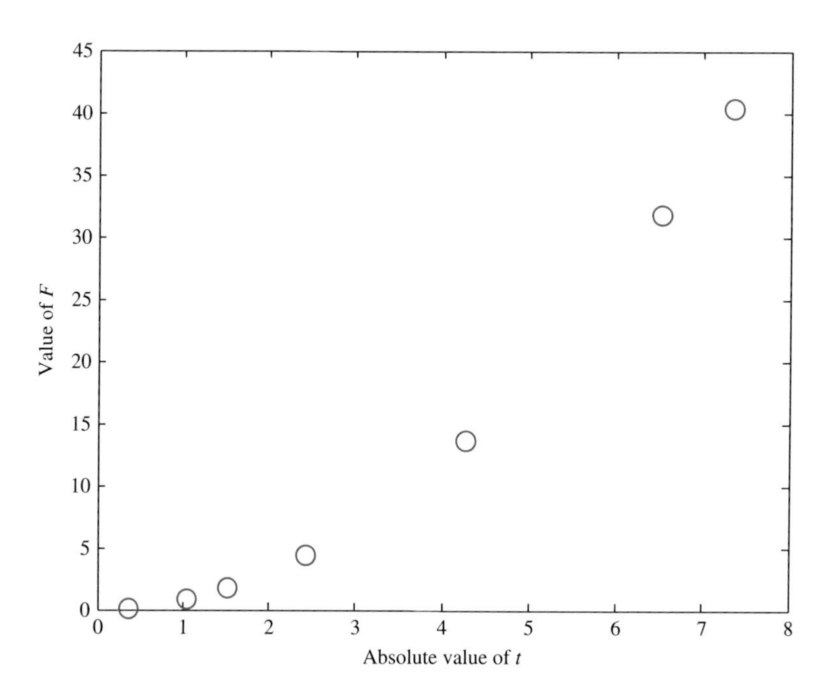

Figure 9.11 Relationship between the t-statistic and the value of b obtained on the standardized data using MLR for the small dataset of Table 9.8

and the higher the ratio of MS_{BET} / MS_{WIT} (or mean between group sum of squares to mean within group sum of squares), the more significant the variable is. This ratio is called the F-ratio.

We will illustrate this with a small example in Table 9.9, in which one variable is measured over two classes, each consisting of 10 samples. There are 19 degrees of freedom overall of which 1 is due to 'between group variance' and 18 due to 'within group

Table 9.9 *ANOVA example*

Class	Value		
A	4.101	x_j	4.075
A	3.851	$\bar{x}_{jA}$	4.164
A	4.045	$\bar{x}_{jB}$	3.987
A	4.616		
A	4.238		
A	4.462	SS_{TOT}	1.474
A	4.448	SS_{WIT}	1.319
A	3.803	SS_{BET}	0.155
A	4.008		
A	4.063		
B	3.994		
B	3.906	MS_{WIT}	0.155
B	4.120	MS_{BET}	0.073
B	4.071		
B	4.497		
B	4.082		
B	3.965		
B	4.092		
B	3.532		
B	3.614		

variance'. The mean sum of squares due to 'between group variation' is 0.155 whereas that due to 'within group variation' 0.073, giving an F-ratio of 2.121. Typically the data are presented as an ANOVA table as illustrated in Table 9.10

Table 9.10 *ANOVA table for the example of Table 9.9*

Source	SS	df	MS	F-ratio
'Between'	0.155	1	0.155	2.121
'Within'	1.319	18	0.073	—
Total	1.474	19	—	—

We can see that the F-ratio is proportional to the square of the t-statistic. Consider our simple example of two equally sized classes, each consisting of $I_g = I/2$ samples. We can see that:

$$SS_{BET} = 2I_g(\delta/2)^2 = I\delta^2/4$$

and:

$$SS_{WIT} = 2I_g p^2 = Ip^2$$

where p is the pooled variance (remembering that the class sizes are equal in this example – the derivation can be extended to unequal class sizes). For two classes the 'between class' sum of squares has one degree of freedom and the 'within class' sum of squares I–1 degrees of freedom. Hence:

$$F = \mathrm{MS}_{\mathrm{BET}}/\mathrm{MS}_{\mathrm{WIT}} = \frac{I\delta^2/4}{Ip^2/(I-2)} = \frac{(I-2)\delta^2}{4p^2}$$

but we know that:

$$t = \delta/(p\sqrt{4/I})$$

so that:

$$t^2 = \frac{I\delta^2}{4p^2}$$

Hence

$$F = t^2(I-2)/I$$

We can calculate F for the variables in our example of Table 9.8 and show this relationship in Figure 9.11. Hence ranking variables using the value of the F-statistic will result in identical ranks as for the t-statistic, and the MLR b coefficient (if the data are appropriately scaled).

Many statisticians then interpret the F-ratio further and attach a probability that it is significant (in chemometrics terms we may be asking whether a variable which may represent a compound is a significant candidate as a marker; in statistical terms we can see whether we can reject the null hypothesis that there is no relationship between the level of a factor which could be a classifier and the distribution of a variable). However the probability depends on the data being normally distributed and this is not often the case as we have shown above; we will discuss alternative approaches for obtaining empirical probabilities in Section 9.3. In addition the aim in variable selection for chemometrics is somewhat different. We are not primarily interested in which variable *per se* is significant, but to compare the significance of different variables; the larger the F-ratio, the more they appear to be significant, and they can be ranked accordingly. F-ratios can often be interpreted as probabilities using tables such as Table 9.11 and Table 9.12. The critical value of the F-ratio is tabulated for two probability levels. A 5 % probability level means that we are 95 % certain that we can reject the null hypothesis (in statistical terms). To use these tables, first determine the number of degrees of freedom (df) of the sum of squares with the highest F-value (between groups in our case), and which is interpreted as the as the df along the top, and then the df of the sum of squares with the lowest variance (along the side). The value in the table is the critical value; if exceeded we can be more that 95 % certain (using a one tailed test and Table 9.12) that our variable is significant. Hence the critical value of F (4,6) is 4.534. If the F-ratio exceeds this value then it is considered significant at greater than 95 % level for 4 and 6 dfs. We have introduced other applications of the F-statistic in Chapter 6 in the context of one class classifiers. However it is important to recognize that these probabilities are based on the assumption of normality. When we compare distributions of variables, some may be normally distributed and others are not, and so it may not necessarily be safe to translate the F values into probabilities. We will see that there are other empirical ways of establishing probabilities (Section 9.3.2).

Table 9.11 *Critical values of F-statistic: 1 % one-tailed; 2 % two-tailed*

	1	2	3	4	5	6	7	8	9	10	11	12	13	14	15	20	30	40	50	75	100	250	500	∞
1	4052	4999	5404	5624	5764	5859	5928	5981	6022	6056	6083	6107	6126	6143	6157	6209	6260	6286	6302	6324	6334	6353	6360	6366
2	98.50	99.00	99.16	99.25	99.30	99.33	99.36	99.38	99.39	99.40	99.41	99.42	99.42	99.43	99.43	99.45	99.47	99.48	99.48	99.48	99.49	99.50	99.50	99.499
3	34.12	30.82	29.46	28.71	28.24	27.91	27.67	27.49	27.34	27.23	27.13	27.05	26.98	26.92	26.87	26.69	26.50	26.41	26.35	26.28	26.24	26.17	26.15	26.125
4	21.20	18.00	16.69	15.98	15.52	15.21	14.98	14.80	14.66	14.55	14.45	14.37	14.31	14.25	14.20	14.02	13.84	13.75	13.69	13.61	13.58	13.51	13.49	13.463
5	16.26	13.27	12.06	11.39	10.97	10.67	10.46	10.29	10.16	10.05	9.963	9.888	9.825	9.770	9.722	9.553	9.379	9.291	9.238	9.166	9.130	9.064	9.042	9.020
6	13.75	10.92	9.780	9.148	8.746	8.466	8.260	8.102	7.976	7.874	7.790	7.718	7.657	7.605	7.559	7.396	7.229	7.143	7.091	7.022	6.987	6.923	6.901	6.880
7	12.25	9.547	8.451	7.847	7.460	7.191	6.993	6.840	6.719	6.620	6.538	6.469	6.410	6.359	6.314	6.155	5.992	5.908	5.858	5.789	5.755	5.692	5.671	5.650
8	11.26	8.649	7.591	7.006	6.632	6.371	6.178	6.029	5.911	5.814	5.734	5.667	5.609	5.559	5.515	5.359	5.198	5.116	5.065	4.998	4.963	4.901	4.880	4.859
9	10.56	8.022	6.992	6.422	6.057	5.802	5.613	5.467	5.351	5.257	5.178	5.111	5.055	5.005	4.962	4.808	4.649	4.567	4.517	4.449	4.415	4.353	4.332	4.311
10	10.04	7.559	6.552	5.994	5.636	5.386	5.200	5.057	4.942	4.849	4.772	4.706	4.650	4.601	4.558	4.405	4.247	4.165	4.115	4.048	4.014	3.951	3.930	3.909
11	9.646	7.206	6.217	5.668	5.316	5.069	4.886	4.744	4.632	4.539	4.462	4.397	4.342	4.293	4.251	4.099	3.941	3.860	3.810	3.742	3.708	3.645	3.624	3.602
12	9.330	6.927	5.953	5.412	5.064	4.821	4.640	4.499	4.388	4.296	4.220	4.155	4.100	4.052	4.010	3.858	3.701	3.619	3.569	3.501	3.467	3.404	3.382	3.361
13	9.074	6.701	5.739	5.205	4.862	4.620	4.441	4.302	4.191	4.100	4.025	3.960	3.905	3.857	3.815	3.665	3.507	3.425	3.375	3.307	3.272	3.209	3.187	3.165
14	8.862	6.515	5.564	5.035	4.695	4.456	4.278	4.140	4.030	3.939	3.864	3.800	3.745	3.698	3.656	3.505	3.348	3.266	3.215	3.147	3.112	3.048	3.026	3.004
15	8.683	6.359	5.417	4.893	4.556	4.318	4.142	4.004	3.895	3.805	3.730	3.666	3.612	3.564	3.522	3.372	3.214	3.132	3.081	3.012	2.977	2.913	2.891	2.868
20	8.096	5.849	4.938	4.431	4.103	3.871	3.699	3.564	3.457	3.368	3.294	3.231	3.177	3.130	3.088	2.938	2.778	2.695	2.643	2.572	2.535	2.468	2.445	2.421
30	7.562	5.390	4.510	4.018	3.699	3.473	3.305	3.173	3.067	2.979	2.906	2.843	2.789	2.742	2.700	2.549	2.386	2.299	2.245	2.170	2.131	2.057	2.032	2.006
40	7.314	5.178	4.313	3.828	3.514	3.291	3.124	2.993	2.888	2.801	2.727	2.665	2.611	2.563	2.522	2.369	2.203	2.114	2.058	1.980	1.938	1.860	1.833	1.805
50	7.171	5.057	4.199	3.720	3.408	3.186	3.020	2.890	2.785	2.698	2.625	2.563	2.508	2.461	2.419	2.265	2.098	2.007	1.949	1.868	1.825	1.742	1.713	1.683
75	6.985	4.900	4.054	3.580	3.272	3.052	2.887	2.758	2.653	2.567	2.494	2.431	2.377	2.329	2.287	2.132	1.960	1.866	1.806	1.720	1.674	1.583	1.551	1.516
100	6.895	4.824	3.984	3.513	3.206	2.988	2.823	2.694	2.590	2.503	2.430	2.368	2.313	2.265	2.223	2.067	1.893	1.797	1.735	1.646	1.598	1.501	1.466	1.427
250	6.737	4.691	3.861	3.395	3.091	2.875	2.711	2.583	2.479	2.392	2.319	2.256	2.202	2.154	2.111	1.953	1.774	1.674	1.608	1.511	1.457	1.343	1.297	1.244
500	6.686	4.648	3.821	3.357	3.054	2.838	2.675	2.547	2.443	2.356	2.283	2.220	2.166	2.117	2.075	1.915	1.735	1.633	1.566	1.465	1.408	1.285	1.232	1.164
∞	6.635	4.605	3.782	3.319	3.017	2.802	2.639	2.511	2.407	2.321	2.248	2.185	2.130	2.082	2.039	1.878	1.696	1.592	1.523	1.419	1.358	1.220	1.153	1.000

Table 9.12 *Critical values of F-statistic: 5 % one-tailed; 10 % two-tailed*

	1	2	3	4	5	6	7	8	9	10	11	12	13	14	15	20	30	40	50	75	100	250	500	∞
1	161.4	199.5	215.7	224.6	230.2	234.0	236.8	238.9	240.5	241.9	243.0	243.9	244.7	245.4	245.9	248.0	250.1	251.1	251.8	252.6	253.0	253.8	254.1	254.3
2	18.51	19.00	19.16	19.25	19.30	19.33	19.35	19.37	19.38	19.40	19.40	19.41	19.42	19.42	19.43	19.45	19.46	19.47	19.48	19.48	19.49	19.49	19.49	19.50
3	10.13	9.552	9.277	9.117	9.013	8.941	8.887	8.845	8.812	8.785	8.763	8.745	8.729	8.715	8.703	8.660	8.617	8.594	8.581	8.563	8.554	8.537	8.532	8.526
4	7.709	6.944	6.591	6.388	6.256	6.163	6.094	6.041	5.999	5.964	5.936	5.912	5.891	5.873	5.858	5.803	5.746	5.717	5.699	5.676	5.664	5.643	5.635	5.628
5	6.608	5.786	5.409	5.192	5.050	4.950	4.876	4.818	4.772	4.735	4.704	4.678	4.655	4.636	4.619	4.558	4.496	4.464	4.444	4.418	4.405	4.381	4.373	4.365
6	5.987	5.143	4.757	4.534	4.387	4.284	4.207	4.147	4.099	4.060	4.027	4.000	3.976	3.956	3.938	3.874	3.808	3.774	3.754	3.726	3.712	3.686	3.678	3.669
7	5.591	4.737	4.347	4.120	3.972	3.866	3.787	3.726	3.677	3.637	3.603	3.575	3.550	3.529	3.511	3.445	3.376	3.340	3.319	3.290	3.275	3.248	3.239	3.230
8	5.318	4.459	4.066	3.838	3.688	3.581	3.500	3.438	3.388	3.347	3.313	3.284	3.259	3.237	3.218	3.150	3.079	3.043	3.020	2.990	2.975	2.947	2.937	2.928
9	5.117	4.256	3.863	3.633	3.482	3.374	3.293	3.230	3.179	3.137	3.102	3.073	3.048	3.025	3.006	2.936	2.864	2.826	2.803	2.771	2.756	2.726	2.717	2.707
10	4.965	4.103	3.708	3.478	3.326	3.217	3.135	3.072	3.020	2.978	2.943	2.913	2.887	2.865	2.845	2.774	2.700	2.661	2.637	2.605	2.588	2.558	2.548	2.538
11	4.844	3.982	3.587	3.357	3.204	3.095	3.012	2.948	2.896	2.854	2.818	2.788	2.761	2.739	2.719	2.646	2.570	2.531	2.507	2.473	2.457	2.426	2.415	2.404
12	4.747	3.885	3.490	3.259	3.106	2.996	2.913	2.849	2.796	2.753	2.717	2.687	2.660	2.637	2.617	2.544	2.466	2.426	2.401	2.367	2.350	2.318	2.307	2.296
13	4.667	3.806	3.411	3.179	3.025	2.915	2.832	2.767	2.714	2.671	2.635	2.604	2.577	2.554	2.533	2.459	2.380	2.339	2.314	2.279	2.261	2.229	2.218	2.206
14	4.600	3.739	3.344	3.112	2.958	2.848	2.764	2.699	2.646	2.602	2.565	2.534	2.507	2.484	2.463	2.388	2.308	2.266	2.241	2.205	2.187	2.154	2.142	2.131
15	4.543	3.682	3.287	3.056	2.901	2.790	2.707	2.641	2.588	2.544	2.507	2.475	2.448	2.424	2.403	2.328	2.247	2.204	2.178	2.142	2.123	2.089	2.078	2.066
20	4.351	3.493	3.098	2.866	2.711	2.599	2.514	2.447	2.393	2.348	2.310	2.278	2.250	2.225	2.203	2.124	2.039	1.994	1.966	1.927	1.907	1.869	1.856	1.843
30	4.171	3.316	2.922	2.690	2.534	2.421	2.334	2.266	2.211	2.165	2.126	2.092	2.063	2.037	2.015	1.932	1.841	1.792	1.761	1.718	1.695	1.652	1.637	1.622
40	4.085	3.232	2.839	2.606	2.449	2.336	2.249	2.180	2.124	2.077	2.038	2.003	1.974	1.948	1.924	1.839	1.744	1.693	1.660	1.614	1.589	1.542	1.526	1.509
50	4.034	3.183	2.790	2.557	2.400	2.286	2.199	2.130	2.073	2.026	1.986	1.952	1.921	1.895	1.871	1.784	1.687	1.634	1.599	1.551	1.525	1.475	1.457	1.438
75	3.968	3.119	2.727	2.494	2.337	2.222	2.134	2.064	2.007	1.959	1.919	1.884	1.853	1.826	1.802	1.712	1.611	1.555	1.518	1.466	1.437	1.381	1.360	1.338
100	3.936	3.087	2.696	2.463	2.305	2.191	2.103	2.032	1.975	1.927	1.886	1.850	1.819	1.792	1.768	1.676	1.573	1.515	1.477	1.422	1.392	1.331	1.308	1.283
250	3.879	3.032	2.641	2.408	2.250	2.135	2.046	1.976	1.917	1.869	1.827	1.791	1.759	1.732	1.707	1.613	1.505	1.443	1.402	1.341	1.306	1.232	1.202	1.166
500	3.860	3.014	2.623	2.390	2.232	2.117	2.028	1.957	1.899	1.850	1.808	1.772	1.740	1.712	1.686	1.592	1.482	1.419	1.376	1.312	1.275	1.194	1.159	1.113
∞	3.841	2.996	2.605	2.372	2.214	2.099	2.010	1.938	1.880	1.831	1.789	1.752	1.720	1.692	1.666	1.571	1.459	1.394	1.350	1.283	1.243	1.152	1.106	1.000

However, ANOVA based approaches have an advantage over the t-statistic (for example) in that like the Fisher weight they can be extended to multiclass datasets. The description of one way ANOVA is included in this text is because many readers will have encountered this in other contexts but it offers no advantages in terms of ranking of variables over the t-statistic or Fisher weight or the MLR b coefficient for one way problems where there is a single classifier, apart from the fact that in classical statistics a probability can be attached to the F-ratio on the assumption that errors are normally distributed. We have shown that MLR can also cope with such situations, but only easily if there are two groups for each classifier. In some circumstances each sample may be influenced by more than one factor and so could be classified in two distinct ways. Case Study 6 is such an example, where the LCMS depends on the origins of the tablets and on the instrument or time it was recorded. There are 3 origins and 5 instrumental conditions. How can we distinguish between instrumental condition and origin? One way of course would be to treat each independently and to calculate Fisher weights or PLS-DA regression coefficients (Section 9.2.4) or whatever we choose, both for the origins and instruments separately, but this ignores the fact that there may be some interactions between origins and instrumental conditions – for example, samples from one of the origins may exhibit ions that are detected better on one instrument or at one sampling time than another. Under these circumstances we could employ a two-way ANOVA. There are in fact a large number of designs each of which require specific analyses, but a common design is illustrated in Figure 9.12. An unfortunate limitation is that ideally

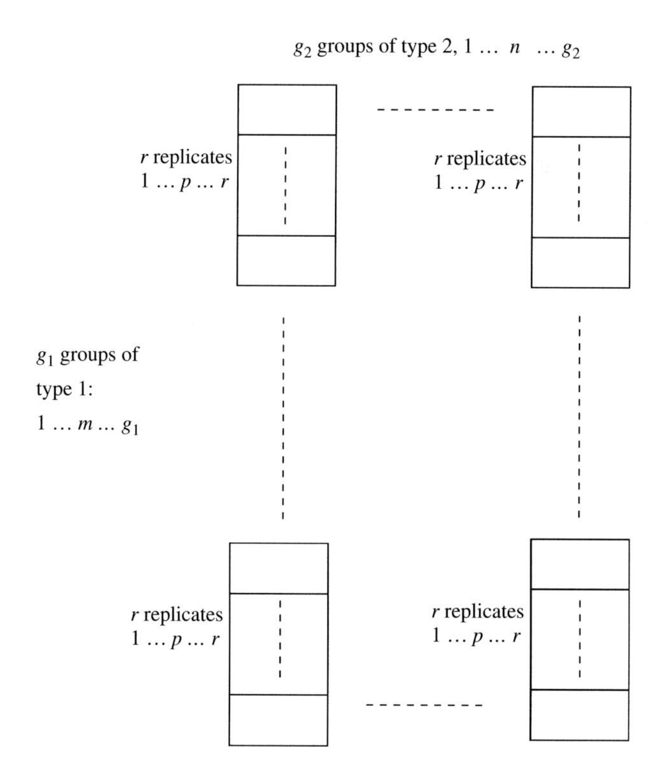

Figure 9.12 Design for two way ANOVA

the design should be balanced, that is, all possible combinations of factor levels must be included in the analysis and there should be replicates at each combination of factors. In simple terms that means that if there are 10 classes or groups of type 1 (e.g. 10 individuals) and 2 of type 2 (e.g. treatments) then all 20 possible combinations must be studied, and ideally this should be balanced, that is, there should be the same number of combinations for every situation if there are repeats. In many situations this is quite difficult, and when dealing with analytical measurements there may be instrumental and sampling problems which mean that a few of the samples are not of sufficient quality to reliably include in the analysis. This does introduce some dilemmas and although there are ways around this, the interpretation is rather complex and we will restrict our discussion to balanced sample designs in this text. The use of two-way ANOVA, whilst quite common in areas such as engineering and clinical studies, is rarer in chemometrics, although should be introduced as a potential method for determining discriminatory variables in certain circumstances, and so the main equations will be listed.

We will express the common ANOVA equations in terms of notation typically used in chemometrics pattern recognition. We will consider an example in which we are interested both in the influence of the origin of a sample and the instrument it is run on, on its chromatographic signal:

1. We are interested in each variable j, which may be a chromatographic peak.
2. We assume that there are g_1 groups of type 1 (e.g. origins of a sample) and g_2 groups of type 2 (e.g. instrument), and each unique set of conditions is replicated r times.
3. The mean of variable j over all samples is given by $\bar{x}_j = \sum_{m=1}^{g_1} \sum_{n=1}^{g_2} \sum_{p=1}^{r} x_{jnmp}/(g_1 g_2 r)$
4. The mean of variable j in class m of type 1 is given by $\bar{x}_{jm} = \sum_{n=1}^{g_2} \sum_{p=1}^{r} x_{jnmp}/(g_2 r)$
5. The mean of variable j in class n of type 2 is given by

$$\bar{x}_{jn} = \sum_{m=1}^{g_1} \sum_{p=1}^{r} x_{jnmp}/(g_1 r)$$

6. The mean for variable j over each replicate is given by

$$\bar{x}_{jmn} = \sum_{p=1}^{r} x_{jnmp}/r$$

The algebraic definitions of the F-ratios are presented in Table 9.13 and the degrees of freedom given in Figure 9.13. In the context of variable selection, the higher the F-ratio for a particular variable, the more significant the variable for a particular classifier. Two-way ANOVA and F-ratios are useful if there is suspected to be an interaction between two classifiers. However such types of experiments are comparatively elaborate in the context of chemometrics and so these approaches are rare although it is important to evaluate in advance if there is a real need for such experimental design and so whether such types of experiment are desirable. Experimental design and the associated ANOVA can be extended to more complex situations, including those where there are several factors (or

Table 9.13 *Calculations of F-ratios for two way ANOVA, for two types of class*

Variation	SS	Equation	df	Mean sum of squares	*F*-ratio
'Within'	SS_{WIT}	$\sum_{m=1}^{g_1}\sum_{n=1}^{g_2}\sum_{p=1}^{k}(x_{jnmp} - \bar{x}_{jnm})^2$	$I - g_1g_2$	$MS_{WIT} = SS_{WIT}/(I - g_1g_2)$	—
'Between'	SS_{BET}	$\sum_{m=1}^{g_1}\sum_{n=1}^{g_2}\sum_{p=1}^{k}(\bar{x}_{jnm} - \bar{x}_j)^2$	$g_1g_2 - 1$	$MS_{BET} = SS_{BET}/(g_1g_2 - 1)$	MS_{BET}/MS_{WIT}
Class type 1	SS_1	$\sum_{m=1}^{g_1}\sum_{n=1}^{g_2}\sum_{p=1}^{k}(\bar{x}_{jm} - \bar{x}_j)^2$	$g_1 - 1$	$MS_1 = SS_1/(g_1 - 1)$	MS_1/MS_{WIT}
Class type 2	SS_2	$\sum_{m=1}^{g_1}\sum_{n=1}^{g_2}\sum_{p=1}^{k}(\bar{x}_{jn} - \bar{x}_j)^2$	$g_2 - 1$	$MS_2 = SS_2/(g_2 - 1)$	MS_2/MS_{WIT}
Interaction	SS_{12}	$SS_{BET} - SS_1 - SS_2$	$(g_1 - 1)(g_2 - 1)$	$MS_{12} = SS_{12}/[(g_1 - 1)(g_2 - 1)]$	MS_{12}/MS_{WIT}

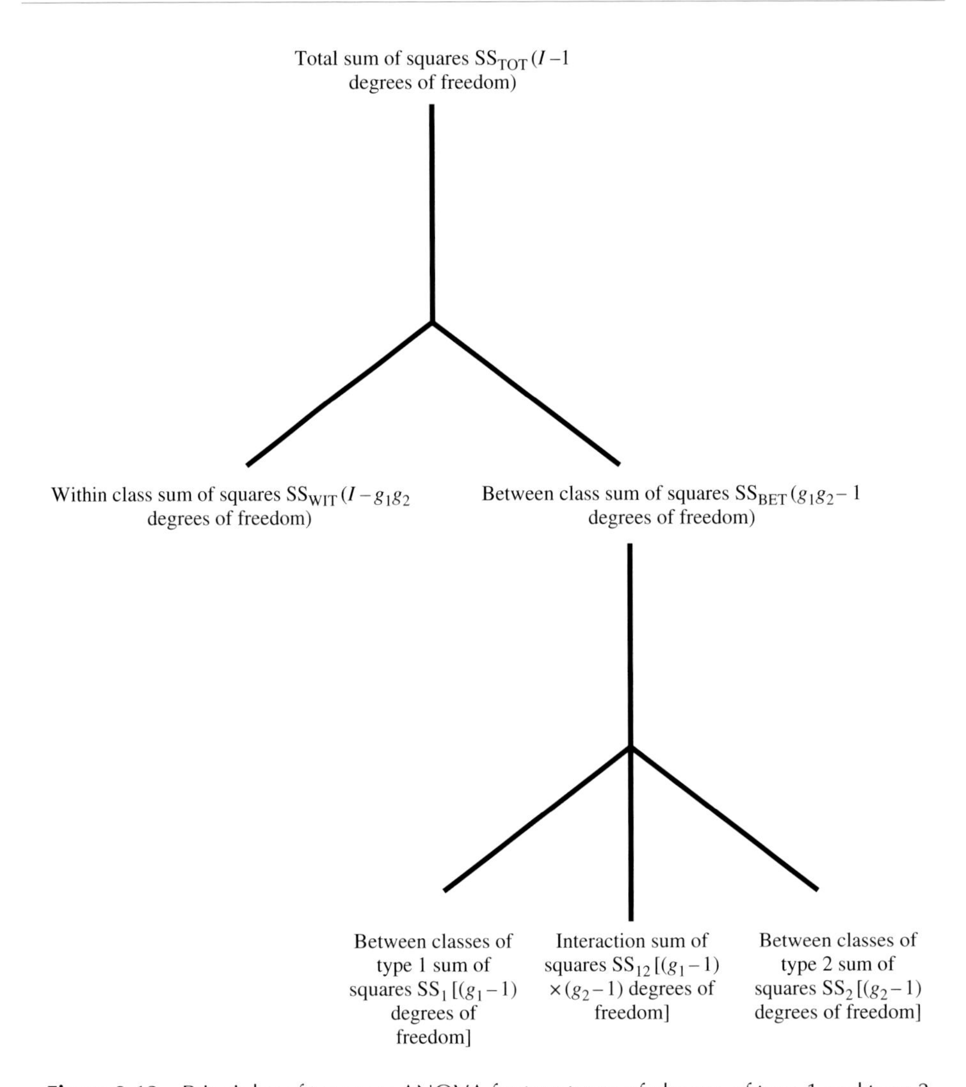

Figure 9.13 Principles of two way ANOVA for two types of classes, of type 1 and type 2

potentially interacting classifiers), but is beyond this text; however there are many such methods presented in mainstream texts on statistical experimental design and regression analysis.

There are several multivariate enhancements of ANOVA that take into account the interaction between the variables as well as the classifiers (or factors), which are an active area of research in the interface between chemometrics and biomedical studies. However such methods are out of the scope of this introductory text but are of potential interest to leading researchers and certainly should not be ignored, especially in clinical applications.

9.2.4 Partial Least Squares

A weakness of the methods of Sections 9.2.2 and 9.2.3 is that they are univariate, that is, each variable is considered separately and the relationship between variables is not taken into account. Whereas many traditional statisticians often prefer such models, particularly because equations of significance, probabilities and confidence levels are well established, in many practical situations we might expect some relationship between variables, and so we cannot assess the significance of each variable independently.

PLS-DA is one of the most widespread multivariate methods for classification and has been introduced previously in the context of classifiers (Section 5.5). PLS however can also be used to determine which variables contribute most to the classifier and so are best potential discriminators. PLS has an advantage over multivariate methods such as Mahalanobis distance approaches such as LDA and QDA in that all the original variables can be employed in the model even if the number of variables far exceeds the number of samples, and so the influence of each variable on the model can be directly interpreted and quantified; for Mahalonobis distance based approaches, if the number of variables exceeds the number of samples, they have to be reduced either by removing some of the original variables or by combining them into new variables such as PCs – the latter being most common in the context of chemometrics. In the case of LVQ and SVMs it is very hard to obtain indicators of influence of variables on the model. Hence, for multivariate models, we will be placing emphasis on PLS based approaches for variable selection.

It is most usual to use the PLS1 algorithm in which samples are assigned to one class or another using a single calculation. In the case that there are more than two classes a series of one against all comparisons can be made or else PLS2 could be employed as an alternative, as discussed in Chapter 7. In this chapter we will restrict the discussion to two class PLS1 models although the principles can easily be extended. There are several algorithms for PLS but we base the description in this text on the NIPALS algorithm where matrices $\boldsymbol{T}$ of dimensions $I \times A$, $\boldsymbol{P}$ of dimensions $A \times J$, a vector $\boldsymbol{q}$ of dimensions $A \times 1$, a weights matrix $\boldsymbol{W}$ of dimensions $J \times A$ where I is the number of samples, J is the number of variables and A is the number of PLS components used in the model, often determined by cross-validation or the bootstrap (Section 8.5.1). For illustration in this chapter, when we want to optimize the number of PLS components we will employ the bootstrap, although this is not prescriptive. In order for the significance of each variable to be compared it is essential to standardize the data prior to performing PLS, even if this was not done for model building; otherwise the values of the PLS coefficients will not be comparable. Note an important issue, in this chapter to exemplify the results of PLS, we standardize the $\boldsymbol{X}$ matrix so it is exactly centred, even if the numbers of variables in each class are different, unlike when we use PLS-DA for a classifier. We showed in Figure 5.10 that using weighted adjustment of means can change position the class boundary considerably if the number of samples in each class is different. However it has a much less serious influence on the relative significance of variables, although the numerical value and central point may be shifted, the order stays approximately the same (dependent slightly on how different the class means are and on the relative class sizes). The reason we employ overall standardization for this purpose is that the significance of variables is then directly comparable to the results of other statistical indicators such as the t-statistic, MLR b value and the F-statistic. Of course, whereas it is important to put each variable on a comparable scale if the original units are very different, alternative approaches to mean centring could be envisaged, and the aim of this text is to describe and exemplify different approaches rather than be highly prescriptive.

There are two principal indicators that can be computed as described below.

PLS Weights

PLS weights use the weights matrix ($\boldsymbol{W}$) to determine which variables are good discriminators. The value of each element in this matrix relates to how important a variable is for defining each PLS component. The greater the magnitude, the more influential the variable is in a specific PLS component; the sign relates to the class the variable is most diagnostic of. The root mean square PLS weight ($\overline{w}$) for each variable j (see Section 5.5.2 for definition) over all A PLS components used in the model can be calculated as follows:

$$\overline{w}_j = \sqrt{\frac{\sum_{a=1}^{A} w_{ja}^2}{A}}$$

The variable with the highest PLS weight is ranked number 1 as the best potential marker for the dataset and so on. The lower the value of the PLS weight, the less significant the variable. PLS weights are always positive and so do not contain information about which class the variable is a marker for.

PLS Regression Coefficients

An alternative method is the PLS regression coefficients ($\boldsymbol{b}$). In its simplest form, the regression model specifies the relationship between an experimental data matrix and a classifier; it can be expressed by:

$$\boldsymbol{c} = \boldsymbol{X}\boldsymbol{b} + \boldsymbol{f} = \boldsymbol{T}\boldsymbol{q} + \boldsymbol{f}$$

where $\boldsymbol{b}$ is the regression coefficient vector, estimated as follows:

$$\boldsymbol{b} = \boldsymbol{W}\boldsymbol{q}$$

and $\boldsymbol{W}$ is the PLS weight matrix.

The magnitudes of the coefficients $\boldsymbol{b}$ can be used to determine which variables are significant or most influential in the dataset. The coefficient b_j corresponds to the magnitude of each variable j. The greater the magnitude, the more likely it is to be a good discriminator. The sign of the regression coefficient can also be employed to determine which group the variable is a marker for. Variables are ranked from 1 (best) to J (worst) according to the magnitude of b_j. Note that the PLS regression coefficient differs from the MLR regression coefficient as described in Section 9.2.3, even though we use 'b' in both cases, to be consistent with generally accepted terminology.

Relationship to other Methods and Number of Components

There are some interesting relationships between the rank of variables using both PLS indicators and the t-statistic. When there are two classes, both with equal numbers of samples and 1 PLS component, employed, the rank of variables using PLS weights, PLS regression coefficients, the t-statistic, MLR b coefficient and F-ratio are the same as the same as shown as follows.

Because the value of $c = +1$ for class A and –1 for class B, and since $\mathbf{w} = \mathbf{X}'\ \mathbf{c}$, the weight for the first PLS component for variable j is given by:

$$w_j = \sum_{i=1}^{I} x_{ij}c_i = \sum_{i \in A} x_{ij} - \sum_{i \in B} x_{ij}$$

where $\sum_{i \in g} x_{ij}$ is the sum of the values of x_{ij} for all samples that are a member of class g. But the data are standardised, and so:

$$\sum_{i \in A} x_{ij} = -\sum_{i \in B} x_{ij}$$

if there are equal numbers in each class as the sum of the columns must come to 0. Therefore:

$$\sum_{i \in A} x_{ij} - \sum_{i \in B} x_{ij} = 2\sum_{i \in A} x_{ij} = I\delta/2s$$

if δ is the difference in means between each class, before standardization, and there are $I/2$ samples in each class. Hence the PLS weights are linearly related to the MLR regression coefficient b by a factor of I (total number of samples). The regression coefficients can be calculated, for one component as the product of w_j and q and so will maintain the same rank.

Since the t-statistic and F-ratio also are monotonically related to the MLR b coefficient, all these measures are related and lead to the same relative ranks of variables providing the data are scaled appropriately. Hence the PLS weights, PLS regression coefficients and all the univariate statistics result in the same ranking even though they are not linearly related.

When there are more PLS components, the relationships are more complex, but despite PLS being a multivariate method and the others not, they are not so radically different. Table 9.14 lists the values of the t-statistic, PLS weights and PLS regression coefficients for Case Study 8a (mouse urine – diet study) as both 1 and 5 PLS components are computed. We can see that whereas for the 1 PLS component, the ranks of the variables obtained using the t-statistic, PLS weight and PLS regression coefficient all agree, for 5 PLS components there are now some differences in order of the variables, especially using PLS regression coefficients. The change using PLS weights is somewhat limited as these are an average over each component and larger for the earlier components, and so the magnitude decreases as more smaller numbers are being averaged with quite considerable change in order of the variables. We illustrate how changing the number of PLS components changes the relative ranks of

Table 9.14 *Values of the t-statistic, PLS weights (PLSW) and PLS regression coefficients (PLSRC) for 1 and 5 PLS components for Case Study 8a, and 1 and 5 PLS components, for the top 20 most significant variables*

Rank	*t*-statistic		1 PLS component				5 PLS components			
	Var	*t*	Var	PLSW	Var	PLSRC	Var	PLSW	Var	PLSRC
1	197	9.76	197	79.35	197	14.13	197	35.71	197	14.31
2	55	7.83	55	69.40	55	12.36	55	31.60	110	11.50
3	110	6.68	110	62.15	110	11.07	110	28.12	262	−10.74
4	263	6.56	263	61.31	263	10.92	263	27.48	55	10.71
5	124	6.22	124	58.93	124	10.49	124	26.56	263	10.56
6	238	5.89	238	56.57	238	10.07	238	26.33	62	−10.39
7	175	5.44	175	53.10	175	9.46	149	25.00	271	−10.07
8	149	5.41	149	52.89	149	9.42	175	24.22	124	10.02
9	262	−5.31	262	52.05	262	−9.27	262	23.89	4	9.26
10	270	5.20	270	51.21	270	9.12	270	23.52	175	9.16
11	201	5.14	201	50.71	201	9.03	271	23.07	207	−9.02
12	271	−5.13	271	50.62	271	−9.01	185	23.02	145	8.92
13	185	4.96	185	49.22	185	8.77	201	22.86	270	8.82
14	218	−4.90	218	48.75	218	−8.68	285	22.54	129	8.73
15	4	4.90	4	48.74	4	8.68	4	22.35	186	8.67
16	285	−4.90	285	48.72	285	−8.68	218	22.04	201	8.61
17	77	−4.62	77	46.41	77	−8.26	77	21.52	218	−8.59
18	69	4.58	69	46.05	69	8.20	69	21.08	294	−8.54
19	212	4.58	212	46.04	212	8.20	212	20.71	212	8.43
20	209	4.31	209	43.69	209	7.78	62	20.21	82	8.41

the variables in Figure 9.14 using the two PLS indicators and the *t*-statistic, for Case Study 8a (mouse urine – diet study) and Case Study 10c (simulation – maximum overlap).

9.2.5 Relationship between the Indicator Functions

We have demonstrated that there are a number of mathematical results above between the ranks and values of the indicator functions. These are illustrated in Figure 9.15 for Case Study 8a (mouse urine – diet study) for both 1 PLS component and 5 component models (found as the optimum from the bootstrap) and for Case Study 4 (pollution) for 2 PLS components. The MLR *b* value, *t*-statistic, *F*-ratio, PLS weights and PLS regression coefficients are compared.

It can be seen that the rank of all variables is the same when 1 PLS component is used in the model. This result stands in all cases providing the data are preprocessed as discussed in this chapter. The values for 1 PLS component (see Case Study 8a – mouse urine – diet study) show appropriate curvilinear relationships with each other as expected, but they are monotonic and so the ranks are always the same. When 5 PLS components are computed (Case Study 8a) the values for the PLS weights and regression coefficients against the univariate statistics start to deviate, which has some influence on the relative ranks of the least significant variables (higher ranks – to the

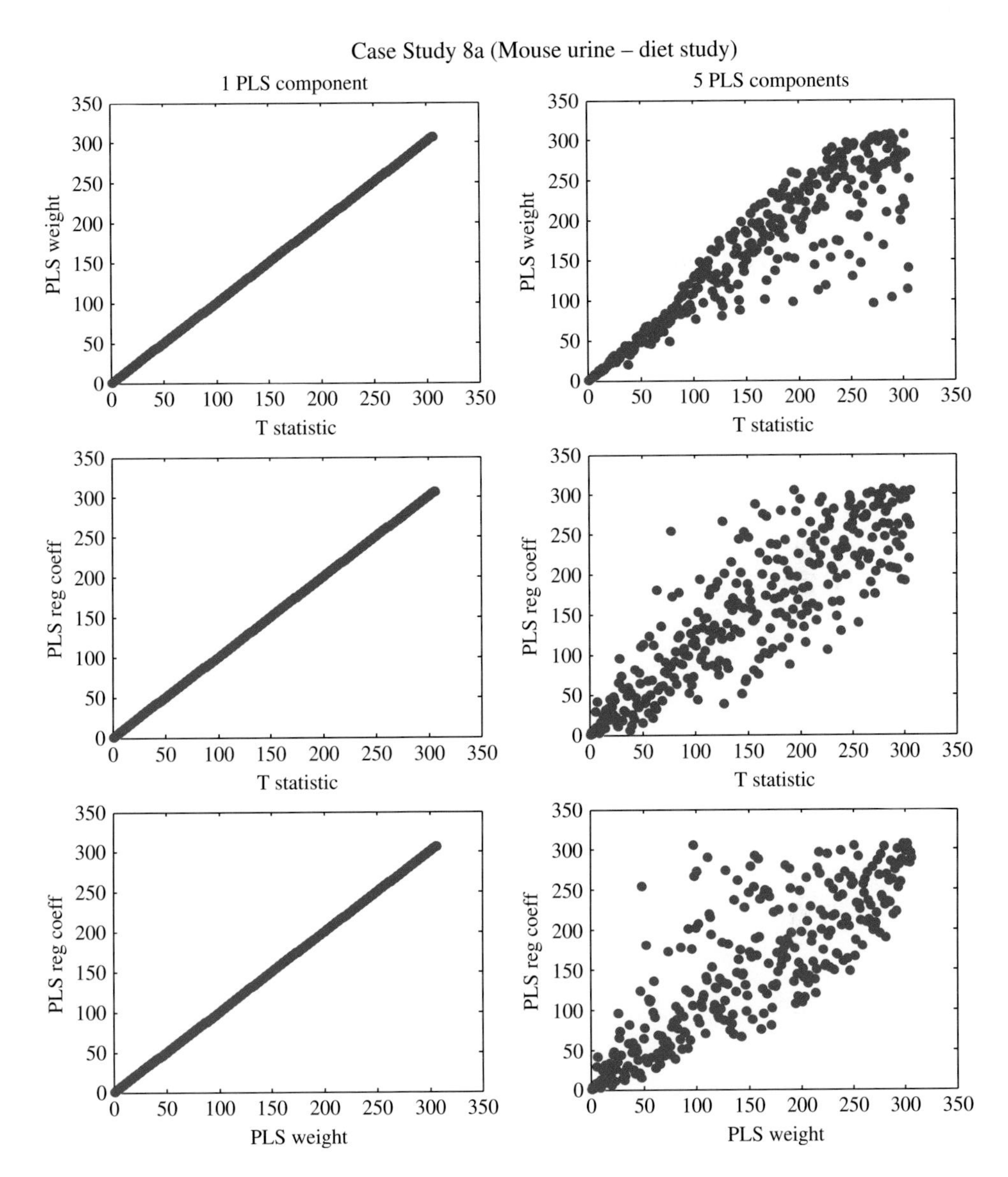

Figure 9.14 Rank of variables for Case Studies 8a (mouse urine – diet study) and 10c (simulation – maximum overlap) for the *t*-statistic, PLS weights and PLS regression coefficients, for 1 PLS components and the optimal number of components (5 for Case Study 8a and 2 for Case Study 10c)

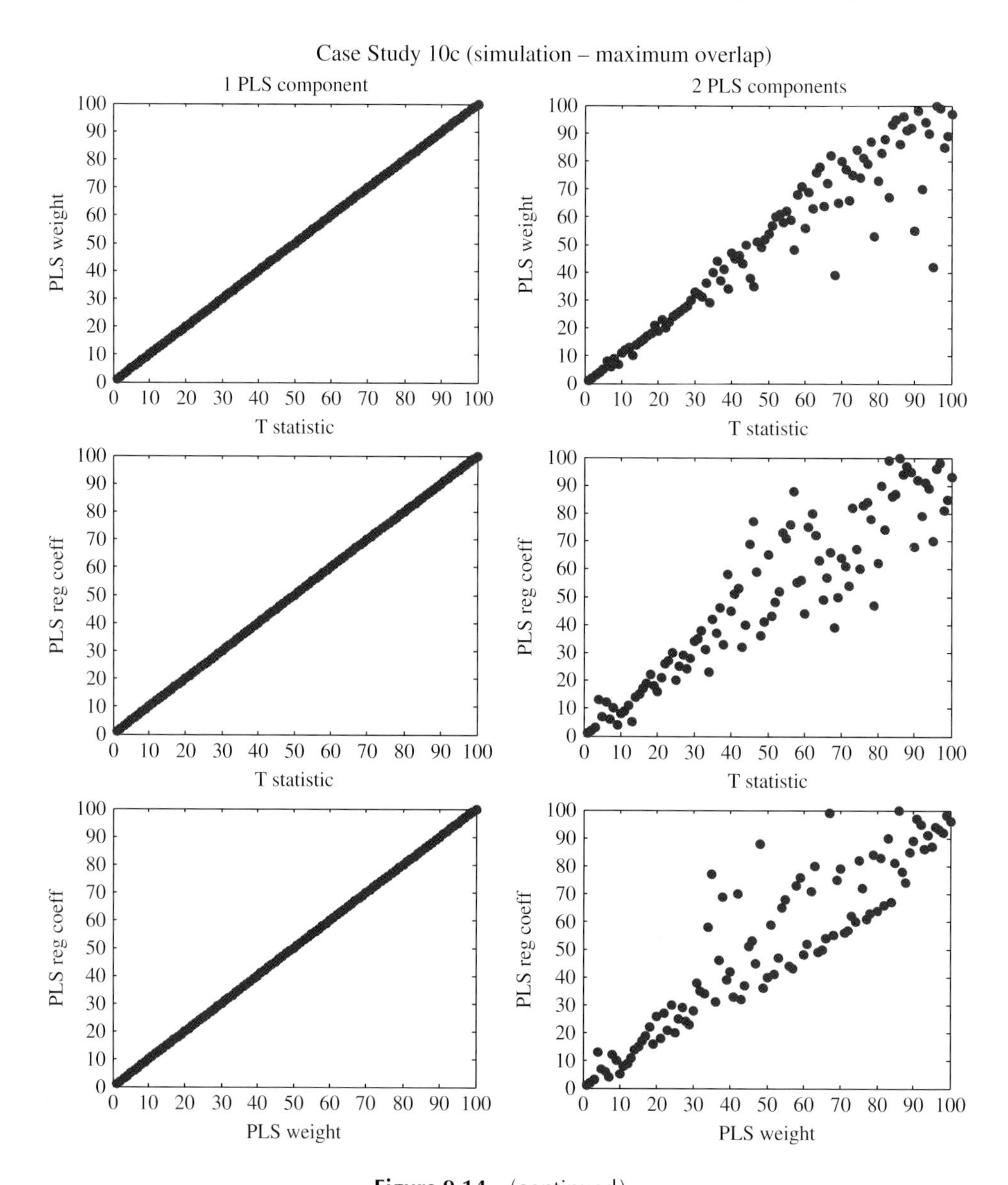

Figure 9.14 (continued)

right of each graph), but the most significant variables especially for PLS weights retain similar ranks to the univariate ones. For Case Study 4 (pollution) we illustrate a 2 component model: for this case study the relationship between t and b is noticeably curved because there is a wider range of b values. These relationships are reflected in the graphs of the values of PLS weights and regression coefficients. Note that although the relative ranks of the least significant variables appear to deviate with each indicator function, those of the more important ones do not change much.

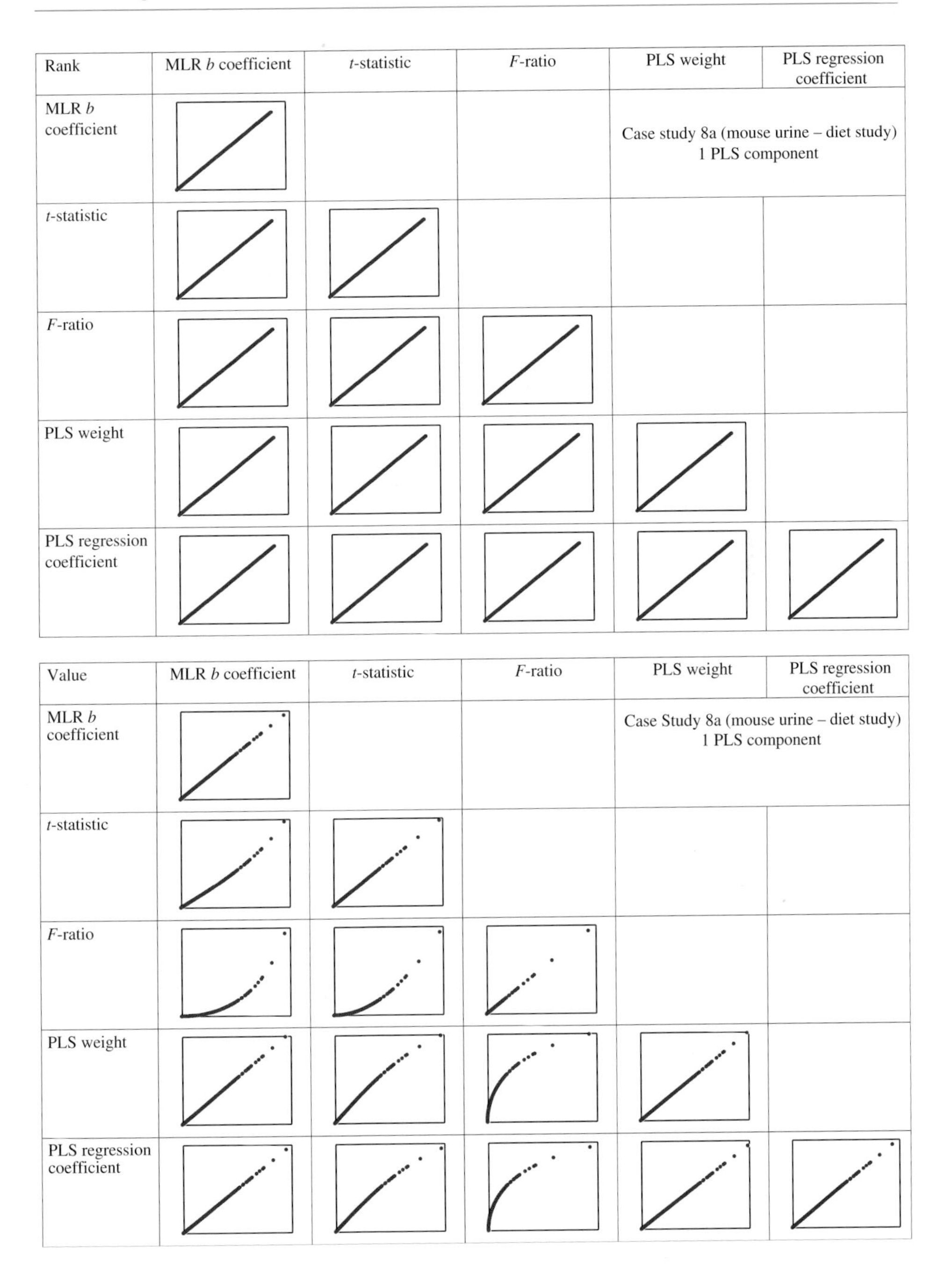

Figure 9.15 Ranks and absolute values of five indicator functions for Case Studies 8a (mouse urine – diet study) and 4 (pollution), using different numbers of PLS components. The column headings correspond to the horizontal axes and the row headings to the vertical axes

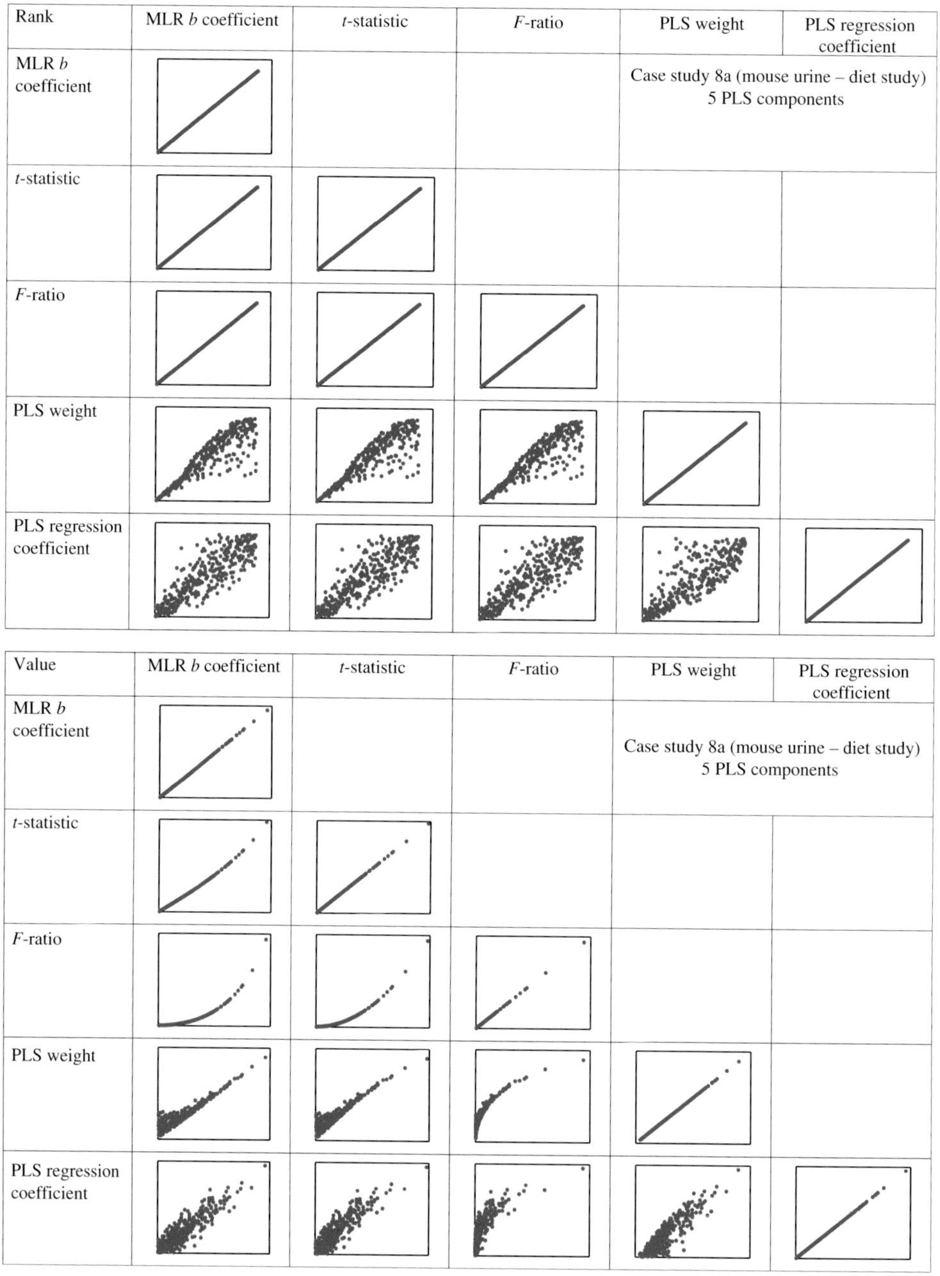

Rank	MLR *b* coefficient	*t*-statistic	*F*-ratio	PLS weight	PLS regression coefficient
MLR *b* coefficient				Case study 8a (mouse urine – diet study) 5 PLS components	
t-statistic					
F-ratio					
PLS weight					
PLS regression coefficient					

Value	MLR *b* coefficient	*t*-statistic	*F*-ratio	PLS weight	PLS regression coefficient
MLR *b* coefficient				Case study 8a (mouse urine – diet study) 5 PLS components	
t-statistic					
F-ratio					
PLS weight					
PLS regression coefficient					

Figure 9.15 (continued)

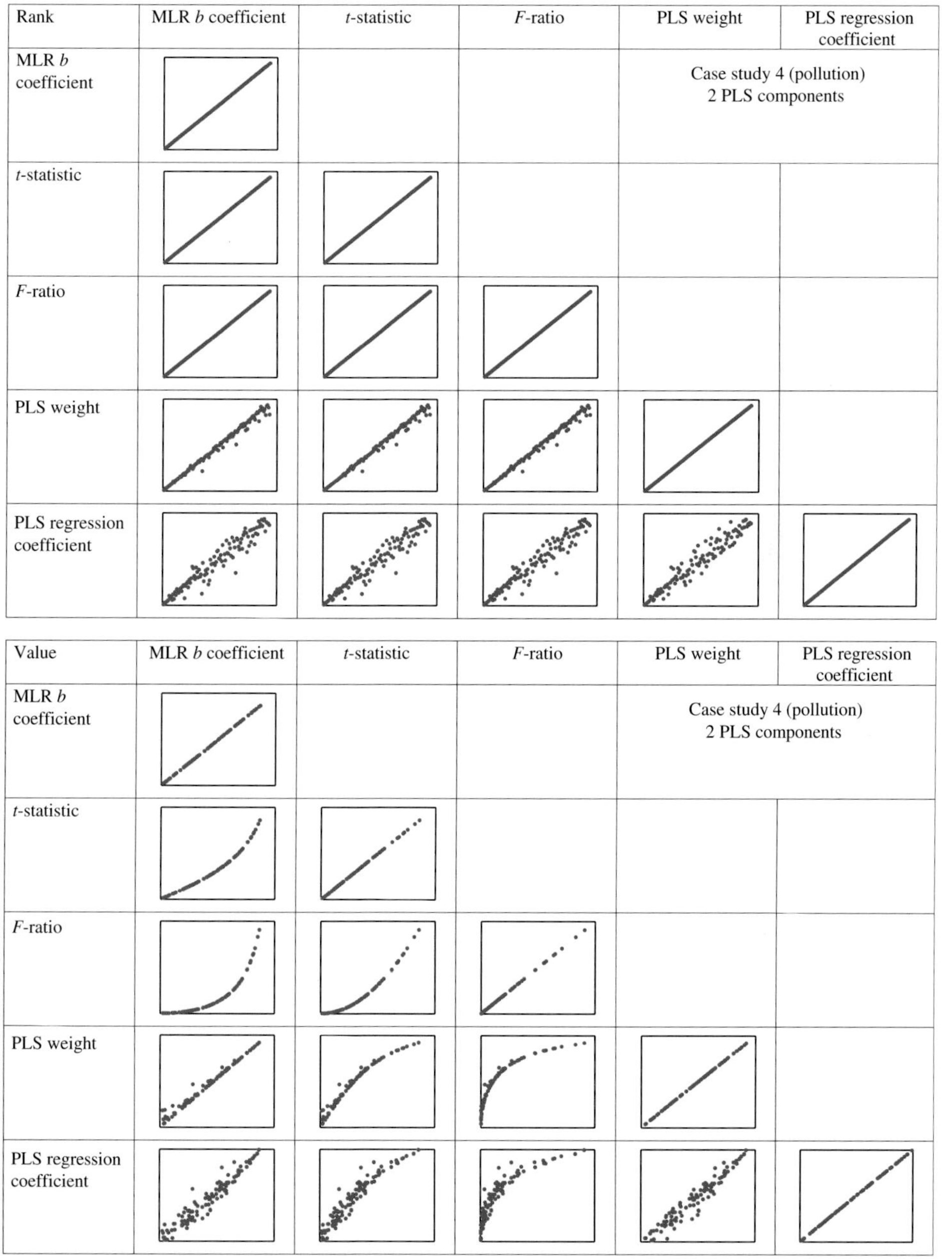

Rank	MLR b coefficient	t-statistic	F-ratio	PLS weight	PLS regression coefficient
MLR b coefficient				Case study 4 (pollution) 2 PLS components	
t-statistic					
F-ratio					
PLS weight					
PLS regression coefficient					

Value	MLR b coefficient	t-statistic	F-ratio	PLS weight	PLS regression coefficient
MLR b coefficient				Case study 4 (pollution) 2 PLS components	
t-statistic					
F-ratio					
PLS weight					
PLS regression coefficient					

Figure 9.15 (continued)

9.3 How Many Variables are Significant?

Up to this point in the text we have described a number of methods for obtaining statistical indicators of significance for each variable as a potential discriminator and then ranking them in order of significance. Once we have determined the relative order of significance, the next step is to decide how many markers are significant. There are a large number of potential methods in the literature and we will summarize just a few in this chapter.

9.3.1 Probabilistic Approaches

In biology in particular, there is a tendency to attach probabilities to the values of parameters. Many biologists use p values, a p of 0.01 implying 99 % confidence that a variable is significant at a level greater than 99 %, a p value of 0.05 implying 95 % confidence and so on. For many of the methods described above, a probability can be assigned, usually either via the t-statistic or F-ratio, using standard tables or calculations (e.g. in Excel or Matlab). The critical values of the t-statistic are presented in Table 9.7 and the F-ratio in Table 9.11 and Table 9.12. In order to interpret the numbers obtained it is necessary to know the degrees of freedom associated with the statistic. We have discussed the probabilistic interpretation of the t-statistic in Section 9.2.2. For the F-ratio, it is only considered significant if the numerator is greater than the denominator: take the number of degrees of freedom of the numerator along the columns and the denominator along the rows and find the critical values in Table 9.11 and Table 9.12; if it exceeds this value it is of greater significance, for example, if there are 12 dfs for the numerator and 10 for the denominator, an F-statistic exceeding 4.706 has a p value of less than 0.01, or a significance value of >99 %; a one tailed test is used as we are interested in whether the numerator is significantly larger than the denominator rather than whether it is significantly different. Using standard software such as Excel it is possible to convert any value of the F-ratio and t-statistic into a probability knowing the dfs, although conventionally, prior to the application of user friendly computing, people relied primarily on tables of critical values that were precalculated, and asked questions such as 'does the significance level exceed 99 %' rather than calculate precise significance levels.

However, as discussed above, the interpretation of these numbers in the form of probabilities is critically dependent on the raw data being approximately normally distributed. It is recommended always to use a normal distribution test first as described in Section 9.2.2. In most cases of chemometrics this fails for individual variables.

If however probabilities are to be used, it is essential to understand their meaning in context. A p value of 0.01 or significance of 99 % means that, assuming an underlying normal distribution, the value of the statistic is expected to be at or above this level one time in a hundred. However, if we have a chromatographic dataset in which we identify 1000 peaks, we would anticipate exceeding this level for 10 out of the 1000 peaks even if there were no underlying markers. This takes us back to the illustration of the toss of coins in Section 9.1.4 but in a different context, if we toss a coin a sufficient number of times we definitely expect some extreme distributions, from time to time, and so having a few peaks whose t-statistic or F-ratio is large which suggests they have a p value of less than 0.01 is expected even if there is no underlying discrimination between groups.

However, when using the t-statistic we can rank all variables and assign a probability assuming normality. A p value is often the reverse of what we expect; a p value of 0.01 implies 99 % confidence that the null hypothesis can be rejected. For Case Studies 11 (null) and 4 (pollution) this is done in Figure 9.16. We see that there are quite different distributions in each situation. For Case Study 11 some variables do appear to have low p values, and so we have apparently high confidence that these are markers. However the highest ranked marker has a p value of 1.6 % and indeed we expect one variable, by chance, out of the 100 to fall with a 1 % limit. A p value of 1 % means that we expect the

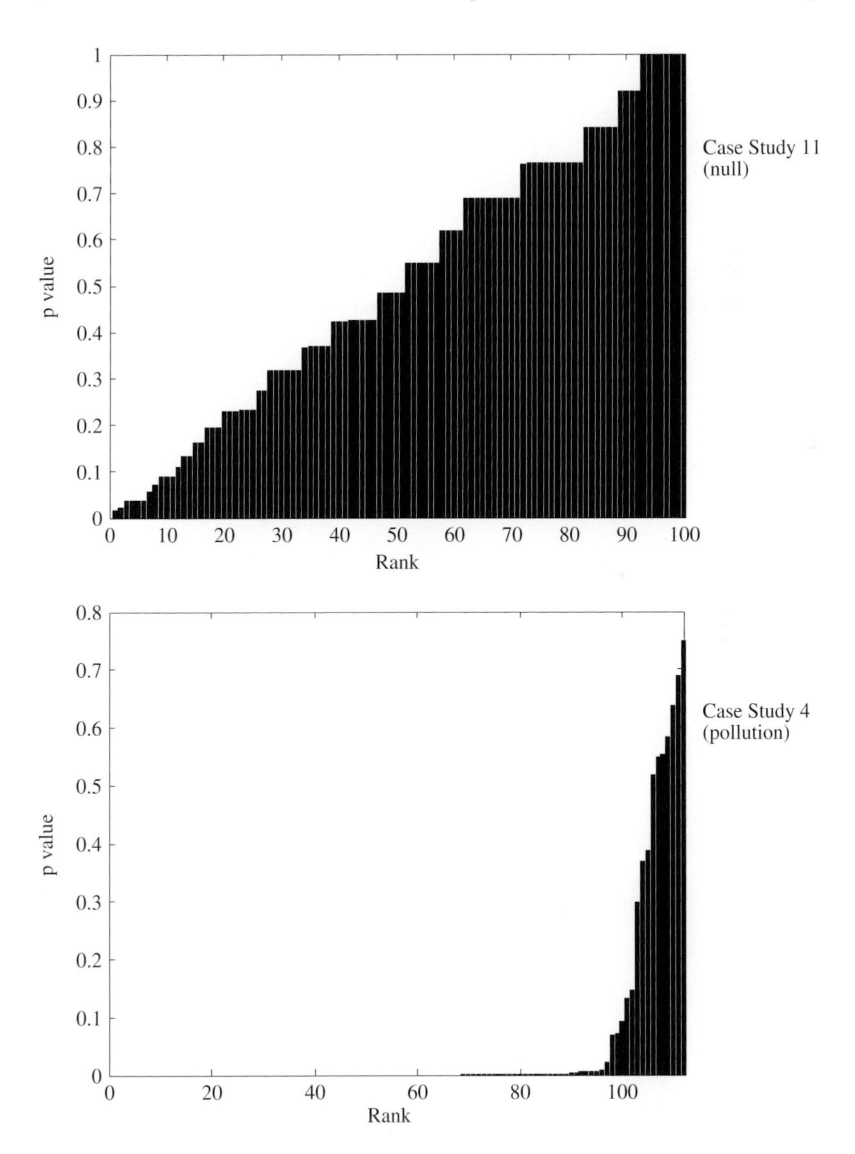

Figure 9.16 Distribution of p values as assessed using the t-statistic for variables in Case Studies 11 (null) and 4 (pollution) assuming normality

t-statistic to be exceeded 1 time in 100, and so is nothing remarkable. The p values change almost linearly with rank of marker which is a good indication of randomness (and suggests that a model of approximate normality is not completely inappropriate), as anticipated. For Case Study 4, there are quite different trends, with many markers having some significance, and there are certainly many more markers at low p (high t) values than we would expect by chance, suggesting that there really are some significant trends.

For PLS-DA it is harder to convert the values of the statistical indicator function into probabilities although empirical methods as discussed below are alternatives.

9.3.2 Empirical Methods: Monte Carlo

Empirical methods do not assume that there is a specific underlying normal distribution in the intensities of the variables. A popular approach involves using Monte Carlo methods which compare the value of an observed statistic with a set of null distributions. In the case of classifiers, Monte Carlo methods can be used to compare the observed data with randomly generated permutations of the sample class membership. The comparison is between the test criteria, for example, a t-statistic or PLS regression coefficient, which are calculated for the value of each variable on the observed data and values generated using a null distribution to determine which of the observed variables are truly significant, and are therefore potential discriminators.

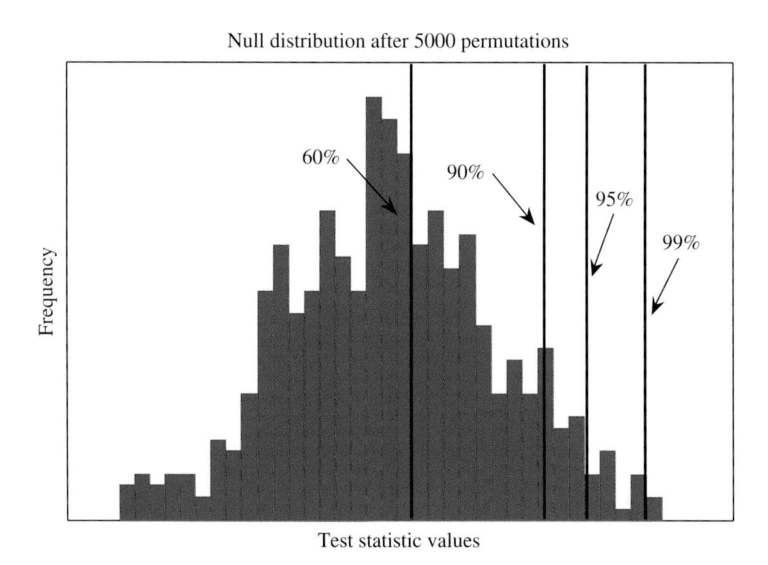

Figure 9.17 Level of significance of an observed value compared with null distribution (distribution of random class assignment at 99 % (highly significant), 95 %, 90 % (moderately significant) and 60 % (not significant))

Permutation is used to create a null distribution. The significance of the observed values of each variable is determined by comparing to this distribution. How empirical significance levels can be obtained is illustrated in Figure 9.17. The method for permutation is as follows:

1. Numerical indicators of significance using one or more variable selection methods for all variables to be compared are calculated for each dataset using the known class information. The Monte Carlo method can be used with any statistical indicator.
2. The calculation of the observed test statistics involves the random assignment of a class label for each sample in vector $\mathbf{c}$. The permutation in the Monte Carlo method involves randomly assigning a class label to each sample, in other words, randomly permuting the class labels. The numeric indicators of significance are calculated for this randomly permuted dataset using a new class list and the chosen variable selection method(s). This represents a null distribution for the class labels permuted and so the assignment of samples to classes is random.
3. The procedure in step 2 is repeated a number of times ($= T$ – typically several thousand) to produce a null distribution of test statistics for each variable. The values of the significance indicators that are found for these permutations form a null distribution.
4. The value of the statistical indicator function is compared to the null distribution on the permuted data. It is important to establish whether the distribution is two tailed (e.g. t-statistic and PLS regression coefficients) in which case the magnitudes of the coefficients are used, or one tailed (e.g. PLS weights) in which case the raw numbers can be employed.
5. The empirical significance relates to how large the indicator statistic obtained from the real data is compared to the series of values obtained during Monte Carlo permutations. If there are T permutations where class information is randomly assigned to each sample, then a threshold of γ for variable j represents the value of the indicator that is exceeded $\gamma \times T$ times. For example, if there are $T = 5000$ permutations, a threshold of 95 % or 0.95 represents the value of the indicator obtained with the true class information that exceeds the value from that obtained from the null permutations 4750 out of 5000 times: if a variable has a value higher than this, it is considered to be a significant variable with an empirical significance of 95 % or a p value of 0.05. Note that these are estimated probabilities, for example, a value exceeding that obtained for T permutations would have an empirical significance of 100 % or a p of 0.0.
6. Variables can then be ranked according to their empirical significance, the higher the significance, the higher the rank. Note that this ranking approximates to the original ranks of the variables (the greater the magnitude, the higher the rank) but may not be exactly the same, because the significance is assessed against a background distribution, which is randomly generated and may differ slightly in shape according to the variable and also the nature of the permutations.

The number of variables exceeding a threshold can be obtained for different approaches, and we compare the t-statistic, PLS-DA weights and PLS-DA regression coefficients for Case Studies 8a (mouse urine – diet study), 4 (pollution) and 11 (null) in Figure 9.18. The distribution of values of each statistic is indicated. PLS regression coefficients and the t-statistic have a symmetric distribution whereas PLS weights are asymmetric. The value of the indicator for the 1st, 20th and least significant variables are indicated. Note that for the two tailed statistics, these can be either side of the mean. For the one tailed statistics, the least significant variable is always at a low value, whereas for the two tailed ones it is in

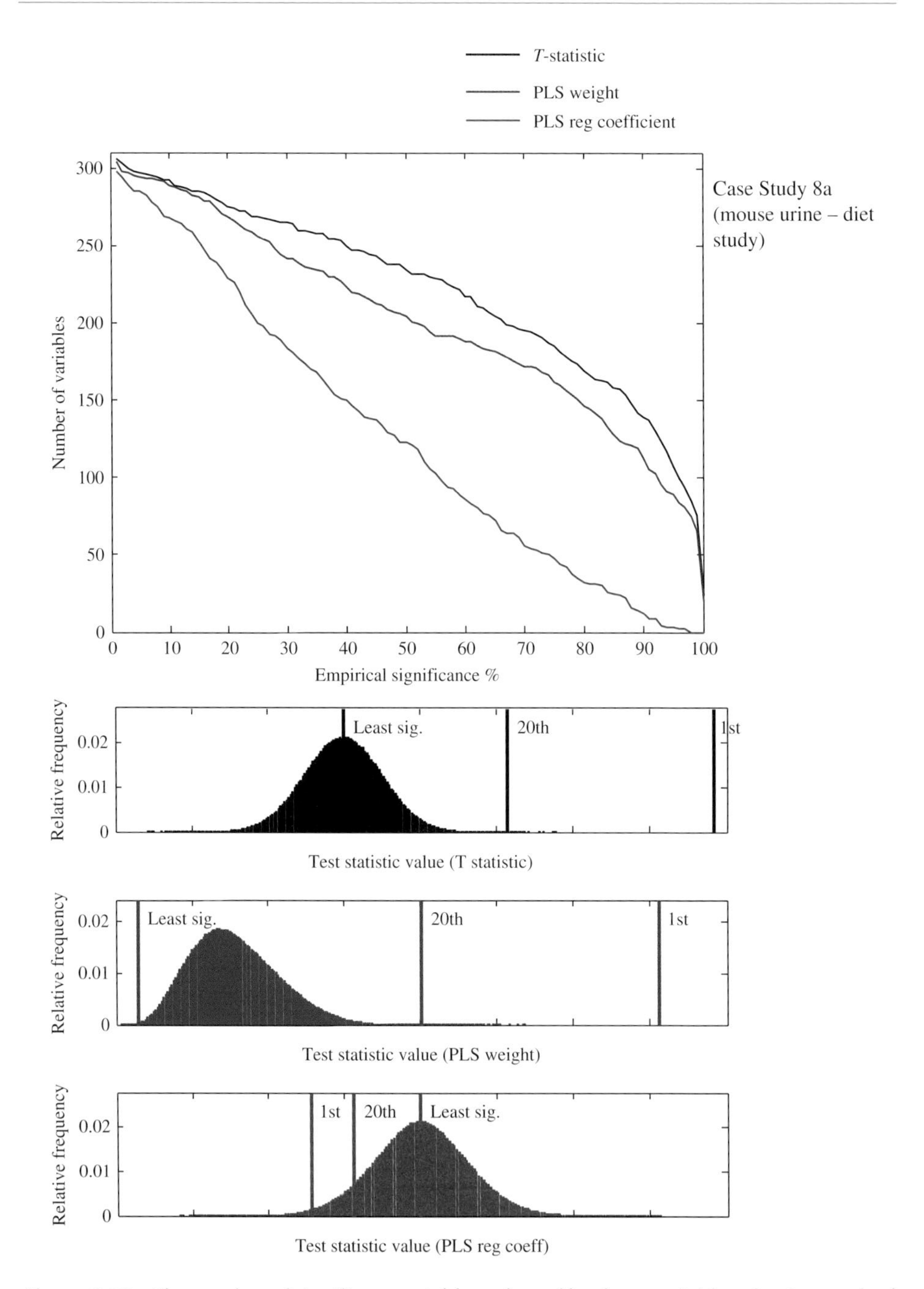

Figure 9.18 The number of significant variables selected by three variable selection methods at different empirical significance thresholds for Case Studies 8a (mouse urine – diet study), 4 (pollution) and 11 (null) using Monte Carlo methods, together with an indication of where they fall in the null distribution. The background distributions are for all J variables together for illustration, although the recommendation is to do this on the basis of individual variables

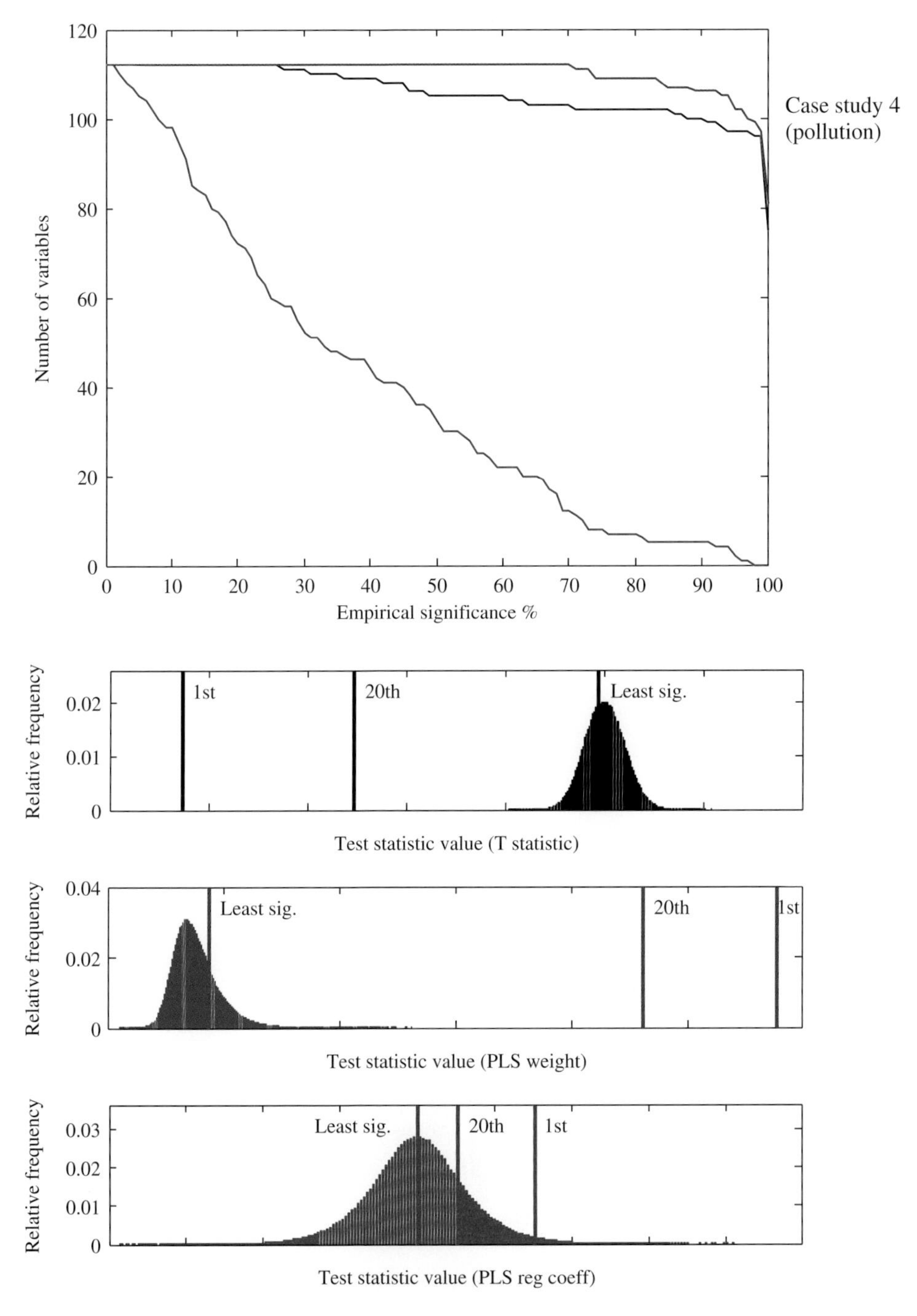

Figure 9.18 (continued)

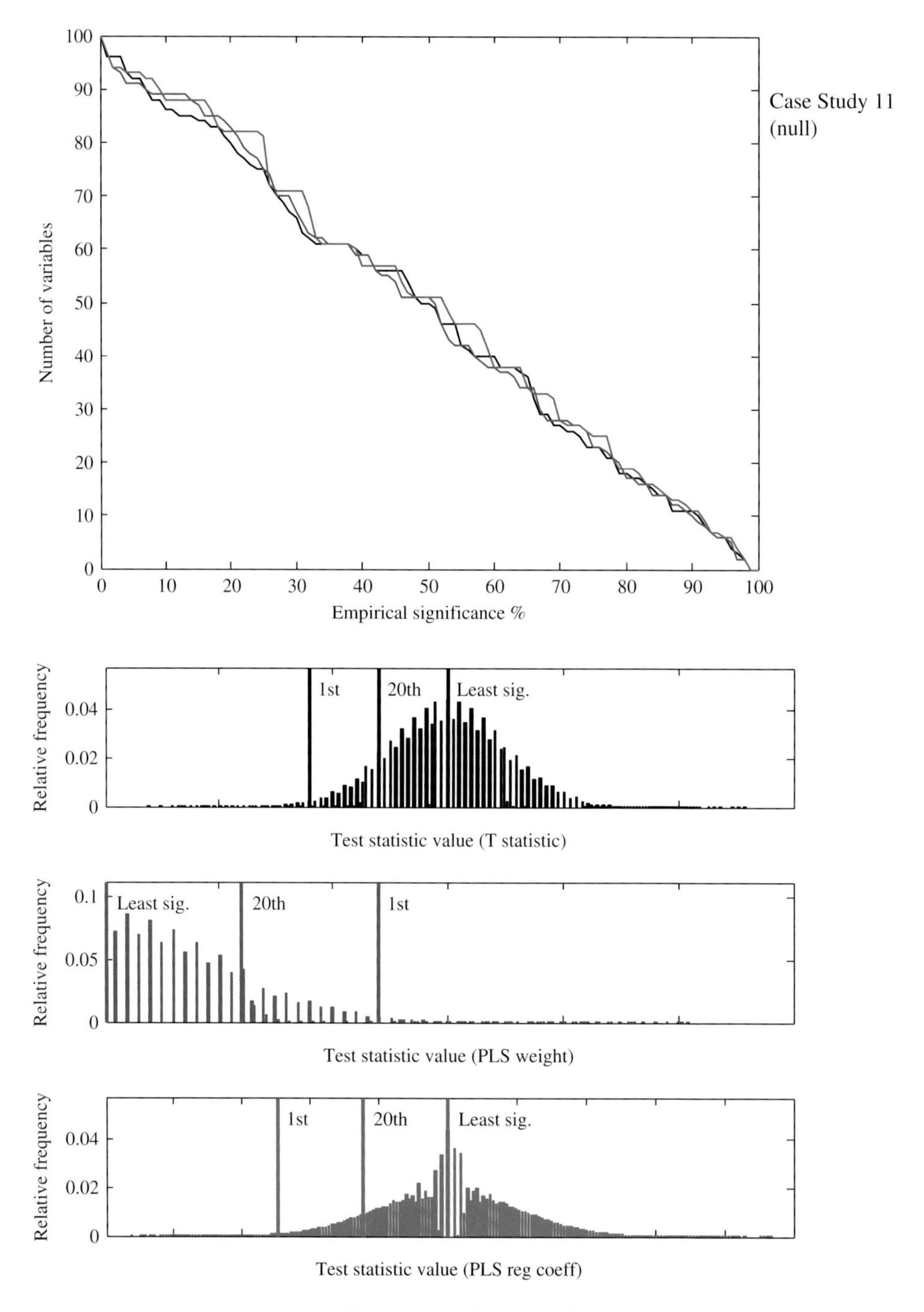

Figure 9.18 (continued)

the mean. Case Study 11 (null) looks somewhat quantized: this is because the dataset consists of '1s' and '0s', and so the distinction between variables is discrete. We can see that the empirical significance distribution is fairly similar for PLS weights and the *t*-statistic, but PLS regression coefficients diverge quite strongly for Case Study 8a and 4 (mouse urine – diet study). For Case Study 11 (null) the graph of number of variables against empirical significance looks almost linear using all three statistics, similar to the probabilistic graph of Figure 9.16.

Monte Carlo methods are valuable methods for determining the empirical significance of a variable and can then be used to determine a cut off level below which the variable is not significant. No assumptions of underlying distributions are needed.

9.3.3 Cost/Benefit of Increasing the Number of Variables

In Sections 9.3.1 and 9.3.2 we discussed how we can attach a numerical significance level to a variable, and so provide some type of empirical probability that it is an important discriminator. Of many thousand possible variables, only a few may be useful for forming a specific classification model. The methods above will allow us to assign variables in rank order and, if we wish, to assign a cut-off threshold for which variables we consider most significant. Each method may come up with a different list, but variables that appear frequently in each list are likely to be important in forming a classification model between two or more groups. For multivariate models, what often happens is that there need to be several variables included in a model before an acceptable quality of classification can be achieved; 1 or 2 variables is inadequate, and so increasing the number of variables usually improves classification ability. However once an optimum number of variables has been obtained, the remainder usually degrade the quality of the classifier as less significant variables may contribute noise or be influenced by other, undesirable, factors. So an alternative to the methods discussed above is to see how increasing the number of variables in the model influences the classifier. This is best done using test and training set splits, and assessing classification ability on the test set, as autopredictive ability is likely to increase the more the variables in the model, but this will not necessarily be so for the test set.

Using the methods of Section 9.2, variables can be ranked according to the size of an indicator of significance. Models can then be constructed as the number of variables is increased, for example, a model can be formed on the top 10 ranked variables, and performance than can be assessed by one or more methods discussed in Chapter 8. The number of variables can be increased and the model performance is assessed on the new subset of variables and so on. Ideally the model performance should increase until an optimal number of variables is found, and then decrease afterwards. However sometimes the improvement of the model is not very great as the number of variables is increased above a certain number. If variables are expensive to measure (e.g. by constructing specific sensors, each of which has to monitor additional compounds), the cost of, for example, using 30 rather than 5 variables could be a six fold increase, yet the benefit of including these additional variables may be limited, perhaps improving classification rates by a few percent. So it is important not only to select the optimal number of variables, but also determine the benefit of including them.

As discussed in Chapter 8, we recommend repeatedly splitting data into test and training sets (we recommend a default of 100 times but for specific datasets this number could be varied). For every training/test set split, variables can be ranked according to the size of the chosen statistical indicator function. For brevity we will illustrate the methods using PLS weights, but this does not advocate or restrict the analysis to PLS weights: the number of PLS components is determined on the training set using the bootstrap as discussed in Section 8.5.1; PLS-DA weights are calculated to determine which variables are most significant in the training set and subsequently PLS-DA classification models are performed using the subsets of the most significant variables. It is important to realize that the approaches in this chapter are generic and could be applied equally well to the use of other indicator functions or other methods for determining the optimal number of components or other methods for modelling the data once the variables are determined. One advantage of this procedure is that variable selection is performed on the training set alone, and not the test set, thus removing the possibility of overfitting if variables are selected on all data together as we discussed in Section 9.1.5.

A recommended procedure is as follows:

1. The dataset of I samples is split into a training set and test set. The numbers of samples picked for the training set for each class are equal, according to the criteria of Section 8.3.2. This should be performed several times, typically 100.
2. The top ranked variables are selected from the training set for each iteration. Note that for each of the training and test set splits, variables have slightly different ranks because the training set is different each time. The variables are ranked according to relevant indicator function. A predictive model, e.g. PLS-DA, is built from the training set using only the top variables.
3. In the validation step, the variables in the test set are also reduced to the top variables selected from the training set in step 2. The predictive model from step 2 is used to predict the classification performance on the test set using this reduced subset of variables. It is important that the variables are selected from the training set and the test set has no influence on this in order to avoid the pitfalls discussed in Section 9.1.5.
4. This is repeated for every iteration. Once this has been done a statistic such as the test set %CC can be plotted as a function of the number of variables. The AUC for the test set can also be plotted as a function of number of variables using the average test set ROC curve (see Section 8.5.2) as differing number of parameters are computed.

The effect of increasing variables on the %CC and AUC using PLS-DA weights to rank the variables is presented in Figure 9.19. This allows us to see what the benefit is of including more variables in the model. In some cases it may be expensive to measure additional variables, and we can assess from these graphs if it is worth increasing the number of variables. Consider, for example, a sensor for which each measurement has an equal cost associated with it: if one found that one could obtain 95% CC with 5 variables, is it worth increasing to 98 % by including an extra 15 variables and so tripling the cost?

A separate issue involves selecting those variables that appear most diagnostic; so far in this chapter we have found these variables using autoprediction, that is, on the entire training set. If we are convinced there is a relationship between the classifier and the analytical data,

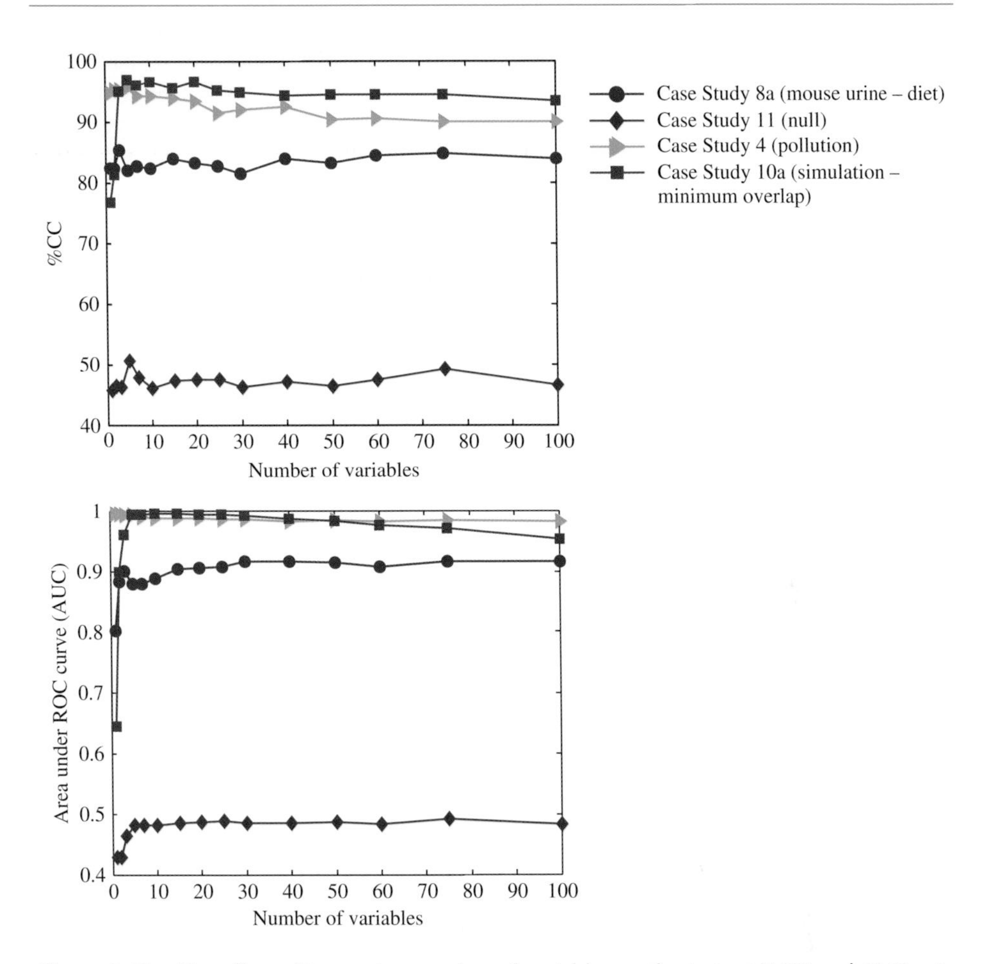

Figure 9.19 The effect of increasing number of variables on the test set %CC and AUC using PLS-DA weights to order the variables for Case Studies 8a, 11, 4 and 10a

the approaches discussed above may well be suited for purpose; however, the three steps above allow us to determine the influence of including more variables in the model, but because the training set differs in each iteration, the nature of the variables may vary slightly according to which samples are included. This causes a dilemma because although we may decide, for example, to include 20 variables in a model, the specific variables may differ in each test and training set iteration. There are various ways around it, as follows:

1. Once the optimum number of variables is found, if by careful validation we are convinced that there is a relationship between the classifier and the measured data, select these on the autoprediction dataset.
2. For each iteration (test/training set split), use a cut-off for the number of variables selected (e.g. via Monte Carlo, or just the top 10 or via a probabilistic threshold), and score how many times within all the iterations a variable is selected. For example, if a variable is selected in 90 out of a 100 iterations, it is more significant than a variable is

selected 80 times. This threshold could be similar for all iterations or be calculated separately at each iteration – the latter however could be very computationally intensive for limited benefit.

3. Over all iterations simply compute an average rank for each variable, the higher the average rank, the greater its significance.

Strategies 2 and 3 overcome the dilemma that each test and training set iterations may select slightly different subsets of variables. Furthermore if the same variables are selected regularly this suggests a stable model, whereas if not, this suggests that there is little meaning to these variables as they are highly influenced by the samples included in the training set. Using a cut off of the top 10 variables found in each of 100 iterations, and PLS-DA weights, the variables found more than 50 times in the top 10 of the training set are listed according to frequency in Table 9.15 for Case Studies 8a (mouse urine – diet study) and 11 (null). It can be seen that there is a good correlation with the overall rank in this case even for the null dataset.

Table 9.15 *Frequency with which variables are selected in Case Studies 8a and 11 using PLS weights and 100 iterative selection of variables over the training set, taking the top 10 each time. Only variables that are selected 50 times or more are listed, together with the rank they have when using the straight PLS weight over all samples*

Case Study 8a (mouse urine – diet study)	Variable	PLS weight (rank)	Frequency
	55	69.11 (2)	100
	197	79.02 (1)	100
	110	61.89 (3)	97
	263	61.06 (4)	91
	124	58.68 (5)	85
	238	56.33 (6)	76
	149	52.67 (8)	52
	262	51.83 (9)	51
Case Study 11 (null)	**Variable**	**PLS weight (Rank)**	**Frequency**
	19	17.15 (1)	69
	74	16.43 (2)	63
	47	14.96 (6)	61
	24	14.99 (5)	58
	98	15.04 (4)	56

Bibliography

Basic Statistical Definitions

NIST Engineering Statistics Handbook [http://www.itl.nist.gov/div898/handbook/].

R. Caulcutt, R. Boddy, *Statistics for Analytical Chemists*, Chapman & Hall, London, UK, 1983.

J.N. Miller, J.C. Miller, *Statistics and Chemometrics for Analytical Chemistry*, Fifth Edition, Pearson, Harlow, UK, 2005.

W.P.Gardiner, *Statistical Analysis Methods for Chemists: A Software-Based Approach*, The Royal Society of Chemistry, Cambridge, UK, 1997.

Partial Least Squares Coefficients

H. Martens, T. Næs, *Multivariate Calibration*, John Wiley & Sons, Ltd, Chichester, 1989.
B.K. Alsberg, D.B. Kell, R. Goodacre, Variable selection in discriminant partial least-squares analysis, *Anal. Chem.*, **70**, 4126–4133 (1998).
A. Hoskuldsson, Variable and Subset Selection in PLS Regression, *Chemometrics Intell. Lab. Systems*, **55**, 23–58 (2001).
S.J. Dixon, Y. Xu, R.G. Brereton, A. Soini , M.V.Novotny, E. Oberzaucher, K. Grammer, D.J. Penn, Pattern Recognition of Gas Chromatography Mass Spectrometry of Human Volatiles in Sweat to Distinguish the Sex of Subjects and Determine Potential Discriminatory Marker Peaks, *Chemometrics Intell. Lab. Systems*, **87**, 161–172 (2007).

ANOVA Based Methods and Designs

G.P. Quinn, M.J. Keough, *Experimental Design and Data Analysis for Biologists*, Cambridge University Press, Cambridge, UK, 2002.
A.J. Hayter, *Probability and Statistics for Engineers and Scientists*, Second Edition, Duxbury, Pacific Grove, CA, USA, 2002.

Monte Carlo Methods

A.C.A. Hope, A Simplified Monte Carlo Significance Test Procedure, *J. R. Statistical Soc., Ser. B – Statistical Methodology*, **30**, 582–598 (1968).
L. Ståhle and S. Wold, Partial least squares analysis with cross-validation for the two-class problem: A Monte Carlo study, *J. Chemometrics*, **1**, 185–196 (1987).
K. Wongravee, G.R. Lloyd, J. Hall, M.E. Holmboe, M.L. Schaefer, R.R. Reed, J. Trevejo, R.G. Brereton, Monte-Carlo methods for determining optimal number of significant variables. Application to mouse urinary profiles, *Metabolomics*, (2009) doi 10.1007/s11306-009-0164-4.

10

Bayesian Methods and Unequal Class Sizes

10.1 Introduction

In many real world classification problems there may be unequal number of experimental samples in each group. This is common in clinical chemistry where it is often much easier to obtain control samples from subjects as opposed to samples with a specific disease. When studying populations it is often difficult to control the number of samples in each subgroup, e.g. by age, gender, genetics and so on. Many of the classification methods we have introduced so far can be modified to take this into account, but in most cases this involves using the classifier to determine the probability that a sample is a member of a specified class, and adjusting this for prior information about class membership.

In addition, most methods described in this chapter can be employed also to allow for any prior information about the probability that a sample is a member of a specific class into algorithms, which is often called Bayesian statistics. Probabilistic methods can only realistically be employed for binary or multiclass classifiers, when there are several defined groups, each of which has a defined probability of membership, such that the total probability of being a member of each group adds up to 1. When adapting to one class classifiers, the 'in group' and 'out group' can be assigned probabilities, but all samples not found to be part of the 'in group' should be assigned to the 'out group' without ambiguities.

10.2 Contingency Tables and Bayes' Theorem

In univariate statistics often Bayesian methods are employed to allow for this imbalance. In Chapter 8 we introduced the concept of contingency tables for binary classifiers. In Table 10.1(a) we illustrate a simple 2×2 contingency table which is the same as that of Table 8.1. We will use this to demonstrate the application of Bayesian methods.

Chemometrics for Pattern Recognition Richard G. Brereton

Table 10.1 *Examples of contingency tables using different prior probabilities*

(a) Raw data assuming the relative number of measurements equals the relative prior probabilities

	Group A (true class)	Group B (true class)	$p(g\|\hat{g})$
Group A (predicted class)	60	10	0.857
Group B (predicted class)	20	90	0.818

(b) *Raw data assuming the relative number of measurements equals the relative prior probabilities*

	Group A (true class)	Group B (true class)	$p(g\|\hat{g})$
Group A (predicted class)	60	100	0.375
Group B (predicted class)	20	900	0.978

(c) *Assuming the data of (a) and a prior probability of belonging to class A of 0.5*

	Group A (true class)	Group B (true class)	$p(g\|\hat{g})$
Group A (predicted class)	37.5	5	0.882
Group B (predicted class)	12.5	45	0.783

(d) *Assuming the data of (b) and a prior probability of belonging to class A of 0.5*

	Group A (true class)	Group B (true class)	$p(g\|\hat{g})$
Group A (predicted class)	37.5	5	0.882
Group B (predicted class)	12.5	45	0.783

(e) *Assuming the data of (a) and a prior probability of belonging to class A of 0.1*

	Group A (true class)	Group B (true class)	$p(g\|\hat{g})$
Group A (predicted class)	7.5	9	0.455
Group B (predicted class)	2.5	81	0.970

Bayesian analysis assigns a prior probability of belonging to a specific group before incorporating the results of experimental measurements, and mathematically formulates the effect of the measurement process (e.g. extracting a sample and recording a GCMS) as adjusting this prior probability. We need to define a probability $p(q|r)$ which is the probability that the underlying result is of type q given observation r. In chemometrics terms, we may be interested in whether a person is diseased (q) given that a specific peak is found in the GCMS (r). In our terminology we will define $p(g|\hat{g})$ as the probability a sample that belongs to class g is predicted to belong to class g (= A or B if there are two classes), $\hat{g}$ using our experimental observations. In Table 10.1 (a) we define the following:

- $p(A|\hat{A}) = (\hat{A}|A)/A = 60/70 = 0.857$
- $p(B|\hat{B}) = (\hat{B}|B)/B = 90/110 = 0.783$

where $(\hat{A}|A)$ is the total number of samples predicted to be in class A that are genuinely members of class A (sometimes called TPs if A is the 'in group') and A the total number of samples in class A that have been analysed, with similar definitions for class B. This may appear straightforward but now consider a different situation, that of Table 10.1(b). In this case we have available ten times more samples in class B than in Table 10.1(a). This situation is quite common, for example, we may want to sample a number of diseased patients but have a much larger background population, or we may be analysing patients in a hospital and want to single out only those with one specific disease amongst others. Even in our examples in this book for Case Study 3 (polymers), there are 9 groups, and so what would be our approach for distinguishing one polymer group from all the rest? If we then calculate the probabilities we find:

- $p(A|\hat{A}) = 0.375$
- $p(B|\hat{B}) = 0.978$

This means that if a test is positive (or our classifier assigns a sample into group A) we now assess the chance of the patient having a disease as only 0.375, compared to 0.857 previously. There must be something wrong here.

The problem with this logic is that it assumes that the prior probability of a sample belonging to each class is equal to the proportion of samples obtained for analysis, and so in Table 10.1(a) this prior probability, which we will define, as $\pi(g)$ for group g, is 80/180 for class A (= 0.444) and 100/180 for class B (= 0.556). In Table 10.1 (b) these become 0.074 and 0.926, because these are the proportions of samples in the entire population from each group. The probability $\pi(\hat{g})$ is the probability that a sample is estimated to be a member of group g which for Table 10.1 (a) is 70/180 for class A (= 0.388) and 110/180 for class B (= 0.611).

The calculations can be reformulated using Bayes' theorem. This can be expressed in a variety of ways, one of which states:

$$p(g|\hat{g}) = \frac{p(\hat{g}|g)\pi(g)}{\pi(\hat{g})}$$

where $p(\hat{g}|g)$ is the probability that a sample is predicted to be a member of group g if it is genuinely a member of the group or:

- $p(\hat{A}|A) = (\hat{A}|A)/A = 60/80 = 0.75$
- $p(\hat{B}|B) = (\hat{B}|B)/B = 90/100 = 0.90$

for both the examples of Table 10.1(a) and Table 10.1(b) (the latter involving multiplying top and bottom by 10) so that in Table 10.1(a):

$$p(A|\hat{A}) = \frac{0.75(0.444)}{(0.388)} = 0.857$$

and so on as tabulated.

In Bayesian terms $p(g|\hat{g})$ is called a posterior probability, that is, one that is obtained after the experiment (e.g. recording some analytical data and then performing pattern recognition on it), has been performed.

The posterior probabilities of Table 10.1(a) and Table 10.1(b) are based on assuming that the prior probabilities of being a member of each class equal the proportion of samples in the original population. In some cases this may be a perfectly reasonable assumption, for example, our samples may be representative of the underlying population. As an example, we may be entering a remote village and screening all inhabitants for a disease, which is rare (see Table 10.1(b)) and the underlying hypothesis is that not many people have this disease and so before the person enters the room their probability of having this disease is determined by the proportion in our underlying sample as we are doing a full population survey.

But in other cases the assumption that the relative probability that a sample is a member of a specific group according to its occurrence in our population, may be invalid or misleading, for example, we may have many more controls (class B) than subjects with a disease, and if a person is up for testing we assume they have a 50 % chance of this disease. So we would like the prior probabilities to be 50 %. In the cases of Table 10.1(a) and Table 10.1(b) $\pi(g)$ becomes 0.5 for both classes. The value of $\pi(\hat{g})$ is slightly harder to estimate, but is given by:

$$\pi(\hat{A}) = (\pi(A) \times (\hat{A}|A)/A + \pi(B) \times (\hat{A}|B)/B)$$

and

$$\pi(\hat{B}) = (\pi(B) \times (\hat{B}|B)/B + \pi(A) \times (\hat{B}|A)/A)$$

and just involves scaling each column so that the relative totals equal the relative probabilities, and so in our case for the data of Table 10.1(a):

$$\pi(\hat{A}) = (0.5 \times 60/80 + 0.5 \times 10/90) = 0.425$$

$$\pi(\hat{B}) = (0.5 \times 90/100 + 0.5 \times 20/80) = 0.575$$

which calculation is illustrated in Table 10.1(c) (a simple alternative is to scale each column according to the relative prior probability), to give:

$$\mathrm{p(A|\hat{A})} = \frac{0.75(0.5)}{(0.425)} = 0.882 \text{ and } \mathrm{p(B|\hat{B})} = \frac{0.90(0.5)}{(0.575)} = 0.783$$

We can now also calculate the posterior probabilities for the data of Table 10.1(b) as illustrated in Table 10.1(d) assuming equal prior probabilities of 0.5 for membership of each class, and come up with the same posterior probabilities as in Table 10.1(c). Hence unequal class sizes have no influence on the resultant prediction.

We can of course use any Bayesian prior probabilities we like. Table 10.1(e) illustrates the case where the prior probability of belonging to class A is set to 0.1 whereas for class B it is 0.9, giving the posterior probabilities of 0.455 and 0.970 of belonging to these two classes, respectively, if the test is positive (or in our terminology the classifier predicts that the sample belongs to a specific class). In other words if the prior probability of belonging to class A is 0.1, and if the classifier predicts it belongs to this class, the posterior probability (after the chemical and computational analysis) becomes 0.455. What level we decide to set the prior probability at depends on why we are performing our analysis. Take, as an example, a forensic study. We may be interested in whether a person is guilty of a crime. If we were to draw a randomly chosen person from the street, the prior probability will be very low, so even if they do test positive, the posterior probability is still quite low (a classic example is DNA matching); however if there is other evidence that has led to a person being a suspect (e.g. present close to the scene of the crime at the time it was committed), then the prior probability is much higher.

Bayesian analysis can be incorporated into chemometrics in a variety of ways. In Section 10.3 we will look at how algorithms can be adapted to include Bayesian terms. This section will focus exclusively on how contingency tables can be interpreted and used for Bayesian analysis. Note that many statisticians like to talk about the probability that a sample arises from a specific class, whereas chemometricians tend to prefer to talk about overall %CCs and look at the qualities of the models, and then look at the quality of prediction of individual samples, for example, is it close to a boundary or is its PA low or is it ambiguously classified? These are two different philosophies and the Bayesian approach fits a bit awkwardly into the chemometrics scheme.

However, in order to relate to the overall philosophy of this text it is important to remind the reader that the result of most classification algorithms is to determine the %CC or percentage correctly classified, and this is often used to determine how well the classifier performs, which can be used for validation or for optimization as discussed in Chapter 8. The default calculation of %CC as used elsewhere in this text is to divide the number of correctly classified samples in each group by the overall number of samples in the group. If there are g groups this can be expressed numerically by:

$$\%\mathrm{CC} = \sum_{g=1}^{G} I_g \%\mathrm{CC}_g / I$$

where $\%\mathrm{CC}_g$ is the percentage correctly classified in group g. So for the example of Table 10.1(a), this results in %CC = (80 × 75 + 100 × 90) /180 = 83.33 %. However a Bayesian assessment of the %CC will take the prior probabilities into account and the %CC can be redefined by:

$$\%\mathrm{CC} = \sum_{g=1}^{G} \pi(g)\ \%\mathrm{CC}_g$$

assuming that the probabilities add up to 1 (i.e. all groups are included and there are no outliers, etc.). The default in this text is to assume $\pi(g) = I_g/I$. The %CC can now be recalculated as in Table 10.2 taking these prior probabilities into account that relate to the relative proportion of samples from each class and compared %CC assuming a 50 % prior probability of belonging to each class. The changes are not dramatic but this is because the %CC of the two classes is approximately similar (75 % and 90 %). If these differ substantially the influence of the prior probabilities on the overall %CC may be dramatic. Note that the overall %PA and %MS can similarly be adjusted according to prior probabilities.

Table 10.2 *%CCs for the five examples of Table 10.1 taking prior probabilities into account*

	(a)	(b)	(c)	(d)	(e)
Prior probabilities relating to relative proportions in each class	83.33	88.89	82.50	82.50	88.50
Prior probabilities of 50 %	82.50	82.50	82.50	82.50	82.50

The prior probability can be thought of as a type of weighting – which class is considered to be more significant or important for the overall assessment of model quality. In many cases in this text when the number of samples in each class is approximately equal and there is no particular reason to be prejudiced in advance of experimentation including prior probabilities makes little difference. However it can make a significant difference if the class sizes are unequal, not only on the overall assessment of %CC but also on the validation of models. If, for example, there are unequal sample sizes in two classes then the recommendation is that the training set includes equal number of representatives of each class (Chapter 8), but this means that the sample distribution in the test set will be biased towards the larger class, but if equal prior probabilities are included this bias is removed. Another example is when we use one class classifiers, e.g. when the 'in group' and 'out group' are heavily unbalanced such as for the Case Study 3 (polymers) when we may want to distinguish one group, consisting of just a few samples, from the remainder of the samples. We protect against this somewhat in optimization if it is performed on a training set consisting of equal representatives of each group, but care must always be taken.

Table 10.3 is of the two class classification results for Case Studies 3 (polymers – 2 types) and 4 (pollution) using the Bayesian and non-Bayesian adjusted %CCs for equal prior probabilities. In both cases the groups sizes are quite different. In Table 10.4 we present the %CCs for Case Study 3 (polymers – 9 types) using a 2 PC QDA model assuming the classification ability relates to the proportion of samples in the dataset, and adjusting the %CC for equal prior probabilities of belonging to the 'in group' and 'out group'. Because there are nine groups, the proportion of samples in each 'in group' out of the full population is quite small, and there are some significant changes, for example, for class A the %CC using the population weighted %CC (default) is 76.45 % compared to 85.39 % on the Bayesian model (equal priors).

10.3 Bayesian Extensions to Classifiers

An alternative is to modify classification algorithms to incorporate prior probabilities. In this context there is an incompatibility between traditional chemometrics thinking and statistical (or probabilistic) thinking. Most chemometricians like to visualize samples as points in multidimensional space and as such often like to talk about distances or to visualize points graphically or draw boundaries between groups of samples, as a natural way of thinking. Statisticians tend to think primarily in probabilities and view most graphical representations as ways of visualizing these probabilities. There is of course some cross over in thinking, for example, when describing one class classifiers (Chapter 6) we usually compute confidence limits based on a probabilistic model and as such the two ways of thinking are not mutually exclusive. However many statisticians go farther and compute numerical values that incorporate probabilities which cannot necessarily be interpreted directly as distances, and so some of the modifications of the classifiers result in numbers, that can indeed be used to say whether a sample is a member of a class and to develop models, but cannot necessarily be visualized as a class distance from a centroid. However these numerical values can be used for model building just as the normal non Bayesian equations and can still be used to show boundaries between groups.

Incorporating prior probabilities is only really sensible for binary or multiclass classifiers and so that the sum of all prior probabilities equals 1. For one class classifiers although this is conceivable, if each one class model is developed independently, with its own prior probability, there is no guarantee that the probabilities of all possible one class models add up to 1; take a simple example, when there are 2 one class classifiers each having a prior probability of 0.5, then a third (independent) class is added – the sum of the

Table 10.3 *Two class %CCs for Case Studies 3 (polymers – 2 types) and 4 (pollution) weighting the %CC together with the Bayesian adjusted %CCs, both assuming equal group sizes (priors of 50 % for 'in group' and 'out group') and the unadjusted probabilities*

Case Study 3 (polymers)

	Training				Test			
	A	B	Normal Overall	Bayesian Overall	A	B	Normal Overall	Bayesian Overall
EDC	97.83	82.09	87.03	89.96	93.48	82.59	86.01	88.04
LDA/PLS-DA	100	82.09	87.71	91.05	100	82.09	87.71	91.05
QDA	96.74	96.02	96.25	96.38	96.74	95.02	95.56	95.88
LVQ	98.91	97.01	97.61	97.96	96.74	96.02	96.25	96.38
SVM	96.74	98.01	97.61	97.38	96.74	97.01	96.93	96.88

Case Study 4 (pollution)

	Training				Test			
	A	B	Normal Overall	Bayesian Overall	A	B	Normal Overall	Bayesian Overall
EDC	89.94	91.18	90.14	90.56	88.83	91.18	89.20	90.01
LDA/PLS-DA	89.94	91.18	90.14	90.56	88.83	91.18	89.20	90.01
QDA	88.83	91.18	89.20	90.00	88.27	91.18	88.73	89.73
LVQ	96.65	91.18	95.77	93.91	93.30	91.18	92.96	92.24
SVM	93.30	91.18	92.96	92.24	91.06	91.18	91.08	91.12

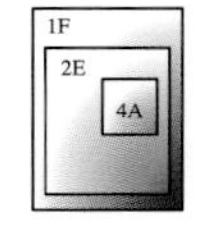

Table 10.4 *One class test set QDA using two PCs and 95 % confidence limit for polymer groups of Case Study 3, together with the Bayesian adjusted %CCs both assuming equal group sizes (priors of 50 % for 'in group' and 'out group') and the unadjusted probabilities*

QDA model using 2 PCs – test set

	'In group' %CC	'Out group' %CC	Unadjusted %CC	Adjusted %CC (equal priors)
Class A ('in group')				
'In group' (35 samples)	97.14	26.36	76.45	85.39
'Out group' (258 samples)	2.86	73.64	—	—
Class B ('in group')				
'In group' (47 samples)	100.00	11.38	90.44	94.31
'Out group' (246 samples)	0.00	88.62	—	—
Class C ('in group')				
'In group' (10 samples)	100.00	0.00	100.00	100.00
'Out group' (283 samples)	0.00	100.00	—	—
Class D ('in group')				
'In group' (56 samples)	96.43	6.75	93.86	94.84
'Out group' (237 samples)	3.57	93.25	—	—
Class E ('in group')				
'In group' (45 samples)	100.00	5.65	95.22	97.18
'Out group' (248 samples)	0.00	94.35	—	—
Class F ('in group')				
'In group' (30 samples)	96.67	4.18	95.90	96.24
'Out group' (263 samples)	3.33	95.82	—	—
Class G ('in group')				
'In group' (20 samples)	95.00	0.37	99.32	97.32
'Out group' (273 samples)	5.00	99.63	—	—
Class H ('in group')				
'In group' (10 samples)	100.00	1.77	98.29	99.12
'Out group' (283 samples)	0.00	98.23	—	—
Class I ('in-group')				
'In group' (40 samples)	97.50	18.18	83.96	89.66
'Out group' (253 samples)	2.50	81.82	—	—

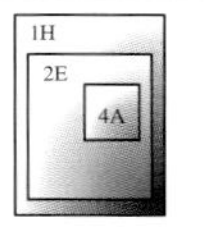

probabilities will now be more than 1, so long as the chance of belonging to the third class is non zero. This means that the one class classifiers cannot really be modelled independently if probabilities are to be attached to each model, and so we are no longer employing truly independent one class classifiers. Therefore our formulation of the Bayesian model will be restricted to the binary classifiers of Chapter 5 although easily extended to multi-class classifiers. Below we discuss Bayesian extensions of two common classifiers. They are not defined for EDC (Euclidean Distance to Centroids).

The equations for both LDA and QDA can be extended to include an additional term which takes into account the prior probability that an unknown sample has of belonging to either of the classes. The equation for calculating the discriminant score between sample i and class g in LDA, assuming unequal class probabilities, is:

$$d_{ig}^2 = (\boldsymbol{x}_i - \bar{\boldsymbol{x}}_g)\boldsymbol{S}_p^{-1}(\boldsymbol{x}_i - \bar{\boldsymbol{x}}_g)' - 2\ln(\pi(g)) + \ln(|\boldsymbol{S}_p|)$$

Although this is based on the squared Mahalanobis distance, it can no longer be directly interpreted as a squared distance from a centroid as the term $\ln(|\boldsymbol{S}_p|)$, where $\boldsymbol{S}_p$ is the pooled variance covariance matrix and '|' stands for determinant, may be negative. However since the pooled variance covariance matrix is the same for each class this term is often in practice be omitted, and, for two classes, a decision function of the form:

$$g(\boldsymbol{x}_i) = \mathrm{sgn}\left\{\left[(\boldsymbol{x}_i - \bar{\boldsymbol{x}}_B)\boldsymbol{S}_p^{-1}(\boldsymbol{x}_i - \bar{\boldsymbol{x}}_B)'\right] - \left[(\boldsymbol{x}_i - \bar{\boldsymbol{x}}_A)\boldsymbol{S}_p^{-1}(\boldsymbol{x}_i - \bar{\boldsymbol{x}}_A)'\right] + 2\ln[\pi(\mathrm{A})/\pi(\mathrm{B})]\right\}$$

can be used in which a positive value suggests membership of class A and a negative class B. The term $\ln[(\pi(A)/\pi(B)]$ is a function of the relative prior probabilities. If they are equal then this term is 0 and so the equation reduces to the normal one for LDA (Section 5.3), but if they differ a value is added (if $\pi(\mathrm{A}) > \pi(\mathrm{B})$) to the discriminant function (or subtracted if vice versa). It can be shown that the average Mahalanobis distance relates to the number of variables, being larger, the more the variables. Therefore the relative importance of the probabilistic term changes according to the number of variables in the training set; the more the variables, the less important this probabilistic term becomes.

QDA has a similar extension but this includes a term relating to the variance covariance matrix, which will be different for each class:

$$d_{ig}^2 = (\boldsymbol{x}_i - \bar{\boldsymbol{x}}_g)\boldsymbol{S}_g^{-1}(\boldsymbol{x}_i - \bar{\boldsymbol{x}}_g)' + \ln(|\boldsymbol{S}_g|) - 2\ln\pi(g)$$

with the two class decision function of a format of:

$$g(\boldsymbol{x}_i) = \mathrm{sgn}\{[(\boldsymbol{x}_i - \bar{\boldsymbol{x}}_B)\boldsymbol{S}_B^{-1}(\boldsymbol{x}_i - \bar{\boldsymbol{x}}_B)'] - [(\boldsymbol{x}_i - \bar{\boldsymbol{x}}_A)\boldsymbol{S}_A^{-1}(\boldsymbol{x}_i - \bar{\boldsymbol{x}}_A)'] \\ + 2\ln[\pi(\mathrm{A})/\pi(\mathrm{B})] + \ln(|\boldsymbol{S}_B|/|\boldsymbol{S}_A|)\}$$

and takes into account the ratio of determinants of the variance covariance matrices for each class. Table 10.5 provides contingency tables for Case Studies 11 (null), 3 (polymers – 2 types) and 4 (pollution) as the relative probabilities are changed . Note that we would only expect one of these values, the %CC when the relative probabilities of belonging to each group, using LDA if the group sizes are identical to be similar to the non-Bayesian %CC; for QDA because the variances are taken into account these values will differ. Since we want to visualize the results (see below) we are using slightly different data scaling that was employed in Table 8.9. However the general trends are clear. The change in position of

Table 10.5 *Contingency tables for application of differing prior probabilities to the classification abilities using (a) LDA and (b) QDA for Case Studies 11, 3 and 4. Note that the QDA %CC at equal probabilities is not necessarily the same as the non-Bayesian form for equal class sizes as there is a variance term; however, LDA will be the same*

(a)

Case Study 11 (null)

2PCs model	A(0.9), B(0.1)		A(0.75), B(0.25)		A(0.5), B(0.5)		A(0.25), B(0.75)		A(0.1), B(0.9)	
Autoprediction	Class A	Class B	Class A	Class B	Class A	Class B	Class A	Class B	Class A	Class B
Class A	100	100	100	100	53	50.5	0	0	0	0
Class B	0	0	0	0	47	49.5	100	100	100	100
Training set										
Class A	100	100	100	100	54.48	44.78	0	0	0	0
Class B	0	0	0	0	45.52	55.22	100	100	100	100
Test set										
Class A	100	100	100	100	45.45	37.88	0	0	0	0
Class B	0	0	0	0	54.55	62.12	100	100	100	100

Case Study 3 (polymers)

2PCs model	A(0.9), B(0.1)		A(0.75), B(0.25)		A(0.5), B(0.5)		A(0.25), B(0.75)		A(0.1), B(0.9)	
Autoprediction	Class A	Class B	Class A	Class B	Class A	Class B	Class A	Class B	Class A	Class B
Class A	100.00	31.84	100.00	24.88	100.00	19.90	98.91	7.46	85.87	1.00
Class B	0.00	68.16	0.00	75.12	0.00	80.10	1.09	92.54	14.13	99.00
Training set										
Class A	100.00	27.87	100.00	21.31	100.00	16.39	100.00	14.75	98.36	1.64
Class B	0.00	72.13	0.00	78.69	0.00	83.61	0.00	85.25	1.64	98.36
Test set										
Class A	100.00	30.00	100.00	26.43	100.00	20.71	100.00	13.57	96.77	7.14
Class B	0.00	70.00	0.00	73.57	0.00	79.29	0.00	86.43	3.23	92.86

Case Study 4 (pollution)

2PCs model	A(0.9), B(0.1)		A(0.75), B(0.25)		A(0.5), B(0.5)		A(0.25), B(0.75)		A(0.1), B(0.9)	
Autoprediction	Class A	Class B	Class A	Class B	Class A	Class B	Class A	Class B	Class A	Class B
Class A	95.53	8.82	93.30	8.82	90.50	8.82	88.83	8.82	87.15	8.82
Class B	4.47	91.18	6.70	91.18	9.50	91.18	11.17	91.18	12.85	91.18
Training set										
Class A	95.45	13.64	95.45	13.64	95.45	13.64	90.91	13.64	81.82	13.64
Class B	4.55	86.36	4.55	86.36	4.55	86.36	9.09	86.36	18.18	86.36
Test set										
Class A	90.45	0.00	88.54	0.00	87.26	0.00	84.08	0.00	82.17	0.00
Class B	9.55	100.00	11.46	100.00	12.74	100.00	15.92	100.00	17.83	100.00

(b)

Case Study 11 (null)

2PCs model	A(0.9), B(0.1)		A(0.75), B(0.25)		A(0.5), B(0.5)		A(0.25), B(0.75)		A(0.1), B(0.9)	
Autoprediction	Class A	Class B	Class A	Class B	Class A	Class B	Class A	Class B	Class A	Class B
Class A	100	100	100	100	88.5	88	1	0	0	0
Class B	0	0	0	0	11.5	12	99	100	100	100
Training set										
Class A	100	100	100	100	95.52	91.79	1.49	0.75	0	0
Class B	0	0	0	0	4.48	8.21	98.51	99.25	100	100
Test set										
Class A	100	100	100	100	92.42	90.91	0	4.55	0	0
Class B	0	0	0	0	7.58	9.09	100	95.45	100	100

Case Study 3 (polymers)

2PCs model	A(0.9), B(0.1)		A(0.75), B(0.25)		A(0.5), B(0.5)		A(0.25), B(0.75)		A(0.1), B(0.9)	
Autoprediction	Class A	Class B	Class A	Class B	Class A	Class B	Class A	Class B	Class A	Class B
Class A	98.91	8.96	95.65	3.98	82.61	1.00	69.57	0.50	39.13	0.00
Class B	1.09	91.04	4.35	96.02	17.39	99.00	30.43	99.50	60.87	100.00

(*continued overleaf*)

Table 10.5 *(continued)*

Training set										
Class A	100.00	6.56	98.36	4.92	83.61	0.00	77.05	0.00	62.30	0.00
Class B	0.00	93.44	1.64	95.08	16.39	100.00	22.95	100.00	37.70	100.00
Test set										
Class A	96.77	8.57	93.55	5.00	87.10	1.43	77.42	1.43	58.06	0.00
Class B	3.23	91.43	6.45	95.00	12.90	98.57	22.58	98.57	41.94	100.00
Case Study 4 (pollution)										
2PCs model	A(0.9), B(0.1)		A(0.75), B(0.25)		A(0.5), B(0.5)		A(0.25), B(0.75)		A(0.1), B(0.9)	
Autoprediction	Class A	Class B	Class A	Class B	Class A	Class B	Class A	Class B	Class A	Class B
Class A	96.09	11.76	93.30	8.82	92.18	8.82	91.62	8.82	89.39	8.82
Class B	3.91	88.24	6.70	91.18	7.82	91.18	8.38	91.18	10.61	91.18
Training set										
Class A	95.45	13.64	95.45	13.64	95.45	13.64	90.91	13.64	86.36	13.64
Class B	4.55	86.36	4.55	86.36	4.55	86.36	9.09	86.36	13.64	86.36
Test set										
Class A	91.72	0.00	90.45	0.00	89.17	0.00	86.62	0.00	86.62	0.00
Class B	8.28	100.00	9.55	100.00	10.83	100.00	13.38	100.00	13.38	100.00

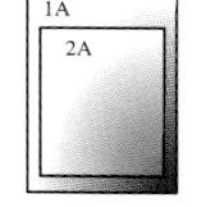

decision boundary is somewhat analogous to the change in position for the PLS-DA boundary when the decision threshold is altered as described in Section 8.5.2 (Figure 5.10). There are, however, no straightforward Bayesian extensions to PLS-DA probably due to the different philosophical user base: many users of LDA and QDA are by nature statisticians and as such comfortable with probabilistic extensions of classifiers, whereas many users of PLS are chemometricians who tend to think in terms of boundaries or classification rates. However the different effects are comparable, and changing the PLS-DA threshold has an analogous effect of changing position of boundaries to changing the prior probability in PLS-DA.

The effect of differing prior probabilities are represented by boundaries in Figure 10.1 and Figure 10.2 for autopredictive models on Case Studies 11 (null), 3 (polymers) and 4 (pollution). Reducing the probability of membership of class A moves the boundary closer

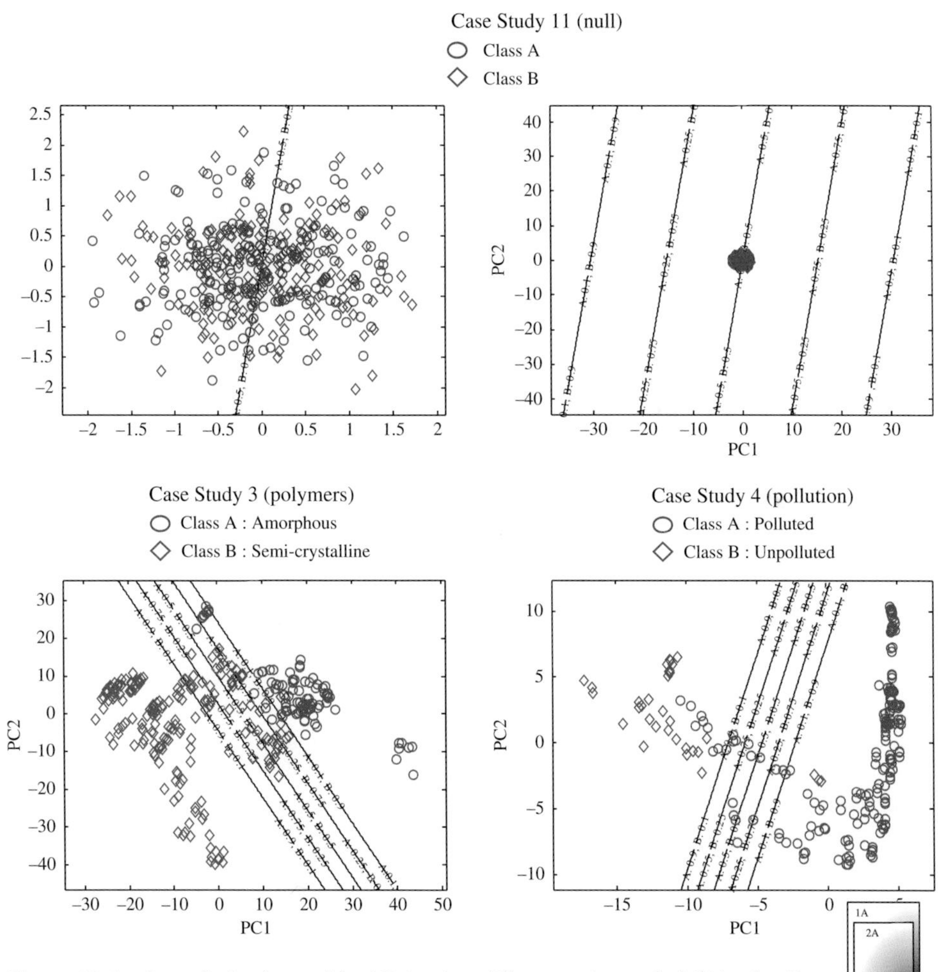

Figure 10.1 Boundaries formed for LDA using different prior probabilities for Case Studies 11, 3 and 4; the Case Study 11 graph is expanded – top right

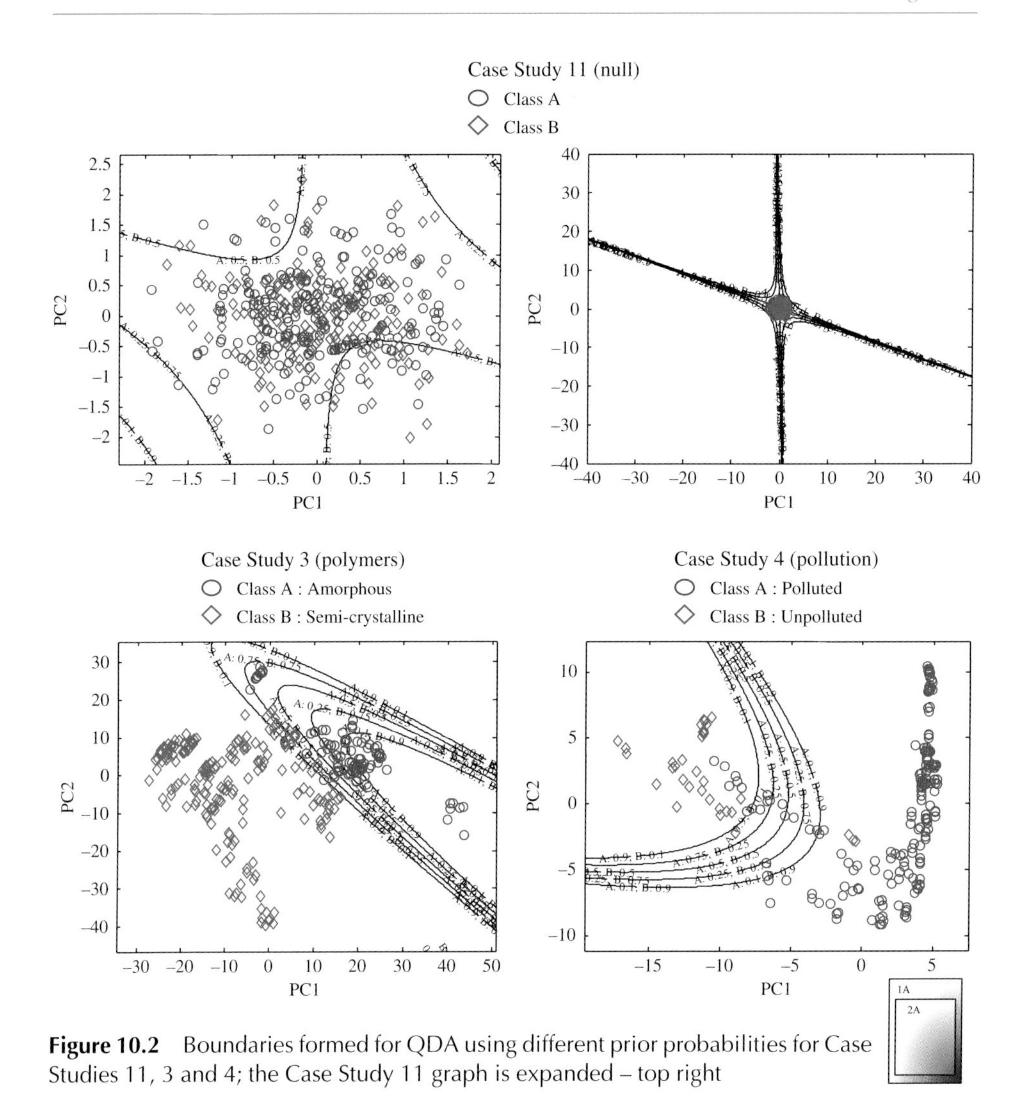

Figure 10.2 Boundaries formed for QDA using different prior probabilities for Case Studies 11, 3 and 4; the Case Study 11 graph is expanded – top right

to the centroid of this class, which results in a greater proportion of samples being rejected as members of this class. For LDA, these boundaries are represented by parallel lines. QDA results in somewhat more complex boundaries, rather analogous to confidence contours of one class classifiers (Chapter 6) but no longer so constrained in shape. For Case Study 11 (null) the LDA boundaries for probabilities differing significantly from 0.5 are very far away from the samples, effectively perfectly classifying samples, if there is a change from the prior probability of 50 % (which equals the relative number of samples in each group).

Prior probabilities can also be employed when class sizes are very different, as in, for example, Case Studies 3 (polymers, class A = 92, B = 201 samples) and 4 (pollution, class A = 175, B = 34 samples). The results of LDA and QDA assuming unequal prior probabilities based on the ratio of samples in each class, are compared to the equal priors in Figure 10.3. The effect of changing from prior probabilities of 50 % (the default) to making

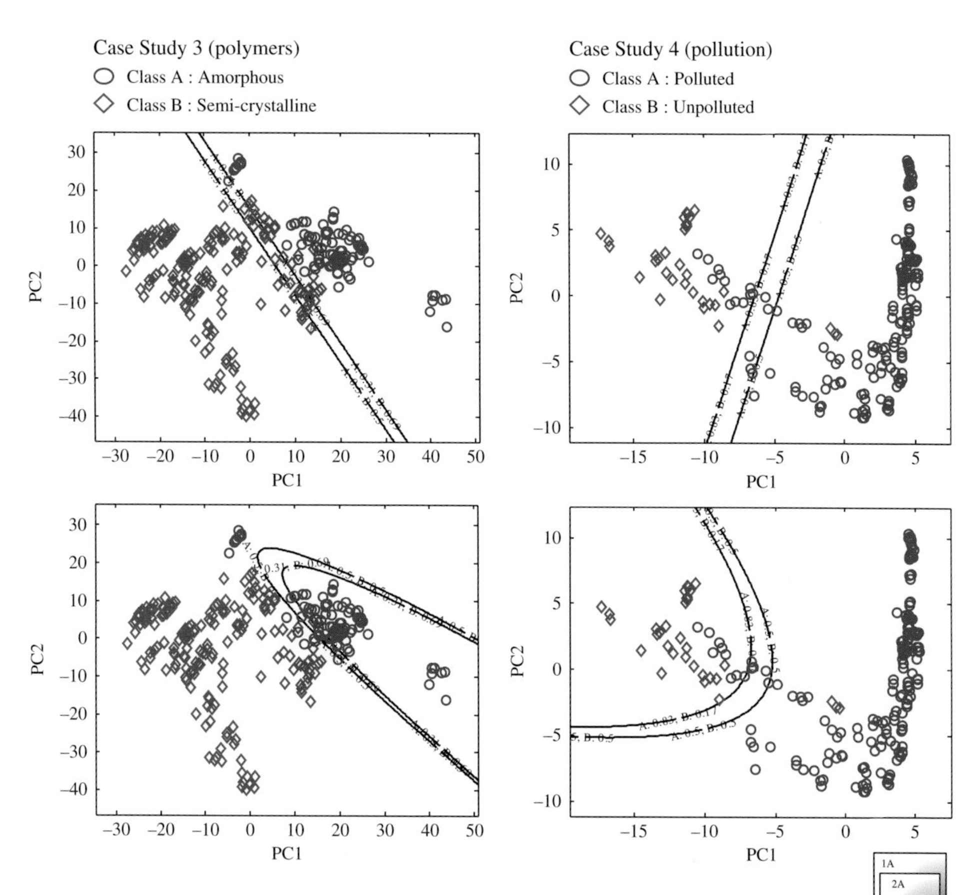

Figure 10.3 Boundaries formed for LDA and QDA both using prior probabilities proportional to the relative number of samples in classes A and B and using equal prior probabilities (A = 0.5, B = 0.5)

these proportional to group size, moves the boundaries towards the centroid of the smaller group. In most cases it improves overall classification, with better classification of the larger group at the expense of a degradation of the smaller group; however if the boundary already fits the border between the two groups well (as QDA of Case Study 3) and there are several samples close to this border, it can lead to poorer classification. In all cases the %CC of the larger group will either stay the same or improve and the smaller group either reduces or gets worse. Note that the Bayesian modification is not only applicable to autopredictive models but all aspects of prediction including test and training sets.

Bibliography

Bayesian Extensions to Classifiers

J.H. Friedman, Regularized Discriminant-Analysis, *J. Am. Statistical Assoc.*, **84**, 165–175(1989).

11

Class Separation Indices

11.1 Introduction

It is useful to obtain a numerical value as to how well different classes can be distinguished, or how separate these classes are, and also to compare separation of different classes to see which are more readily distinguishable. The easier it is to distinguish two or more classes, the more significant the influence of the factor that causes this difference, but it is useful to develop a scale that can compare different datasets and case studies. In Chapter 9, we defined statistical indices which were used to determine which variables are best at separating classes by looking at how distinct the distributions of the intensities of the variables from each class are; in this chapter we will extend these to multivariate indices of class separation. There are quite a number of indices available, but in this chapter we will focus on four such indices. We will illustrate the calculation of two class Cluster Separation Indices (CSIs) that indicate how far apart the two classes lie. Hence the separation observed in different datasets can easily be compared even though the raw data cannot. In the literature there are also many references to Cluster Validation Indices (CVIs) which are used during unsupervized cluster analysis when it is not known how many clusters are present in the data: the data is grouped into $g = 2, 3, \ldots g_{max}$ clusters, where g_{max} is a preset maximum number of clusters to be tested and a CVI is calculated for each value of g, the value which gives the optimal CVI taken as the optimal number of clusters. However, CVIs are often designed only to be used for comparing different methods applied to a single dataset, and it is only the relative value of each CVI that is important. Many CVIs have no upper or lower limits, and so it is difficult to produce meaningful interpretation of the numbers obtained. CSIs, however, can be employed to compare class separation on different datasets, and we will, therefore, restrict discussion to these types of index. In this chapter, four indices are described, the Davies Bouldin index (dbi), the Silhouette width (sw), the modified Silhouette width (msw) and the Overlap Coefficient (oc).

Chemometrics for Pattern Recognition Richard G. Brereton

In Figure 11.1 we illustrate the principle of why it is useful to compute class separation indications. The centroids of the two classes in both groups are equally spaced, but one dataset is more spread out than the other, and so the two classes overlap more; therefore using the distance between centroids alone as an indication of class separation is unlikely to provide sufficient information to assess whether two groups are distinct. There are several criteria of overlap, each of which gives a different viewpoint of class separation, and which are not mathematically equivalent. Many class separation indices differ in how they handle outliers, that is, samples that although assigned to a specific class, may have been assigned by accident or mislabelled. These can have a large, unintended, influence on some indices, as they represent the extremes borders of a class, and so inadvertently make a class look more spread out. However most of the approaches described below have safeguards against this problem.

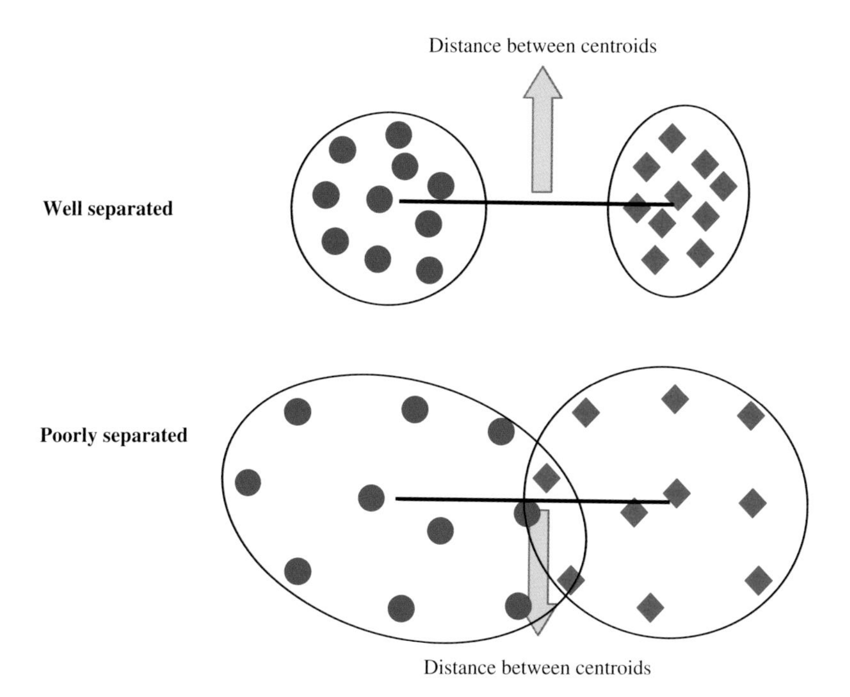

Figure 11.1 Two datasets each consisting of two classes, both with the same distance between centroids, one of which is well separated and one poorly separated

11.2 Davies Bouldin Index

The idea of the Davies Bouldin index (dbi) is that classes are more separated the lower the ratio of the spread of individual samples in each class to the distance between the class centroids. There are various ways of defining the dbi, but one that has a definition for the separation between groups g and h is as follows:

$$dbi_{gh} = 0.5 \frac{\left\{ \left(\sum_{i \in g, i < I_g} \sum_{k \in g, k > i} d_{ik} \right) \Big/ [I_g(I_g - 1)/2] + \left(\sum_{i \in h, i < I_h} \sum_{k \in h, k > i} d_{ik} \right) \Big/ [I_h(I_h - 1)/2] \right\}}{d_{gh}}$$

where d_{ik} is the Euclidean distance between two samples i and k and d_{gh} is the distance between the centroids of the two groups. The numerator involves calculating the average distance between all samples in class g and in class h (average intraclass distances). Hence the dbi is the ratio of the mean within class distances of all samples to the distance between the centroids of each class. The lower the value, the more separated the classes are, as the ratio of within to the 'between class distance' is less. High values suggest high overlap between groups. The dbi is unbounded in that there is no upper limit.

There are some variations of the theme, the main one being to compute the largest distance between the samples in each class rather than the mean distance between all pairs of samples in each class (so for each class the distance between the two extreme samples is computed), but this measure is sensitive to outliers, and takes no note of the 'within class sample distribution'. Finally, distance measures are not restricted to the Euclidean distance, but the Euclidean metric has the advantage that all variables can be employed in the model.

Table 11.1 list the values of the index for several of the case studies in this text. We can see that there is as consistent trend in the three Case Studies 10a to 10c (simulations) as the level of overlap increases. Case Study 11 (null) has a high value of the dbi as there is in effect no significant separation between the classes. Values around or less than 1 imply on the whole good separation, as for Case Studies 1 (forensic), 3 (polymers) and 4 (pollution). We can see that Case Study 8a (mouse urine – diet study) has an overlap that is slightly below the least overlapping simulation (10a). It is important to remember that the index is computed using all variables and so may not necessarily correspond to the separation found in the PC scores plots. However this allows us to calibrate the separation between classes and as such one can have a benchmark set of reference distributions against which it is possible to compare experimental points. A key problem relates to outliers, which will

Table 11.1 *Three of the separation indices discussed in this chapter for several two class case studies*

Case study	Separation Index		
	DBI	SW	MSW
1 (forensic)	0.650	0.447	0.958
3 (polymers)	1.053	0.292	0.952
4 (pollution)	1.038	0.381	0.958
8a (mouse urine – diet study)	3.437	0.031	0.832
10a (simulation – minimum overlap)	5.004	0.015	0.740
10b (simulation – intermediate overlap)	5.391	0.012	0.810
10c (simulation – maximum overlap)	6.500	0.007	0.730
11 (null)	14.493	0.000	0.468

have an undue effect on the overall value of the 'within class distances', although, since every pairwise intraclass distance has been averaged this influence is diluted somewhat.

Table 11.2 *Calculation of dbi for pairwise comparison of classes in Case Study 3 (polymers – 9 groups) for both the scores of the first 2 PCs and when all variables are employed as input to the classifier*

DBI 2 PCs	A	B	C	D	E	F	G	H	I
A									
B	1.129								
C	0.200	0.162							
D	0.222	0.187	0.164						
E	0.273	0.231	0.157	0.691					
F	0.298	0.225	0.121	0.245	0.361				
G	0.359	0.371	0.214	0.246	0.285	0.196			
H	0.343	0.305	0.152	0.354	0.590	0.293	0.428		
I	0.755	0.769	0.260	0.366	0.521	0.438	0.635	0.948	
DBI All variables	A	B	C	D	E	F	G	H	I
A									
B	0.655								
C	0.258	0.244							
D	0.332	0.335	0.194						
E	0.396	0.410	0.216	0.709					
F	0.368	0.342	0.164	0.303	0.429				
G	0.401	0.399	0.179	0.274	0.317	0.239			
H	0.388	0.375	0.168	0.333	0.416	0.333	0.511		
I	0.741	0.649	0.288	0.460	0.568	0.475	0.475	0.559	

It is also useful to compare the class separation when there are several different groups within the same dataset, an example being Case Study 3 (polymers – 9 groups), to see which are more similar to each other. The dbi can be computed both in the original dataspace or in the PC space. In Table 11.2 we calculate the pairwise class separation indices both for the raw data and the first 2 PCs. The results for 2 PCs can be compared to the scores plot. For example, the highest value of dbi (most overlapping) is for classes A and B which definitely exhibit overlap in the scores plot. Because there is a small subgroup of class A that is far away from the centroid of A this reduces the dbi somewhat, and we would anticipate a substantially higher dbi if this subgroup was not present. Class F is on the bottom centre of the scores plot and distinct from all other groups and so has low values of dbi. Its farthest neighbour is class C and as such the dbi is low (0.121). Closer classes are characterized by higher values of dbi. Interestingly when all variables are used, the values of dbi converge towards a mean value, with small ones increasing in magnitude and large ones decreasing. In Figure 11.2 we show the change in dbi as increasing number of PCs are used in the model for six case studies. This is a good visual indicator as to how including additional PCs in a model influences cluster separation. We see that for Case Study 11 (null) adding PCs improves separation, whereas for the others it slightly decreases separation. However in most cases the dbi has largely converged after 3 or 4 PCs. For

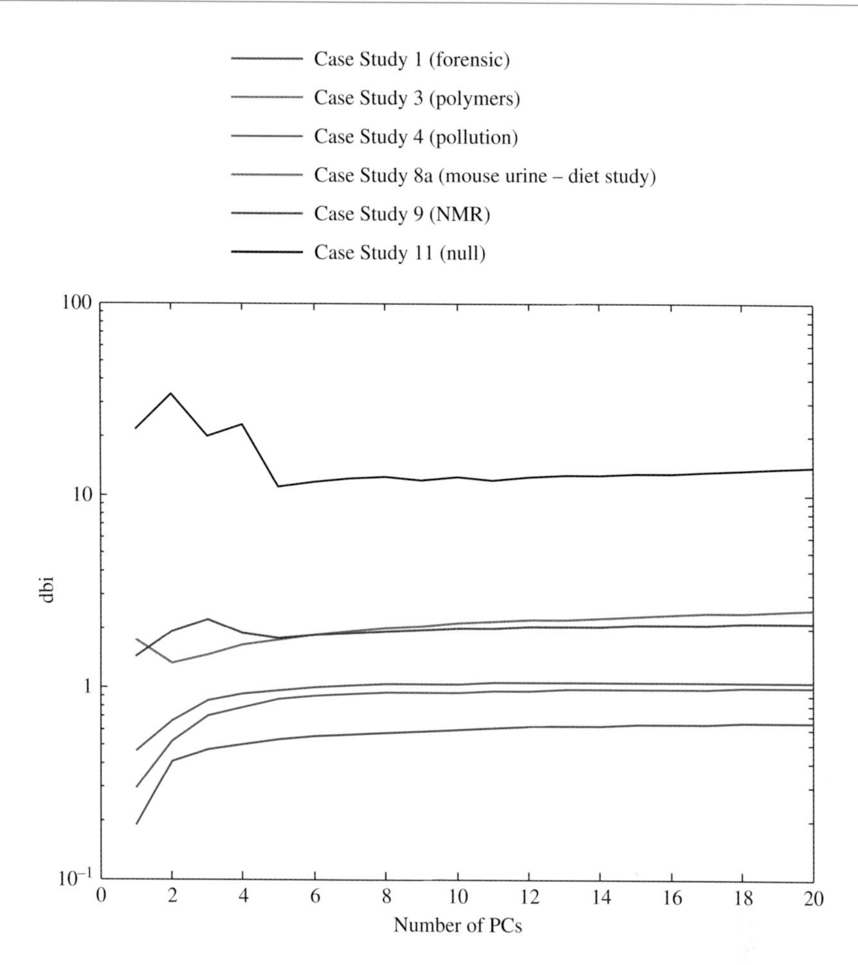

Figure 11.2 Change in value of dbi as the number of PCs is increased for various two class case studies

many real datasets, the class structure has a major influence on the data and so is reflected in the largest PCs, so whereas later PCs do add information they may not be influenced by these major factors and as such the class structure may slightly degrade. Case Study 11 (null) shows a slight increase in separation as more PCs are included in the model, because there are no overriding factors that influence separation and so the more the PCs, the more the information and possibility of chance distributions having a small effect on separation: the value of dbi is very high and as such there is no real separation for any number of PCs.

The value of dbi, as all cluster separation indices, can also be calculated for single variable models, and can be used as an additional (univariate) indicator of which variables are better discriminators as well as those discussed in Chapter 9, but is approximately related to other indicators. As an illustration we calculate the values of dbi for each of the twenty models formed on each of the largest 20 PCs for Case Studies 8a (mouse urine – diet study) and 1 (forensic) against the absolute value of t on a log–log scale (Figure 11.3). The gradient is

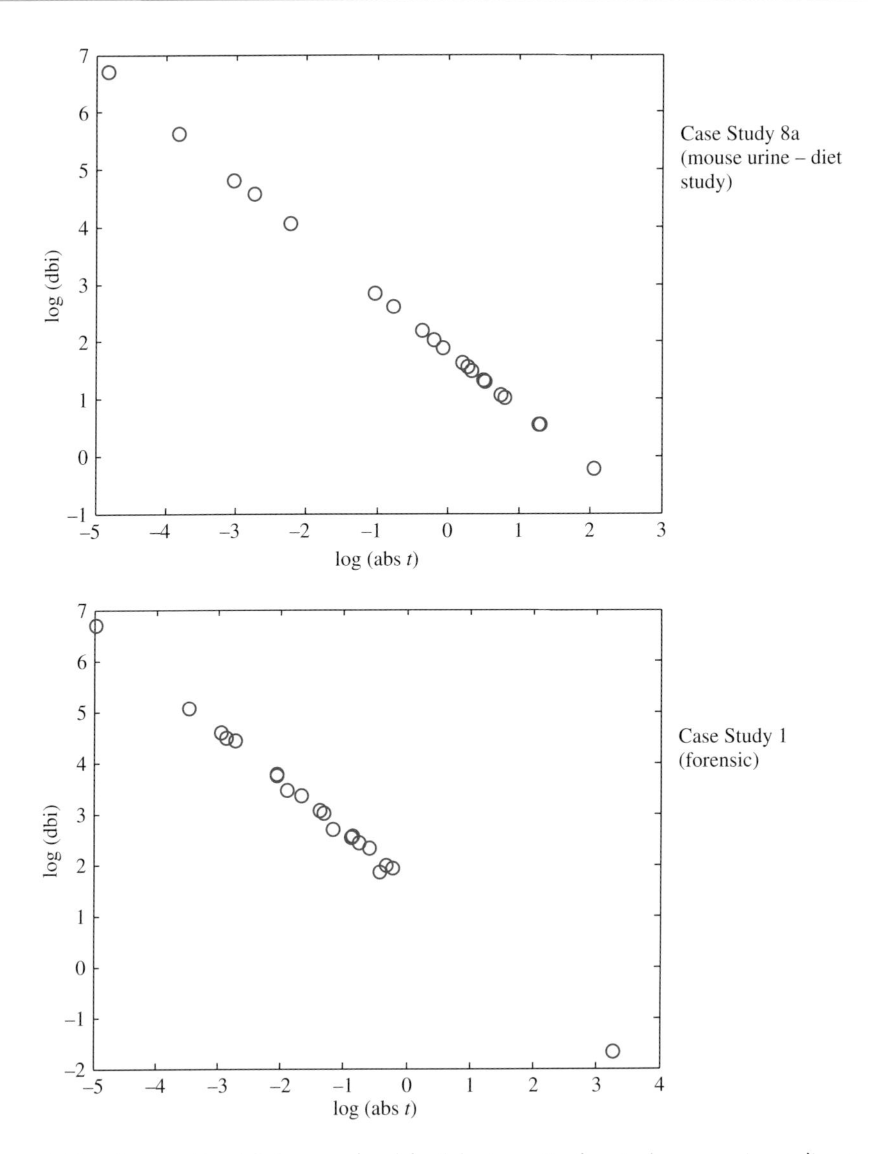

Figure 11.3 Graph of log (dbi) versus log (abs t) for Case Studies 8a (mouse urine – diet study) and 1 (forensic) for twenty 1 PC models formed on each of the largest 20 PCs

approximately −1, implying a reciprocal relationship. This is not exact and will depend on the structure of the data; for the forensic case study there are a small number of outlying samples that lie within each other's group which will be influential on the distance measure and exhibit some deviation from perfect behaviour. However class separation indices such as dbi nevertheless provide alternatives to the t-statistic or other indicators as discussed in Chapter 9, but for brevity we only illustrate general principles in this chapter and could be employed as an alternative univariate indicator of discriminatory power.

11.3 Silhouette Width and Modified Silhouette Width

11.3.1 Silhouette Width

The silhouette width (sw) between two classes is defined as follows:

1. The average distance for each sample i of group g to all other samples k within its cluster is calculated, except to itself:

$$\overline{d(i_g, g)} = \sum_{k \in g,\, k \neq i} d_{ik}/(I_g - 1)$$

where there are I_g samples in this cluster.
2. The average distance to all samples in the other cluster is calculated separately for cluster h. The average distance between sample i and samples of cluster h consisting of I_h samples is then taken as follows:

$$\overline{d(i_g, h)} = \sum_{j \in h} d_{ij}/I_h$$

3. The silhouette width s_i is calculated for sample i (member of class g) as follows:

$$S_{i_g} = [\overline{d(i_g, h)} - \overline{d(i_g, g)}]/\max\ [\overline{d(i_g, h)},\ \overline{d(i_g, g)}]$$

   In simple terms this equals the average of the (mean distance between each sample and samples not in its own class minus the mean distance between each sample and samples in the same class except itself) over the maximum of these two measures.
4. The average sw over all samples in the two groups to be compared is then computed to give sw_{gh}, which is a measure of class separation.

The more positive sw_{gh}, the more separated the classes are, and so this follows an opposite trend to dbi as can be seen in Table 11.1, although this relationship is not exactly a monotonic one as each index is based on slightly different criteria; to illustrate the relationship we plot a graph of silhouette width versus dbi for Case Studies 9 (NMR) and 8a (mouse urine – diet) in Figure 11.4 as the number of PCs in the modelled in increased.

The value of sw for an individual sample can vary between –1 and 1, but for two groups each of size I_g, the minimum possible is $-1/I_g$ which represents complete overlap (this is the value of the sw of a group of samples with itself). When used for individual samples it can give a measure as to how well a sample belongs to its own group (in which case the summations in the equation above are removed), and is a useful approach for identifying outliers or mislabelled samples within a dataset.

11.3.2 Modified Silhouette Width

The modified silhouette width can be calculated by examining the sign of the silhouette width calculated for each individual sample. If both clusters were drawn from the same population then both clusters would have a very similar mean. On average, samples would be a similar distance to the mean of the cluster they are assigned to as the distance to the mean of the other cluster, leading to approximately 50 % of samples obtaining a positive sw. As the clusters become more separated, then samples will be closest to the mean of their

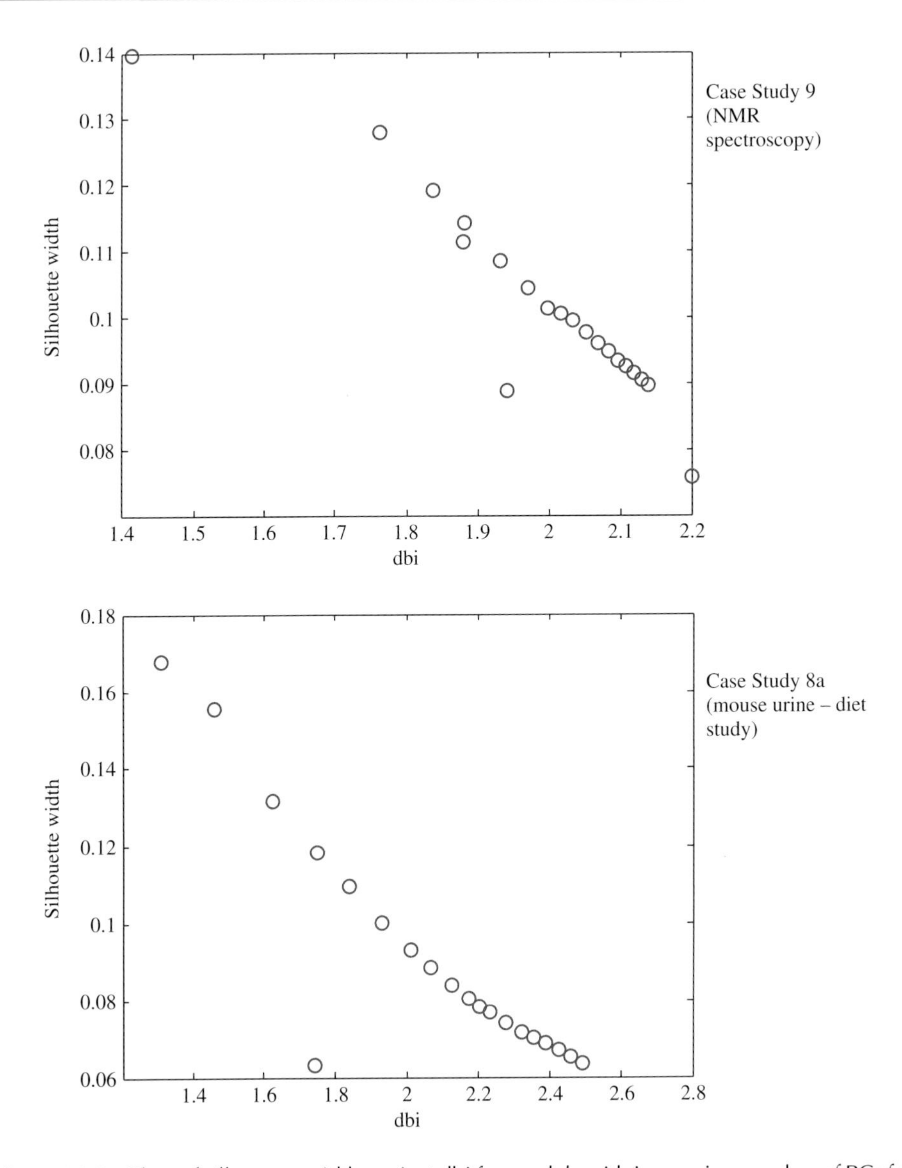

Figure 11.4 Plots of silhouette width against dbi for models with increasing number of PCs for Case Studies 9 (NMR spectroscopy) and 8a (mouse urine – diet study)

own cluster with a greater frequency. The msw (modified silhouette width) is the proportion of samples with a positive sw.

It can be seen that this metric is very similar to simply calculating the % 'correctly classified' obtained using EDC as the classification method. However there is a slight difference in the method of calculation. When classification is performed using EDC, the mean is calculated over all samples in the cluster. In this method, when the distance to the mean of the class to which a sample belongs is calculated, the mean is calculated using all samples excluding the sample of interest. This difference becomes more noticeable when dealing with outliers. In traditional distance-to-means calculations, the outliers will be involved during the calculation of the class mean, as all class members are involved in the computation; however using the silhouette width method, the distance is calculated from an outlier to the remainder of the class, often leading to a larger distance than would be seen in traditional distance-to-means. The msw will vary between approximately 0.5 and 1; in Table 11.1 we see that Case Study 11 (null) has a msw of 0.468, compared to 0.958 for Case Study 1 (forensic). High values (e.g. 1) can be obtained if two classes are completely separate. There is not a direct mathematical relationship between msw and the other indicators discussed in this chapter.

As usual, msw, as per other approaches discussed in this chapter, could also be used to rank PCs and also variables according to their power to differentiate between two classes.

11.4 Overlap Coefficient

Another way of determining the amount of overlap between two clusters is by examining the number of samples that have a high probability of belonging to both of the clusters. For each sample, the distance between that sample and the cluster centroid is calculated using the Hotelling's T^2 statistic:

$$T_{ig}^2 = I_g(\boldsymbol{x}_i - \overline{\boldsymbol{x}}_g)\boldsymbol{S}_g^{-1}(\boldsymbol{x} - \overline{\boldsymbol{x}}_g)'$$

where $\boldsymbol{S}_g$ is the variance covariance matrix of class g. It can be seen this is in fact the squared Mahalanobis distance multiplied by the number of samples in each group, and is calculated using the mean and covariance of the cluster. For a multinormal distribution, the values calculated for all of the samples will be described by a Hotelling's T^2 distribution, dependent on the number of variables (J) and the number of samples (I_g). The distribution of the T^2 is actually robust when the rows of a data matrix are not necessarily independent and also if the variables do not strictly follow a normal distribution, making this distribution appropriate for this method. It is simple to convert the vector T^2 (containing the calculated statistic for all samples) into a vector which forms an F-distribution with parameters as follows, as discussed in Chapter 6:

$$T_{ig}^2 \approx \frac{J(I_g - 1)(I_g + 1)}{I_g(I_g - J)} F(J, I_g - J)$$

The F-statistic for each sample (f_g) can then be compared to critical values obtained from an F-distribution table. For the purposes of calculating a coefficient to describe the amount of overlap between two clusters (Overlap Coefficient (oc)), we recommend a sample is assumed to belong to a given cluster if T^2 is within 95 % confidence limits. This was determined by examining an F-distribution table for a critical value of $\alpha = 0.05$, with the two degrees of freedom equal to J, and $I_g - J$.

Each sample is compared to both of the clusters, and if both f_1 and f_2 are greater than the critical value from the F-distribution table, i.e. the sample has within 95 % confidence limits of belonging to both of the clusters, then that sample can be assumed to be in a region of overlap. The oc is then determined as 1 – the proportion of samples that were calculated as lying in the region of overlap, with a value of 1 indicating no ambiguous samples and 0 that all are ambiguous.

Because this is calculated using the Mahalanobis distance it cannot be computing using all the original variables, unless the number of variables is less than the number of samples in each class, and there first must be variable reduction, usually via PCA. Hence the oc will depend both on the number of PCs used in the model and the value of α. The values of oc for several case studies are calculated in Table 11.3. Note that there is a good and expected gradation between Case Studies 10a (simulation – minimum overlap) to 10c (simulation – maximum overlap), and that Case Study 11 (null) has an oc close to zero. There are some differences in the models using different numbers of PCs, which can be of interest. The oc can rank separations between classes in a similar way that the t-statistic does between variables, both based on similar principles, but as discussed in this chapter the oc allows the possibility of comparing different datasets, whereas the t-statistic is used to compare distributions of variables within a single dataset.

Table 11.3 *Overlap coefficient for several datasets using 2 and 10 PC models, and a 95 % confidence limit*

Case study	Overlap coefficient	
	2 PCs	10 PCs
1 (forensic)	1.000	0.989
3 (polymers)	0.816	1.000
4 (pollution)	0.939	1.000
8a (mouse urine – diet study)	0.319	0.454
10a (simulation – minimum overlap)	0.425	0.315
10b (simulation – intermediate overlap)	0.250	0.190
10c (simulation – maximum overlap)	0.135	0.175
11 (null)	0.058	0.030

Bibliography

Cluster Separation Indices

S.J. Dixon, N. Heinrich, M. Holmboe, M.L. Schaefer, R.R. Reed, J. Trevejo, R.G. Brereton, Use of cluster separation indices and the influence of outliers: application of two new separation indices, the modified silhouette index and the overlap coefficient to simulated data and mouse urine metabolomic profiles, *J. Chemometrics*, **23**, 19–31 (2009).

D.L. Davies, D.W. Bouldin, A Cluster Separation Measure, *IEEE Trans. Pattern Anal. Machine Intell.*, **1**, 224–227 (1979).

P.J. Rousseeuw, Silhouettes: A Graphical Aid to the Interpretation and Validation of Cluster Analysis, *J. Computat. Appl. Math.*, **20**, 53–65 (1987).

12
Comparing Different Patterns

12.1 Introduction

It is often useful to compare information from two different sources. For example, we may be monitoring a reaction using two types of spectroscopy, e.g. IR and UV/Vis. Is there a common or consensual view between these two types of information? Each type of information can be considered a 'block' containing different types of data each containing complementary information about the samples.

There are several questions we might wish to answer:

1. Is there information common to both types of measurement, for example, if we use two spectroscopies, are there common trends that can be found in each spectroscopy?
2. Can we produce a single unified model of the two blocks?
3. What variables are responsible for the common trends between the blocks?

There are a very large number of so-called 'mutliblock' approaches in the chemometrics literature. This is because there are lots of opportunities for the development of algorithms, which often fall into the definition of chemometrics research that gets published in fundamental journals. The topics in this chapter could be extended considerably, almost to fill a book in its own rights, but we restrict this chapter to some of the main approaches. Multiblock data, whilst increasingly available, especially in areas such as process monitoring, is still quite rare in the majority of mainstream chemometrics applications and so the length of this chapter is restricted for this reason, whilst this is certainly a growing topic for the future. Often more theoretical research in algorithm development is applied by a few pioneers and reported first in quite technical papers but then can take several years to become generally accepted.

In order to illustrate the methods we will use Case Study 12, which consists of two blocks of information, one obtained from GCMS and one from microbiology. The samples belong to 4 subgroups, and we are asking whether each technique can distinguish the

Chemometrics for Pattern Recognition Richard G. Brereton

groups but also whether there are common trends between both techniques. We will call the GCMS data block 1 and the microbiology data block 2, each yielding matrices denoted by $\mathbf{X}_b$ where $b = 1$ or 2, with dimensions $I \times J_b$, as illustrated symbolically in Figure 12.1. In our case $I = 34$, $J_1 = 334$ and $J_2 = 30$. Preprocessing each block of data separately by square root scaling the elements of each matrix, summing the rows of each matrix separately to 1, and standardizing the columns. The resultant PC scores and loadings plots are presented in Figure 12.2. As discussed in Section 2.14, the groups can be

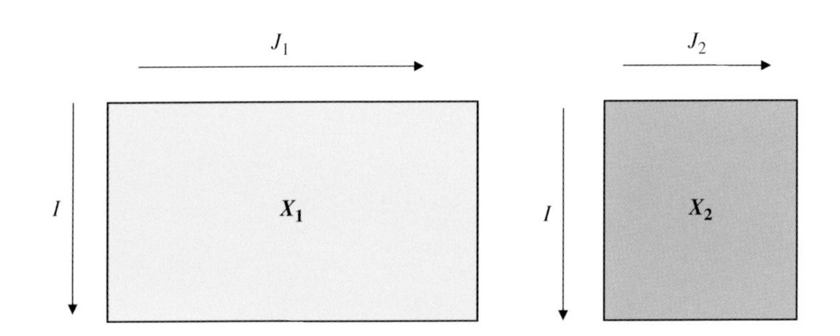

Figure 12.1 Two blocks of data

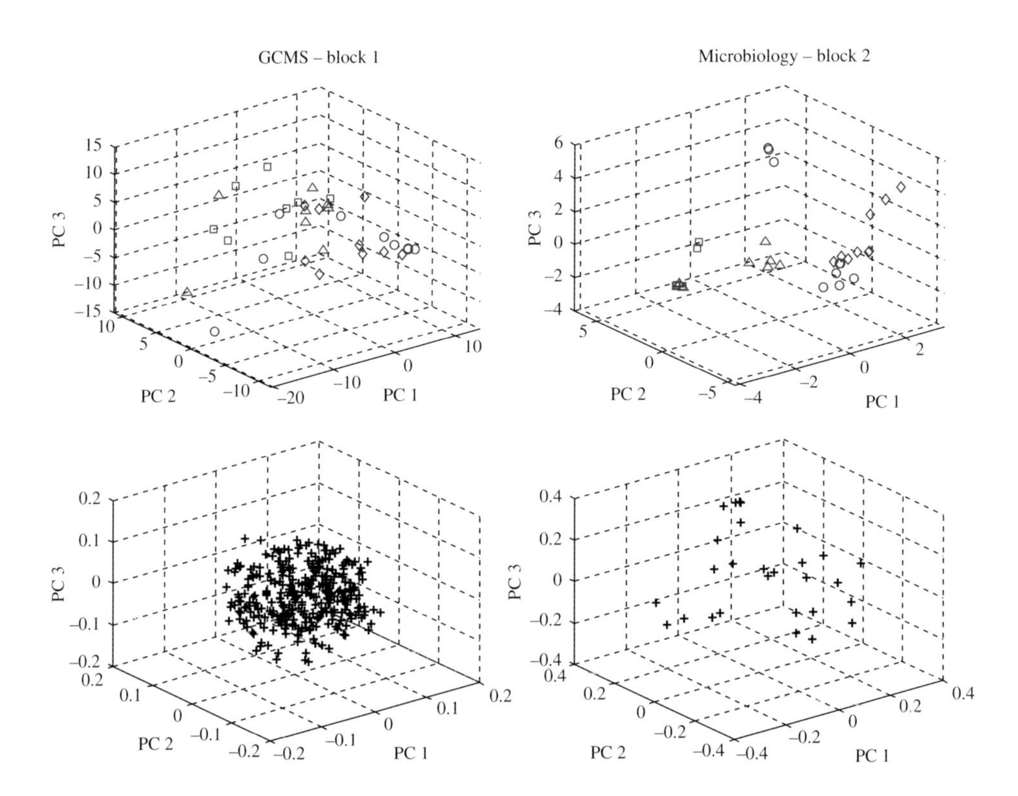

Figure 12.2 Scores and loadings plots for the data of Case Study 12 for the first three PCs

distinguished visually using the first 3 PCs. The parallel loadings plots suggest that there are variables that appear to associate with specific groups, especially using microbiology. Loadings plots are a simple exploratory visual method, complementary to the more formal approaches discussed in Chapter 9, for determining which variables are most likely to be associated with each group. For Case Study 12 the group sizes are quite small and so it is best to stick with visual approaches.

12.2 Correlation Based Methods

The simplest way of answering whether two blocks are similar is by seeing how well they are correlated, the higher the correlation, the better. There are several tests in the literature of which two common ones are described in the following.

12.2.1 Mantel Test

For each of the two techniques a (dis)similarity matrix is computed. Several approaches have been described in Section 3.3.1. To compare two matrices directly, the Mantel test is a commonly used and effective method. Given two (dis)similarity matrices $\boldsymbol{D}_1$ and $\boldsymbol{D}_2$ (from blocks 1 and 2 respectively in our case) of dimensions of $I \times I$, the lower (or equivalent upper) triangular parts of the matrices, consisting of $I \times (I-1)/2$ elements are unfolded into two vectors: $\boldsymbol{d_1} = \{d_{1,11}, d_{1,21}, \ldots d_{1,I1}, d_{1,22}, \ldots d_{1,I2,\ldots}, d_{1,II}\}$ and $\boldsymbol{d_2} = \{d_{2,11}, d_{2,21}, \ldots d_{2,I1}, d_{2,22}, \ldots d_{2,I2,\ldots}, d_{2,II}\}$. The correlation coefficient r between $\boldsymbol{d_1}$ and $\boldsymbol{d_2}$ is then calculated.

The significance of the correlation is determined using Monte Carlo permutations (Section 9.3.2). The order of the elements in one vector is permutated while the other remains unchanged, to give a null distribution, and the correlation coefficient between these two vectors is calculated. This process is repeated a large number of times and the correlation coefficients obtained by these permutations are used to form an empirical null distribution. The empirical significance of the correlation coefficient r calculated above is calculated to provide an indication as to whether this belongs to the null distribution or not. For an estimate of the significance of the correlation coefficient we can calculate the proportion of times the Monte Carlo simulation results in a correlation coefficient greater than the observed value r (one tailed test) or the absolute value is greater than r (two tailed test). These approximate to empirical probabilities and are denoted by the p values below. If there are, for example, 10 000 permutations, and the r value is greater than 9900 of the r values for the null distribution, but less than 9901 then the empirical significance that the two matrices are similar is 99 % which is often reported as a p value of 0.01.

The results of performing the Mantel test on the two matrices using the both Euclidean distance measure and the square root of (1 – the Jaccard index), and a two tailed test for r, are presented in Figure 12.3, using GCMS as the reference technique (that is, the microbiological data are permutated). We find that after 10 000 permutations, the empirical correlation coefficient for the Euclidean distance between the two sets of data is 0.0254, which exceeds 7809 of the absolute values of the correlation coefficients using random permutations, leading to an empirical p value of 0.2191. This

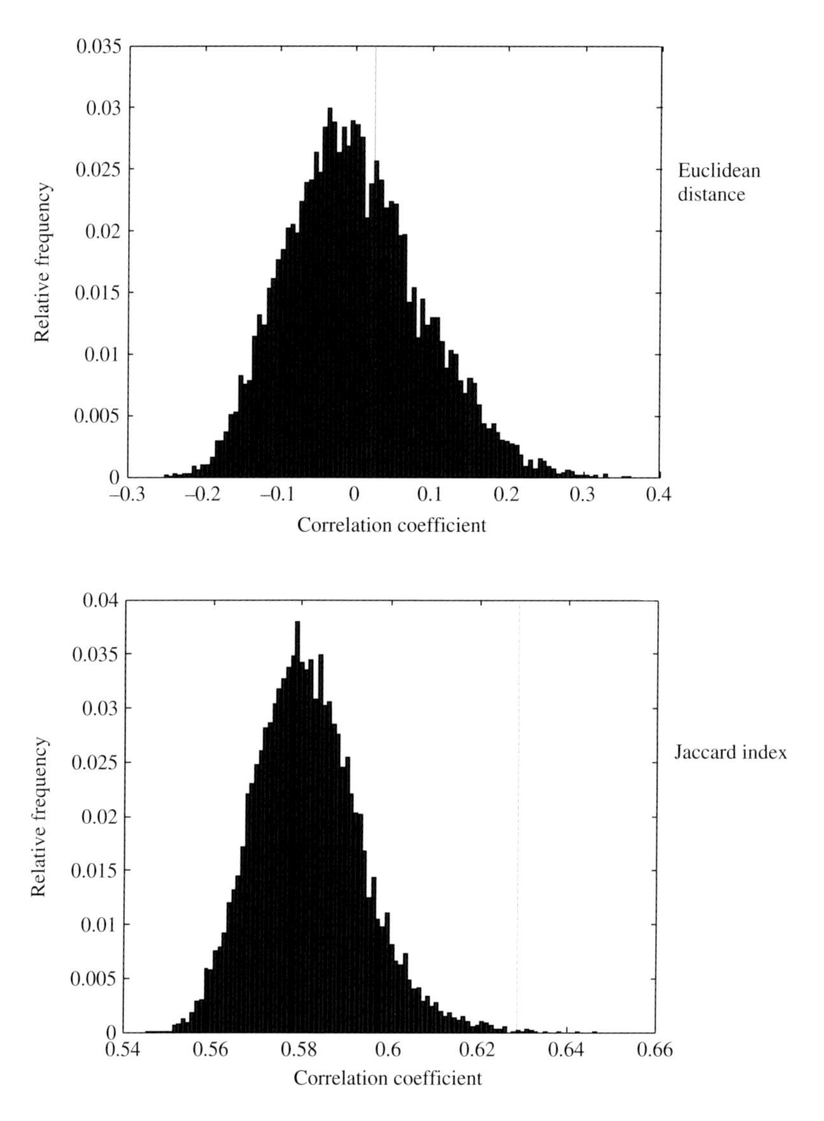

Figure 12.3 Results of Mantel test on Case Study 12, using the D = Euclidean distance, and $D = \sqrt{(1 - S)}$ where S is the Jaccard index with GCMS as the reference technique, with 10 000 Monte Carlo permutations

correlation coefficient appears relatively low, but this is primarily because many variables are not detected in any of the samples which has a distorting effect on the data. Using the Jaccard index, which is calculated from the presence and absence of variables as detected in the samples, we obtain a much more promising result. The empirical correlation coefficient between the two sets of data is 0.6288, which exceeds 9987 of the absolute values of the correlation coefficients using random permutations, leading to an empirical p value of 0.0013. Of course these results are dependent on the data we use, and case studies less dependent on detection or otherwise of variables may perform better using quantitative dissimilarity measures and worse using qualitative ones.

12.2.2 R_V Coefficient

Another way of calculating the amount of correlation between two matrices is the R_V coefficient:

$$R_V(\mathbf{X}_1, \mathbf{X}_2) = \frac{\text{trace}(\mathbf{X}_1\mathbf{X}_1{}'\mathbf{X}_2\mathbf{X}'{}_2)}{\sqrt{\text{trace}(\mathbf{X}_1\mathbf{X}_1{}'\mathbf{X}_1\mathbf{X}'{}_1)\,\text{trace}(\mathbf{X}_2\mathbf{X}_2{}'\mathbf{X}_2\mathbf{X}'{}_2)}}$$

where 'trace' indicates the trace of the matrix (the sum of its diagonal elements). This coefficient does not require any prior computation of similarities. The coefficient spans the interval [0,1] and returns 1 in the case of perfect correlation. In addition its value is invariant to scaling and rotation, and therefore the coefficient can be effectively used also on matrices after PCA if necessary (as well as on the raw data), to measure to what extent the configurations derived give the same view of the samples. Note that the R_V coefficient copes with matrices that have different number of columns (or variables); if the number of columns is identical a simpler approach called congruence coefficients can be used, but the method in this section is of more general use where the number of variables in each block is likely to be very different. For some of the approaches below, such as procrustes analysis, both matrices should have the same dimensions (after PCA).

The significance of the R_V coefficient can be obtained empirically using Monte Carlo simulations, just like the Mantel test, and is given in Figure 12.4. The correlation

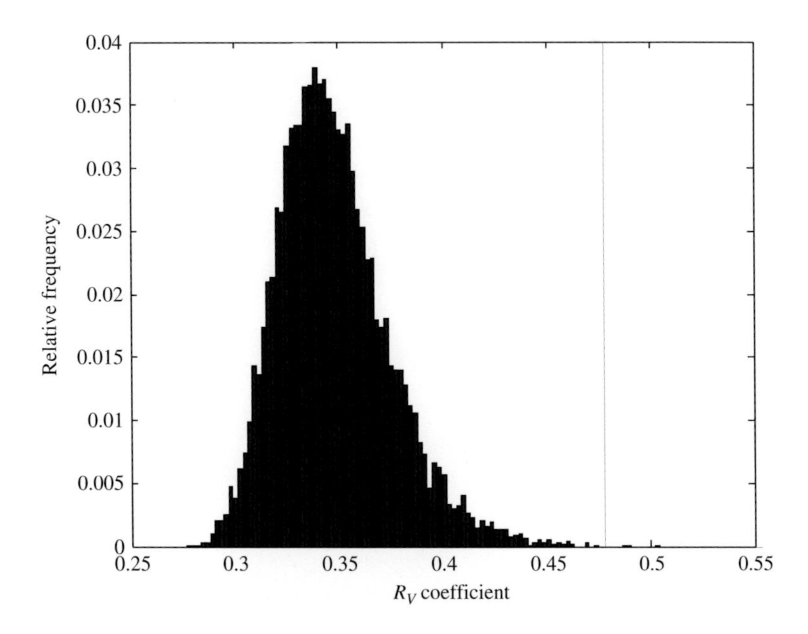

Figure 12.4 R_V for Case Study 12 compared to 10 000 Monte Carlo permutations

coefficient between the two sets of data is 0.478, and is greater than 9994 of the random permutations leading to an empirical p value of 0.0006.

12.3 Consensus PCA

PCA has been used widely in this text to visualize relationships between samples using one type of measurement. Consensus PCA (C-PCA) is an approach employed to look at trends common to more than one block of data simultaneously. In the case discussed in this chapter it is interest to verify what is the 'consensual' view obtained from merging the information available from both blocks and to determine what is the contribution of each of the two profiles to the overall consensual view.

The method starts by selecting a super-score $\boldsymbol{t_T}$ that is multiplied on all blocks to give the block loadings. These block loadings are scaled so that their sum of squares equals 1 and then multiplied through the respective blocks to give the block scores. The block scores are combined in a super-block $\boldsymbol{T}$. Then a PCA iteration on $\boldsymbol{T}$ is performed to give the super-weight $\boldsymbol{w}$ that is again scaled so its sum of squares equals 1 and multiplied onto the super-block $\boldsymbol{T}$ to obtain an improved estimate of the super-score $\boldsymbol{t_T}$. This is repeated until convergence of $\boldsymbol{t_T}$. The super-score $\boldsymbol{T_T}$ summarizes the main trends between all blocks and block scores $\boldsymbol{T_1}$ and $\boldsymbol{T_2}$ represent the individual contribution of each block that are related to the former by means of the weight vector $\boldsymbol{w}$. The algorithm usually incorporates a block scaling factor which usually set to $s_b = 1/\sqrt{J_b}$ where J_b are the number of variables in block b, which we will use in this text, but other scaling can be employed if it is desired to attach different relevant significance to each block.

A formal description of the algorithm for B blocks follows:

1. Guess the initial super-scores for the first or new component ${}^{initial}\hat{\boldsymbol{t}}_T$. A guidance is to select the column with maximum variance from the super-matrix $\boldsymbol{X} = [s_1\ \boldsymbol{X}_1 \ldots\ldots\ldots s_B\ \boldsymbol{X_B}]$ which is the concatenation of the matrices for each block scaled appropriately. The relative weight for each block s_b can be set in a variety of ways but a common one (which we will adopt) is to use $s_b = \sqrt{1/J_b}$ where there are J_b variables in each block.

Steps 2 to 4 are performed on each block separately:

2. For each block b calculate:

$$ {}^{unnorm}\hat{\boldsymbol{p}}_b = \frac{{}^{initial}\hat{\boldsymbol{t}}_T{}'\boldsymbol{X}_b}{\sum \hat{\boldsymbol{t}}_T^2} $$

3. For each block b scale the guess of the loadings, so:

$$ \hat{\boldsymbol{p}}_b = \frac{{}^{unnorm}\hat{\boldsymbol{p}}_b}{\sqrt{\sum {}^{unnorm}\hat{\boldsymbol{p}}_b^2}} $$

4. Now calculate a guess of the scores for block b, $\hat{\boldsymbol{t}}_b = s_b \boldsymbol{X}_b\, \hat{\boldsymbol{p}}_b{}'$ where s_b is the scaling factor as above.

The next steps are performed on all blocks together:

5. The scores for all blocks B are combined into a matrix $\boldsymbol{T}$ of dimensions $I \times B$:

$$\hat{\boldsymbol{T}} = \left[{}^{new}\hat{\boldsymbol{t}}_{1......}{}^{new}\hat{\boldsymbol{t}}_{B}\right]$$

6. Calculate the super-weight, a vector of dimensions $1 \times B$:

$$^{unnorm}\hat{\boldsymbol{w}} = \frac{\hat{\boldsymbol{t}}_T{}'\,\hat{\boldsymbol{T}}}{\sum \hat{t}_T^{\,2}}$$

7. Scale the super-weight so that its sum of squares equals 1:

$$\hat{\boldsymbol{w}} = \frac{^{unnorm}\hat{\boldsymbol{w}}}{\sqrt{\sum {}^{unnorm}\hat{w}^2}}$$

8. Calculate the new guess of the super-scores:

$$\hat{\boldsymbol{t}}_T = \hat{\boldsymbol{T}}\hat{\boldsymbol{w}}'$$

9. Check if this new guess differs from the first guess; a simple approach is to look at the size of the sum of square difference in the old and new super-scores. If this is small, the PCs have been extracted.

The next two steps are performed on each block separately:

10. Set the loadings for each block b to:

$$\boldsymbol{p}_b = \frac{\boldsymbol{t}_T{}'\boldsymbol{X}_b}{\sum t_T^2}$$

11. Calculate the residuals for each block:

$$^{resid}\boldsymbol{X}_b = \boldsymbol{X} - \boldsymbol{t}_T\,\boldsymbol{p}_b$$

A final step involves determining whether to continue computing more PCs:

12. If it is desired to compute further PCs, substitute the residual data matrices and go to step 1.

This algorithm has many steps in common with the NIPALS algorithm for PCA described in Section 3.2.4.

The super-scores can be visually inspected to determine the consensus view provided by both blocks and can be compared to block scores that will contribute proportionally to their weights. The super-scores for Case Study 12 are illustrated in Figure 12.5. It can be seen that the two strains are well separated on PC1, but interestingly the two haplotypes (classes A and C are H1 and classes B and C are H2) are separated well in PC3. It is possible to calculate the relative weights w_b of the block scores for each block and so which are more significant in determining the overall trend as presented in Figure 12.6. It can be seen that the GCMS has most influence over PC2 but microbiology over PC1 and PC3, with a big relative influence on PC3. In Figure 12.7, we illustrate the PC scores (as 1D plots) for both GCMS and microbiology. We see that GCMS can distinguish the strains (classes A & B and C & D) well especially in PC2. Microbiology however can distinguish both

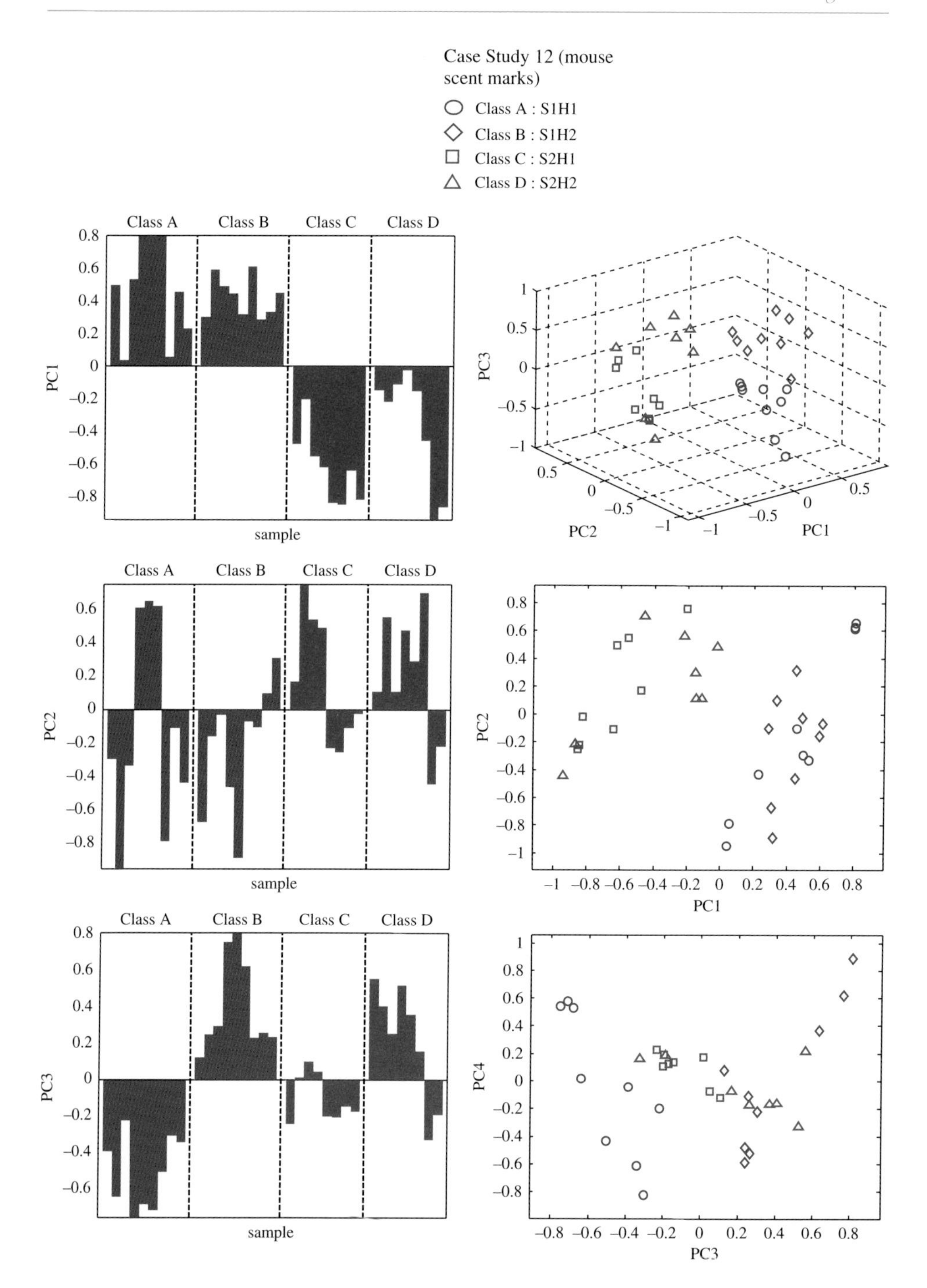

Figure 12.5 Super-scores for Case Study 12

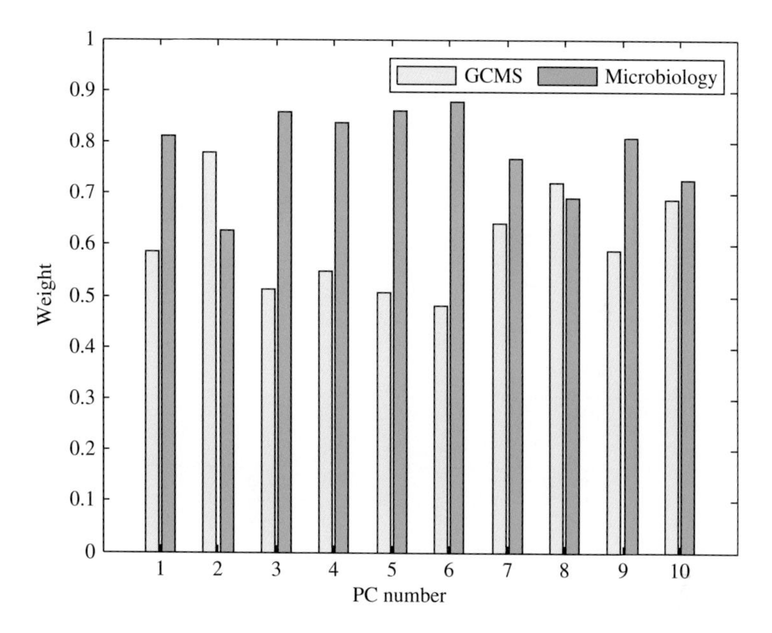

Figure 12.6 Weights of the block scores for block 1 (GCMS) and block 2 (microbiology)

strains (PC1) and haplotypes (A and C and B and D) in PC2. In the consensus PCs, PC3 distinguishes haplotypes well, and we can see that microbiology has a greater influence over PC3 from the relative weights. Both microbiology and GCMS can separate out the strains which are well differentiated on PC1, representing the largest consensus between the two methods.

12.4 Procrustes Analysis

Procrustes analysis is a slightly different approach, although it, too, aims to reach a consensus. In procrustes analysis one of the technique is designated as a reference, and the aim is to see how closely data from other techniques fit this reference. Let us say we are trying to select an analytical method that can be best used to determine pollution. Our reference goal may be a set of samples for which we are sure (from independent measurements) that we know the pattern of pollution, e.g. a set of analyses of elemental composition. These measurements may be costly and slow, and so it is desired to select one technique, e.g. NIR spectroscopy, MS, or HPLC that best reflects the original and known pattern of pollution. We could use procrustes analysis to see how well the information from each analytical technique in turn relates to the known pollution pattern. The one that fits it closest is probably the most suitable (dependent on cost of course). We might however, in contrast want to determine which technique gives complementary information – for example, which are most similar and which are most different, as each technique may be expensive and an aim might be to minimize the number of techniques,

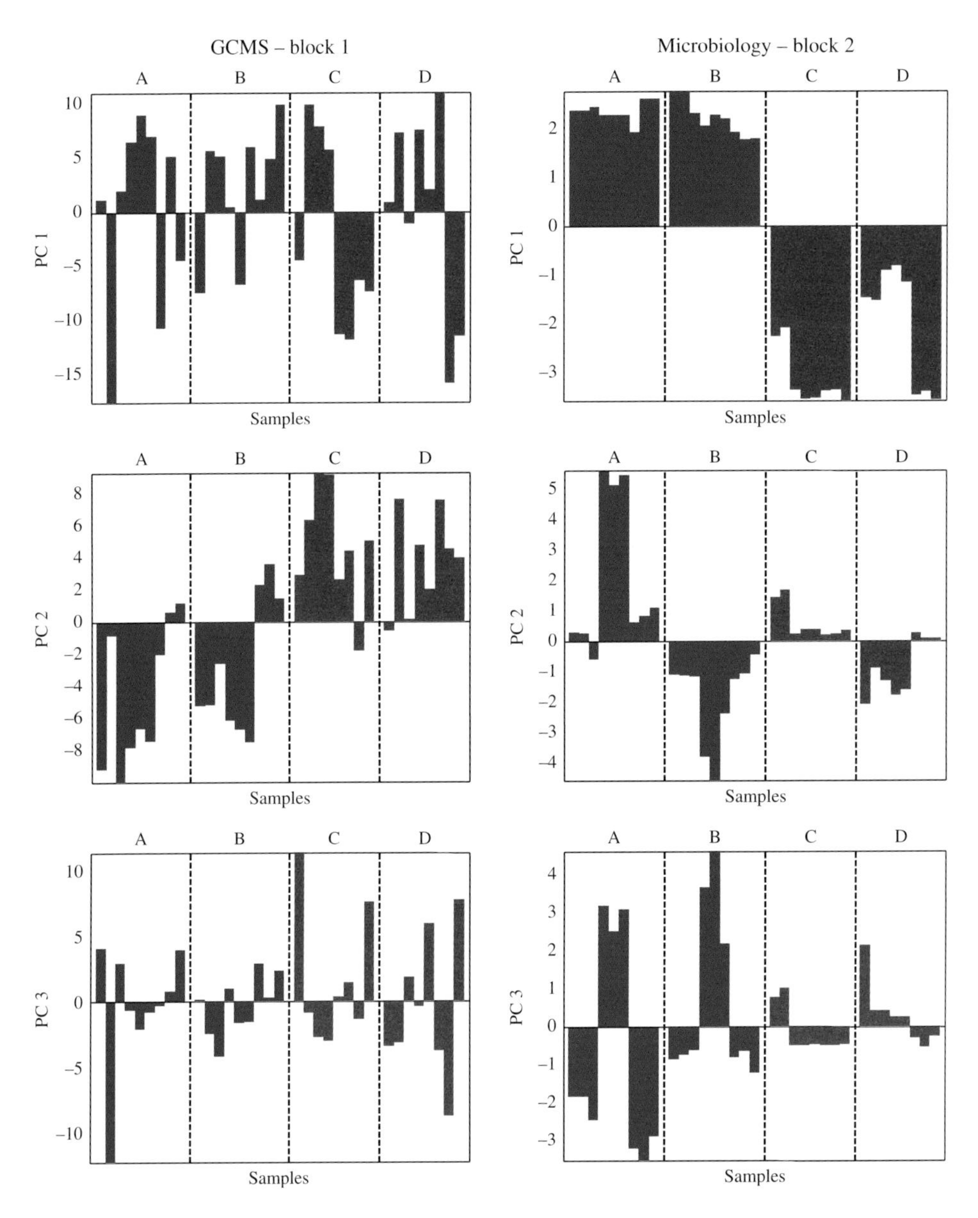

Figure 12.7 Scores of the first 3 PCs for the GCMS and microbiology data, presented as bar graphs

cutting cost, and so only employ techniques that offer different information. Finally a common and related use of procrustes is to reduce the number of variables in a multivariate dataset; each variable may represent a measurement that is expensive and we may be interested in whether we can pick a subset of variables that gives information that as closely as possible fits the full variable set, and if so, which variables, and how many are most appropriate.

In the context of Case Study 12, we have to set one block as a reference block, say for example, the GCMS in our situation, and ask whether we can obtain similar information from the other, in our case from microbiology. This is a slightly different question from choosing the consensus view (Section 12.3). Procrustes analysis starts by comparing the PC scores plots, and asks whether the scores plot obtained by the second technique can be transformed to fit that of the reference technique as closely as possible. Three transformations are employed, namely translation, scaling and finally rotation/reflection. Note that because the sign of the PCs cannot be controlled easily, the reflection step is integral to the rotation.

There are several ways of describing the algorithm, the steps below representing one description; all approaches should obtain the same result except for an overall scaling factor:

1. Perform PCA separately on each of the two blocks, to give a scores matrix $\boldsymbol{T_R}$ for the reference block and $\boldsymbol{T_C}$ for the second or comparison block. Select the top A components for each block using a method for determining the number of significant components (Section 4.3.1) or empirically. In most descriptions of procrustes analysis, the number of components for each block must be the same. There are no generally agreed guidelines if they are different, although it is probably best to use the optimal number of components from each block that is greatest, so as to avoid losing information, keeping redundant PCs from the block that has least significant PCs.
2. The next step involves translation, formally described as mean centring. If the original data matrices have already been mean centred, no further transformation is required. Otherwise create new matrices ${}^{cen}\boldsymbol{T_R}$ and ${}^{cen}\boldsymbol{T_M}$ where ${}^{cen}t_{b,\,ij_b} = t_{b,\,ij_b} - \bar{t}_{b,\,j_b}$ where b denotes the block, i the sample and j_b variable j from block b. In matrix terms this can be expressed by ${}^{cen}\boldsymbol{T}_b = (\boldsymbol{U} - \boldsymbol{Z})\boldsymbol{T}_b$ where $\boldsymbol{U}$ is the $I \times I$ identity matrix, and $\boldsymbol{Z}$ is an $I \times I$ matrix whose elements all equal $1/I$. In the context of translation it does not matter which matrix is the reference. In the case study described in this chapter the two matrices have already been mean centred via standardization.
3. There are a variety of ways of scaling the matrices, some authors preferring to scale each block, others to scale the comparison block to fit the reference block. Both will result in the same answer. A simple and symmetric approach is to transform ${}^{cen}\boldsymbol{T_b}$ to ${}^{scale}\boldsymbol{T_b}$ for each block separately by setting ${}^{scale}\boldsymbol{T_b} = {}^{cen}\boldsymbol{T_b}/s_b$ where $s_b = \sqrt{\mathrm{trace}({}^{cen}\boldsymbol{T}_b\,{}^{cen}\boldsymbol{T}'_b)}$: this results in the sum of squared distances to the centroid of each block to be equal to 1, which we employ.
4. The final rotation and reflection step involves finding a matrix $\boldsymbol{R}$ to transform ${}^{rot}\boldsymbol{T_C} = {}^{scale}\boldsymbol{T_C}\boldsymbol{R}$. This is done as follows:
   - Perform SVD (see Section 3.2.2) on the $A \times A$ matrix ${}^{scale}\boldsymbol{T_R}'\,{}^{scale}\boldsymbol{T_C}$ to give $\boldsymbol{USV}$.
   - The rotation/reflection matrix is defined by $\boldsymbol{R} = \boldsymbol{VU}'$.

   This new rotated matrix should fit the matrix ${}^{scale}\boldsymbol{T_R}$ as closely as possible.

Visually for two or three PCs, the two matrices can be superimposed, as illustrated in Figure 12.8 for Case Study 12 when 2 PCs are employed in the model. Note that because the PC scores plots do exhibit a different configuration, they will never overlap perfectly, but by examining the microbiology scores and comparing to Figure 3.17, it should be evident that the microbiology scores are rotated whereas the GCMS ones are invariant.

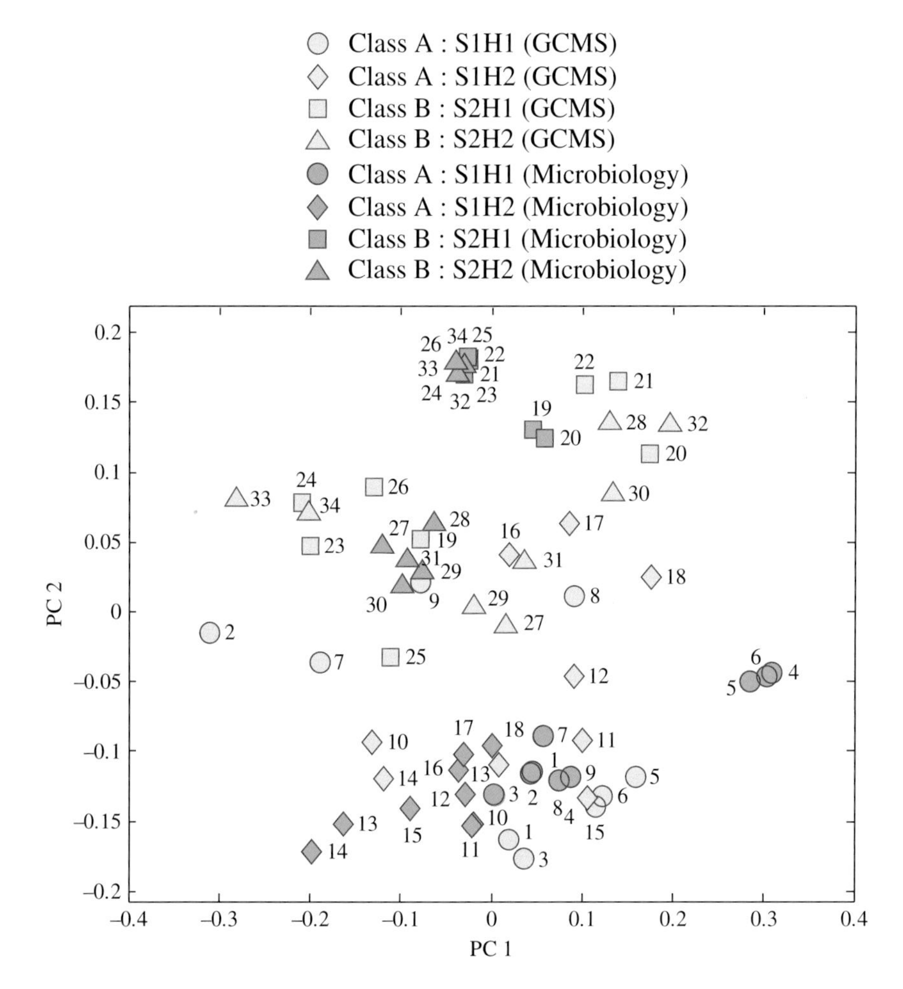

Figure 12.8 Superimposing the two matrices for GCMS and microbiology in Case Study 12, using 2 PCs and procrustes rotation to obtain as close a fit as possible

It is often useful to determine how well the matrices are matched after procrustes analysis. This can be done by computing the root mean square error between the two transformed matrices:

$$e_{RC} = \sqrt{\sum_{i=1}^{I} \sum_{a=1}^{A} \left({}^{scale}t_{R,ia} - {}^{rot}t_{C,ia} \right)^2 / IA}$$

The greater this error, the more different the patterns. Using this equation one could compare, for example, which of the different analytical techniques fits a reference method best, or which method of variable reduction preserves the original data space directly. Note that there are several different equations in the literature for computing this error, most giving equivalent results. Note that this error does not necessarily decrease with the number of PCs in the model, as there may be differing information in later PCs.

The significance of the fit between the two methods can be assessed empirically, where a common approach is to employ Monte Carlo methods. In the context of procrustes analysis, these involve randomly permuting the samples in one of the two blocks (we recommend the comparison block, leaving the reference unchanged), and then calculating the error e_{RC} after procrustes rotation. This is then repeated many times (e.g. 10 000 times) to provide a null distribution of e values. The actual value of e is then compared to the distribution, in a one sided empirical test. If, for example, its value is equal to the 20th highest value obtained by the Monte Carlo distribution, then its empirical probability is $100 \times (10\,000 - 20)/10\,000$ % or 99.8 % or a p value of 0.002. Note that if its value is below the lowest obtained using the null distribution, the empirical probability becomes 100 %, but this, of course, is an approximation. The results for Case Study 12 are presented in Figure 12.9, together with the empirical probability for 2 and 5 PC models.

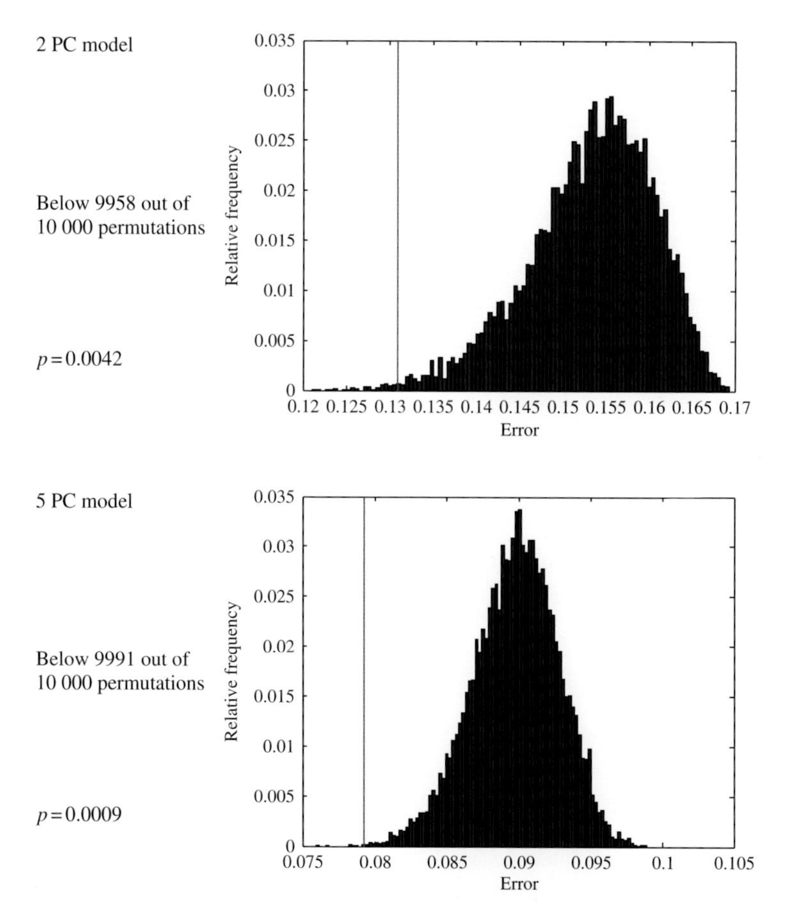

Figure 12.9 Monte Carlo simulations using procrustes analysis for 1 PC and 5 PC models on Case Study 12, showing an empirical error distribution together with the observed error and 10 000 permutations, together with empirical probabilities

Bibliography

Multiblock Methods

S. Zomer, S.J. Dixon, Y. Xu, S.P. Jensen, H. Wang, C.V. Lanyon, A.G. O'Donnell, A.S. Clare, L.M. Gosling, D.J. Penn, R.G. Brereton, Consensus Multivariate methods in Gas Chromatographic Mass Spectrometry and Denaturing Gradient Gel Electrophoresis: MHC-congenic and other strains of mice can be classified according to the profiles of volatiles and microflora in their scent-marks, *Analyst*, **134**, 114–123 (2009).

Y. Xu, S.J. Dixon, R.G. Brereton, H.A. Soini, M.V. Novotny, K. Trebesius, I. Bergmaier, E. Oberzaucher, K. Grammer, D.J. Penn, Comparison of human axillary odour profiles obtained by gas chromatography mass spectrometry and skin microbial profiles obtained by denaturing gradient gel electrophoresis using multivariate pattern recognition, *Metabolomics*, **3**, 427–437 (2007).

J.M. Andrade, M.P. Gomez-Carracedo, W. Krzanowski, M. Kubista, Procrustes rotation in analytical chemistry, a tutorial, *Chemometrics Intell. Lab. Systems*, **72**, 123–132 (2004).

J.C. Gower, Generalized Procrustes Analysis, *Psychometrika*, **40**, 33–51 (1975).

J.A. Westerhuis,T. Kourti, J.F. MacGregor, Analysis of Multiblock and Hierarchical PCA and PLS Models, *J. Chemometrics*, **12**, 301–321 (1998).

P. Robert, Y. Escoufier, Unifying Tool for Linear Multivariate Statistical-Methods – R_v-Coefficient *J. R. Statistical Soc., Ser. C – Appl. Statistics*, **25**, 257–265 (1976).

Index

Note: References to figures are given in italic type. References to tables are given in bold type.

Analysis of variance (ANOVA) 397–8, 420–24
  advantages 427, 431
  F-ratio 421–24
  governing equations 421, **426**
  one way 421–24
  two way 424
    method design *430*
Area under the Curve (AUC)
  Receiver Operator Characteristic curves 279
    Partial Least Squares Discriminatory Analysis 377
Atomic spectroscopy, data preprocessing 34–5
Autoprediction 181–4
  process flow charts *153–6, 162–3*
    data reduction *164*
  test models and 320–1
  see also Classifiers
Autoscaling 125–7

Banknotes
  forensic analysis 20–3
  forgery 319
Bayesian analysis
  Bayes' theorem 453–8
  classifiers 458–67
Best Matching Unit
  Self Organizing Maps 89, 92
    hit histograms 97–9
Bias (of data) 16–17
Binary classifiers, *see* Classifiers, two-class
Binary dissimilarity measures 76–9
Binomial distribution 399–401, *400*
Block scaling
  columns 128–9
  rows 120–4
Blood 34–5
Bootstrap
  for classifier optimization 368–74
    percentage correctly classified 372, **373**
  for PCA data reduction 134–7, *138–40*
  process flow charts
    for PCA *168*
    for PLS *169*
    for Support Vector Machines or Learning Vector Quantization *173*
Box Cox transformation 115–17

Canberra dissimilarity index **79,** 80
Case studies
  1 (forensic analysis of banknotes) 20–3
    class membership plots 275, *275*
    class separation indices **471**
      overlap coefficient **478**
    comparison of autopredictive, test and training sets **325,** *327*, **332**
    dissimilarity rankings 85, *86*
    D-statistic 243, 247, 282, *285*
      disjoint model *252, 260, 283*
    effect of test and training set selection on classifier boundaries 352, *353*
    Euclidean distance to centroids *186,* 190–1
    Learning Vector Quantization 201, *212*
    Linear Discriminant Analysis *189, 190*
    multivariate normality **264**
    one-class, Support Vector Machines *270*
    predictive ability and model stability **339,** *344*, **344**
    Principal Components Analysis
      bootstrap *138*
      column standardization and 126

Chemometrics for Pattern Recognition Richard G. Brereton

Case studies (*continued*)
cross-validation *135*
scores plots 58–9, *58*
most discriminatory PCs 143
standardization and *126*
t-statistics **141**
variance *55, 56,* **131**
Q-statistic, disjoint models *261*
Quadratic Discriminant Analysis *194, 283*
Receiver Operator Characteristic curves *283*
Self Organizing Maps
class structure visualization *102*
hit histograms *98*
Support Vector Machines *229*
optimization 390, *387, 389*
variable selection, t-statistics **409**
2 (NIR spectroscopy of food) 23–4, *25*
data scaling 120
D-statistic *244, 246*
disjoint model *260, 261*
Linear Discriminant Analysis, multiclass *293*
multiclass classification 291
effect of test and training set selection on classifier boundaries *359*
Euclidean distance to centroids *292*
Learning Vector Quantization *296*
multiclass, Support Vector Machines *305*
Partial Least Squares Discriminatory Analysis
PLS1 *301, 302, 303*
PLS2 *299*
Quadratic Discriminant Analysis *294*
multivariate normality **264**
Principal Components Analysis
bootstrap *138*
cross-validation *135*
loadings plots 71–4, *72–3*
scores plots *59*
variance **131**
Q-statistic
disjoint models 263, **264–265**
Self Organizing Maps
class structure visualization *103*
U-matrix 95, *96, 97*
3 (thermal analysis of polymers) 25–6
Bayesian weighting **459**
bootstrap *370,* **373**
class separation indices **471**
overlap coefficient **478**
comparison of autopredictive, test and training sets **325**, *327*, **332–3**
directed acrylic graphs *308*
dissimilarity measures, rankings 84, *86*
D-statistic *244*
disjoint model *250–1, 256*
effect of test and training set selection on classifier boundaries 352, *354, 359*
multiclass 358, *360*
Euclidean distance to centroids *186*
multiclass 292
Learning Vector Quantization 203–5, *212*
multiclass *296, 297, 298*
Linear Discriminant Analysis *189*
multiclass *293*
prior probabilities **460,** *465*
multivariate normality **264**
Partial Least Squares Discriminatory Analysis
cross-validation *367*
PLS1 *301, 302, 303*
PLS2 *299*
predictive ability and model stability **340–1**
preprocessing
power element scaling *116–17*
row scaling 120
Principal Components Analysis
bootstrap *138*
cross-validation *135*
scores plots 59–61, *60*
variance **131**
Q-statistic, disjoint model *257, 285*
Quadratic Discriminant Analysis *194*
multiclass *294*
prior probabilities **462–4**
row scaling *121*
Self Organizing Maps
class structure visualization 101, *104*
component planes 100
hit histograms 98–9
U-matrix 95, *96, 97*
Support Vector Machines *229, 272*
multiclass *305, 307*
4 (sand pollution analysed by headspace mass spectrometry) 27–30
Bayesian weighting **459**
class separation indices **471**
overlap coefficient **478**
Euclidean distance to centroids *186*
Learning Vector Quantization 203, *212*
Linear Discriminant Analysis *189*
prior probabilities **462,** *465*
multivariate normality **264**
Principal Components Analysis
bootstrap *139*

cross-validation *136*
scores plots 61–2, *61*
most discriminatory PCs *142*
stepwise discrimination analysis **144**
t-statistics **141**
variance **131**
Quadratic Discriminant Analysis *194*
prior probabilities **462**
samples **28–9**
Support Vector Machines *227–8, 229*
optimization 380, *381–2, 386, 388*
test and training set boundaries *355*
effect on classifier boundaries 352, 358
variable selection
indicator functions compared *437–8*
Monte Carlo method *444*
t-statistics **409,** 412–13
5 (sweat analysis by GCMS) 30–2
dissimilarity measures 77
Euclidean distance to centroids *186, 190*
Linear Discriminant Analysis *189, 190*
multivariate normality **264**
one-class
D-statistic *244, 283*
disjoint model *252*
predictive ability and model stability **339, 341**
Principal Components Analysis
scores plots 62–3, *62*
variance **131**
Quadratic Discriminant Analysis 193, *194*
Receiver Operator Characteristic (ROC) curves *378*
Support Vector Machines *227–8, 229*
multiclass *305*
6 (liquid chromatography-mass spectrometry of pharmaceuticals) 32–4
Principal Components Analysis
scores plots 63, *64*
variance **131**
samples **33**
Support Vector Machines *270*
7 (atomic spectroscopy for study of hypertension) 34–5
Euclidean distance to centroids *186, 190*
multiclass *292*
Learning Vector Quantization *212*
multiclass *296*
Linear Discriminant Analysis *189, 190*
multiclass *293*
Principal Components Analysis
biplots 74
scores plots 63, *65*
variance **131**
Principal Coordinates Analysis *83*
Quadratic Discriminant Analysis 193, *194*
multiclass *294*
Self Organizing Maps, class structure visualization 104, *105*
Support Vector Machines *227–8, 229*
8 (metabolic profiling of mouse urine) 36
bootstrap *370,* 372, **373**
class separation indices **471**
overlap coefficient **478**
comparison of autopredictive, test and training sets **325–6,** *328*
dissimilarity measures
binary 77
rankings 84, *86*
Euclidean distance to centroids *186, 190*
multiclass 292
Fisher weight **414**
indicator functions compared *438–9*
Learning Vector Quantization *213*
multiclass *296*
Linear Discriminant Analysis *189, 190*
multiclass *293*
Monte Carlo method *444*
one-class
class membership plots 275, *276*
D-statistic 245, 247, 269, 282, *284, 285*
disjoint model *253*
Q-statistic *284*
disjoint model *263, 284*
Receiver Operator Characteristic curves *281, 283*
Support Vector Machines *268*
partial least squares **434**
Partial Least Squares Discriminatory Analysis (PLS-DA)
classification thresholds **375**
cross-validation *367*
optimization, receiver operator characteristic (ROC) curves *378*
preprocessing
column standardization 126–7
logarithmic scaling *112*
power scaling *116–17*
Principal Components Analysis, stepwise discriminant analysis **144**
row scaling 120, *122*
standardization *126*
Principal Components Analysis
column standardization and 126–7
cross-validation *135*
loadings plots 69–70, *72–3*

Case studies (*continued*)
scores plots 63–4, *66*
most discriminatory PCs *142*
standardization and *126*
t-statistics **140–1**
variance *55, 56,* **131**
Principal Co-ordinates Analysis, dissimilarity indices 81–4, *82*
Quadratic Discriminant Analysis *194, 283*
multiclass *294*
Support Vector Machines *229*
multiclass *305*
optimization *386, 388*
t-statistic **409,** 407, *410–11*, **414**
compared with PLS indicators *435*
validation
confusion matrices **347**
iterative, predictive ability and model stability **340, 341, 343,** *345*, **345,** 347
variable selection, 412
9 (NMR spectroscopy of saliva) 37–8
Bayesian weighting **459**
bootstrap *370,* 372, **373**
comparison of autopredictive, test and training sets **325–6,** *329*, **332**
effect of test and training set selection on classifier boundaries *356,* 358
Euclidean distance to centroids *187*
Learning Vector Quantization 203, *204, 213*
Linear Discriminant Analysis *190*
Partial Least Squares Discriminatory Analysis (PLS-DA), cross-validation *367*
preprocessing
logarithmic scaling *112*
row scaling *123*
Principal Components Analysis
scores plots 66–7, *67*
variance *55, 56,* **131**
Quadratic Discriminant Analysis *195*
Support Vector Machines *230*
10 (simulations) 38–40
class separation indices **471**
overlap coefficient **478**
Euclidean distance to centroids *187*
Learning Vector Quantization *213*
Linear Discriminant Analysis *190*
Principal Components Analysis
scores plots 67, *68*
variance **131**
Quadratic Discriminant Analysis *195*
Support Vector Machines *230*
variable selection
t-statistic 407, **409**
t-statistic compared with PLS indicators *435*
11 (null data) 40–1, *42*
class membership plots *276*
class separation indices **471**
overlap coefficients **478**
comparison of autopredictive, test and training set data **326,** *329–30*, **332**
D-statistic *245, 284, 285*
disjoint model *252, 256*
effect of test and training set selection on classifier boundaries *357,* 358
Euclidean distance to centroids *187*
Learning Vector Quantization *213*
Linear Discriminant Analysis *190*
prior probabilities **462,** *465*
Partial Least Squares Discriminatory Analysis
bootstrap *370,* 372, **373**
cross-validation *367, 367*
optimization, receiver operator characteristic (ROC) curves *378*
preprocessing, Principal Components Analysis, stepwise Discriminant Analysis **144**
Principal Components Analysis
bootstrap *139*
cross-validation *136*
for exploratory data analysis, scores plots 67
scores plots *71*
most discriminatory PCs *142*
variance *55, 56,* **131**
Q-statistic *285*
disjoint model *259, 284*
Quadratic Discriminant Analysis *195, 283*
prior probabilities **462**
Receiver Operator Characteristic curves *283, 284*
Support Vector Machines *230, 268*
optimization 390, *387, 389*
validation, predictive ability and model stability **340,** *346*
variable selection **403**
Monte Carlo method *444*
12 (GCMS and microbiology of mouse scent marks) 42–4
Principal Components Analysis 67, *72*
scores and loadings plots *480*
super-scores 484, *486*
variance **131**
procrustes analysis *490, 491*

C block, *see* Class information block
Centroids 184–5
 see also Euclidean distance to centroids
Chebyshev dissimilarity index **79,** 81, *82*
Chemometrics
 approaches 2
 development of field 9–10
 history 1–4
Chromatography, internal standards 117–18
City block dissimilarity index **79**
Classifiers
 multiclass
  cross-class classification rate 337, **338**
  overview 289–94
  success measures 318
   cross-class classification rate 337, **338**
  variable selection 393–4
 one-class
  effect of test and training set selection on classifier boundaries *362–4*
  empirical approaches 266
  incorporation of prior probabilities 456–8
  overview 233–5
  results summaries 275–9
   receiver operator characteristic (ROC) curves 279–86
  samples not approximated by multinormal distribution 263–6
  significance indicators
   chi-squared 239–41
   D-statistic 243–8
   Gaussian 239–41
   Hotelling T-squared 241–2
   squared prediction error 248–9
  Support Vector Machines 266–75
  test set prediction 323
  test and training set boundaries 361
 stability 350–2
 success measures, multiclass 318
 two-class 177–80
  class boundaries 181–4
  compared to one-class 233–5
  Euclidean distance to centroids 184–5
  Learning Vector Quantization 201–5
  Linear Discriminant Analysis 185–92
  nonhomogeneous populations 201–2
  Quadratic Discriminant Analysis 192–5
  success measures 315–17
  Support Vector Machines, *see* Support Vector Machines (SVM)
  test set prediction 323
  variable selection 397–8
 unequal class sizes, contingency tables 453–8
 see also Class separation indices
Class information block 18–19, *19*
 multiclass datasets 24
 Self Organizing Maps 97–105
Class membership plots 275–79
Class separation indices 469–70
 Davies Bouldin index 470–4
 dissimilarity measures and 84
 modified silhouette width 475–7
 multiple classes 477
 overlap coefficient 477–8
 silhouette width 475
Closure 117, 411–12
Cluster Validation Indices (CVI) 469–70
Cocaine 20–3
Codebooks 206–7
 generation 207–8
Column scaling
 block 128–9
 mean centring 124–5
 range 127–8
 standardization 125–7
 weighted centring 125
Component Planes 99–100
Computing power 314
 development 4
 Support Vector Machines and 385
Confusion matrix 316
 multiclass **318**
Consensus PCA 484–7
Contingency tables 316
 incorporating prior probabilities 453–8
 one-class classifiers **319**
Copper, in blood 34–5
Cosine similarity index **79**
C-PCA *see* Consensus PCA
Cross-validation 132–3, *135–6*
 Partial Least Squares Discriminatory Analysis 365–7
  bootstrap flow chart *168*
 Principal Components Analysis
  Leave One Out (LOO) method 132–4, 365, *366*, *367*
  process flow chart *174*
CSI, *see* Class separation indices

Data analysis, errors 6–7
Databases 7–8
Data reduction
 Principal Components Analysis 129–34
 see also Data scaling; Preprocessing; Variable selection

Data scaling
  individual elements 108–9
    Box Cox 115–17
    logarithmic 109–12
    power 113–15
  rows
    block scaling 120–4
    to constant total 118–20
    to landmarks 117–18
Davies Bouldin index (dbi) 470–4
  outlying data 474
Denaturing Gradient Gel Electrophoresis (DGGE) 43, *44*
Diabetes 313
Directed acrylic graph (DAG) tree 306–9
Discriminatory Principal Components 137–45, **146**
Disjoint principal component models 238
  D- and Q-statistic visualization 249–64
  Support Vector Machines 267–8, *268*
Dissimilarity indices 75–6
Dissimilarity measures
  binary 75–9, **78**
  Principal Coordinates Analysis and 80–3
  quantitative 79–80
  rankings 84–7
    Kolgomorov-Smirnov measure 87
  see also Class separation indices
Distances (between points in sample space) 75
Doornik-Hansen Omnibus Test **264–5**
Dot product 17–18
Drug syndicates 20
D-statistic 243–8
  disjoint models 249–64
Dynamic Mechanical Analysis (DMA) 25, 75

EDA, *see* Exploratory Data Analysis (EDA)
EDC, *see* Euclidean distance to centroids (EDC)
Eigenvectors 50
Elements (of a matrix) 18
Euclidean dissimilarity index **79,** 81, *82*, *83*
Euclidean distance to centroids 184–5, *186–7*
  Bayesian weighting **459**
  class separation and *470*
  compared with Linear Discriminant Analysis 190–1, *191*
  effect of test and training set selection on classifier boundaries *353–7*
  multiclass 291–5
    effect of test and training set selection on classifier boundaries *359–60*
  one-class 235–6
  test set prediction 323
Experimental design 16–17
Exploratory Data Analysis (EDA) 47–75
  dissimilarity measures 75–6
    binary 75–9
    Principal Coordinates Analysis and 80–3
    rankings 84–7
  Principal Components Analysis, *see* Principal Components Analysis (PCA)
  Self Organizing Maps, 10, 87

False negatives (FN) 316
  Partial Least Squares Discriminatory Analysis (PLS-DA) 374
False positives (FP) *281*, 316
  Partial Least Squares Discriminatory Analysis (PLS-DA) 374
Fisher Discriminant Analysis (FDA) 188–91
Fisher weight 412–413
  case study 8 (mouse urine) **414**
Flow charts
  autoprediction *153–6*
  overview *151*
  symbols 152
  test/training
    one class, visualization 159
    two class
      no visualization *158*
      visualization *157*
Food, NIR spectroscopy 23–7
FP, *see* False positives (FP)
F-statistic
  calculation **426**
  class separation and 476
  critical values **429**
  t-statistic and 423

Gas chromatography-mass spectrometry 30–2
  mouse scent marks 42–4
  mouse urine 36
Gaussian density estimators 239–41
GC-MS, *see* Gas chromatography-mass spectrometry
Gower and Legendre similarity index **78**
In group 235
Growing SOM 89

Haplotypes 36, 43
Hard margin vector support machines 224
Hit histograms 97–9
Hotelling T^2 statistic 241–2
Hypertension 34–5
Hypothesis confirmation 312–13
  variable selection 401–404

Indicator functions, *see* Variable selection, significance measures
Informaticians 7–8

Intelligent buckets 37–8
Interaction term 420
Internal standards 117–18
Iterative optimization
bootstrap 368–74
cross-validation 365–7
Iterative validation
effectiveness measures 335–47
number of iterations 348–52
Jaccard similarity index **78**

K Early Neighbours 2
optimization 377–8
Kernel functions 218–23
one-class classification 267
Kohnonen maps, *see* Self Organizing maps
Kolomorogov-Smirnov measure 87, **416**
variables 415

Lagrange multipliers 216
Landmark variables 117–18
LC-MS, *see* liquid chromatography-mass spectrometry
LDA, *see* Linear Discriminant Analysis
Learning rate, Self Organizing Maps 90
Learning Vector Quantization (LVQ) 201–5
Bayesian weighting **459**
bootstrap flow chart *168*
boundary stability *205*
codebooks 206–7
cross-validation, process flow chart *174*
effect of test and training set selection on classifier boundaries *353–7*
examples 211–13
LVQ1 algorithm 207–8
LVQ3 algorithm 209–11
multiclass 295–7
optimization 377–9
test set prediction 323
test and training set boundaries, NIR of food 359
training parameters **211**
Voronoi tessellation 206–7
Leave One Out (LOO) cross-validation 132–3, 365, *366*, *367*
Likelihood ratios 317
Limit of Detection, data scaling and 110–111
Linear Discriminant Analysis (LDA) 185–92
Bayesian weighting **459,** 461
effect of differing prior probabilities **462–4,** *465*
classification ability **146**
compared with Euclidean distance to centroids 190–1, *191*
comparison of autopredictive, test and training sets **324–6,** *327–30*
effect of test and training set selection on classifier boundaries *353–7*
multiclass 291–5
Partial Least Squares Discriminatory Analysis (PLA-DA) and 191, 196, *197*
test set prediction 323
test and training set boundaries
case study 3 (polymer mechanical analysis) *360*
NIR of food 359
Liquid chromatography-mass spectrometry
data preprocessing 32–4
instrument performance and data quality 63
Loadings, Principal Components Analysis 50, 52
Logarithmic data element scaling 109–12
LOO, *see* Leave One Out
Loss Modulus 25
LVQ, *see* Learning Vector Quantization

Magnesium, in blood 34–5
Mahalanobis distance 188
density estimation 241
one-class classification 235–6
see also Linear Discriminant Analysis; Quadratic Discriminant Analysis
Majority voting 336–7
Mantel test 481–3
Mass spectrometry
tandem 20–1
see also gas chromatography-mass spectrometry; liquid chromatography-mass spectrometry
Matrices (numerical)
basic principles 17–19
element transformations 108–9
Box Cox 115–17
logarithmic 109–12, *114*
square root 113–15, *114*, *116*
pseudoinverse 18
rank 53
transpose 17–18
see also confusion matrices; U-matrix
MDS, *see* Principal Co-ordinates analysis
Mean centring 124–5
case study 2 (NIR of food) 24
Mean Quantization Error, Self Organizing Maps 93–4
Metabolomic profiling
development of field 6
mouse urine 36

Metabolomic profiling (*continued*)
sample variability 30–2
Microbiological data 42–4
Minkowski dissimilarity index **79,** 80
MLR, *see* multiple linear regression
Model stability 336, **338**
case studies 1, 3 and 8 **343**
Monte Carlo method 442–3, *444–6*
to assess multiblock data correlation 491
multiblock data correlation tests 481
Moore's law 4, 314
MSC (multiplicative scatter correction) 23–4
Msw, *see* modified silhouette width
Multiblock data 42–4, 479–81
consensus PCA 484–7
correlation tests
Mantel 481–3
Rv coefficient 483–4
procrustes analysis 487–91
scaling and weighting 128–9
Multidimensional scaling, *see* Principal Co-ordinates Analysis
Multiple linear regression (MLR) 397, 417–20
interaction term 420
Multiplicative scatter correction (MSC) 24–5
Multivariate normality tests 263–6

Neighbourhood width, Self Organizing Maps 90, *91*
Neural Networks, *see* Self Organizing Maps, Learning Vector Quantization
NIPALS algorithm 2
algorithm 57
for variable selection 431
NIR spectroscopy
development of field 3, 5–6
food 23–7
NMR spectroscopy 37–8
Noise
column standardization and 126–7
as result of irrelevant variables 405
row scaling and 119
see also Data reduction
Normal distribution 115
variables 412
see also Multivariate normality tests
Normalization
data columns 125–7
data rows 118–20
selective 120–4
Null dataset
classification
Euclidean distance to centroids *187*
Linear Discriminant Analysis *190*
class membership plots *275*
preprocessing validation 107
stepwise Discriminant Analysis **144**
t-statistics **140–1**

Ochiai similarity index **78**
One-class classifiers, *see* Classifiers, one-class
One vs all multiclass decision process
partial least squares methods 300
Support Vector Machines 303, *305*
One vs one multiclass decision process 304–9
directed acrylic graphs 306–9
partial least squares methods 300–2, *308*
Support Vector Machines 303
Optimization
general considerations 315
Learning Vector Quantization 377–9
partial least squares models 361–5
bootstrap 368–74
cross-validation 365–7
success measures
multiclass classifiers 318
one-class classifiers 318–20
two-class classifiers 315–17
variables, *see* Variable selection
Out group 235
Outlying data
class separation indexes and, Davies Bouldin index 470
effect on Self Organizing map quality 93–5
logarithmic data scaling and 109–10
Overfitting 6–7
null data and 40–1
as a result of disjoint PCA 282
as a result of PLS-DA data reduction 146–8
as a result of variable selection 48–9, 402
Support Vector Machines 214
Overlap coefficient 477–8

Partial Least Squares algorithm 198–9
for variable selection 431
Partial Least Squares Discriminatory Analysis (PLS-DA)
advantages 196
algorithm 198–9
classification thresholds 374–7, 377
prior probabilities and 465
class size correction 200
comparison of autopredictive, test and training sets **324–6,** *327–30,* **332–3,** 334
component size 198
as data reduction method 145–50
effect of test and training set selection on classifier boundaries *353–7,* 358–61
Linear Discriminant Analysis and 191

multiclass
effect of test and training set selection on classifier boundaries *359–60*
PLS1 300–3
PLS2 198, 298–300
see also Directed acrylic graph (DAG) tree
number of PLS components 361–2
optimization 361–2
bootstrap 368–74
cross-validation 365–7
receiver-operator curves 377, *378*, *379*
preprocessing 180
weighted column centring 125
test set prediction 323
Pattern recognition
development of field 3
overview of methods 2
PCA, *see* Principal Components Analysis (PCA)
PCO, *see* Principal Co-ordinates Analysis (PCO)
Pearson similarity index **78, 79,** *82*, *83*
Penalty error 223–8
one-class 266
optimization 380, *381–4*
Percentage correctly classified 316
bootstrap and 372
training, test and autopredictive sets compared 324–31, **325**
unequal class sizes 456
Pharmaceuticals 32–3
PLS1 algorithm 300–3
PLS2 algorithm, multiclass 298–300
Pollution 27–9
Polymers, thermal analysis 25–6
Posterior probabilities 455
Predictive Ability (PA) 335, **338**, **339–41**
Predictive modeling 15–16
principal component number 334
Preprocessing 107–8
case study 1 (forensic analysis of banknotes) 22–3
case study 2 (NIR of food) 23–6
case study 3 (thermal analysis of polymers) 25
case study 4 (hydrocarbons in sand) 27
case study 5 (sweet analysis by GCMS) 31–32
case study 6 (liquid-chromatography mass-spectrometry of pharmaceuticals) 33
case study 7 (atomic spectroscopy for hypertension) 34–5
case study 8 (metabolic profiling of mouse urine) 36
case study 9 (NMR of saliva) 37–8
case study 11 (null dataset) 41
case study 12 (GCMS and microbiology of mouse scent marks) 43
column scaling 124–9
mean centring 124–5
range scaling 127–9
standardization 125–8
weighted centring 125
gas chromatography-mass spectrometry
human sweat 30–2
mouse scent marks 43
multiblock data, mouse scent marks 42–4
multivariate methods, largest principal components 129–37
one-class classification, principal components analysis 237–9
row scaling 117–24
block 120–4
to constant total 118–20
to landmarks 117–18
variable selection and 410–12
single element transformations 108–9
Box Cox 115–17
logarithmic 109–12, *114*
power transformations 113–15, *116*
strategies 150–6
for test and training sets 323–4
see also variable selection
PRESS (Predicted Residual Error Sum of Squares) 133–4
Principal Components Analysis (PCA)
algorithm 57
basic principles 50–4
for data reduction 129–34
bootstrap 134–7
discriminatory PC selection 137–43
stepwise discriminant analysis 143–5
Leave One Out (LOO) cross-validation 132–4
number of PCs and prediction efficiency 331–4
t-statistics 140–3
disjoint models 238
class separation 254–5
visualization 249–64
D-statistic 243–8
eigenvalues 53–6
error matrix 56
for exploratory data analysis
graphical output, biplots 74
loadings plots 67–74
scores plots 57–8
case studies 58–67, *68–9*
column standardization and 126
multiblock data 484–7

Principal Components Analysis (PCA) (*continued*)
procrustes analysis 487–91
NIPALS algorithm 57
Singular Value Decomposition and 52
number of PCs 132
for one-class classification 236–9
overview 49
Q-statistic 248–9
residual sum of squares (RSS) error 55–6
Singular Value Decomposition method, NIPALS and 52
Principal Co-ordinates Analysis (PCO) 80–4
Probability
Bayes' theorem 453–8
prior probabilities
in classification algorithms 458–61, **462–4**
incorporation into classifiers 458–67
to reflect differing class sizes 466–7
variable ranking and 418–20
variable selection utilising 440–2
Procrustes analysis 487–91
Pseudoinverse (of a matrix) 18
Psychology, factor analysis *50*
P values 415, 440
Q-statistic 248–9
Quadratic Discriminant Analysis (QDA) 192–5
Bayesian extension 461
Bayesian weighting **459, 460**
effect of changing prior probabilities 466
comparison of autopredictive, test and training sets **324–6,** *327–30*, **332–3**
multiclass 291–4, *294*
effect of test and training set selection on classifier boundaries *359–60*
test and training set boundaries *359*
test set prediction 323
Quantitative dissimilarity measures 79–80

Rank (of a matrix) 53
Rank (of a Principal Component by discriminatory ability) **140–1**
Rank (of a sample), dissimilarity measures 84–7
Rank (significance, of a raw variable) 405–7
by Fisher weight 412–413
by multiple linear regression 417
by partial least squares methods 432, **434**
by *t*-statistic 408
Receiver operator characteristic (ROC) curves 279–8, 377, *378*, *379*
Residual sum of squares (RSS) errors 55–6, 133
Rogers and Tanimoto similarity index **78**
Row scaling
block 120–4
case study 4 (environmental pollution) 27
to constant total 118–20
ratioing to landmarks 117–18
variable selection and 410–12
RSS errors 55–6, 134
Russell and Rao similarity index **78**
R_v coefficient 483–4

Saliva 37–8
Sample dissimilarity indices, *see* Dissimilarity indices
Sample size (numerical) 7–8, 399–401
Sample size (physical) 118–19
Sand 25–7
Scalar notation 17
Scores, Principal Components Analysis 50, 52
Selective normalization 120–4
Self Organizing Maps
algorithm 88–9
initialization 89
training
iteration 92–3
learning rate 90–2
neighbourhood width 90–2, *91*
Best Matching Unit 89, 92
Growing 89
map quality 93–5
map shape 89
learning rate and 91
Mean Quantization Error 93–4
overview 87
Topographic Error 94–5
training 89, *94*
recommended parameter values **93**
visualization
class information and 101–5
component planes 99–100
hit histograms 97–9
unified distance matrix 95–7
weight vectors, generation 89
Silhouette width (SW) 475
SIMCA 2, 236–9, 239
Similarity measures, *see* Dissimilarity measures
Simulated data
generation 38–40
null dataset 40–1
see also Case studies, 10 (simulations) 11
Singular Value Decomposition 52
Sockal and Sneith similarity index **78**
Soft margin vector support machines 224

Software companies 4
Software, development costs 8
Sorensen similarity index **78,** 81, *82*, *83*
SPE, *see* Q-statistic
Squared Pearson correlation coefficient **79**
Squared prediction error (SPE) 248–9
Square root element scaling 113–15
Statistical indicators 405–7
Stepwise discriminant analysis 143, **144**
Structural risk minimization 223–8
Sum of squares after mean centring 422
Support Vector Machines (SVM) 213–14, *268–9*
  Bayesian weighting **459**
  bootstrap flow chart *170*
  cross-validation flow chart *173*
  effect of test and training set selection on classifier boundaries *353–7*
  kernel functions 218–23, 219–23
  linear classification 214–17
  multiclass 304
    effect of test and training set selection on classifier boundaries *359–60*
  nonlinear classification 218–23
  one-class 266–9
  optimization 216–17, 380–90
  overfitting 223–8
  penalty error 223–8
    one-class classification 266
    optimization 380–90
      case study 4 (sand pollution) *381–4*
  'soft margin' 224–5
  structural risk minimization 223–8
  test set prediction 323
Sweat 30–2
SW (silhouette width) 475

Test datasets 321–4
  for bootstrap 368–9
    percentage correctly classified 372
  effect on classifier boundaries 352–61
  iterative splitting 337–48
  for variable selection 402
Testing 320–1
Topographic Error, Self Organizing Maps 94–5
TP, *see* True positives (TP)
Training datasets 321–4
  bootstrap 368–9
    percentage correctly classified 372
  effect on classifer boundaries 352–61
  for variable selection 402
Training (of software users) 5
Transpose (of a matrix) 17
True negatives (TN) 316
  Partial Least Squares Discriminatory Analysis (PLS-DA) 374
True positives (TP) 281, 316, 374
T-statistic 140–3, 407–12
  F-ratio and 4212
  Hotelling's T^2 241–2
  multiple linear regression and 417–20, *419*
  partial least squares measures and **434**
  of Principal Components 143–5
  of variables 407–13
    case study 8 (mouse urine) **414**
    compared with PLS indicators *435*
Two-class classifiers, *see* Classifiers, two-class

U-matrix *see* Unified distance matrix
Unified distance matrix 95–7
Urine analysis 36

Validation 311–15
  autoprediction and 320–1
  iterative approaches
    classifier stability 350–2
    number of iterations 348–52
    performance indicators, Predictive Ability 335
  number of principal components and 331–4
  process flow charts 157–61, *157–60*, *166–7*
    two-class visualization *157*
  test sets 161
  test and training datasets 321–4
  unequal class sizes 456
  see also Cross-validation
Validation dataset, *see* Test dataset
Variable selection 7
  Davies Bouldin index and 470–4
  general considerations 391–2
  independent variables 396–7
  multilevel and multiway distributions 396–8
  normality test 413
  number of significant variables
    cost/benefit 445–8
    empirical determination 442–3
    probabilistic determination 440–2
    procedure 448–50
  partial least squares 431–34
  Principal Components Analysis, loadings plots 67–74
  procrustes analysis 487
  to remove irrelevant variables 405
  sample size 399–401

Variable selection (*continued*)
  Self Organizing Maps, Component Planes 99–105
  significance measures 405–7
    analysis of variance 420–4
    ANOVA 420–4
    Fisher weight 412–417
    multiple linear regression 417–20
    partial least squares
      regression coefficients 432–3
      weights matrix 432
    t-statistic 407–12
  two-class distributions 394–5
  see also Data reduction; discriminatory Principal Components; Exploratory Data Analysis (EDA)
Variance
  Principal Components 130–4
    case studies **131**
  see also Analysis of variance
Variance (of a matrix) 54
Vector normalization 119–20
Vector notation 17
Vegetable oils 23–5
Visualization
  disjoint models, D- and Q- statistics 249–64
  multidimensional spaces 331
  see also Case studies
Voronoi tessellation 206–7

Weighted centring 125